MATHEMATICS-I

Dr. Zubair Khan
Lecturer in the Department of Mathematics
Integral University, Lucknow

Shadab Ahmad Khan
Lecturer in the Department of Mathematics
Integral University, Lucknow

Dr. Qazi Shoeb Ahmad
Assistant Professor in the Department of Mathematics
Integral University, Lucknow

Ane Books Pvt. Ltd.

New Delhi ♦ Chennai

Mathematics-I

Dr. Zubair Khan, Shadab Ahmad Khan & Dr. Qazi Shoeb Ahmad

Price: ₹ 595/-
Pages: 590 + xii

Published by

Ane Books Pvt. Ltd.
4821, Parwana Bhawan, 1st Floor, 24 Ansari Road,
Darya Ganj, **New Delhi** - 110 002, India
Tel.: +91(011) 23276843-44, Fax: +91(011) 23276863
e-mail: kapoor@anebooks.com, Website: www.anebooks.com

Branch Office:

- Avantika Niwas, 1st Floor, 19 Doraiswamy Road, T. Nagar,
  Chennai - 600 017, Tel.: +91(044) 28141554, 28141209
  email: anebookschennai@gmail.com, rathinam@anebooks.com

ISBN : 978-81-8052-294-9

Printed at : R.K. Offset Process, New Delhi.

Preface

The authors feel great pleasure in introducing the book **"Mathematics-I"**. This book is designed to meet the requirements of B.C.A. students of Integral University and its study centres (UPTEC).

The subject matter has been discussed in such a way that the students will find no difficulty to understand it. Each chapter of this book contains complete self-explanatory theory and a large number of solved examples, followed by a collection of good exercises.

The language of the book, on the whole, is lucid and easy to understand. The authors hope that the students, teachers and other readers will find the book interesting and to the point. It covers the whole course. We hope that the students will receive the book warmly.

We have taken great care in eliminating the misprints, but if there are still any, we shall be highly obliged to those who will take trouble of pointing them out. Suggestions for the improvement of the book will be gratefully acknowledged.

We wish to place in record our sincere appreciation & thanks for the help we received from our respected teachers, colleagues and friends in presenting the manuscript of the book.

The last but not the least we are also thankful to Mr. Sunil Saxena, Mr. Jai Raj Kapoor, Mr. H. Rahman and their team of "Ane Books Pvt. Ltd." for their kind cooperation at every stage, without which it would not have been possible to bring out this book in such a fine format.

Zubair Khan
Shadab Ahmad Khan
Qazi Shoeb Ahmad

Contents

1 Trigonometry and Complex Numbers

1.1 TRIGONOMETRY

"Trigonometry" is the science of measuring triangles. It is "that branch of mathematics which deals with the measurement of angle and the problems allied with angles." It is used in surveying, Physics, Astronomy, Navigation, etc.

1.2 ANGLES

Angles in geometry are always positive and lie between 0° and 360°. But angles in trigonometry may be positive or negative and their measurements are not restricted to any fixed limit.

The angles are formed by the rotation of a straight line about a fixed point lying on it starting from a fixed position.

Sign of Angles: Let a revolving ray starting from OX, rotate about O in a plane and stop at any position OP. Then it is said to generate and angle XOP, OX is called initial position, OP is final or terminal position and O is the vertex of the angle. If rotation is anti-clockwise the angle is positive, if rotation is clockwise, the angle is negative.

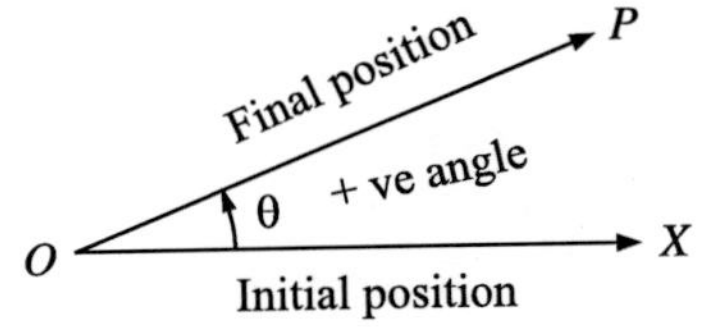

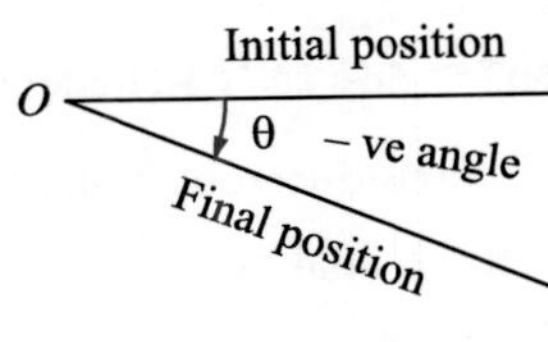

Lines at Right Angle: Two lines are said to be at right angles if the revolving ray starting from initial position to the final position describes one quarter of a circle.

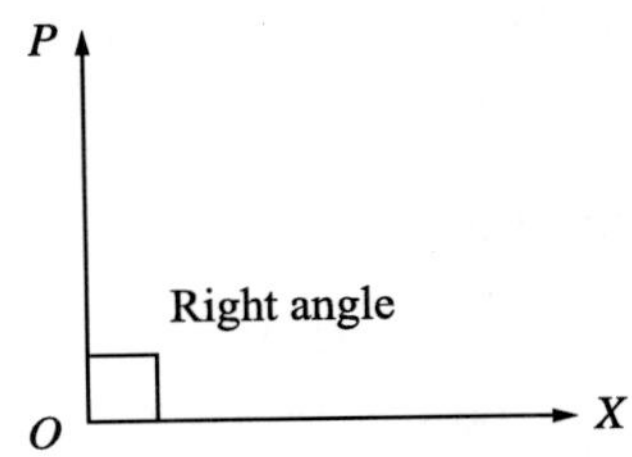

Quadrants: Let X′OX and YOY′ be two coplanar lines at right angles to each other. X′OX is called the *x*-axis and YOY′ is *y*-axis

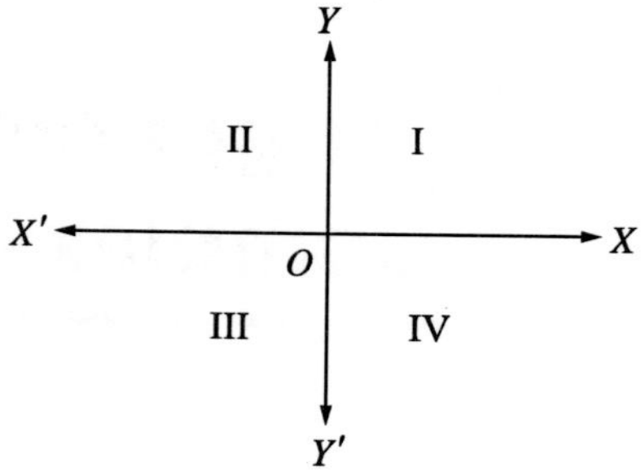

Clearly, these lines divide the plane into four equal parts, called quadrants.

The parts XOY, YOX′, X′OY′ and Y′OX are known as the first, the second, the third and the fourth quadrant respectively.

Co - Terminal Angles: Two angles are called Co - terminal angles if they have the same initial sides and the same terminal sides but different measures.

Example: (*i*) The angles with measures 90° and – 270 are co - terminal.

(*ii*) The angles with measures 45° and – 315° are co - terminal

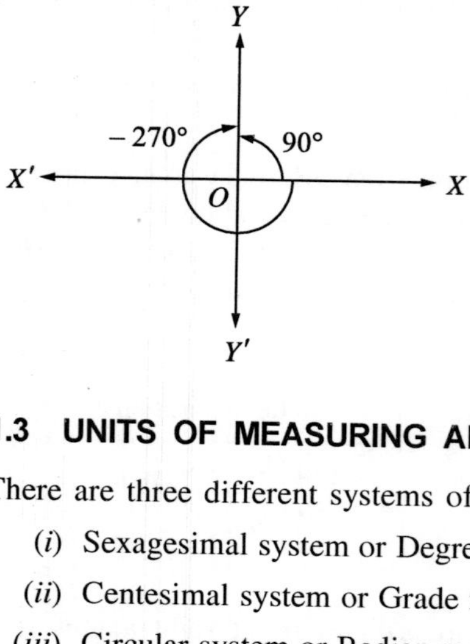

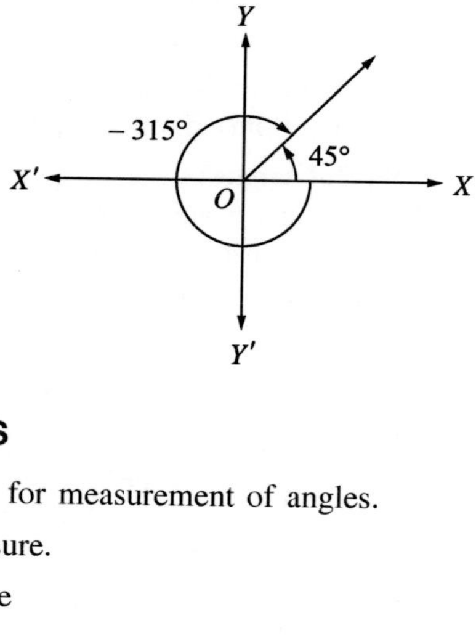

1.3 UNITS OF MEASURING ANGLES

There are three different systems of units for measurement of angles.

(*i*) Sexagesimal system or Degree measure.

(*ii*) Centesimal system or Grade measure

(*iii*) Circular system or Radian measure.

(i) Sexagesimal system or Degree measure: The most common unit of measurement of angles is the degree. In this system an angle is measured in degrees, minutes and seconds. A complete rotation describes 360º.

$\therefore$ 1 right angle = 90° (since right angle is $\frac{1}{4}$ the of full rotation)

1 degree i.e.. 1° = 60 minutes ie. 60′

and 1 minute i.e; 1′ = 60 seconds i.e; 60″

Example: 45° 16′ 25″ means 45 degree, 16 minutes and 25 seconds.

Remark: This system is also called the English system of measurement of angles.

Various Type of Angles: An angle θ is said to be

(*i*) an acute angle, if $0° \leq \theta < 90°$
(*ii*) an obtuse angle, if $90° < \theta < 180°$
(*iii*) a reflex angle, if $180° < \theta < 360°$
(*iv*) a right angle , if $\theta = 90°$
(*v*) a straight line, if $\theta = 180°$
(*vi*) a complete angle, if $\theta = 360°$

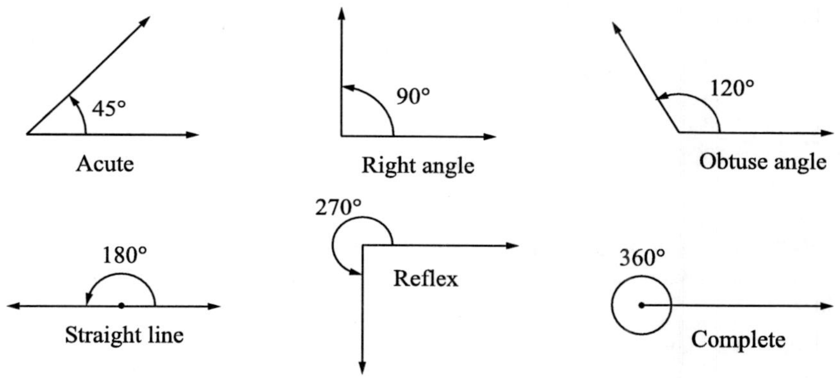

(*ii*) **Centesimal System or Grade measure:** In this system an angle is measured in grades, minutes and seconds.

Here, 1 right angle = 100 grades i.e; 100^g
1 grade i.e; 1^g = 100 minutes i.e; 100′
and 1 minute i.e; 1′ = 100 seconds 1.e; 100″

Example: 45^g 16′ means 45 grades and 16 minutes
or 45.16^g means 45 grades and 16 minutes

(*iii*) **Circular System or Radio Measure:** The unit of angle in this system is radian.

1 radian, written as 1^c is the measure of angle subtended at the centre of a circle of radius r by an arc of length r.

Remark: A radian is a constant angle

Theorem: The number of radians is an angle subtended by an arc of a circle at the centre = $\frac{\text{arc}}{\text{radius}}$, i.e; $\theta = \frac{s}{r}$, where θ^c is the angle subtended at the centre of a circle of radius r by an arc of length.

1.4 RELATION BETWEEN DEGREES AND RADIANS

Suppose a ray OA revolves about its end point O and generates a circle of radius r. Then the radian measure of the angle formed by one complete revolution of the revolving ray is

$$\frac{2\pi r}{r} = 2\pi \qquad \left(\because \theta = \frac{s}{r} = \frac{2\pi r}{r}\right)$$

But one complete revolution of the revolving ray generates an angle of 360°

$\therefore$ 2π radians = 360°

or π radians = 180°

i.e; $\pi^c = 180°$

Corollary:

1. $1 \text{ radian} = \frac{180°}{\pi} = \left(\frac{180}{\frac{22}{7}}\right)^0 = 57°16'$ (approx)

2. $1° = \frac{\pi}{180}$ radians $= \left(\frac{22}{7 \times 180}\right)$ radians = 0.01746 radians.

3. π radians = 180° = 200 grades

Remark : If an angle is given without mentioning units, it is assumed to be in radians or when an angle is expressed in radians, the word radian is often omitted.

Thus, we may write $\pi = 180°$ instead of π radians = 180°.

Circular measure of some standard angles are:

Angle in degrees	0°	30°	45°	60°	75°	90°	120°	150°	180°	270°	360°
Circular measure	0	$\frac{\pi}{6}$	$\frac{\pi}{4}$	$\frac{\pi}{3}$	$\frac{5\pi}{12}$	$\frac{\pi}{2}$	$\frac{2\pi}{3}$	$\frac{5\pi}{6}$	π	$\frac{3\pi}{2}$	2π

ILLUSTRATIVE EXAMPLES

Example 1. Express the following in radian measure:

(*i*) 105° (*ii*) 40° 37′ 30″ (*iii*) – 22° 30′

Solution:

(*i*) $180° = \pi^c$

$\therefore$ $105° = \frac{\pi}{180} \times 105 = \left(\frac{7\pi}{12}\right)^c$

(*ii*) $40° \, 37' \, 30'' = 40° \left(37\frac{30}{60}\right)' = 40° \left(\frac{75}{2}\right)'$

$$= \left(40 \, . \, \frac{75}{2 \times 60}\right)^° = \left(\frac{325}{8}\right)^°$$

$180° = \pi^c$

$$\therefore \qquad \left(\frac{325}{8}\right)^{\circ} = \frac{\pi}{180}\times\frac{325}{8} = \left(\frac{65\pi}{288}\right)^{c}$$

$$(iii) \qquad -22^{\circ}\ 30' = -22^{\circ}\frac{1}{2}^{\circ} = \left(\frac{45}{2}\right)^{\circ}$$

$$180^{\circ} = \pi^{c}$$

$$\therefore \qquad -\left(\frac{45}{2}\right)^{\circ} = \frac{\pi}{180}\times\frac{45}{2} = \left(-\frac{\pi}{8}\right)^{c}$$

Example 2: Express the following is degree measure:

(*i*) $\left(\frac{18\pi}{5}\right)^{c}$ (ii) -3^{c} (iii) $\left(\frac{3}{4}\right)^{c}$

Solution.

(*i*) $\pi^{c} = 180^{\circ}$

$$\therefore \qquad \left(\frac{18\pi}{5}\right)^{c} = \frac{180}{\pi}\times\frac{18\pi}{5} = 648^{\circ}$$

(*ii*) $\pi^{c} = 180^{\circ}$

$$\therefore \qquad -3^{c} = \frac{180}{\pi}\times(-3) = -\left(\frac{180}{22}\times 7\times 3\right)^{\circ} = -(171^{\circ}49'5')$$

(*iii*) $\pi^{c} = 180^{\circ}$

$$\therefore \qquad \left(\frac{3}{4}\right)^{c} = \frac{180}{\pi}\times\frac{3}{4} = \left(\frac{180}{22}\times 7\times\frac{3}{4}\right)^{\circ} = 42^{\circ}\ 57'\ 16''$$

Example 3: Express $\frac{\pi}{4}$ in sexagesimal and centesimal system

Solution. $\pi^{2} = 180^{\circ}$

$$\therefore \qquad \frac{\pi}{4} = \frac{180}{\pi}\times\frac{\pi}{4} = 45^{\circ}$$

Again, $\pi^{c} = 200g \quad \therefore \quad \frac{\pi}{4} = \frac{200}{\pi}\times\frac{\pi}{4} = 50g$

Example 4: Find the degree measure of the angle subtended at the centre of a circle of diameter 200cm by are arc of length 22cm ($\pi = 3.14$ approx)

Solution. Here radius of circle,

$$r = \frac{\text{diameter}}{2} = \frac{200}{2} = 100\text{cm}$$

and length of arc, $s = 22\text{cm}$

$$\therefore \text{ angle} \qquad = \frac{\text{arc}}{\text{radius}} \Rightarrow \theta = \frac{s}{r}$$

$$\Rightarrow \qquad \theta = \frac{22}{100} = \frac{11}{50} \text{ radians} = \frac{11}{50}\times\frac{180}{\pi}\text{ degree}$$

$$= \frac{11}{50} \times \frac{180 \times 7}{22} = \left(\frac{63}{5}\right)^{\circ}$$

$$= \left(12\frac{3}{5}\right)^{\circ} = 12^{\circ} \left(\frac{3}{5} \times 60\right)' = 12^{\circ}36'$$

Example 5: The large hand of a clock is 35cm long. How much distance (in cm.) does its extremity move in 9 minutes?

Solution. The angle traced by the large hand in 60 minutes = 360° = 2π radians

∴ The angle traced by the large hand in 9 minutes

$$= \frac{2\pi}{60} \times 9 = \frac{3\pi}{10} \text{ radians}$$

Let s be the length of the arc moved by the extremity of the large hand, then

$$s = r\theta = 35 \times \frac{3\pi}{10} \text{ cm} = 35 \times \frac{3}{10} \times \frac{22}{7}$$

$$= 33\text{cm}$$

Example 6: The difference between two acute angles of a right angled triangle is $\frac{\pi}{5}$. Find these angles in degrees.

Solution. Let two acute angles be x and y, x > y. Since, in a right angled triangle ; sum of two acute angles is 90°,

∴ $x + y = 90^{\circ}$ ...(i)

Also $x - y = \frac{\pi}{5}$ radians $= \left(\frac{\pi}{5} \times \frac{180}{\pi}\right)^{\circ}$

i.e, $x - y = 36^{\circ}$ ...(ii)

Solving (i) and (ii) simultaneously, we get

$$x = 63^{\circ} \text{ and } y = 27^{\circ}$$

Example 7: The angles of a triangle are in *A.P.* and the greatest is double the least. Find all the angles in radians and degrees.

Solution. Let the angles of a triangle be $(a - d)^{\circ}$, a°, $(a + d)^{\circ}$.

Then, $(a - d) + a + (a + d) = 180^{\circ} \Rightarrow a = 60$

Hence, the angles are $(60 - d)^{\circ}$, 60° and $(60 + d)^{\circ}$, the least angle = $(60 - d)^{\circ}$ and the greatest = $(60 + d)^{\circ}$. By given condition,

$$(60 + d) = 2\ (60 - d)$$

$$\Rightarrow \quad 3d = 60 \Rightarrow d = 20$$

Thus the angles are (60 – 20)°, 60°, (60 + 20)° i.e; 40°, 60°, 80°

The same angles in radians are

$$40 \times \frac{\pi}{180}, 60 \times \frac{\pi}{180}, 80 \times \frac{\pi}{180} \text{ i.e; } \frac{2\pi}{9}, \frac{\pi}{3}, \frac{4\pi}{9} \text{ radians}$$

Example 8: A horse moves uniformly along a circular path of radius 21mt. The angle subtended at the centre of the track by the arc described by the horse in 2 seconds is 60°. What distance will the horse pass over in a minute

Solution. Let P be the centre of the circular path and A be the initial position of the horse. Suppose, The horse moves along the arc AB in 2 seconds making angle of 60° as it comes at B then, the distance moved by the horse in 2 seconds,

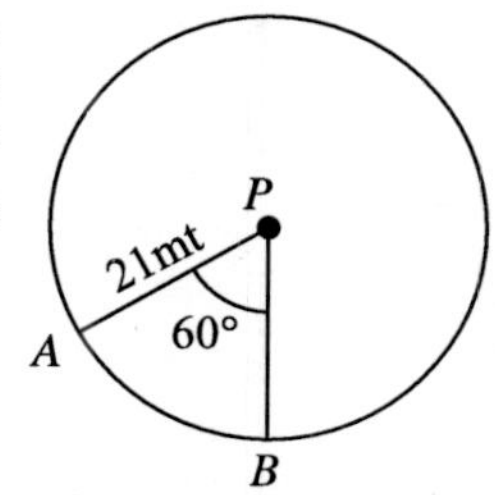

$$AB = r.\theta = 21 \times 60° = 21 \times \frac{\pi}{3}$$

$$= 7 \times \frac{22}{7} = 22\text{mt}$$

The horse moves in 2 seconds, distance of 22mt.

∴ The distance moved by the horse in 1 minute

$$= \frac{22}{2} \times 60 = 660 \text{ meters.}$$

Example 9: The difference of two angles is 1° and their sum is 1 in curricular measure. Find the angles in degrees in terms of π.

Solution. Let the two angles be x and y, then

$$x - y = 1° \qquad ...(i)$$

and $$x + y = 1 \text{ radian} = \left(\frac{\pi}{180}\right)^{\circ}$$

⇒ $$x + y = \left(\frac{\pi}{180}\right)^{\circ} \qquad ...(ii)$$

Solving (i) and (ii), we get

$$x = \frac{180 + \pi}{2 \times 180} = \left(\frac{180 + \pi}{2\pi}\right)^{\circ}$$

and $$y = \left(\frac{180 - \pi}{2\pi}\right)^{\circ}$$

Example 10: If G, D and R be the number of grades, degrees and radius in an angle, prove that

(i) $\frac{G}{100} = \frac{D}{90} = \frac{2R}{\pi}$ and (ii) $G - D = \frac{20R}{\pi}$

Solution. Let the measure of the given angle be x right angles.

Then 1 right angle = 100g ⇒ x $rt.$ ∠s = (100 x)g;

1 right angle = 90° ⇒ x $rt.$ ∠s = (90 x)°;

$$2\ rt \text{ angle} = \pi^c \Rightarrow x\ rt.\ \angle s = \left(\frac{\pi}{2}x\right)^c$$

Now, (*i*) $100\,x = G;\ 90x = D$ and $\frac{x}{2}\ x = R$

Hence $\frac{G}{100} = \frac{D}{90} = \frac{2R}{\pi}$

And (*ii*) $G = 100x,\ D = 90x$ and $x = \frac{2R}{\pi}$

$\Rightarrow$ $G - D = 10x = (10 \times 2R)/\pi = \frac{20R}{\pi}$

EXERCISE 1.1

1. Express the following angles in radian measure:
(*i*) 530° (*ii*) 7°30′ (*iii*) – 270° (*iv*) – 37° 30′

2. Find the degree measures corresponding to the following radian measures:

(*i*) $(-4)^c$ (*ii*) $\left(\frac{-18\pi}{5}\right)^c$ (*iii*) $\left(\frac{\pi}{8}\right)^c$ (*iv*) $\left(\frac{1}{4}\right)^c$

3 In a circle of diameter 40cm, the length of a chord is 20cm. Find the length of the minor arc of the chord

4. The angles of a triangle are in *A.P.* and the number of degrees in the least to the number of radians in the greatest is 60 : π. Find the angles in degrees and radians.

5. The minute hand of a watch is 1.5cm long. How for does its tip move in 50 minutes? Use π = 3.14

6. Find the angles in degrees through which a pendulum swings if its length is 50cm and the tip describes an arc of length 16cm. $\left(\text{use } \pi = \frac{22}{7}\right)$

7. A wire 121cm long is bent so as to lie along the arc of a circle of radius 180cm. Find in degrees the angle subtended at the centre by the arc.

8. The difference of two angles is 20° and their sum is 2 radians. Find each angle in degrees.

9. A railway train is travelling on a circular curve of 1500 meters radius at the rate of 66km per hour. Through what angle had it turned in 10 seconds

10. If the arc of the same length in two circles subtend angles of 75° and 120° at their respective centre, find the ratio of their radii.

ANSWERS

1. (*i*) $\frac{53\pi}{8}$ (*ii*) $\frac{\pi}{24}$ (*iii*) $\frac{-3\pi}{2}$ (*iv*) $\frac{-5\pi}{24}$

2. (*i*) – 229° 5′ 25″ (*ii*) – 648° (*iii*) 22° 30′ (*iv*) 14° 19′ 15″

3. $\frac{20}{3}\ \pi$cm **4.** 30°, 60°, 90°; $\frac{\pi}{6}, \frac{\pi}{3}, \frac{\pi}{2}$ radians

5. 7.85cm **6.** 18° 19′ 38″

7. 7 . 38° 30′ **8.** 67° 16′ 22″ and 47° 16′ 22″

9 $\left(\frac{11}{90}\right)$ radian **10.** 8 : 5

1.5 TRIGONOMETRIC FUNCTIONS OR TRIGONOMETRIC RATIOS

Let $X'OX$ and $Y'OY$ be two mutually perpendicular lines intersecting at point O. Let a revolving line starting from OX move about O in the anti- clockwise direction and generate out an angle $\angle XOA = \theta$. Let $P(x, y)$ be any point on the line OA other than O. OX is called the initial line and OA, the final line or terminal line. Let $OP = r > 0$. From P draw a perpendicular on the x - axis.

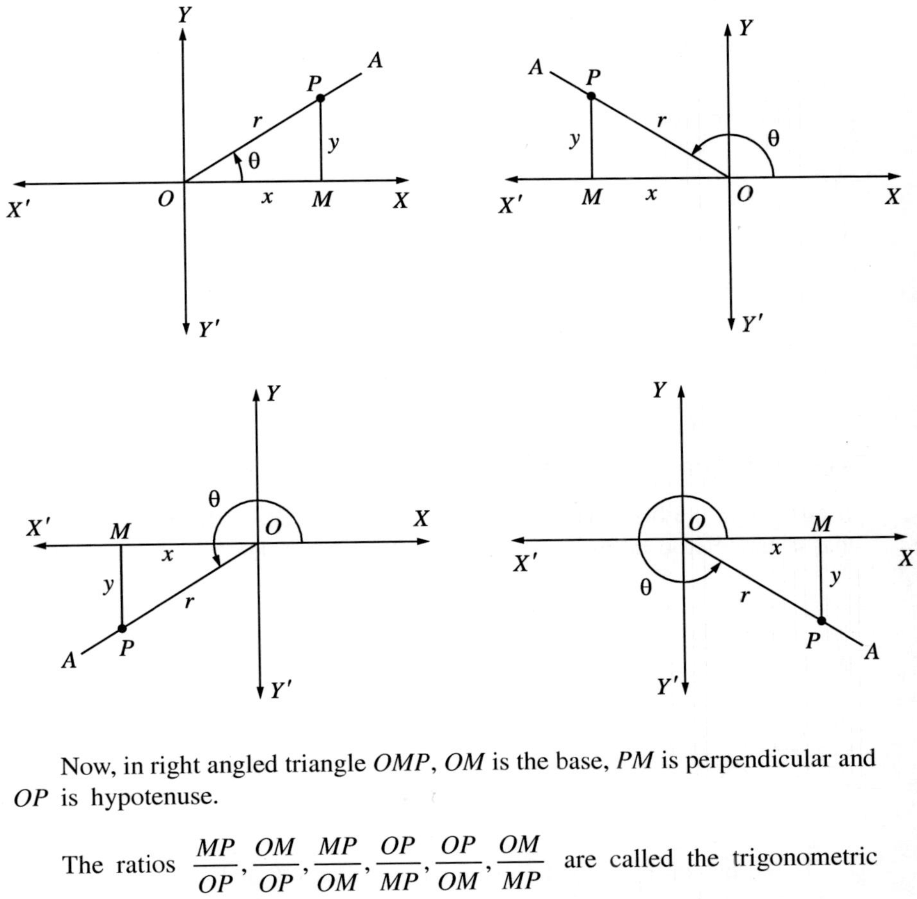

Now, in right angled triangle OMP, OM is the base, PM is perpendicular and OP is hypotenuse.

The ratios $\frac{MP}{OP}, \frac{OM}{OP}, \frac{MP}{OM}, \frac{OP}{MP}, \frac{OP}{OM}, \frac{OM}{MP}$ are called the trigonometric ratios of the angle θ and defined as follows:

(*i*) Since θ (or $\sin\theta$) $= \dfrac{MP}{OP} = \dfrac{\text{perpendicilar}}{\text{hypotenuse}} = \dfrac{y}{r}$

(*ii*) Cosine θ (or $\cos\theta$) $= \dfrac{OM}{OP} = \dfrac{\text{base}}{\text{hypotenuse}} = \dfrac{x}{r}$

(*iii*) tangent θ (or $\tan\theta$) $= \dfrac{MP}{OM} = \dfrac{\text{perpendicular}}{\text{Base}} = \dfrac{y}{x}$

(*iv*) Cosecant θ (or tan θ) = $\dfrac{OP}{MP} = \dfrac{\text{hypotenuse}}{\text{perpendicular}} = \dfrac{r}{y}$

(*v*) Secant θ (or sec θ) = $\dfrac{OP}{OM} = \dfrac{\text{hypotenuse}}{\text{base}} = \dfrac{r}{x}$

(*vi*) Cotangent θ (or cot θ) = $\dfrac{OM}{MP} = \dfrac{\text{base}}{\text{perpendicular}} = \dfrac{x}{y}$

The above six trigonometric ratios are also known as, Trigonometric functions.

Also we have

$$\tan\theta = \frac{\sin\theta}{\cos\theta}$$

$$\sec\theta = \frac{1}{\cos\theta}$$

$$\text{cosec}\,\theta = \frac{1}{\sin\theta}$$

$$\cot\theta = \frac{1}{\tan\theta} = \frac{\cos\theta}{\sin\theta}$$

Sign of Trigonometric Ratios

If *OP* lies in the first quadrant *XOY*, then *OP*, *OM* and *PM* are all positive, so that trigonometric ratios sin θ, cosθ, sec θ, cosec θ, tan θ, cot θ are positive

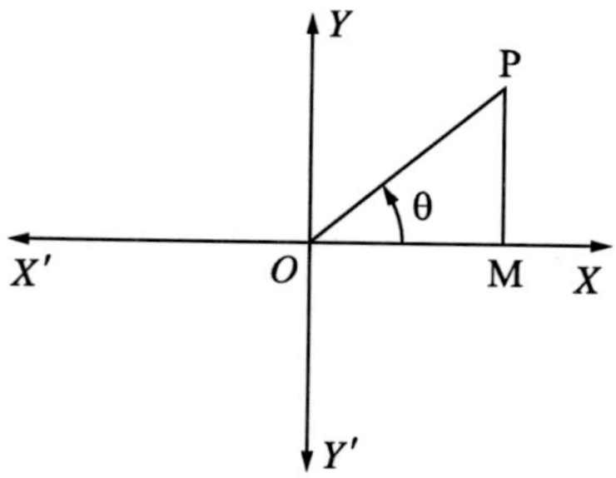

If *OP* lies in the second quadrant *YOX'*, then *OP* and *MP* are positive but *OM* is negative and hence sin θ and cosec θ are positive but cos θ, sec θ, tan θ, cot θ, are negative

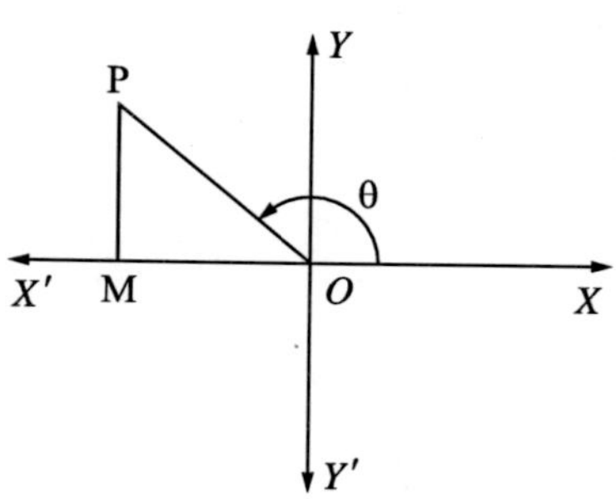

If *OP* lies in the third quadrant *X′OY′* then *OP* is positive but *OM* and *MP* are both negative, so that tan θ and cot θ are positive but sin θ, cosec θ, cos θ, sec θ are all negative

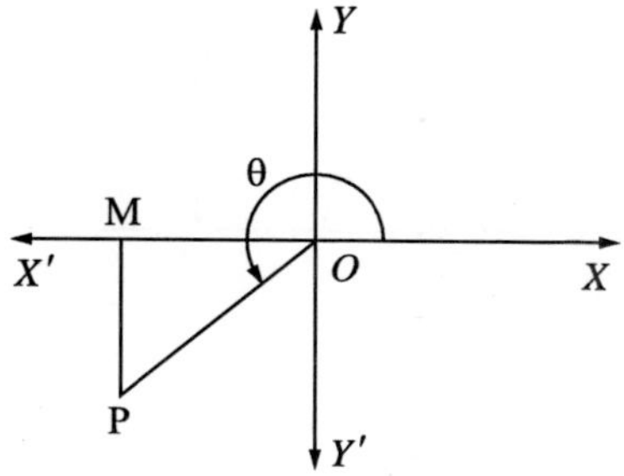

Similarly, in the fourth quadrant cos θ and sec θ are positive but sin θ, tan θ, cosec θ, cot θ are all negative

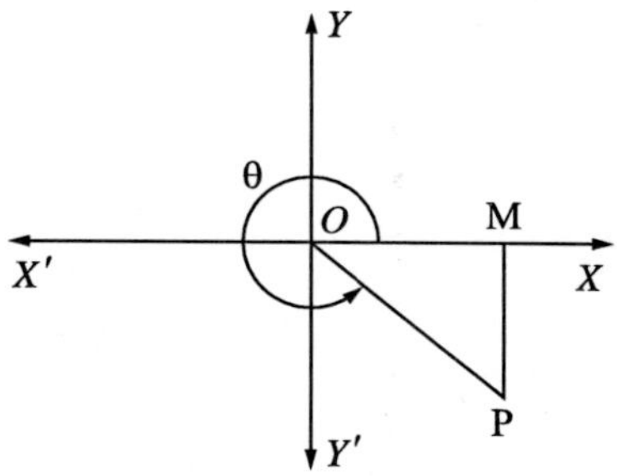

Hence, it clear from the following figure

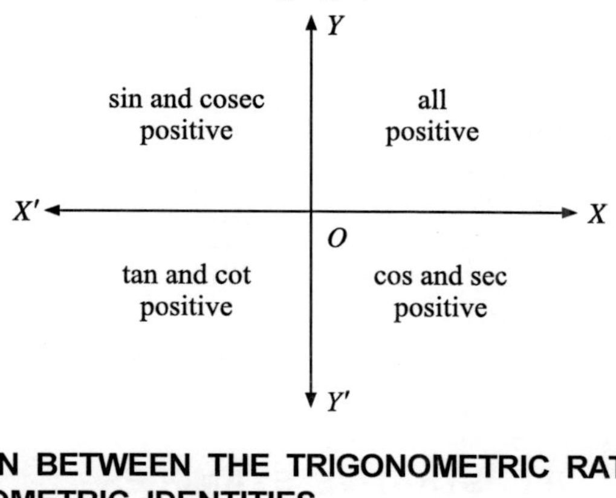

1.6 RELATION BETWEEN THE TRIGONOMETRIC RATIOS: TRIGONOMETRIC IDENTITIES

An equation involving trigonometric functions, which is true for all values of θ, for which the functions are defined is called a trigonometric Identity.

(*i*) $\sin^2\theta + \cos^2\theta = 1$

(*ii*) $\sec^2\theta - \tan^2\theta = 1$

(*iii*) $\operatorname{cosec}^2\theta - \cot^2\theta = 1$

(*iv*) $\tan\theta \,.\, \cot\theta = 1$

Proof: In right angled triangle OMP, we have

$$OP^2 = OM^2 + PM^2 \qquad \text{...(i)}$$

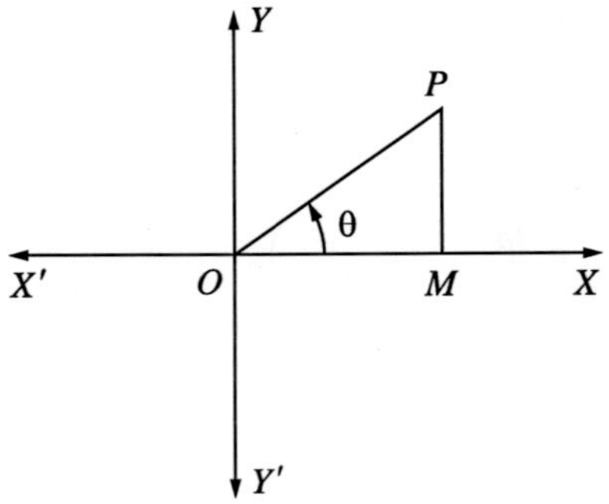

(*i*) $\sin^2\theta + \cos^2\theta = \left(\frac{PM}{OP}\right)^2 + \left(\frac{OM}{OP}\right)^2$

$$= \frac{PM^2 + OM^2}{OP^2} = \frac{OP^2}{OP^2} \qquad \text{(by (i))}$$

$\Rightarrow$ $\sin^2\theta + \cos^2\theta = 1$

(*ii*) $\sec^2\theta - \tan^2\theta = \left(\frac{OP}{OM}\right)^2 - \left(\frac{MP}{OM}\right)^2$

$$= \frac{OP^2 - MP^2}{OM^2} = \frac{OM^2}{OM^2} \qquad \text{(by (i))}$$

$\Rightarrow$ $\sec^2\theta - \tan^2\theta = 1$

(iii) $\text{cosec}^2\theta - \cot^2\theta = \left(\frac{OP}{MP}\right)^2 - \left(\frac{OM}{PM}\right)^2$

$$= \frac{OP^2 - OM^2}{MP^2} = \frac{PM^2}{PM^2} \qquad \text{(by (i))}$$

$\Rightarrow$ $\text{cosec}^2\theta - \cot^2\theta = 1$

(*iv*) $\tan\theta . \cot\theta = \frac{MP}{OM} . \frac{OM}{PM} = 1$

ILLUSTRATIVE EXAMPLES

Example 1. (*i*) If $\cos\theta = \frac{4}{5}$ and θ lies in the first quadrant, find $\sin\theta$ and $\cot\theta$

(*ii*) If $\tan\theta = \frac{3}{4}$, find the value of $3\sin\theta + 4\cos\theta$, taking θ in the first quadrant.

Solution : (*i*) $\cos\theta = \frac{4}{5}$

$\therefore \quad \sin^2 \theta = 1 - \cos^2 \theta$

$\Rightarrow \quad \sin \theta = \pm\sqrt{1-\cos^2\theta} = \pm\sqrt{1-\frac{16}{25}} = \pm\frac{3}{5}$

$\therefore \quad \sin \theta = \frac{3}{4}$, since θ lies in first quadrant

Also $\quad \cot \theta = \frac{\cos\theta}{\sin\theta} = \frac{\frac{4}{5}}{\frac{3}{5}} = \frac{4}{3}.$

(*ii*) $\because \quad \sec^2\theta - \tan^2\theta = 1$

$\Rightarrow \quad \sec^2 \theta = 1 + \tan^2 \theta$

$\Rightarrow \quad \sec \theta = +\sqrt{1+\tan^2\theta}$ (In first quadrant)

$$= \sqrt{1+\left(\frac{3}{4}\right)^2} = \sqrt{\frac{25}{16}}$$

$\Rightarrow \quad \sec \theta = \frac{5}{4}$

$\therefore \quad \cos \theta = \frac{1}{\sec\theta} = \frac{4}{5}$

Also $\quad \sin \theta = +\sqrt{1-\cos^2\theta} = \sqrt{1-\frac{16}{25}} = \frac{3}{5}$

$\therefore \quad 3 \sin \theta + 4 \cos \theta = 3.\frac{3}{5} + 4.\frac{4}{5}$

$$= \frac{9}{5} + \frac{16}{5}$$

$$= \frac{25}{5}$$

$$= 5$$

Example 2. Prove that:

(*i*) $\tan^2\theta - \sin^2\theta = \tan^2\theta . \sin^2\theta$

(*ii*) $\sin^4\theta + \cos^4\theta = 1 - 2\sin^2\theta\cos^2\theta$

Solution :

(*i*) L.H.S $= \tan^2\theta - \sin^2\theta$

$$= \left(\frac{\sin^2\theta}{\cos^2\theta} - \sin^2\theta\right) = \frac{\sin^2\theta - \sin^2\theta\cos^2\theta}{\cos^2\theta}$$

$$= \frac{\sin^2\theta\,(1-\cos^2\theta)}{\cos^2\theta} = \frac{\sin^2\theta}{\cos^2\theta}.\sin^2\theta$$

$$= \tan^2\theta\,.\,\sin^2\theta = \text{R.H.S.}$$

(ii) $$\text{L.H.S} = \sin^4\theta + \cos^4\theta$$

$$= (\sin^4\theta + \cos^4\theta + 2\sin^2\theta\cos^2\theta - 2\sin^2\theta\cos^2\theta)$$

$$= (\sin^2\theta + \cos^2\theta)^2 - 2\sin^2\theta\cos^2\theta$$

$$= 1 - 2\sin^2\theta\cos^2\theta = \text{R.H.S}$$

Example 3. Prove that: $\cos^8 x + \sec^8 x = 2$, if $\cos x + \sec x = 2$

Solution: we have $\cos x + \sec x = 2$

$\Rightarrow \quad \cos x + \dfrac{1}{\cos x} = 2$

$\Rightarrow \quad \cos^2 x + 1 = 2\cos x$

$\Rightarrow \quad \cos^2 x - 2\cos x + 1 = 0$

$\Rightarrow \quad (\cos x - 1)^2 = 0$

$\Rightarrow \quad \cos x = 1 \quad \therefore \quad \sec x = \dfrac{1}{\cos x} = 1$

Hence, $\cos^8 x + \sec^8 x = 1^8 + 1^8 = 1 + 1 = 2$

Example 4. Prove that

$$\sin^8\theta - \cos^8\theta = (\sin^2\theta - \cos^2\theta)(1 - 2\sin^2\theta\cos^2\theta)$$

Solution: L.H.S. $= \sin^8\theta - \cos^8\theta$

$$= (\sin^4\theta)^2 - (\cos^4\theta)^2$$

$$= (\sin^4\theta - \cos^4\theta)(\sin^4\theta + \cos^4\theta)$$

$$= (\sin^4\theta - \cos^4\theta)\ [(\sin^2\theta + \cos^2\theta)^2 - 2\sin^2\theta\cos^2\theta)$$

$$= (\sin^2\theta - \cos^2\theta)(\sin^2\theta + \cos^2\theta)\ (1 - 2\sin^2\theta\cos^2\theta)$$

$$= (\sin^2\theta - \cos^2\theta)(1 - 2\sin^2\theta\cos^2\theta)\ (\because \sin^2\theta + \cos^2\theta = 1)$$

Example 5. It $\sin\theta = x$ and $\tan\theta = y$, prove that $\dfrac{1}{x^2} - \dfrac{1}{y^2} = 1$

Solution:

$$\text{L.H.S.} = \frac{1}{x^2} - \frac{1}{y^2} = \frac{1}{\sin^2\theta} - \frac{1}{\tan^2\theta}$$

$$= \frac{1}{\sin^2\theta} - \frac{\cos^2\theta}{\sin^2\theta} = \frac{1-\cos^2\theta}{\sin^2\theta}$$

$$= \frac{\sin^2\theta}{\sin^2\theta} = 1$$

Example 6. Prove that

(*i*) $\dfrac{\sin\theta}{1+\cos\theta} + \dfrac{1+\cos\theta}{\sin\theta} = \quad 2\ \text{cosec}\ \theta$

(*ii*) $\dfrac{\text{cosec}\ \theta}{\cot\theta + \tan\theta} = \quad \cos\ \theta$

Solution:

(*i*) L.H.S. $= \dfrac{\sin\theta}{1+\cos\theta} + \dfrac{1+\cos\theta}{\sin\theta} = \dfrac{\sin^2\theta + (1+\cos\theta)^2}{\sin\theta(1+\cos\theta)}$

$$= \frac{\sin^2\theta + \cos^2\theta + 1 + 2\cos\theta}{\sin\theta(1+\cos\theta)}$$

$$= \frac{1+1+2\cos\theta}{\sin\theta\,(1+\cos\theta)} = \frac{2}{\sin\theta}$$

$= 2\ \text{cosec}\ \theta =$ R.H.S

(*ii*) L.H.S $= \dfrac{\text{cosec}\,\theta}{\cot\theta + \tan\theta} = \dfrac{\dfrac{1}{\sin\theta}}{\dfrac{\cos\theta}{\sin\theta} + \dfrac{\sin\theta}{\cos\theta}}$

$$= \frac{1}{\sin\theta} \times \frac{\sin\theta\cos\theta}{\cos^2\theta + \sin^2\theta} = \frac{1}{\sin\theta} \times \frac{\sin\theta\cos\theta}{1}$$

$= \cos\ \theta =$ R.H.S

Example 7. If $\sin\ \theta = \dfrac{2pq}{p^2+q^2}$, prove that $\sec\ \theta + \tan\ \theta = \dfrac{p+q}{p-q}$

Solution: we have, $\cos\ \theta = \sqrt{1-\sin^2\theta}$

$$= \sqrt{1 - \frac{4p^2q^2}{(p^2+q^2)^2}} = \sqrt{\frac{(p^2-q^2)^2}{(p^2+q^2)^2}}$$

$\Rightarrow$ $\cos\ \theta = \dfrac{p^2-q^2}{p^2+q^2}$

$\therefore$ $\sec\ \theta = \dfrac{1}{\cos\theta} = \dfrac{p^2+q^2}{p^2-q^2}$

and $$\tan\theta = \frac{\sin\theta}{\cos\theta} = \frac{2pq}{p^2 - q^2}$$

Now $$\sec\theta + \tan\theta = \frac{p^2+q^2}{p^2-q^2} + \frac{2pq}{p^2-q^2}$$

$$= \frac{p^2+q^2+2pq}{p^2-q^2} = \frac{(p+q)^2}{(p-q)(p+q)}$$

$\Rightarrow$ $$\sec\theta + \tan\theta = \frac{p+q}{p-q}$$

Example 8: Prove that

(*i*) $\frac{1-\cos\theta}{\sin\theta} = \frac{\sin\theta}{1+\cos\theta}$

(*ii*) $\frac{1}{\sec\theta+\tan\theta} - \frac{1}{\cos\theta} = \frac{1}{\cos\theta} - \frac{1}{\sec\theta-\tan\theta}$

Solution:

(*i*) L.H.S $= \frac{1-\cos\theta}{\sin\theta}$

$$= \frac{1-\cos\theta}{\sin\theta}\cdot\frac{\sin\theta}{\sin\theta}$$

$$= \frac{(1-\cos\theta)\sin\theta}{\sin^2\theta} = \frac{(1-\cos\theta)\sin\theta}{1-\cos^2\theta}$$

$$= \frac{(1-\cos\theta)\sin\theta}{(1-\cos\theta)(1+\cos\theta)} = \frac{\sin\theta}{1+\cos\theta} = \text{R.H.S}$$

(*ii*) L.H.S $= \frac{1}{\sec\theta+\tan\theta} - \frac{1}{\cos\theta}$

$$= \frac{\sec^2\theta-\tan^2\theta}{\sec\theta+\tan\theta} - \frac{1}{\cos\theta} \qquad (\because 1 = \sec^2\theta - \tan^2\theta)$$

$$= (\sec\theta - \tan\theta) - \sec\theta = -\tan\theta$$

Similarly, R.H.S $= \frac{1}{\cos\theta} - \frac{1}{\sec\theta-\tan\theta}$

$$= \sec\theta - \frac{\sec^2\theta-\tan^2\theta}{\sec\theta-\tan\theta}$$

$$= \sec\theta - (\sec + \tan\theta)$$

$$= -\tan\theta$$

$\Rightarrow$ R.H.S = L.H.S

Example 9. If tan θ + sin θ = *p* and tan θ – sin θ = *q* then prove that $p^2 - q^2 = 4\sqrt{pq}$

Solution: we are given that,

$$\tan\theta + \sin\theta = p \quad ...(i)$$

$$\tan\theta - \sin\theta = q \quad ...(ii)$$

Adding (*i*) and (*ii*) we get

$$2\tan\theta = p + q$$

$$\Rightarrow \quad \tan\theta = \frac{p+q}{2}$$

$$\Rightarrow \quad \cot\theta = \frac{2}{p+q} \quad ...(iii)$$

Subtracting (*ii*) from (*i*), we get

$$2\sin\theta = p - q$$

$$\Rightarrow \quad \sin\theta = \frac{p-q}{2}$$

$$\Rightarrow \quad \operatorname{cosec}\theta = \frac{2}{p-q} \quad ...(iv)$$

Now, since $\operatorname{cosec}^2\theta - \cot^2\theta = 1$

$$\Rightarrow \quad \left(\frac{2}{p-q}\right)^2 - \left(\frac{2}{p+q}\right)^2 = 1 \text{ (Frour (iii) and (iv))}$$

$$\Rightarrow \quad 4(p+q)^2 - 4(p-q)^2 = (p-q)^2\,(p+q)^2$$

$$\Rightarrow \quad 16pq = (p^2 - q^2)^2$$

$$\Rightarrow \quad p^2 - q^2 = 4\sqrt{pq}$$

Example 10. Prove that

(*i*) $\tan\theta + \sqrt{\dfrac{1-\sin\theta}{1+\sin\theta}} = \sec\theta$

(*ii*) $\sqrt{\dfrac{1+\sin\theta}{1-\sin\theta}} = \sec\theta + \tan\theta$

Solution:

(*i*) L.H.S. $= \tan\theta + \sqrt{\dfrac{1-\sin\theta}{1+\sin\theta}}$

$$= \tan\theta + \sqrt{\frac{(1-\sin\theta)\,(1-\sin\theta)}{(1+\sin\theta)\,(1-\sin\theta)}}$$

$$= \tan\theta + \sqrt{\frac{(1-\sin\theta)^2}{1-\sin^2\theta}}$$

$$= \tan\theta + \sqrt{\frac{(1-\sin\theta)^2}{\cos^2\theta}} = \tan\theta + \frac{1-\sin\theta}{\cos\theta}$$

$$= \frac{\sin\theta}{\cos\theta} + \frac{1-\sin\theta}{\cos\theta} = \frac{1}{\cos\theta} = \sec\theta = \text{R.H.S}$$

(*ii*) L.H.S $= \sqrt{\frac{1+\sin\theta}{1-\sin\theta}} = \sqrt{\frac{(1+\sin\theta)(1+\sin\theta)}{(1-\sin\theta)(1+\sin\theta)}} = \sqrt{\frac{(1+\sin\theta)^2}{1-\sin^2\theta}}$

$$= \sqrt{\frac{(1+\sin\theta)^2}{\cos^2\theta}} = \frac{1+\sin\theta}{\cos\theta} = \frac{1}{\cos\theta} + \frac{\sin\theta}{\cos\theta}$$

$= \sec\theta + \tan\theta =$ R.H.S

EXERCISE 1.2

1. If $\sin\theta = \frac{3}{5}$ find the value of $\cos\theta$ and $\tan\theta$, if θ lies in the first quadrant.

Prove the following:

2. $(\cos A + \sin A)^2 + (\sin A - \cos A)^2 = 2$

3. $(\tan\theta + \sec\theta)^2 = \frac{1+\sin\theta}{1-\sin\theta}$

4. $\sin^6\theta + \cos^6\theta = 1 - 3\sin^2\theta\cos^2\theta$

5. $(1 + \cot\theta - \operatorname{cosec}\theta)(1 + \tan\theta + \sec\theta) = 2$

6 $\cot\theta + \sqrt{\frac{1-\cos\theta}{1+\cos\theta}} = \operatorname{cosec}\theta$

7. $\sqrt{\frac{1+\cos\theta}{1-\cos\theta}} + \sqrt{\frac{1-\cos\theta}{1+\cos\theta}} = 2\operatorname{cosec}\theta$

8. $\frac{1+\tan^2\theta}{1+\cos^2\theta} = \left(\frac{1-\tan\theta}{1-\cot\theta}\right)^2$

9. $\frac{\tan\theta}{1-\cot\theta} + \frac{\cot\theta}{1-\tan\theta} = (\sec\theta\operatorname{cosec}\theta + 1)$

10. $\sqrt{\frac{1+\cos\theta}{1-\cos\theta}} = \operatorname{cosec}\theta + \cot\theta\ ;\ \cos\theta \neq 1$

11. $\cot^2\theta\left(\frac{\sec\theta-1}{1+\sin\theta}\right) + \sec^2\theta\left(\frac{\sin\theta-1}{1+\sec\theta}\right) = 0$

12. $\tan^2 A \sec^2 B - \sec^2 A \tan^2 B = \tan^2 A - \tan^2 B$

13. $\sqrt{\sec^2\theta + \operatorname{cosec}^2\theta} = (\tan\theta + \cot\theta)$

14. $\dfrac{\operatorname{cosec}\theta + \cot\theta}{\operatorname{cosec}\theta - \cot\theta} = (\operatorname{cosec}\theta + \cot\theta)^2$

15. $\sqrt{\dfrac{2}{1+\cos\theta} + \dfrac{2}{1-\cos\theta}} = \dfrac{2}{|\sin\theta|}$

16. $\dfrac{\sin^3 A + \cos^3 A}{\sin A + \cos A} + \dfrac{\sin^3 A - \cos^3 A}{\sin A - \cos A} = 2$

17. If $\sin A + \cos A = m$ and $\sec A + \operatorname{cosec} A = n$, prove that $n(m^2 - 1) = 2m$

18. If $\cos\theta + \sec\theta = \sqrt{3}$, prove that $\cos^3\theta + \sec^3\theta = 0$

19. If $\tan A = \dfrac{1}{\sqrt{3}}$, find the value of $\dfrac{\operatorname{cosec}^2 A - \sec^2 A}{\operatorname{cosec}^2 A + \sec^2 A}$

20. If $x = r\cos\theta\cos\phi$, $y = r\cos\theta\sin\phi$, $z = r\sin\theta$, slow that $x^2 + y^2 + z^2 = r^2$

ANSWERS

1. $\dfrac{4}{5}, \dfrac{3}{4}$ **19.** $\dfrac{1}{2}$

1.7 TRIGONOMETRIC FUNCTIONS OF SOME STANDARD ANGLES

(*i*) Trigonometric Functions of 30°

Let a revolving line starting from OX rotate about O and generate an angle $\angle XOA = 30°$. From any point P on OA draw perpendicular PM on OX. Then $\angle OPM = 60°$. Produce PM to P' such that $PM = MP'$. Join OP'. Then in triangles OPM and $OP'M$, $MP = MP'$, OM is a common side and $\angle OMP = \angle OMP'$

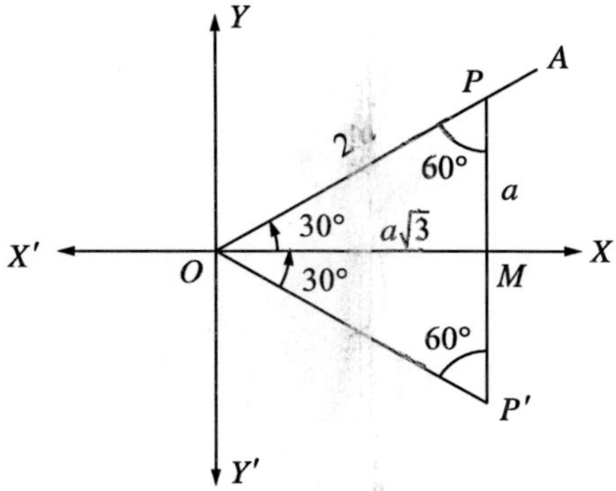

$\therefore$ The two triangles are congruent.

$\therefore \qquad \angle OP'M = \angle OPM = 60°$

$\therefore \qquad \angle POP' = 60°$

$\Delta\, OPP'$ is equilateral

$\therefore \qquad OP = PP'$

Let the sides of this equilateral triangles is 2a.

Then $\qquad PM = a,\ PP' = 2a$ and $OP = 2a$

Now, $\qquad OM^2 = OP^2 - PM^2 = (2a)^2 - a^2 = 3a^2$

$\Rightarrow \qquad OM = a\sqrt{3}$

Since $\angle$MOP = 30° is acute, MP, OP and OM are all positive

$\therefore$ From the right angled triangle OPM, we have

$$\sin 30° = \frac{PM}{OP} = \frac{a}{2a} = \frac{1}{2}$$

$$\cos 30° = \frac{OM}{OP} = \frac{a\sqrt{3}}{2a} = \frac{\sqrt{3}}{2}$$

$$\tan 30° = \frac{PM}{OM} = \frac{a}{a\sqrt{3}} = \frac{1}{\sqrt{3}}$$

$$\text{cosec } 30° = \frac{OP}{PM} = \frac{2a}{a} = 2$$

$$\sec 30° = \frac{OP}{OM} = \frac{2a}{a\sqrt{3}} = \frac{2}{\sqrt{3}}$$

$$\cot 30° = \frac{OM}{PM} = \frac{a\sqrt{3}}{a} = \sqrt{3}$$

(*ii*) Trigonometric Functions of 45°

Similarly if the angle generated is 45° i.e; $\angle POM = 45°$. Then in right angled triangle PMO,

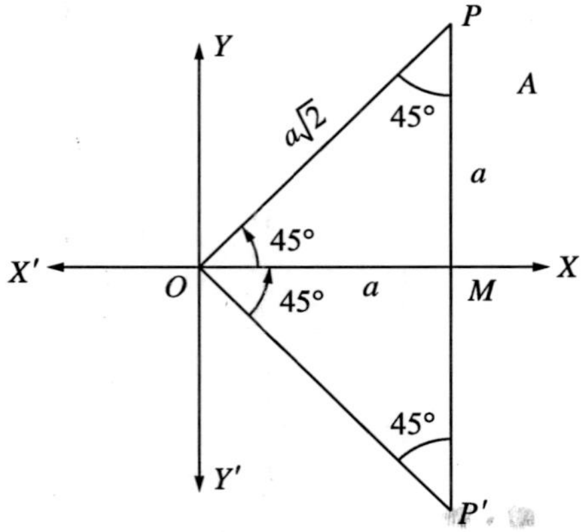

$$\angle OPM = 45°$$

$\therefore \qquad PM = OM = a$ (let)

$$\therefore \qquad OP = \sqrt{OM^2 + PM^2} = \sqrt{a^2 + a^2} = a\sqrt{2}$$

Hence, we have

$$\sin 45° = \frac{PM}{OP} = \frac{a}{a\sqrt{2}} = \frac{1}{\sqrt{2}}$$

$$\cos 45° = \frac{OM}{OP} = \frac{a}{a\sqrt{2}} = \frac{1}{\sqrt{2}}$$

$$\tan 45° = \frac{PM}{OM} = \frac{a}{a} = 1$$

$$\operatorname{cosec} 45° = \frac{OP}{PM} = \frac{a\sqrt{2}}{a} = \sqrt{2}$$

$$\sec 45° = \frac{OP}{OM} = \frac{a\sqrt{2}}{a} = \sqrt{2}$$

$$\cot 45° = \frac{OM}{PM} = \frac{a}{a} = 1$$

(*iii*) Trigonometric Functions of 60°

Fig. is same as for 30° angles. But is this case *PM* is base, *OM* is perpendicular and *OP* is hypotenuse.

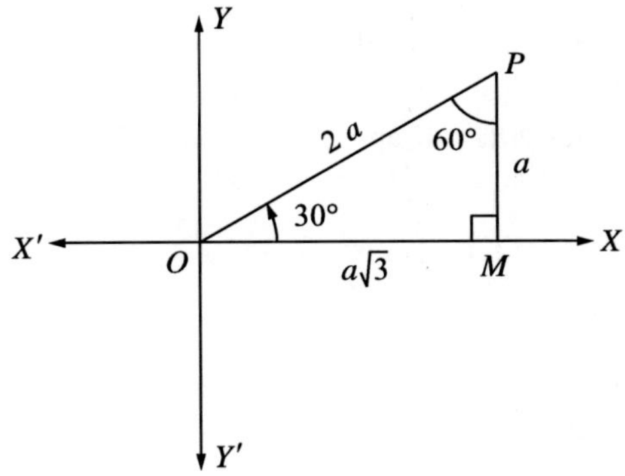

Hence,

$$\sin 60° = \frac{OM}{OP} = \frac{a\sqrt{3}}{2a} = \frac{\sqrt{3}}{2}$$

$$\cos 60° = \frac{OM}{OP} = \frac{a}{2a} = \frac{1}{2}$$

$$\operatorname{cosec} 60° = \frac{OP}{OM} = \frac{2a}{a\sqrt{3}} = \frac{2}{\sqrt{3}}; \ \sec 60° = \frac{OP}{PM} = \frac{2a}{a} = 2$$

$$\cot 60° = \frac{PM}{OM} = \frac{a}{a\sqrt{3}} = \frac{1}{\sqrt{3}}; \ \tan 60° = \frac{OM}{PM} = \frac{a\sqrt{3}}{a} = \sqrt{3}$$

(iv) Trigonometric Functions of 0°

Let $\angle XOA = \theta$ be very small positive angle and let PM be perpendicular to OX. Then as θ becomes smaller; PM also becomes smaller and smaller and OP tends to coincide with OM and ultimately when $\theta = 0°$, $PM = 0$ and $OP = OM$

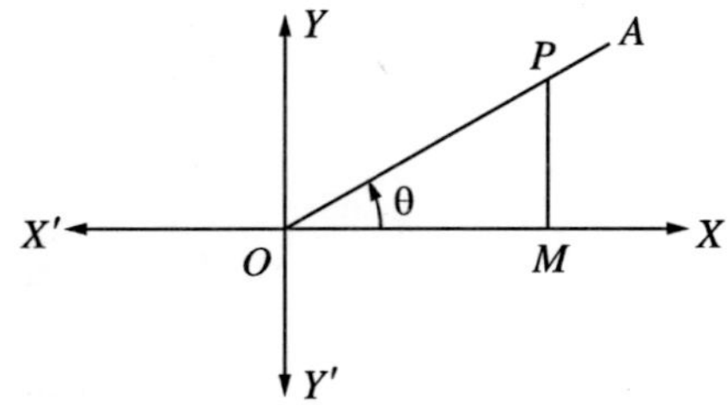

$$\therefore \quad \sin 0° = \frac{PM}{OP} = \frac{0}{OM} = 0$$

$$\cos 0° = \frac{OM}{OP} = \frac{OM}{OM} = 1$$

$$\tan 0° = \frac{OM}{OM} = \frac{0}{OM} = 0$$

$$\sec 0° = \frac{OP}{OM} = \frac{OM}{OM} = 1$$

$$\text{cosec } 0° = \frac{OM}{PM} = \frac{OM}{0} = \text{undefined}$$

$$\cot 0° = \frac{OM}{PM} = \frac{OM}{0} = \text{undefined}$$

(v) Trigonometric Function of 90°

Let $\angle XOP = \theta$ be very nearly equal to 90°, where $\theta \not> 90°$ and let PM be perpendicularly to OX.

Then as θ approaches 90°, OM approaches O and OP tends to coincide with PM and ultimately when $\theta = 90°$, $OM = 0$, $OP = PM$

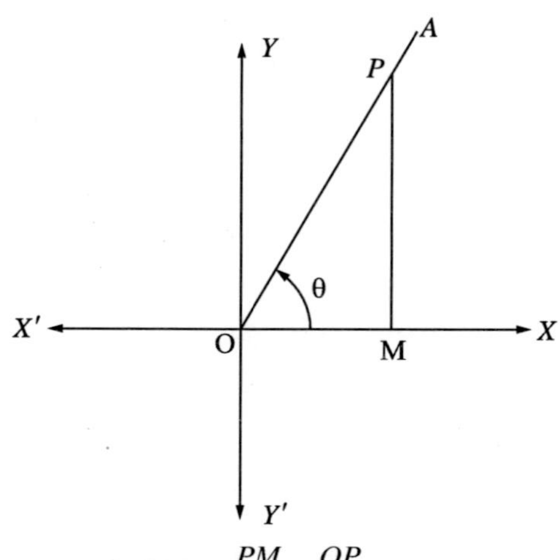

$$\therefore \quad \sin 90° = \frac{PM}{OP} = \frac{OP}{OP} = 1$$

$$\cos 90° = \frac{OM}{OP} = \frac{0}{OP} = 0$$

$$\text{cosec } 90^\circ = \frac{OP}{PM} = 1$$

$$\cot 90^\circ = \frac{OM}{PM} = \frac{0}{PM} = 0$$

$$\sec 90^\circ = \frac{OP}{OM} = \frac{OP}{0} = \text{undefined}$$

$$\tan 90^\circ = \frac{PM}{OM} = \frac{PM}{0} = \text{undefined}$$

Summary Table

θ	0°	30°	45°	60°	90°
$\sin\theta$	0	$\frac{1}{2}$	$\frac{1}{\sqrt{2}}$	$\frac{\sqrt{3}}{2}$	1
$\cos\theta$	1	$\frac{\sqrt{3}}{2}$	$\frac{1}{\sqrt{2}}$	$\frac{1}{2}$	0
$\tan\theta$	0	$\frac{1}{\sqrt{3}}$	1	$\sqrt{3}$	undefined
$\cot\theta$	undefined	$\sqrt{3}$	1	$\frac{1}{\sqrt{3}}$	0
$\sec\theta$	1	$\frac{2}{\sqrt{3}}$	$\sqrt{2}$	2	undefined
$\text{cosec }\theta$	undefined	2	$\sqrt{2}$	$\frac{2}{\sqrt{3}}$	1

Note: (1) Since 0° and 90° are complimentary, we have

$$\sin 0^\circ = \cos 90^\circ = 0$$

$$\cos 0^\circ = \sin 90^\circ = 1 \text{ etc.}$$

(2) Again 30° and 60° are complimentary, we have

$$\sin 30^\circ = \cos 60^\circ = \frac{1}{2}$$

$$\cos 30^\circ = \sin 60^\circ = \frac{\sqrt{3}}{2}, \text{ etc.}$$

ILLUSTRATIVE EXAMPLES

Example 1. Prove that:

$$\text{cosec}^2 45^\circ . \sec^2 30 \ (\cos^2 60^\circ + 4\tan^2 45^\circ - \sec^2 60^\circ) = \frac{2}{3}$$

Solution: L.H.S = cosec^{2}45°. sec^{2}30° $(\cos^2 60° + 4\tan^2 45° - \sec^2 60°)$

$$= (\sqrt{2})^2 . \left(\frac{2}{\sqrt{3}}\right)^2 \left(\left(\frac{1}{2}\right)^2 + 4.(1)^2 - (2)^2\right)$$

$$= 2.\frac{4}{3}\left(\frac{1}{4} + 4 - 4\right) = 2.\frac{4}{3}.\frac{1}{4} = \frac{2}{3} = \text{R.H.S}$$

Example 2. Find the value of 3 tan^2 45° – sin^2 60° $-\frac{1}{2}\cot^2 30° + \frac{1}{8}\sec^2 45°$

Solution: $3\tan^2 45° - \sin^2 60° - \frac{1}{2}\cot^2 30° + \frac{1}{8}\sec^2 45°$

$$= 3.(1)^2 - \left(\frac{\sqrt{3}}{2}\right)^2 - \frac{1}{2}(\sqrt{3})^2 + \frac{1}{8}.(\sqrt{2})^2$$

$$= 3 - \frac{3}{4} - \frac{3}{2} + \frac{2}{8}$$

$$= \frac{4}{4} = 1$$

Example 3: Prove that

(*i*) $\sqrt{\dfrac{1+\cos 30°}{1-\cos 30°}} - \tan 60° = \text{cosec } 30°$

(*ii*) $\sin 30° \cos° + \sin 45° \cos 45° + \sin 60° \cos 30° = \dfrac{7}{4}$

Solution:

(*i*)
$$\text{L.H.S.} = \sqrt{\frac{1+\cos 30°}{1-\cos 30°}} - \tan 60°$$

$$= \sqrt{\frac{1+\frac{\sqrt{3}}{2}}{1-\frac{\sqrt{3}}{2}}} - \sqrt{3} = \sqrt{\frac{2+\sqrt{3}}{2-\sqrt{3}}} - \sqrt{3}$$

$$= \sqrt{\frac{2+\sqrt{3}}{2-\sqrt{3}}} \times \sqrt{\frac{2+\sqrt{3}}{\sqrt{2+\sqrt{3}}}} - \sqrt{3}$$

$$= \frac{(2+\sqrt{3})}{4-3} - \sqrt{3} = (2+\sqrt{3}) - \sqrt{3} = 2$$

R.H.S = cosec 30° = 2

Hence, L.H.S = R.H.S

(*ii*) $$\text{L.H.S} = \text{L.H.S}= \sin 30^\circ \cos^\circ + \sin 45^\circ \cos 45^\circ + \sin 60^\circ \cos 30^\circ$$

$$= \frac{1}{2}.1 + \frac{1}{\sqrt{2}}.\frac{1}{\sqrt{2}} + \frac{\sqrt{3}}{2}.\frac{\sqrt{3}}{2}$$

$$= \frac{1}{2} + \frac{1}{2} + \frac{3}{4} = \frac{7}{4} = \text{R.H.S}$$

Example 4. Solve (i) 3 tan θ = cot θ (θ is positive acute)

(ii) 2 sinθ = cosec θ (θ is positive acute)

Solution: (i) we have 3 tan θ = cot θ

$$\Rightarrow \quad 3 \tan \theta = \frac{1}{\tan\theta} \Rightarrow 3\tan^2 \theta = 1$$

$$\Rightarrow \quad \tan \theta = \pm\frac{1}{\sqrt{3}}$$

But since θ is acute; ∴ tan θ is positive

$$\therefore \quad \tan \theta = \frac{1}{\sqrt{3}} = \tan 30^\circ$$

Hence, $\theta = 30^\circ$

(ii) we have $2 \sin \theta = \text{cosec } \theta = \dfrac{1}{\sin\theta}$

$$\Rightarrow \quad 2 \sin^2 \theta = 1 \Rightarrow \sin \theta = \pm\frac{1}{\sqrt{2}}$$

But since θ is acute, sin θ is positive

$$\therefore \quad \sin \theta = \frac{1}{\sqrt{2}} = \sin 45^\circ$$

Hence $\theta = 45^\circ$

EXERCISE 1.3

1. Find the value of $\sin^2 45^\circ + \cos^2 45^\circ + \tan^2 30^\circ$

2. Show that $\cos 60^\circ = \cos^2 30^\circ - \sin^2 30 = \dfrac{1 - \tan^2 30^\circ}{1 + \tan^2 30^\circ}$

3. If $x \sin 45^\circ \cdot \cos 45^\circ \cdot \tan 60^\circ = \tan^2 45^\circ - \cos^2 60^\circ$, find the value of x

4. Prove that: $\dfrac{4}{3}\cot^2 30^\circ + 3\sin^2 60^\circ - 2\text{cosec}^2 60^\circ - \dfrac{3}{4}\tan^2 30^\circ = \dfrac{10}{3}$

5. Prove that:

$$\sqrt{\frac{1 + \sin 60^\circ}{1 - \sin 60^\circ}} - \sec 60^\circ = \sec 60^\circ - \sqrt{\frac{1 - \sin 60^\circ}{1 + \sin 60^\circ}}$$

6. Show that $\cos\frac{\pi}{3}.\cos\frac{\pi}{4}+\sin\frac{\pi}{3}.\sin\frac{\pi}{4} = \sin\frac{\pi}{4}.\cos\frac{\pi}{6}+\cos\frac{\pi}{4}\sin\frac{\pi}{6}$

7. Taking $A = 30°$ verity that

(i) $\sin^2 A = \dfrac{2\tan A}{1+\tan^2 A}$ (ii) $\cos^2 A = \dfrac{1-\tan^2 A}{1+\tan^2 A}$

8. Taking $A = 60°$, $B = 60°$, verity that

$\cos (A + B) \neq \cos A + \cos B$

9. Solve : $4 \sin \theta + \text{cosec}\, \theta = 4$

10. Solve : $2 \sin \theta \cdot \tan \theta + 1 = \tan \theta + 2 \sin \theta$

11. Solve : $3 \sec^2 \theta = 2\, \text{cosec}\, \theta$

12. Solve : $4 \cos^2 \theta + 2 \sin^2 \theta = 3$

ANSWERS

1. $\frac{4}{3}$ **3.** $\frac{\sqrt{3}}{2}$ **9.** 30°

10. 30°, 45° **11.** 30° **12.** 45°

1.8 TRIGONOMETRIC FUNCTIONS OF ALLIED ANGLES

Two angles are said to be allied angles iff their sum or difference is a multiple of 90°

For example 30° and 60° are allied, 90°– θ and θ are allied

(i) Trigonometric Functions of (– θ)

Let a revolving line *OA* starting from *OX* rotate in the anti - clockwise direction about *O* and generate an angle $\angle XOA = \theta$ in any of the four quadrants (Here we are taking the first quadrant). Again, Let another revolving line *OA′* starting from *OX* rotate in the clockwise direction and form the angle $\angle OXA'$ such that $\angle XOA' = \angle XOA$

Clearly, $\angle XOA' = -\theta$

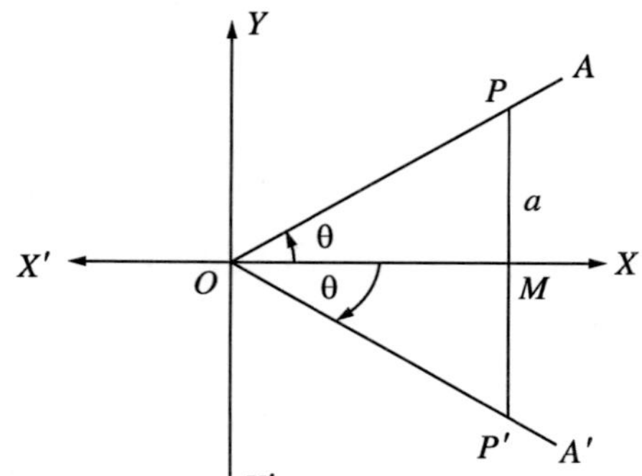

From any point *P* on *OA* draw a perpendicular *PM* on *OX* and produce it to meet *OA′* at P′

In right – angled triangle OPM and $OP'M$, $\angle MOP = \angle MOP'$, $\angle OMP = \angle OMP' = 90°$ and OM is common.

$\therefore$ Triangles are congruent

$\therefore$ $$OP = OP' \text{ and } MP = -MP'$$

$\therefore$ $$\sin(-\theta) = \frac{MP'}{OP'} = -\frac{MP}{OP} = -\sin\theta$$

$$\cos(-\theta) = \frac{OM}{OP'} = \frac{OM}{OP} = \cos\theta$$

$$\tan(-\theta) = \frac{MP'}{OM} = \frac{-MP}{OM} = -\tan\theta$$

$$\text{cosec}(-\theta) = \frac{OP'}{MP'} = \frac{OP}{-MP} = -\cot\theta$$

$$\sec(-\theta) = \frac{OP'}{OM} = \frac{OP}{OM} = \sec\theta$$

$$\cot(-\theta) = \frac{OM}{MP'} = \frac{OM}{-MP} = -\cot\theta$$

(ii) Trigonometric Functions of (90° – θ)

Let a revolving line OA starting from OX rotate in the anti - clock wise direction about O and generate an angle $\angle XOA = \theta$.

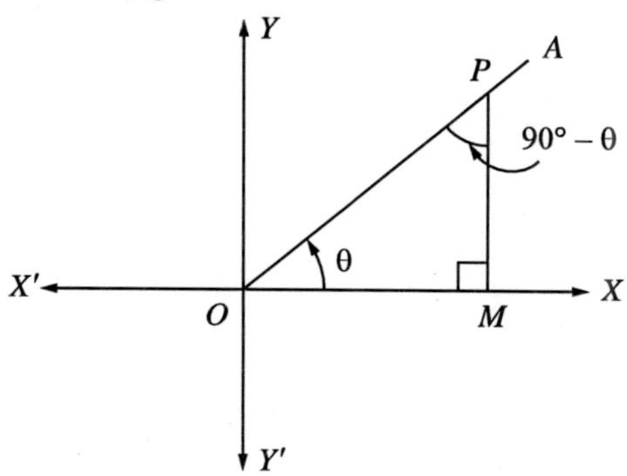

From any point P on OA draw a perpendicular PM on OX, then $\angle OPM = 90° - \theta$.

Now

$$\sin(90° - \theta) = \frac{OM}{OP} = \cos\theta$$

$$\cos(90° - \theta) = \frac{MP}{OP} = \sin\theta$$

$$\tan(90° - \theta) = \frac{OM}{PM} = \cot\theta$$

$$\operatorname{cosec}(90° - \theta) = \frac{OP}{OM} = \sec\theta$$

$$\sec(90° - \theta) = \frac{OP}{PM} = \operatorname{cosec}\theta$$

$$\cot(90° - \theta) = \frac{PM}{OM} = \tan\theta$$

(iii) Trigonometric Functions of (90° + θ)

we have,
$$\sin(90° + \theta) = \sin[90° - (-\theta)] = \cos(-\theta) = \cos\theta$$
$$\cos(90° + \theta) = \cos[90° - (-\theta)] = \sin(-\theta) = -\sin\theta$$
$$\tan(90° + \theta) = \tan[90° - (-\theta)] = \cot(-\theta) = -\cot\theta$$
$$\operatorname{cosec}(90° + \theta) = \operatorname{cosec}[90° - (-\theta)] = \sec(-\theta) = \sec\theta$$
$$\sec(90° + \theta) = \sec[90° - (-\theta)] = \operatorname{cosec}(-\theta) = -\operatorname{cosec}\theta$$
$$\cot(90° + \theta) = \cot[90° - (-\theta)] = \tan(-\theta) = -\tan\theta$$

(iv) Trigonometric Functions of (180° – θ)

Let a revolving line OA starting from OX generate an angle $\angle XOA = \theta$ and let another revolving line starting from OX generate an angle 180° and then revolve back and trace out $\angle X'OA' = \theta$. Then $\angle XOA' = 180° - \theta$

Take any point P on OA and P' on OA' such that $OP = OP'$. Draw perpendicular PM and $P'M'$ on $X'OX$. Then since $\angle POM = \angle P'OM'$ and $OP = OP'$, the right angled triangles OPM and $OP'M'$ are congruent and $M'P' = PM$, $OM' = -OM$

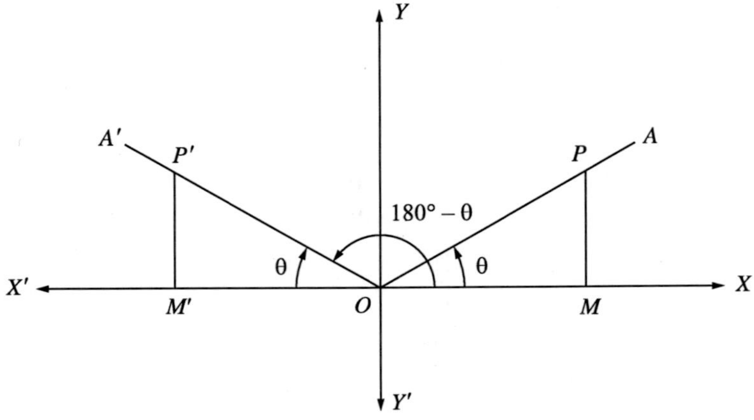

$$\therefore \qquad \sin(180° - \theta) = \frac{M'P'}{OP'} = \frac{PM}{OP} = \sin\theta$$

$$\cos(180° - \theta) = \frac{OM'}{OP'} = \frac{-OM}{OP} = -\cos\theta$$

$$\tan(180° - \theta) = \frac{MP'}{OM'} = \frac{PM}{-OM} = -\tan\theta$$

$$\text{cosec}\,(180^\circ - \theta) = \frac{OP'}{M'P'} = \frac{OP}{MP} = \text{cosec}\,\theta$$

$$\sec\,(180^\circ - \theta) = \frac{OP'}{OM'} = \frac{OP}{-OM} = -\sec\,\theta$$

$$\cot\,(180^\circ - \theta) = \frac{OM'}{MP'} = \frac{-OM}{MP} = -\cot\,\theta$$

Alternate Method.

$$\sin\,(180^\circ - \theta) = \sin\,[90^\circ + (90^\circ - \theta)] = \cos\,(90^\circ - \theta) = \sin\,\theta$$

$$\cos\,(180^\circ - \theta) = \cos\,[90^\circ + (90^\circ - \theta)] = -\sin\,(90^\circ - \theta) = -\cos\,\theta$$

$$\tan\,(180^\circ - \theta) = \tan\,[90^\circ + (90^\circ - \theta)] = -\cot\,(90^\circ - \theta) = -\tan\theta$$

$$\text{cosec}\,(180^\circ - \theta) = \text{cosec}\,[90^\circ + (90^\circ - \theta)] = \sec\,(90^\circ - \theta) = \text{cosec}\theta$$

$$\sec\,(180^\circ - \theta) = \sec\,[90^\circ + (90^\circ - \theta)] = -\text{cosec}\,(90^\circ - \theta) = -\sec\,\theta$$

$$\cot\,(180^\circ - \theta) = \cot\,[90^\circ + (90^\circ - \theta)] = -\tan\,(90^\circ - \theta) = -\cot\,\theta$$

(v) Trigonometric Functions of (180° + θ):

$$\sin\,(180^\circ + \theta) = \sin\,[180^\circ - (-\theta)] = \sin\,(-\theta) = -\sin\,\theta$$

$$\cos\,(180^\circ + \theta) = \cos\,[180^\circ - (-\theta)] = -\cos\,(-\theta) = -\cos\,\theta$$

$$\tan\,(180^\circ + \theta) = \tan\,[180^\circ - (-\theta)] = -\tan\,(-\theta) = -(-\tan\theta) = \tan\,\theta$$

Similarly,

$$\text{cosec}\,(180^\circ + \theta) = -\text{cosec}\,\theta$$

$$\sec\,(180^\circ + \theta) = -\sec\,\theta$$

$$\cot\,(180^\circ + \theta) = \cot\,\theta$$

(vi) Trigonometric Functions of (270° – θ):

$$\sin\,(270^\circ - \theta) = \sin\,[180^\circ + (90^\circ - \theta)] = -\sin\,(90^\circ - \theta) = -\cos\,\theta$$

$$\cos\,(270^\circ - \theta) = \cos\,[180^\circ + (90^\circ - \theta)] = -\cos\,(90^\circ - \theta) = -\sin\,\theta$$

$$\tan\,(270^\circ - \theta) = \tan\,[180^\circ + (90^\circ - \theta)] = \tan\,(90^\circ - \theta) = \cot\,\theta$$

$$\text{cosec}\,(270^\circ - \theta) = \text{cosec}\,[180^\circ + (90^\circ - \theta)] = -\text{cosec}\,(90^\circ - \theta) = -\sec\,\theta$$

$$\sec\,(270^\circ - \theta) = \sec\,[180^\circ + (90^\circ - \theta)] = -\sec\,(90^\circ - \theta) = -\text{cosec}\,\theta$$

$$\cot\,(270^\circ - \theta) = \cot\,[180^\circ + (90^\circ - \theta)] = \cot\,(90^\circ - \theta) = \tan\,\theta$$

(vii) Trigonometric Functions of (270° + θ):

In the above manner we can show that:

$$\sin\,(270^\circ + \theta) = -\cos\,\theta$$

$$\cos\,(270^\circ + \theta) = \sin\,\theta$$

$$\tan\,(270^\circ + \theta) = -\cot\,\theta$$

$$\text{cosec}\,(270^\circ + \theta) = -\sec\,\theta$$

$$\sec\,(270^\circ + \theta) = \text{cosec}\,\theta$$

$$\cot\,(270^\circ + \theta) = -\tan\,\theta$$

(viii) Trigonometric Functions of (360° – θ):

$$\sin(360° - \theta) = \sin[180° + (180° - \theta)] = -\sin(180° - \theta) = -\sin\theta$$

$$\cos(360° - \theta) = \cos\theta$$

$$\tan(360° - \theta) = -\tan\theta$$

$$\operatorname{cosec}(360° - \theta) = -\operatorname{cosec}\theta$$

$$\sec(360° - \theta) = \sec\theta$$

$$\cot(360° - \theta) = -\cot\theta$$

(ix) Trigonometric Functions of (360° + θ)

$$\sin(360° + \theta) = \sin[180° + (180° + \theta)]$$

$$= -\sin(180° - \theta) = -(-\sin\theta) = \sin\theta$$ Similarly,

$$\cos(360° + \theta) = \cos\theta$$

$$\tan(360° + \theta) = \tan\theta$$

$$\operatorname{cosec}(360° + \theta) = \operatorname{cosec}\theta$$

$$\sec(360° + \theta) = \sec\theta$$

$$\cot(360° + \theta) = \cot\theta$$

(x) Trigonometric Functions of (*n*. 360° ± θ), *n* being any integer

Since increasing or decreasing an angle by an identical multiple of 360° amounts to complete revolutions of the revolving line, therefore trigonometric functions of (n. 360° ± θ) are the same as that of (± θ).

Thus, $\sin(n \,.\, 360° + \theta) = \sin\theta,\ \sin(n \,.\, 360° - \theta) = -\sin\theta$

$$\cos(n \,.\, 360° + \theta) = \cos\theta,\ \cos(n \,.\, 360° - \theta) = \cos\theta$$

$$\tan(n \,.\, 360° + \theta) = \tan\theta,\ \tan(n \,.\, 360° - \theta) = -\tan\theta$$

$$\operatorname{cosec}(n \,.\, 360° + \theta) = \operatorname{cosec}\theta,\ \operatorname{cosec}(n \,.\, 360° - \theta) = -\operatorname{cosec}\theta$$

$$\sec(n \,.\, 360° + \theta) = \sec\theta,\ \sec(n \,.\, 360° - \theta) = \sec\theta$$

$$\cot(n \,.\, 360° + \theta) = \cot\theta,\ \cot(n \,.\, 360° - \theta) = -\cot\theta$$

Summary Table

	– θ	90° – θ	90° + θ	180° – θ	180° + θ	270° – θ	270° + θ	360° – θ	360° + θ
sin	– sin θ	cos θ	cos θ	sin θ	– sin θ	– cos θ	– cos θ	– sin θ	sin θ
cos	cos θ	sin θ	– sin θ	– cos θ	– cos θ	– sin θ	sin θ	cos θ	cos θ
tan	– tan θ	cot θ	– cot θ	– tan θ	tan θ	cot θ	– cot θ	– tan θ	tan θ
cosec	– cosec θ	sec θ	– sec θ	cosec θ	– cosec θ	– sec θ	– sec θ	– cosec θ	cosec θ
sec	sec θ	cosec θ	– cosec θ	– cosec θ	– sec θ	– cosec θ	cosec θ	sec θ	sec θ
cot	– cot θ	tan θ	– tan θ	– cot θ	cot θ	tan θ	– tan θ	– cot θ	cot θ

ILLUSTRATIVE EXAMPLES

Example 1. Evaluate

(i) cot 1140° (ii) sec (– 300°) (iii) sin (930°)

(iv) cos 1230° (v) cosec $\left(\frac{35\pi}{6}\right)$

Solution:(i) $\cot 1140° = \cot(3 \times 360° + 60°) = \cot 60° = \frac{1}{\sqrt{3}}$

$$(ii) \quad \sec(-300°) = \sec 300° = \sec(180° + 120°)$$
$$= -\sec 120° = -\sec(90° + 30°)$$
$$= -(-\operatorname{cosec} 30°) = \operatorname{cosec} 30°$$
$$= 2$$

$$(iii) \quad \sin(930°) = \sin(2 \times 360° + 210°) = \sin 210°$$
$$= \sin(270° - 60°) = -\cos 60°$$
$$= -\frac{1}{2}$$

$$(iv) \quad \cos 1230° = \cos(3 \times 360° + 150°) = \cos 150°$$
$$= \cos(90° + 60°) = -\sin 60° = \frac{-\sqrt{3}}{2}$$

$$(v) \quad \operatorname{cosec}\left(\frac{35\pi}{6}\right) = \operatorname{cosec}\left(\frac{35 \times 180}{6}\right) = \operatorname{cosec} 1050°$$
$$= (\operatorname{cosec}(3 \times 360° - 30°)$$
$$= -\operatorname{cosec} 30° = -2$$

Example 2. If $\sin\theta = \frac{1}{\sqrt{2}}$ and θ lies in the second quadrant, find all other trigonometric functions.

Solution: In the second quadrant, $\cos\theta$ is negative

$$\therefore \quad \cos\theta = -\sqrt{1-\sin^2\theta} = -\sqrt{1-\frac{1}{2}} = \frac{-1}{\sqrt{2}}$$

So,
$$\tan\theta = \frac{\sin\theta}{\cos\theta} = \frac{\frac{1}{\sqrt{2}}}{\frac{-1}{\sqrt{2}}} = -1$$

$$\cot\theta = \frac{1}{\tan\theta} = \frac{1}{-1} = -1$$

$$\sec\theta = \frac{1}{\cos\theta} = -\sqrt{2}$$

$$\operatorname{cosec}\theta = \frac{1}{\sin\theta} = \sqrt{2}$$

Example 3. Evaluate

(i) $\frac{4}{3}\cot^2 30^\circ + 3\sin^2 60^\circ - 2\,\text{cosec}^2 45^\circ + \frac{1}{3}\tan^2 60^\circ - \frac{1}{4}\cos 0^\circ$

(ii) $\sqrt{2}\sin 135^\circ \cos 210^\circ \tan 240^\circ \cot 300^\circ \sec 330^\circ$

(iii) $\tan 36^\circ \tan 54^\circ$

Solution: (i) $\frac{4}{3}\cot^2 30^\circ + 3\sin^2 60^\circ - 2\,\text{cosec}^2 45^\circ + \frac{1}{3}\tan^2 60^\circ - \frac{1}{4}\cos 0^\circ$

$$= \frac{4}{3}.(\sqrt{3})^2 + 3.\left(\frac{\sqrt{3}}{2}\right)^2 - 2.(\sqrt{2})^2 + \frac{1}{3}.(\sqrt{3})^2 - \frac{1}{4}.1$$

$$= 4 + \frac{9}{4} - 4 + 1 - \frac{1}{4} = 3$$

(ii) $\sqrt{2}\sin 135^\circ \cos 210^\circ \tan 240^\circ \cot 300^\circ \sec 330^\circ$

$= \sqrt{2}\sin(90^\circ + 45^\circ).\cos(180^\circ + 30^\circ).\tan(180^\circ + 60^\circ).\cot(360^\circ - 60^\circ).$
$\sec(360 - 30^\circ) = \sqrt{2}(\cos 45^\circ)(-\cos 30^\circ)(\tan 60^\circ)(-\cot 60^\circ)(\sec 30^\circ)$

$$= \sqrt{2}\cdot\frac{1}{\sqrt{2}}\cdot\left(\frac{-\sqrt{3}}{2}\right)\cdot\sqrt{3}\cdot\left(\frac{-1}{\sqrt{3}}\right)\cdot\frac{2}{\sqrt{3}} = 1$$

(iii) $\tan 36^\circ \tan 54^\circ = \tan(90^\circ - 54^\circ).\tan 54^\circ$

$= \cot 54^\circ, \ \tan 54^\circ$

$= \frac{1}{\tan 54^\circ}, \tan 54^\circ = 1$

Example 4. Find all the values of θ between 0° and 360° for which $\sin\theta = \frac{1}{2}$

Solution: sin θ is positive in I and II quadrants

∴ $\sin\theta = \frac{1}{2} = \sin 30^\circ \Rightarrow \theta = 30^\circ$

Also, $\sin 30^\circ = \sin(180 - 30^\circ) = \sin 150^\circ \Rightarrow \theta = 150^\circ$

$[\because \sin(180 - \theta) = \sin\theta]$

∴ $\theta = 30^\circ, 150^\circ$

Example 5. For what values of θ, sec θ and cosec θ are equal?

Solution: $\sec\theta = \text{cosec}\,\theta \Rightarrow \frac{1}{\cos\theta} = \frac{1}{\sin\theta} \Rightarrow \tan\theta = 1$

$\Rightarrow \tan\theta = \tan 45^\circ$

$\Rightarrow \theta = 45^\circ$

Example 6. Prove that

(*i*) $$\frac{\cos(90^\circ + \theta)\sec(-\theta)\tan(180^\circ - \theta)}{\sec(360^\circ - \theta)\sin(180^\circ + \theta)\cot(90^\circ + \theta)} = 1$$

(*ii*) $\sin^2 \frac{\pi}{8} + \sin^2 \frac{3\pi}{8} + \sin^2 \frac{5\pi}{8} + \sin^2 \frac{7\pi}{8} = 2$

Solution: (i) L.H.S. $= \dfrac{\cos(90^\circ + \theta)\sec(-\theta)\tan(180^\circ - \theta)}{\sec(360^\circ - \theta)\sin(180^\circ + \theta)\cot(90^\circ + \theta)}$

$$= \frac{(-\sin\theta).\sec\theta.(-\tan\theta)}{\sec\theta.(-\sin\theta).(-\tan\theta)} = 1 = \text{R.H.S}$$

(*ii*) L.H.S $= \sin^2 \frac{\pi}{8} + \sin^2 \frac{3\pi}{8} + \sin^2 \frac{5\pi}{8} + \sin^2 \frac{7\pi}{8}$

$$= \sin^2 \frac{\pi}{8} + \sin^2 \frac{3\pi}{8} + \left(\sin \frac{4\pi + \pi}{8}\right)^2 + \left(\sin \frac{4\pi + 3\pi}{8}\right)^2$$

$$= \sin^2 \frac{\pi}{8} + \sin^2 \frac{3\pi}{8} + \left[\sin\left(\frac{\pi}{2} + \frac{\pi}{8}\right)\right]^2 + \left[\left(\sin \frac{\pi}{2} + \frac{3\pi}{8}\right)\right]^2$$

$$= \sin^2 \frac{\pi}{8} + \sin^2 \frac{3\pi}{8} + \left(\cos \frac{\pi}{8}\right)^2 + \left(\cos \frac{3\pi}{8}\right)^2$$

$$= \left(\sin^2 \frac{\pi}{8} + \cos^2 \frac{\pi}{8}\right) + \left(\sin^2 \frac{3\pi}{8} + \cos^2 \frac{3\pi}{8}\right)$$

$= 1 + 1 = 2$ R.H.S

Example 7. Solve for θ when $0^\circ \le \theta < 360^\circ$:

(i) $\sin\theta - \cos\theta = 0$

(ii) $\cot\theta + \tan\theta = 2\,\text{cosec}\,\theta$

Solution. (i) $\sin\theta - \cos\theta = 0 \Rightarrow \sin\theta = \cos\theta$

$\Rightarrow \quad \tan\theta = 1 = \tan 45^\circ$

$\therefore \quad \theta = 45^\circ$

Again $\quad \tan\theta = \tan 45^\circ = \tan(180^\circ + 45^\circ)$

$= \tan 225^\circ$

$\therefore \quad \theta = 225^\circ$

Hence $\quad \theta = 45^\circ, 225^\circ$

(ii) $\cot\theta + \tan\theta = 2\,\text{cosec}\,\theta$

$\Rightarrow \quad \dfrac{1}{\tan\theta} + \tan\theta = 2\,\text{cosec}\,\theta$

$\Rightarrow \quad 1 + \tan^2\theta = 2\,\text{cosec}\,\theta\,.\,\tan\theta$

$\Rightarrow \quad \sec^2\theta = 2\sec\theta$

$\Rightarrow \quad \sec\theta(\sec\theta - 2) = 0 \Rightarrow \sec\theta = 0$ or $\sec\theta = 2$

$\therefore \quad \sec\theta = 0 = \sec 0^\circ \Rightarrow \theta = 0^\circ$

or $\quad \sec\theta = 2 = \sec 60^\circ \Rightarrow \theta = 60^\circ$

Also, $\sec\theta = \sec 60^\circ = \sec(360^\circ - 60^\circ) = \sec 300^\circ$

$\Rightarrow \quad \theta = \theta = 300^\circ$

Hence $\quad \theta = 0^\circ, 60^\circ, 300^\circ$

Example 8. If A, B, C, are angles of a triangle, prove that

(*i*) $\tan\frac{A}{2} = \cot\left(\frac{B+C}{2}\right)$

(*ii*) $\frac{\text{cosec}\,(B+C) + \text{cosec}\,(C+A) + \text{cosec}\,(A+B)}{\sec(270^\circ + A) + \sec(270^\circ + B) + \sec(270^\circ + C)} = 1$

Solution: (i) In a triangle ABC

$$A + B + C = \pi$$

$$\Rightarrow \quad B + C = \pi - A$$

$$\Rightarrow \quad \frac{B+C}{2} = \frac{\pi}{2} - \frac{A}{2}$$

$$\therefore \quad \text{R.H.S} = \cot\left(\frac{B+C}{2}\right) = \cot\left(\frac{\pi}{2} - \frac{A}{2}\right) = \cot\left(90^\circ - \frac{A}{2}\right)$$

$$= \tan\frac{A}{2} = \text{LHS}$$

(*ii*) $\therefore A + B + C = \pi$

$$\text{L.H.S} = \frac{\text{cosec}\,(B+C) + \text{cosec}\,(C+A) + \text{cosec}\,(A+B)}{\sec(270^\circ + A) + \sec(270^\circ + B) + \sec(270^\circ + C)}$$

$$= \frac{\text{cosec}\,(\pi - A) + \text{cosec}\,(\pi - B) + \text{cosec}\,(\pi - c)}{\text{cosec}\,A + \text{cosec}\,B + \text{cosec}\,C}$$

$[\because \sec(270^\circ + \theta) = \text{cosec}\,\theta]$

$$= \frac{\text{cosec}\,A + \text{cosec}\,B + \text{cosec}\,C}{\text{cosec}\,A + \text{cosec}\,B + \text{cosec}\,C} \quad [\because \text{cosec}\,(\pi - \theta) = \text{cosec}\,\theta]$$

$= 1 = \text{R.H.S.}$

Example 9. In a cyclic quadrilateral $ABCD$, show that

$\cot A + \cot B + \cot C + \cot D = 0$

Solution: Since $ABCD$ is a cyclic quadrilateral

$$\therefore \quad A + C = \pi,\ B + D = \pi$$

$$\therefore \cot A + \cot B + \cot C + \cot D = \cot A + \cot B + \cot(\pi - A) + \cot(\pi - B)$$

$$= \cot A + \cot B - \cot A - \cot B$$

$$= 0$$

EXERCISE 1.4

1. Find the values of

(*i*) $\cos(-870^\circ)$ (*ii*) $\tan(-2025^\circ)$ (*iii*) $\sin 1845^\circ$

(*iv*) $\text{cosec}\,(-1200^\circ)$ (*v*) $\sec(-150^\circ)$ (*vi*) $\cot(-1590^\circ)$

2. Find the values of θ between 0° and 360°, when

(*i*) $\sec\theta = -2$ (*ii*) $\tan\theta = -1$

3. Prove that: $\tan 720^\circ - \cos 270^\circ - \sin 150^\circ \,.\, \cos 120^\circ = \frac{1}{4}$

4. Prove that: $\dfrac{\sin 300^\circ \tan 330^\circ \sec 420^\circ}{\cot 135^\circ \cos 210^\circ \operatorname{cosec} 315^\circ} = -\sqrt{\dfrac{2}{3}}$

5. Prove that: $\tan 225^\circ \cot 405^\circ + \tan 765^\circ \cot 675^\circ = 0$

6. If $\cos\theta = \frac{5}{13}$ and θ lies in the fourth quadrant, find the value of $\dfrac{2 - 3\cot\theta}{4 + 9\sqrt{\sec^2\theta - 1}}$

7. If A, B, C, be the angles of a triangle, prove that

(i) $\dfrac{\sin(B+C) + \sin(C+A) + \sin(A+B)}{\sin(\pi + A) + \sin(3\pi + B) + \sin(5\pi + C)} = 1$

(ii) $\cos(A + B) + \cos C = 0$

8. In a cyclic quadrilateral *ABCD* prove that.

$\cos A + \cos B + \cos C + \cos D = 0$

9. If $\tan 25^\circ = a$, prove that $\dfrac{\tan 155^\circ - \tan 115^\circ}{1 + \tan 155^\circ \cdot \tan 115^\circ} = \dfrac{1 - a^2}{2a}$

10. Solve for θ when $0^\circ \le \theta < 360^\circ$;

$\cos^2\theta - \sin\theta = \frac{1}{4}$

11. Prove that $\sin^2(-300^\circ) \,.\, \cos^2(120^\circ) + \cos^2(-240^\circ) \,.\, \sin^2(390^\circ) = -\frac{1}{16}$

12. Find the simplest value of

(i) $\sin 140^\circ \,.\, \cos 250^\circ + \cos(-290^\circ) \,.\, \sin 400^\circ$

(ii) $\cos\dfrac{5\pi}{3} \,.\, \cos\dfrac{37\pi}{6} + \sin\dfrac{43\pi}{6} \,.\, \sin\dfrac{55\pi}{3}$

13 Prove that: $\cos^2\dfrac{\pi}{8} + \cos^2\dfrac{3\pi}{8} + \cos^2\dfrac{5\pi}{8} + \cos^2\dfrac{7\pi}{8} = 2$

14. Prove that: $\dfrac{\cos(90^\circ + \theta) \,.\, \sec(270^\circ + \theta) \cdot \sin(180^\circ + \theta)}{\operatorname{cosec}(-\theta)\cos(270^\circ - \theta)\tan(180 + \theta)} = \cos\theta$

ANSWERS

1. (*i*) $-\dfrac{\sqrt{3}}{2}$ (*ii*) -1 (*iii*) $\dfrac{1}{\sqrt{2}}$ (*iv*) $\dfrac{-2}{\sqrt{3}}$ (*v*) $\dfrac{-2}{\sqrt{3}}$ (*vi*) $\sqrt{3}$

2. (*i*) 120°, 240° (*ii*) 135°, 315°

6. $\dfrac{-65}{352}$

10. 30° , 150°

12. (*i*) 0 (*ii*) 0

1.9 COMPOUND ANGLES

The algebraic sum or difference of two or more angles is called compound angle.

If A, B, C, are three angles, then $A + B$, $B + C$, $C + A$, $A - B$, $B - C$, $A - C$, $A + B + C$ etc. are compound angles

Sum Formulae

For any angles A and B, we have

(*i*) $\text{Sin}\ (A + B) = \sin A \cos B + \cos A \sin B$

(*ii*) $\cos (A + B) = \cos A \cos B - \sin A \sin B$

(*iii*) $\tan (A + B) = \dfrac{\tan A + \tan B}{1 - \tan A \tan B}$

Theorem: For any angles A and B, we have

(*i*) $\sin (A + B) \sin (A - B) = \sin^2 A - \sin^2 B$

(*ii*) $\cos (A + B) \cos (A - B) = \cos^2 A - \sin^2 B$

COROLLARY: $\tan (A + B + C) = \dfrac{\tan A + \tan B + \tan C - \tan A \tan B \tan C}{1 - \tan A \tan B - \tan B \tan C - \tan C \tan A}$

ILLUSTRATIVE EXAMPLES

Example 1. Find the value of sin 15°, cos 15°, sin 75°, cos 75°

Solution:

(*i*) $\sin 15^\circ = \sin (45^\circ - 30^\circ)$

$$= \sin 45^\circ \cos 30^\circ - \cos 45^\circ \sin 30^\circ$$

$$= \left(\frac{1}{\sqrt{2}} \cdot \frac{\sqrt{3}}{2} - \frac{1}{\sqrt{2}} \cdot \frac{1}{2}\right) = \left(\frac{\sqrt{3} - 1}{2\sqrt{2}}\right)$$

(*ii*) $\cos 15^\circ = \cos (45^\circ - 30^\circ)$

$$= \cos 45^\circ \cos 30^\circ + \sin 45^\circ \sin 30^\circ$$

$$= \left(\frac{1}{\sqrt{2}} \cdot \frac{\sqrt{3}}{2} + \frac{1}{\sqrt{2}} \cdot \frac{1}{2}\right) = \left(\frac{\sqrt{3} + 1}{2\sqrt{2}}\right)$$

(*iii*) $\sin 75^\circ = \sin (45^\circ + 30^\circ)$

$$= \sin 45^\circ \cos 30^\circ + \cos 45^\circ \sin 30^\circ$$

$$= \left(\frac{1}{\sqrt{2}} \cdot \frac{\sqrt{3}}{2} + \frac{1}{\sqrt{2}} \cdot \frac{1}{2}\right) = \left(\frac{\sqrt{3} + 1}{2\sqrt{2}}\right)$$

(*iv*) $\cos 75^\circ = \cos (45^\circ + 30^\circ)$

$$= \cos 45^\circ \cos 30^\circ - \sin 45^\circ \sin 30^\circ$$

$$= \left(\frac{1}{\sqrt{2}} \cdot \frac{\sqrt{3}}{2} - \frac{1}{\sqrt{2}} \cdot \frac{1}{2}\right) = \left(\frac{\sqrt{3} - 1}{2\sqrt{2}}\right)$$

Alternate method:

$$\sin 75^\circ = \sin (90^\circ - 15^\circ) = \cos 15^\circ = \left(\frac{\sqrt{3}+1}{2\sqrt{2}}\right)$$

and
$$\cos 75^\circ = \cos (90^\circ - 15^\circ) = \sin 15^\circ = \left(\frac{\sqrt{3}-1}{2\sqrt{2}}\right)$$

Example 2. If $\tan A = \frac{5}{6}$ and $\tan B = \frac{1}{11}$, prove that $A + B = 45^\circ$

Solution: $\tan (A + B) = \frac{\tan A + \tan B}{1 - \tan A \tan B} = \frac{\frac{5}{6}+\frac{1}{11}}{1-\frac{5}{6}.\frac{1}{11}} = 1$

$$= \tan 45^\circ$$

$\therefore$ $A + B = 45^\circ$

Example 3. Find the value of $\cos^2 45^\circ - \sin {}^2 45^\circ$

Solution: $\cos^2 45^\circ - \sin^2 15^\circ = \cos (45^\circ + 15^\circ) \cos (45^\circ - 15^\circ)$

$$= \cos 60^\circ \,.\, \cos 30^\circ$$

$$= \frac{1}{2}.\frac{\sqrt{3}}{2} = \frac{\sqrt{3}}{4}$$

Example 4. Prove that: $\frac{\cos 11^\circ + \sin 11^\circ}{\cos 11^\circ - \sin 11^\circ} = \tan 56^\circ$

Solution: L.H.S $= \frac{\cos 11^\circ + \sin 11^\circ}{\cos 11^\circ - \sin 11^\circ} = \frac{1 + \tan 11^\circ}{1 - \tan 11^\circ}$

$$= \frac{\tan 45^\circ + \tan 11^\circ}{1 - \tan 45^\circ + \tan 11^\circ} = \tan (45^\circ + 11^\circ)$$

$$= \tan 56^\circ = \text{R.H.S}$$

Example 5. Find the value of $\tan (\alpha + \beta)$, given that $\cot \alpha = \frac{1}{2}$, $\sec \beta = \frac{-5}{3}$, where $\pi < \alpha < \frac{3\pi}{2}$ and $\frac{\pi}{2} < \beta < \pi$

Solution: $\sin \beta$ lies in II quadrant, $\tan \beta$ is negative

$\therefore$ $\tan \beta = -\sqrt{\sec^2 \beta - 1} = -\sqrt{\frac{25}{9} - 1} = \frac{-4}{3}$

Also $\cot \alpha = \frac{1}{2} \Rightarrow \tan \alpha = 2$

$\therefore$ $\tan (\alpha + \beta) = \frac{\tan \alpha + \tan \beta}{1 - \tan \alpha \tan \beta} = \frac{2 - \frac{4}{3}}{1 - 2 \times \left(\frac{-4}{3}\right)} = \frac{2}{11}$

Example 6. If $\cos\theta = \frac{1}{7}$ and $\cos\phi = \frac{13}{14}$, prove that $\theta - \phi = 60$

Solution: $\sin\theta = \sqrt{1-\cos^2\theta} = \sqrt{1-\frac{1}{49}} = \frac{\sqrt{48}}{7}$

and $\sin\phi = \sqrt{1-\cos^2\phi} = \sqrt{1-\frac{169}{196}} = \frac{\sqrt{27}}{14}$

$\therefore$ $\cos(\theta-\phi) = \cos\theta\cos\phi + \sin\theta\sin\phi$

$$= \frac{1}{7}\cdot\frac{13}{14} + \frac{\sqrt{48}}{7}\cdot\frac{\sqrt{27}}{14}$$

$$= \frac{13}{98} + \frac{36}{98} = \frac{49}{98}$$

$$= \frac{1}{2} = \cos 60^\circ$$

$\Rightarrow$ $\theta - \phi = 60^\circ$

Example 7. If $A + B + C = \pi$ and $\cos A = \cos B.\cos C$, show that $\tan A = \tan B + \tan C$

Solution:

$$\text{R.H.S} = \tan B + \tan C = \frac{\sin B}{\cos B} + \frac{\sin C}{\cos C}$$

$$= \frac{\sin B\cos C + \sin C\cos B}{\cos B\cos C} = \frac{\sin(B+C)}{\cos A} \quad [\because \cos B\cos C = \cos A]$$

$$= \frac{\sin(\pi - A)}{\cos A} \quad [\because A + B + C = \pi]$$

$$= \frac{\sin A}{\cos A} = \tan A = \text{L.H.S}$$

Example 8. Find the value of tan 15° and tan 75°

Solution: (i) $\tan 15^\circ = \tan(45^\circ - 30^\circ) = \frac{\tan 45^\circ - \tan 30^\circ}{1 + \tan 45^\circ + \tan 30^\circ}$

$$= \frac{1-\frac{1}{\sqrt{3}}}{1+1\frac{1}{\sqrt{3}}} = \frac{\sqrt{3}-1}{\sqrt{3}+1} \times \frac{\sqrt{3}-1}{\sqrt{3}-1}$$

$$= \frac{3+1-2\sqrt{3}}{3-1} = \frac{4-2\sqrt{3}}{2} = 2-\sqrt{3}$$

(*ii*) $\tan 75^\circ = \tan(45^\circ + 30^\circ) = \frac{\tan 45^\circ + \tan 30^\circ}{1 - \tan 45^\circ + \tan 30^\circ}$

$$= \frac{1+\frac{1}{\sqrt{3}}}{1-1.\frac{1}{\sqrt{3}}} = \frac{\sqrt{3}+1}{\sqrt{3}-1} = \frac{\sqrt{3}+1}{\sqrt{3}-1} \times \frac{\sqrt{3}+1}{\sqrt{3}+1}$$

$$= \frac{3+1+2\sqrt{3}}{3-1}$$

$$= 2+\sqrt{3}$$

Example 9. Show that $\tan x + \tan 2x - \tan 3x = -\tan x \,.\, \tan 2x \,.\, \tan 3x$

Solution: we have $\tan 3x = \tan (x + 2x)$

$$\Rightarrow \quad \tan 3x = \frac{\tan x + \tan 2x}{1 - \tan x \tan 2x}$$

$$\Rightarrow \quad \tan 3x - \tan x \tan 2x \tan 3x = \tan x + \tan 2x$$

$$\Rightarrow \quad \tan x + \tan 2x - \tan 3x = -\tan x \tan 2x \tan 3x$$

Example 10. If $\tan x - \tan y = a$ and $\cot y - \cot x = b$, prove that $\frac{1}{a}+\frac{1}{b} = \cot (x - y)$

Solution: we have $\frac{1}{a}+\frac{1}{b} = \frac{1}{\tan x - \tan y} + \frac{1}{\cot y - \cot x}$

$$= \frac{\cos x \cos y}{\sin x \cos y - \cos x \sin y} + \frac{\sin x \sin y}{\sin x \cos y - \cos x \sin y}$$

$$= \frac{\cos x \cos y + \sin x \sin y}{\sin x \cos y - \cos x \sin y}$$

$$= \frac{\cos (x-y)}{\sin (x-y)} = \cot (x - y)$$

Example 11. If $\tan A = m \tan B$, prove that

$$\frac{\sin (A-B)}{\sin (A+B)} = \left(\frac{m-1}{m+1}\right)$$

Solution: Given, $\tan A = m \tan B$

$$\Rightarrow \quad \frac{\tan A}{\tan B} = m$$

$$\Rightarrow \quad \frac{\sin A \cos B}{\sin B \cos A} = \frac{m}{1}$$

Applying Componendo 0 and dividendo rule, we get

$$\frac{\sin A \cos B - \sin B \cos A}{\sin A \cos B + \cos A \sin B} = \frac{m-1}{m+1}$$

$$\Rightarrow \qquad \frac{\sin(A-B)}{\sin(A+B)} = \frac{m-1}{m+1}$$

Example 12. Prove that: $\cos^2(A - 120^\circ) + \cos^2 A + \cos^2(A + 120^\circ) = \frac{3}{2}$

Solution: L.H.S = $[\cos A . \cos 120^\circ + \sin A . \sin 120^\circ]^2 + \cos^2 A + [\cos A . \cos 120^\circ - \sin A \sin 120^\circ]^2$

$$= \left(\frac{-1}{2}\cos A + \frac{\sqrt{3}}{2}\sin A\right)^2 + \cos^2 A + \left(\frac{-1}{2}\cos A - \frac{\sqrt{3}}{2}\sin A\right)^2$$

$$= \frac{1}{4}\cos^2 A + \frac{3}{4}\sin^2 A - \frac{\sqrt{3}}{2}\sin A \cos A + \cos^2 A$$

$$+ \frac{1}{4}\cos^2 A + \frac{1}{4}\sin^2 A + \frac{\sqrt{3}}{2}\sin A . \cos A$$

$$= \frac{1}{4}\cos^2 A + \frac{3}{4}\sin^2 A + \cos^2 A + \frac{1}{4}\cos^2 A + \frac{3}{4}\sin^2 A$$

$$= \frac{3}{2}\cos^2 A + \frac{3}{2}\sin^2 A$$

$$= \frac{3}{2}(\cos^2 A + \sin^2 A) = \frac{3}{2} . 1 = \frac{3}{2} \text{ R.H.S}$$

EXERCISE 1.5

1. Find the value of sin 105°, cos 105°, tan 135°
2. Evaluate:
   (*i*) sin 76° cos 16° – cos 76° sin 16°
   (*ii*) sin 36° cos9° + cos 36° sin 9°
   (*iii*) cos 130° cos 40° + sin 130° sin 40°
3. Find the value of sin (α + β), cos (α + β), tan (α + β), it $\cos\alpha = \frac{-12}{13}$ and $\cot\beta = \frac{24}{7}$, where α lies in second quadrant and β in third quadrant.
4 Prove that: $\sin\theta = \frac{1}{\sqrt{5}}$ and $\sin\phi = \frac{1}{\sqrt{10}} \Rightarrow \theta + \phi = 45^\circ$

Prove that:

5. tan 70° = tan 20° + 2 tan 50°
6. sin (60° – *A*) cos (30° + *A*) + cos (60° – *A*) sin (30° + *A*) = 1

7 $\frac{\cos 8^\circ - \sin 8^\circ}{\cos 8^\circ + \sin 8^\circ} = \tan 37^\circ$

8. $\dfrac{\sin B}{\sin A} = \dfrac{\sin(2A+B)}{\sin A} - 2\cos(A+B)$

9. $\tan 20^\circ + \tan 25^\circ + \tan 20^\circ \tan 25^\circ = 1$

10. $\tan A + \tan B = 1 - \tan A \tan B$ if $A + B + C = \pi$

11. $\sin^2\left(\dfrac{\pi}{12}+\dfrac{\theta}{2}\right) - \sin^2\left(\dfrac{\pi}{12}-\dfrac{\theta}{2}\right) = \dfrac{1}{2}\sin\theta$

12. $\dfrac{1+\tan A}{1+\tan B} = 2$ if $A + B = \dfrac{\pi}{2}$

13. $\dfrac{\tan\left(\dfrac{\pi}{4}+x\right)}{\tan\left(\dfrac{\pi}{4}-x\right)} = \left(\dfrac{1+\tan x}{1-\tan x}\right)^2$

14. $\cos A \cos(A-B) + \sin A \sin(A-B) = \cos B$

15. $\tan 13\theta - \tan 9\theta - \tan 4\theta = \tan 4\theta \tan 9\theta \tan 13\theta$

16. $\sin(n+1)A . \sin(n-1)A + \cos(n+1)A . \cos(n-1)A = \cos 2A$

17. $\sin(\theta+\phi) = \dfrac{1}{2}(m^2+n^2-2)$ if $\cos\theta + \sin\phi = m$ and $\sin\theta + \cos\phi = n$

ANSWERS

1. $\dfrac{\sqrt{3}+1}{2\sqrt{2}}, \dfrac{1-\sqrt{3}}{2\sqrt{2}}, \dfrac{\sqrt{3}+1}{\sqrt{3}-1}$

2. (*i*) $\dfrac{\sqrt{3}}{2}$ (*ii*) $\dfrac{1}{\sqrt{2}}$ (*iii*) 0

3. $\dfrac{-36}{325}, \dfrac{323}{325}, \dfrac{-36}{323}$

1.10 TRANSFORMATION OF PRODUCTS INTO SUM OR DIFFERENCE

I $\quad 2\sin A\cos B = \sin(A+B) + \sin(A-B)$

II $\quad 2\cos A\sin B = \sin(A+B) - \sin(A-B)$

III $\quad 2\cos A\cos B = \cos(A+B) + \cos(A-B)$

IV $\quad 2\sin A\sin B = \cos(A-B) - \cos(A+B)$

Proof: We know that

$$\sin(A+B) = \sin A\cos B + \cos A\sin B \quad ...(i)$$

$$\sin(A-B) = \sin A\cos B - \cos A\sin B \quad ...(ii)$$

$$\cos(A+B) = \cos A\cos B - \sin A\sin B \quad ...(iii)$$

$$\cos(A-B) = \cos A\cos B + \sin A\sin B \quad ...(iv)$$

Adding (*i*) and (*ii*), we get I

subtracting (*ii*) from (*i*) we get II

Adding (*iii*) and (*iv*), we get III

Subtracting (*iii*) from (*iv*) get IV

1.11 TERAUSFORMATION OF SUM OF DIFFERENEC INTO PRODUCTS

I. $\sin C + \sin D = 2 \sin \frac{C+D}{2}.\cos\frac{C-D}{2}$

II. $\sin C - \sin D = 2 \cos \frac{C+D}{2}.\sin\frac{C-D}{2}$

III. $\cos C + \cos D = 2 \cos \frac{C+D}{2}.\cos\frac{C-D}{2}$

IV. $\cos C - \cos D = 2 \sin \frac{C+D}{2}.\sin\frac{D-C}{2}$

ILLUSTRATIVE EXAMPLES

Example 1. Prove that: (i) $\sin 105^\circ + \cos 105^\circ = \cos 45^\circ$

(ii) $2 \sin 45^\circ \cos 15^\circ = \frac{\sqrt{3}}{2} + \frac{1}{2}$

Solution: (i) L.H.S $= \sin 105^\circ + \cos 105^\circ$

$= \sin (90^\circ + 15^\circ) + \cos 105^\circ$ [$\because \sin (90^\circ + \theta) = \cos \theta$]

$= 2 \cos 60^\circ, \cos (-45^\circ)$

$= 2 \cos 60^\circ \cos 45^\circ$

$= 2. \frac{1}{2}; \cos 45^\circ$

$= \cos 45^\circ =$ R.H.S

(ii) LHS $= \sin (45^\circ + 15^\circ) + \sin (45^\circ - 15^\circ)$

$= \sin 60^\circ + \sin 30^\circ$

$= \frac{\sqrt{3}}{2} + \frac{1}{2} =$ R.H.S

Example 2. Prove that: $\frac{\sin A + \sin 3A}{\cos 3A + \cos A} = \tan 2A$

Solution: LHS $= \frac{\sin A + \sin 3A}{\cos 3A + \cos A} = \frac{2 \sin\left(\frac{A+3A}{2}\right)\cos\left(\frac{A-3A}{2}\right)}{2 \cos\left(\frac{3A+A}{2}\right)\cos\left(\frac{3A-A}{2}\right)}$

$= \frac{2 \sin 2A \cos (-A)}{2 \cos 2A \cos A} = \frac{\sin 2A \,.\, \cos A}{\cos 2A \,.\, \cos A} = \frac{\sin 2A}{\cos 2A}$

$= \tan 2A =$ RHS

Example 3. Prove that:

(*i*) $\cos 20^\circ . \cos 40^\circ . \cos 80^\circ = \frac{1}{8}$

(*ii*) $\cos 20^\circ . \cos 40^\circ . \cos 60^\circ . \cos 80^\circ = \frac{1}{16}$

(*iii*) $\sin \alpha + \sin\left(\alpha + \frac{2\pi}{3}\right) + \sin\left(\alpha + \frac{4\pi}{3}\right) = 0$

Solution:

(*i*) $\text{L.H.S} = \frac{1}{2}(2 \cos 40^\circ . \cos 20^\circ) . \cos 80^\circ$

$$= \frac{1}{2}\{\cos (40^\circ + 20^\circ) + \cos (40^\circ - 20^\circ)\} \cos 80^\circ$$

$$= \frac{1}{2}(\cos 60^\circ + \cos 20^\circ) . \cos 80^\circ$$

$$= \frac{1}{2}\left(\frac{1}{2} + \cos 20^\circ\right)\cos 80^\circ = \frac{1}{4}\cos 80^\circ + \frac{1}{2}\cos 20^\circ \cos 80^\circ$$

$$= \frac{1}{4}\cos 80^\circ + \frac{1}{4}\cdot 2 \cos 20^\circ \cos 80^\circ$$

$$= \frac{1}{4}\cos 80^\circ + \frac{1}{4}[\cos (20^\circ + 80^\circ) + \cos (20^\circ - 80^\circ)]$$

$$= \frac{1}{4}\cos (180^\circ - 100^\circ) + \frac{1}{4}[\cos 100^\circ + \cos 60^\circ]$$

$$= -\frac{1}{4}\cos 100^\circ + \frac{1}{4}\cos 100^\circ + \frac{1}{8} = \frac{1}{8}$$

(*ii*) $\text{L.H.S} = \frac{1}{2}.(2 \cos 20^\circ \cos 40^\circ).\frac{1}{2}.\cos 80^\circ$

$$= \frac{1}{2}\{\cos (20^\circ + 40^\circ) + \cos (20^\circ - 40^\circ)\}.\frac{1}{2}.\cos 80^\circ$$

$$= \frac{1}{4}(\cos 60^\circ + \cos 20^\circ).\cos 80^\circ = \frac{1}{4}\left(\frac{1}{2} + \cos 20^\circ\right)\cdot \cos 80^\circ$$

$$= \frac{1}{8}\cos 80^\circ + \frac{1}{8}(2 \cos 20^\circ \cos 80^\circ)$$

$$= \frac{1}{8}\cos 80^\circ + \frac{1}{8}(\cos 100^\circ + \cos 60^\circ)$$

$$= \frac{1}{8}\cos 80^\circ + \frac{1}{8}\cos 100^\circ + \frac{1}{8}.\frac{1}{2}$$

$$= \frac{1}{8}\cos 80^\circ - \frac{1}{8}\cos 80^\circ + \frac{1}{16}$$

[$\because \cos 100^\circ = \cos (180 - 80^\circ) = - \cos 80^\circ$)

$$= \frac{1}{16} = \text{RHS}$$

(iii) L.H.S. $= \sin\alpha + \left[\sin\left(\alpha + \frac{2\pi}{3}\right) + \sin\left(\alpha + \frac{4\pi}{3}\right)\right]$

$= \sin\alpha + \left[2\sin(\alpha + \pi) \cdot \frac{\pi}{3}\right]$

$= \sin\alpha + 2\sin(\alpha + \pi) \cdot \frac{1}{2} = \sin\alpha + \sin(\pi + \alpha)$

$= \sin\alpha - \sin\alpha = 0 =$ R.H.S

Example 4. Prove that: $\dfrac{\sin 6\theta \cdot \cos 3\theta - \sin 8\theta \cdot \cos\theta}{\sin 3\theta \sin 4\theta - \cos 2\theta \cos\theta}$ $\tan 2\theta$

Solution: L.H.S $= \dfrac{2\sin 6\theta \cdot \cos 3\theta - 2\sin 8\theta \cdot \cos\theta}{2\sin 3\theta \sin 4\theta - 2\cos 2\theta \cos\theta}$

$$= \frac{\sin(6\theta + 3\theta) + \sin(6\theta - 3\theta) - \{\sin(8\theta + \theta) + \sin(8\theta - \theta)\}}{\cos(3\theta - 4\theta) - \cos(4\theta + 3\theta) - \{\cos(2\theta + \theta) + \cos(2\theta - \theta)\}}$$

$$= \frac{\sin 9\theta + \sin 3\theta - \sin 9\theta - \sin 7\theta}{\cos\theta - \cos 7\theta - \cos 3\theta - \cos\theta}$$

$$= \frac{\sin 3\theta - \sin 7\theta}{-(\cos 7\theta + \cos 3\theta)} = \frac{\sin 7\theta - \sin 3\theta}{\cos 7\theta + \cos 3\theta}$$

$$= \frac{2\cos\dfrac{7\theta + 3\theta}{2} \sin\dfrac{7\theta - 3\theta}{2}}{2\cos\dfrac{7\theta + 3\theta}{2} \cdot \cos\dfrac{7\theta - 3\theta}{2}}$$

$$= \frac{2\cos 5\theta \cdot \sin 2\theta}{2\cos 5\theta \cdot \cos 2\theta}$$

$= \tan 2\theta$

$=$ R.H.S

Example 5. Prove that:

$$\frac{\sin\theta + \sin 3\theta + \sin 5\theta + \sin 7\theta}{\cos\theta + \cos 3\theta + \cos 5\theta + \cos 7\theta} = \tan 4\theta$$

Solution:

L.H.S $= \dfrac{(\sin 7\theta + \sin\theta) + (\sin 5\theta + \sin 3\theta)}{(\cos 7\theta + \cos\theta) + (\cos 5\theta + \cos 3\theta)}$

$$= \frac{2\sin\theta \cdot \cos 3\theta + 2\sin 4\theta \cdot \cos\theta}{2\cos 4\theta \cdot \cos 3\theta + 2\cos 4\theta \cdot \cos\theta}$$

$$= \frac{\sin 4\theta(\cos 3\theta + \cos\theta)}{\cos 4\theta(\cos 3\theta + \cos\theta)} = \tan 4\theta = \text{R.H.S}$$

Example 6. Prove that:

$$(\cos\alpha + \cos\beta)^2 + (\sin\alpha + \sin\beta)^2 = 4\cos^2\left(\frac{\alpha - \beta}{2}\right)$$

Solution: L.H.S. = $(\cos\alpha + \cos\beta)^2 + (\sin\alpha + \sin\beta)^2$

$$= \left[2\cos\left(\frac{\alpha+\beta}{2}\right)\cos\left(\frac{\alpha-\beta}{2}\right)\right]^2 + \left[2\sin\left(\frac{\alpha+\beta}{2}\right)\cos\left(\frac{\alpha-\beta}{2}\right)\right]^2$$

$$= 4\cos^2\left(\frac{\alpha+\beta}{2}\right)\cos^2\left(\frac{\alpha-\beta}{2}\right) + 4\sin^2\left(\frac{\alpha+\beta}{2}\right)\cos^2\left(\frac{\alpha-\beta}{2}\right)$$

$$= 4\cos^2\left(\frac{\alpha-\beta}{2}\right)\left[\cos^2\left(\frac{\alpha+\beta}{2}\right) + \sin^2\left(\frac{\alpha+\beta}{2}\right)\right]$$

$$= 4\cos^2\left(\frac{\alpha-\beta}{2}\right).1 \quad [\because \cos^2 x + \sin^2 x = 1]$$

$$= 4\cos^2\left(\frac{\alpha-\beta}{2}\right) = \text{R.H.S}$$

Example 7. If $\sin\alpha = 2\sin\beta$, prove that

$$\tan\left(\frac{\alpha-\beta}{2}\right) = \frac{1}{3}\tan\left(\frac{\alpha+\beta}{2}\right)$$

Solution: we have $\sin\alpha = 2\sin\beta \Rightarrow \frac{\sin\alpha}{\sin\beta} = \frac{2}{1}$. By componendo and dividendo rule, we have

$$\frac{\sin\alpha + \sin\beta}{\sin\alpha - \sin\beta} = \frac{2+1}{2-1}$$

$$\Rightarrow \frac{2\sin\left(\frac{\alpha+\beta}{2}\right)\cos\left(\frac{\alpha-\beta}{2}\right)}{2\cos\left(\frac{\alpha+\beta}{2}\right)\sin\left(\frac{\alpha-\beta}{2}\right)} = \frac{3}{1}$$

$$\Rightarrow \tan\left(\frac{\alpha+\beta}{2}\right).\cot\left(\frac{\alpha-\beta}{2}\right) = \frac{3}{1}$$

$$\Rightarrow \quad \tan\left(\frac{\alpha+\beta}{2}\right) = \frac{3}{1}.\tan\left(\frac{\alpha-\beta}{2}\right)$$

$$\Rightarrow \quad \tan\left(\frac{\alpha-\beta}{2}\right) = \frac{1}{3}.\tan\left(\frac{\alpha+\beta}{2}\right)$$

Example 8. Prove that

$$\sin\alpha + \sin\beta + \sin\gamma - \sin(\alpha+\beta+\gamma) = 4\sin\left(\frac{\alpha+\beta}{2}\right).\sin\left(\frac{\beta+\gamma}{2}\right).\sin\left(\frac{\gamma+\alpha}{2}\right)$$

Solution:

L.H.S $= (\sin\alpha + \sin\beta) - \{\sin(\alpha+\beta+\gamma) - \sin\gamma\}$

$$= 2.\sin\frac{\alpha+\beta}{2}.\cos\frac{\alpha-\beta}{2} - 2\cos\frac{\alpha+\beta+2\gamma}{2}\sin\frac{\alpha+\beta}{2}$$

$$= 2\sin\frac{\alpha+\beta}{2}\left[\cos\frac{\alpha-\beta}{2} - \cos\frac{\alpha+\beta+2\gamma}{2}\right]$$

$$= 2\sin\frac{\alpha+\beta}{2}\left[2\,.\sin\frac{2\alpha+2\gamma}{4}\sin\frac{2\beta+2\gamma}{4}\right]$$

$$= 4\sin\frac{\alpha+\beta}{2}.\sin\frac{\beta+\gamma}{2}.\sin\frac{\alpha+\gamma}{2} = \text{RHS}$$

EXERCISE 1.6

Prove that:

1. $\cos 5^\circ - \sin 25^\circ = \sin 35^\circ$

2. $\sin 115^\circ + \cos 115^\circ = \sqrt{2}\cos 70^\circ$

3. $\cos 80^\circ + \cos 40^\circ - \cos 20^\circ = 0$

4. $2\sin 75^\circ \cos 15^\circ = \left(1+\frac{\sqrt{3}}{2}\right)$

5. $\frac{\sin 50^\circ - \sin 40^\circ}{\cos 50^\circ - \cos 40^\circ} = -1$

6. $\frac{\sin A + \sin 2A}{\cos A - \cos 2A} = \cot\frac{A}{2}$

7. $\frac{\sin 5A - \sin 3A}{\cos 3A + \cos 5A} = \tan A$

8. $\tan\left(\frac{\pi}{3}+A\right)\tan\left(\frac{\pi}{3}-A\right) = \frac{2\cos 2A+1}{2\cos 2A-1}$

9. $\frac{\sin\alpha+\sin\beta}{\cos\alpha+\cos\beta} = \tan\left(\frac{\alpha+\beta}{2}\right)$

10. $\cos\alpha + \cos\beta + \cos\gamma + \cos(\alpha+\beta+\gamma) = 4\cos\left(\frac{\alpha+\beta}{2}\right).\cos\left(\frac{\beta+\gamma}{2}\right).\cos\left(\frac{\gamma+\alpha}{2}\right)$

11. $\frac{\sin\theta+\sin 3\theta+\sin 5\theta}{\cos\theta+\cos 3\theta+\cos 5\theta} = \tan 3\theta.$

12. $\sin 20^\circ \sin 40^\circ \sin 60^\circ \sin 80^\circ = \frac{3}{16}$

13. $\frac{\cos\alpha+\cos\beta}{\cos\beta-\cos\alpha} = \cot\left(\frac{\alpha+\beta}{2}\right).\cot\left(\frac{\alpha-\beta}{2}\right)$

14. $\tan(45^\circ+\theta) - \tan(45^\circ-\theta) = 2\tan 2\theta$

15. If $\cos A + \cos B = \frac{1}{3}$. and $\sin A + \sin B = \frac{1}{4}$, prove that.

$$\tan\left(\frac{A+B}{2}\right) = \frac{3}{4}$$

1.12 TRIGONOMETRIC FUNCTIONS OF MULTIPLE ANGLES

I. $$\sin 2A = 2\sin A\cos A = \frac{2\tan A}{1+\tan^2 A}$$

II. $$\cos 2A = \cos^2 A - \sin^2 A = 1 - 2\sin^2 A = 2\cos^2 A - 1 = \frac{1-\tan^2 A}{1+\tan^2 A}$$

III. $$\tan 2A = \frac{2\tan A}{1-\tan^2 A}$$

IV. $$\sin 3A = 3\sin A - 4\sin^3 A$$

V. $$\cos 3A = 4\cos^3 A - 3\cos A$$

VI. $$\tan 3A = \frac{3\tan A - \tan^3 A}{1-3\tan^2 A}$$

1.13 TRIGONOMETRIC FUNCTIONS OF SUB-MULTIPLE ANGLES

I. $$\sin A = 2\sin\frac{A}{2}\cos\frac{A}{2} = \frac{2\tan\frac{A}{2}}{1+\tan^2\frac{A}{2}}$$

II. $$\cos A = \cos^2\frac{A}{2} - \sin^2\frac{A}{2}$$

$$= 1 - 2\sin^2\frac{A}{2}$$

$$= 2\cos^2\frac{A}{2} - 1$$

$$= \frac{1-\tan^2\frac{A}{2}}{1+\tan^2\frac{A}{2}}$$

III. $$\tan A = \frac{2\tan\frac{A}{2}}{1-\tan^2\frac{A}{2}}$$

IV. $$1 - \cos A = 2\sin^2\frac{A}{2} \Rightarrow \sin\frac{A}{2} = \sqrt{\frac{1-\cos A}{2}}$$

V. $$1 + \cos A = 2\cos^2\frac{A}{2} \Rightarrow \cos\frac{A}{2} = \sqrt{\frac{1+\cos A}{2}}$$

VI. $$\tan\frac{A}{2} = \sqrt{\frac{1-\cos A}{1+\cos A}}$$

Corollary:

1. $$\cos^2 A = \frac{1+\cos 2A}{2} \Rightarrow \cos A = \sqrt{\frac{1+\cos 2A}{2}}$$

2. $$\sin 2A = \frac{1-\cos 2A}{2} \Rightarrow \sin A = \sqrt{\frac{1-\cos 2A}{2}}$$

3. $$\tan^2 A = \frac{1-\cos 2A}{1+\cos 2A} \Rightarrow \tan A = \sqrt{\frac{1-\cos 2A}{1+\cos 2A}}$$

1.14 TRIGONOMETRIC FUNCTIONS OF SOME SPECIAL ANGLES

1. Sin 18°

Let $\theta = 18^\circ$, then $5\theta = 90^\circ$

$\Rightarrow \quad 2\theta = 90^\circ - 3\theta$

$\Rightarrow \quad \sin 2\theta = \sin(90^\circ - 3\theta) = \cos 3\theta$

$\Rightarrow \quad 2\sin\theta\cos\theta = 4\cos^3\theta - 3\cos\theta$

$\Rightarrow \quad 2\sin\theta = 1 - 4\sin^2\theta$

$\Rightarrow \quad 4\sin\theta = 4\cos^2\theta - 3 \quad [\because \cos\theta \neq 0]$

$\Rightarrow \quad 4\sin^2\theta + 2\sin\theta - 1 = 0$

$$\Rightarrow \quad \sin\theta = \frac{-2 \pm \sqrt{4+16}}{8} = \frac{-1\pm\sqrt{5}}{4}$$

$$\Rightarrow \quad \sin 18^\circ = \frac{-1+\sqrt{5}}{4} = \left(\frac{\sqrt{5}-1}{4}\right)$$

$[\because \theta = 18^\circ$ and $\sin 18^\circ$ is $+ve]$

2. cos 18°

$$\cos 18^\circ = \sqrt{1-\sin^2 18^\circ} = \sqrt{1-\left(\frac{\sqrt{5}-1}{4}\right)^2} = \frac{\sqrt{10+2\sqrt{5}}}{4}$$

3. cos 36°

$\cos 36^\circ = \cos(2 \times 18^\circ) = (1 - 2\sin^2 18^\circ)$

$$= \left[1 - 2\left(\frac{\sqrt{5}-1}{4}\right)^2\right] = \frac{2+2\sqrt{5}}{8} = \left(\frac{\sqrt{5}+1}{4}\right)$$

4. sin 36°

$$\sin 36^\circ = \sqrt{1 - 2\cos^2 36} = \sqrt{1 - \left(\frac{\sqrt{5}+1}{4}\right)^2}$$

$$= \frac{\sqrt{10 - 2\sqrt{5}}}{4}$$

5. sin 54° $= \sin(90^\circ - 36^\circ) = \cos 36^\circ = \left(\frac{\sqrt{5}+1}{4}\right)$

6. cos 54°

$$\cos 54^\circ = \cos(90^\circ - 36^\circ) = \sin 36^\circ = \frac{\sqrt{10 - 2\sqrt{5}}}{4}$$

7. sin 72°

$$\sin 72^\circ = \sin(90^\circ - 18^\circ) = \cos 18^\circ = \frac{\sqrt{10 + 2\sqrt{5}}}{4}$$

8. cos 72°

$$\cos 72^\circ = \cos(90^\circ - 18^\circ) = \sin 18^\circ = \left(\frac{\sqrt{5}-1}{4}\right)$$

Remark: It is clear that

$\sin 18^\circ = \cos 72^\circ$; $\cos 18^\circ = \sin 72^\circ$; $\sin 36^\circ = \cos 54^\circ$; $\cos 36^\circ = \sin 54^\circ$

9. sin 22° 30′.

we know that,

$$\sin^2 A = \left(\frac{1 - \cos 2A}{2}\right)$$

$$\Rightarrow \quad \sin^2(22^\circ 30') = \left(\frac{1 - \cos 45^\circ}{2}\right) \qquad \text{[Putting } 2A = 45^\circ\text{]}$$

$$\Rightarrow \quad \sin^2(22^\circ 30') = \frac{1 - \frac{1}{\sqrt{2}}}{2} = \frac{\sqrt{2}-1}{2\sqrt{2}}$$

$$\Rightarrow \quad \sin(20^\circ 30') = \sqrt{\left(\frac{\sqrt{2}-1}{2\sqrt{2}}\right)} \qquad [\because \sin(22^\circ 30') \text{ is } + ve]$$

10. cos 22°30′.

We know that,

$$\cos^2 A = \left(\frac{1-\cos 2A}{2}\right)$$

$$\Rightarrow \qquad \cos^2(22°30') = \left(\frac{1+\cos 45°}{2}\right) = \left(\frac{1+\frac{1}{\sqrt{2}}}{2}\right) \quad [\text{Putting } 2A = 45°]$$

$$\Rightarrow \qquad \cos^2(22°30') = \frac{\sqrt{2}+1}{2\sqrt{2}}$$

$$\Rightarrow \qquad \cos(22°30') = \sqrt{\left(\frac{\sqrt{2}+1}{2\sqrt{2}}\right)} \qquad [\because \ \cos(22°30') \text{ is +ve}]$$

11. tan $22\frac{1°}{2}$.

$$\tan 22\frac{1°}{2} = \frac{\sin(22°\,30')}{\cos(22°\,30')} = \sqrt{\left(\frac{\sqrt{2}-1}{\sqrt{2}+1}\right)}$$

$$= \sqrt{\frac{(\sqrt{2}-1)}{(\sqrt{2}+1)} \times \frac{(\sqrt{2}-1)}{(\sqrt{2}-1)}} = (\sqrt{2}-1)$$

12. sin $7\frac{1°}{2}$

$$\sin^2 A = \left(\frac{1-\cos 2A}{2}\right)$$

$$\sin^2 7\frac{1°}{2} = \left(\frac{1-\cos 15°}{2}\right) = \left(\frac{1}{2} - \frac{\sqrt{3}+1}{4\sqrt{2}}\right)$$

$$= \left(\frac{2\sqrt{2}-\sqrt{3}-1}{4\sqrt{2}}\right)$$

$$\Rightarrow \qquad \sin^2 7\frac{1°}{2} = \left(\frac{2\sqrt{2}-\sqrt{3}-1}{4\sqrt{2}}\right) \times \frac{\sqrt{2}}{\sqrt{2}} = \left(\frac{4-\sqrt{6}-\sqrt{2}}{8}\right)$$

$$\Rightarrow \qquad \sin 7\frac{1°}{2} = \frac{\sqrt{4-\sqrt{6}-\sqrt{2}}}{2\sqrt{2}} \qquad \left[\because \ \sin 7\frac{1°}{2} \text{ is + } ve\right]$$

13. cos $7\frac{1°}{2}$

$$\cos^2 A = \left(\frac{1+\cos 2A}{2}\right)$$

$$\Rightarrow \quad \cos^2 7\frac{1^\circ}{2} = \left(\frac{1}{2}+\frac{\cos 15^\circ}{2}\right) = \left(\frac{1}{2}+\frac{\sqrt{3}+1}{4\sqrt{2}}\right)$$

$$\Rightarrow \quad \cos^2 7\frac{1^\circ}{2} = \frac{2\sqrt{2}+\sqrt{3}+1}{4\sqrt{2}}$$

$$= \left(\frac{2\sqrt{2}+\sqrt{3}+1}{4\sqrt{2}}\right)\times\frac{\sqrt{2}}{\sqrt{2}} = \frac{4+\sqrt{6}+\sqrt{2}}{8}$$

$$\Rightarrow \quad \cos 7\frac{1^\circ}{2} = \frac{\sqrt{4+\sqrt{6}+\sqrt{2}}}{2\sqrt{2}} \qquad \left[\because \cos 7\frac{1^\circ}{2} \text{ is } + ve\right]$$

14. tan $11\frac{1^\circ}{4}$.

$$\tan A = \frac{2\tan\frac{A}{2}}{1-\tan^2\frac{A}{2}}$$

$$\Rightarrow \quad \tan 22\frac{1^\circ}{2} = \frac{2\tan 11\frac{1^\circ}{4}}{1-\tan^2 11\frac{1^\circ}{4}}$$

Let $\tan 11\frac{1^\circ}{4} = x$

$$\therefore \quad (\sqrt{2}-1) = \frac{2x}{1-x^2} \qquad \left[\because \tan 22\frac{1^\circ}{2} = (\sqrt{2}-1)\right]$$

$$\Rightarrow \quad (\sqrt{2}-1)-(\sqrt{2}-1)x^2-2x = 0$$

$$\Rightarrow \quad (\sqrt{2}-1)x^2+2x-(\sqrt{2}-1) = 0$$

$$\Rightarrow \quad x = \frac{-2\pm\sqrt{4+4(\sqrt{2}-1)^2}}{2(\sqrt{2}-1)}$$

$$= \frac{-1\pm\sqrt{1+(\sqrt{2}-1)^2}}{(\sqrt{2}-1)}$$

$$\Rightarrow \quad \tan 11\frac{1^\circ}{4} = \frac{-1+\sqrt{1+2+1-2\sqrt{2}}}{\sqrt{2}-1}$$

$$\Rightarrow \quad \tan 11\frac{1^\circ}{4} = \frac{-1+\sqrt{(4-2\sqrt{2})}}{\sqrt{2}-1} \quad \left[\because \tan 11\frac{1^\circ}{4} \text{ is +ve}\right]$$

ILLUSTRATIVE EXAMPLE

Example 1. Prove that

(*i*) $\sin^2 72° - \sin^2 60° = \dfrac{\sqrt{5}-1}{8}$

(*ii*) $\cos 4x = 1 - 8\sin^2 x \cos^2 x$

(*iii*) $\cos^2 A + \cos^2 (A + 120°) + \cos^2 (A - 120°) = \dfrac{3}{2}$

Solution: (*i*) L.H.S $= \sin^2 72° - \sin^2 60° = \left[\dfrac{\sqrt{10+2\sqrt{5}}}{4}\right]^2 - \left(\dfrac{\sqrt{3}}{2}\right)^2$

$$= \frac{10+2\sqrt{5}}{16} - \frac{3}{4} = \frac{(8\sqrt{5}-8)}{16\times4} = \left(\frac{\sqrt{5}-1}{8}\right) = \text{R.H.S}$$

(*ii*) L.H.S $= \cos 4x = \cos (2 \times 2x) = 1 - 2\sin^2 2x$

$= 1 - 2(\sin x \cos x)^2 = 1 - 8\sin^2 x \cos^2 x =$ RHS

(*iii*) L.H.S $= \cos^2 A + \cos^2 (A + 120°) + 1 - \sin^2 (A - 120°)$

$= \cos^2 A + 1 + [\cos^2 (A + 120°) - \sin^2 (A - 120°)]$

$= \cos^2 A + 1 + \cos 2A \cos 240°$

$[\because \cos^2 A - \sin^2 B = \cos (A + B) \cos (A - B)]$

$= \cos^2 A + 1 + (\cos^2 A - 1) \cos (180 + 60°)$

$= \cos^2 A + 1 + (2\cos^2 A - 1)(-\cos 60°)$

$= \cos^2 A + 1 + (2\cos^2 A - 1)\left(\dfrac{-1}{2}\right)$

$= \cos^2 A + 1 - \cos^2 A + \dfrac{1}{2} = \dfrac{3}{2} =$ RHS.

Example 2. (*i*) If $\cos \alpha = \dfrac{1}{2}$, find the value of $\cos 3\alpha$

(*ii*) If $\tan A = \dfrac{1}{2}$ and $\tan B = \dfrac{1}{3}$, show that $\cos 2A = \sin 4B$

Solution: (i) $\cos 3\alpha = 4\cos^3\alpha - 3\cos\alpha$

$$= \left[4.\left(\frac{1}{2}\right)^3 - 3.\frac{1}{2}\right] = \frac{4}{8} - \frac{3}{2} = -1$$

(*ii*) $$\cos 2A = \frac{1-\tan^2 A}{1+\tan^2 A} = \frac{\left(1-\frac{1}{49}\right)}{\left(1+\frac{1}{49}\right)} = \frac{48}{49}\times\frac{49}{50} = \frac{24}{25}$$

and $\sin 4B = 2\sin 2B \cos 2B$

$$= 2\left(\frac{2\tan B}{1+\tan^2 B}\right).\left(\frac{1-\tan^2 B}{1+\tan^2 B}\right)$$

$$= 2\left(\frac{2.\frac{1}{3}}{1+\frac{1}{9}}\right).\left(\frac{1-\frac{1}{9}}{1+\frac{1}{9}}\right) = 2\times\frac{3}{5}\times\frac{8}{10} = \frac{24}{25}$$

$\cos 2A = \sin 4B$

Example 3. If $135^\circ = \frac{-1}{\sqrt{2}}$, find the value of $\sin 67\frac{1^\circ}{2}$ and $\cos 67\frac{1^\circ}{2}$

Solution: $\therefore \quad \sin A = \pm\sqrt{\frac{1-\cos 2A}{2}}$

$$\Rightarrow \quad \sin 67\frac{1^\circ}{2} = \pm\sqrt{\frac{1-\cos 135^\circ}{2}} = \pm\sqrt{\frac{1+\frac{1}{\sqrt{2}}}{2}} = \pm\sqrt{\frac{\sqrt{2}+1}{2\sqrt{2}}}$$

As $67\frac{1^\circ}{2}$ lies in second quadrant, $\sin 67\frac{1^\circ}{2}$ is + *ve*

$$\therefore \quad \sin 67\frac{1^\circ}{2} = \sqrt{\frac{\sqrt{2}+1}{2\sqrt{2}}}$$

also, since $\cos A = \pm\sqrt{\frac{1+\cos 2A}{2}}$

$$\Rightarrow \quad \cos 67\frac{1^\circ}{2} = \pm\sqrt{\frac{1+\cos 135^\circ}{2}} = \pm\sqrt{\frac{1-\frac{1}{\sqrt{2}}}{2}} = \pm\sqrt{\frac{\sqrt{2}-1}{2\sqrt{2}}}$$

As $67\frac{1^\circ}{2}$ lies in second quadrant, $\cos 67\frac{1^\circ}{2}$ is – *ve*

$$\therefore \quad \cos 67\frac{1^\circ}{2} = -\sqrt{\frac{\sqrt{2}-1}{2\sqrt{2}}}$$

Example 4. Prove that:

$$16\cos\frac{2\pi}{15}.\cos\frac{4\pi}{15}.\cos\frac{8\pi}{15}.\cos\frac{14\pi}{15} = 1$$

Solution: L.H.S $= 16\cos 24^\circ . \cos 48^\circ . \cos 96^\circ . \cos 168^\circ$

$= 4\,(2\cos 24^\circ \cos 96^\circ) . (2\cos 48^\circ \cos 168^\circ)$

$= 4\,(\cos 120^\circ + \cos 72^\circ)(\cos 216^\circ + \cos 120^\circ)$

$$= 4\left[-\frac{1}{2}+\sin 18^o\right]\left[-\cos 36^o-\frac{1}{2}\right] \quad (\because \cos 72^o = \sin 18^o$$

and $\cos 216^o = \cos(180^o + 36^o) = -\cos 36^o)$

$$= 4\left[\frac{1}{2}-\sin 18^o\right]\left[\cos 36^o+\frac{1}{2}\right]$$

$$= 4\left[\frac{1}{2}-\frac{(\sqrt{5}-1)}{4}\right]\left[\frac{(\sqrt{5}+1)}{4}+\frac{1}{2}\right]$$

$$= 4\left(\frac{3-\sqrt{5}}{4}\right)\left(\frac{3+\sqrt{5}}{4}\right) = 1 = \text{RHS.}$$

Example 5. Prove that:

$\tan 6^o . \tan 42^o . \tan 66^o . \tan 78^o = 1$

Solution: $\text{LHS} = \dfrac{\sin 6^o . \sin 42^o . \sin 66^o . \sin 78^o}{\cos 6^o . \cos 42^o . \cos 66^o . \cos 78^o}$

$$= \frac{(2\sin 6^o \sin 66^o)(2\sin 42^o \sin 78^o)}{(2\cos 6^o \cos 66^o)(2\cos 42^o \cos 78^o)}$$

$$= \frac{(\cos 60^o - \cos 72^o)(\cos 36^o - \cos 120^o)}{(\cos 72^o + \cos 60^o)(\cos 120^o + \cos 36^o)}$$

$$= \frac{(\cos 60^o - \sin 18^o)(\cos 36^o + \sin 30^o)}{(\sin 18^o + \cos 60^o)(\cos 36^o - \sin 30^o)}$$

$$= \frac{\left[\frac{1}{2}-\left(\frac{\sqrt{5}-1}{4}\right)\right]\left[\left(\frac{\sqrt{5}+1}{4}\right)+\frac{1}{2}\right]}{\left[\left(\frac{\sqrt{5}-1}{4}\right)+\frac{1}{2}\right]\left[\left(\frac{\sqrt{5}+1}{4}\right)-\frac{1}{2}\right]} = 1 = \text{RH.S}$$

Example 6. Prove that: $\sqrt{2+\sqrt{2+2\cos 4\theta}} = 2\cos\theta$

Solution: $\text{L.H.S} = \sqrt{2+\sqrt{2+2\cos 4\theta}} = \sqrt{2+\sqrt{2(1+\cos 4\theta)}}$

$$= \sqrt{2+\sqrt{2(2\cos^2 2\theta}} = \sqrt{2+2\cos 2\theta}$$

$$= \sqrt{2(1+\cos 2\theta)} = \sqrt{2(2\cos^2\theta)} = 2\cos\theta = \text{R.H.S}$$

Example 7. Prove that: $\dfrac{\sin 2\theta}{1+\cos 2\theta} = \tan\theta$

Solution: $\text{LHS} = \dfrac{\sin 2\theta}{1+\cos 2\theta} = \dfrac{2\sin\theta\cos\theta}{1+(2\cos^2\theta-1)} = \dfrac{2\sin\theta\cos\theta}{2\cos^2\theta} = \tan\theta = \text{RHS}$

Example 8. If $\tan^2\alpha = 1 + 2\tan^2\beta$, prove that $\cos 2\beta = 1 + 2\cos 2\alpha$

Solution: $\therefore\ \tan^2\alpha = 1 + 2\tan^2\beta$

$$\Rightarrow \quad \frac{1-\cos 2\alpha}{1+\cos 2\alpha} = \frac{1+2\,(1-\cos 2\beta)}{(1+\cos 2\beta)} \qquad \left(\because \tan^2\theta = \frac{1-\cos 2\theta}{1+\cos 2\theta}\right)$$

$$\Rightarrow \quad \frac{1-\cos 2\alpha}{1+\cos 2\alpha} = \frac{3-\cos 2\beta}{1+\cos 2\beta}$$

Applying componendo and dividend rule, we get

$$\frac{2}{-2\cos 2\alpha} = \frac{4}{2-2\cos 2\beta}$$

$$\Rightarrow \quad \frac{1}{\cos 2\alpha} = \frac{2}{\cos 2\beta - 1}$$

$$\Rightarrow \quad \cos 2\beta - 1 = 2\cos 2\alpha$$

$$\Rightarrow \quad \cos 2\beta = 1 + 2\cos 2\alpha.$$

Example 9. Prove that: $\dfrac{1+\sin 2\theta+\cos 2\theta}{1+\sin 2\theta-\cos 2\theta} = \cot\theta$

Solution: L.H.S. $= \dfrac{1+2\sin\theta\cos\theta+(2\cos^2\theta-1)}{1+2\sin\theta\cos\theta-(1-2\sin^2\theta)}$

$$= \frac{2\sin\theta\cos\theta+2\cos^2\theta}{2\sin\theta\cos+2\sin^2\theta} = \frac{2\cos\theta\,(\sin\theta+\cos\theta)}{2\sin\theta\,(\cos\theta+\sin\theta)}$$

$$= \frac{\cos\theta}{\sin\theta} = \cot\theta = \text{R.HS.}$$

EXERCISE 1.7

1. If $\cos 315^\circ = \dfrac{1}{\sqrt{2}}$, find the value of $\sin 157\dfrac{1^\circ}{2}$ and $\cos 157\dfrac{1^\circ}{2}$
2. Find the value of tan 36° and cot 36°
3. If $\sin\theta = \dfrac{4}{5}$, find the value of $\sin 2\theta$, $\cos 2\theta$, and $\tan 2\theta$
4. If $\tan\dfrac{8}{2} = (2-\sqrt{3})$, find the value of $\sin\theta$
5. Express $\cot^2 A$ in terms of $\cos 2A$

   Prove that:
6. $\cos^2 48^\circ - \sin^2 12^\circ = \dfrac{\sqrt{5}+1}{8}$

7. $\sin^2 24^\circ - \sin^2 6^\circ = \dfrac{\sqrt{5}-1}{8}$

8. $\dfrac{\sin 2\theta}{1+\cos 2\theta} = \tan\theta$

9. $\dfrac{1+\cos 2\theta+\sin 2\theta}{1-\cos 2\theta+\sin 2\theta} = \cot\theta$

10. $\sin^4 x + \cos^4 x = 1-\dfrac{1}{2}\sin^2 2x$

11. $\dfrac{\cot\theta-\tan\theta}{\cot\theta+\tan\theta} = \cos 2\theta$

12. $8\cos^3\dfrac{\pi}{9} - 6\cos\dfrac{\pi}{9} = 1$

13. $\sqrt{\dfrac{1-\cos 2x}{1+\cos 2x}} = \tan x$

14. $\tan\left(45^\circ+\dfrac{\theta}{2}\right) = \dfrac{\cos\theta}{1-\sin\theta}$

15. $\dfrac{\sec 8\theta-1}{\sec 4\theta-1} = \dfrac{\tan 8\theta}{\tan 2\theta}$

16. $\sin 36^\circ \sin 72^\circ \sin 108^\circ \sin 144^\circ = \dfrac{5}{16}$

17. $\dfrac{\tan^2\theta}{\tan^2\theta-1} + \dfrac{\operatorname{cosec}^2\theta}{\sec^2\theta-\operatorname{cosec}^2\theta} = -\sec 2\theta$

18. $\cos^2\left(\dfrac{\pi}{4}-\theta\right) - \sin^2\left(\dfrac{\pi}{4}-\theta\right) = \sin 2\theta$

ANSWERS

1. $\sqrt{\dfrac{\sqrt{2}-1}{2\sqrt{2}}};-\sqrt{\dfrac{\sqrt{2}+1}{2\sqrt{2}}}$

2. $\dfrac{\sqrt{10-2\sqrt{5}}}{\sqrt{5}+1};\dfrac{\sqrt{5}+1}{\sqrt{10-2\sqrt{5}}}$

3. $\dfrac{24}{25},\dfrac{-7}{25},\dfrac{-24}{7}$

4. $\dfrac{1}{2}$

5. $\dfrac{1+\cos 2A}{1-\cos 2A}$

1.15 INVERSE TRIGONOMETRIC FUNCTIONS.

The function $\sin^{-1}x$, $\cos^{-1}x$, $\tan^{-1}x$, $\cot^{-1}x$, $\sec^{-1}x$ $\operatorname{cosec}^{-1}x$ are called inverse trigonometric (or inverse circular) functions.

Note: $\sin^{-1}x \neq (\sin x)^{-1} = \dfrac{1}{\sin x}$; $\tan^{-1}x \neq (\tan x)^{-1} = \dfrac{1}{\tan x}$ etc.

DOMAIN AND RANGE OF INVERSE TRIGONOMETRIC FUNCTION

Function	Domain	Range
$y = \sin^{-1} x$	$\{x : -1 \le x \le 1\}$	$\left\{y : -\frac{\pi}{2} \le y \le \frac{\pi}{2}\right\}$
$y = \cos^{-1} x$	$\{x : -1 \le x \le 1\}$	$\{y : 0 \le y \le \pi\}$
$y = \tan^{-1} x$	set of read numbers	$\left\{y : \frac{\pi}{2} < y < \frac{\pi}{2}\right\}$
$y = \text{cosec}^{-1} x$	set of real numbers except $\|x\| < 1$ i.e; $x \ge 1$ or $x \le -1$	$\left\{y : -\frac{\pi}{2} \le y \le \frac{\pi}{2}, y \ne 0\right\}$
$y = \sec^{-1} x$	set of the numbers except $\|x\| < 1$ i.e; $x \ge 1$ or $x \le -1$	$\left\{y : 0 < y < \pi : y \ne \frac{\pi}{2}\right\}$
$y = \cot^{-1} x$	set of real numbers	$\{y : 0 < y < \pi)$

Principal value of an inverse Trigonometric function

If $y = \sin^{-1}x$, then y is called the principal value of $\sin^{-1}x$.

ILLUSTRATIVE EXAMPLES

Example 1: Find the principal values of:

(*i*) $\sin^{-1}\left(\frac{-\sqrt{3}}{2}\right)$ (*ii*) $\cot^{-1}\left(\frac{-1}{\sqrt{3}}\right)$ (*iii*) $\cos^{-1}\left(\frac{1}{2}\right)$

(*iv*) $\text{cosec}^{-1}(-2)$ (*v*) $\tan^{-1}\left(\frac{1}{\sqrt{3}}\right)$ (*vi*) $\sec^{-1}\left(\frac{2}{\sqrt{3}}\right)$

(*vii*) $\cos^{-1}\left(\frac{-1}{\sqrt{2}}\right)$ (*viii*) $\cot^{-1}(1)$

Solution: (*i*) Let $\sin^{-1}\left(\frac{-\sqrt{3}}{2}\right) = y$ where $\frac{-\pi}{2} \le y \le \frac{\pi}{2}$

$$\Rightarrow \quad \sin y = \frac{-\sqrt{3}}{2} = -\sin\frac{\pi}{3} = \sin\left(\frac{-\pi}{3}\right)$$

$$\Rightarrow \quad y = \frac{-\pi}{3}$$

$\therefore$ The principal value of $\sin^{-1}\left(\frac{-\sqrt{3}}{2}\right) = \frac{-\pi}{3}$

(*ii*) Let $\cot^{-1}\left(\frac{-1}{\sqrt{3}}\right) = y$ where $0 < y < \pi$

$$\Rightarrow \quad \cot y = \frac{-1}{\sqrt{3}} = -\cot\frac{\pi}{3} = \cot\left(\pi - \frac{\pi}{3}\right) = \cot\frac{2\pi}{3}$$

$$\Rightarrow \quad y = \frac{2\pi}{3}$$

$\therefore$ The principle value of $\cot^{-1}\left(\frac{-1}{\sqrt{3}}\right) = \frac{2\pi}{3}$

(*iii*) Let $\cos^{-1}\left(\frac{1}{2}\right) = y$ where $0 \le y \le \pi$

$$\Rightarrow \quad \cos y = \frac{1}{2} = \cos\frac{\pi}{3}$$

$$\Rightarrow \quad y = \frac{\pi}{3}$$

$\therefore$ The principal value of $\cos^{-1}\left(\frac{1}{2}\right) = \frac{\pi}{3}$

(*iv*) Let $\text{cosec}^{-1}(-2) = y$ where $\frac{-\pi}{2} \le y < 0$

$$\Rightarrow \quad \text{cosec } y = -2 = \text{cosec}\left(\frac{-\pi}{6}\right) \Rightarrow y = -\frac{\pi}{6}$$

$\therefore$ The principal value of $\text{cosec}^{-1}(-2) = \frac{-\pi}{6}$

(*v*) Let $\tan^{-1}\left(\frac{1}{\sqrt{3}}\right) = y$ where $\frac{-\pi}{2} < y < \frac{\pi}{2}$

$$\Rightarrow \quad \tan y = \frac{1}{\sqrt{3}} = \tan\frac{\pi}{6}$$

$$\Rightarrow \quad y = \frac{\pi}{6}$$

$\therefore$ The The principal value of $\tan^{-1}\left(\frac{1}{\sqrt{3}}\right) = \frac{\pi}{6}$

(*vi*) Let $\sec^{-1}\left(\frac{2}{\sqrt{3}}\right) = y$ where $0 < y < \frac{\pi}{2}$

$$\Rightarrow \quad \sec y = \frac{2}{\sqrt{3}} = \sec\frac{\pi}{6}$$

$$\Rightarrow \quad y = \frac{\pi}{6}$$

$\therefore$ The principal value of $\sec^{-1}\left(\frac{2}{\sqrt{3}}\right) = \frac{\pi}{6}$

(*vii*) Let $\cos^{-1}\left(\frac{-1}{\sqrt{2}}\right) = y$ where $0 \le y \le \pi$

$\Rightarrow \quad \cos y = \frac{-1}{\sqrt{2}} = -\cos\left(\frac{\pi}{4}\right) = \cos\left(\pi - \frac{\pi}{4}\right) = \cos\frac{3\pi}{4}$

$\Rightarrow \quad y = \frac{3\pi}{4}$

$\therefore$ The principal value of $\cos^{-1}\left(\frac{-1}{\sqrt{2}}\right) = \frac{3\pi}{4}$

(*viii*) Let $y = \cot^{-1}(1)$ where $0 < y < \pi$

$\Rightarrow \quad \cot y = 1 = \cot\frac{\pi}{4}$

$\Rightarrow \quad y = \frac{\pi}{4}$

$\therefore$ The principal value of $\cot^{-1}(1) = \frac{\pi}{4}$.

Example 2:(*i*) If $\tan^{-1}\left(\frac{3}{4}\right) = \theta$, find the value of $\sin\theta$

(*ii*) If $\cot^{-1}\left(\frac{1}{7}\right) = \theta$, find the value of $\cos\theta$

(*iii*) If $\sin^{-1}\left(\frac{1}{2}\right) = \tan^{-1}x$, find the value of x

Solution: (*i*) $\tan^{-1}\left(\frac{3}{4}\right) = \theta \Rightarrow \tan\theta = \frac{3}{4}$

$\Rightarrow \quad \cot\theta = \frac{4}{3}$

$\therefore \quad \text{cosec}\,\theta = \sqrt{1+\cot^2\theta} = \sqrt{1+\frac{16}{9}} = \frac{5}{3}$

$\Rightarrow \quad \sin\theta = \frac{3}{5}$

(*ii*) $\cot^{-1}\left(\frac{1}{7}\right) = \theta \Rightarrow \cot\theta = \frac{1}{7} \Rightarrow \tan\theta = 7$

$\Rightarrow \quad \sec\theta = \sqrt{1+\tan^2\theta} = 5\sqrt{2}$

$\Rightarrow \quad \cos\theta = \frac{1}{5\sqrt{2}}$

(*iii*) $\tan^{-1}x = \sin^{-1}\left(\frac{1}{2}\right) = \frac{\pi}{6} \Rightarrow x = \tan\frac{\pi}{6} = \frac{1}{\sqrt{3}}$

$\Rightarrow \quad x = \frac{1}{\sqrt{3}}$.

1.16 PROPERTIES OF INVERSE TRIGONOMETRIC FUNCTIONS.

Theorem 1.(*i*) $\sin^{-1}(\sin x) = x$ (*ii*) $\cos^{-1}(\cos x) = x$

(*iii*) $\tan^{-1}(\tan x) = x$ (*iv*) $\text{cosec}^{-1}(\text{cosec } x) = x$

(*v*) $\sec^{-1}(\sec x) = x$ (*vi*) $\cot^{-1}(\cot x) = x$

Proof: (i) Let $\sin x = y$. Then,

$$x = \sin^{-1} y = \sin^{-1}(\sin x) \qquad [\because y = \sin x]$$

$\therefore \quad \sin^{-1}(\sin x) = x$

Similarly, the other results may be proved

Theorem 2.(*i*) $\sin(\sin^{-1}x) = x$ (*ii*) $\cos(\cos^{-1}x) = x$

(*iii*) $\tan(\tan^{-1}x) = x$ (*iv*) $\text{cosec}(\text{cosec}^{-1}x) = x$

(*v*) $\sec(\sec^{-1}x) = x$ (*vi*) $\cot(\cot^{-1}x) = x$

Proof: (*ii*) Let $\cos^{-1}x = y$ Then,

$$x = \cos y = \cos(\cos^{-1}x) \ [\because y = \cos^{-1}x]$$

$\therefore \quad x = \cos(\cos^{-1}x)$

similarly, the other results may be proved

Theorem 3.(*i*) $\sin^{-1}\left(\frac{1}{x}\right) = \text{cosec}^{-1}x$ i.e; $\sin^{-1}x = \text{cosec}^{-1}\left(\frac{1}{x}\right)$

(*ii*) $\cos^{-1}\left(\frac{1}{x}\right) = \sec^{-1}x$ i.e; $\cos^{-1}x = \sec^{-1}\left(\frac{1}{x}\right)$

(*iii*) $\tan^{-1}\left(\frac{1}{x}\right) = \cot^{-1}x$ i.e; $\tan^{-1}x = \cot^{-1}\left(\frac{1}{x}\right)$

Proof: (i) Let $\sin^{-1}\left(\frac{1}{x}\right) = y \Rightarrow \sin y = \frac{1}{x}$

$\Rightarrow \quad \text{cosec } y = x$

$\Rightarrow \quad y = \text{cosec}^{-1}\text{x}$

$\therefore \quad \sin{-1}\left(\frac{1}{x}\right) = \text{cosec}^{-1}x$

Similarly, the other results may be proved

Theorem 4: (*i*) $\sin^{-1}(-x) = -\sin^{-1}x$ (*ii*) $\cos^{-1}(-x) = \pi - \cos^{-1}x$.

(*iii*) $\tan^{-1}(-x) = -\tan^{-1}x$ (*iv*) $\text{cosec}^{-1}(-x) = -\text{cosec}^{-1}x$.

(*v*) $\sec^{-1}(-x) = \pi - \sec^{-1}x$ (*vi*) $\cot^{-1}(-x) = -\cot^{-1}x$.

Proof: (i) let $\sin^{-1}(-x) = y \Rightarrow -x = \sin y$

$\Rightarrow \quad x = -\sin y = \sin(-y)$

$\Rightarrow \quad -y = -\sin^{-1}x$

$\Rightarrow \quad y = -\sin^{-1}x$

$\Rightarrow \quad \sin^{-1}(-x) = -\sin^{-1}x$

(ii) Let $\cos^{-1}(-x) = y \Rightarrow -x = \cos y$

$\Rightarrow \quad x = -\cos y = \cos(\pi - y)$

$\Rightarrow \quad \pi - y = \cos^{-1}x$

$\Rightarrow \quad y = \pi - \cos^{-1}x.$

$\Rightarrow \quad \cos^{-1}(-x) = \pi - \cos^{-1}x.$

Similarly, the other results may be proved.

Theorem 5. (i) $\sin^{-1}x + \cos^{-1}x = \dfrac{\pi}{2}$ (ii) $\tan^{-1}x + \cot^{-1}x = \dfrac{\pi}{2}$

(*iii*) $\sec^{-1}x + \text{cosec}^{-1}x = \dfrac{\pi}{2}$

Proof: (i) Let $\sin^{-1}x = y \Rightarrow x = \sin y = \cos\left(\dfrac{\pi}{2} - y\right)$

$$\Rightarrow \quad \cos^{-1}x = \frac{\pi}{2} - y = \frac{\pi}{2} - \sin^{-1}x$$

$$\Rightarrow \cos^{-1}x + \sin^{-1x} = \frac{\pi}{2}$$

Similarly, the other results may be proved.

Theorem 6.

(*i*) $\tan^{-1} + \tan^{-1}y = \tan^{-1}\left(\dfrac{x+y}{1-xy}\right)$

(*ii*) $\tan^{-1}x - \tan^{-1}y = \tan^{-1}\left(\dfrac{x-y}{1+xy}\right)$

(*iii*) $\sin^{-1}x + \sin^{-1}y = \sin^{-1}(x\sqrt{1-y^2} + y\sqrt{1-x^2})$

(*iv*) $\sin^{-1}x - \sin^{-1}y = \sin^{-1}(x\sqrt{1-y^2}) - y\sqrt{1-x^2})$

(*v*) $\cos^{-1}x + \cos^{-1}y = \cos^{-1}(xy - \sqrt{1-x^2}\,.\sqrt{1-y^2})$

(*vi*) $\cos^{-1}x - \cos^{-1}y = \cos^{-1}(xy + \sqrt{1-x^2}\sqrt{1-y^2})$; $x, y > 0$; $xy < 1$ and $x > y$

Proof:

(*i*) Let $\tan^{-1}x = \theta_1$ and $\tan^{-1}y = \theta_2$

Then $x = \tan\theta_1$ and $y = \tan\theta_2$

we have $\tan(\theta_1 + \theta_2) = \dfrac{\tan\theta_1 + \tan\theta_2}{1 - \tan\theta_1 \tan\theta_2}$

$$\Rightarrow \qquad \theta_1 + \theta_2 = \tan^{-1}\left(\frac{\tan\theta_1 + \tan\theta_2}{1 - \tan\theta_1 \tan\theta_2}\right)$$

$$\Rightarrow \qquad \tan^{-1}x + \tan^{-1}y = \tan^{-1}\left(\frac{x+y}{1-xy}\right)$$

(*ii*) Same as (*i*)

(*iii*) Let $\sin^{-1}(x) = \theta_1$ and $\sin^{-1}y = \theta_2$. Then

$$x = \sin\theta_1 \text{ and } y = \sin\theta_2$$

we have $\sin(\theta_1 + \theta_2) = \sin\theta_1 \cos\theta_2 + \cos\theta_1 \sin\theta_2$

$$= \sin\theta_1 \sqrt{1 - \sin^2\theta_2} + \sqrt{(1 - \sin^2\theta_1)}\sin\theta_2$$

$$\Rightarrow \qquad \sin(\theta_1 + \theta_2) = \sin(\theta_1 + \theta_2) = [x\sqrt{1-y^2} + y\sqrt{1-x^2}]$$

$$\Rightarrow \qquad \theta_1 + \theta_2 = \sin^{-1}[x\sqrt{1-y^2} + y\sqrt{1-x^2}]$$

$$\Rightarrow \qquad \sin^{-1}x + \sin^{-1}y = \sin^{-1}[x\sqrt{1-y^2} + y\sqrt{1-x^2}]$$

(*iv*) Same as (*iii*)

(*v*) Let $\cos^{-1}x = \theta_1$ and $\cos^{-1}y = \theta_2$. Then

$$x = \cos\theta_1 \text{ and } y = \cos\theta_2$$

we have $\cos(\theta_1 + \theta_2) = \cos\theta_1 \cos\theta_2 - \sin\theta_1 \sin\theta_2$

$$= \cos\theta_1 \cos\theta_2 - \sqrt{1-\cos^2\theta_1}\sqrt{1-\cos^2\theta_2}$$

$$\Rightarrow \qquad \cos(\theta_1 + \theta_2) = x.\,y - \sqrt{1-x^2}\sqrt{1-y^2}$$

$$\Rightarrow \qquad \theta_1 + \theta_2 = \cos^{-1}[xy - \sqrt{1-x^2}\sqrt{1-y^2}]$$

$$\Rightarrow \qquad \cos^{-1}x + \cos^{-1}y = \cos^{-1}[xy - \sqrt{1-x^2}\sqrt{1-y^2}]$$

(*vi*) Same as (*v*)

Theorem 7. $2\tan^{-1}x = \sin^{-1}\left(\frac{2x}{1+x^2}\right)$

$$= \cos^{-1}\left(\frac{1-x^2}{1+x^2}\right)$$

$$= \tan^{-1}\left(\frac{2x}{1-x^2}\right) \quad (\text{if } x^2 < 1)$$

Proof: Let $\tan^{-1}x = \theta$. Then $x = \tan\theta$

$$\therefore \qquad \sin 2\theta = \frac{2\tan\theta}{1+\tan^2\theta} = \frac{2x}{1+x^2} \Rightarrow 2\theta = \sin^{-1}\left(\frac{2x}{1+x^2}\right)$$

$$\Rightarrow \qquad 2\tan^{-1}x = \sin^{-1}\left(\frac{2x}{1+x^2}\right)$$

Again, $\tan 2\theta = \dfrac{2\tan\theta}{1-\tan^2\theta} = \dfrac{2x}{1-x^2} \Rightarrow 2\theta = \tan^{-1}\left(\dfrac{2x}{1-x^2}\right)$

$\Rightarrow \quad 2\tan^{-1}x = \tan^{-1}\left(\dfrac{2x}{1-x^2}\right)$

Again, $\cos 2\theta = \dfrac{1-\tan^2\theta}{1+\tan^2\theta} = \dfrac{1-x^2}{1+x^2} \Rightarrow 2\theta = \cos^{-1}\left(\dfrac{1-x^2}{1+x^2}\right)$

$\Rightarrow \quad 2\tan^{-1}x = \cos^{-1}\left(\dfrac{1-x^2}{1+x^2}\right)$

Theorem 8. $\tan^{-1}\left[\dfrac{x}{\sqrt{a^2-x^2}}\right] = \sin^{-1}\left(\dfrac{x}{a}\right)$

Proof: Let $\sin^{-1}\left(\dfrac{x}{a}\right) = y$, Then

$$\therefore \quad \frac{x}{\sqrt{a^2-x^2}} = \frac{a\sin y}{\sqrt{a^2-a^2\sin^2 y}} = \frac{a\sin y}{a\sqrt{1-\sin^2 y}}$$

$$= \frac{\sin y}{\sqrt{\cos^2 y}} = \frac{\sin y}{\cos y}$$

$$\Rightarrow \quad \frac{x}{\sqrt{a^2-x^2}} = \tan y$$

$$\Rightarrow \tan^{-1}\left[\frac{x}{\sqrt{a^2-x^2}}\right] = y$$

$$\Rightarrow \tan^{-1}\left[\frac{x}{\sqrt{a^2-x^2}}\right] = \sin^{-1}\left(\frac{x}{a}\right).$$

ILLUSTRATIVE EXAMPLES

Example 1. Find the values of;

(*i*) $\sin^{-1}\left(\sin\dfrac{2\pi}{3}\right)$ (*ii*) $\tan^{-1}\left(\tan\dfrac{3\pi}{4}\right)$

(*iii*) $\sin\left(\cos^{-1}\dfrac{3}{5}\right)$ (*iv*) $\tan\dfrac{1}{2}\left(\cos^{-1}\dfrac{\sqrt{5}}{3}.\right)$

(*v*) $\sin(\cot^{-1}x)$ (*vi*) $\cos\left(\tan^{-1}\frac{3}{4}\right)$

Solution: (i) Let $y = \sin^{-1}\left(\sin\frac{2\pi}{3}\right)$. Then, $-\frac{\pi}{2} \le y \le \frac{\pi}{2}$

Now, $$\sin\frac{2\pi}{3} = \sin\left(\pi - \frac{\pi}{3}\right) = \sin\frac{\pi}{3}$$

$$\therefore \quad \sin^{-1}\left(\sin\frac{2\pi}{3}\right) = \sin^{-1}\left(\sin\frac{\pi}{3}\right) = \frac{\pi}{3}$$

(*ii*) Let $y = \tan^{-1}\left(\tan\frac{3\pi}{4}\right)$. Then, $-\frac{\pi}{2} < y < \frac{\pi}{2}$

Now, $$\tan\left(\frac{3\pi}{4}\right) = \tan\left(\pi - \frac{\pi}{4}\right) = -\tan\frac{\pi}{4} = \tan\left(-\frac{\pi}{4}\right)$$

$$\therefore \quad \tan^{-1}\left(\tan\frac{3\pi}{4}\right) = \tan^{-1}\left[\tan\left(\frac{-\pi}{4}\right)\right] = -\frac{\pi}{4}$$

(*iii*) Let $\cos^{-1}\frac{3}{5} = y$. Then $\cos y = \frac{3}{5}$

$$\therefore \quad \sin\left[\cos^{-1}\frac{3}{5}\right] = \sin y = \sqrt{1-\cos^2 y} = \sqrt{1-\frac{9}{25}} = \frac{4}{5}$$

(*iv*) Let $\cos^{-1}\frac{\sqrt{5}}{3} = y$. Then, $\cos y = \frac{\sqrt{5}}{3}$

$$\therefore \quad \tan\frac{1}{2}\left[\cos^{-1}\frac{\sqrt{5}}{3}\right] = \tan\frac{y}{2} = \sqrt{\frac{1-\cos\theta}{1+\cos\theta}}$$

$$= \sqrt{\frac{1-(\sqrt{5}/3)}{1+(\sqrt{5}/3)}} = \sqrt{\frac{(3-\sqrt{5})}{(3+\sqrt{5})} \times \frac{(3-\sqrt{5})}{(3-\sqrt{5})}} = \frac{(3-\sqrt{5})}{2}$$

(*v*) Let $\cot^{-1} x = y$. Then, $\cot y = x$.

$$\therefore \quad \operatorname{cosec}\theta = \sqrt{1+\cot^2\theta} = \sqrt{1+x^2}$$

$$\Rightarrow \quad \sin\theta = \frac{1}{\sqrt{1+x^2}}$$

$$\therefore \quad \sin(\cot^{-1}x) = \sin y = \frac{1}{\sqrt{1+x^2}}$$

(*vi*) Let $\tan^{-1}\frac{3}{4} = y$, Then, $\tan y = \frac{3}{4}$

$$\therefore \quad \cos\left(\tan^{-1}\frac{3}{4}\right) = \cos\theta = \frac{1}{\sec\theta} = \frac{1}{\sqrt{1+\tan^2\theta}} = \frac{4}{5} \quad \left[\because \tan^2\theta = \frac{3}{4}\right]$$

Example 2. Evaluate: $\cos\left[\sin^{-1}\frac{3}{5}+\sin^{-1}\frac{5}{13}\right]$

Solution. Let $\sin^{-1}\frac{3}{5} = A$ and $\sin^{-1}\frac{5}{13} = B$

Then, $\quad \sin A = \frac{3}{5}$ and $\sin B = \frac{5}{13}$

$$\therefore \cos\left[\sin^{-1}\frac{3}{5}+\sin^{-1}\frac{5}{13}\right] = \cos(\text{A} + \text{B})$$

$$= \cos \text{A} \,.\, \cos \text{B} - \sin \text{A} \,.\, \sin \text{B}$$

$$= \left(\frac{4}{5}\cdot\frac{12}{13}-\frac{3}{5}\frac{5}{13}\right)=\frac{33}{65}$$

$$\left[\because \cos A=\sqrt{1-\sin^2 A}=\frac{4}{5} \text{ and } \cos B=\sqrt{1-\sin^2 B}=\frac{12}{13}\right]$$

Example 3. Prove that:

$$\sin^{-1}\frac{12}{13}+\cos^{-1}\frac{4}{5}+\tan^{-1}\frac{63}{16} = \pi$$

Solution: $\because \sin^{-1}x = \tan^{-1}\left(\frac{x}{\sqrt{1-x^2}}\right)$ and $\cos^{-1}x=\tan^{-1}\left(\frac{\sqrt{1-x^2}}{x}\right)$

$$\therefore \quad \sin^{-1}\frac{12}{13} = \tan^{-1}\left[\frac{(12/13)}{\sqrt{\left(1-\frac{144}{169}\right)}}\right]=\tan^{-1}\left[\frac{12/13}{(5/13)}\right]$$

$$= \tan^{-1}\frac{12}{5}$$

And, $$\cos^{-1}\frac{4}{5} = \tan^{-1}\left(\frac{\sqrt{1-\frac{16}{25}}}{4/5}\right)=\tan^{-1}\left[\frac{(3/5)}{(4/5)}\right] = \tan^{-1}\frac{3}{4}$$

$$\therefore \quad \text{L.H.S.} = \left[\tan^{-1}\frac{12}{5}+\tan^{-1}\frac{3}{4}\right]+\tan^{-1}\frac{63}{16}$$

$$=\pi+\tan^{-1}\left[\frac{\left[\frac{12}{5}+\frac{3}{4}\right]}{\left(1-\frac{12}{5}\times\frac{3}{4}\right)}\right]+\tan^{-1}\frac{63}{16}\left[\because\left(\frac{12}{5}\times\frac{3}{4}\right)>1\right]$$

$$= \pi + \tan^{-1}\left(\frac{-63}{16}\right) + \tan^{-1}\frac{63}{16}$$

$$= \left[\pi - \tan^{-1}\frac{63}{16} + \tan^{-1}\frac{63}{16}\right]$$

$$= \pi = \text{RHS.}$$

Example 4. Prove that:

$$\tan^{-1}\left(\frac{1}{3}\right) + \tan^{-1}\left(\frac{1}{5}\right) + \tan^{-1}\left(\frac{1}{7}\right) + \tan^{-1}\left(\frac{1}{8}\right) = \frac{\pi}{4}$$

Solution: $\left[\tan^{-1}\left(\frac{1}{3}\right) + \tan^{-1}\left(\frac{1}{5}\right)\right] + \left[\tan^{-1}\left(\frac{1}{7}\right) + \tan^{-1}\left(\frac{1}{8}\right)\right]$

$$= \tan^{-1}\left[\frac{\frac{1}{3}+\frac{1}{5}}{1-\frac{1}{3}\cdot\frac{1}{5}}\right] + \tan^{-1}\left[\frac{\frac{1}{7}+\frac{1}{8}}{1-\frac{1}{7}\cdot\frac{1}{8}}\right] = \tan^{-1}\left(\frac{8}{14}\right) + \tan^{-1}\left(\frac{15}{55}\right)$$

$$= \tan^{-1}\left(\frac{4}{7}\right) + \tan^{-1}\left(\frac{3}{11}\right) = \tan^{-1}\left[\frac{\frac{4}{7}+\frac{3}{11}}{1-\frac{4}{7}\cdot\frac{3}{11}}\right]$$

$$= \tan^{-1}\left(\frac{65}{65}\right) = \tan^{-1}(1) = \frac{\pi}{4} = \text{RHS.}$$

Example 5. Prove that:

(*i*) $\sin^{-1}\frac{3}{5} + \sin^{-1}\frac{8}{17} = \sin^{-1}\frac{77}{85}$

(*ii*) $\tan^{-1}\left[\frac{\cos x - \sin x}{\cos x + \sin x}\right] = \left(\frac{\pi}{4} - x\right)$

Solution:

(*i*) $\text{L.H.S} = \sin^{-1}\frac{3}{4} + \sin^{-1}\frac{8}{17}$

$$= \sin^{-1}\left[\frac{3}{5}\cdot\sqrt{1-\left(\frac{8}{17}\right)^2} + \frac{8}{17}\cdot\sqrt{1-\left(\frac{3}{5}\right)^2}\right]$$

$$= \sin^{-1}\frac{77}{85} = \text{RHS.}$$

(*ii*) $\text{LHS.} = \tan^{-1}\left(\frac{\cos x - \sin x}{\cos x + \sin x}\right) = \tan^{-1}\left(\frac{1-\tan x}{1+\tan x}\right)$

$$= \tan^{-1}\left[\frac{\tan\frac{\pi}{4} - \tan x}{1 + \tan\frac{\pi}{4}.\tan x}\right] = \tan^{-1}\left[\tan\left(\frac{\pi}{4} - x\right)\right]$$

$$= \left(\frac{\pi}{4} - x\right) = \text{RHS.}$$

Example 6. If $\cos^{-1}\frac{x}{a} + \cos^{-1}\frac{y}{b} = \alpha$, prove that

$$\frac{x^2}{a^2} - \left(\frac{2xy}{ab}\right)\cos\alpha + \frac{y^2}{b^2} = \sin^2\alpha.$$

Solution: Given, $\cos^{-1}\frac{x}{a} + \cos^{-1}\frac{y}{b} = \alpha.$

$$\Rightarrow \quad \cos^{-1}\left[\frac{x}{a}.\frac{y}{b} - \sqrt{\left(1 - \frac{x^2}{a^2}\right)\left(1 - \frac{y^2}{b^2}\right)}\right] = \alpha$$

$$\Rightarrow \quad \frac{xy}{ab} - \sqrt{\left(1 - \frac{x^2}{a^2}\right)\left(1 - \frac{y^2}{b^2}\right)} = \cos\alpha$$

$$\Rightarrow \quad \left(\frac{xy}{ab} - \cos\alpha\right) = \sqrt{\left(1 - \frac{x^2}{a^2}\right)\left(1 - \frac{y^2}{b^2}\right)}$$

$$\Rightarrow \quad \left(\frac{xy}{ab} - \cos\alpha\right)^2 = \left(1 - \frac{x^2}{a^2}\right)\left(1 - \frac{y^2}{b^2}\right)$$

$$\Rightarrow \quad \frac{x^2y^2}{a^2b^2} + \cos^2\alpha - \frac{2xy}{ab}\cos\alpha = 1 + \frac{x^2}{a^2}.\frac{y^2}{b^2} - \frac{x^2}{a^2} - \frac{y^2}{b^2}$$

$$\Rightarrow \quad \frac{x^2}{a^2} - \frac{2xy}{ab}\cos\alpha + \frac{y^2}{b^2} = 1 - \cos^2\alpha = \sin^2\alpha$$

Example 7. Solve:

(*i*) $\tan^{-1} 2x + \tan^{-1} 3x = \frac{\pi}{4}$

(*ii*) $2\tan^{-1}(\cos x) = \tan^{-1}(2\,\text{cosec x})$

Solution. (i) $\tan^{-1}2x + \tan^{-1}3x = \frac{\pi}{4}$

$$\Rightarrow \quad \tan^{-1}\left(\frac{2x + 3x}{1 - 2x\,.\,3x}\right) = \frac{\pi}{4}$$

$$\Rightarrow \qquad \frac{5x}{1-6x^2} = \tan\frac{\pi}{4} = 1$$

$$\Rightarrow \qquad 5x = 1 - 6x^2$$

$$\Rightarrow \qquad 6x^2 + 5x - 1 = 0$$

$$\Rightarrow \qquad (6x - 1)(x + 1) = 0$$

$$\Rightarrow \qquad x = \frac{1}{6} \text{ or } x = -1$$

But, on putting $x = -1$ in the given equation,

$$\text{L.H.S.} = \tan^{-1}(-2) + \tan^{-1}(-3) = \pi + \tan^{-1}\left[\frac{(-2)+(-3)}{1-(-2)(-3)}\right]$$

$$= \pi + \tan^{-1} 1 = \left(\pi + \frac{\pi}{4}\right) \neq \frac{\pi}{4}$$

$\therefore$ $x = -1$ does not satisfy the given equation

Hence, $x = \dfrac{1}{6}$ is the required solution.

(*ii*) let $2\tan^{-1}(\cos x) = y$. Then, $\cos x = \tan\dfrac{y}{2}$

$$\therefore \qquad 2\operatorname{cosec} x = \frac{2}{\sin x} = \frac{2}{\sqrt{1-\cos^2 x}} = \frac{2}{\sqrt{1-\tan^2\frac{y}{2}}}$$

Now, $2\tan^{-1}(\cos x) = \tan^{-1}(2\operatorname{cosec}(x)$

$$\Rightarrow \qquad y = \tan^{-1}\left[\frac{2}{\sqrt{1-\tan^2\frac{y}{2}}}\right]$$

$$\Rightarrow \qquad \tan y = \frac{2}{\sqrt{1-\tan^2\frac{y}{2}}}$$

$$\Rightarrow \qquad \frac{2\tan\frac{y}{2}}{1-\tan^2\frac{y}{2}} = \frac{2}{\sqrt{1-\tan^2\frac{y}{2}}}$$

$$\Rightarrow \qquad \tan\frac{y}{2} = \sqrt{1-\tan^2\frac{y}{2}}$$

$$\Rightarrow \qquad \tan^2\frac{y}{2} = 1-\tan^2\frac{y}{2}$$

$$\Rightarrow \qquad \tan^2\frac{y}{2} = \frac{1}{2}$$

$$\Rightarrow \qquad \tan\frac{y}{2} = \frac{1}{\sqrt{2}}$$

$$\Rightarrow \qquad \cos x = \frac{1}{\sqrt{2}} \left[\because \tan\frac{y}{2} = \cos x\right]$$

$$\Rightarrow \qquad x = \frac{\pi}{4}$$

Example 8. Prove that

(*i*) $\sin^{-1}(2x\sqrt{1-x^2}) = 2\sin^{-1}x$

(*ii*) $\cos^{-1}(2x^2 - 1) = 2\cos^{-1}x$

Solution:

(*i*) $\text{LHS} = \sin^{-1}(2x\sqrt{1-x^2})$; Put $x = \sin\theta \Rightarrow \theta = \sin^{-1}x$

$= \sin^{-1}(2\sin\theta\sqrt{1-\sin^2\theta})$

$= \sin^{-1}(2\sin\theta\cos\theta) = \sin^{-1}(\sin 2\theta)$

$= 2\theta = 2\sin^{-1}x = \text{RHS}$

(*ii*) $\text{L.H.S} = \cos^{-1}(2x^2 - 1)$; Put $x = \cos\theta \Rightarrow \theta = \cos^{-1}x$

$= \cos^{-1}(2\cos^2\theta - 1)$

$= \cos^{-1}(\cos 2\theta) = 2\theta = 2\cos^{-1}x = \text{RHS}$

Example 9. Solve:

$$\sin^{-1}(1-x) - 2\sin^{-1}x = \frac{\pi}{2}$$

Solution: $\sin^{-1}(1-x) - 2\sin^{-1}x = \frac{\pi}{2}$ ...(i)

$$\Rightarrow \qquad \sin^{-1}(1-x) = \frac{\pi}{2} + 2\sin^{-1}x$$

$$\Rightarrow \qquad (1-x) = \sin\left(\frac{\pi}{2} + 2\sin^{-1}x\right)$$

$$\Rightarrow \qquad 1 - x = \cos(2\sin^{-1}x) = \cos[\cos^{-1}(1 - 2x^2)]$$

$$\Rightarrow \qquad 1 - x = 1 - 2x^2 \Rightarrow 2x^2 - x = 0$$

$$\Rightarrow \qquad x(2x - 1) = 0$$

$$\Rightarrow \qquad x = 0 \text{ or } x = \frac{1}{2}$$

But $x = \frac{1}{2}$ does not satisfy the equations (i). Hence, $x = 0$.

Example 10. Prove that: $\tan^{-1}\dfrac{m}{n} - \tan^{-1}\left(\dfrac{m-n}{m+n}\right) = \dfrac{\pi}{4}$

Solution:

$$\text{L.H.S.} = \tan^{-1}\left[\frac{\frac{m}{n} - \left(\frac{m-n}{m+n}\right)}{1+\frac{m}{n}\left(\frac{m-n}{m+n}\right)}\right]$$

$$= \tan^{-1}\left[\frac{m^2+mn-mn+n^2}{mn+n^2+m^2-mn}\right] = \tan^{-1}\left[\frac{m^2+n^2}{m^2+n^2}\right]$$

$$= \tan^{-1}(1)$$

$$= \frac{\pi}{4} = \text{R.H.S.}$$

EXERCISE 1.8

1. Find the principal values of:

(*i*) $\cos^{-1}\left(\cos\dfrac{7\pi}{6}\right)$ (*ii*) $\cos^{-1}\left(\dfrac{1}{\sqrt{2}}\right)$ (*iii*) $\cot^{-1}(\sqrt{3})$ (*iv*) $\sec^{-1}(\sqrt{2})$

(*v*) $\tan^{-1}\left(\dfrac{-1}{\sqrt{3}}\right)$ (*vi*) $\sin^{-1}(-1)$ (*vii*) $\tan^{-1}(-1)$ (*viii*) $\sin^{-1}\left(\sin\dfrac{2\pi}{3}\right)$

(*ix*) $\tan^{-1}\left(\tan\dfrac{3\pi}{4}\right)$ (*x*) $\sin\left(\dfrac{1}{2}\cos^{-1}\dfrac{4}{5}\right)$ (*xi*) $\sin\left(\dfrac{\pi}{3} - \sin^{-1}\left(\dfrac{-1}{2}\right)\right)$

2. Prove that

(*i*) $\tan^{-1}\left(\dfrac{1}{2}\right) + \tan^{-1}\left(\dfrac{1}{3}\right) = \dfrac{\pi}{4}$

(*ii*) $\cos^{-1}\left(\dfrac{4}{5}\right) + \cos^{-1}\left(\dfrac{12}{13}\right) = \cos^{-1}\left(\dfrac{33}{65}\right)$

(*iii*) $\sin^{-1}\left(\dfrac{12}{13}\right) + \sin^{-1}\left(\dfrac{4}{5}\right) = \sin^{-1}\left(\dfrac{63}{65}\right)$

Prove that:

3. $2\tan^{-1}\left(\dfrac{1}{x}\right) = \sin^{-1}\left(\dfrac{2x}{1+x^2}\right)$

4. $\tan^{-1}\left(\dfrac{\cos x}{1+\sin x}\right) = \left(\dfrac{\pi}{4} - \dfrac{x}{2}\right)$

5. $\cot^{-1}\left(\dfrac{ab+1}{a-b}\right) + \cot^{-1}\left(\dfrac{bc+1}{b-c}\right) + \cot^{-1}\left(\dfrac{ac+1}{c-a}\right) = 0$

6. $\tan^{-1}x + \cot^{-1}(x+1) = \tan^{-1}(1+x+x^2)$.

7. $\tan^{-1}\sqrt{\dfrac{a-x}{a+x}} = \dfrac{1}{2}\cos^{-1}\left(\dfrac{x}{a}\right)$

8. $\tan^{-1}\left(\dfrac{2x}{1-x^2}\right) + \cot^{-1}\left(\dfrac{1-x^2}{2x}\right) = \dfrac{\pi}{3}$

9. $2\tan^{-1}\dfrac{1}{2} + \tan^{-1}\dfrac{1}{7} = \tan^{-1}\dfrac{31}{17}$

10. $\sec^2(\tan^{-1}2) + \text{cosec}^2(\cot^{-1}3) = 15$

Solve for x, the following functions:

11. $\tan^{-1}\left(\dfrac{x-1}{x-2}\right) + \tan^{-1}\left(\dfrac{x+1}{x+2}\right) = \dfrac{\pi}{4}$

12. $\cot^{-1}x - \cot^{-1}(x+2) = \dfrac{\pi}{12}$, $x > 0$

13. $\sin^{-1}x + \sin^{-1}2x = \dfrac{\pi}{3}$

14. $\cos^{-1}x + \sin^{-1}\dfrac{x}{2} = \dfrac{\pi}{6}$

15. If $\tan^{-1}a + \tan^{-1}b + \tan^{-1}c = \pi$; prove that a + b + c = abc

16. If $\sin^{-1}a + \sin^{-1}b + \sin^{-1}c = \pi$, prove that

$$a\sqrt{1-a^2} + b\sqrt{1-b^2} + c\sqrt{1-c^2} = 2abc$$

17. If $\cos^{-1}a + \cos^{-1}b + \cos^{-1}c = \pi$, prove that

$a^2 + b^2 + c^2 + 2abc = 1$

18. Prove that

$$\tan^{-1}\left(\frac{c_1x-y}{c_1y-x}\right) + \tan^{-1}\left(\frac{c_2-c_1}{1+c_2c_1}\right) + \tan^{-1}\left(\frac{c_2-c_2}{1+c_3c_2}\right) + .. + \tan^{-1}\left(\frac{1}{c_n}\right) = \tan^{-1}\left(\frac{x}{y}\right)$$

ANSWERS

1. (*i*) $\dfrac{5\pi}{6}$ (*ii*) $\dfrac{\pi}{4}$ (*iii*) $\dfrac{\pi}{6}$ (*iv*) $\dfrac{\pi}{4}$ (*v*) $\dfrac{-\pi}{4}$ (*vi*) $\dfrac{-\pi}{2}$

(*vii*) $\dfrac{-\pi}{4}$ (*viii*) $\dfrac{\pi}{3}$ (*ix*) $\dfrac{-\pi}{4}$ (*x*) $\dfrac{1}{\sqrt{10}}$ (*xi*) $\dfrac{1}{2}$

11. $\pm\dfrac{1}{\sqrt{2}}$ **12.** $\sqrt{3}$ **13.** $\dfrac{1}{2}\sqrt{\dfrac{3}{7}}$

14. ± 1

COMPLEX NUMBERS

1.17 INTRODUCTION

If a, b, are natural numbers such that $a > b$ then the equation $x + a = b$ is not solvable in N, the set of natural numbers i.e; there is no natural number satisfying the equation $x + a = b$. So the set of natural numbers is extended to form the set I of integers in which every equation of the form $x + a = b$; $a, b, \in N$ is solvable. But equations of the form $xa = b$, where $a, b, \in I$, $a \neq 0$ are not solvable in I also. There fore the set I of integers is extended to obtain the set Q of all rational numbers in which every equation of the form $xa = b$, $a \neq 0$ $a, b \in I$ is uniquely solvable. The equations of the form $x^2 = 2$, $x^2 = 3$ etc. are not solvable in Q because there is no rational number whose square is 2. Such numbers are known as irrational numbers. The set Q of all rational numbers is extended to obtain the set R which includes both rational and irrational numbers. This set is known as the set of real numbers. The equations of the form $x^2 + 1 = 0$, $x^2 + 4 = 0$ etc. are not solvable in R i.e; there is no real number whose square is a negative real number. Euler was the first mathematician to introduce the symbol i (iota) for the square root of –1 with the property that $i^2 = -1$. He called this symbol as the imaginary unit.

1.18 INTEGRAL POWERS OF IOTA (I)

Positive integral powers of *i*.

we have $\quad i = \sqrt{-1}$

$$\therefore \quad i^2 = -1$$

$$i^3 = i^2 \times i = -i$$

$$i^4 = (i^2)^2 = (-1)^2 = 1$$

In order to compute i^n for $n > 4$, we divide n by 4 and obtain the remainder r. Let m be the quotient when n is divided by 4. Then .

$$n = 4\,m + r, \text{ where } 0 \le r < 4$$

$$\Rightarrow \quad i^n = (i)^{4m + r} = (i^4)^m\, i^r = i^r$$

Thus, the value of i^n for $n > 4$ is i, where r is the remainder when n is divided by 4

Negative integral powers of *i*:

By the law of indices, we have

$$i^{-1} = \frac{1}{i} = \frac{i^3}{i^4} = i^3 = -i$$

$$i^{-2} = \frac{1}{i^2} = \frac{1}{-1} = -1$$

$$i^{-3} = \frac{1}{i^3} = \frac{i}{i^4} = i$$

$$i^{-4} = \frac{1}{i^4} = \frac{1}{1} = 1$$

of $n > 4$, then

$i^{-n} = \frac{1}{i^n} = \frac{1}{i^r}$, where r is the remainder when n is divided by 4.

ILLUSTRATIVE EXAMPLES

Example 1. Evaluate the following

(*i*) i^{17} (*ii*) i^{135} (*iii*) i^{-998}

Solution: (i) the remainder is 1 when 17 is divided by 4, there fore $i^{17} = i$

(*ii*) The remainder is 3 when 135 is divided by 4, therefore $i^{135} = i^3 = -i$

(*iii*) $i^{-998} = \frac{1}{i^{998}} = \frac{1}{i^2} = \frac{1}{-1} = -1$

Example 2. Show that $1 + i^{10} + i^{20} + i^{30}$ is a real number.

Solution: $1 + i^{10} + i^{20} + i^{30} = 1 + i^2 + 1 + i^2 = 1 - 1 + 1 - 1 = 0$

Example 3. Find the values of the following expressions.

(*i*) $\left[i^{19} + \left(\frac{1}{i}\right)^{25}\right]^2$ (*ii*) $\left[i^{17} - \left(\frac{1}{i}\right)^{34}\right]^2$ (*iii*) $i^n + i^{n+1} + i^{n+2} + i^{n+3}$, $n \in N$

Solution:

(*i*) $\left[i^{19} + \left(\frac{1}{i}\right)^{25}\right]^2 = \left[i^{19} + \frac{1}{i^{25}}\right]^2 = \left[i^3 + \frac{1}{i}\right]^2 = [-i + i^3]^2$

$= (-i - i)^2 = 4i^2 = -4$

(*ii*) $\left[i^{17} - \left(\frac{1}{i}\right)^{34}\right]^2 = \left[i^{17} - \frac{1}{i^{34}}\right]^2 = \left[i - \frac{1}{i^2}\right]^2 = \left[i - \frac{1}{(-1)}\right]^2 = (i + 1)^2$

$= i^2 + 2i + 1 = -1 + 2i + 1 = 2i$

(*iii*) $i^n + i^{n+1} + i^{n+2} + i^{n+3} = i^n + i^n \times i + i^n \times i^2 + i^n \times i^3$

$= i^n [1 + i + i^2 + i^3] = i^n [1 + i - 1 - i] = i^n \times 0 = 0$

1.19 IMAGINARY QUANTITIES

The square root of a negative real number is called an imaginary qnantity or an imaginary number.

For example $\sqrt{-3}$, $\sqrt{-5}$, $\sqrt{\frac{-13}{4}}$ etc. are imaginary quantities.

Note: 1. If a, b, are positive real numbers, then $\sqrt{-a} \times \sqrt{-b} = -\sqrt{ab}$

2. For any two real numbers $\sqrt{a} \times \sqrt{b} = \sqrt{ab}$ is true only when atleast one of a and b is either positive or zero. In other words $\sqrt{a} \times \sqrt{b} = \sqrt{ab}$ is not valid if a and b both are negative.

ILLUSTRATIVE EXAMPLES

Example 1. Compute the following (i) $\sqrt{-121}$ (ii) $\sqrt{-4} \times \sqrt{\frac{-9}{4}}$

Solution: $\sqrt{-121} = \sqrt{-1 \times 121} = \sqrt{-1} \times \sqrt{121} = i \times 11 = 11i$

(*ii*) $\sqrt{-4} \times \sqrt{\frac{-9}{4}} = (2i)\left(\frac{3i}{2}\right) = 3i^2 = -3.$

1.20 COMPLEX NUMBER

If a, b, are two real numbers then a number of the form $a + ib$ is called a complex number.

eg. $2 + 3i$, $-1 + 2i$, $0 + 3i$, $1 + 0i$, etc are complex numbers

Real and Imaginary Parts of a Complex Numbers

If $z = a + ib$ is a complex number then a is called real part of z and 'b' is known as imaginary part of z. The real part of z is denoted by Re (z) and the imaginary part of Im(z).

e.g. If $z = 2 + 3i$, then Re $(z) = 2$ and Im$(z) = 3$

Set of Complex Numbers

The set of all complex numbers is denoted by C. i.e $C = \{a + ib \mid a, b \in R\}$

Since a real number 'a' can be written as $a + 0.i$, therefore every real number is a complex number. Hence $R \subset C$, where R is the set of all real numbers.

Equality of Complex Numbers

Two complex numbers $z_1 = a_1 + ib_1$ and $z_2 = a_2 + ib_2$ are equal it $a_1 = a_2$ and $b_1, = b_2$ i.e; Re (z_1) = Re (z_2) and Im (z_1) = Im (z_2). Thus $z_1 = z_2 \Leftrightarrow$ Re (z_1) = Re (z_2) and Im (z_1) = Im (z_2).

e.g. If $z_1 = 2 + iy$ and $z_2 = x + 4i$ are equal, find x and y.

Solution: we have $z_1 = z_2$

$2 + iy = x + 4i \Rightarrow 2 = x$ and $y = 4 \Rightarrow x = 2$ and $y = 4$

Addition of complex numbers

Let $z_1 = a_1 + ib_1$ and $z_2 = a_2 + ib_2$ be two complex numbers. Then their sum $z_1 + z_2$ is defined as the complex number $(a_1 + a_2) + i\,(b_1 + b_2)$. If follows from this definition that the sum $z_1 + z_2$ is a complex number such that Re $(z_1 + z_2)$ = Re (z_1) + Re (z_2) and Im $(z_1 + z_2)$ = Im (z_1) + *Im* (z_2)

eg. If $z_1 = 1 + 2i$ and $z_2 = 2 + 3i$

then $z_1 + z_2 = (1 + 2) + i\,(2 + 3) = 3 + 5i$

1.21 PROPERTIES OF ADDITION OF COMPLEX NUMBERS

(*i*) **Addition is commutative.** For any two complex numbers z_1 and z_2 we have $z_1 + z_2 = z_2 + z_1$

(*ii*) **Addition is Associative.** For any three complex numbers z_1, z_2, z_3 we have $(z_1 + z_2) + z_3 = z_1 + (z_2 + z_3)$.

(*iii*) **Existence of Additive Identity.** The complex numbers $0 = 0 + i0$ is the identity element for addition i.e; $z + 0 = z = 0 + z$ for all $z \in c$

(*iv*) **Existence of Additive Inverse.** For any complex number $z = a + ib$, there exists $-z = (-a) + i(-b)$ such that $z + (-z) = 0 = (-z) + z$.

1.22 SUBTRACTION OF COMPLEX NUMBERS

Let $z_1 = a_1 + ib_1$ and $z_2 = a_2 + ib_2$ be two complex numbers. Then the subtraction of z_2 from z_1 is dendied by $z_1 - z_2$ and is defined as the addition of z_1 and $-z_2$.

Thus $z_1 - z_2 = z_1 + (-z_2) = (a_1 + ib_1) + (-a_2 - ib_2) = (a_1 - a_2) + i(b_1 - b_2)$

eg. If $z_1 = -4 + 5i$ and $z_2 = 6 + 7i$, then

$$z_1 - z_2 = (-4 + 5i) + (-6 - 7i) = (-4 - 6) = (-4 - 6) + i(5 - 7)$$
$$= -10 - 2i.$$

1.23 MULTIPLICATION OF COMPLEX NUMBERS

Let $z_1 = a_1 + ib_1$ and $z_2 = a_2 + ib_2$ be two complex numbers. Then the multiplication of z_1 with z_2 is denoted by $z_1 z_2$ and is defined as the complex number $(a_1 a_2 - b_1 b_2) + \text{i}(a_1 b_2 + a_2 b_1)$

eg. If $z_1 = 2 + 3i$ and $z_2 = 4 - 2i$, then

$$z_1 z_2 = (2 + 3i)(4 - 2\text{i}) = (2 \times 4 - 3 \times -2) + i(2 \times -2 + 4 \times 3)$$
$$= (8 + 6) + i(-4 + 12)$$
$$= 14 + 8i$$

Properties of Multiplication

(*i*) **Multiplication is commutative.** For any two complex numbers z_1 and z_2, we have $z_1 z_2 = z_2 z_1$.

(*ii*) **Multiplication is Associative.** For any three complex numbers z_1, z_2, z_3, we have $(z_1 z_2) z_3 = z_1 (z_2 z_3)$.

(*iii*) **Existence of identity element for multiplication.** The complex number $1 = 1 + i0$ is the identity element for multiplication i.e; for every complex number z, we have $z . 1 = z = 1 . z$

(*iv*) **Existence of Multiplicative Inverse.** Corresponding to every non zero complex number $z = a + ib$ there exist a complex number $z_1 = x + iy$ such that $z . z_1 = 1 = z_1 . z$

Every non - zero complex number $z = a + ib$ possesses multiplicative inverse given by $\left(\frac{a}{a^2+b^2}\right)+i\left(\frac{-b}{a^2+b^2}\right)$

Note: The multiplicative inverse of z is denoted by z^{-1} or $\frac{1}{z}$.

Example. Find the multiplicative inverse of $z = 2 + 3i$

Solution: $z^{-1} = \left(\frac{2}{2^2+3^2}\right)+i\left(\frac{-3}{2^2+3^2}\right)=\frac{2}{13}-\frac{3i}{13}$

(*v*) Multiplication of complex numbers is distributive over addition of complex numbers: For any three complex numbers z_1, z_2, z_3 we have

(*i*) $z_1 (z_2 + z_3)= z_1z_2 + z_1z_3$

(*ii*) $(z_2 + z_3) z_1= z_2z_1 + z_3z_1$

1.24 DIVISION OF COMPLEX NUMBER

The division of a complex number z_1 by a non-zero complex number z_2 is defined as the multiplication of z_1 by the multiplicative inverse of z_2 and is denoted by $\frac{z_1}{z_2}$.

Thus $$\frac{z_1}{z_2} = z_1.z_2^{-1} = z_1.\left(\frac{1}{z_2}\right)$$

Example. If $z_1 = 2 + 3i$ and $z_2 = 2 + 2i$, then

$$\frac{z_1}{z_2} = z_1 .\frac{1}{z_2}= (2+3i)\left(\frac{1}{2+2i}\right) = (2+3i)\left(\frac{2}{8}-\frac{2}{8}i\right)$$

$$= (2 + 3i) \left(\frac{1}{4}-\frac{1}{4}i\right)=\left(2+\frac{1}{4}\right)+i\left(3-\frac{1}{4}\right)$$

$$= \frac{9}{4}+i\frac{11}{4}=\frac{9}{4}+\frac{11}{4}i$$

1.25 CONJUGATE OF A COMPLEX NUMBER

Let $z = a + ib$ be a complex number then the conjugate of z is denoted by $\overline{z}$ and is equal to $a - ib$.

Thus $$z = a + ib \Rightarrow \overline{z} = a - ib$$

It follows from this definition that the conjugate of a complex number is obtained by replacing i by $- i$

Example. If $z = 2 + 3i$, then $\overline{z} = 2 - 3i$

Properties of conjugate

If z, z_1, z_2 are complex numbers, then

(*i*) $\overline{(\overline{z})} = z$

(*ii*) $z + \overline{z} = 2 \text{ Re } (z)$

(*iii*) $z - \overline{z} = 2 \text{ Im } (z)$

(*iv*) $z = \overline{z} \Leftrightarrow z$ is purely real

(*v*) $z + \overline{z} = 0 \Rightarrow z$ is purely imaginary

(*vi*) $z\,\overline{z} = \{\text{Re}(z)\}^2 + \{\text{I}_\text{m}(z)\}^2$

(*vii*) $\overline{z_1 + z_2} = \overline{z}_1 + \overline{z}_2$

(*viii*) $\overline{z_1 - z_2} = \overline{z}_1 - \overline{z}_2$

(*ix*) $\overline{z_1 z_2} = \overline{z}_1\,\overline{z}_2$.

(*x*) $\overline{\left(\dfrac{z_1}{z_2}\right)} = \dfrac{\overline{z}_1}{\overline{z}_2}$, $z_2 \neq 0$.

1.26 MODULUS OF A COMPLEX NUMBERS

The modulus of a complex number $z = a + ib$ is denoted by $|\,z\,|$ and is defined as $|\,z\,| = \sqrt{a^2 + b^2} = \sqrt{\{\text{Re}\,(z)\}^2 + \{\text{Im}\,(z)\}^2}$. Clearly $|\,z\,| \geq 0$ for all $z \in c$.

Example. If $z_1 = 2 + 4i$, $z_2 = -2 + 3i$ then

$$|\,z_1\,| = \sqrt{2^2 + 4^2} = \sqrt{4 + 16} = \sqrt{20}, |\,z_2\,| = \sqrt{(-2)^2 + 3^2}$$

$$= \sqrt{4 + 9} = \sqrt{13}$$

Note: In the set C of all complex numbers, the order relation is not defined. As such $z_1 > z_2$ or $z_1 < z_2$ has no meaning but $|\,z_1| > |\,z_2|$ or $|\,z_1| < |\,z_2|$ has got its meaning since $|z_1|$ and $|z_2|$ are real numbers.

Properties of Modulus

If z, z_1, $z_2 \in C$ then

(*i*) $|\,z| = 0 \Leftrightarrow z = 0$ i.e; $\text{Re}\,(z) = \text{Im}\,(z) = 0$

(*ii*) $|\,z| = |\,\overline{z}\,| = |-z|$

(*iii*) $z\overline{z} = |\,z\,|^2$

(*iv*) $|\,z_1 z_2| = |\,z_1|\,|\,z_2|$

(*v*) $\left|\dfrac{z_1}{z_2}\right| = \dfrac{|z_1|}{|z_2|}$, $z_2 \neq 0$

(*vi*) $|\,z_1 + z_2\,|^2 = |\,z_1|^2 + |\,z_2|^2 + 2 \text{ Re } (z_1\,\overline{z}_2)$

(vii) $| z_1 - z_2|^2 = | z |^2 + | z_2|^2 - 2 \text{ Re } (z_1 \, \overline{z}_2)$

(viii) $| z_1 + z_2|^2 + |z_1 - z_2|^2 = 2 \,(| z_1|^2 + |z_2|^2)$

1.27 RECIPROCAL OF A COMPLEX NUMBER

Let $z = a + ib$ be a non zero complex number. Then,

$$\frac{1}{z} = \frac{1}{a+ib} = \frac{1}{a+ib} \times \frac{a-ib}{a-ib} = \frac{a-ib}{a^2 - i^2 b^2} = \frac{a-ib}{a^2+b^2}$$

$$\frac{1}{z} = \frac{a}{a^2+b^2} + \frac{i\,(-b)}{a^2+b^2}$$

Clearly, $\frac{1}{z}$ is equal to the multiplicative inverse of z.

Also , $$\frac{1}{z} = \frac{a-ib}{a^2+b^2} = \frac{\overline{z}}{|z|^2}$$

Thus the multiplicative inverse of a non - zero complex number z is same as its reciprocal and is given by

$$\frac{\text{Re}(z)}{|z|^2} + i\,\frac{(-\text{Im}\,(z))}{|z|^2} = \frac{\overline{z}}{|z|^2}$$

1.28 SQUARE ROOTS OF A COMPLEX NUMBER

Let $a + ib$ be a complex number such that $\sqrt{a+ib} = x + iy$, where x and y are real numbers. Then,

$$\sqrt{a+ib} = x + iy$$

$$(a + ib) = (x + iy)^2$$

$$\Rightarrow \quad (a + ib) = (x^2 - y^2) + 2ixy$$

Equating real and imaginary parts, we get

$$x^2 - y^2 = a \qquad \text{...(1)}$$

and $$2xy = b \qquad \text{...(2)}$$

Now $$(x^2 + y^2)^2 = (x^2 - y^2)^2 + 4x^2y^2$$

$$(x^2 + y^2)^2 = a^2 + b^2$$

$$(x^2 + y^2) = \sqrt{a^2+b^2} \quad [\because x^2 + y^2 \geq 0] \qquad \text{...(3)}$$

Solving equations (1) and (3), we get

$$x^2 = \frac{1}{2}[\sqrt{a^2+b^2} + a] \text{ and } y^2 = \frac{1}{2}[\sqrt{a^2+b^2} - a]$$

$$\Rightarrow \quad x = \pm\sqrt{\frac{1}{2}[\sqrt{a^2+b^2} + a]} \text{ and } y = \pm\sqrt{\frac{1}{2}[\sqrt{a^2+b^2} - a]}$$

If b is positive, then by equation (2), x and y are of the same signs.

Hence $$\sqrt{a+ib} = \pm\left\{\sqrt{\frac{1}{2}[\sqrt{a^2+b^2}+a]}+i\sqrt{\frac{1}{2}[\sqrt{a^2+b^2}-a}\right\}$$

If b is negative then by equation (2) x and y are of different signs.

Hence $$\sqrt{a+ib} = \pm\left\{\sqrt{\frac{1}{2}[\sqrt{a^2+b^2}+a]}-i\sqrt{\frac{1}{2}[\sqrt{a^2+b^2}-a}\right\}$$

Note: To find the square root of a – ib replace i by – i in the above results.

ILLUSTRATIVE EXAMPLES

Example 1. Find the conjugate of $\frac{1}{2+3i}$

Solution: Let $z = \frac{1}{2+3i}$, then

$$z = \frac{1}{2+3i}\cdot\frac{2-3i}{2-3i} = \frac{2-3i}{4+9} = \frac{2}{13}-\frac{3}{13}i$$

$$\therefore \qquad \bar{z} = \frac{2}{13}+\frac{3}{13}i$$

Example 2. Express the following complex numbers in the standard form. Also, find their conjugate.

(*i*) $\frac{1-i}{1+i}$ (*ii*) $\frac{(2+3i)^2}{2-i}$

Solution: (*i*) Let $z = \frac{1-i}{1+i} = \frac{1-i}{1+i}\times\frac{1-i}{1-i} = \frac{(1-i)^2}{1^2-i^2} = \frac{1-2i+i^2}{1-i^2}$

$$z = \frac{1-2i-1}{1+1} = 0 - i$$

$$\bar{z} = 0 + i$$

(*ii*) Let $z = \frac{(2+3i)^2}{2-i} = \frac{4+12i+9i^2}{2-i} = \frac{4+12i-9}{2-i}\times\frac{2+i}{2+i}$

$$= \frac{-5+12i}{2-i}\times\frac{2+i}{2+i} = \frac{-22+19i}{4-i^2} = \frac{-22+19i}{4+1}$$

$$z = \frac{-22}{5}+\frac{19}{5}i$$

$$\bar{z} = \frac{-22}{5}-\frac{19}{5}i$$

Example 3. Find the multiplicative inverse of the following complex numbers:

(*i*) $2 + 3i$ (*ii*) $3 + 4i$ (*iii*) $(1+\sqrt{2}i)^2$

Solution:

(*i*) Let $z = 2 + 3i$ then

$$\frac{1}{z} = \frac{1}{2+3i} = \frac{2-3i}{(2+3i)(2-3i)} = \frac{2-3i}{4+9} = \frac{2-3i}{13}$$

$$\frac{1}{z} = \frac{2}{13} - \frac{3}{13}i$$

(*ii*) Let $z = 3 + 4i$, then

$$\frac{1}{z} = \frac{1}{3+4i} = \frac{3-4i}{(3+4i)(3-4i)} \frac{(3-4i)}{9+16} = \frac{(3-4i)}{25}$$

$$\frac{1}{z} = \frac{3}{25} - \frac{4}{25}i$$

(*iii*) Let $z = (1+\sqrt{2}\,i)^2$, then

$$z = 1 + 2\sqrt{2}i + 2i^2 = 1 + 2\sqrt{2}i - 2 = -1 + 2\sqrt{2}i$$

$$\therefore \quad \frac{1}{z} = \frac{1}{-1+2\sqrt{2}i} = \frac{(-1-2\sqrt{2}i)}{(-1+2\sqrt{2}i)(-1-2\sqrt{2}i)}$$

$$= \frac{(-1-2\sqrt{2}i)}{-1+8} = \frac{(-1-2\sqrt{2}i)}{7}$$

$$\frac{1}{z} = \frac{-1}{7} - \frac{-2\sqrt{2}}{7}i$$

Example 4. Find the real values of x and y for which the following equality hold. $(1 + i)y^2 + (6 + i) = (2 + i)x$

Solution: $(1 + i)y^2 + (6 + i) = (2 + i)x$

$\Rightarrow (y^2 + 6) + i(y^2 + 1) = 2x + ix$

$\Rightarrow \quad y^2 + 6 = 2x$ ...(1)

and $\quad y^2 + 1 = x$ ...(2)

From (1) and (2), we get

$$y^2 + 6 = 2(y^2 + 1) \Rightarrow y^2 = 4 \Rightarrow y = \pm 2$$

Now $\quad x = y^2 + 1 \Rightarrow x = 5$ when $y = \pm 2$

Thus $\quad x = 5$ and $y = 2$ or, $x = 5$ and $y = -2$

Example 5. Find the square root of $7 - 24i$

Solution: Let $\sqrt{7-24i} = x + iy$, then

$$7 - 24i = (x + iy)^2$$

$\Rightarrow \qquad 7 - 24i = (x^2 - y^2) + 2i\,xy$

$\Rightarrow \qquad x^2 - y^2 = 7 \qquad ...(1)$

$2xy = -24 \qquad ...(2)$

Now $\quad (x^2 + y^2)^2 = (x^2 - y^2)^2 + 4x^2y^2$

$\Rightarrow \qquad (x^2 + y^2)^2 = 49 + 576 = 625$

$x^2 + y^2 = 25 \; [\because x^2 + y^2 > 0] \qquad ...(3)$

Solving (1) and (3) we get

$$x^2 = 16 \text{ and } y^2 = 9 \Rightarrow x = \pm 4 \text{ and } y = \pm 3$$

From (2) , $2xy$ is negative. So x and y are of oppositive signs.

$\therefore$ ($x = 4$ and $y = -3$) or ($x = -4$ and $y = 3$)

Hence $\quad \sqrt{7 - 24i} = \pm (4 - 3i)$

Example 6. Find the square root of i

Solution: Let $\sqrt{i} = x + iy$, then $\sqrt{i} = x + iy$

$\Rightarrow \qquad i = (x + iy)^2 = (x^2 - y^2) + 2ixy$

$\Rightarrow \qquad 0 + i = (x^2 - y^2) + 2ixy$

$\Rightarrow \qquad x^2 - y^2 = 0 \qquad ...(1)$

and $\qquad 2xy = 1 \qquad ...(2)$

Now $\quad (x^2 + y^2)^2 = (x^2 - y^2)^2 + 4x^2y^2$

$(x^2 + y^2) = 0 + 1 = 1$

$\Rightarrow \qquad x^2 + y^2 = 1 \; [\because x^2 + y^2 > 0] \qquad ...(3)$

Solving (1) and (3) , we get

$$x^2 = \frac{1}{2} \text{ and } y^2 = 1/2 \Rightarrow x = \pm\frac{1}{\sqrt{2}} \text{ and } y = \pm\frac{1}{\sqrt{2}}$$

From (2), $2xy$ is positive so x and y are of same sign.

$$\therefore \qquad \left(x = \frac{1}{\sqrt{2}} \text{ and } y = \frac{1}{\sqrt{2}}\right) \text{or} \left(x = -\frac{1}{\sqrt{2}} \text{ and } y = -\frac{1}{\sqrt{2}}\right)$$

$$\text{Hence} \qquad \sqrt{i} = \pm\left(\frac{1}{\sqrt{2}} + \frac{1}{\sqrt{2}}i\right) = \pm\frac{1}{\sqrt{2}}(1 + i)$$

Example 7. Find the square root of $-15 - 8i$

Solution: let $\sqrt{-15 - 8i} = x + iy$, then

$-15 - 8i = (x + iy)^2$

$\Rightarrow \qquad -15 - 8i = (x^2 - y^2) + 2ixy$

$\Rightarrow \qquad -15 = x^2 - y^2 \qquad ...(1)$

and $\qquad 2xy = -8 \qquad ...(2)$

Now $\quad (x^2 + y^2)^2 = (x^2 - y^2) + 4x^2y^2$

$$\Rightarrow \qquad (x^2 + y^2)^2 = (-15)2 + 64 = 289$$

$$\Rightarrow \qquad x^2 + y^2 = 17 \qquad \text{...(3)}$$

Solving (1) and (3), we get

$$x^2 = 1 \text{ and } y^2 = 16 \Rightarrow x = \pm 1 \text{ and } y = \pm 4$$

From (2), $2xy$ is negative. So x and y are of opposite signs

$\therefore (x = 1$ and $y = -4)$or $(x = -1$ and $y = 4)$

Hence $\sqrt{-15 - 8i} = (1 - 4i)$

EXERCISE 1.9

1. Express the following complex numbers in the standard form $a + ib$

(i) $\dfrac{1-i}{1+i}$ (ii) $(1 + 2i)^{-3}$ (iii) $\dfrac{1}{(2+i)^2}$ (iv) $\dfrac{3-4i}{(4-2i)(1+i)}$

2. Find the real values of x and y if

(i) $(x + iy)(2 - 3i) = 4 + i$ (ii) $(1 + i)(x + iy) = 2 - 5i$

3. Find the conjugates of the following complex numbers:

(i) $\dfrac{1}{3+5i}$ (ii) $4 - 5i$ (iii) $\dfrac{(3-i)^2}{2+i}$

4. Find the multiplicative inverse of the following complex number:

(i) $1 - i$ (ii) $(1 + i\sqrt{3})^2$

5. Find the square root of the following complex numbers:

(i) $4i$ (ii) $5 + 12i$ (iii) $-8 - 6i$ (iv) $-i$ (v) $-7 - 24i$

ANSWERS

1. (i) $-i$ (ii) $\dfrac{-11}{125} - \dfrac{2i}{125}$ (iii) $\dfrac{3}{25} - \dfrac{4}{25}i$ (iv) $\dfrac{1}{4} - \dfrac{3}{4}i$

2. (i) $x = \dfrac{5}{13}, y = \dfrac{14}{13}$ (ii) $x = \dfrac{-3}{2}, y = \dfrac{-7}{2}$

3. (i) $\dfrac{1}{34}(3 + 5i)$ (ii) $4 + 5i$ (iii) $2 + 4i$

4. (i) $\dfrac{1}{2} + \dfrac{1}{2}i$ (ii) $\dfrac{-1}{8} - i\dfrac{\sqrt{3}}{8}$

5. (i) $\pm\sqrt{2}(1 + i)$ (ii) $\pm(3 + 2i)$ (iii) $\pm(1 - 3i)$

(iv) $\pm\dfrac{1}{\sqrt{2}}(1 - i)$ (v) $\pm(3 - 4i)$

1.29 CUBE ROOTS OF UNITY

let $z = (1)^{1/3}$ then

$$\Rightarrow \quad z^3 = 1$$

$$\Rightarrow \quad z^3 - 1 = 0$$

$$\Rightarrow \quad (z-1)(z^2+z+1) = 0$$

$$\Rightarrow \quad z - 1 = 0 \text{ or } z^2 + z + 1 = 0$$

$$\Rightarrow \quad z = 1 \text{ or, } z = \frac{-1 \pm \sqrt{1-4}}{2}$$

$$\Rightarrow \quad z = 1, \text{ or } z = \frac{-1 \pm i\sqrt{3}}{2}$$

So that cube roots of unity are: $1, \dfrac{-1+i\sqrt{3}}{2}$ and $\dfrac{-1-i\sqrt{3}}{2}$

1.30 PROPERTIES OF CUBE ROOTS OF UNITY

1. One of the cube roots of unity is real and the other two are conjugate complex numbers.
2. Each complex cube root of unity is the square of the other. If we denote one of the complex cube roots of unity by w (omega), then the other complex cube root of unity is w^2.

   Let $w = \dfrac{-1+i\sqrt{3}}{2}$ then $w^2 = \dfrac{-1-i\sqrt{3}}{2}$

   clearly $\overline{w} = w^2$ and $(\overline{w})^2 = w$

   Thus, complex cube roots of unity are conjugate of each other.
3. The sum of three cube roots of unity is zero, ie; $1 + w + w^2 = 0$
4. $w^3 = 1$
5. The product of three cube roots of unity is 1.
6. Each complex cube root of unity is the reciprocal of the other we have w. $w^2 = w^3 = 1$

   $\Rightarrow w = \dfrac{1}{w^2}$ and $w^2 = \dfrac{1}{w}$.
7. Cube roots of – 1 are – 1, – w, w^2

ILLUSTRATIVE EXAMPLES

Example 1. If 1, w, w^2 are three cube roots of unity, show that

(i) $(1 - w + w^2)^3 = (1 + w - w^2)^3 = -8$

(ii) $(1 - w + w^2)(1 + w - w^2) = 4$

(iii) $(a + bw + cw^2)(a + bw^2 + cw) = (a^2 + b^2 + c^2 - ab - bc - ca)$

Solution:

(*i*) we have

$$(1 - w + w^2)^3 = (1 + w^2 - w)^3 = (-w - w)^3 = -8w^3 = -8$$

and $(1 + w - w^2)^3 = (-w^2 - w^2)^3 = (-2w^2)^3 = -8\,(w^3)^2 = -8$

$\therefore \quad (1 - w + w^2)^3 = (1 + w - w^2)^3 = -8$

(*ii*) we have

$$\begin{aligned}(1 - w + w^2)(1 + w - w^2) &= (1 + w^2 - w)(1 + w - w^2) \\ &= (-w - w)(-w^2 - w^2) \\ &= (-2w)(-2w^2) = 4w^3 = 4\end{aligned}$$

(*iii*) we have

$$\begin{aligned}&(a + bw + cw^2)(a + bw^2 + cw) \\ &= a^2 + b^2w^3 + c^2w^3 + ab\,(w + w^2) + bc\,(w^2 + w^4) + ca\,(w + w^2) \\ &= a^2 + b^2 + c^2 + ab\,(w + w^2) + bc\,(w^2 + w) + ca\,(w + w^2) \quad [\because w^3 = 1] \\ &= a^2 + b^2 + c^2 + ab\,(-1) + bc\,(-1) + ca\,(-1) \\ &= a^2 + b^2 + c^2 - ab - bc - ca \quad [\because w + w^2 = -1]\end{aligned}$$

Example 2. If 1, w, w^2 are three cube roots of unity, show that

$$(1 + w)(1 + w^2)(1 + w^4)(1 + w^8) = 1$$

Solution: we have

$$\begin{aligned}(1 + w)(1 + w^2)(1 + w^4)(1 + w^8) &= (1 + w)(1 + w^2)(1 + w)(1 + w^2)[\because w^4 = w, w^8 = w^2] \\ &= (1 + w)^2 (1 + w^2))^2 \\ &= [(1 + w)(1 + w^2)]^2 = [(-w)(-w^2)]^2 = [w^3]^2 = (1)^2 = 1\end{aligned}$$

Example 3. If $x = a + b$, $y = aw + bw^2$ and $z = aw^2 + bw$, where w is a cube root of unity, prove that (i) $xyz = a^3 + b^3$ (ii) $x^2 + y^2 + z^2 = 6ab$ (iii) $x^3 + y^3 + z^3 = 3xyz$ (iv) $x^3 + y^3 + z^3 = 3\,(a^3 + b^3)$.

Solution: we have

$$\begin{aligned}(i) \quad xyz &= (a + b)(aw + bw^2)(aw^2 + bw) \\ &= (a + b)\{a^2 w^3 + b^2w^3 + ab\,(w^2 + w^4\} \\ &= (a + b)\{a^2 + b^2 + ab\,(w^2 + w)\} \\ &= (a + b)(a^2 + b^2 - ab) = a^3 + b^3 \quad [\because w + w^2 = -1]\end{aligned}$$

$$\begin{aligned}(ii)\, x^2 + y^2 + z^2 &= (a + b)^2 + (aw + bw^2)^2 + (aw^2 + bw)^2 \\ &= (a^2 + b^2 + 2ab) + (a^2w^2 + b^2w^4 + 2abw^3) + a^2w^4 \\ &\qquad + b^2w^2 + 2abw^3) \\ &= (a^2 + b^2 + 2ab) + a^2w^2 + b^2w + 2ab) + (a^2w + b^2w^2 \\ &\qquad + 2ab) \\ &= a^2\,(1 + w^2 + w) + b^2\,(1 + w + w^2) + 6ab \\ &= 6ab \quad [\because 1 + w + w^2 = 0]\end{aligned}$$

(*iii*) $x + y + z = (a + b) + (aw + bw^2) + (aw^2 + bw)$

$= a(1 + w + w^2) + b(1 + w^2 + w) = a.0 + b.0 = 0$

$\therefore\ x^3 + y^3 + z^3 = 3xyz \qquad \left[\begin{array}{l} \because x^3 + y^3 + z^3 - 3xyz \\ = (x + y + z)(x^2 + y^2 + z^2 - xy - yz - zx) \end{array}\right]$

(*iv*) From, (*iii*) we have $z^3 + y^3 + z^3 = 3xyz$

$\Rightarrow x^3 + y^3 + z^3 = 3(a^3 + b^3)$ [using (i)]

Example 4. If 1, w, w^2 be the cube roots of unity, prove that $(1 - w + w^2)(1 - w^2 + w^4)(1 - w^4 + w^8)$ ... to $2n$ factors $= 2^{2n}$.

Solution: $(1 - w + w^2)(1 - w^2 + w^4)(1 - w^4 + w^8)$ ... to $2n$ factors

$= (1 - w + w^2)(1 - w^2 + w)(1 - w + w^2)$ ... to $2n$ factors

$= (-2w)(-2w^2)(-2w)(-2w^2)$ ... to $2n$ factors

$= (2^2 w^3)(2^2 w^3)$ ... to n factors.

$= (2^2)^n = 2^{2n}$.

Example 5. Prove that $\left(\dfrac{i + \sqrt{3}}{-i + \sqrt{3}}\right)^{200} + \left(\dfrac{i - \sqrt{3}}{i + \sqrt{3}}\right)^{200} = -1$

Solution: we have

$$\left(\frac{i + \sqrt{3}}{-i + \sqrt{3}}\right)^{200} + \left(\frac{i - \sqrt{3}}{i + \sqrt{3}}\right)^{200} = \left[\frac{i(i + \sqrt{3})}{i(-i + \sqrt{3})}\right]^{200} + \left[\frac{i(i - \sqrt{3})}{i(i + \sqrt{3})}\right]^{200}$$

$$= \left(\frac{-1 + i\sqrt{3}}{1 + i\sqrt{3}}\right)^{200} + \left(\frac{-1 - i\sqrt{3}}{-1 + i\sqrt{3}}\right)^{200}$$

$$= \left(\frac{2w}{-2w^2}\right)^{200} + \left(\frac{2w^2}{2w}\right)^{200} \qquad \left[\begin{array}{l} \because w = \dfrac{-1 + i\sqrt{3}}{2} \\ \text{and } w^2 = \dfrac{-1 - i\sqrt{3}}{2} \end{array}\right]$$

$$= \left(-\frac{1}{w}\right)^{200} + w^{200}$$

$$= (-w^2)^{200} + w^{200} \qquad \left[\because \frac{1}{w} = w^2\right]$$

$$= w^{400} + w^{200} = (w^3)^{133}\, w + (w^3)^{66} \,.\, w^2$$

$$= w + w^2 = -1$$

Example 6. If α and β are the complex cube roots of unity, show that $\alpha^4 + \beta^4 + \alpha^{-1}\beta^{-1} = 0$

Solution: Since the complex cube roots of unity are w and w^2 where $w = \dfrac{-1 + i\sqrt{3}}{2}$. so let $\alpha = w$ and $\beta = w^2$, then

$$\alpha^4 + \beta^4 + \alpha^{-1}\beta^{-1} = \alpha^4 + \beta^4 + \frac{1}{\alpha\beta}$$

$$= (w)^4 + (w^2)^4 + \frac{1}{w.\,w^2}$$

$$= w^4 + w^8 + \frac{1}{w^3} = w^3.\,w + (w^3)^2\,.\,w^2 + \frac{1}{w^3}$$

$$= w + w^2 + 1 = 0$$

EXERCISE 1.10

1. If 1, w, w^2 are the cube roots of unity prove that

(*i*) $(3 + 5w + 3w^2)^6 = 64$

(*ii*) $(1 + w)^3 - (1 + w^2)^3 = 0$

(*iii*) $(1 - w + w^2)\,(1 + w - w^2) = 4$

(*iv*) $(1 - w)\,(1 - w^2)\,(1 - w^4)\,(1 - w^5) = 9.$

2. Prove that $\left(\frac{-1+i\sqrt{3}}{2}\right)^{17} + \left(\frac{-1-i\sqrt{3}}{2}\right)^{17} = -1$

3. Prove that $\left(\frac{i+\sqrt{3}}{2}\right)^{100} + \left(\frac{i-\sqrt{3}}{2}\right)^{100} = -1$.

4. If $\left(-\frac{1}{2}+\frac{i\sqrt{3}}{2}\right)^{1000} = a + ib$, then find the values of a and b.

5. Evaluate $\frac{(-1+i\sqrt{3})^{15}}{(1-i)^{20}} + \frac{(-1-i\sqrt{3})^{15}}{(1+i)^{20}}$

6. If $x^2 + x + 1 = 0$, then prove that

$$\left(x+\frac{1}{x}\right)^2 + \left(x^2+\frac{1}{x^2}\right)^2 + \ldots + \left(x^{27}+\frac{1}{x^{27}}\right)^2 = 54.$$

ANSWERS

4. $a = -\frac{1}{2}$, $b = \frac{\sqrt{3}}{2}$ **5.** -16

1.31 REPRESENTATION OF A COMPLEX NUMBER

A complex number can be represented in the following forms:

1. Geometrical form
2. Vectorial form
3. Trigonometrical form or polar form

1.32 GEOMETRICAL REPRESENTATION OF A COMPLEX NUMBER

A complex number $z = x + iy$ can be represented by a point (x, y) on the plane which is known as the Argand plane. To represent $z = x + iy$ geometrically we take two mutually perpendicular straight lines $X'OX$ and $Y'OY$. Now plot a point whose x and y coordinates are respectively the real and imaginary parts of z. This point P (x, y) represent the complex number $Z = x + iy$.

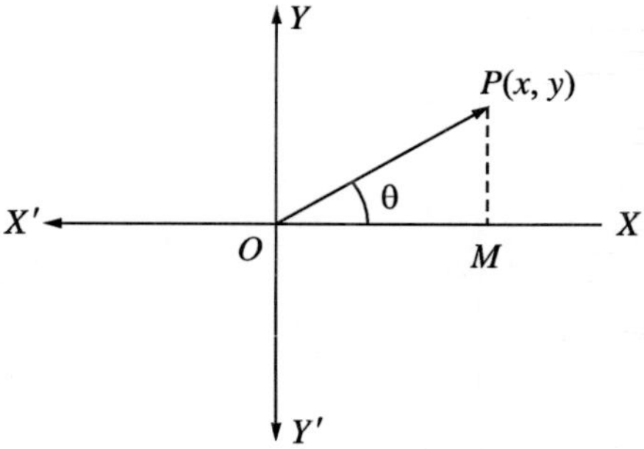

If a complex number is purely real, then its imaginary part is zero. Therefore, a purely real number is represented by a point on x-axis. A purely imaginary complex number is represented by a point on y-axis. That is why x-axis is known as the real axis and y-axis, as the imaginary axis.

The plane in which we represent a complex number geometrically is known as the complex plane or Argand plane. The point P, plotted on the Argand plane is called the Argand diagram.

The length of the line segment OP is called the modulus of z and is denoted by $|z|$.

From figure, we have

$$OP^2 = OM^2 + MP^2$$

$$\Rightarrow \quad OP^2 = x^2 + y^2$$

$$\Rightarrow \quad OP = \sqrt{x^2 + y^2}$$

Thus
$$|z| = \sqrt{x^2 + y^2} = \sqrt{\{Re(z)\}^2 + \{Im(z)\}^2}$$

The angle θ which OP makes with positive direction of x-axis in anticlock wise sense is called the argument or amplitude of z and is denoted by arg (z) or amp (z).

From figure, we have

$$\tan\theta = \frac{PM}{OM} = \frac{y}{x} \Rightarrow \theta = \tan^{-1}\left(\frac{y}{x}\right)$$

This angle θ has infinitely many values differing by multiples of 2π. The unique value of θ such that $-\pi < \theta \le \pi$ is called the principle value of the amplitude or principle argument.

Argument or Amplitude of a complex number $z = x + iy$ for different signs of x and y

I. **Argument of $z = x + iy$ when $x > 0$ and $y > 0$**

Since x and y both are positive, therefore the point P (x, y) representing $z = x + iy$ in the Argand plane lies in the first quadrant. Let θ be the argument of z and α be the acute angle satisfying $\tan\alpha = |y/x|$. Then from figure 1.1 that $\theta = \alpha$.

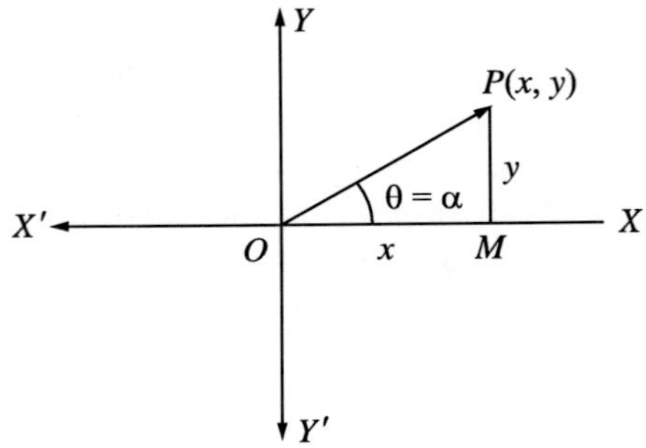

Fig. 1.1

II. Argument of $z = x + iy$ when $x < 0$ and $y > 0$.

Since $x < 0$ and $y > 0$, therefore the point $P(x, y)$ representing $z = x + iy$ in the Argand plane lies in the second quadrant. Let θ be the argument of z and let α be the acute angle satisfying $\tan \alpha = | y/x |$. Then from figure 1.2 that $\theta = \pi - \alpha$.

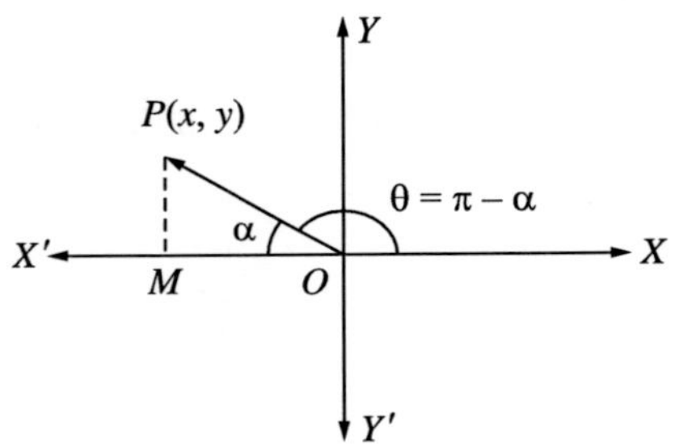

Fig. 1.2

III. Argument of $z = x + iy$ when $x < 0$ and $y < 0$.

Since $x < 0$ and $y < 0$, therefore the point $P(x, y)$ representing $z = x + iy$ lies in the third quadrant. Let θ be the argument of z and α be the acute angle given by $\tan \alpha = | y/x |$. Then from figure 1.3, we have $\theta = -(\pi - \alpha) = \alpha - \pi$.

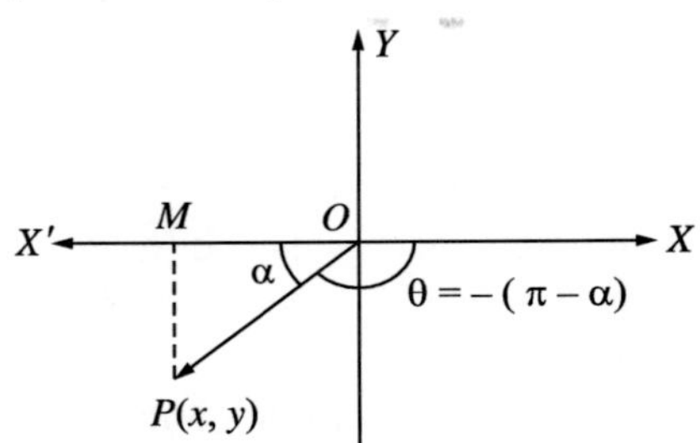

Fig. 1.3

IV. Argument of $z = x + iy$ when $x > 0$ and $y < 0$.

Since $x > 0$ and $y < 0$, therefore the point $P(x, y)$ representing $z = x + iy$ lies in the fourth quadrant. Let θ be the argument of z and let α be the acute angle given by $\tan \alpha = | y/x |$.

Then from figure 1.4, we have $\theta = -\alpha$.

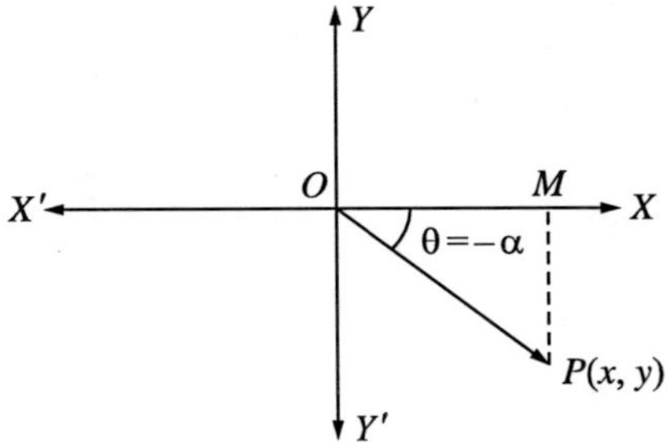

Fig. 1.4

Steps for finding Argument of $z = x + iy$

Step I: Find the value of $\tan^{-1} |y/x|$ lying between 0 and $\frac{\pi}{2}$. Let it be α.

Step II: Determine in which quadrant the point P(x, y) belongs

If $P(x, y)$ belongs to the first quadrant, then $\arg(z) = \alpha$.

If $P(x, y)$ belongs to the second quadrant, then $\arg(z) = \pi - \alpha$.

If $P(x, y)$ belongs to the third quadrant, then $\arg(z) = -(\pi - \alpha) = \alpha - \pi$ or $\pi + \alpha$

If $P(x, y)$ belongs to the fourth quadrant, then $\arg(z) = -\alpha$ or $2\pi - \alpha$.

ILLUSTRATIVE EXAMPLES

Example 1. Find the modulus and argument of each of the following complex numbers:

(*i*) $1 + i\sqrt{3}$ (*ii*) $-1 + i\sqrt{3}$ (*iii*) $-\sqrt{3} - i$ (*iv*) $\sqrt{3} - i$

Solution:

(*i*) Let $z = 1 + i\sqrt{3}$, then $|z| = \sqrt{(1)^2 + (\sqrt{3})^2} = \sqrt{1+3} = 2$

Let α be the acute angle given by

$$\alpha = \tan^{-1} |y/x| = \tan^{-1} \left|\frac{Im(z)}{Re(z)}\right| = \tan^{-1} \left|\frac{\sqrt{3}}{1}\right| = \tan^{-1} \sqrt{3}$$

$$\tan \alpha = \sqrt{3} \Rightarrow \alpha = \frac{\pi}{3}$$

Since $Re(z) > 0$ and $Im(z) > 0$. Therefore, $\arg(z) = \alpha = \pi/3$.

(*ii*) Let $z = -1 + i\sqrt{3}$, then $|z| = \sqrt{(-1)^2 + (\sqrt{3})^2} = 2$

Let α be the acute angle given by

$$\alpha = \tan^{-1} \left|\frac{Im(z)}{Re(z)}\right| = \tan^{-1} \left|\frac{\sqrt{3}}{-1}\right| = \tan^{-1}(\sqrt{3})$$

$$\tan \alpha = \sqrt{3} \Rightarrow \alpha = \frac{\pi}{3}$$

Since $Re(z) < 0$ and $I_m(z) > 0$. Therefore,

$$\arg(z) = \pi - \alpha = \pi - \frac{\pi}{3} = \frac{2\pi}{3}.$$

(*iii*) Let $z = -\sqrt{3} - i$, then $|z| = \sqrt{(-\sqrt{3})^2 + (-1)^2} = 2$

Let α be the acute angle given by

$$\alpha = \tan^{-1}\left|\frac{Im(z)}{Re(z)}\right| = \tan^{-1}\left|\frac{-1}{-\sqrt{3}}\right| = \tan^{-1}\frac{1}{\sqrt{3}}$$

$$\tan \alpha = \frac{1}{\sqrt{3}} \Rightarrow \alpha = \pi/6$$

Since $Re(z) < 0$ and $Im(z) < 0$. Therefore

$$\arg(z) = -(\pi - \alpha) = -(\pi - \pi/6) = -\frac{5\pi}{6}$$

(*iv*) Let $z = \sqrt{3} - i$, then $|z| = \sqrt{(\sqrt{3})^2 + (-1)^2} = 2$

Let α be the acute angle given by $\alpha = \tan^{-1}\left|\frac{Im(z)}{Re(z)}\right| = \tan^{-1}$

$$\left|\frac{-1}{\sqrt{3}}\right| = \tan^{-1}\frac{1}{\sqrt{3}}$$

$$\Rightarrow \qquad \tan \alpha = \frac{1}{\sqrt{3}} \Rightarrow \alpha = \pi/6.$$

since $Re(z) > 0$ and $Im(z) < 0$, therefore

$$\arg(z) = -\alpha = -\pi/6.$$

Example 2. Find the modulus and argument of each of the following complex number:

(*i*) $1 + i$ (*ii*) $-1 + i$ (*iii*) $-1 - i$ (*iv*) $1 - i$

Solution:

(*i*) Let $z = 1 + i$, then $|z| = \sqrt{(1)^2 + (1)^2} = \sqrt{2}$

Let α be the acute angle given by

$$\alpha = \tan^{-1}\left|\frac{Im(z)}{Re(z)}\right| = \tan^{-1}\left|\frac{1}{1}\right| = \tan^{1} 1$$

$$\tan \alpha = 1 \Rightarrow \alpha = \frac{\pi}{4}$$

Since $Re(z) > 0$ and $Im(z) > 0$, therefore $\arg(z) = \alpha = \frac{\pi}{4}$.

(*ii*) Let $z = -1 + i$, then $|z| = \sqrt{(-1)^2 + (1)^2} = \sqrt{2}$

Let α be the acute angle given by

$$\alpha = \tan^{-1}\left|\frac{Im\,(z)}{Re\,(z)}\right| = \tan^{-1}\left|\frac{1}{-1}\right| = \tan^{-1} 1$$

$$\tan\alpha = 1 \Rightarrow \alpha = \frac{\pi}{4}$$

Since $Re\,(z) < 0$ and $Im\,(z) > 0$, therefore $\arg(z) = \pi - \alpha = \pi - \frac{\pi}{4} = \frac{3\pi}{4}$

(*iii*) Let $z = -1 - i$, then $|z| = \sqrt{(-1)^2 + (-1)^2} = \sqrt{2}$

Let α be the acute angle given by

$$\alpha = \tan^{-1}\left|\frac{\text{Im}\,(z)}{\text{Re}\,(z)}\right| = \tan^{-1}\left|\frac{1}{-1}\right| = \tan^{-1} 1$$

$$\tan\alpha = 1 \Rightarrow \alpha = \frac{\pi}{4}$$

since $Re\,(z) < 0$ and $Im\,(z) < 0$, therefore $\arg(z) = -(\pi - \alpha)$

$$\Rightarrow \quad \arg(z) = -\left(\pi - \frac{\pi}{4}\right) = -\frac{3\pi}{4}.$$

(*iv*) Let $z = 1 - i$, then $|z| = \sqrt{(1)^2 + (-1)^2} = \sqrt{2}$.

Let α be the acute angle given by

$$\alpha = \tan^{-1}\left|\frac{Im\,(z)}{Re\,(z)}\right| = \tan^{-1}\left|\frac{-1}{1}\right| = \tan^{-1} 1$$

$$\tan\alpha = 1 \Rightarrow \alpha = \frac{\pi}{4}$$

Since $Re\,(z) > 0$ and $Im\,(z) < 0$, therefore $\arg(z) = -\alpha = -\frac{\pi}{4}$.

1.33 VECTORIAL REPRESENTATION OF A COMPLEX NUMBER

A complex number $z = x + iy$ can be represented by the position vector OP of point $P(x, y)$ in a two dimensional plane because a complex number depends on two things e.g.

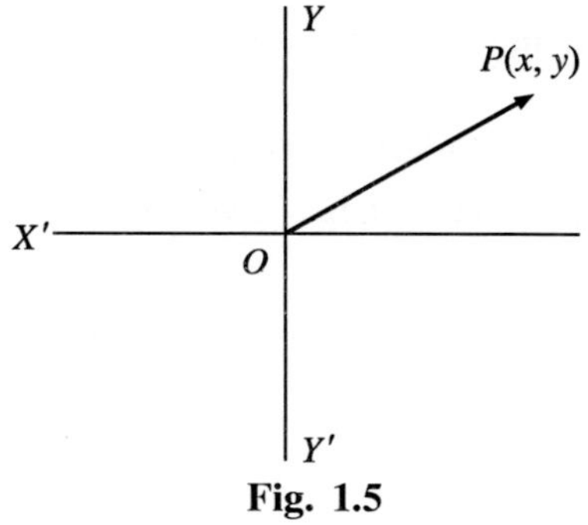

Fig. 1.5

(*i*) its modulus and (*ii*) its argument which are also the requirements of a vector on a plane.

In figure 1.5 the complex number $z = x + iy$ is represented by the vector $\overrightarrow{OP}$ and in such a case $|z|$ is the length OP and arg (z) is the angle which the directed line OP makes with the positive direction of x-axis.

1.34 POLAR FORM OF A COMPLEX NUMBER

Let $z = x + iy$ be a complex number represented by a point $P(x, y)$ in the Argand plane. Then by the geometrical representation of $z = x + iy$, we have $OP = |z|$ and $\angle POX = \theta = \arg(z)$

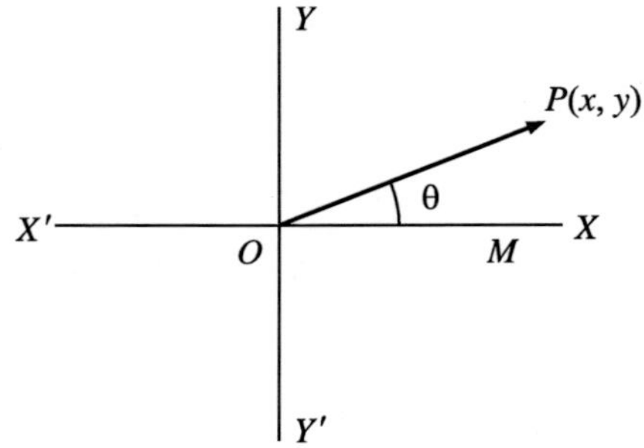

Fig. 1.6

In ΔPOM, we have

$$\cos\theta = \frac{OM}{OP} = \frac{y}{|z|} \Rightarrow x = |z|\cos\theta$$

and $$\sin\theta = \frac{PM}{OP} = \frac{y}{|z|} \Rightarrow y = |z|\sin\theta$$

since $$z = x + iy$$

$\Rightarrow$ $$z = |z|\cos\theta + i\,|z|\sin\theta$$

$$z = |z|(\cos\theta + i\sin\theta)$$

$$z = \gamma(\cos\theta + i\sin\theta), \text{ where } \gamma = |z| \text{ and } \theta = \arg(z)$$

This form of z is called a polar form of z.

If we use the general value of the argument of θ, then the polar form of z is given by

$z = \gamma$ (cos $(2n\pi + \theta)$ i sin $(2n\pi + \theta)$], where $\gamma = |z|$, $\theta = \arg(z)$ and n is an integer.

Polar form of $z = x + iy$ for different signs of x and y:

Let $|z| = \gamma$ and α be the acute angle given by $\tan^{-1}(y/x)$

Let θ be the argument of z.

Case I: Polar form of $z = x + iy$, when $x > 0$ and $y > 0$:

In this case, $\theta = \alpha$. So, the polar form of $z = x + iy$ is given by

$$z = r(\cos\alpha + i\sin\alpha)$$

Case II: Polar form of $z = x + iy$, when $x < 0$ and $y > 0$:

In this case, $\theta = \pi - \alpha$. So, the polar form of $z = x + iy$ is given by

$$z = \gamma\,[\cos(\pi - \alpha) + i \sin(\pi - \alpha)] = \gamma\,(-\cos\alpha + i \sin\alpha)$$

Case III: Polar form of $z = x + iy$, when $x < 0$ and $y < 0$:

In this case, $\theta = -(\pi - \alpha)$. So, the polar form of $z = x + iy$ is given by

$$z = \gamma\,[\cos(\pi - \alpha) + i \sin\{-(\pi - \alpha)\}]$$

$$\Rightarrow \quad z = \gamma\,(-\cos - i \sin\alpha)$$

Case IV: Polar form of $z = x + iy$, when $x > 0$ and $y < 0$:

In this case, $\theta = -\alpha$. So, the polar form of z is given by

$$z = \gamma\,[\cos(-\alpha) + i \sin(-\alpha)] = \gamma\,[\cos\alpha - i \sin\alpha].$$

ILLUSTRATIVE EXAMPLES

Example 1. Write the following complex numbers in the polar form:

(*i*) 1 + i (*ii*) $1 - i$ (*iii*) $-1 - i$

Solution:

(*i*) let $z = 1 + i$ then

$$r = |z| = \sqrt{(1)^2 + (1)^2} = \sqrt{2}$$

let
$$\tan\alpha = \left|\frac{\text{Im}(z)}{\text{Re}(z)}\right|, \text{ then}$$

$$\tan\alpha = \left|\frac{1}{1}\right| = 1 \Rightarrow \alpha = \frac{\pi}{4}$$

Since Re $(z) > 0$ and Im $(z) > 0$, therefore $\theta = \arg(z) = \alpha = \frac{\pi}{4}$. The polar form of $z = 1 + i$ is given by.

$$z = r(\cos\theta + i \sin\theta) = \sqrt{2}\left(\cos\frac{\pi}{4} + i \sin\frac{\pi}{4}\right)$$

(*ii*) let $z = 1 - i$, then

$$r = \sqrt{(1)^2 + (-1)^2} = \sqrt{2}$$

let
$$\tan\alpha = \left|\frac{\text{Im}(z)}{\text{Re}(z)}\right| = \left|\frac{-1}{1}\right|$$

$$\alpha = \frac{\pi}{4}$$

Since Re $(z) > 0$ and Im $(z) < 0$, So $\theta = \arg(z) = -\alpha = \frac{-\pi}{4}$. The polar form of $z = 1 - i$ is given by

$$z = r(\cos\theta + i\sin\theta) = \sqrt{2}\left[\cos\left(\frac{-\pi}{4}\right) + i\sin\left(\frac{-\pi}{4}\right)\right]$$

$$= \sqrt{2}\left[\cos\frac{\pi}{4} - i\sin\frac{\pi}{4}\right]$$

(*iii*) let $z = -1 - i$, then

$$r = |z| = \sqrt{(-1)^2 + (-1)^2} = \sqrt{2}$$

let $$\tan\alpha = \left|\frac{\text{Im}(z)}{\text{Re}(z)}\right| = \left|\frac{-1}{-1}\right| = 1$$

$\Rightarrow$ $$\alpha = \frac{\pi}{4}$$

Since Re $(z) < 0$ and Im $(z) < 0$. So the argument of z is given by

$$\theta = -(\pi - \alpha) = -(\pi - \pi/4) = \frac{-3\pi}{4}$$

The polar form of $z = -1 - i$ is given by

$$z = r(\cos\theta + i\sin\theta) = \sqrt{2}\left[\cos\left(\frac{-3\pi}{4}\right) + i\sin\left(\frac{-3\pi}{4}\right)\right]$$

$$z = \sqrt{2}\left[\cos\frac{3\pi}{4} - i\sin\frac{3\pi}{4}\right]$$

Example 2. Express $\frac{1+i}{1-i}$ in the polar form.

Solution: Let $z = \frac{1+i}{1-i}$

Now $$z = \frac{1+i}{1-i} = \frac{(1+i)(1+i)}{(1-i)(1+i)} = \frac{1+i^2+2i}{1-i^2}$$

$$z = \frac{2i}{2} = i = 0 + i$$

$\therefore$ $$r = |z| = \sqrt{0^2 + 1^2} = 1$$

Since the point (0, 1) representing $z = 0 + \text{i}$ lies on positive direction of imaginary axis therefore,

$$\theta = \arg(z) = \frac{\pi}{2}$$

Hence $$z = r(\cos\theta + i\sin\theta)$$

$$= 1\left(\cos\frac{\pi}{2} + i\sin\pi/2\right) = \cos\frac{\pi}{2} + i\sin\pi/2$$

Example 3. Find the modulus and principle argument of (1 + i) and hence express it in the polar form.

Solution: Let $z = 1 + i$, then

$$r = |z| = \sqrt{1^2 + 1^2} = \sqrt{2}$$

let α be the acute angle given by $\tan \alpha = \left|\frac{\text{Im}(z)}{\text{Re}(z)}\right|$, then

$$\tan \alpha = \left|\frac{1}{1}\right| = 1 \Rightarrow \alpha = \pi / 4$$

Since the point (1, 1) representing $z = 1 + i$ lies in the fist quadrant, therefore $\theta = \arg(z) = \pi / 4$

Hence polar form of $z = 1 + i$ is given by

$$z = r(\cos \theta + i \sin \theta) = \sqrt{2}\left(\cos \frac{\pi}{4} + i \sin \pi / 4\right)$$

1.35 DE - MOIVRE'S THEOREM

Statement.

(*i*) If n is a positive or negative integer, then
$(\cos \theta + i \sin \theta)^n = \cos n\theta + i \sin n\theta$

(*ii*) If $n \in Q$ (the set of rational numbers), then $\cos n\theta + i \sin n\theta$ is one of the values of $(\cos \theta + i \sin \theta)^n$.

Proof. (*i*) we divide the proof into the following three cases:

Case I. When $n = 0$

In this case,

$$(\cos \theta + i \sin \theta)^n = (\cos \theta + i \sin \theta)\ 0 = 1 = \cos (0.\ \theta) + i \sin (0\ .\ \theta)$$
$$= \cos n\ \theta + i \sin n\theta\ [\because\ n = 0]$$

so, the theorem holds for $n = 0$

Case II. When $n \in N$ i.e; n is a positive integer.

In this case, we shall prove the theorem by induction the theorem is true for $n = 1$ because

$$(\cos \theta + i \sin \theta)^1 = \cos\theta + i \sin\theta = \cos(1.\theta) + i \sin(1.\theta) \quad ...(1)$$

Let us assume that the theorem is true for $n = m$, then

$$(\cos\theta + i \sin \theta)^m = \cos m\theta + i \sin m\theta \quad ...(1)$$

Now we shall show that theorem is true for $n = m + 1$

we have $(\cos \theta + i \sin \theta)^{m+1}$

$$= (\cos \theta + i \sin \theta)^m (\cos \theta + i \sin \theta)$$
$$= (\cos m\theta + i \sin m\theta)(\cos \theta + i \sin \theta) \quad \text{[using (1)]}$$
$$= (\cos m\ \theta \cos \theta - \sin m\theta \sin \theta) + i (\cos m\theta \sin \theta + \sin m\theta \cos \theta)$$
$$= \cos (m\theta + \theta) + i \sin (m\theta + \theta)$$
$$= \cos (m + 1)\ \theta + i \sin (m + 1)\ \theta$$

So, the theorem holds for $n = m + 1$

Hence by induction the theorem holds good for all $n \in N$.

Case III. When n is a negative integer.

let $n = -m$, where $m \in N$, then

$$(\cos\theta + i\sin\theta)^n = (\cos\theta + i\sin\theta)^{-m}$$

$$= \frac{1}{(\cos\theta + i\sin\theta)^m}$$

$$= \frac{1}{\cos m\theta + i\sin m\theta} \qquad \text{[By case II]}$$

$$= \frac{1}{\cos m\theta + i\sin m\theta} \times \frac{\cos m\theta - i\sin m\theta}{\cos m\theta - i\sin m\theta}$$

$$= \frac{\cos m\theta - i\sin m\theta}{\cos^2 m\theta + \sin^2 m\theta} = \cos m\theta - i\sin m\theta$$

$$= \cos(-m)\theta + i\sin(-m)\theta \qquad \left[\begin{array}{c}\because \cos(-\theta) = \cos\theta \\ \text{and } \sin(-\theta) = -\sin\theta\end{array}\right]$$

$$= \cos n\theta + i\sin n\theta$$

Hence $(\cos\theta + i\sin\theta)^n = \cos n\theta + i\sin n\theta$, for all $n \in Z$ (set of integers).

(ii) let $n = \frac{p}{q}$, where p, q are integers and $q > 0$

From proof (i) we have

$$\left(\cos\frac{p\theta}{q} + i\sin\frac{p\theta}{q}\right)^q = \left[\cos\frac{p\theta}{q}q + i\sin\left(\frac{p\theta}{q}\right)q\right] = \cos p\,\theta + i\sin p\,\theta$$

$\Rightarrow \cos\frac{p\theta}{q} + i\sin\frac{p\theta}{q}$ is one of the values of $(\cos p\theta + \text{i}\sin p\theta)^{1/q}$

$\Rightarrow \cos\frac{p\theta}{q} + i\sin\frac{p\theta}{q}$ is one of the values of $[(\cos\theta + i\sin\theta)^{1/q}$

$\Rightarrow \cos\frac{p\theta}{q} + i\sin\frac{p\theta}{q}$ is one of the values of $(\cos\theta + i\sin\theta)^{p/q}$

Note: 1. The theorem is also true for $(\cos\theta - i\sin\theta)$ i.e;

$(\cos\theta - i\sin\theta)^n = \cos\ n\theta - i\sin \text{n}\,\theta$, because

$$(\cos\theta - i\sin\theta)^n = [\cos(-\theta) + i\sin(-\theta)^n] = \cos n(-\theta) + i\sin n(-\theta)$$

$$= \cos(-n\theta) + i\sin(-n\theta)$$

$$= \cos n\theta - i\sin n\theta$$

2. $$\frac{1}{\cos\theta + i\sin\theta} = (\cos\theta + i\sin\theta)^{-1} = \cos\theta - i\sin\theta.$$

ILLUSTRATIVE EXAMPLES

Example 1. Simplify

(*i*) $\dfrac{(\cos\theta + i\sin\theta)^4}{(\cos\theta - i\sin\theta)^5}$

(*ii*) $\dfrac{\cos 3\theta + i\sin 3\theta)^4\,(\cos 4\theta - i\sin 4\theta)^5}{(\cos 4\theta + i\sin 4\theta)^3\,(\cos 5\theta + i\sin 5\theta)^{-4}}$

(*iii*) $(\cos\theta + i\sin\theta)^4\,(\cos\theta - i\sin\theta)^2$

Solution: (i) $\dfrac{(\cos\theta + i\sin\theta)^4}{(\cos\theta - \sin\theta)^5} = \dfrac{(\cos\theta + i\sin\theta)^4}{(\cos\theta + i\sin\theta)^5}$

$[\because \cos\theta - i\sin\theta = (\cos\theta + i\sin\theta)^{-1}]$

$= (\cos\theta + i\sin\theta)4\;\;(\cos\theta + i\sin\theta)^5$

$= (\cos\theta + i\sin\theta)^9 = \cos 9\theta + i\sin 9\theta$

(*ii*) $\dfrac{(\cos 3\theta + i\sin 3\theta)^4\,(\cos 4\theta - i\sin 4\theta)^5}{(\cos 4\theta + i\sin 4\theta)^3\,(\cos 5\theta + i\sin 5\theta)^{-4}}$

$$= \frac{[(\cos\theta + i\sin\theta)^3]^4\,[(\cos\theta + i\sin\theta)^{-4}]^5}{[(\cos\theta + i\sin\theta)^4]^3\,[(\cos\theta + i\sin\theta)^5]^{-4}}$$

$$= \frac{(\cos\theta + i\sin\theta)^{12}\,(\cos\theta + i\sin\theta)^{-20}}{(\cos\theta + i\sin\theta)^{12}\,(\cos\theta + i\sin\theta)^{-20}} = 1$$

(*iii*) $(\cos\theta + i\sin\theta)^4\,(\cos\theta - i\sin\theta)^2$

$= (\cos\theta + i\sin\theta)^4\,(\cos\theta + i\sin\theta)^{-2}$

$= (\cos\theta + i\sin\theta)^{4-2} = (\cos\theta + i\sin\theta)^2$

$= \cos 2\theta + i\sin 2\theta$

Example 2. Prove that

$$(1 + \cos\theta + i\sin\theta)^n + (1 + i\sin\theta)^n = 2^{n+1}\cos^n\frac{\theta}{2}\cos\frac{n\theta}{2} \text{ or}$$

Prove that $(1 + \cos\theta + i\sin\theta)^n + (1 + \cos\theta - \text{i}\sin\theta)^n$ is real and find its value.

Solution: We have

(*ii*) L.H.S $= (1 + \cos\theta + i\sin\theta)^n + (1 + \cos\theta - i\sin\theta)^n$

$$= \left(2\cos^2\frac{\theta}{2} + 2i\sin\frac{\theta}{2}\cos\theta/2\right)^n$$

$$+ \left(2\cos^2\frac{\theta}{2} - 2\,i\sin\theta/2\cos\theta/2\right)^n$$

$$= 2^n \left[\cos\theta/2\left(\cos\frac{\theta}{2} + i\sin\theta/2\right)\right]^n$$

$$+ 2^n\left[\cos\theta/2\,(\cos\theta/2 - i\sin\theta/2\right]^n$$

$$= 2^n \cos{}^n\frac{\theta}{2}\left[\left(\cos\frac{\theta}{2} + i\sin\theta/2\right)^n + \left(\cos\frac{\theta}{2} - i\sin\theta/2\right)^n\right]$$

$$= 2^n \cos^n\frac{\theta}{2}\left[\cos\frac{n\theta}{2} + i\sin\frac{n\theta}{2} + \cos\frac{n\theta}{2} - i\sin\frac{n\theta}{2}\right]$$

$$= 2^n \cos^n\frac{\theta}{2}\left(2\cos\frac{n\theta}{2}\right) = 2^{n+1}\cos{}^n\frac{\theta}{2}\cos\frac{n\theta}{2}$$

Since $2^{n+1}\cos^n\frac{\theta}{2}\cos\frac{n\theta}{2}$ is real, therefore it follows that the value of $(1 + \cos\theta + i\sin\theta)^n + (1 + \cos\theta - \mathrm{i}\sin\theta)^n$ is real and its value is $2^{n+1}\cos^n\frac{\theta}{2}\cos\frac{n\theta}{2}$.

Example 3. Prove that $\left(\frac{1+\cos\theta + i\sin\theta}{1+\cos\theta - i\sin\theta}\right)^n = \cos n\theta + i\sin n\theta.$

Solution:

$$\left(\frac{1+\cos\theta + i\sin\theta}{1+\cos\theta - i\sin\theta}\right)^n = \left(\frac{2\cos^2\frac{\theta}{2} + 2i\sin\theta/2\cos\theta/2}{2\cos^2\frac{\theta}{2} - 2i\sin\theta/2\cos\theta/2}\right)^n$$

$$= \left(\frac{\cos\frac{\theta}{2} + i\sin\frac{\theta}{2}}{\cos\frac{\theta}{2} - i\sin\theta/2}\right)^n$$

$$= \left(\cos\frac{\theta}{2} + i\sin\theta/2\right)^n\left(\cos\frac{\theta}{2} - i\sin\theta/2\right)^{-n}$$

$$= \left(\cos\frac{\theta}{2} + i\sin\theta/2\right)^n\left(\cos\frac{\theta}{2} + i\sin\theta/2\right)^n$$

$$= \left(\cos\frac{\theta}{2} + i\sin\frac{\theta}{2}\right)^{2n} = \cos n\theta + i\sin n\,\theta$$

Example 4. If $x + \frac{1}{x} = 2\cos\theta$, $y + \frac{1}{y} = 2\cos\phi$, prove that

$$\frac{x^m}{y^n} + \frac{y^n}{x^m} = 2\cos(\mathrm{m}\theta - n\phi).$$

Solution: we have

$$x+\frac{1}{x} = 2\cos\theta$$

$$\Rightarrow \quad x^2 - 2x\cos\theta + 1 = 0$$

$$\Rightarrow \quad x = \frac{2\cos\theta \pm \sqrt{4\cos^2\theta - 4}}{2}$$

$$\Rightarrow \quad x = \cos\theta \pm i\sin\theta$$

$$\Rightarrow \quad x = \cos\theta + i\sin\theta \quad \text{(taking positive sign only)}$$

similarly $y = \cos\phi + i\sin\phi$

Now
$$\frac{x^m}{y^n} = \left(\frac{(\cos\theta + i\sin\theta)^m}{(\cos\phi + i\sin\phi)^n}\right)$$

$$= (\cos\theta + i\sin\theta)^m (\cos\phi + i\sin\phi)^{-n}$$

$$\frac{x^m}{y^n} = (\cos m\theta + i\sin m\theta)\,[\cos(-n\phi) + i\sin(-n\phi)]$$

$$= \cos(m\theta - n\phi) + i\sin(m\theta - n\phi) \quad \text{...(1)}$$

$$\therefore \quad \frac{y^n}{x^m} = [\cos(m\theta - n\phi) + i\sin(m\theta - n\phi)]^{-1}$$

$$= \cos(m\theta - n\phi) - i\sin(m\theta - n\phi) \quad \text{...(2)}$$

From equation (1) and (2). we get

$$\frac{x^m}{y^n} + \frac{y^n}{x^m} = \cos(m\theta - n\phi) + i\sin(m\theta - n\phi) + \cos(m\theta - n\phi) - i\sin(m\theta - n\phi)$$

$$= 2\cos(m\theta - n\phi)$$

Example 5. If $x_n = \cos\left(\frac{\pi}{2^n}\right) + i\sin\left(\frac{\pi}{2^n}\right)$, prove that $x_1 \,.\, x_2 \,.\, x_3 \ldots \infty = -1$

Solution: we have $x_1, x_2, x_3 \ldots \infty$

$$= \left(\cos\frac{\pi}{2} + i\sin\frac{\pi}{2}\right)\left(\cos\frac{\pi}{2^2} + i\sin\frac{\pi}{2^2}\right)\left(\cos\frac{\pi}{2^3} + i\sin\frac{\pi}{2^3}\right) \ldots \infty$$

$$= \cos\left(\frac{\pi}{2} + \frac{\pi}{2^2} + \frac{\pi}{2^3} + \ldots\right) + i\sin\left(\frac{\pi}{2} + \frac{\pi}{2^2} + \frac{\pi}{2^3} + \ldots\right)$$

$$= \cos\left(\frac{\pi/2}{1-1/2}\right) + i\sin\left(\frac{\pi/2}{1-1/2}\right) \qquad \left[\because a + ar + ar^2 + \ldots \infty = \frac{a}{1-r}\right]$$

$= \cos\pi + i\sin\pi = -1.$

Example 6. If n is a positive integer, prove that

$$(\sqrt{3}+i)^n + (\sqrt{3}-i)^n = 2^{n+1}\cos\frac{n\pi}{6}.$$

Solution: let r (cos θ + *i* sin θ) be the polar form of $\sqrt{3}+i$, then

$$r = \sqrt{(\sqrt{3}) + (1)^2} = 2$$

$$\tan\theta = \frac{1}{\sqrt{3}} \Rightarrow \theta = \pi/6$$

$$\therefore \quad \sqrt{3}+i = 2\left(\cos\frac{\pi}{6} + i\sin\pi/6\right)$$

$$\Rightarrow \quad (\sqrt{3}+i)^n = 2^n\left(\cos\frac{\pi}{6} + i\sin\pi/6\right)^n$$

$$\Rightarrow \quad (\sqrt{3}+i)^n = 2^n\left(\cos\frac{n\pi}{6} + i\sin\frac{n\pi}{6}\right) \qquad ...(1)$$

The polar form of $(\sqrt{3}-i)$ is $2\left[\cos(-\pi/6) + i\sin(-\pi/6)\right]$ = $2\left[\cos\frac{\pi}{6} - i\sin\pi/6\right]$

$$\therefore \quad (\sqrt{3}-i)^n = 2^n\left(\cos\frac{\pi}{6} - i\sin\frac{\pi}{6}\right)^n$$

$$(\sqrt{3}-i)^n = 2^n\left(\cos\frac{n\pi}{6} - i\sin\frac{n\pi}{6}\right) \qquad ...(2)$$

Adding (1) and (2), we get

$$(\sqrt{3}+i)^n + (\sqrt{3}-i)^n = 2^n\left(\cos\frac{n\pi}{6} + i\sin\frac{n\pi}{6}\right) + 2^n\left(\cos\frac{n\pi}{6} - i\sin\frac{n\pi}{6}\right)$$

$$= 2^n\left(2\cos\frac{n\pi}{6}\right) = 2^{n+1}\cos\frac{n\pi}{6}.$$

Example 7. Prove that

$$(1+i)^n + (1-i)^n = 2^{\frac{n}{2}+1}\cos\frac{n\pi}{4}$$

Solution: let $1 + i = r(\cos\theta + i\sin\theta)$

then $\quad 1 - i = r(\cos\theta - i\sin\theta)$

In both cases

$$r\cos\theta = 1 \text{ and } r\sin\theta = 1$$

$$\therefore \quad r^2 = 2 \Rightarrow r = \sqrt{2}$$

and $\quad \sin\theta = \frac{1}{\sqrt{2}}, \cos\theta = \frac{1}{\sqrt{2}}$

$$\Rightarrow \qquad \theta = \frac{\pi}{4}$$

$$\therefore \qquad 1 + i = \sqrt{2}\left(\cos\frac{\pi}{4} + i\sin\frac{\pi}{4}\right)$$

and
$$1 - i = \sqrt{2}\left(\cos\frac{\pi}{4} - i\sin\frac{\pi}{4}\right)$$

$$\therefore \; (1 + i)^n + (1 - i)^n = \left[\sqrt{2}\left\{\cos\frac{\pi}{4} + i\sin\frac{\pi}{4}\right\}\right]^n + \left[\sqrt{2}\left\{\cos\frac{\pi}{4} - i\sin\frac{\pi}{4}\right\}\right]^n$$

$$= 2^{n/2}\left[\cos\frac{n\pi}{4} + i\sin\frac{n\pi}{4}\right] + 2^{n/2}\left[\cos\frac{n\pi}{4} - i\sin\frac{n\pi}{4}\right]$$

$$= 2^{n/2}\,.\,2\cos\frac{n\pi}{4}$$

$$= 2^{\frac{n}{2}+1}\cos\frac{n\pi}{4}$$

Example 8. If $\sin\alpha + \sin\beta + \sin\gamma = \cos\alpha + \cos\beta + \cos\gamma = 0$ then

Prove that $\cos 3\alpha + \cos 3\beta + \cos 3\gamma = 3\cos(\alpha + \beta + \gamma)$

and $\sin 3\alpha + \sin 3\beta + \sin 3\gamma = 3\sin(\alpha + \beta + \gamma)$

Solution:

Let
$$a = \cos\alpha + i\sin\alpha$$
$$b = \cos\beta + i\sin\beta$$
$$c = \cos\gamma + i\sin\gamma$$

$$\therefore \qquad a + b + c = (\cos\alpha + i\sin\alpha) + (\cos\beta + i\sin\beta) + (\cos\gamma + i\sin\gamma)$$
$$a + b + c = (\cos\alpha + \cos\beta + \cos\gamma) + i(\sin\alpha + \sin\beta + \sin\gamma)$$
$$\Rightarrow \qquad a + b + c = 0$$
$$\Rightarrow \qquad a + b = -c$$
$$\Rightarrow \qquad (a + b)^3 = -c^3$$
$$\Rightarrow \qquad a^3 + b^3 + 3ab(a + b) = -c^3$$
$$\Rightarrow \qquad a^3 + b^3 + 3ab(-c) = -c^3$$
$$\Rightarrow \qquad a^3 + b^3 + c^3 = 3abc$$

Substituting the values of a, b, and c, we have

$$(\cos\alpha + i\sin\alpha)^3 + (\cos\beta + i\sin\beta)^3 + (\cos\gamma + i\sin\gamma)^3$$
$$= 3(\cos\alpha + i\sin\alpha)(\cos\beta + i\sin\beta)(\cos\gamma + i\sin\gamma)$$

$\Rightarrow (\cos 3\alpha + i \sin 3\alpha) + (\cos 3\beta + i \sin 3\beta) + (\cos 3\gamma + i \sin \gamma)$
$= 3 [\cos (\alpha + \beta + \gamma) + i \sin (\alpha + \beta + \gamma)]$

Equating real and imaginary parts, we get

$\cos 3\alpha + \cos 3\beta + \cos 3\gamma = 3 \cos (\alpha + \beta + \gamma)$

$\sin 3\alpha + \sin 3\beta + \sin 3\gamma = 3 \sin (\alpha + \beta + \gamma)$

Example 9. If $x = \cos \theta + i \sin \theta$ and $\sqrt{1-c^2} = n\,c - 1$, show that

$$1 + c \cos \theta = \frac{c}{2n}(1+nx)\left(1+\frac{n}{x}\right)$$

Solution: Given $x = \cos \theta + i \sin \theta$

$$\frac{1}{x} = (\cos \theta + i \sin \theta)^{-1} = \cos \theta - i \sin \theta$$

$$\therefore \quad x+\frac{1}{x} = (\cos \theta + i \sin \theta) + (\cos \theta - i \sin \theta) = 2 \cos \theta \quad ...(1)$$

Also $\sqrt{1-c^2} = nc - 1$

$\Rightarrow \quad 1 - c^2 = n^2c^2 - 2nc + 1$

$\Rightarrow \quad 2nc = c^2 (1 + n^2)$

$2n = c (1 + n^2) \quad ...(2)$

It is not equal to zero

$$\therefore \frac{c}{2n}(1+nx)\left(1+\frac{n}{x}\right) = \frac{c}{2n}\left(1+nx+\frac{n}{x}+n^2\right)$$

$$= \frac{c}{2n}\left\{1+n^2+n\left(x+\frac{1}{x}\right)\right\}$$

$$= \frac{c}{2n}\{1+n^2+n.2\cos\theta\} \text{ [From (1)]}$$

$$= \frac{c(1+n^2)}{2n} + c \cos \theta$$

$$= \frac{2n}{2n} + c \cos \theta$$

$$= 1 + c \cos \theta$$

Example 10. If α and β are the roots of the equation $x^2 - 2x + 4 = 0$,

Prove that $\alpha^n + \beta^n = 2^{n+1} \cos \dfrac{n\pi}{3}$

Solution: $x^2 - 2x + 4 = 0$

$$\therefore \quad x = \frac{2 \pm \sqrt{4-16}}{2} = 1 \pm i\sqrt{3}$$

since α and β are the roots of the equation, therefore, we have

$$\alpha = 1+i\sqrt{3} = 2\left(\frac{1}{2}+i\frac{\sqrt{3}}{2}\right) = 2\left(\cos\frac{\pi}{3}+i\sin\frac{\pi}{3}\right)$$

$$\beta = 1-i\sqrt{3}\ \ 2\left(\frac{1}{2}-i\frac{\sqrt{3}}{2}\right) = 2\left(\cos\frac{\pi}{3}-i\sin\pi/3\right)$$

$$\therefore \qquad \alpha^n + \beta^n = \left[2\left(\cos\frac{\pi}{3}+i\sin\frac{\pi}{3}\right)\right]^n + \left[2\left(\cos\frac{\pi}{3}-i\sin\pi/3\right)\right]^n$$

$$= 2^n\left[\left(\cos\frac{n\pi}{3}+i\sin\frac{n\pi}{3}\right)+\left(\cos\frac{\pi}{3}-i\sin\frac{n\pi}{3}\right)\right]$$

$$= 2^n.2\cos\frac{n\pi}{3} = 2^{n+1}\cos\frac{n\pi}{3}$$

EXERCISE 1.11

1. Simplify the following:

(*i*) $\dfrac{(\cos\theta+i\sin\theta)^4}{(\sin\theta-i\cos\theta)^5}$ (*ii*) $\left(\sin\dfrac{\pi}{6}+i\cos\dfrac{\pi}{6}\right)^7$ (*iii*) $[2\ (\cos 40^\circ + i\sin 40^\circ)]^6$

2. Prove that $(\sin\theta + i\cos\theta)^n = \cos\left(\dfrac{n\pi}{2}-n\theta\right)+i\sin\left(\dfrac{n\pi}{2}-n\theta\right)$

3. Prove that $(1+\cos\theta+i\sin\theta)^n + (1+\cos\theta-i\sin\theta)^n$

$$= 2^{n+1}\cos^n\left(\frac{\theta}{2}\right)\cos\frac{n\theta}{2}\ \left(\frac{\theta}{2}\right)\cos\frac{n\theta}{2}\ \left(\frac{\theta}{2}\right)\cos\left(\frac{n\theta}{2}\right)$$

4. If $x^2 - 2x\cos\theta + 1 = 0$, prove that $x^{2n} + 1 = 2x^n \cos n\theta$

5. If $x + \dfrac{1}{x} = 2\cos\theta$, prove that $x^n + x^{-n} = 2\cos n\theta$.

6. If $x_n = \cos\dfrac{\pi}{3^n}+i\sin\dfrac{\pi}{3^n}$ prove that $x_1\, x_2\, x_3 \ldots \infty = i$.

7. If $a = \cos 2\alpha + i\sin 2\alpha$ and $b = \cos 2\beta + i\sin 2\beta$, prove that

(*i*) $\sqrt{ab}+\dfrac{1}{\sqrt{ab}} = 2\cos(\alpha+\beta)$

(*ii*) $\dfrac{a-b}{a+b} = \text{i}\tan(\alpha-\beta)$

8. If $\cos\alpha + \cos\beta + \cos\gamma = \sin\alpha + \sin\beta + \sin\gamma = 0$ prove that

(*i*) $\cos 2\alpha + \cos 2\beta + \cos 2\gamma = \sin 2\alpha + \sin 2\beta + \sin 2\gamma = 0$

(*ii*) $\cos^2\alpha + \cos^2\beta + \cos^2\gamma = \sin^2\alpha + \sin^2\beta + \sin^2\gamma = \dfrac{3}{2}$

9. simplify $\left(\dfrac{1+\sin\theta+i\cos\theta}{1+\sin\theta-i\cos\theta}\right)^n$

10. Prove that $(1 + i)^8 + (1 - i)^8 = 32$

ANSWERS

1. (*i*) $\sin\theta + i\cos\theta$ (*ii*) $\dfrac{1}{2}(1+i\sqrt{3})$ (*iii*) $-2^5(1+i\sqrt{3})$

9. $\cos\left(\dfrac{n\pi}{2}-n\theta\right)+i\sin\left(\dfrac{n\pi}{2}-n\theta\right)$

2

Matrices and Determinant

2.1 MATRIX

A set of mn numbers (real or imaginary) arranged in the form of a rectangular array of m rows and n columns is called a matrix of order m by n written as $m \times n$ matrix. Matrices are usually written as by capital letters. A $m \times n$ matrix is usually written as

$$A = \begin{bmatrix} a_{11} & a_{12} & \cdots & a_{1n} \\ a_{21} & a_{22} & \cdots & a_{2n} \\ \vdots & \vdots & & \vdots \\ a_{m1} & a_{m2} & \cdots & a_{mn} \end{bmatrix}$$

The above matrix can be represented by $A = [a_{ij}]_{m \times n}$ or $A = [a_{ij}]$, where $i = 1,2, ..., m$ and $j = 1, 2,..., n$.

The numbers $a_{11}, a_{12},..., a_{mn}$ are known as the elements of the matrix A. In general, a_{ij} is called the element of the ith row and i^{th} column or the $(i, j)^{\text{th}}$ element.

For example:

$A = \begin{bmatrix} 3 & 2 & -2 \\ 2 & 4 & 3 \end{bmatrix}$ is a matrix having two rows and three columns i.e., the order of the matrix is 2×3. Here $a_{11} = 3, a_{12} = 2, a_{13} = -2, a_{21} = 2, a_{22} = 4, a_{23} = 3$.

Example. Construct a 2×3 matrix $A = [a_{ij}]$ whose elements are given by $a_{ij} = i + j$

Solution: We have $A = \begin{bmatrix} a_{11} & a_{12} & a_{13} \\ a_{21} & a_{22} & a_{23} \end{bmatrix}$

Here, we have $a_{11} = 1 + 1 = 2, a_{12} = 1 + 2 = 3, a_{13} = 1 + 3 = 4$

$a_{21} = 2 + 1 = 3, a_{22} = 2 + 2 = 4, a_{23} = 2 + 3 = 5$

$$\therefore \quad A = \begin{bmatrix} 2 & 3 & 4 \\ 3 & 4 & 5 \end{bmatrix}$$

2.2 DIAGONAL ELEMENTS OF A MATRIX

An element a_{ij} of the matrix $A = [a_{ij}]_{m \times n}$ is said to be diagonal element if $i = j$; i.e., the elements a_{11}, a_{22}, a_{33}, etc. are the diagonal elements.

Principal diagonal: The line along which the diagonal elements lie is called the principal or leading diagonal of a matrix.

For example: $$A = \begin{bmatrix} 3 & 2 & 4 \\ 2 & 5 & 6 \\ 4 & 1 & 2 \end{bmatrix}_{3\times 3}$$

Here 3, 5, 2 are diagonal elements.

2.3 TYPES OF MATRICES

1. Row matrix: A matrix having only one row and any finite number of columns is called a row matrix; i.e., a matrix of the order $1 \times n$ is a row matrix.

For example: $[2 \quad 3 \quad 4 \quad 7]$ is a row matrix of order 1×4.

2. Column matrix: A matrix having only one column and any finite number of rows is called a column matrix; i.e., a matrix of the order $m \times 1$ is a column matrix.

For example: $\begin{bmatrix} 2 \\ 5 \\ 8 \end{bmatrix}$ is a column matrix of order 3×1.

3. Null or zero matrix: A matrix, each of whose element is zero is called a null matrix or a zero matrix. A null matrix of order $m \times n$ is denoted by $O_{m\times n}$ or simply by O.

For example: $\begin{bmatrix} 0 & 0 & 0 \\ 0 & 0 & 0 \end{bmatrix} = O$ is a null matrix of order 2×3.

4. Square matrix: A matrix in which the number of rows and columns are equal is called a square matrix, i.e., a matrix $A = [a_{ij}]_{m\times n}$ is said to be a square matrix if $m = n$.

For example: $A = \begin{bmatrix} 2 & 3 \\ 4 & 5 \end{bmatrix}_{2\times 2}$ is a square matrix of order 2.

and $B = \begin{bmatrix} 1 & 2 & 3 \\ 5 & 6 & 7 \\ 2 & 3 & 4 \end{bmatrix}_{3\times 3}$ is a square matrix of order 3.

5. Diagonal matrix: A square matrix is called a diagonal matrix, if all its non-diagonal elements are zero, i.e., a matrix $A = [a_{ij}]_{m\times n}$ is called a diagonal matrix if $a_{ij} = 0$ for $i \neq j$.

For example: $\begin{bmatrix} 1 & 0 & 0 \\ 0 & 4 & 0 \\ 0 & 0 & 2 \end{bmatrix}$ is a diagonal matrix of order 3×3 or simply '3'.

6. Unit or identity matrix: A diagonal matrix is said to be a unit or identity matrix if its diagonal elements are unity, i.e., a matrix $A = [a_{ij}]_{m\times n}$ is a unit matrix if $a_{ij} = 0$ for $i \neq j$

and $a_{ij} = 1$ for $i = j$.

For example: $I_{2\times 2}$ (or I_2) $= \begin{bmatrix} 1 & 0 \\ 0 & 1 \end{bmatrix}$

$I_{3\times 3}$ (or I_3) $= \begin{bmatrix} 1 & 0 & 0 \\ 0 & 1 & 0 \\ 0 & 0 & 1 \end{bmatrix}$

7. Equal matrices: Let $A = [a_{ij}]$ and $B = [b_{ij}]$ be two matrices. Then A and B are said to equal matrices if

(*i*) they are of the same order, i.e., numbers of rows and columns in both the matrices are same.

(*ii*) the elements in the corresponding positions of the two matrices are equal, i.e., $a_{ij} = b_{ij}$ for all values of i and j.

For example: The following two matrices are equal:

$$\begin{bmatrix} 1 & 2 & \frac{3}{5} \\ \frac{2}{3} & 1 & 4 \\ \frac{1}{2} & 0 & 5 \end{bmatrix} = \begin{bmatrix} \frac{4}{4} & 2 & \frac{9}{15} \\ \frac{6}{9} & 1 & \frac{16}{4} \\ \frac{5}{10} & 0 & 5 \end{bmatrix}$$

8. Scalar matrix: A diagonal matrix is said to be a scalar matrix if its diagonal elements are equal, i.e., a square matrix $A = [a_{ij}]_{m\times n}$ is said to be a scalar matrix if

$a_{ij} = 0$ for $i \neq j$

$a_{ij} = C$ for $i = j$, C is constant.

For example: $A = \begin{bmatrix} 5 & 0 & 0 \\ 0 & 5 & 0 \\ 0 & 0 & 5 \end{bmatrix}$

is a scalar matrix of order 3. Also $A = 5I$, where I is identity matrix of order 3.

9. Transpose of matrix: If we interchange the rows and the corresponding columns of a matrix say A, then the new matrix obtained is called the transpose of the given matrix and is denoted by A' or A^T.

Symbolically if, $A = [a_{ij}]_{m\times n}$ then $A' = [b_{ji}]_{n\times m}$

where $b_{ji} = a_{ij}$ i.e., the $(j, i)^{th}$ element of A' is the $(i, j)^{th}$ element of A.

For Example: If $A \begin{bmatrix} 1 & 3 & 5 & 7 \\ 2 & 4 & 6 & 8 \\ 3 & 6 & 7 & 5 \end{bmatrix}$

then A' or $A^T = \begin{bmatrix} 1 & 2 & 3 \\ 3 & 4 & 6 \\ 5 & 6 & 7 \\ 7 & 8 & 5 \end{bmatrix}$ is transpose matrix of A.

Properties: If A' and B' are the transposes of matrix A and B respectively, then

(*i*) $(A')' = A$

(*ii*) $(A+B)' = A' + B'$, A and B being of the same order.

(*iii*) $(kA)' = kA'$, k being any scalar.

(*iv*) $(AB)' = B'A'$

10. Triangular matrix: A square matrix, in which all the elements either above or below the principal diagonal are zero is called a triangular matrix. A triangular matrix is either upper triangular or lower triangular.

A square matrix, in which all the elements below the principal diagonal are zero is called an upper triangular matrix.

For example: $A = \begin{bmatrix} 1 & 3 & 5 \\ 0 & 4 & 2 \\ 0 & 0 & 6 \end{bmatrix}$ is an upper triangular matrix.

A square matrix, in which all the elements above the principal diagonal are zero is called a lower triangular matrix.

For example: $A = \begin{bmatrix} 1 & 0 & 0 \\ 2 & 3 & 0 \\ 4 & 6 & 5 \end{bmatrix}$ is a lower triangular matrix.

11. Symmetric matrix: A square matrix $A = [a_{ij}]$ is said to be a symmetric matrix if $a_{ij} = a_{ji}$ for all values of i and j. In other words, a square matrix A is said to be a symmetric matrix if $A' = A$.

For example: $\begin{bmatrix} 1 & 2 & 3 \\ 2 & 4 & 6 \\ 3 & 6 & 8 \end{bmatrix}$ and $\begin{bmatrix} a & b & c \\ b & d & e \\ c & e & f \end{bmatrix}$ are symmetric matrices of order 3.

12. Skew-symmetric matrix: A square matrix $A = [a_{ij}]$ is said to be a skew symmetric matrix if $a_{ij} = -a_{ji}$, for all values of $i \neq j$ and all its diagonal elements are zero. In otherwords, a square matrix is said to be a skew-symmetric matrix if $A' = -A$

For example: $\begin{bmatrix} 0 & 3 & -5 \\ -3 & 0 & 7 \\ 5 & -7 & 0 \end{bmatrix}$ and $\begin{bmatrix} 0 & a & -b \\ -a & 0 & -c \\ b & c & 0 \end{bmatrix}$ are skew-symmetric matrices of order 3.

13. Sub-matrix of a matrix: A matrix obtained from a given matrix A, by deleting some rows or columns or both is called a sub-matrix of A.

For example: If $A = \begin{bmatrix} 2 & 3 & 5 \\ 3 & 6 & 7 \\ 8 & 9 & 10 \end{bmatrix}$

then, $\begin{bmatrix} 2 & 3 \\ 3 & 6 \end{bmatrix}$, $\begin{bmatrix} 2 & 3 & 5 \\ 8 & 9 & 10 \end{bmatrix}$ and $\begin{bmatrix} 3 & 5 \\ 6 & 7 \\ 9 & 10 \end{bmatrix}$ are some of the sub-matrices of the matrix A.

2.4 MATRIX ADDITION (ADDITION OF MATRICES)

Two matrices are added only if they are of same order. If A and B are two matrices of same order $m \times n$ then their sum (denoted by $A + B$) is defined to be the matrix of the same order $m \times n$ obtained by adding the corresponding elements of A and B.

If $$A = [a_{ij}]_{m \times n} \text{ and } B = [b_{ij}]_{m \times n} \text{ then}$$
$$A + B = [a_{ij} + b_{ij}]_{m \times n}$$

Note: If two matrices are not of same order, then their sum is not defined.

2.5 PROPERTIES OF MATRIX ADDITION

(*i*) Matrix addition is commutative,
i.e., $A + B = B + A$

(*ii*) Matrix addition is associative,
i.e., $(A + B) + C = A + (B + C)$

(*iii*) Existence of additive identity:
For every matrix A, there exists a null matrix O of the same order as A such that
$A + O = O + A = A$

(*iv*) For any matrix A, there exists a matrix $-A$ of the same order as A such that
$A + (-A) = (-A) + (A) = 0$

(*v*) Cancellation law:
$A + B = A + C \Rightarrow B = C$ (Left cancellation law)
$B + A = C + A \Rightarrow B = C$ (Right cancellation law)

2.6 SUBTRACTION OF MATRICES

If A and B are two matrices of same order $m \times n$, then we define
$$A - B = A + (-B)$$

Thus the difference $A - B$ is obtained by subtracting from each element of A the corresponding element of B.

Example. If $A = \begin{bmatrix} 1 & 2 & 3 \\ 3 & 4 & -2 \\ -3 & 2 & 4 \end{bmatrix}$ and $B = \begin{bmatrix} 1 & 0 & 2 \\ 2 & 3 & 4 \\ 4 & 3 & 2 \end{bmatrix}$

Find the sum and difference of A and B.

Solution: Here, the order of both the matrices A and B is same as 3×3. So, we can find the sum and difference of A and B.

$$\therefore \quad A + B = \begin{bmatrix} 1 & 2 & 3 \\ 3 & 4 & -2 \\ -3 & 2 & 4 \end{bmatrix} + \begin{bmatrix} 1 & 0 & 2 \\ 2 & 3 & 4 \\ 4 & 3 & 2 \end{bmatrix}$$

$$= \begin{bmatrix} 1+1 & 2+0 & 3+2 \\ 3+2 & 4+3 & -2+4 \\ -3+4 & 2+3 & 4+2 \end{bmatrix} = \begin{bmatrix} 2 & 2 & 5 \\ 5 & 7 & 2 \\ 1 & 5 & 6 \end{bmatrix}$$

Also, $$A - B = A + (-B)$$

$$= \begin{bmatrix} 1 & 2 & 3 \\ 3 & 4 & -2 \\ -3 & 2 & 4 \end{bmatrix} + (-1) \begin{bmatrix} 1 & 0 & 2 \\ 2 & 3 & 4 \\ 4 & 3 & 2 \end{bmatrix}$$

$$= \begin{bmatrix} 1 & 2 & 3 \\ 3 & 4 & -2 \\ -3 & 2 & 4 \end{bmatrix} + \begin{bmatrix} -1 & 0 & -2 \\ -2 & -3 & -4 \\ -4 & -3 & -2 \end{bmatrix}$$

$$= \begin{bmatrix} 1-1 & 2+0 & 3-2 \\ 3-2 & 4-3 & -2-4 \\ -3-4 & 2-3 & 4-2 \end{bmatrix}$$

$$\therefore \qquad A - B = \begin{bmatrix} 0 & 2 & 1 \\ 1 & 1 & -6 \\ -7 & -1 & 2 \end{bmatrix}$$

2.7 MULTIPLICATION OF MATRICES

Let $A = [a_{ij}]_{m \times n}$ and $B = [b_{jk}]_{n \times p}$ be two matrices such that the number of columns in A is equal to the number of rows in B. Then the $m \times p$ matrix $C = [C_{ik}]_{m \times p}$ such that

$$C_{ik} = \sum_{j=1}^{n} a_{ij}\, b_{jk} \; ; i = 1, 2, ..., m$$

$$k = 1, 2, ..., p$$

is called the product of the matrices A and B in that order and we write $C = AB$.

In the product AB, the matrix A is called the pre-factor and the matrix B is called the post-factor.

Note: Two matrices are said to be conformable for multiplication, if the number of columns in first is equal to the number of rows in second.

Example: If $A = \begin{bmatrix} 1 & 2 & 0 \\ 3 & 1 & 2 \\ 1 & 0 & 3 \end{bmatrix}$ and $B = \begin{bmatrix} 1 & 2 & 3 & 4 \\ 0 & 1 & 2 & 1 \\ 2 & 1 & 0 & 6 \end{bmatrix}$

Find AB.

Solution: The order of A is 3×3 and order of B is 3×4. Here, the number of columns in A is equal to the number of rows in B. Therefore, AB is possible.

$$\text{Now, } AB = A \times B = \begin{bmatrix} 1 & 2 & 0 \\ 3 & 1 & 2 \\ 1 & 0 & 3 \end{bmatrix} \times \begin{bmatrix} 1 & 2 & 3 & 4 \\ 0 & 1 & 2 & 1 \\ 2 & 1 & 0 & 6 \end{bmatrix}$$

$$= \begin{bmatrix} (1\times1+2\times0+0\times2) & (1\times2+2\times1+0\times1) & (1\times3+2\times2+0\times0) & (1\times4+2\times1+0\times6) \\ (3\times1+1\times0+2\times2) & (3\times2+1\times1+2\times1) & (3\times3+1\times2+2\times0) & (3\times4+1\times1+2\times6) \\ (1\times1+0\times0+3\times2) & (1\times2+0\times1+3\times1) & (1\times3+0\times2+3\times0) & (1\times4+0\times1+3\times6) \end{bmatrix}$$

$$AB = \begin{bmatrix} 1 & 4 & 7 & 6 \\ 7 & 9 & 11 & 25 \\ 7 & 5 & 3 & 22 \end{bmatrix}_{3 \times 4}$$

2.8 PROPERTIES OF MATRIX MULTIPLICATION

Assuming that A, B and C are conformable to multiplication, we have

(*i*) $A(B + C) = AB + AC$ ⎤
(*ii*) $(A + B)C = AC + BC$ ⎦ Distributive property

(*iii*) $(AB)C = A(BC)$ Associative property

(*iv*) The multiplication of matrices is not always commutative, i.e., if $AB = BA$ we say that A and B are commutative otherwise not.

(*v*) $AB = O$ does not necessarily mean $A = O$ or $B = O$. That is the product of two non-zero matrices can be a zero matrix.

(*vi*) $AB = AC$ and $BC = CA$ does not necessarily mean $B = C$, that is cancellation laws, in general do not hold for matrix multiplication.

(*vii*) $AO = OA = O$, where O is a null matrix.

(*viii*) $AI = A = IA$, where I is a unit matrix.

2.9 ELEMENTARY TRANSFORMATIONS ON A MATRIX

An elementary transformation (or an E – transformation) is an operation of any one of the following type:

1. Interchanging of any two rows or two columns. The operation is denoted by R_{ij} or $R_i \leftrightarrow R_j$ for rows and C_{ij} or $C_i \leftrightarrow C_j$ for columns.
2. The multiplication of any row or column by a non-zero number or scalar k, denoted by kR_i or $R_i \to kR_i$ for row and kC_i or $C_i \to kC_i$ for column.
3. The addition of k times of jth row to i^{th} row or k times of jth column to i^{th} column, denoted by $R_i + kR_j$ or $R_i \to R_i + kR_j$ for row and $C_i + kC_j$ or $C_i \to C_i + kC_j$ for column.

If a matrix B is obtained from a matrix A by one or more elementary transformations, then B is said to be equivalent to A. The symbol ~ is used for equivalence. i.e., $A \sim B$.

Example: Apply $R_2 \to R_2 + 2R_1$ and $R_3 \to R_3 + R_1$ on the matrix $\begin{bmatrix} 1 & 2 & 3 \\ 2 & 3 & 4 \\ 5 & 4 & 3 \end{bmatrix}$.

Solution: Let $A = \begin{bmatrix} 1 & 2 & 3 \\ 2 & 3 & 4 \\ 5 & 4 & 3 \end{bmatrix}$

$$\sim \begin{bmatrix} 1 & 2 & 3 \\ 4 & 7 & 10 \\ 6 & 6 & 6 \end{bmatrix} \text{ (by } R_2 \to R_2 + 2R_1, R_3 \to R_3 + R_1)$$

Example: Apply $R_2 \to R_2 - 2R_1$ and $R_3 \to \frac{1}{3} R_3$ on the matrix $\begin{bmatrix} 1 & 2 & 4 \\ 3 & 6 & 9 \\ 9 & 12 & 18 \end{bmatrix}$.

Solution: Let $A = \begin{bmatrix} 1 & 2 & 4 \\ 3 & 6 & 9 \\ 9 & 12 & 18 \end{bmatrix}$

$$\sim \begin{bmatrix} 1 & 2 & 4 \\ 1 & 2 & 1 \\ 3 & 4 & 6 \end{bmatrix} \text{ (by } R_2 \to R_2 - 2R_1, R_3 \to \frac{1}{3} R_3)$$

2.10 ELEMENTARY MATRIX

A matrix obtained from a unit matrix by a single elementary transformation is called an elementary matrix. For example, $\begin{bmatrix} 0 & 0 & 1 \\ 0 & 1 & 0 \\ 1 & 0 & 0 \end{bmatrix}$ is the elementary matrix obtained from I_3 by the elementary operation $C_1 \leftrightarrow C_3$.

$\begin{bmatrix} 1 & 2 & 0 \\ 0 & 1 & 0 \\ 0 & 0 & 1 \end{bmatrix}$ is the elementary matrix obtained from I_3 by the elementary operation $R_1 \to R_1 + 2R_2$.

2.11 ORTHOGONAL MATRIX

A square matrix A is called an orthogonal matrix if $AA' = I$ or $A'A = I$

For example, $\frac{1}{3}\begin{bmatrix} 1 & 2 & 2 \\ 2 & 1 & -2 \\ -2 & 2 & -1 \end{bmatrix}$ is orthogonal.

ILLUSTRATIVE EXAMPLES

Example 1. Find the values of x, y, z and t satisfying the matrix equation $\begin{bmatrix} x+3 & 2y+x \\ z-2 & 2t-3 \end{bmatrix} = \begin{bmatrix} 0 & 7 \\ 8 & t \end{bmatrix}$

Solution: As the given matrices are equal, their corresponding elements will be equal.

$$x + 3 = 0 \quad \text{or} \quad x = -3$$
$$2y + x = 7 \quad \text{or} \quad 2y = 7 - x = 7 - (-3) + 7 + 3 = 10$$
$$\text{or} \quad y = 5$$
$$z - 2 = 8 \quad \text{or} \quad z = 8 + 2 = 10$$

$$2t-3=t \quad \text{or} \quad 2t-t=3$$
$$\text{or} \quad t=3$$

$\therefore \quad x=-3, y=5, z=10, t=3$

Example 2. If $A=\begin{bmatrix} 3 & 4 & 2 \\ 1 & 0 & 6 \end{bmatrix}$ and $B=\begin{bmatrix} 2 & 3 & 0 \\ 1 & 0 & 4 \end{bmatrix}$. Find $3A-2B$.

Solution: $3A=\begin{bmatrix} 3\times3 & 3\times4 & 3\times2 \\ 3\times1 & 3\times0 & 3\times6 \end{bmatrix}=\begin{bmatrix} 9 & 12 & 6 \\ 3 & 0 & 18 \end{bmatrix}$

$$2B=\begin{bmatrix} 2\times2 & 2\times3 & 2\times0 \\ 2\times1 & 2\times0 & 2\times4 \end{bmatrix}=\begin{bmatrix} 4 & 6 & 0 \\ 2 & 0 & 8 \end{bmatrix}$$

$$\therefore \quad 3A-2B=\begin{bmatrix} 9 & 12 & 6 \\ 3 & 0 & 18 \end{bmatrix}-\begin{bmatrix} 4 & 6 & 0 \\ 2 & 0 & 8 \end{bmatrix}$$

$$=\begin{bmatrix} 9-4 & 12-6 & 6-0 \\ 3-2 & 0-0 & 18-8 \end{bmatrix}=\begin{bmatrix} 5 & 6 & 6 \\ 1 & 0 & 10 \end{bmatrix}$$

Example 3. Find the matrix A, if $\begin{bmatrix} 3 & 5 & -9 \\ -1 & 4 & -7 \end{bmatrix}+A=\begin{bmatrix} 6 & 2 & 3 \\ 4 & 8 & 6 \end{bmatrix}$

Solution: We have,

$$\begin{bmatrix} 3 & 5 & -9 \\ -1 & 4 & -7 \end{bmatrix}+A=\begin{bmatrix} 6 & 2 & 3 \\ 4 & 8 & 6 \end{bmatrix}$$

Now,
$$A=\begin{bmatrix} 6 & 2 & 3 \\ 4 & 8 & 6 \end{bmatrix}-\begin{bmatrix} 3 & 5 & -9 \\ -1 & 4 & -7 \end{bmatrix}$$

$$=\begin{bmatrix} 6-3 & 2-5 & 3+9 \\ 4+1 & 8-4 & 6+7 \end{bmatrix}$$

$$=\begin{bmatrix} 3 & -3 & 12 \\ 5 & 4 & 13 \end{bmatrix}$$

Example 4. Find the matrices A and B if $A+B=\begin{bmatrix} 1 & -1 \\ 3 & 0 \end{bmatrix}$ and $A-B=\begin{bmatrix} 3 & 1 \\ 1 & 4 \end{bmatrix}$.

Solution: On adding the given matrices, we have

$$(A+B)+(A-B)=\begin{bmatrix} 1 & -1 \\ 3 & 0 \end{bmatrix}+\begin{bmatrix} 3 & 1 \\ 1 & 4 \end{bmatrix}$$

$$\Rightarrow \quad 2A=\begin{bmatrix} 4 & 0 \\ 4 & 4 \end{bmatrix} \Rightarrow A=\frac{1}{2}\begin{bmatrix} 4 & 0 \\ 4 & 4 \end{bmatrix}=\begin{bmatrix} 2 & 0 \\ 2 & 2 \end{bmatrix}$$

On subtraiting the given matrices, we have

$$(A+B)-(A-B)=\begin{bmatrix} 1 & -1 \\ 3 & 0 \end{bmatrix}-\begin{bmatrix} 3 & 1 \\ 1 & 4 \end{bmatrix}$$

$$\Rightarrow \qquad 2B = \begin{bmatrix} -2 & -2 \\ 2 & -4 \end{bmatrix} \Rightarrow B = \frac{1}{2}\begin{bmatrix} -2 & -2 \\ 2 & -4 \end{bmatrix} = \begin{bmatrix} -1 & -1 \\ 1 & -2 \end{bmatrix}$$

Example 5. If $A = \begin{bmatrix} 1 & 2 \\ 3 & 4 \end{bmatrix}, B = \begin{bmatrix} 2 & 1 \\ 4 & 2 \end{bmatrix}, C = \begin{bmatrix} 5 & 1 \\ 7 & 4 \end{bmatrix}$ verity that $A\,(B + C) = AB + AC$.

Solution: $B + C = \begin{bmatrix} 2 & 1 \\ 4 & 2 \end{bmatrix} + \begin{bmatrix} 5 & 1 \\ 7 & 4 \end{bmatrix} = \begin{bmatrix} 7 & 2 \\ 11 & 6 \end{bmatrix}$

$$A\,(B + C) = \begin{bmatrix} 1 & 2 \\ 3 & 4 \end{bmatrix}\begin{bmatrix} 7 & 2 \\ 11 & 6 \end{bmatrix} = \begin{bmatrix} (1\times 7 + 2\times 11) & (1\times 2 + 2\times 6) \\ (3\times 7 + 4\times 11) & (3\times 2 + 4\times 6) \end{bmatrix} = \begin{bmatrix} 29 & 14 \\ 65 & 30 \end{bmatrix}$$

$$\text{Now, } AB = \begin{bmatrix} 1 & 2 \\ 3 & 4 \end{bmatrix}\begin{bmatrix} 2 & 1 \\ 4 & 2 \end{bmatrix} = \begin{bmatrix} (1\times 2 + 2\times 4) & (1\times 1 + 2\times 2) \\ (3\times 2 + 4\times 4) & (3\times 1 + 4\times 2) \end{bmatrix} = \begin{bmatrix} 10 & 5 \\ 22 & 11 \end{bmatrix}$$

$$\text{and } AC = \begin{bmatrix} 1 & 2 \\ 3 & 4 \end{bmatrix}\begin{bmatrix} 5 & 1 \\ 7 & 4 \end{bmatrix} = \begin{bmatrix} (1\times 5 + 2\times 7) & (1\times 1 + 2\times 4) \\ (3\times 5 + 4\times 7) & (3\times 1 + 4\times 4) \end{bmatrix} = \begin{bmatrix} 19 & 9 \\ 43 & 19 \end{bmatrix}$$

$$AB + AC = \begin{bmatrix} 10 & 5 \\ 22 & 11 \end{bmatrix} + \begin{bmatrix} 19 & 9 \\ 43 & 19 \end{bmatrix} = \begin{bmatrix} 29 & 14 \\ 65 & 30 \end{bmatrix}$$

$\therefore A\,(B + C) = AB + BC$

Example 6. If $A = \begin{bmatrix} 4 & 1 \\ 3 & 2 \end{bmatrix}$, then prove that $A^2 - 6A + 5I = \begin{bmatrix} 0 & 0 \\ 0 & 0 \end{bmatrix}$, where $I = \begin{bmatrix} 1 & 0 \\ 0 & 1 \end{bmatrix}$.

Solution: $A^2 = A \cdot A = \begin{bmatrix} 4 & 1 \\ 3 & 2 \end{bmatrix}\begin{bmatrix} 4 & 1 \\ 3 & 2 \end{bmatrix}$

$$= \begin{bmatrix} (4\times 4 + 1\times 3 & (4\times 1 + 1\times 2) \\ (3\times 4 + 2\times 3) & (3\times 1 + 2\times 2) \end{bmatrix} = \begin{bmatrix} 19 & 6 \\ 18 & 7 \end{bmatrix}$$

$$6A = 6\begin{bmatrix} 4 & 1 \\ 3 & 2 \end{bmatrix} = \begin{bmatrix} 24 & 6 \\ 18 & 12 \end{bmatrix}$$

$$5I = 5\begin{bmatrix} 1 & 0 \\ 0 & 1 \end{bmatrix} = \begin{bmatrix} 5 & 0 \\ 0 & 5 \end{bmatrix}$$

$$\therefore \qquad A^2 - 6A + 5I = \begin{bmatrix} 19 & 6 \\ 18 & 7 \end{bmatrix} - \begin{bmatrix} 24 & 6 \\ 18 & 12 \end{bmatrix} + \begin{bmatrix} 5 & 0 \\ 0 & 5 \end{bmatrix}$$

$$= \begin{bmatrix} 19 - 24 + 5 & 6 - 6 + 0 \\ 18 - 18 + 0 & 7 - 12 + 5 \end{bmatrix} = \begin{bmatrix} 0 & 0 \\ 0 & 0 \end{bmatrix} \text{ Proved.}$$

Example 7. If $A = \begin{bmatrix} 1 & 2 & 2 \\ 2 & 1 & 2 \\ 2 & 2 & 1 \end{bmatrix}$, then prove that $A^2 - 4A - 5I = 0$

Solution: $A^2 = \begin{bmatrix} 1 & 2 & 2 \\ 2 & 1 & 2 \\ 2 & 2 & 1 \end{bmatrix} \begin{bmatrix} 1 & 2 & 2 \\ 2 & 1 & 2 \\ 2 & 2 & 1 \end{bmatrix}$

$$= \begin{bmatrix} (1\times1+2\times2+2\times2) & (1\times2+2\times1+2\times2) & (1\times2+2\times2+2\times1) \\ (2\times1+1\times2+2\times2) & (2\times2+1\times1+2\times2) & (2\times2+1\times2+2\times1) \\ (2\times1+2\times2+1\times2) & (2\times2+2\times1+1\times2) & (2\times2+2\times2+1\times1) \end{bmatrix}$$

$$A^2 = \begin{bmatrix} 9 & 8 & 8 \\ 8 & 9 & 8 \\ 8 & 8 & 9 \end{bmatrix}, 4A = 4\begin{bmatrix} 1 & 2 & 2 \\ 2 & 1 & 2 \\ 2 & 2 & 1 \end{bmatrix} = \begin{bmatrix} 4 & 8 & 8 \\ 8 & 4 & 8 \\ 8 & 8 & 4 \end{bmatrix}$$

$$5I = 5I_{3\times3} = 5\begin{bmatrix} 1 & 0 & 0 \\ 0 & 1 & 0 \\ 0 & 0 & 1 \end{bmatrix} = \begin{bmatrix} 5 & 0 & 0 \\ 0 & 5 & 0 \\ 0 & 0 & 5 \end{bmatrix}$$

$$\therefore \quad A^2 - 4A - 5I = \begin{bmatrix} 9 & 8 & 8 \\ 8 & 9 & 8 \\ 8 & 8 & 9 \end{bmatrix} - \begin{bmatrix} 4 & 8 & 8 \\ 8 & 4 & 8 \\ 8 & 8 & 4 \end{bmatrix} - \begin{bmatrix} 5 & 0 & 0 \\ 0 & 5 & 0 \\ 0 & 0 & 5 \end{bmatrix}$$

$$= \begin{bmatrix} 9-4-5 & 8-8-0 & 8-8-0 \\ 8-8-0 & 9-4-5 & 8-8-0 \\ 8-8-0 & 8-8-0 & 9-4-5 \end{bmatrix} = \begin{bmatrix} 0 & 0 & 0 \\ 0 & 0 & 0 \\ 0 & 0 & 0 \end{bmatrix}$$

$= O_{3\times3} = O$ Proved.

Example 8. If $A = \begin{bmatrix} 1 & 3 & 0 \\ -1 & 2 & 1 \\ 0 & 0 & 2 \end{bmatrix}$ and $B = \begin{bmatrix} 2 & 3 & 4 \\ 1 & 2 & 3 \\ -1 & 1 & 2 \end{bmatrix}$ then, show that $AB \neq BA$.

Solution:

$$AB = \begin{bmatrix} 1 & 3 & 0 \\ -1 & 2 & 1 \\ 0 & 0 & 2 \end{bmatrix} \begin{bmatrix} 2 & 3 & 4 \\ 1 & 2 & 3 \\ -1 & 1 & 2 \end{bmatrix}$$

$$= \begin{bmatrix} (1\times2+3\times1+0\times-1) & (1\times3+3\times2+0\times1) & (1\times4+3\times3+0\times2) \\ (-1\times2+2\times1+1\times-1) & (-1\times3+2\times2+1\times1) & (-1\times4+2\times3+1\times2) \\ (0\times2+0\times1+2\times-1) & (0\times3+0\times2+2\times1) & (0\times4+0\times3+2\times2) \end{bmatrix}$$

$$= \begin{bmatrix} 5 & 9 & 13 \\ -1 & 2 & 4 \\ -2 & 2 & 4 \end{bmatrix}$$

and $$BA = \begin{bmatrix} 2 & 3 & 4 \\ 1 & 2 & 3 \\ -1 & 1 & 2 \end{bmatrix} \begin{bmatrix} 1 & 3 & 0 \\ -1 & 2 & 1 \\ 0 & 0 & 2 \end{bmatrix}$$

$$= \begin{bmatrix} (2\times1+3\times-1+4\times0) & (2\times3+3\times2+4\times0) & (2\times0+3\times1+4\times2) \\ (1\times1+2\times-1+3\times0) & (1\times3+2\times2+3\times0) & (1\times0+2\times1+3\times2) \\ (-1\times1+1\times-1+2\times0) & (-1\times3+1\times2+2\times0) & (-1\times0+1\times1+2\times2) \end{bmatrix}$$

$$BA = \begin{bmatrix} -1 & 12 & 11 \\ -1 & 7 & 8 \\ -2 & -1 & 5 \end{bmatrix}$$

$\therefore AB \neq BA$ Proved.

Example 9. If $\begin{bmatrix} 1 & 1 & x \end{bmatrix} \begin{bmatrix} 1 & 0 & 2 \\ 0 & 2 & 1 \\ 2 & 1 & 0 \end{bmatrix} \begin{bmatrix} 1 \\ 1 \\ 1 \end{bmatrix} = 0$. Find x.

Solution: We have,

$$\begin{bmatrix} 1 & 1 & x \end{bmatrix} \begin{bmatrix} 1 & 0 & 2 \\ 0 & 2 & 1 \\ 2 & 1 & 0 \end{bmatrix} \begin{bmatrix} 1 \\ 1 \\ 1 \end{bmatrix} = 0$$

$$\Rightarrow \begin{bmatrix} 1\times1+1\times0+x\times2 & 1\times0+1\times2+x\times1 & 1\times2+1\times1+x\times0 \end{bmatrix} \begin{bmatrix} 1 \\ 1 \\ 1 \end{bmatrix} = 0$$

$$\Rightarrow \quad [1+2x \quad 2+x \quad 3] \begin{bmatrix} 1 \\ 1 \\ 1 \end{bmatrix} = 0$$

$$\Rightarrow \quad [(1+2x)\times1+(2+x)\times1+3\times1] = 0$$

$$\Rightarrow \quad 3x+6 = 0$$

$$\Rightarrow \quad x = -2$$

Example 10. If $A = \begin{bmatrix} 1 & 1 & -1 \\ 2 & 0 & 2 \\ 3 & -1 & 3 \end{bmatrix}$, $B = \begin{bmatrix} 1 & 3 \\ 0 & 2 \\ -1 & 4 \end{bmatrix}$ and $C = \begin{bmatrix} 1 & 2 & 3 & -4 \\ 2 & 0 & -2 & 1 \end{bmatrix}$

show that $(AB)C = A(BC)$

Solution. It $$AB = \begin{bmatrix} 1 & 1 & -1 \\ 2 & 0 & 2 \\ 3 & -1 & 3 \end{bmatrix} \begin{bmatrix} 1 & 3 \\ 0 & 2 \\ -1 & 4 \end{bmatrix}$$

$$= \begin{bmatrix} 1\times1+1\times0+(-1)\times(-1) & 1\times3+1\times2+(-1)\times4 \\ 2\times1+0\times0+2\times(-1) & 2\times3+0\times2+2\times4 \\ 3\times1+(-1)\times0+3\times(-1) & 3\times3+(-1)\times2+3\times4 \end{bmatrix}$$

$$= \begin{bmatrix} 2 & 1 \\ 0 & 14 \\ 0 & 19 \end{bmatrix}$$

$$(AB)\,C = \begin{bmatrix} 2 & 1 \\ 0 & 14 \\ 0 & 19 \end{bmatrix} \begin{bmatrix} 1 & 2 & 3 & -4 \\ 2 & 0 & -2 & 1 \end{bmatrix}$$

$$= \begin{bmatrix} 2\times 1+1\times 2 & 2\times 2+1\times 0 & 2\times 3+1\times(-2) & 2\times(-4)+1\times 1 \\ 0\times 1+14\times 2 & 0\times 2+14\times 0 & 0\times 3+14\times(-2) & 0\times(-4)+14\times 1 \\ 0\times 1+19\times 2 & 0\times 2+19\times 0 & 0\times 3+19\times(-2) & 0\times(-4)+19\times 1 \end{bmatrix}$$

$$= \begin{bmatrix} 4 & 4 & 4 & -7 \\ 28 & 0 & -28 & 14 \\ 38 & 0 & -38 & 19 \end{bmatrix}$$

$$\text{Now,} BC = \begin{bmatrix} 1 & 3 \\ 0 & 2 \\ -1 & 4 \end{bmatrix} \begin{bmatrix} 1 & 2 & 3 & -4 \\ 2 & 0 & -2 & 1 \end{bmatrix}$$

$$= \begin{bmatrix} 1\times 1+3\times 2 & 1\times 2+3\times 0 & 1\times 3+3\times(-2) & 1\times(-4)+3\times 1 \\ 0\times 1+2\times 2 & 0\times 2+2\times 0 & 0\times 3+2\times(-2) & 0\times(-4)+2\times 1 \\ (-1)\times 1+4\times 2 & (-1)\times 2+4\times 0 & (-1)\times 3+4\times(-2) & (-1)\times(-4)+4\times 1 \end{bmatrix}$$

$$= \begin{bmatrix} 7 & 2 & -3 & -1 \\ 4 & 0 & -4 & 2 \\ 7 & -2 & -11 & 8 \end{bmatrix}$$

$$A\,(BC) = \begin{bmatrix} 1 & 1 & -1 \\ 2 & 0 & 2 \\ 3 & -1 & 3 \end{bmatrix} \begin{bmatrix} 7 & 2 & -3 & -1 \\ 4 & 0 & -4 & 2 \\ 7 & -2 & -11 & 8 \end{bmatrix}$$

$$= \begin{bmatrix} 1\times 7+1\times 4+(-1)\times 7 & 1\times 2+1\times 0+(-1)\times(-2) & 1\times(-3)+1\times(-4)+(-1)\times(-11) & 1\times(-1)+1\times 2+(-1)\times 8 \\ 2\times 7+0\times 4+2\times 7 & 2\times 2+0\times 0+2\times(-2) & 2\times(-3)+0\times(-4)+2\times(-11) & 2\times(-1)+0\times 2+2\times 8 \\ 3\times 7+(-1)\times 4+3\times 7 & 3\times 2+(-1)\times 0+3\times(-2) & 3\times(-3)+(-1)\times(-4)+3\times(-11) & 3\times(-1)+(-1)\times 2+3\times 8 \end{bmatrix}$$

$$= \begin{bmatrix} 4 & 4 & 4 & -7 \\ 28 & 0 & -28 & 14 \\ 38 & 0 & -38 & 19 \end{bmatrix}$$

$\therefore \quad (AB)\,C = A\,(BC)$ Proved.

Example 11. If $A \begin{bmatrix} 1 & 2 & -1 \\ 3 & 0 & 2 \\ 4 & 5 & 0 \end{bmatrix}$ and $B = \begin{bmatrix} 1 & 0 & 0 \\ 2 & 1 & 0 \\ 0 & 1 & 3 \end{bmatrix}$, verity that $(AB)' = B'A'$.

Solution: $A' = \text{Transpose of } A = \begin{bmatrix} 1 & 3 & 4 \\ 2 & 0 & 5 \\ -1 & 2 & 0 \end{bmatrix}$

$$B' = \text{Transpose of } B = \begin{bmatrix} 1 & 2 & 0 \\ 0 & 1 & 1 \\ 0 & 0 & 3 \end{bmatrix}$$

Now,
$$AB = \begin{bmatrix} 1 & 2 & -1 \\ 3 & 0 & 2 \\ 4 & 5 & 0 \end{bmatrix} \begin{bmatrix} 1 & 0 & 0 \\ 2 & 1 & 0 \\ 0 & 1 & 3 \end{bmatrix}$$

$$AB = \begin{bmatrix} 1+4+0 & 0+2-1 & 0+0-3 \\ 3+0+0 & 0+0+2 & 0+0+6 \\ 4+10+0 & 0+5+0 & 0+0+0 \end{bmatrix}$$

$$= \begin{bmatrix} 5 & 1 & -3 \\ 3 & 2 & 6 \\ 14 & 5 & 0 \end{bmatrix}$$

$$(AB)' = \text{Transpose of } AB = \begin{bmatrix} 5 & 3 & 14 \\ 1 & 2 & 5 \\ -3 & 6 & 0 \end{bmatrix}$$

Now,
$$B'A' = \begin{bmatrix} 1 & 2 & 0 \\ 0 & 1 & 1 \\ 0 & 0 & 3 \end{bmatrix} \begin{bmatrix} 1 & 3 & 4 \\ 2 & 0 & 5 \\ -1 & 2 & 0 \end{bmatrix}$$

$$= \begin{bmatrix} 1+4+0 & 3+0+0 & 4+10+0 \\ 0+2-1 & 0+0+2 & 0+5+0 \\ 0+0-3 & 0+0+6 & 0+0+0 \end{bmatrix}$$

$$= \begin{bmatrix} 5 & 3 & 14 \\ 1 & 2 & 5 \\ -3 & 6 & 0 \end{bmatrix}$$

$\therefore$ $(AB)' = B'A'$ **Proved.**

Example 12. If $A = \begin{bmatrix} 3 & -4 \\ 1 & 1 \\ 1 & 0 \end{bmatrix}$ and $B = \begin{bmatrix} 2 & 1 & 2 \\ 1 & 3 & 4 \end{bmatrix}$, verity that $(AB)' = B'A'$

Solution: $A' = \text{Transpose of } A = \begin{bmatrix} 3 & 1 & 1 \\ -4 & 1 & 0 \end{bmatrix}$

$$B' = \text{Transpose of } B = \begin{bmatrix} 2 & 1 \\ 1 & 3 \\ 2 & 4 \end{bmatrix}$$

$$AB = \begin{bmatrix} 3 & -4 \\ 1 & 1 \\ 1 & 0 \end{bmatrix} \begin{bmatrix} 2 & 1 & 2 \\ 1 & 3 & 4 \end{bmatrix}$$

$$= \begin{bmatrix} 6-4 & 3-12 & 6-16 \\ 2+1 & 1+3 & 2+4 \\ 2+0 & 1+0 & 2+0 \end{bmatrix} = \begin{bmatrix} 2 & -9 & -10 \\ 3 & 4 & 6 \\ 2 & 1 & 2 \end{bmatrix}$$

$$(AB)' \text{ Transpose of } AB = \begin{bmatrix} 2 & 3 & 2 \\ -9 & 4 & 1 \\ -10 & 6 & 2 \end{bmatrix}$$

Now, $$B'A' = \begin{bmatrix} 2 & 1 \\ 1 & 3 \\ 2 & 4 \end{bmatrix} \begin{bmatrix} 3 & 1 & 1 \\ -4 & 1 & 1 \end{bmatrix}$$

$$= \begin{bmatrix} 6-4 & 2+1 & 2+0 \\ 3-12 & 1+3 & 1+0 \\ 6-16 & 2+4 & 2+0 \end{bmatrix}$$

$$= \begin{bmatrix} 2 & 3 & 2 \\ -9 & 4 & 1 \\ -10 & 6 & 2 \end{bmatrix}$$

$\therefore \quad (AB)' = B'A'$ Proved.

Example 13. If $A = \begin{bmatrix} \cos\alpha & \sin\alpha \\ -\sin\alpha & \cos\alpha \end{bmatrix}$, verity that $AA' = I = A'A$

Solution: A' = Transpose of $A = \begin{bmatrix} \cos\alpha & -\sin\alpha \\ \sin\alpha & \cos\alpha \end{bmatrix}$

Now, $$AA' = \begin{bmatrix} \cos\alpha & \sin\alpha \\ -\sin\alpha & \cos\alpha \end{bmatrix} \begin{bmatrix} \cos\alpha & -\sin\alpha \\ \sin\alpha & \cos\alpha \end{bmatrix}$$

$$= \begin{bmatrix} \cos^2\alpha + \sin^2\alpha & -\cos\alpha\sin\alpha + \cos\alpha\sin\alpha \\ -\sin\alpha\cos\alpha + \cos\alpha\sin\alpha & \sin^2\alpha + \cos^2\alpha \end{bmatrix}$$

$$= \begin{bmatrix} 1 & 0 \\ 0 & 1 \end{bmatrix} = I$$

Now, $$A'A = \begin{bmatrix} \cos\alpha & -\sin\alpha \\ \sin\alpha & \cos\alpha \end{bmatrix} \begin{bmatrix} \cos\alpha & \sin\alpha \\ -\sin\alpha & \cos\alpha \end{bmatrix}$$

$$= \begin{bmatrix} \cos^2\alpha + \sin^2\alpha & \cos\alpha\sin\alpha - \sin\alpha\cos\alpha \\ \sin\alpha\cos\alpha - \sin\alpha\cos\alpha & \sin^2\alpha + \cos^2\alpha \end{bmatrix}$$

$$= \begin{bmatrix} 1 & 0 \\ 0 & 1 \end{bmatrix} = I$$

$\therefore$ $$AA' = I = A'A$$ Proved.

Example 14. Verity that $\frac{1}{3}\begin{bmatrix} 1 & 2 & 2 \\ 2 & 1 & -2 \\ -2 & 2 & -1 \end{bmatrix}$ is an orthogonal matrix.

Solution: Let $A = \frac{1}{3}\begin{bmatrix} 1 & 2 & 2 \\ 2 & 1 & -2 \\ -2 & 2 & -1 \end{bmatrix}$

$$A' = \frac{1}{3}\begin{bmatrix} 1 & 2 & -2 \\ 2 & 1 & 2 \\ 2 & -2 & -1 \end{bmatrix}$$

Now, $$AA' = \frac{1}{9}\begin{bmatrix} 1 & 2 & 2 \\ 2 & 1 & -2 \\ -2 & 2 & -1 \end{bmatrix}\begin{bmatrix} 1 & 2 & -2 \\ 2 & 1 & 2 \\ 2 & -2 & -1 \end{bmatrix}$$

$$= \frac{1}{9}\begin{bmatrix} 1+4+4 & 2+2-4 & -2+4-2 \\ 2+2-4 & 4+1+4 & -4+2+2 \\ -2+4-2 & -4+2+2 & 4+4+1 \end{bmatrix}$$

$$= \frac{1}{9}\begin{bmatrix} 9 & 0 & 0 \\ 0 & 9 & 0 \\ 0 & 0 & 9 \end{bmatrix}$$

$$= \begin{bmatrix} 1 & 0 & 0 \\ 0 & 1 & 0 \\ 0 & 0 & 1 \end{bmatrix} = I$$

Hence, the given matrix is orthogonal.

Example 15. Verity that $\begin{bmatrix} \cos\theta & 0 & \sin\theta \\ 0 & 1 & 0 \\ -\sin\theta & 0 & \cos\theta \end{bmatrix}$ is an orthogonal matrix.

Solution: Let $A = \begin{bmatrix} \cos\theta & 0 & \sin\theta \\ 0 & 1 & 0 \\ -\sin\theta & 0 & \cos\theta \end{bmatrix}$

$$A' = \begin{bmatrix} \cos\theta & 0 & -\sin\theta \\ 0 & 1 & 0 \\ \sin\theta & 0 & \cos\theta \end{bmatrix}$$

$$\text{Now, } AA' = \begin{bmatrix} \cos\theta & 0 & \sin\theta \\ 0 & 1 & 0 \\ -\sin\theta & 0 & \cos\theta \end{bmatrix} \begin{bmatrix} \cos\theta & 0 & -\sin\theta \\ 0 & 1 & 0 \\ \sin\theta & 0 & \cos\theta \end{bmatrix}$$

$$= \begin{bmatrix} \cos^2\theta + 0 + \sin^2\theta & 0+0+0 & -\sin\theta\cos\theta + 0 + \sin\theta\cos\theta \\ 0+0+0 & 0+1+0 & 0+0+0 \\ -\sin\theta\cos\theta + 0 + \sin\theta\cos\theta & 0+0+0 & \sin^2\theta + 0 + \cos^2\theta \end{bmatrix}$$

$$= \begin{bmatrix} 1 & 0 & 0 \\ 0 & 1 & 0 \\ 0 & 0 & 1 \end{bmatrix} = \text{I}$$

Hence, the given matrix is orthogonal.

Example 16. Determine the values of a, b and c when $\begin{bmatrix} 0 & 2b & c \\ a & b & -c \\ a & -b & c \end{bmatrix}$ is orthogonal.

Solution: We know that, a square matrix is orthogonal if $AA' = I$.

Let $$A = \begin{bmatrix} 0 & 2b & c \\ a & b & -c \\ a & -b & c \end{bmatrix}$$

$$A' = \begin{bmatrix} 0 & a & a \\ 2b & b & -b \\ c & -c & c \end{bmatrix}$$

We have, $AA' = I$

$$\begin{bmatrix} 0 & 2b & c \\ a & b & -c \\ a & -b & c \end{bmatrix} \begin{bmatrix} 0 & a & a \\ 2b & b & -b \\ c & -c & c \end{bmatrix} = \begin{bmatrix} 1 & 0 & 0 \\ 0 & 1 & 0 \\ 0 & 0 & 1 \end{bmatrix}$$

$$\begin{bmatrix} 4b^2 + c^2 & 2b^2 - c^2 & -2b^2 + c^2 \\ 2b^2 - c^2 & a^2 + b^2 + c^2 & a^2 - b^2 - c^2 \\ -2b^2 + c^2 & a^2 - b^2 - c^2 & a^2 + b^2 + c^2 \end{bmatrix} = \begin{bmatrix} 1 & 0 & 0 \\ 0 & 1 & 0 \\ 0 & 0 & 1 \end{bmatrix}$$

On equating the corresponding elements, we have

$$4b^2 + c^2 = 1$$

$$2b^2 - c^2 = 1$$

and $$a^2 - b^2 - c^2 = 0$$

On solving above equations, we have

$$a = \pm\frac{1}{\sqrt{2}}, b = \pm\frac{1}{\sqrt{6}} \text{ and } c = \pm\frac{1}{\sqrt{3}}$$

EXERCISE 2.1

1. Construct a 2×3 matrix whose elements a_{ij} are given by

(*i*) $a_{ij} = i \cdot j$ (*ii*) $a_{ij} = 2i - j$

2. If $\begin{bmatrix} x+2 & 3 \\ 1 & y+1 \end{bmatrix} = \begin{bmatrix} 8 & 3 \\ 1 & -2 \end{bmatrix}$, find the values of x and y.

3. If $A = \begin{bmatrix} 2 & 0 \\ 4 & -2 \end{bmatrix}$ and $B = \begin{bmatrix} 6 & 14 \\ 8 & 16 \end{bmatrix}$, prove that $A + B = B + A$.

4. If $A = \begin{bmatrix} 2 & 3 & 1 \\ 0 & -1 & 5 \end{bmatrix}$ and $B = \begin{bmatrix} 1 & 2 & -6 \\ 0 & -1 & 3 \end{bmatrix}$, find $3A - 4B$.

5. If $A = \begin{bmatrix} 3 & 8 & 1 \\ 2 & 6 & 3 \\ 7 & 4 & -5 \end{bmatrix}$ and $B = \begin{bmatrix} 4 & 0 & 2 \\ 6 & 2 & 3 \\ 1 & 3 & 2 \end{bmatrix}$, prove that $4(A - B) = 4A - 4B$.

6. If $A = \begin{bmatrix} 4 & 0 & 3 \\ 2 & 1 & 2 \\ 1 & 3 & 0 \end{bmatrix}$, $B = \begin{bmatrix} 1 & 3 & 2 \\ 2 & 0 & 4 \end{bmatrix}$, find AB and BA, if defined.

7. If $A = \begin{bmatrix} 1 & 0 \\ 1 & 1 \end{bmatrix}$, $B = \begin{bmatrix} 2 & 0 \\ 1 & 1 \end{bmatrix}$ and $C = \begin{bmatrix} -1 & 2 \\ 3 & 1 \end{bmatrix}$, show that

(*i*) $A(B + C) = AB + AC$ (*ii*) $(AB)C = A(BC)$

8. If $A = \begin{bmatrix} 1 & -1 \\ -1 & 1 \end{bmatrix}$, then prove that $A^3 = 4A$.

9. If $A = \begin{bmatrix} 1 & 2 & 3 \\ 2 & 3 & 4 \\ 1 & 1 & 2 \end{bmatrix}$ and $B = \begin{bmatrix} 0 & 2 & -1 \\ 1 & 2 & 4 \\ 0 & 2 & -3 \end{bmatrix}$, then show that $AB \neq BA$.

10. If $A = \begin{bmatrix} 1 & 0 & 5 \\ 2 & -1 & 3 \\ 4 & 1 & 0 \end{bmatrix}$ and $B = \begin{bmatrix} 3 & 2 & 1 \\ 0 & 2 & 1 \\ 3 & 2 & 5 \end{bmatrix}$, show that $(A + B)' = A' + B'$.

11. If $A \begin{bmatrix} 1 & -1 & 0 \\ 2 & 1 & 3 \\ 4 & 1 & 8 \end{bmatrix}$ and $B = \begin{bmatrix} 4 & 1 & 0 \\ 2 & -3 & 1 \\ 1 & 1 & -1 \end{bmatrix}$, prove that $(AB)' = B'A'$.

12. Find the value of $A^3 - 2A^2 + A - I$, if $A = \begin{bmatrix} 1 & 3 & 2 \\ 2 & 0 & 3 \\ 1 & -1 & 1 \end{bmatrix}$, where I is 3×3 unit matrix.

13. If $A = \begin{bmatrix} 2 & -3 & -5 \\ -1 & 4 & 5 \\ 1 & -3 & -4 \end{bmatrix}$ and $B = \begin{bmatrix} 2 & -2 & -4 \\ -1 & 3 & 4 \\ 1 & -2 & -3 \end{bmatrix}$, verity that $AB = A$ and $BA = B$.

14. If $A = \begin{bmatrix} 4 & -1 & -4 \\ 3 & 0 & -4 \\ 3 & -1 & -3 \end{bmatrix}$, show that $A^2 = I_3$.

15. verify that $\frac{1}{3}\begin{bmatrix} 1 & -2 & 2 \\ -2 & 1 & 2 \\ -2 & -2 & -1 \end{bmatrix}$ is an orthogonal matrix.

ANSWERS

1. (*i*) $\begin{bmatrix} 1 & 2 & 3 \\ 2 & 4 & 6 \end{bmatrix}$ (*ii*) $\begin{bmatrix} 1 & 0 & -1 \\ 3 & 2 & 1 \end{bmatrix}$ **2.** $x = 6, \quad y = -3$

4. $\begin{bmatrix} 2 & 1 & 27 \\ 0 & 1 & 3 \end{bmatrix}$

6. AB not defined, $BA = \begin{bmatrix} 12 & 9 & 9 \\ 12 & 12 & 6 \end{bmatrix}$

12. $\begin{bmatrix} 6 & 15 & 10 \\ 10 & 1 & 15 \\ 5 & -5 & 6 \end{bmatrix}$

2.12 DETERMINANTS

The determinant of a square matrix A is the determinant whose elements are the corresponding elements of the matrix A, and is denoted by ΔA or $|A|$ or det (A).

If $\quad A = \begin{bmatrix} a_{11} & a_{12} & \cdots & a_{1n} \\ a_{21} & a_{22} & \cdots & a_{2n} \\ \vdots & \vdots & & \vdots \\ a_{n1} & a_{n2} & \cdots & a_{nn} \end{bmatrix}$, then determinant of A is

$$|A| = \begin{vmatrix} a_{11} & a_{12} & \cdots & a_{1n} \\ a_{21} & a_{22} & \cdots & a_{2n} \\ \vdots & \vdots & & \vdots \\ a_{n1} & a_{n2} & \cdots & a_{nn} \end{vmatrix}$$

Let $A = \begin{bmatrix} a_{11} & a_{12} \\ a_{21} & a_{22} \end{bmatrix}$, then $|A| = \begin{vmatrix} a_{11} & a_{12} \\ a_{21} & a_{22} \end{vmatrix} = a_{11}a_{22} - a_{12}a_{21}$

It is called a determinant of second order. The numbers $a_{11}, a_{12}, a_{21}, a_{22}$ are called the elements of the determinant and $(a_{11}\,a_{22}-a_{12}\,a_{21})$ is called the value of the determinant.

$$\text{Let } A=\begin{bmatrix} a_{11} & a_{12} & a_{13} \\ a_{21} & a_{22} & a_{23} \\ a_{31} & a_{32} & a_{33} \end{bmatrix},\ \text{then } |A|=\begin{vmatrix} a_{11} & a_{12} & a_{13} \\ a_{21} & a_{22} & a_{23} \\ a_{31} & a_{32} & a_{33} \end{vmatrix}$$

It is called a determinant of third order.

Note: (1) A matrix is an arrangement of numbers, has no value, Its determinant is not an array of numbers, enclosed between two vertical lines. It is a slalar quantity having a definite value.

(2) Only square matrices have their determinants.

(3) It the determinant of a square matrix is zero, the matrix is called singular, i.e., a square matrix A is singular if $|A|=0$. Otherwise, matrix is called non-singular, i.e., a square matrix A is non-singular if $|A|\neq 0$.

2.13 VALUE OF A DETERMINANT OF ORDER 2

$$\text{Let } A=\begin{bmatrix} a_{11} & a_{12} \\ a_{21} & a_{22} \end{bmatrix},\ \text{then } |A|=\begin{vmatrix} a_{11} & a_{12} \\ a_{21} & a_{22} \end{vmatrix}=a_{11}\,a_{22}-a_{12}\,a_{21}$$

2.14 VALUE OF A DETERMINANT OF ORDER 3

$$\text{Let } A=\begin{bmatrix} a_{11} & a_{12} & a_{13} \\ a_{21} & a_{22} & a_{23} \\ a_{31} & a_{32} & a_{33} \end{bmatrix},\ \text{then}$$

$$|A|=a_{11}\,(a_{22}\,a_{33}-a_{23}\,a_{32})-a_{12}\,(a_{21}\,a_{33}-a_{23}\,a_{31})+a_{13}\,(a_{21}\,a_{32}-a_{22}\,a_{31})$$

Note: Elementary transformations can be used to evaluate the value of a determinant.

2.15 MINORS AND COFACTORS

Minors: The minor of any element of a determinant is the determinant obtained by omitting the row and column in which that element occurs. The minor of an element a_{ij} in $|A|$ is denoted by M_{ij}.

Cofactors: The cofactor of any element of a determinant is its minor with proper sign, plus or minus. The cofacor of an element a_{ij} in $|A|$ is defined as

$$C_{ij}=(-1)^{i+j}\cdot M_{ij}$$

For example: If $A\begin{bmatrix} a_{11} & a_{12} \\ a_{21} & a_{22} \end{bmatrix}$, then $|A|=\begin{vmatrix} a_{11} & a_{12} \\ a_{21} & a_{22} \end{vmatrix}$ and

$$M_{11}=a_{22},\quad M_{12}=a_{21}$$

$$M_{21}=a_{12},\quad M_{22}=a_{11}$$

and $$C_{11}=(-1)^{1+1}M_{11}=a_{22},\quad C_{12}=(-1)^{1+2}M_{12}=-a_{21}$$

$$C_{21}=(-1)^{2+1}M_{21}=-a_{12},\quad C_{22}=(-1)^{2+2}M_{22}=a_{11}$$

For example: Find the minor and cofactor of the elements of the determinant.

$$|A| = \begin{vmatrix} a_{11} & a_{12} & a_{13} \\ a_{21} & a_{22} & a_{23} \\ a_{31} & a_{32} & a_{33} \end{vmatrix}$$

Solution: $M_{11} = \begin{vmatrix} a_{22} & a_{23} \\ a_{32} & a_{33} \end{vmatrix} = (a_{22}\,a_{33} - a_{23}\,a_{32})$

$$M_{12} = \begin{vmatrix} a_{21} & a_{23} \\ a_{31} & a_{33} \end{vmatrix} = (a_{21}\,a_{33} - a_{23}\,a_{31})$$

$$\vdots$$

$$M_{33} = \begin{vmatrix} a_{11} & a_{12} \\ a_{21} & a_{22} \end{vmatrix} = (a_{11}\,a_{22} - a_{12}\,a_{21})$$

and $C_{11} = (-1)_{1+1}\,M_{11} = (a_{22}\,a_{33} - a_{23}\,a_{32})$

$$C_{12} = (-1)^{1+2}\,M_{12} = -(a_{21}\,a_{33} - a_{23}\,a_{31})$$

$$\vdots$$

$$C_{33} = (-1)^{3+3}\,M_{33} = (a_{11}\,a_{22} - a_{12}\,a_{21})$$

2.16 EXPANSION OF DETERMINANT OF THIRD ORDER IN TERMS OF MINORS AND COFACTORS

Let $|A| = \begin{vmatrix} a_{11} & a_{12} & a_{13} \\ a_{21} & a_{22} & a_{23} \\ a_{31} & a_{32} & a_{33} \end{vmatrix}$

Now, expanding the determinant in terms of the elements of the first row, we have

$$|A| = a_{11}\,(a_{22}\,a_{33} - a_{23}\,a_{32}) - a_{12}\,(a_{21}\,a_{33} - a_{23}\,a_{31}) + a_{13}\,(a_{21}\,a_{32} - a_{22}\,a_{31})$$

$$= a_{11}\begin{vmatrix} a_{22} & a_{23} \\ a_{32} & a_{33} \end{vmatrix} - a_{12}\begin{vmatrix} a_{21} & a_{23} \\ a_{31} & a_{33} \end{vmatrix} + a_{13}\begin{vmatrix} a_{21} & a_{22} \\ a_{31} & a_{32} \end{vmatrix}$$

$$= a_{11}\,M_{11} - a_{12}\,M_{12} + a_{13}\,M_{13}$$

$$= a_{11}\,C_{11} + a_{12}\,C_{12} + a_{13}\,C_{13}$$

Note: Any determinant can be expanded in terms of the elements of any row or of any column.

2.17 PROPERTIES OF DETERMINANTS

1. If each element in a row or in a column of a determinant is zero, then the value of the determinant is zero.

Verification: Let $\Delta = \begin{vmatrix} a_1 & b_1 & c_1 \\ a_2 & b_2 & c_2 \\ 0 & 0 & 0 \end{vmatrix}$

Now, expanding the determinant in terms of the elements of the third row, we have

$$\Delta = 0\cdot\begin{vmatrix} b_1 & c_1 \\ b_2 & c_2 \end{vmatrix} - 0\cdot\begin{vmatrix} a_1 & c_1 \\ a_2 & c_2 \end{vmatrix} + 0\cdot\begin{vmatrix} a_1 & b_1 \\ a_2 & b_2 \end{vmatrix}$$

$$= 0 - 0 + 0 = 0$$

2. The value of a determinant remains unchanged if its rows and columns are interchanged.

Verification: Let $\Delta = \begin{vmatrix} a_1 & b_1 & c_1 \\ a_2 & b_2 & c_2 \\ a_3 & b_3 & c_3 \end{vmatrix}$

Now, expanding it in terms of the elements of the first row, we have

$$\Delta = a_1 (b_2c_3 - b_3c_2) - b_1 (a_2c_3 - a_3c_2) + c_1 (a_2 b_3 - a_3 b_2)$$

Now let Δ' be the determinant obtained by interchanging the rows and columns of Δ.

then $\Delta' = \begin{vmatrix} a_1 & a_2 & a_3 \\ b_1 & b_2 & b_3 \\ c_1 & c_2 & c_3 \end{vmatrix}$

Now, expanding it in terms of the elements of the first row, we have

$$\Delta' = a_1 (b_2c_3 - b_3c_2) - a_2 (b_1c_3 - c_1b_3) + a_3 (b_1c_2 - b_2c_1)$$
$$= a_1 (b_2c_3 - b_3c_2) - b_1 (a_2c_3 - a_3c_2) + c_1 (a_2b_3 - a_3b_2)$$
$$= \Delta$$

3. If each element on one side of the principal diagonal of a determinant is zero, then the value of the determinant is the product of the diagonal elements.

Verification: Let $\Delta = \begin{vmatrix} a_1 & b_1 & c_1 \\ 0 & b_2 & c_2 \\ 0 & 0 & c_3 \end{vmatrix}$

Now, expanding the determinant in terms of the elements of the third row, we have

$$\Delta = 0\cdot\begin{vmatrix} b_1 & c_1 \\ b_2 & c_2 \end{vmatrix} - 0\cdot\begin{vmatrix} a_1 & c_1 \\ 0 & c_2 \end{vmatrix} + c_3\cdot\begin{vmatrix} a_1 & b_1 \\ 0 & b_2 \end{vmatrix}$$

$$= 0 - 0 + a_1b_2c_3 = a_1b_2c_3$$

4. If any two adjacent rows or columns of a determinant are interchanged, then the sign of the determinant is changed, but the value remains unchanged.

Verification: Let $\Delta = \begin{vmatrix} a_1 & b_1 & c_1 \\ a_2 & b_2 & c_2 \\ a_3 & b_3 & c_3 \end{vmatrix}$

Now, expanding the determinant in terms of the elements of the first row, we have

$$\Delta = a_1 (b_2c_3 - b_3c_2) - b_1 (a_2c_3 - a_3c_2) + c_1 (a_2b_3 - a_3b_2)$$

Let Δ' be the determinant obtained by interchanging the first and second rows of Δ.

Then $$\Delta' = \begin{vmatrix} a_2 & b_2 & c_2 \\ a_1 & b_1 & c_1 \\ a_3 & b_3 & c_3 \end{vmatrix}$$

Now, expanding it in terms of the elements of the first row, we have

$$\begin{aligned} \Delta' &= a_2 (b_1c_3 - b_3c_1) - b_2 (a_1c_3 - a_3c_1) + c_2 (a_1b_3 - a_3b_1) \\ &= -[a_2 (b_3c_1 - b_1c_3) + b_2 (a_1c_3 - a_3c_1) + c_2 (a_3b_1 - a_1b_3)] \\ &= -[a_1 (b_2c_3 - b_3c_2) - b_1 (a_2c_3 - a_3c_2) + c_1 (a_2b_3 - a_3b_2)] \\ &= -\Delta \end{aligned}$$

Hence, $$\begin{vmatrix} a_1 & b_1 & c_1 \\ a_2 & b_2 & c_2 \\ a_3 & b_3 & c_3 \end{vmatrix} = -\begin{vmatrix} a_2 & b_2 & c_2 \\ a_1 & b_1 & c_1 \\ a_3 & b_3 & c_3 \end{vmatrix}$$

Note: If any row (or column) of a determinant Δ be passed over m rows (or columns), then the resulting determinant

$$\Delta' = (-1)^m \Delta.$$

5. If two rows or columns of a determinant are identical, then the value of the determinant is zero, i.e.,

$$\Delta = \begin{vmatrix} a_1 & b_1 & c_1 \\ a_1 & b_1 & c_1 \\ a_3 & b_3 & c_3 \end{vmatrix} = 0$$

Verification: If we interchange the identical rows or columns of the determinant Δ then there will be no change in Δ. But, interchanging any two rows or columns of the determinant Δ, the sign of the determinant is changed.

So, $$\Delta = -\Delta \Rightarrow 2\Delta = 0$$

$$\Delta = 0 \qquad \text{(by property 3)}$$

6. If all the elements of any row or any column of a determinant is multiplied by the same quantity k (say), then the value of the new determinant is k times the value of the original determinant.

Verification: Let $$\Delta = \begin{vmatrix} a_1 & b_1 & c_1 \\ a_2 & b_2 & c_2 \\ a_3 & b_3 & c_3 \end{vmatrix}$$

Also, let Δ' be the determinant obtained by multiplying first column of Δ by k, then

$$\Delta' = \begin{vmatrix} ka_1 & b_1 & c_1 \\ ka_2 & b_2 & c_2 \\ ka_3 & b_3 & c_3 \end{vmatrix}$$

$$= ka_1 (b_2c_3 - b_3c_2) - b_1 (ka_2c_3 - ka_3c_2) + c_1 (ka_2b_3 - ka_3b_2)$$
$$= ka_1 (b_2c_3 - b_3c_2) - kb_1 (a_2c_3 - a_3c_2) + kc_1 (a_2b_3 - a_3b_2)$$
$$= k [a_1 (b_2c_3 - b_3c_2) - b_1 (a_2c_3 - a_3c_2) + c_1 (a_2b_3 - a_3b_2)]$$
$$= k\Delta$$

7. If each element of a row (or a column) or a determinant consists of sum of two or more terms, then the determinant can be expressed as the sum of two or more determinants whose other rows (or columns) are not altered.

Verification: Let $\Delta = \begin{vmatrix} a_1 + p & b_1 + q & c_1 + r \\ a_2 & b_2 & c_2 \\ a_3 & b_3 & c_3 \end{vmatrix}$

Now, expanding the determinant in terms of the elements of the first row, we have

$$\Delta = (a_1 + p)(b_2c_3 - b_3c_2) - (b_1 + q)(a_2c_3 - a_3c_2) + (c_1 + r)(a_2b_3 - a_3b_2)$$
$$= [a_1 (b_2c_3 - b_3c_2) - b_1 (a_2c_3 - a_3c_2) + c_1 (a_2b_3 - a_3b_2)]$$
$$+ [p (b_2c_3 - b_3c_2) - q (a_2c_3 - a_3c_2) + r (a_2b_3 - a_3b_2)]$$

$$= \begin{vmatrix} a_1 & b_1 & c_1 \\ a_2 & b_2 & c_2 \\ a_3 & b_3 & c_3 \end{vmatrix} + \begin{vmatrix} p & q & r \\ a_2 & b_2 & c_2 \\ a_3 & b_3 & c_3 \end{vmatrix}$$

8. If to each element of a row (or a column) of a determinant be added the equimultiples of the corresponding elements of one or more rows (or columns), the value of the determinant remains unchanged.

Verification: Let $\Delta = \begin{vmatrix} a_1 & b_1 & c_1 \\ a_2 & b_2 & c_2 \\ a_3 & b_3 & c_3 \end{vmatrix}$ and $\Delta' = \begin{vmatrix} a_1 + pb_1 & b_1 & c_1 \\ a_2 + pb_2 & b_2 & c_2 \\ a_3 + pb_3 & b_3 & c_3 \end{vmatrix}$

then, $\Delta' = \begin{vmatrix} a_1 & b_1 & c_1 \\ a_2 & b_2 & c_2 \\ a_3 & b_3 & c_3 \end{vmatrix} + \begin{vmatrix} pb_1 & b_1 & c_1 \\ pb_2 & b_2 & c_2 \\ pb_3 & b_3 & c_3 \end{vmatrix}$

$$= \begin{vmatrix} a_1 & b_1 & c_1 \\ a_2 & b_2 & c_2 \\ a_3 & b_3 & c_3 \end{vmatrix} + p \begin{vmatrix} b_1 & b_1 & c_1 \\ b_2 & b_2 & c_2 \\ b_3 & b_3 & c_3 \end{vmatrix} = \begin{vmatrix} a_1 & b_1 & c_1 \\ a_2 & b_2 & c_2 \\ a_3 & b_3 & c_3 \end{vmatrix}$$

($\because$ the value of the second determinant is zero by property 5)

9. The sum of the products of elements of any row (or column) with the cofactors of the corresponding elements of some other row (or column) is zero.

Verification: Let $\Delta = \begin{vmatrix} a_1 & b_1 & c_1 \\ a_2 & b_2 & c_2 \\ a_3 & b_3 & c_3 \end{vmatrix}$

Now, the co-factors corresponding to first row elements are:

So, $C_{11} = b_2c_3 - b_3c_2,\ C_{12} = -(a_2c_3 - a_3c_2),\ C_{13} = a_2b_3 - a_3b_2)$

$$\begin{aligned} a_2C_{11} + b_2C_{12} + c_2C_{13} &= a_2(b_2c_3 - b_3c_2) - b_2(a_2c_3 - a_3c_2) + c_2(a_2b_3 - a_3b_2) \\ &= a_2b_2c_3 - a_2b_3c_2 - a_2b_2c_3 + a_3b_2c_2 + a_2b_3c_2 - a_3b_2c_2 \\ &= 0 \end{aligned}$$

ILLUSTRATIVE EXAMPLES

Example 1. Find the minors and cofactors of all the elements of $\Delta = \begin{vmatrix} 1 & -2 & 3 \\ 3 & 1 & 2 \\ 2 & 4 & 5 \end{vmatrix}$ and expand Δ to find its value.

Solution: $M_{11} = \begin{vmatrix} 1 & 2 \\ 4 & 5 \end{vmatrix} = 5 - 8 = -3,\ M_{12} = \begin{vmatrix} 3 & 2 \\ 2 & 5 \end{vmatrix} = 15 - 4 = 11,$

$$M_{13} = \begin{vmatrix} 3 & 1 \\ 2 & 4 \end{vmatrix} = 12 - 2 = 10$$

$$M_{21} = \begin{vmatrix} -2 & 3 \\ 4 & 5 \end{vmatrix} = -10 - 12 = -22,$$

$$M_{22} = \begin{vmatrix} 1 & 3 \\ 2 & 5 \end{vmatrix} = 5 - 6 = -1,\ M_{23} = \begin{vmatrix} 1 & -2 \\ 2 & 4 \end{vmatrix} = 4 + 4 = 8$$

$$M_{31} = \begin{vmatrix} -2 & 3 \\ 1 & 2 \end{vmatrix} = -4 - 3 = -7,$$

$$M_{32} = \begin{vmatrix} 1 & 3 \\ 3 & 2 \end{vmatrix} = 2 - 9 = -7,\ M_{33} = \begin{vmatrix} 1 & -2 \\ 3 & 1 \end{vmatrix} = 1 + 6 = 7$$

Now, $C_{11} = (-1)^{1+1} M_{11} = -3,\ C_{12} = (-1)^{1+2} M_{12} = -11,\ C_{13} = (-1)^{1+3} M_{13} = 10$

$C_{21} = (-1)^{2+1} M_{21} = 22,\ C_{22} = (-1)^{2+2} M_{22} = -1,\ C_{23} = (-1)^{2+3} M_{23} = -8$

$c_{31} = (-1)^{3+1} M_{31} = -7,\ c_{32} = (-1)^{3+2} M_{32} = 7,\ C_{33} = (-1)^{3+3} M_{33} = 7$

and $\Delta = 1(1 \times 5 - 2 \times 4) - (-2)(3 \times 5 - 2 \times 2) + 3(3 \times 4 - 2 \times 1)$

$= 1(5 - 8) + 2(15 - 4) + 3(12 - 2) = -3 + 22 + 30$

$= 49$

Example 2. Find the value of the determinant, $|A| = \begin{vmatrix} a & h & g \\ h & b & f \\ g & f & c \end{vmatrix}$

Solution: $|A| = \begin{vmatrix} a & h & g \\ h & b & f \\ g & f & c \end{vmatrix}$

Now, expanding the determinant in terms of the elements of the first row, we have

$$\begin{aligned} |A| &= a(bc - f^2) - h\,(hc - fg) + g\,(hf - bg) \\ &= abc - af^2 - h^2c + fgh + fgh - bg^2 \\ &= abc - af^2 - bg^2 - ch^2 + 2fgh \end{aligned}$$

Example 3. Find the value of the following determinants:

(*i*) $\begin{vmatrix} \cos x & -\sin x \\ \sin x & \cos x \end{vmatrix}$ (*ii*) $\begin{vmatrix} 1 & \log_a b \\ \log_a a & 1 \end{vmatrix}$

Solution: (*i*) $\begin{vmatrix} \cos x & -\sin x \\ \sin x & \cos x \end{vmatrix} = \cos^2 x - (-\sin^2 x) = \cos^2 x + \sin^2 x = 1$

(*ii*) $\begin{vmatrix} 1 & \log_a b \\ \log_b a & 1 \end{vmatrix} = 1 - \log_a b \cdot \log_b a = 1 - 1 = 0 \quad (\because \log_a b \cdot \log_b a = 1)$

Example 4. If $\begin{vmatrix} 3 & m \\ 4 & 5 \end{vmatrix} = 7$, find the value of m.

Solution: $\begin{vmatrix} 3 & m \\ 4 & 5 \end{vmatrix} = 7$

$$\Rightarrow \qquad 15 - 4m = 7 \text{ or } 4m = 8$$

$$\Rightarrow \qquad m = 2$$

Example 5. Find the value of the determinant $\begin{vmatrix} 1 & 2 & 3 \\ 4 & 5 & 6 \\ 7 & 8 & 9 \end{vmatrix}$

Solution: $\begin{vmatrix} 1 & 2 & 3 \\ 4 & 5 & 6 \\ 7 & 8 & 9 \end{vmatrix} = 1\,(45 - 48) - 2\,(36 - 42) + 3\,(32 - 35)$

$$= -3 + 12 - 9 = 0$$

Example 6. Find the value of the determinant $\begin{vmatrix} 0 & -n & -m \\ n & 0 & -l \\ m & l & 0 \end{vmatrix}$

Solution: $\begin{vmatrix} 0 & -n & -m \\ n & 0 & -l \\ m & l & 0 \end{vmatrix} = 0\,(0+l^2) + n\,(0+lm) - m\,(ln-0)$

$$= 0 + lmn - lmn$$
$$= 0$$

Example 7. Without expanding the determinant, prove that $\begin{vmatrix} 15 & 18 & 21 \\ 16 & 19 & 22 \\ 17 & 20 & 23 \end{vmatrix} = 0$

Solution: $\Delta = \begin{vmatrix} 15 & 18 & 21 \\ 16 & 19 & 22 \\ 17 & 20 & 23 \end{vmatrix}$

$$= \begin{vmatrix} 15 & 3 & 6 \\ 16 & 3 & 6 \\ 17 & 3 & 6 \end{vmatrix} \quad (\text{by } C_2 \to C_2 - C_1,\ C_3 \to C_3 - C_1)$$

$$= 3\times 6 \begin{vmatrix} 15 & 1 & 1 \\ 16 & 1 & 1 \\ 17 & 1 & 1 \end{vmatrix}$$

(taking 3 common from second column and 6 from third column)

$= 18 \times 0$ (by property 5)

$= 0$

Example 8. Without expanding the determinant, prove that $\begin{vmatrix} a-b & b-c & c-a \\ b-c & c-a & a-b \\ c-a & a-b & b-c \end{vmatrix} = 0$

Solution: We have,

$$\Delta = \begin{vmatrix} a-b & b-c & c-a \\ b-c & c-a & a-b \\ c-a & a-b & b-c \end{vmatrix}$$

$$= \begin{vmatrix} 0 & b-c & c-a \\ 0 & c-a & a-b \\ 0 & a-b & b-c \end{vmatrix} \quad (\text{by } C_1 \to C_1 + C_2 + C_3)$$

$= 0$ (by property 1)

Example 9. Prove that $\begin{vmatrix} 1 & a & bc \\ 1 & b & ca \\ 1 & c & ab \end{vmatrix} = (a-b)\,(b-c)\,(c-a)$

Solution: We have

$$\Delta = \begin{vmatrix} 1 & a & bc \\ 1 & b & ca \\ 1 & c & ab \end{vmatrix}$$

$$= \begin{vmatrix} 1 & a & bc \\ 0 & b-a & c(a-b) \\ 0 & c-a & b(a-c) \end{vmatrix} \text{ (by } R_2 \to R_2 - R_1, R_3 \to R_3 - R_1)$$

$$= (a-b)\,(c-a) \begin{vmatrix} 1 & a & bc \\ 0 & -1 & c \\ 0 & 1 & -b \end{vmatrix}$$

(taking $(a-b)$ common from R_2 and $(c-a)$ from R_3)

$$= (a-b)\,(c-a) \begin{vmatrix} -1 & c \\ 1 & -b \end{vmatrix} \text{ (on expanding along } C_1)$$

$$= (a-b)\,(c-a)\,[-1 \times -b - c]$$

$$= (a-b)\,(c-a)\,(b-c)$$

$$= (a-b)\,(b-c)\,(c-a)$$

Proved.

Example 10. Prove that $\begin{vmatrix} 1 & a & a^3 \\ 1 & b & b^3 \\ 1 & c & c^3 \end{vmatrix} = (a-b)\,(b-c)\,(c-a)\,(a+b+c)$

Solution: We have

$$\Delta = \begin{vmatrix} 1 & a & a^3 \\ 1 & b & b^3 \\ 1 & c & c^3 \end{vmatrix}$$

$$= \begin{vmatrix} 1 & a & a^3 \\ 0 & b-a & b^3-a^3 \\ 0 & c-a & c^3-a^3 \end{vmatrix} \text{ (by } R_2 \to R_2 - R_1; R_3 \to R_3 - R_1)$$

$$= (b-a)\,(c-a) \begin{vmatrix} 1 & a & a^3 \\ 0 & 1 & b^2+ba+a^2 \\ 0 & 1 & c^2+ca+a^2 \end{vmatrix}$$

(taking $(b-a)$ common from R_2 and $(c-a)$ from R_3)

$$\Delta = (b-a)\,(c-a) \begin{vmatrix} 1 & b^2+ba+a^2 \\ 1 & c^2+ca+a^2 \end{vmatrix}$$

(on expanding along C_1)

$$= (b-a)(c-a)[(c^2+ca+a^2)-(b^2+ba+a^2)]$$
$$= (b-a)(c-a)[(c^2-b^2)+a(c-b)]$$
$$= (b-a)(c-a)(c-b)(c+b+a)$$
$$= (a-b)(b-c)(c-a)(a+b+c)$$ **Proved.**

Example 11. Proved that $\begin{vmatrix} 1+a & 1 & 1 \\ 1 & 1+b & 1 \\ 1 & 1 & 1+c \end{vmatrix} = (abc)\left(\frac{1}{a}+\frac{1}{b}+\frac{1}{c}+1\right)$

Solution: We have, $\Delta = \begin{vmatrix} 1+a & 1 & 1 \\ 1 & 1+b & 1 \\ 1 & 1 & 1+c \end{vmatrix}$

$$= (abc)\begin{vmatrix} \frac{1}{a}+1 & \frac{1}{a} & \frac{1}{a} \\ \frac{1}{b} & \frac{1}{b}+1 & \frac{1}{b} \\ \frac{1}{c} & \frac{1}{c} & \frac{1}{c}+1 \end{vmatrix}$$

(taking a, b, c common from R_1, R_2 and R_3 respectively)

$$= (abc)\begin{vmatrix} \frac{1}{a}+\frac{1}{b}+\frac{1}{c}+1 & \frac{1}{a}+\frac{1}{b}+\frac{1}{c}++1 & \frac{1}{a}+\frac{1}{b}+\frac{1}{c}+1 \\ \frac{1}{b} & \frac{1}{b}+1 & \frac{1}{b} \\ \frac{1}{c} & \frac{1}{c} & \frac{1}{c}+1 \end{vmatrix} \quad (\text{by } R_1 \to R_1+R_2+R_3)$$

$$= (abc)\left(\frac{1}{a}+\frac{1}{b}+\frac{1}{c}+1\right)\begin{vmatrix} 1 & 1 & 1 \\ \frac{1}{b} & \frac{1}{b}+1 & \frac{1}{b} \\ \frac{1}{c} & \frac{1}{c} & \frac{1}{c}+1 \end{vmatrix}$$

(taking $\left(\frac{1}{a}+\frac{1}{b}+\frac{1}{c}+1\right)$ common from R_1)

$$= (abc)\left(\frac{1}{a}+\frac{1}{b}+\frac{1}{c}+1\right)\begin{vmatrix} 1 & 0 & 0 \\ \frac{1}{b} & 1 & 0 \\ \frac{1}{c} & 0 & 1 \end{vmatrix} \quad (\text{by } C_2 \to C_2 - C_1;\ C_3 \to C_3 - C_1)$$

$$\Delta = (abc)\left(\frac{1}{a}+\frac{1}{b}+\frac{1}{c}+1\right)\begin{vmatrix} 1 & 0 \\ 0 & 1 \end{vmatrix} \text{ (on expanding along } R_1\text{)}$$

$$= (abc)\left(\frac{1}{a}+\frac{1}{b}+\frac{1}{c}+1\right)(1)$$

$$= (abc)\left(\frac{1}{a}+\frac{1}{b}+\frac{1}{c}+1\right)$$ **Proved.**

Example 12. Prove that $\begin{vmatrix} a-b-c & 2a & 2a \\ 2b & b-c-a & 2b \\ 2c & 2c & c-a-b \end{vmatrix} = (a+b+c)^3$

Solution: We have,

$$\Delta = \begin{vmatrix} a-b-c & 2a & 2a \\ 2b & b-c-a & 2b \\ 2c & 2c & c-a-b \end{vmatrix}$$

$$= \begin{vmatrix} a+b+c & a+b+c & a+b+c \\ 2b & b-c-a & 2b \\ 2c & 2c & c-a-b \end{vmatrix} \text{ (by } R_1 \to R_1 + R_2 + R_3\text{)}$$

$$= (a+b+c)\begin{vmatrix} 1 & 1 & 1 \\ 2b & b-c-a & 2b \\ 2c & 2c & c-a-b \end{vmatrix}$$

(taking $(a+b+c)$ common from R_1)

$$= (a+b+c)\begin{vmatrix} 1 & 0 & 0 \\ 2b & -(a+b+c) & 0 \\ 2c & 0 & -(a+b+c) \end{vmatrix}$$

(by $C_2 \to C_2 - C_1$; $C_3 \to C_3 - C_1$)

$$= (a+b+c)\begin{vmatrix} -(a+b+c) & 0 \\ 0 & -(a+b+c) \end{vmatrix} \text{ (on expanding along } R_1\text{)}$$

$$= (a+b+c)\,[(a+b+c)^2 - 0]$$

$$= (a+b+c)^3$$ **Proved.**

Example 13. Prove that $\begin{vmatrix} b+c & a & a \\ b & c+a & b \\ c & c & a+b \end{vmatrix} = 4abc$

Solution: We have,

$$\Delta = \begin{vmatrix} b+c & a & a \\ b & c+a & b \\ c & c & a+b \end{vmatrix}$$

$$= \begin{vmatrix} c & -c & a-b \\ b-c & a & -a \\ c & c & a+b \end{vmatrix} \text{(by } R_1 \to R_1 - R_2; R_2 \to R_2 - R_3)$$

$$= \begin{vmatrix} 2c & 0 & 2a \\ b & a+c & b \\ c & c & a+b \end{vmatrix} \text{(by } R_1 \to R_1 + R_3; R_2 \to R_2 + R_3)$$

$$= 2\begin{vmatrix} c & 0 & a \\ b & a+c & b \\ c & c & a+b \end{vmatrix} \text{(taking 2 common from } R_1)$$

$$= 2\begin{vmatrix} c & 0 & a \\ b & a+c & b \\ 0 & c & b \end{vmatrix} \text{(by } R_3 \to R_3 - R_1)$$

$$= 2\begin{vmatrix} c & 0 & a \\ b & a & 0 \\ 0 & c & b \end{vmatrix} \text{(by } R_2 \to R_2 - R_3)$$

$$= 2\,[\,c\,(ab-0) - 0 + a\,(bc-0)]$$

$$= 2\,[abc + abc]$$

$$= 4abc$$

Proved.

Example 14. Solve $\begin{vmatrix} x+a & x+b & x+c \\ x+b & x+c & x+a \\ x+c & x+a & x+b \end{vmatrix} = 0$

Solution: We have, $\begin{vmatrix} x+a & x+b & x+c \\ x+b & x+c & x+a \\ x+c & x+a & x+b \end{vmatrix} = 0$

$$\begin{vmatrix} 3x+a+b+c & x+b & x+c \\ 3x+a+b+c & x+c & x+a \\ 3x+a+b+c & x+a & x+b \end{vmatrix} = 0 \text{ (by } C_1 \to C_1 + C_2 + C_3)$$

$$(3x+a+b+c)\begin{vmatrix} 1 & x+b & x+c \\ 1 & x+c & x+a \\ 1 & x+a & x+b \end{vmatrix} = 0 \text{ (taking } (3x+a+b+c) \text{ common from } C_1)$$

$$(3x+a+b+c)\begin{vmatrix} 1 & x+b & x+c \\ 0 & c-b & a-c \\ 0 & a-b & b-c \end{vmatrix} = 0 \text{ (by } R_2 \to R_2 - R_1; R_3 \to R_3 - R_1)$$

$$(3x+a+b+c)\begin{vmatrix} c-b & a-c \\ a-b & b-c \end{vmatrix} = 0 \text{ (on expanding along } C_1)$$

$$3x + a + b + c = 0$$

$$x = \frac{-(a+b+c)}{3}$$ **Ans.**

Example 15. Solve $\begin{vmatrix} x+a & b & c \\ a & x+b & c \\ a & b & x+c \end{vmatrix} = 0$

Solution: We have $\begin{vmatrix} x+a & b & c \\ a & x+b & c \\ a & b & x+c \end{vmatrix} = 0$

$$\begin{vmatrix} x+a+b+c & b & c \\ x+a+b+c & x+b & c \\ x+a+b+c & b & x+c \end{vmatrix} = 0 \text{ (by } C_1 \to C_1 + C_2 + C_3\text{)}$$

$$(x+a+b+c)\begin{vmatrix} 1 & b & c \\ 1 & x+b & c \\ 1 & b & x+c \end{vmatrix} = 0 \text{ (taking } (x+a+b+c) \text{ common from } C_1\text{)}$$

$$(x+a+b+c)\begin{vmatrix} 1 & b & c \\ 0 & x & 0 \\ 0 & 0 & x \end{vmatrix} \text{ (by } R_2 \to R_2 - R_1; R_3 \to R_3 - R_1\text{)}$$

$$(x+a+b+c)\begin{vmatrix} x & 0 \\ 0 & x \end{vmatrix} = 0 \text{ (on expanding along } C_1\text{)}$$

$$(x+a+b+c)(x^2 - 0) = 0$$

$$x^2(x+a+b+c) = 0$$

$$x = 0,\ x = -(a+b+c)$$ **Ans.**

Example 16. Solve $\begin{vmatrix} a+x & a-x & a-x \\ a-x & a+x & a-x \\ a-x & a-x & a+x \end{vmatrix} = 0$

Solution: We have, $\begin{vmatrix} a+x & a-x & a-x \\ a-x & a+x & a-x \\ a-x & a-x & a+x \end{vmatrix} = 0$

$$\begin{vmatrix} 3a+x & a-x & a-x \\ 3a-x & a+x & a-x \\ 3a-x & a-x & a+x \end{vmatrix} = 0 \text{ (by } C_1 \to C_1 + C_2 + C_3\text{)}$$

$$(3a-x)\begin{vmatrix} 1 & a-x & a-x \\ 1 & a+x & a-x \\ 1 & a-x & a+x \end{vmatrix} = 0 \text{ (taking } (3a-x) \text{ common from } C_1\text{)}$$

$$(3a-x)\begin{vmatrix} 1 & a-x & a-x \\ 0 & 2x & 0 \\ 0 & 0 & 2x \end{vmatrix} = 0 \text{ (by ;} R_2 \to R_2 - R_1; R_3 \to R_3 - R_1)$$

$$(3a-x)\begin{vmatrix} 2x & 0 \\ 0 & 2x \end{vmatrix} = 0 \text{ (on expanding along } C_1)$$

$$(3a-x)(4x^2-0) = 0$$

$$4x^2(3a-x) = 0$$

$$x = 0, x = 3a$$

Ans.

EXERCISE 2.2

1. Find the cofactors of all elements of $\Delta = \begin{vmatrix} 2 & -1 & 3 \\ 1 & 2 & -1 \\ 0 & 5 & -5 \end{vmatrix}$ and expand Δ to find its value.

2. Find the values of the determinant $\begin{vmatrix} 1 & -3 & 2 \\ 4 & -1 & 2 \\ 3 & 5 & 2 \end{vmatrix}$.

3. Evaluate $\begin{vmatrix} x-1 & x+2 \\ x-2 & x+1 \end{vmatrix}$

4. Without expanding the determinant, prove that $\begin{vmatrix} 265 & 240 & 219 \\ 240 & 225 & 198 \\ 219 & 198 & 181 \end{vmatrix} = 0$.

5. Without expanding the determinant, prove that $\begin{vmatrix} 26 & 32 & 38 \\ 28 & 34 & 40 \\ 30 & 36 & 42 \end{vmatrix} = 0$.

6. Without expanding the determinant, prove that

$$\begin{vmatrix} x+a & x+2a & x+3a \\ x+2a & x+3a & x+4a \\ x+4a & x+5a & x+6a \end{vmatrix} = 0.$$

7. Prove that $\begin{vmatrix} 1 & 1 & 1 \\ x & y & z \\ x^3 & y^3 & z^3 \end{vmatrix} = (x-y)(y-z)(z-x)(x+y+z)$.

8. Prove that $\begin{vmatrix} x & y+z & x^2 \\ y & z+x & y^2 \\ z & x+y & z^2 \end{vmatrix} = -(x+y+z)(x-y)(y-z)(z-x)$.

9. Prove that $\begin{vmatrix} a+b & b & c \\ b+c & c & a \\ c+a & a & b \end{vmatrix} = 3abc - a^3 - b^3 - c^3.$

10. Prove that $\begin{vmatrix} -a^2 & ab & ac \\ ba & -b^2 & bc \\ ac & bc & -c^2 \end{vmatrix} = 4a^2b^2c^2.$

11. If x, y, z are different, and $\begin{vmatrix} x & x^2 & 1+x^3 \\ y & y^2 & 1+y^3 \\ z & z^2 & 1+z^3 \end{vmatrix} = 0$, prove that $xyz = -1$.

12. Prove that $\begin{vmatrix} 1 & 1 & 1 \\ 1 & 1+x & 1 \\ 1 & 1 & 1+y \end{vmatrix} = xy$

13. Prove that $\begin{vmatrix} a & b & c \\ a^2 & b^2 & c^2 \\ bc & ca & ab \end{vmatrix} = (a-b)(b-c)(c-a)(ab+bc+ca)$

14. Solve $\begin{vmatrix} x-2 & 2x-3 & 3x-4 \\ x-4 & 2x-9 & 3x-16 \\ x-8 & 2x-27 & 3x-64 \end{vmatrix} = 0$

15. Solve $\begin{vmatrix} 3+x & 5 & 2 \\ 1 & 7+x & 6 \\ 2 & 5 & 3+x \end{vmatrix} = 0$

16. Solve $\begin{vmatrix} x & -6 & -1 \\ 2 & -3x & x-3 \\ -3 & 2x & x+2 \end{vmatrix} = 0$

ANSWERS

1. $C_{11} = -5, C_{12} = 5, C_{13} = 5, C_{21} = 10, C_{22} = -10, C_{23} = -10, C_{31} = -5, C_{32} = 5, C_{33} = 5; \Delta = 0$

2. 40

3. 3

14. 4

15. $0, 3a$

16. $1, 2, -3$

2.18 APPLICATIONS OF DETERMINANTS

Determinants can be used to find the solution of linear equations, area of a triangle, test for collinear points etc.

2.19 AREA OF A TRIANGLE IN DETERMINANT FORM

We know that the area of a triangle whose vertices are (x_1, y_1), (x_2, y_2) and (x_3, y_3) is given by

$$\Delta = \frac{1}{2}\,[x_1(y_2 - y_3) - x_2(y_1 - y_3) + x_3(y_1 - y_2)]$$

But the area of a triangle whose vertices are (x_1, y_1), (x_2, y_2) and (x_3, y_3) using determinants can be obtained by

$$\Delta = \frac{1}{2}\begin{vmatrix} x_1 & y_1 & 1 \\ x_2 & y_2 & 1 \\ x_3 & y_3 & 1 \end{vmatrix}$$

2.20 TEST FOR COLLINEAR POINTS

It the area of a triangle becomes equal to zero, then there would be no triangle, the points would be collinear (on a line). We can use that fact to test collinear points. If the area of a triangle is zero, the points are collinear. Three points (x_1, y_1), (x_2, y_2) and (x_3, y_3) are collinear if

$$\frac{1}{2}\begin{vmatrix} x_1 & y_1 & 1 \\ x_2 & y_2 & 1 \\ x_3 & y_3 & 1 \end{vmatrix} = 0 \text{ or } \begin{vmatrix} x_1 & y_1 & 1 \\ x_2 & y_2 & 1 \\ x_3 & y_3 & 1 \end{vmatrix} = 0$$

ILLUSTRATIVE EXAMPLES

Example 1. Find the area of the triangle whose vertices are (0, 5), (2, 3), (4, 5).

Solution: Area of triangle $\Delta = \frac{1}{2}\begin{vmatrix} x_1 & y_1 & 1 \\ x_2 & y_2 & 1 \\ x_3 & y_3 & 1 \end{vmatrix}$

$$= \frac{1}{2}\begin{vmatrix} 0 & 5 & 1 \\ 2 & 3 & 1 \\ 4 & 5 & 1 \end{vmatrix}$$

$$= \frac{1}{2}[0(3-5) - 5(2-4) + 1(10-12)]$$

$$= \frac{1}{2} \times 8 = 4$$

$\therefore$ Area of triangle $\Delta = 4$ square units

Example 2. Find the area of the triangle whose vertices are (–3, –5), (5, 2), (–9, –3).

Solution: Area of the triangle $\Delta = \frac{1}{2}\begin{vmatrix} x_1 & y_1 & 1 \\ x_2 & y_2 & 1 \\ x_3 & y_3 & 1 \end{vmatrix}$

$$= \frac{1}{2}\begin{vmatrix} -3 & -5 & 1 \\ 5 & 2 & 1 \\ -9 & -3 & 1 \end{vmatrix}$$

$$= \frac{1}{2}[-3(2+3) + 5(5+9) + 1(-15+18)]$$

$$= \frac{1}{2} \times 58 = 29$$

$\therefore$ Area of triangle $\Delta = 29$ square units

Example 3. Prove that the points (–5, 1), (5, 5) and (10, 7) are collinear.

Solution: The given points are collinear, if $\begin{vmatrix} x_1 & y_1 & 1 \\ x_2 & y_2 & 1 \\ x_3 & y_3 & 1 \end{vmatrix} = 0$

Now, $\begin{vmatrix} -5 & 1 & 1 \\ 5 & 5 & 1 \\ 10 & 7 & 1 \end{vmatrix} = -5\,(5-7) - 1\,(5-10) + 1\,(35-50)$

$$= 0$$

$\therefore$ The given points are collinear.

Example 4. Find the value of k when points (k, –1), (2, 1) and (4, 5) are collinear.

Solution: If the points are collinear, we have $\begin{vmatrix} x_1 & y_1 & 1 \\ x_2 & y_2 & 1 \\ x_3 & y_3 & 1 \end{vmatrix} = 0$

Now, $\begin{vmatrix} k & -1 & 1 \\ 2 & 1 & 1 \\ 4 & 5 & 1 \end{vmatrix} = 0$

$$k\,(1-5) + 1\,(2-4) + 1\,(10-4) = 0$$

$$-4k + 4 = 0$$

$$k = 1$$

Ans.

Example 5. Find the value of k when points (k, 1), (5, 5) and (10, 7) are collinear.

Solution: If the points are collinear, we have $\begin{vmatrix} x_1 & y_1 & 1 \\ x_2 & y_2 & 1 \\ x_3 & y_3 & 1 \end{vmatrix} = 0$

Now, $$\begin{vmatrix} k & 1 & 1 \\ 5 & 5 & 1 \\ 10 & 7 & 1 \end{vmatrix} = 0$$

$$k(5-7) - 1(5-10) + 1(35-50) = 0$$
$$-2k - 10 = 0$$
$$k = -5$$ **Ans.**

Example 6. Prove that $(a, 0)$, $(0, b)$ and (x, y) are collinear when $\frac{x}{a} + \frac{y}{b} = 1$.

Solution: If the points are collinear, we have $\begin{vmatrix} x_1 & y_1 & 1 \\ x_2 & y_2 & 1 \\ x_3 & y_3 & 1 \end{vmatrix} = 0$

Now, $$\begin{vmatrix} a & 0 & 1 \\ 0 & b & 1 \\ x & y & 1 \end{vmatrix} = 0$$

$$a(b-y) - 0(0-x) + 1(0-xb) = 0$$
$$ab - ay - bx = 0$$
$$bx + ay = ab$$

On dividing, both the sides by ab we have,

$$\frac{x}{a} + \frac{y}{b} = 1$$ **Proved.**

Example 7. Prove that (a, b), (a', b') and $a - a', b - b')$ are collinear when $ab' = a'b$.

Solution: If the points are collinear, we have $\begin{vmatrix} x_1 & y_1 & 1 \\ x_2 & y_2 & 1 \\ x_3 & y_3 & 1 \end{vmatrix} = 0$

Now, $$\begin{vmatrix} a & b & 1 \\ a' & b' & 1 \\ a-a' & b-b' & 1 \end{vmatrix} = 0$$

$$\begin{vmatrix} a & b & 1 \\ a & b & 2 \\ a-a' & b-b' & 1 \end{vmatrix} = 0 \qquad \text{(by } R_2 \to R_2 + R_3\text{)}$$

$$\begin{vmatrix} 0 & 0 & -1 \\ a & b & 2 \\ a-a' & b-b' & 1 \end{vmatrix} = 0 \qquad \text{(by } R_1 \to R_1 - R_2\text{)}$$

$$0(b - 2b + 2b') - 0(a - 2a + 2a') - 1(ab - ab' - ab + a'b) = 0$$
$$ab' - a'b = 0$$
$$ab' = a'b$$ **Proved.**

EXERCISE 2.3

1. Find the area of the triangle whose vertices are (–2, 4), (2, –6), (5, 4).
2. Find the area of the triangle whose vertices are (3, 4), (2, –1), (4, –6).
3. Find the area of the triangle whose vertices are (0, 0), 6, 0), (4, 3).
4. Prove that the points $(a, b + c)$, $(b, c + a)$ and $(c, a + b)$ are collinear.
5. Prove that the points (2, 3), (–1, –2) and (5, 8) are collinear.
6. For what value of k are the points (1, 5), $(k, 1)$ and (11, 7) collinear? (Use determinants).
7. Prove that $(a, 0)$, $(0, b)$ and (1, 1) are collinear when $\frac{1}{a} + \frac{1}{b} = 1$.
8. Prove that $(a, 0)$, $(0, b)$ and $(3a, y)$ are collinear when $y = -2b$.
9. Find the value of k when points (–1, 4), (–3, 8) and $(-k + 1, 3k)$ lie on a straight line.

ANSWERS

1. 35 square units
2. 7·5 square units
3. 9 square units
6. $k = -19$
9. $k = 0$

2.21 ADJOINT OF A SQUARE MATRIX

The adjoint of a square matrix, A (say) is the transpose of the matrix obtained by the cofactors of A.

Let $A = [a_{ij}]_{n \times n}$ be any square matrix and let C_{ij} denote the co-factor of a_{ij} in $|A|$. Then, the adjoint of A, denoted by *adj A* is defined as

$$adj\ A = [Cji]_{n \times n}$$

If $A = \begin{bmatrix} a_{11} & a_{12} & \cdots & a_{1n} \\ a_{21} & a_{22} & \cdots & a_{2n} \\ \cdots & \cdots & \cdots & \cdots \\ a_{n1} & a_{n2} & \cdots & a_{nn} \end{bmatrix}$, then

$$adj\ A = \text{the transpose of the matrix } \begin{bmatrix} C_{11} & C_{12} & \cdots & C_{1n} \\ C_{21} & C_{22} & \cdots & C_{2n} \\ \cdots & \cdots & \cdots & \cdots \\ C_{n1} & C_{n2} & \cdots & C_{nn} \end{bmatrix}$$

$$= \begin{bmatrix} C_{11} & C_{21} & \cdots & C_{n_1} \\ C_{12} & C_{22} & \cdots & C_{n_2} \\ \cdots & \cdots & \cdots & \cdots \\ C_{1n} & C_{2n} & \cdots & C_{nn} \end{bmatrix}$$

where C_{ij} denoted the co-factors of a_{ij} in $|A|$.

Sometimes the adjoint of a matrix is also called the adjugate of that matrix.

2.22 PROPERTIES OF ADJOINTS

1. If A is a square matrix of order n, then

$$A \cdot (adj\,A) = (adj\,A) \cdot A = |A|\, I_n.$$

2. If A is a square matrix of order n, then

$$adj\,A' = (adj\,A)'$$

3. If A and B are two square matrices, then

$$adj\,(AB) = (adj\,B)\,(adj\,A)$$

2.23 INVERSE OF RECIPROCAL OF A SQUARE MATRIX

Let A be any square matrix of order n. If there exists another matrix B of same order such that

$$AB = BA = I_n$$

then B is called the inverse of the square matrix A and the matrix A is known as invertible.

We can write $\quad B = A^{-1}$

then $\quad AA^{-1} = A^{-1}A = I_n$

We have already seen

$$A\,(adj\,A) = |A|\, I_n$$

or $\quad \dfrac{A\,(adj\,A)}{|A|} = I_n \quad$ provided $|A| \neq 0$

Hence, we get $\quad \dfrac{A^{-1}\,A\,(adj\,A)}{|A|} = A^{-1} I_n$

or $\quad \dfrac{adj\,A}{|A|} = A^{-1} \qquad (\because A^{-1}A = I$ and $A^{-1}I_n = A^{-1})$

2.24 PROPERTIES OF INVERSE

1. Inverse of every square matrix, if it exists, is unique.
2. If A and B are two non-singular matrices of the same order then $(AB)^{-1} = B^{-1}A^{-1}$
3. If A is a non-singular matrix, then

$$(A^{-1})^{-1} = A$$

4. $(A')^{-1} = (A^{-1})'$

   i.e., Inverse of transpose of A is equal to the transpose of inverse of A.
5. A square matrix is invertible if and only if A is non-singular, i.e., A is invertible $\Leftrightarrow |A| \neq 0$

ILLUSTRATIVE EXAMPLES

Example 1. Find the adjoint of the matrix $A = \begin{bmatrix} 0 & 1 & 2 \\ 1 & 2 & 3 \\ 3 & 1 & 1 \end{bmatrix}$

Solution: Let $C = \begin{bmatrix} C_{11} & C_{12} & C_{13} \\ C_{21} & C_{22} & C_{23} \\ C_{31} & C_{32} & C_{33} \end{bmatrix}$ be the co-factor matrix of A.

Then, $C_{11} = \begin{vmatrix} 2 & 3 \\ 1 & 1 \end{vmatrix} = -1, C_{12} = -\begin{vmatrix} 1 & 3 \\ 3 & 1 \end{vmatrix} = 8, C_{13} = \begin{vmatrix} 1 & 2 \\ 3 & 1 \end{vmatrix} = -5$

$C_{21} = -\begin{vmatrix} 1 & 2 \\ 1 & 1 \end{vmatrix} = 1, C_{22} = \begin{vmatrix} 0 & 2 \\ 3 & 1 \end{vmatrix} = -6, C_{23} = -\begin{vmatrix} 0 & 1 \\ 3 & 1 \end{vmatrix} = 3$

$C_{31} = \begin{vmatrix} 1 & 2 \\ 2 & 3 \end{vmatrix} = -1, C_{32} = -\begin{vmatrix} 0 & 2 \\ 1 & 3 \end{vmatrix} = 2, C_{33} = \begin{vmatrix} 0 & 1 \\ 1 & 2 \end{vmatrix} = -1$

$\therefore$ co-factor matrix of A is

$$C = \begin{bmatrix} -1 & 8 & -5 \\ 1 & -6 & 3 \\ -1 & 2 & -1 \end{bmatrix}$$

$\therefore$ $adj\, A = C'$ (transpose of C) $= \begin{bmatrix} -1 & 1 & -1 \\ 8 & -6 & 2 \\ -5 & 3 & -1 \end{bmatrix}$

Example 2. If $A = \begin{bmatrix} 1 & 1 & 1 \\ 1 & 2 & -3 \\ 2 & -1 & 3 \end{bmatrix}$, verity that $A\,(adj\,A) = (adj\,A)\,A = |A|\,I_3$

Solution: Let $C = \begin{bmatrix} C_{11} & C_{12} & C_{13} \\ C_{21} & C_{22} & C_{23} \\ C_{31} & C_{32} & C_{33} \end{bmatrix}$ be the co-factor matrix of A.

Then, $C_{11} = \begin{vmatrix} 2 & -3 \\ -1 & 3 \end{vmatrix} = 3, C_{12} = -\begin{vmatrix} 1 & -3 \\ 2 & 3 \end{vmatrix} = -9, C_{13} = \begin{vmatrix} 1 & 2 \\ 2 & -1 \end{vmatrix} = -5$

$C_{21} = -\begin{vmatrix} 1 & 1 \\ -1 & 3 \end{vmatrix} = -4, C_{22} = \begin{vmatrix} 1 & 1 \\ 2 & 3 \end{vmatrix} = 1, C_{23} = -\begin{vmatrix} 1 & 1 \\ 2 & -1 \end{vmatrix} = 3$

$C_{31} = \begin{vmatrix} 1 & 1 \\ 2 & -3 \end{vmatrix} = -5, C_{32} = -\begin{vmatrix} 1 & 1 \\ 1 & -3 \end{vmatrix} = 4, C_{33} = \begin{vmatrix} 1 & 1 \\ 1 & 2 \end{vmatrix} = 1$

$$\therefore \quad C = \begin{bmatrix} 3 & -9 & -5 \\ -4 & 1 & 3 \\ -5 & 4 & 1 \end{bmatrix}$$

Now, $adj\,A = C' = \begin{bmatrix} 3 & -4 & -5 \\ -9 & 1 & 4 \\ -5 & 3 & 1 \end{bmatrix}$

Also, $|A| = 1(6-3) - 1(3+6) + 1(-1-4) = -11$

Now, $$A\,(adj\,A) = \begin{bmatrix} 1 & 1 & 1 \\ 1 & 2 & -3 \\ 2 & -1 & 3 \end{bmatrix} \begin{bmatrix} 3 & -4 & -5 \\ -9 & 1 & 4 \\ -5 & 3 & 1 \end{bmatrix}$$

$$= \begin{bmatrix} 3-9-5 & -4+1+3 & -5+4+1 \\ 3-18+15 & -4+2-9 & -5+8-3 \\ 6+9-15 & -8-1+9 & -10-4+3 \end{bmatrix}$$

$$= \begin{bmatrix} -11 & 0 & 0 \\ 0 & -11 & 0 \\ 0 & 0 & -11 \end{bmatrix} = -11 \begin{bmatrix} 1 & 0 & 0 \\ 0 & 1 & 0 \\ 0 & 0 & 1 \end{bmatrix} = |A|\,I_3 \qquad ...(1)$$

$$(adj\,A)\,A = \begin{bmatrix} 3 & -4 & -5 \\ -9 & 1 & 4 \\ -5 & 3 & 1 \end{bmatrix} \begin{bmatrix} 1 & 1 & 1 \\ 1 & 2 & -3 \\ 2 & -1 & 3 \end{bmatrix}$$

$$= \begin{bmatrix} 3-4-10 & 3-8+5 & 3+12-15 \\ -9+1+8 & -9+2-4 & -9-3+12 \\ -5+3+2 & -5+6-1 & -5-9+3 \end{bmatrix}$$

$$(adj\,A)\,A = \begin{bmatrix} -11 & 0 & 0 \\ 0 & -11 & 0 \\ 0 & 0 & -11 \end{bmatrix} = -11 \begin{bmatrix} 1 & 0 & 0 \\ 0 & 1 & 0 \\ 0 & 0 & 1 \end{bmatrix} = |A|\,I_3 \qquad ...(2)$$

From (1) and (2) we have,

$$A\,(adj\,A) = (adj\,A)\,A = |A|\,I_3$$ **Proved.**

Example 3. Find the inverse of the matrix $A = \begin{bmatrix} 1 & 3 & 3 \\ 1 & 4 & 3 \\ 1 & 3 & 4 \end{bmatrix}$.

Solution: $|A| = \begin{vmatrix} 1 & 3 & 3 \\ 1 & 4 & 3 \\ 1 & 3 & 4 \end{vmatrix} = 1(16-9) - 3(4-3) + 3(3-4)$

$$= 7 - 3 - 3 = 1$$

$\because \quad |A| \neq 0 \;\therefore\; A^{-1}$ exists

The co-factors of $|A|$ are as fallows:

$$C_{11} = \begin{vmatrix} 4 & 3 \\ 3 & 4 \end{vmatrix} = 7,\; C_{12} = -\begin{vmatrix} 1 & 3 \\ 1 & 4 \end{vmatrix} = -1,\; C_{13} = \begin{vmatrix} 1 & 4 \\ 1 & 3 \end{vmatrix} = -1$$

$$C_{21} = -\begin{vmatrix} 3 & 3 \\ 3 & 4 \end{vmatrix} = -3,\ C_{22} = \begin{vmatrix} 1 & 3 \\ 1 & 4 \end{vmatrix} = 1,\ C_{23} = -\begin{vmatrix} 1 & 3 \\ 1 & 3 \end{vmatrix} = 0$$

$$C_{31} = \begin{vmatrix} 3 & 3 \\ 4 & 3 \end{vmatrix} = -3,\ C_{32} = -\begin{vmatrix} 1 & 3 \\ 1 & 3 \end{vmatrix} = 0,\ C_{33} = \begin{vmatrix} 1 & 3 \\ 1 & 4 \end{vmatrix} = 1$$

$$\text{The co-factor matrix of } A = \begin{bmatrix} 7 & -1 & -1 \\ -3 & 1 & 0 \\ -3 & 0 & 1 \end{bmatrix}$$

$$\therefore\ adj\,A = \text{Transpose of co-factor matrix of } A = \begin{bmatrix} 7 & -3 & -3 \\ -1 & 1 & 0 \\ -1 & 0 & 1 \end{bmatrix}$$

Now,
$$A^{-1} = \frac{adj\,A}{|A|}$$

$$= \begin{bmatrix} 7 & -3 & -3 \\ -1 & 1 & 0 \\ -1 & 0 & 1 \end{bmatrix}$$

Example 4. Find the inverse of the matrix $A = \begin{bmatrix} 2 & 3 & 4 \\ 4 & 3 & 1 \\ 1 & 2 & 4 \end{bmatrix}$

Solution: $|A| = \begin{bmatrix} 2 & 3 & 4 \\ 4 & 3 & 1 \\ 1 & 2 & 4 \end{bmatrix} = 2\,(12-2) - 3\,(16-1) + 4\,(8-3)$

$$= 20 - 45 + 20 = -5$$

$\because$ $|A| \neq 0\ \therefore\ A^{-1}$ exists

The co-factors of $|A|$ are as follows:

$$C_{11} = \begin{vmatrix} 3 & 1 \\ 2 & 4 \end{vmatrix} = 10,\ C_{12} = -\begin{vmatrix} 4 & 1 \\ 1 & 4 \end{vmatrix} = -15,\ C_{13} = \begin{vmatrix} 4 & 3 \\ 1 & 2 \end{vmatrix} = 5$$

$$C_{21} = -\begin{vmatrix} 3 & 4 \\ 2 & 4 \end{vmatrix} = -4,\ C_{22} = \begin{vmatrix} 2 & 4 \\ 1 & 4 \end{vmatrix} = 4,\ C_{23} = -\begin{vmatrix} 2 & 3 \\ 1 & 2 \end{vmatrix} = -1$$

$$C_{31} = \begin{vmatrix} 3 & 4 \\ 3 & 1 \end{vmatrix} = -9,\ C_{32} = -\begin{vmatrix} 2 & 4 \\ 4 & 1 \end{vmatrix} = 14,\ C_{33} = \begin{vmatrix} 2 & 3 \\ 4 & 3 \end{vmatrix} = -6$$

$$\text{The co-factor matrix of } A = \begin{bmatrix} 10 & -15 & 5 \\ -4 & 4 & -1 \\ -9 & 14 & -6 \end{bmatrix}$$

$$\therefore\ adj\,A = \text{Transpose of co-factor matrix of } A = \begin{bmatrix} 10 & -4 & -9 \\ -15 & 4 & 14 \\ 5 & -1 & -6 \end{bmatrix}$$

Now, $$A^{-1} = \frac{adj\,A}{|A|}$$

$$= -\frac{1}{5}\begin{bmatrix} 10 & -4 & -9 \\ -15 & 4 & 14 \\ 5 & -1 & -6 \end{bmatrix}$$

Example 5. Find the inverse of the matrix $A = \begin{bmatrix} 2 & 1 & 2 \\ 2 & 2 & 1 \\ 1 & 2 & 2 \end{bmatrix}$.

Solution: $|A| = \begin{vmatrix} 2 & 1 & 2 \\ 2 & 2 & 1 \\ 1 & 2 & 2 \end{vmatrix} = 2(4-2) - 1(4-1) + 2(4-2)$

$$= 4 - 3 + 4 = 5$$

$\because$ $|A| \neq 0 \ \therefore\ A^{-1}$ exists.

The co-factors of $|A|$ are as follows:

$$c_{11} = \begin{vmatrix} 2 & 1 \\ 2 & 2 \end{vmatrix} = 2,\ c_{12} = -\begin{vmatrix} 2 & 1 \\ 1 & 2 \end{vmatrix} = -3,\ c_{13} = \begin{vmatrix} 2 & 2 \\ 1 & 2 \end{vmatrix} = 2$$

$$c_{21} = -\begin{vmatrix} 1 & 2 \\ 2 & 2 \end{vmatrix} = 2,\ c_{22} = \begin{vmatrix} 2 & 2 \\ 1 & 2 \end{vmatrix} = 2,\ c_{23} = -\begin{vmatrix} 2 & 1 \\ 1 & 2 \end{vmatrix} = -3$$

$$c_{31} = \begin{vmatrix} 1 & 2 \\ 2 & 1 \end{vmatrix} = -3,\ c_{32} = -\begin{vmatrix} 2 & 2 \\ 2 & 1 \end{vmatrix} = 2,\ c_{33} = \begin{vmatrix} 2 & 1 \\ 2 & 2 \end{vmatrix} = 2$$

$$\text{The co-factor matrix of } A = \begin{bmatrix} 2 & -3 & 2 \\ 2 & 2 & -3 \\ -3 & 2 & 2 \end{bmatrix}$$

$$\therefore\ adj\,A = \text{Transpose of co-factor matrix of } A = \begin{bmatrix} 2 & 2 & -3 \\ -3 & 2 & 2 \\ 2 & -3 & 2 \end{bmatrix}$$

Now, $$A^{-1} = \frac{adj\,A}{|A|}$$

$$= \frac{1}{5}\begin{bmatrix} 2 & 2 & -3 \\ -3 & 2 & 2 \\ 2 & -3 & 2 \end{bmatrix}$$

Example 6. If $A = \begin{bmatrix} 3 & 1 \\ 4 & 0 \end{bmatrix}$ and $B = \begin{bmatrix} 4 & 0 \\ 2 & 5 \end{bmatrix}$, verify that $(AB)^{-1} = B^{-1}A^{-1}$.

Solution: $AB = \begin{bmatrix} 3 & 1 \\ 4 & 0 \end{bmatrix}\begin{bmatrix} 4 & 0 \\ 2 & 5 \end{bmatrix}$

$$= \begin{bmatrix} 12+2 & 0+5 \\ 16+0 & 0+0 \end{bmatrix} = \begin{bmatrix} 14 & 5 \\ 16 & 0 \end{bmatrix}$$

$$|AB| = \begin{bmatrix} 14 & 5 \\ 16 & 0 \end{bmatrix} = 0 - 80 = -80$$

$\because$ $|AB| \neq 0 \;\therefore\; (AB)^{-1}$ exists

The co-factors of $|AB|$ are as follows:

$$C_{11} = 0, C_{12} = -16$$
$$C_{21} = -5, C_{22} = 14$$

The co-factor matrix of $AB = \begin{bmatrix} 0 & -16 \\ -5 & 14 \end{bmatrix}$

$\therefore$ $adj\,(AB)$ = Transpose of co-factor matrix of $AB = \begin{bmatrix} 0 & -5 \\ -16 & 14 \end{bmatrix}$

Now, $(AB)^{-1} = \dfrac{adj\,(AB)}{|AB|}$

$$= -\frac{1}{80}\begin{bmatrix} 0 & -5 \\ -16 & 14 \end{bmatrix} \quad ...(1)$$

$$|A| = \begin{vmatrix} 3 & 1 \\ 4 & 0 \end{vmatrix} = -4$$

$$|A| \neq 0 \;\therefore\; A^{-1} \text{ exists}$$

The co-factors of $|A|$ are as follow:

$$C_{11} = 0, C_{12} = -4$$
$$C_{21} = -1, C_{22} = 3$$

The co-factor matrix of $A = \begin{bmatrix} 0 & -4 \\ -1 & 3 \end{bmatrix}$

$\therefore$ $adj\,A$ = Transpose of co-factor matrix of $A = \begin{bmatrix} 0 & -1 \\ -4 & 3 \end{bmatrix}$

Now, $A^{-1} = \dfrac{adj\,A}{|A|}$

$$= -\frac{1}{4}\begin{bmatrix} 0 & -1 \\ -4 & 3 \end{bmatrix}$$

$$|B| = \begin{vmatrix} 4 & 0 \\ 2 & 5 \end{vmatrix} = 20 - 0 = 20$$

$\because$ $|B| \neq 0 \;\therefore\; B^{-1}$ exists

The co-factors of $|B|$ are as follows:

$$C_{11} = 5, C_{12} = -2$$
$$C_{21} = 0, C_{22} = 4$$

The co-factor matrix of $B = \begin{bmatrix} 5 & -2 \\ 0 & 4 \end{bmatrix}$

$\therefore$ *adj* B = Transpose of co-factor matrix of $B = \begin{bmatrix} 5 & 0 \\ -2 & 4 \end{bmatrix}$

Now, $$B^{-1} = \frac{adj\, B}{|B|}$$
$$= \frac{1}{20}\begin{bmatrix} 5 & 0 \\ -2 & 4 \end{bmatrix}$$

Now, $$B^{-1}A^{-1} = \frac{1}{20}\begin{bmatrix} 5 & 0 \\ -2 & 4 \end{bmatrix} \times \frac{-1}{4}\begin{bmatrix} 0 & -1 \\ -4 & 3 \end{bmatrix}$$
$$= -\frac{1}{80}\begin{bmatrix} 0+0 & -5+0 \\ 0-16 & 2+12 \end{bmatrix}$$
$$= -\frac{1}{80}\begin{bmatrix} 0 & -5 \\ -16 & 14 \end{bmatrix} \quad ...(2)$$

From (1) and (2) we have,

$$(AB)^{-1} = B^{-1}A^{-1}$$ **Proved.**

EXERCISE 2.4

1. Find the adjoint of the matrix $A = \begin{bmatrix} 1 & 4 & 5 \\ 3 & 2 & 6 \\ 0 & 1 & 0 \end{bmatrix}$

2. Find the adjoint of the matrix $A = \begin{bmatrix} 1 & 2 & 3 \\ 4 & 5 & 6 \\ 7 & 9 & 8 \end{bmatrix}$

3. Find the adjoint of the transpose of the matrix $\begin{bmatrix} 1 & 1 & 2 \\ 1 & 2 & -1 \\ 1 & -3 & 3 \end{bmatrix}$.

4. If $A = \begin{bmatrix} -4 & -3 & -3 \\ 1 & 0 & 1 \\ 4 & 4 & 3 \end{bmatrix}$, show that *adj* $A = A$.

5. If $A = \begin{bmatrix} \cos\theta & -\sin\theta & 0 \\ \sin\theta & \cos\theta & 0 \\ 0 & 0 & 1 \end{bmatrix}$, verify that $A\,(adj\,A) = (adj\,A)\,A = |A|\,I_3$.

6. If $A = \begin{bmatrix} 1 & 0 & -1 \\ 3 & 4 & 5 \\ 0 & -6 & -7 \end{bmatrix}$, verify that $A\,(adj\,A) = |A|\,I_3$.

7. Find the inverse of the matrix $A = \begin{bmatrix} 1 & 2 & 5 \\ 2 & 3 & 1 \\ -1 & 1 & 1 \end{bmatrix}$

8. If $A = \begin{bmatrix} 1 & -1 & 1 \\ 2 & -1 & 0 \\ 1 & 0 & 0 \end{bmatrix}$, show that $A^{-1} = A^2$.

9. Find the inverse of the matrix $A = \begin{bmatrix} 7 & 7 & -7 \\ -1 & 11 & 7 \\ 11 & 5 & 7 \end{bmatrix}$

10. If $A = \begin{bmatrix} 3 & 2 \\ 7 & 5 \end{bmatrix}$ and $B = \begin{bmatrix} 6 & 7 \\ 8 & 9 \end{bmatrix}$, verity that $(AB)^{-1} = B^{-1}A^{-1}$.

11. Find the inverse of the transpose of the matrix $\begin{bmatrix} 1 & 2 & 1 \\ 1 & -1 & -2 \\ 1 & 2 & -1 \end{bmatrix}$.

12. Find the inverse of the matrix $\begin{bmatrix} \cos\alpha & -\sin\alpha & 0 \\ \sin\alpha & \cos\alpha & 0 \\ 0 & 0 & 1 \end{bmatrix}$.

ANSWERS

1. $\begin{bmatrix} -6 & 5 & 14 \\ 0 & 0 & 9 \\ 3 & -1 & -10 \end{bmatrix}$ 2. $\begin{bmatrix} -14 & 11 & -3 \\ 10 & -13 & 6 \\ 1 & 5 & 3 \end{bmatrix}$ 3. $\begin{bmatrix} 3 & -4 & -5 \\ -9 & 1 & 4 \\ -5 & 3 & 1 \end{bmatrix}$

7. $\dfrac{1}{21}\begin{bmatrix} 2 & 3 & -13 \\ -3 & 6 & 9 \\ 5 & -3 & -1 \end{bmatrix}$ 9. $\dfrac{1}{42}\begin{bmatrix} 1 & -2 & 3 \\ 2 & 3 & -1 \\ -3 & 1 & 2 \end{bmatrix}$ 11. $\dfrac{1}{6}\begin{bmatrix} 6 & 0 & 0 \\ -4 & 2 & 0 \\ 3 & -2 & 3 \end{bmatrix}$

12. $\begin{bmatrix} \cos\alpha & \sin\alpha & 0 \\ -\sin\alpha & \cos\alpha & 0 \\ 0 & 0 & 1 \end{bmatrix}$

2.25 RANK OF A MATRIX

Let A be any $m \times n$ matrix. If has square sub matrices of different orders. The determinants of these square sub matrices are called minors of A. In a square matrix of order n, the largest order minor is the determinant of the matrix itself.

The number r is said to be the rank of a matrix A if

(*i*) The matrix A possesses atleast one non-zero minor of order r.

(*ii*) The matrix A possesses every minor of order $(r + 1)$ as zero.

In other words, the rank of a matrix is the highest of orders of all the non-zero minors of the matrix.

The rank of a matrix A is usually denoted by rank of $A = r$ or $\rho(A) = r$.

There are different methods to determine the rank of a matrix:

1. Minor Method: In this method, we take the highest order minor of A (or minors of A). Let their order be r. If any ore of them is non-zero, then the rank of the matrix A is r, i.e., $\rho(A) = r$.

If all of them are zero, then we have to take the next lower order minors of A and so on till we get a non-zero minor. The order of that minor is the rank of the matrix A.

2. Echelon from Method: In this method, a matrix is reduced in echelon from as

(*i*) The first non-zero element in a row should be unity (if possible).

(*ii*) All the non-zero rows, if any, precede the zero rows.

(*iii*) The number of zeros before the first non-zero element in a row should be less than the number of such zeros in the next row.

The rank of the matrix in echelon from is equal to the number of non-zero rows of the matrix.

3. Normal form Method: In this method, a matrix is reduced by a sequence of elementary transformations (rows or columns or both) into one of the following forms.

$$\begin{bmatrix} I_r & \vdots & 0 \\ \cdots & \cdots & \cdots \\ 0 & \vdots & 0 \end{bmatrix}, \begin{bmatrix} I_r \\ \cdots \\ 0 \end{bmatrix}, [I_r \quad \vdots \quad 0], [I_r]$$

where I_r is the unit matrix of order r.

All these forms are called normal forms or canonical forms. The rank of the matrix in normal form is equal to the order of the unit matrix, i.e., r.

Note: (1) If A is a null matrix, then $\rho(A) = 0$

(2) If A is not a null matrix, then $\rho(A) \geq 1$.

(3) If A is a non-singular $n \times n$ matrix, then $\rho(A) = n$

(4) If I_n is the $n \times n$ unit matrix, then $\rho(I_n) = n$.

(5) If A is an $m \times n$ matrix, then $\rho(A) \leq$ minimum of m and n.

ILLUSTRATIVE EXAMPLES

Example 1. Find the rank of each of the following matrices.

$$(i)\begin{bmatrix}0 & 0 & 0\\0 & 0 & 0\\0 & 0 & 0\end{bmatrix}\quad(ii)\begin{bmatrix}1 & 0 & 0\\0 & 1 & 0\\0 & 0 & 1\end{bmatrix}\quad(iii)\begin{bmatrix}1 & 2 & 3\\2 & 1 & 4\\3 & 2 & 6\end{bmatrix}$$

Solution: (*i*) Let $A=\begin{bmatrix}0 & 0 & 0\\0 & 0 & 0\\0 & 0 & 0\end{bmatrix}=0_{3\times 3}$ (null matrix of order 3)

Hence, rank of $A=\rho(A)=0$

(*ii*) Let $A=\begin{bmatrix}1 & 0 & 0\\0 & 1 & 0\\0 & 0 & 1\end{bmatrix}=I_3$ (unit matrix of order 3)

We have, $|A|=1\neq 0$.

Hence, rank of $A=\rho(A)=3$

(*iii*) Let $A=\begin{bmatrix}1 & 2 & 3\\2 & 1 & 4\\3 & 2 & 6\end{bmatrix}$

We have, $|A|=1(6-8)-2\,(12-12)+3(4-3)=1\neq 0$

Therefore, A is non-singular matirx.

Hence, $\rho(A\,)=3$

Example 2: Find the rank of each of the following matrices:

$$(i)\begin{bmatrix}1 & 2 & 3 & 4\\4 & 3 & 5 & 2\end{bmatrix}\quad(ii)\begin{bmatrix}1 & 3\\4 & 2\\3 & 0\end{bmatrix}\quad(iii)\begin{bmatrix}1 & 1 & 2\\2 & 2 & 4\\3 & 5 & 8\end{bmatrix}$$

Solution: (*i*) Let $A=\begin{bmatrix}1 & 2 & 3 & 4\\4 & 3 & 5 & 2\end{bmatrix}_{2\times 4}$

Here, rank of $A\leq 2$, the smaller of 2 and 4.

Now there is at least one non-zero minor of A of order 2.

i.e., $$\begin{vmatrix}1 & 2\\4 & 3\end{vmatrix}=3-8=-5\neq 0$$

Hence, $$\rho(A)\;=2$$

(*ii*) Let $A=\begin{bmatrix}1 & 3\\4 & 2\\3 & 0\end{bmatrix}_{3\times 2}$

Here, rank of $A\leq 2$, the smaller of 3 and 2.

Now there is least one non-zero minor of A of order 2.

i.e., $\begin{vmatrix} 1 & 3 \\ 4 & 2 \end{vmatrix} = 2 - 12 = -10 \neq 0$

Hence, $\rho(A) = 2$

(*iii*) Let $A = \begin{bmatrix} 1 & 1 & 2 \\ 2 & 2 & 4 \\ 3 & 5 & 8 \end{bmatrix}$

we have, $|A| = 1(16-20) - 1(16-12) + 2(10-6) = -4 - 4 + 8 = 0$

Therefore, the rank of $A < 3$.

Now there is at least one non-zero minor of A order 2.

i.e., $\begin{vmatrix} 2 & 2 \\ 3 & 5 \end{vmatrix} = 10 - 6 = 4 \neq 0$

Hence, $\rho(A) = 2$ **Ans.**

Example 3. Find the rank of the matrix $\begin{bmatrix} 1 & 2 & 1 & 2 \\ 1 & 3 & 2 & 2 \\ 2 & 4 & 3 & 4 \\ 3 & 7 & 4 & 6 \end{bmatrix}$ by echelon form.

Solution: Let $A = \begin{bmatrix} 1 & 2 & 1 & 2 \\ 1 & 3 & 2 & 2 \\ 2 & 4 & 3 & 4 \\ 3 & 7 & 4 & 6 \end{bmatrix}$

$$\sim \begin{bmatrix} 1 & 2 & 1 & 2 \\ 0 & 1 & 1 & 0 \\ 0 & 0 & 1 & 0 \\ 0 & 1 & 1 & 0 \end{bmatrix} \text{(by } R_2 \to R_2 - R_1, R_3 \to R_3 - 2R_1, R_4 \to R_4 - 3R_1)$$

$$\sim \begin{bmatrix} 1 & 2 & 1 & 2 \\ 0 & 1 & 1 & 0 \\ 0 & 0 & 1 & 0 \\ 0 & 0 & 0 & 0 \end{bmatrix} \text{(by } R_4 \to R_4 - R_2)$$

The last equivalent matrix is in echelon form having three non-zero rows.

Hence, $\rho(A) = 3$ **Ans.**

Example 4. Find the rank of the matrix $\begin{bmatrix} 1 & 2 & 3 & 4 & 5 \\ 2 & 3 & 4 & 5 & 1 \\ 5 & 8 & 11 & 14 & 7 \end{bmatrix}$ by echelon form.

Solution: Let $A = \begin{bmatrix} 1 & 2 & 3 & 4 & 5 \\ 2 & 3 & 4 & 5 & 1 \\ 5 & 8 & 11 & 14 & 7 \end{bmatrix}$

$$\sim \begin{bmatrix} 1 & 2 & 3 & 4 & 5 \\ 0 & -1 & -2 & -3 & -9 \\ 0 & -2 & -4 & -6 & -18 \end{bmatrix} \text{(by } R_2 \to R_2 - 2R_1, R_3 \to R_3 - 5R_1\text{)}$$

$$\sim \begin{bmatrix} 1 & 2 & 3 & 4 & 5 \\ 0 & 1 & 2 & 3 & 9 \\ 0 & 1 & 2 & 3 & 9 \end{bmatrix} \left(\text{by } R_2 \to -R_2, R_3 \to -\frac{1}{2}R_3\right)$$

$$\sim \begin{bmatrix} 1 & 2 & 3 & 4 & 5 \\ 0 & 1 & 2 & 3 & 9 \\ 0 & 0 & 0 & 0 & 0 \end{bmatrix} \text{(by } R_3 \to R_3 - R_2\text{)}$$

The last equivalent matrix is in echelon form having two non-zero rows.

Hence, $\rho(A) = 2$ **Ans.**

Example 5. Find the rank of the matrix $\begin{bmatrix} 3 & 4 & 1 & 1 \\ 2 & 4 & 3 & 6 \\ -1 & -2 & 6 & 4 \\ 1 & -1 & 2 & -3 \end{bmatrix}$ using elementary row transformations.

Solution: Let $A = \begin{bmatrix} 3 & 4 & 1 & 1 \\ 2 & 4 & 3 & 6 \\ -1 & -2 & 6 & 4 \\ 1 & -1 & 2 & -3 \end{bmatrix}$

$$\sim \begin{bmatrix} 1 & -1 & 2 & -3 \\ 2 & 4 & 3 & 6 \\ -1 & -2 & 6 & 4 \\ 3 & 4 & 1 & 1 \end{bmatrix} \text{(by } R_1 \leftrightarrow R_4\text{)}$$

$$\sim \begin{bmatrix} 1 & -1 & 2 & -3 \\ 0 & 6 & -1 & 12 \\ 0 & -3 & 8 & 1 \\ 0 & 7 & -5 & 10 \end{bmatrix} \text{(by } R_2 \to R_2 - 2R_1, R_3 \to R_3 + R_1, R_4 \to R_4 - 3R_1\text{)}$$

$$\sim \begin{bmatrix} 1 & -1 & 2 & -3 \\ 0 & 1 & \frac{-1}{6} & 2 \\ 0 & -3 & 8 & 1 \\ 0 & 7 & -5 & 10 \end{bmatrix} \left(\text{by } R_2 \to \frac{1}{6}R_2\right)$$

$$\sim \begin{bmatrix} 1 & -1 & 2 & -3 \\ 0 & 1 & \frac{-1}{6} & 2 \\ 0 & 0 & \frac{15}{2} & 7 \\ 0 & 0 & \frac{-23}{6} & -4 \end{bmatrix} \text{(by } R_3 \to R_3 + 3R_2, R_4 \to R_4 - 7R_2\text{)}$$

$$\sim \begin{bmatrix} 1 & -1 & 2 & -3 \\ 0 & 1 & \frac{-1}{6} & 2 \\ 0 & 0 & 1 & \frac{14}{15} \\ 0 & 0 & \frac{23}{6} & 4 \end{bmatrix} \left(\text{by } R_3 \to \frac{2}{15} R_3, R_4 \to -R_4\right)$$

$$\sim \begin{bmatrix} 1 & -1 & 2 & -3 \\ 0 & 1 & \frac{-1}{6} & 2 \\ 0 & 0 & 1 & \frac{14}{15} \\ 0 & 0 & 0 & \frac{19}{45} \end{bmatrix} \left(\text{by } R_4 \to R_4 - \frac{23}{6} R_3\right)$$

The last equivalent matrix is in echelon form having four non-zero rows.

Hence, $\rho(A) = 4$

Example 6. Find the rank of the matrix $\begin{bmatrix} 1 & 2 & -1 & 3 \\ 4 & 1 & 2 & 1 \\ 3 & -1 & 1 & 2 \\ 1 & 2 & 0 & 1 \end{bmatrix}$ using elementary row transformations.

Solution: Let $A = \begin{bmatrix} 1 & 2 & -1 & 3 \\ 4 & 1 & 2 & 1 \\ 3 & -1 & 1 & 2 \\ 1 & 2 & 0 & 1 \end{bmatrix}$

$$\sim \begin{bmatrix} 1 & 2 & -1 & 3 \\ 0 & -7 & 6 & -11 \\ 0 & -7 & 4 & -7 \\ 0 & 0 & 1 & -2 \end{bmatrix} \text{(by } R_2 \to R_2 - 4R_1, R_3 \to R_3 - 3R_1, R_4 \to R_4 - R_1\text{)}$$

$$\sim \begin{bmatrix} 1 & 2 & -1 & 3 \\ 0 & -7 & 6 & -11 \\ 0 & 0 & -2 & 4 \\ 0 & 0 & 1 & -2 \end{bmatrix} \text{(by } R_3 \to R_3 - R_2\text{)}$$

$$\sim \begin{bmatrix} 1 & 2 & -1 & 3 \\ 0 & 1 & \frac{-6}{7} & \frac{11}{7} \\ 0 & 0 & 1 & -2 \\ 0 & 0 & 1 & -2 \end{bmatrix} \left(\text{by } R_2 \to -\frac{1}{7}R_2,\ R_3 \to -\frac{1}{2}R_3\right)$$

$$\sim \begin{bmatrix} 1 & 2 & -1 & 3 \\ 0 & 1 & \frac{-6}{7} & \frac{11}{7} \\ 0 & 0 & 1 & -2 \\ 0 & 0 & 0 & 0 \end{bmatrix} \text{(by } R_4 \to R_4 - R_3\text{)}$$

The last equivalent matrix is in echelon form having three non-zero rows.

Hence, $\rho(A) = 3$ **Ans.**

Example 7. Find the rank of the matrix $\begin{bmatrix} 1^2 & 2^2 & 3^2 & 4^2 \\ 2^2 & 3^2 & 4^2 & 5^2 \\ 3^2 & 4^2 & 5^2 & 6^2 \\ 4^2 & 5^2 & 6^2 & 7^2 \end{bmatrix}$ using elementary row transformations.

Solution: Let $A = \begin{bmatrix} 1^2 & 2^2 & 3^2 & 4^2 \\ 2^2 & 3^2 & 4^2 & 5^2 \\ 3^2 & 4^2 & 5^2 & 6^2 \\ 4^2 & 5^2 & 6^2 & 7^2 \end{bmatrix}$

$$= \begin{bmatrix} 1 & 4 & 9 & 16 \\ 4 & 9 & 16 & 25 \\ 9 & 16 & 25 & 36 \\ 16 & 25 & 36 & 49 \end{bmatrix}$$

$$\sim \begin{bmatrix} 1 & 4 & 9 & 16 \\ 3 & 5 & 7 & 9 \\ 5 & 7 & 9 & 11 \\ 7 & 9 & 11 & 13 \end{bmatrix} \text{(by } R_2 \to R_2 - R_1, R_3 \to R_3 - R_2, R_4 \to R_4 - R_3\text{)}$$

$$\sim \begin{bmatrix} 1 & 4 & 9 & 16 \\ 3 & 5 & 7 & 9 \\ 2 & 2 & 2 & 2 \\ 2 & 2 & 2 & 2 \end{bmatrix} \text{(by } R_3 \to R_3 - R_2, R_4 \to R_4 - R_3)$$

$$\sim \begin{bmatrix} 1 & 4 & 9 & 16 \\ 3 & 5 & 7 & 9 \\ 1 & 1 & 1 & 1 \\ 0 & 0 & 0 & 0 \end{bmatrix} \left(\text{by } R_3 \to \frac{1}{2}R_3, R_4 \to R_4 - R_3\right)$$

$$\sim \begin{bmatrix} 1 & 4 & 9 & 16 \\ 0 & -7 & -20 & -39 \\ 0 & -3 & -8 & -15 \\ 0 & 0 & 0 & 0 \end{bmatrix} \text{(by } R_2 \to R_2 - 3R_1, R_3 \to R_3 - R_1)$$

$$\sim \begin{bmatrix} 1 & 4 & 9 & 16 \\ 0 & 1 & \frac{20}{7} & \frac{39}{7} \\ 0 & 1 & \frac{8}{3} & 5 \\ 0 & 0 & 0 & 0 \end{bmatrix} \left(\text{by } R_2 \to \frac{-1}{7}R_2, R_3 \to \frac{-1}{3}R_3\right)$$

$$\sim \begin{bmatrix} 1 & 4 & 9 & 16 \\ 0 & 1 & \frac{20}{7} & \frac{39}{7} \\ 0 & 1 & \frac{-4}{21} & \frac{-4}{7} \\ 0 & 0 & 0 & 0 \end{bmatrix} \text{(by } R_3 \to R_3 - R_2)$$

$$\sim \begin{bmatrix} 1 & 4 & 9 & 16 \\ 0 & 1 & \frac{20}{7} & \frac{39}{7} \\ 0 & 0 & 1 & 3 \\ 0 & 0 & 0 & 0 \end{bmatrix} \left(by\ R_3 \to \frac{-21}{4}R_3\right)$$

The last equivalent matrix is in echelon form having three non-zero rows.

Hence, $\rho(A) = 3$

Example 8. Find the rank of the matrix $\begin{bmatrix} 1 & 2 & 0 & -1 \\ 3 & 4 & 1 & 2 \\ -2 & 3 & 2 & 5 \end{bmatrix}$ by normal form.

Solution: Let $A = \begin{bmatrix} 1 & 2 & 0 & -1 \\ 3 & 4 & 1 & 2 \\ -2 & 3 & 2 & 5 \end{bmatrix}$

$$\sim \begin{bmatrix} 1 & 2 & 0 & -1 \\ 0 & -2 & 1 & 5 \\ 0 & 7 & 2 & 3 \end{bmatrix} \text{(by } R_2 \to R_2 - 3R_1, R_3 \to R_3 + 2R_1\text{)}$$

$$\sim \begin{bmatrix} 1 & 0 & 0 & 0 \\ 0 & -2 & 1 & 5 \\ 0 & 7 & 2 & 3 \end{bmatrix} \text{(by } C_2 \to C_2 - 2C_1, c_4 \to C_4 + C_1\text{)}$$

$$\sim \begin{bmatrix} 1 & 0 & 0 & 0 \\ 0 & 1 & -2 & 5 \\ 0 & 2 & 7 & 3 \end{bmatrix} \text{(by } C_2 \leftrightarrow C_3\text{)}$$

$$\sim \begin{bmatrix} 1 & 0 & 0 & 0 \\ 0 & 1 & -2 & 5 \\ 0 & 0 & 11 & -7 \end{bmatrix} \text{(by } R_3 \to R_3 - 2R_2\text{)}$$

$$\sim \begin{bmatrix} 1 & 0 & 0 & 0 \\ 0 & 1 & 0 & 0 \\ 0 & 0 & 11 & -7 \end{bmatrix} \text{(by } C_3 \to C_3 + 2C_2, C_4 \to C_4 - 5C_2\text{)}$$

$$\sim \begin{bmatrix} 1 & 0 & 0 & 0 \\ 0 & 1 & 0 & 0 \\ 0 & 0 & 1 & -7 \end{bmatrix} \left(\text{by } C_3 \to \frac{1}{11} C_3\right)$$

$$\sim \begin{bmatrix} 1 & 0 & 0 & 0 \\ 0 & 1 & 0 & 0 \\ 0 & 0 & 1 & 0 \end{bmatrix} \text{(by } C_4 \to C_4 + 7C_3\text{)}$$

$$\sim [I_3 \ \vdots \ 0]$$

Hence, $\rho(A) = 3$

Example 9. Find the rank of the matrix $\begin{bmatrix} 1 & 2 & 3 & 2 \\ 2 & 3 & 5 & 1 \\ 1 & 3 & 4 & 5 \end{bmatrix}$ by reducing into normal form.

Solution: Let $A = \begin{bmatrix} 1 & 2 & 3 & 2 \\ 2 & 3 & 5 & 1 \\ 1 & 3 & 4 & 5 \end{bmatrix}$

$$\sim \begin{bmatrix} 1 & 2 & 3 & 2 \\ 0 & -1 & -1 & -3 \\ 0 & 1 & 1 & 3 \end{bmatrix} \text{(by } R_2 \to R_2 - 2R_1, R_3 \to R_3 - R_1\text{)}$$

$$\sim \begin{bmatrix} 1 & 2 & 3 & 2 \\ 0 & -1 & -1 & -3 \\ 0 & 0 & 0 & 0 \end{bmatrix} \text{(by } R_3 \to R_3 + R_2\text{)}$$

$$\sim \begin{bmatrix} 1 & 2 & 3 & 2 \\ 0 & 1 & 1 & 3 \\ 0 & 0 & 0 & 0 \end{bmatrix} \text{(by } R_2 \to -R_2\text{)}$$

$$\sim \begin{bmatrix} 1 & 2 & 1 & 2 \\ 0 & 1 & 0 & 3 \\ 0 & 0 & 0 & 0 \end{bmatrix} \text{(by } C_3 \to C_3 - C_2\text{)}$$

$$\sim \begin{bmatrix} 1 & 0 & 0 & 0 \\ 0 & 1 & 0 & 2 \\ 0 & 0 & 0 & 0 \end{bmatrix} \text{(by } C_2 \to C_2 - 2C_1, C_3 \to C_3 - C_1, C_4 \to C_4 - C_2\text{)}$$

$$\sim \begin{bmatrix} 1 & 0 & 0 & 0 \\ 0 & 1 & 0 & 0 \\ 0 & 0 & 0 & 0 \end{bmatrix} \text{(by } C_4 \to C_4 - 2C_2\text{)}$$

$$\sim \begin{bmatrix} I_2 & \vdots & 0 \\ \cdots & & \cdots \\ 0 & \vdots & 0 \end{bmatrix}$$

Hence, $\rho(A) = 2$ **Ans.**

Example 10. Find the rank of the matrix $\begin{bmatrix} 2 & -2 & 0 & 6 \\ 4 & 2 & 0 & 2 \\ 1 & -1 & 0 & 3 \\ 1 & -2 & 1 & 2 \end{bmatrix}$ by reducing into normal form.

Solution: Let $A = \begin{bmatrix} 2 & -2 & 0 & 6 \\ 4 & 2 & 0 & 2 \\ 1 & -1 & 0 & 3 \\ 1 & -2 & 1 & 2 \end{bmatrix}$

$$\sim \begin{bmatrix} 1 & -1 & 0 & 3 \\ 2 & 1 & 0 & 1 \\ 1 & -1 & 0 & 3 \\ 1 & -2 & 1 & 2 \end{bmatrix} \left(\text{by } R_1 \to \frac{1}{2}R_1, R_2 \to \frac{1}{2}R_2\right)$$

$$\sim \begin{bmatrix} 1 & -1 & 0 & 3 \\ 0 & 3 & 0 & -5 \\ 0 & 0 & 0 & 0 \\ 0 & -1 & 1 & -1 \end{bmatrix} (\text{by } R_2 \to R_2 - 2R_1, R_3 \to R_3 - R_1, R_4 \to R_4 - R_1)$$

$$\sim \begin{bmatrix} 1 & 0 & 0 & 0 \\ 0 & 3 & 0 & -5 \\ 0 & 0 & 0 & 0 \\ 0 & -1 & 1 & -1 \end{bmatrix} (\text{by } C_2 \to C_2 + C_1, C_4 \to C_4 - 3C_1)$$

$$\sim \begin{bmatrix} 1 & 0 & 0 & 0 \\ 0 & -1 & 1 & -1 \\ 0 & 0 & 0 & 0 \\ 0 & 3 & 0 & -5 \end{bmatrix} (\text{by } R_2 \leftrightarrow -R_4)$$

$$\sim \begin{bmatrix} 1 & 0 & 0 & 0 \\ 0 & 1 & -1 & 1 \\ 0 & 0 & 0 & 0 \\ 0 & 3 & 0 & -5 \end{bmatrix} (\text{by } R_2 \to -R_2)$$

$$\sim \begin{bmatrix} 1 & 0 & 0 & 0 \\ 0 & 1 & -1 & 1 \\ 0 & 0 & 0 & 0 \\ 0 & 0 & 3 & -8 \end{bmatrix} (\text{by } R_4 \to R_4 - 3R_2)$$

$$\sim \begin{bmatrix} 1 & 0 & 0 & 0 \\ 0 & 1 & 0 & 0 \\ 0 & 0 & 0 & 0 \\ 0 & 0 & 3 & -8 \end{bmatrix} (\text{by } C_3 \to C_3 + C_2, C_4 \to C_4 - C_2)$$

$$\sim \begin{bmatrix} 1 & 0 & 0 & 0 \\ 0 & 1 & 0 & 0 \\ 0 & 0 & 3 & -8 \\ 0 & 0 & 0 & 0 \end{bmatrix} (\text{by } R_4 \leftrightarrow R_3)$$

$$\sim \begin{bmatrix} 1 & 0 & 0 & 0 \\ 0 & 1 & 0 & 0 \\ 0 & 0 & 1 & 1 \\ 0 & 0 & 0 & 0 \end{bmatrix} \left(by\, C_3 \to \frac{1}{3}C_3, C_4 \to -\frac{1}{8}C_4 \right)$$

$$\sim \begin{bmatrix} 1 & 0 & 0 & 0 \\ 0 & 1 & 0 & 0 \\ 0 & 0 & 1 & 0 \\ 0 & 0 & 0 & 0 \end{bmatrix} (\text{by } C_4 \to C_4 - C_3)$$

$$\sim \begin{bmatrix} I_3 & \vdots & 0 \\ \cdots & & \cdots \\ 0 & \vdots & 0 \end{bmatrix}$$

Hence, $\rho(A) = 3$

Example 11. Find the rank of the matrix $\begin{bmatrix} 0 & 1 & 2 & -1 \\ 1 & 0 & 1 & 1 \\ 3 & 1 & 0 & 2 \\ 1 & 1 & -2 & 0 \end{bmatrix}$ by reducing into normal form.

Solution: Let $A = \begin{bmatrix} 0 & 1 & 2 & -1 \\ 1 & 0 & 1 & 1 \\ 3 & 1 & 0 & 2 \\ 1 & 1 & -2 & 0 \end{bmatrix}$

$$\sim \begin{bmatrix} 1 & 0 & 1 & 1 \\ 0 & 1 & 2 & -1 \\ 3 & 1 & 0 & 2 \\ 1 & 1 & -2 & 0 \end{bmatrix} \text{(by } R_1 \leftrightarrow R_2\text{)}$$

$$\sim \begin{bmatrix} 1 & 0 & 1 & 1 \\ 0 & 1 & 2 & -1 \\ 0 & 1 & -3 & -1 \\ 0 & 1 & -3 & -1 \end{bmatrix} \text{(by } R_3 \to R_3 - 3R_1, R_4 \to R_4 - R_1\text{)}$$

$$\sim \begin{bmatrix} 1 & 0 & 1 & 1 \\ 0 & 1 & 2 & -1 \\ 0 & 1 & -3 & -1 \\ 0 & 0 & 0 & 0 \end{bmatrix} \text{(by } R_4 \to R_4 - R_3\text{)}$$

$$\sim \begin{bmatrix} 1 & 0 & 0 & 0 \\ 0 & 1 & 2 & -1 \\ 0 & 1 & -3 & -1 \\ 0 & 0 & 0 & 0 \end{bmatrix} \text{(by } C_3 \to C_3 - C_1, C_4 \to C_4 - C_1\text{)}$$

$$\sim \begin{bmatrix} 1 & 0 & 0 & 0 \\ 0 & 1 & 2 & -1 \\ 0 & 0 & -5 & 0 \\ 0 & 0 & 0 & 0 \end{bmatrix} \text{(by } R_3 \to R_3 - R_2\text{)}$$

$$\sim \begin{bmatrix} 1 & 0 & 0 & 0 \\ 0 & 1 & 0 & 0 \\ 0 & 0 & -5 & 0 \\ 0 & 0 & 0 & 0 \end{bmatrix} \text{(by } C_3 \to C_3 - 2C_2, C_4 \to C_4 + C_2\text{)}$$

$$\sim \begin{bmatrix} 1 & 0 & 0 & 0 \\ 0 & 1 & 0 & 0 \\ 0 & 0 & 1 & 0 \\ 0 & 0 & 0 & 0 \end{bmatrix} \left(\text{by } C_3 \to -\frac{1}{5}C_3\right)$$

$$\sim \begin{bmatrix} I_3 & \vdots & 0 \\ \cdots & & \cdots \\ 0 & \vdots & 0 \end{bmatrix}$$

Hence, $\rho(A) = 3$ **Ans.**

EXERCISE 2.5

1. Using minor property find the rank of the matrix.

$$\begin{bmatrix} 2 & 4 & 6 \\ 2 & 4 & 10 \\ 4 & 8 & 16 \end{bmatrix}$$

2. Using minor property find the rank of the matrix.

$$\begin{bmatrix} 6 & 2 & 4 \\ 12 & 4 & 8 \\ 6 & 2 & 4 \end{bmatrix}$$

3. Using minor property find the rank of the matrix.

$$\begin{bmatrix} 1 & -1 & 3 & 6 \\ 1 & 3 & -3 & -4 \\ 5 & 3 & 3 & 11 \end{bmatrix}$$

4. Find the rank of the following matrix by echelon form.

$$\begin{bmatrix} 2 & 3 & 4 & -1 \\ 5 & 2 & 0 & -1 \\ -4 & 5 & 12 & -1 \end{bmatrix}$$

5. Find the rank of the following matrix by echelon form.

$$\begin{bmatrix} 2 & 3 & -1 & -1 \\ 1 & -1 & -2 & -4 \\ 3 & 1 & 3 & -2 \\ 6 & 3 & 0 & -7 \end{bmatrix}$$

6. Find the rank of the following matrix by echelon form.

$$\begin{bmatrix} 3 & -2 & 0 & -1 \\ 0 & 2 & 2 & 1 \\ 1 & -2 & -3 & 2 \\ 0 & 1 & 2 & 1 \end{bmatrix}$$

7. Find the rank of the following matrix by normal from.

$$\begin{bmatrix} 1 & 2 & -1 & 4 \\ 2 & 4 & 3 & 4 \\ 1 & 2 & 3 & 4 \\ -1 & -2 & 6 & -7 \end{bmatrix}$$

8. Find the rank of the following matrix by normal form.

$$\begin{bmatrix} 2 & 1 & 3 \\ 4 & 7 & 13 \\ 4 & -3 & -1 \end{bmatrix}$$

9. Find the rank of the following matrix by normal form.

$$\begin{bmatrix} 9 & 7 & 3 & 6 \\ 5 & -1 & 4 & 1 \\ 6 & 8 & 2 & 4 \end{bmatrix}$$

10. Find the rank of the following matrix by normal form.

$$\begin{bmatrix} 0 & 1 & 2 & -2 \\ 4 & 0 & 2 & 6 \\ 2 & 1 & 3 & 1 \end{bmatrix}$$

11. Find the rank of the following matrix by normal form.

$$\begin{bmatrix} 1 & 2 & 3 & 0 \\ 2 & 4 & 3 & 2 \\ 3 & 2 & 1 & 3 \\ 6 & 8 & 7 & 5 \end{bmatrix}$$

ANSWERS

1. 2 **2.** 1 **3.** 3 **4.** 2 **5.** 3 **6.** 4 **7.** 3 **8.** 2
9. 3 **10.** 2 **11.** 3

2.26 SOLUTION OF A SYSTEM OF LINEAR EQUATIONS

Consider the system of n equations having n unknown $x_1, x_2, \ldots, x_n$ as

$$a_{11}x_1 + a_{12}x_2 + \ldots + a_{1n}x_n = b_1$$
$$a_{21}x_1 + a_{22}x_2 + \ldots + a_{2n}x_n = b_2$$
$$\ldots\ldots\ldots\ldots\ldots\ldots\ldots\ldots\ldots\ldots\ldots\ldots$$
$$\ldots\ldots\ldots\ldots\ldots\ldots\ldots\ldots\ldots\ldots\ldots\ldots$$
$$a_{n1}x_1 + a_{n2}x_2 + \ldots + a_{nn}x_n = b_n$$

In matrix notation, these equations can be written as

$$\begin{bmatrix} a_{11} & a_{12} & \cdots & a_{1n} \\ a_{21} & a_{22} & \cdots & a_{2n} \\ \cdots & \cdots & \cdots & \cdots \\ \cdots & \cdots & \cdots & \cdots \\ a_{n1} & a_{n2} & \cdots & a_{nn} \end{bmatrix} \begin{bmatrix} x_1 \\ x_2 \\ \vdots \\ x_n \end{bmatrix} = \begin{bmatrix} b_1 \\ b_2 \\ \vdots \\ b_n \end{bmatrix}$$

or $$AX = B$$

where $A = \begin{bmatrix} a_{11} & a_{12} & \cdots & a_{1n} \\ a_{21} & a_{22} & \cdots & a_{2n} \\ \cdots & \cdots & \cdots & \cdots \\ \cdots & \cdots & \cdots & \cdots \\ a_{n1} & a_{n2} & \cdots & a_{nn} \end{bmatrix}$ is known as the coefficient matrix,

$X = \begin{bmatrix} x_1 \\ x_2 \\ \vdots \\ x_n \end{bmatrix}$ is the column matrix of unknowns and

$B = \begin{bmatrix} b_1 \\ b_2 \\ \vdots \\ b_n \end{bmatrix}$ is the column matrix of constants.

If $b_1 = b_2 = \ldots = b_n = 0$, then $B = 0$ and the matrix equation $AX = B$ reduces to AX = 0. Such a system of equations is called a system of homogeneous linear equations. If at least one of $b_1, b_2, \ldots, b_n$ is non-zero, then $B \neq 0$. Such a system of equations is called a system of non-homogeneous linear equations.

The matrix equation $AX = B$ may have no solution or a unique solution or an infinite number of solutions. When the system of equations has one or more solution, the equations are said to be consistent.

When the system of equations has no solution, the equations are said to be inconsistent.

2.27 SOLUTION OF SYSTEM OF LINEAR EQUATIONS BY CRAMER'S RULE

System of Linear Equations in Two Unknowns:

The solution of the system of linear equations

$$a_{11}x + a_{12}y = b_1$$
$$a_{21}x + a_{22}y = b_2$$

is given by

$$x = \frac{D_1}{D} \text{ and } y = \frac{D_2}{D},$$

where, $$D = \begin{vmatrix} a_{11} & a_{12} \\ a_{21} & a_{22} \end{vmatrix}, D_1 = \begin{vmatrix} b_1 & a_{12} \\ b_2 & a_{22} \end{vmatrix} \text{ and } D_2 = \begin{vmatrix} a_{11} & b_1 \\ a_{21} & b_2 \end{vmatrix}$$

Conditions for Consistency:

In case of two unknowns, the following cases arise

1. If $D \neq = 0$, then the system of equations is consistent and has a unique solution.

2. If $D = 0$ and $D_1 \neq 0$, then the system of equations is inconsistent and has no solution.

3. If $D = 0$ and $D_1 = 0$, then the system of equations is consistent and has an infinite number of solution.

System of Linear Equations in Three Unknowns:

The solution of the system of linear equations

$$a_{11}x + a_{12}y + a_{13}z = b_1$$
$$a_{21}x + a_{22}y + a_{23}z = b_2$$
$$a_{31}x + a_{32}y + a_{33}z = b_3$$

is given by

$$x = \frac{D_1}{D}, y = \frac{D_2}{D} \text{ and } z = \frac{D_3}{D},$$

where, $$D = \begin{vmatrix} a_{11} & a_{12} & a_{13} \\ a_{21} & a_{22} & a_{23} \\ a_{31} & a_{32} & a_{33} \end{vmatrix}, D_1 = \begin{vmatrix} b_1 & a_{12} & a_{13} \\ b_2 & a_{22} & a_{23} \\ b_3 & a_{32} & a_{33} \end{vmatrix}$$

$$D_2 = \begin{vmatrix} a_{11} & b_1 & a_{13} \\ a_{21} & b_2 & a_{23} \\ a_{31} & b_3 & a_{33} \end{vmatrix} \text{ and } D_3 = \begin{vmatrix} a_{11} & a_{12} & b_1 \\ a_{21} & a_{22} & b_2 \\ a_{31} & a_{32} & b_3 \end{vmatrix}$$

Conditions for Consistency:

In case of three unknowns, the following cases arise

1. If $D \neq 0$, then the system of equations is consistent and has a unique solution.

2. If $D = 0$ and at least one of D_1, D_2, D_3 is non-zero, then the system of equations is inconsistent and has no solution.

3. If $D = D_1 = D_2 = D_3 = 0$, then the system of equations may or may not be consistent.

2.28 SOLUTION OF SYSTEM OF LINEAR EQUATIONS BY MATRIX METHOD

When A is a non-singular matrix, then A^{-1} exists.

Take the matrix equation

$$AX = B$$

On pre-multiplying both sides by A^{-1}, we have

$$A^{-1}AX = A^{-1}B$$

or $$(A^{-1}A)X = A^{-1}B$$

or $$IX = A^{-1}B$$

$$X = A^{-1}B, \text{ where } A^{-1} = \frac{adj\,A}{|A|}$$

From here, we obtain the required unique solution.

Conditions of Consistency:

1. If $|A| \neq 0$, then the system of equations is consistent and has a unique solution.

2. If $|A| = 0$ and $(adj\,A)B \neq 0$, then the system of equations is inconsistent and has no solution.

3. If $|A| = 0$ and $(adj\,A)B = 0$, then the system of equations is consistent and has an infinite number of solutions.

2.29 SOLUTION OF SYSTEM OF LINEAR EQUATIONS BY GAUSS-ELIMINATION METHOD

This method is based on the elimination of the unknowns by combining equations such that the n linear equations in n unknowns are reduced to an equivalent upper triangular system which could be solved by back substitution.

Consider the system of n equations having n unknowns $x_1, x_2, \ldots, x_n$ as

$$a_{11}x_1 + a_{12}x_2 + \ldots + a_{1n}x_n = b_1$$
$$a_{21}x_1 + a_{22}x_2 + \ldots + a_{2n}x_n = b_2$$
$$\ldots\ldots\ldots\ldots\ldots\ldots\ldots\ldots\ldots\ldots$$
$$\ldots\ldots\ldots\ldots\ldots\ldots\ldots\ldots\ldots\ldots$$
$$a_{n1}x_1 + a_{n2}x_2 + \ldots + a_{nn}x_n = b_n$$

In matrix notation, these equations can be written as

$$\begin{bmatrix} a_{11} & a_{12} & \cdots & a_{1n} \\ a_{21} & a_{22} & \cdots & a_{2n} \\ \cdots & \cdots & \cdots & \cdots \\ \cdots & \cdots & \cdots & \cdots \\ a_{n1} & a_{n2} & \cdots & a_{nn} \end{bmatrix} \begin{bmatrix} x_1 \\ x_2 \\ \vdots \\ x_n \end{bmatrix} = \begin{bmatrix} b_1 \\ b_2 \\ \vdots \\ b_n \end{bmatrix}$$

or $$AX = B$$

Now, our aim is to reduce the augmented matrix $[A \vdots B]$ to upper triangular matrix by using elementary row transformations.

$$[A \vdots B] = \begin{bmatrix} a_{11} & a_{12} & \cdots & a_{1n} & \vdots & b_1 \\ a_{21} & a_{22} & \cdots & a_{2n} & \vdots & b_2 \\ \cdots & \cdots & \cdots & \cdots & \vdots & \cdots \\ \cdots & \cdots & \cdots & \cdots & \vdots & \cdots \\ a_{n1} & a_{n2} & \cdots & a_{nn} & \vdots & b_n \end{bmatrix}$$

After applying the elementary row transformations, we have the following upper triangular matrix.

$$[A \vdots B] = \begin{bmatrix} a_{11} & a_{12} & \cdots & a_{1n} & \vdots & b_1 \\ 0 & a_{22}' & \cdots & a_{2n}' & \vdots & b_2' \\ 0 & 0 & \cdots & a_{3n}'' & \vdots & b_3'' \\ \cdots & \cdots & \cdots & \cdots & \vdots & \vdots \\ 0 & 0 & 0 & \cdots & \vdots & b_n''' \end{bmatrix}$$

Now, the given system of linear equations is equivalent to

$$a_{11}x_1 + a_{12}x_2 + \ldots + a_{1n}x_n = b_1$$

$$a_{22}'x_2 + \ldots + a_{2n}'x_n = b_2'$$

....

....

$$a_{nn}'''x_n = b_n'''$$

Now, starting from the last equation of these equations, we solve for $x_n = \dfrac{b_n'''}{a_{nn}'''}$

Using this in the penultimate equation, we get x_{n-1} and so on. By this back substitution, we get $x_n, x_{n-1}, \ldots, x_2, x_1$.

ILLUSTRATIVE EXAMPLES

Example 1. Solve the following system of equations by Cramer's rule

$$2x + y = 4$$

$$x + 3y = 7$$

Solution: Here, $a_{11} = 2, a_{12} = 1, a_{21} = 1, a_{22} = 3, b_1 = 4$ and $b_2 = 7$

Now, $$D = \begin{vmatrix} a_{11} & a_{12} \\ a_{21} & a_{22} \end{vmatrix} = \begin{vmatrix} 2 & 1 \\ 1 & 3 \end{vmatrix} = 6 - 1 = 5 \neq 0$$

$$D_1 = \begin{vmatrix} b_1 & a_{12} \\ b_2 & a_{22} \end{vmatrix} = \begin{vmatrix} 4 & 1 \\ 7 & 3 \end{vmatrix} = 12 - 7 = 5$$

$$D_2 = \begin{vmatrix} a_{11} & b_1 \\ a_{21} & b_2 \end{vmatrix} = \begin{vmatrix} 2 & 4 \\ 1 & 7 \end{vmatrix} = 14 - 4 = 10$$

By Cramer's rule, we have

$$x = \frac{D_1}{D} = \frac{5}{5} = 1$$

and $$y = \frac{D_2}{D} = \frac{10}{5} = 2$$

Hence, $x = 1, y = 2$ **Ans.**

Example 2. Show that the following system of equations is inconsistent:

$$x + 2y = 3$$
$$4x + 8y = 13$$

Solution: Here, $D = \begin{vmatrix} 1 & 2 \\ 4 & 8 \end{vmatrix} = 8 - 8 = 0$

and $$D_1 = \begin{vmatrix} 3 & 2 \\ 13 & 8 \end{vmatrix} = 24 - 26 = -2 \neq 0$$

$\because$ $D = 0$ and $D_1 = \neq = 0$

$\therefore$ The given system of equations is inconsistent.

Example 3. Solve the following system of equations by Cramer's rule:

$$x + 3y = 5$$
$$2x + 6y = 10$$

Solution: Here,

$$D = \begin{vmatrix} 1 & 3 \\ 12 & 6 \end{vmatrix} = 6 - 6 = 0$$

and $$D_1 = \begin{vmatrix} 5 & 3 \\ 10 & 6 \end{vmatrix} = 30 - 30 = 0$$

$\because$ $D = 0$ and $D_1 = 0$

$\therefore$ The given system of equations has an infinite number of solutions.

Let $y = k$, we have

$$x + 3k = 5 \text{ or } x = 5 - 3k$$

Hence, $x = 5 - 3k, y = k$ (k is arbitrary constant) **Ans.**

Example 4. Solve the following system of equations by Cramer's rule:

$$x + y + z = 6$$
$$x - y + z = 2$$
$$2x + y - z = 1$$

Solution: Here,

$$D = \begin{vmatrix} 1 & 1 & 1 \\ 1 & -1 & 1 \\ 2 & 1 & -1 \end{vmatrix} = 1(1-1) - 1(-1-2) + 1(1+2) = 6 \neq 0$$

$$D_1 = \begin{vmatrix} 6 & 1 & 1 \\ 2 & -1 & 1 \\ 1 & 1 & -1 \end{vmatrix} = 6(1-1) - 1(-2-1) + 1(2+1) = 6$$

$$D_2 = \begin{vmatrix} 1 & 6 & 1 \\ 1 & 2 & 1 \\ 2 & 1 & -1 \end{vmatrix} = 1(-2-1) - 6(-1-2) + 1(1-4) = 12$$

$$D_3 = \begin{vmatrix} 1 & 1 & 6 \\ 1 & -1 & 2 \\ 2 & 1 & 1 \end{vmatrix} = 1(-1-2) - 1(1-4) + 6(1+2) = 18$$

By Cramer's rule, we have

$$x = \frac{D_1}{D} = \frac{6}{6} = 1$$

$$y = \frac{D_2}{D} = \frac{12}{6} = 2$$

and $$z = \frac{D_3}{D} = \frac{18}{6} = 3$$

Hence, $x = 1, y = 2, z = 3$ **Ans.**

Example 5. Show that the following system of equations is inconsistent:

$$x + y + z = -3$$
$$3x + y - 2z = -2$$
$$2x + 4y + 7z = 7$$

Solution: Here,

$$D = \begin{vmatrix} 1 & 1 & 1 \\ 3 & 1 & -2 \\ 2 & 4 & 7 \end{vmatrix} = 1(7+8) - 1(21+4) + 1(12-2) = 0$$

$$D_1 = \begin{vmatrix} -3 & 1 & 1 \\ -2 & 1 & -2 \\ 7 & 4 & 7 \end{vmatrix} = -3(7+8) - 1(-14+14) + 1(-8-7) = -60 \neq 0$$

$\because$ $D = 0$ and $D_1 \neq 0$

$\therefore$ The given system of equations is inconsistent.

Example 6. Solve the following system of equations by Cramer's rule:

$$\begin{aligned} 2x-y+3z &= 1 \\ x+2y-z &= 2 \\ 5y-5z &= 3 \end{aligned}$$

Solution: Here,

$$D = \begin{vmatrix} 2 & -1 & 3 \\ 1 & 2 & -1 \\ 0 & 5 & -5 \end{vmatrix} = 2(-10+5)+1((-5-0)+3(5-0)=0$$

$$D_1 = \begin{vmatrix} 1 & -1 & 3 \\ 2 & 2 & -1 \\ 3 & 5 & -5 \end{vmatrix} = 1(-10+5)+1(-10+3)+3(10-6)=0$$

$$D_2 = \begin{vmatrix} 2 & 1 & 3 \\ 1 & 2 & -1 \\ 0 & 3 & -5 \end{vmatrix} = 2(-10+3)-1(-5-0)+3(3-0)=0$$

$$D_3 = \begin{vmatrix} 2 & -1 & 1 \\ 1 & 2 & 2 \\ 0 & 5 & 3 \end{vmatrix} = 2(6-10)+1(3-0)+1(5-0)=0$$

$\because \quad D = D_1 = D_2 = D_3 = 0$

$\therefore$ The given system of equations may be consistent or inconsistent.

Now consider the equations

$$\begin{aligned} 2x-y+3z &= 1 \\ x+2y-z &= 2 \end{aligned}$$

Let $z = k$, then we have

$$\begin{aligned} 2x-y &= 1-3k \\ x+2y &= 2+k \end{aligned}$$

Again,
$$D = \begin{vmatrix} 2 & -1 \\ 1 & 2 \end{vmatrix} = 4+1 = 5 \neq 0$$

$$D_1 = \begin{vmatrix} 1-3k & -1 \\ 2+k & 2 \end{vmatrix} = 2-6k+2+k = 4-5k$$

$$D_2 = \begin{vmatrix} 2 & 1-3k \\ 1 & 2+k \end{vmatrix} = 4+2k-1+3k = 3+5k$$

By Cramer's rule, we have

$$x = \frac{D_1}{D} = \frac{4-5k}{5}$$

$$y = \frac{D_2}{D} = \frac{3+5k}{5}$$

Hence,
$$x = \frac{4-5k}{5},\ y = \frac{3+5k}{5},\ z = k$$

Note that these values satisfies the third equation, i.e., $5y-5z=3$ of the given system. Hence the system is consistent and has infinitely many solutions. **Ans.**

Example 7: Solve the following system of equations by matrix method:

$$2x+y=4$$
$$x+3y=7$$

Solution: The above equations, in matrix form can be written as

$$\begin{bmatrix} 2 & 1 \\ 1 & 3 \end{bmatrix} \begin{bmatrix} x \\ y \end{bmatrix} = \begin{bmatrix} 4 \\ 7 \end{bmatrix}$$

$$AX = B,$$

where, $$A = \begin{bmatrix} 2 & 1 \\ 1 & 3 \end{bmatrix}, X = \begin{bmatrix} x \\ y \end{bmatrix} \text{ and } B = \begin{bmatrix} 4 \\ 7 \end{bmatrix}$$

Now, $$|A| = \begin{vmatrix} 2 & 1 \\ 1 & 3 \end{vmatrix} = 6-1=5 \neq 0$$

The co-factors of the elements of $|A|$ are as follows:

$$C_{11} = 3, C_{12} = -1$$
$$C_{21} = -1, C_{22} = 2$$

The co-factor matrix of $A = \begin{bmatrix} 3 & -1 \\ -1 & 2 \end{bmatrix}$

$\therefore$ $adj\, A$ = Transpose of co-factor matrix of $A = \begin{bmatrix} 3 & -1 \\ -1 & 2 \end{bmatrix}$

Now, $$A^{-1} = \frac{adj\, A}{|A|} = \frac{1}{5}\begin{bmatrix} 3 & -1 \\ -1 & 2 \end{bmatrix}$$

$\therefore$ $$X = A^{-1}B = \frac{1}{5}\begin{bmatrix} 3 & -1 \\ -1 & 2 \end{bmatrix}\begin{bmatrix} 4 \\ 7 \end{bmatrix}$$

$$= \frac{1}{5}\begin{bmatrix} 12-7 \\ -4+14 \end{bmatrix} = \frac{1}{5}\begin{bmatrix} 5 \\ 10 \end{bmatrix} = \begin{bmatrix} 1 \\ 2 \end{bmatrix}$$

$$\begin{bmatrix} x \\ y \end{bmatrix} = \begin{bmatrix} 1 \\ 2 \end{bmatrix}$$

Hence, $x = 1, y = 2$ **Ans.**

Example 8: Solve the following system of equations by matrix method:

$$5x+3y+7z=4$$
$$3x+26y+2z=9$$
$$7x+2y+11z=5$$

Solution: The above equations, in the matrix form can be written as

$$\begin{bmatrix} 5 & 3 & 7 \\ 3 & 26 & 2 \\ 7 & 2 & 11 \end{bmatrix} \begin{bmatrix} x \\ y \\ z \end{bmatrix} = \begin{bmatrix} 4 \\ 9 \\ 5 \end{bmatrix}$$

$$AX = B$$

$$\text{where, } A = \begin{bmatrix} 5 & 3 & 7 \\ 3 & 26 & 2 \\ 7 & 2 & 11 \end{bmatrix}, X = \begin{bmatrix} x \\ y \\ z \end{bmatrix} \text{ and } B = \begin{bmatrix} 4 \\ 9 \\ 5 \end{bmatrix}$$

$$\text{Now, } |A| = \begin{vmatrix} 5 & 3 & 7 \\ 3 & 26 & 2 \\ 7 & 2 & 11 \end{vmatrix} = 5(286-4) - 3(33-14) + 7(6-182)$$

$$= 121 \neq 0$$

$$\text{The co-factor matrix of } A = \begin{bmatrix} 282 & -19 & -176 \\ -19 & 6 & 11 \\ -176 & 11 & 121 \end{bmatrix}$$

$$\therefore \; adj\,A = \text{Transpose of co-factor of } A = \begin{bmatrix} 282 & -19 & -176 \\ -19 & 6 & 11 \\ -176 & 11 & 121 \end{bmatrix}$$

Now,
$$A^{-1} = \frac{adj\,A}{|A|} = \frac{1}{121}\begin{bmatrix} 282 & -19 & -176 \\ -19 & 6 & 11 \\ -176 & 11 & 121 \end{bmatrix}$$

$\therefore$
$$X = A^{-1}B = \frac{1}{121}\begin{bmatrix} 282 & -19 & -176 \\ -19 & 6 & 11 \\ -176 & 11 & 121 \end{bmatrix}\begin{bmatrix} 4 \\ 9 \\ 5 \end{bmatrix}$$

$$\begin{bmatrix} x \\ y \\ z \end{bmatrix} = \begin{bmatrix} 7/11 \\ 3/11 \\ 0 \end{bmatrix}$$

Hence,
$$x = \frac{7}{11}, y = \frac{3}{11}, z = 0$$
Ans.

Example 9: Solve the following system of equations by matrix method:

$$5x + 3y + z = 16$$
$$2x + y + 3z = 19$$
$$x + 2y + 4z = 25$$

Solution: The above equations, in the matrix form can be written as

$$\begin{bmatrix} 5 & 3 & 1 \\ 2 & 1 & 3 \\ 1 & 2 & 4 \end{bmatrix} \begin{bmatrix} x \\ y \\ z \end{bmatrix} = \begin{bmatrix} 16 \\ 19 \\ 25 \end{bmatrix}$$

$$AX = B$$

where, $A = \begin{bmatrix} 5 & 3 & 1 \\ 2 & 1 & 3 \\ 1 & 2 & 4 \end{bmatrix}, X = \begin{bmatrix} x \\ y \\ z \end{bmatrix}$ and $B = \begin{bmatrix} 16 \\ 19 \\ 25 \end{bmatrix}$

Now, $|A| = \begin{vmatrix} 5 & 3 & 1 \\ 2 & 1 & 3 \\ 1 & 2 & 4 \end{vmatrix} = 5(4-6) - 3(8-3) + 1(4-1) = -22 \neq 0$

The co-factor matrix of $A = \begin{bmatrix} -2 & -5 & 3 \\ -10 & 19 & -7 \\ 8 & -13 & -1 \end{bmatrix}$

$\therefore \qquad adj\ A = \begin{bmatrix} -2 & -10 & 8 \\ -5 & 19 & -13 \\ 3 & -7 & -1 \end{bmatrix}$

Now, $\qquad A^{-1} = \dfrac{adj\ A}{|A|} = -\dfrac{1}{22}\begin{bmatrix} -2 & -10 & 8 \\ -5 & 19 & -13 \\ 3 & -7 & -1 \end{bmatrix}$

$\therefore \qquad X = A^{-1}B = -\dfrac{1}{22}\begin{bmatrix} -2 & -10 & 8 \\ -5 & 19 & -13 \\ 3 & -7 & -1 \end{bmatrix}\begin{bmatrix} 16 \\ 19 \\ 25 \end{bmatrix}$

$$\begin{bmatrix} x \\ y \\ z \end{bmatrix} = \begin{bmatrix} 1 \\ 2 \\ 5 \end{bmatrix}$$

Hence, $\qquad x = 1, y = 2, z = 5$ **Ans.**

Example 10: Solve the following system of equations by matrix method:

$$4x + 2y + 3z = 2$$
$$x + y + z = 1$$
$$3x + y - 2z = 5$$

Solution: The above equations, in the matrix form can be written as

$$\begin{bmatrix} 4 & 2 & 3 \\ 1 & 1 & 1 \\ 3 & 1 & -2 \end{bmatrix}\begin{bmatrix} x \\ y \\ z \end{bmatrix} = \begin{bmatrix} 2 \\ 1 \\ 5 \end{bmatrix}$$

$$AX = B$$

where, $A = \begin{bmatrix} 4 & 2 & 3 \\ 1 & 1 & 1 \\ 3 & 1 & -2 \end{bmatrix}, X = \begin{bmatrix} x \\ y \\ z \end{bmatrix}$ and $B = \begin{bmatrix} 2 \\ 1 \\ 5 \end{bmatrix}$

$$\text{Now, } |A| = \begin{vmatrix} 4 & 2 & 3 \\ 1 & 1 & 1 \\ 3 & 1 & -2 \end{vmatrix} = 4(-2-1) - 2(-2-3) + 3(1-3) = -8 \neq 0$$

$$\text{The cofactor matrix of } A = \begin{bmatrix} -3 & 5 & -2 \\ 7 & -17 & 2 \\ -1 & -1 & 2 \end{bmatrix}$$

$$\therefore \qquad adj\ A = \begin{bmatrix} -3 & 7 & -1 \\ 5 & -17 & -1 \\ -2 & 2 & 2 \end{bmatrix}$$

$$\text{Now, } \qquad A^{-1} = \frac{adj\ A}{|A|} = -\frac{1}{8}\begin{bmatrix} -3 & 7 & -1 \\ 5 & -17 & -1 \\ -2 & 2 & 2 \end{bmatrix}$$

$$\therefore \qquad X = A^{-1}B = -\frac{1}{8}\begin{bmatrix} -3 & 7 & -1 \\ 5 & -17 & -1 \\ -2 & 2 & 2 \end{bmatrix}\begin{bmatrix} 2 \\ 1 \\ 5 \end{bmatrix}$$

$$\begin{bmatrix} x \\ y \\ z \end{bmatrix} = \begin{bmatrix} 1/2 \\ 3/2 \\ -1 \end{bmatrix}$$

$$\text{Hence, } \qquad x = \frac{1}{2}, y = \frac{3}{2}, z = -1$$ **Ans.**

Example 11. Solve the following system of equations by matrix method:

$$x + y = 0$$
$$y + z = 1$$
$$z + x = 3$$

Solution: The above equations, in the matrix form can be written as

$$\begin{bmatrix} 1 & 1 & 0 \\ 0 & 1 & 1 \\ 1 & 0 & 1 \end{bmatrix}\begin{bmatrix} x \\ y \\ z \end{bmatrix} = \begin{bmatrix} 0 \\ 1 \\ 3 \end{bmatrix}$$

$$AX = B$$

$$\text{where, } A = \begin{bmatrix} 1 & 1 & 0 \\ 0 & 1 & 1 \\ 1 & 0 & 1 \end{bmatrix}, X = \begin{bmatrix} x \\ y \\ z \end{bmatrix} \text{ and } B = \begin{bmatrix} 0 \\ 1 \\ 3 \end{bmatrix}$$

$$\text{Now, } |A| = \begin{vmatrix} 1 & 1 & 0 \\ 0 & 1 & 1 \\ 1 & 0 & 1 \end{vmatrix} = 1(1-0) - 1(0-1) + 0(0-1) = 2 \neq 0$$

$$\text{The co-factor matrix of } A = \begin{bmatrix} 1 & 1 & -1 \\ -1 & 1 & 1 \\ 1 & -1 & 1 \end{bmatrix}$$

$$\therefore \qquad adj\ A = \begin{bmatrix} 1 & -1 & 1 \\ 1 & 1 & -1 \\ -1 & 1 & 1 \end{bmatrix}$$

Now,
$$A^{-1} = \frac{adj\ A}{|A|} = \frac{1}{2}\begin{bmatrix} 1 & -1 & 1 \\ 1 & 1 & -1 \\ -1 & 1 & 1 \end{bmatrix}$$

$$\therefore \qquad X = A^{-1}B = \frac{1}{2}\begin{bmatrix} 1 & -1 & 1 \\ 1 & 1 & -1 \\ -1 & 1 & 1 \end{bmatrix}\begin{bmatrix} 0 \\ 1 \\ 3 \end{bmatrix}$$

$$\begin{bmatrix} x \\ y \\ z \end{bmatrix} = \begin{bmatrix} 1 \\ -1 \\ 2 \end{bmatrix}$$

Hence, $x = 1, y = -1, z = 2$ **Ans.**

Example 12. Solve the following system of equations by matrix method:

$$x - 2y + z = 1$$
$$-2x + y + z = 1$$
$$x + y - 2z = -2$$

Solution: The above equations in the matrix form can be written as

$$\begin{bmatrix} 1 & -2 & 1 \\ -2 & 1 & 1 \\ 1 & 1 & -2 \end{bmatrix}\begin{bmatrix} x \\ y \\ z \end{bmatrix} = \begin{bmatrix} 1 \\ 1 \\ -2 \end{bmatrix}$$

$$AX = B$$

$$\text{where, } A = \begin{bmatrix} 1 & -2 & 1 \\ -2 & 1 & 1 \\ 1 & 1 & -2 \end{bmatrix}, X = \begin{bmatrix} x \\ y \\ z \end{bmatrix} \text{ and } B = \begin{bmatrix} 1 \\ 1 \\ -2 \end{bmatrix}$$

$$\text{Now, } |A| = \begin{vmatrix} 1 & -2 & 1 \\ -2 & 1 & 1 \\ 1 & 1 & -2 \end{vmatrix} = 1(-2-1) + 2(4-1) + 1(-2-1) = 0$$

$\because \quad |A| = 0$

$\therefore$ The given system of equations may be consistent or inconsistent according as $(adj\ A)B = 0$ or $(adj\ A)B \neq 0$ respectively.

$$\text{The co-factor matrix of } A = \begin{bmatrix} -3 & -3 & -3 \\ -3 & -3 & -3 \\ -3 & -3 & -3 \end{bmatrix}$$

$$\therefore \qquad adj\ A = \begin{bmatrix} -3 & -3 & -3 \\ -3 & -3 & -3 \\ -3 & -3 & -3 \end{bmatrix}$$

$$\text{Now,} \qquad (adj\ A)\ B = \begin{bmatrix} -3 & -3 & -3 \\ -3 & -3 & -3 \\ -3 & -3 & -3 \end{bmatrix} \begin{bmatrix} 1 \\ 1 \\ -2 \end{bmatrix} = \begin{bmatrix} 0 \\ 0 \\ 0 \end{bmatrix}$$

$$\because \qquad (adj\ A)B = 0$$

$\therefore$ The given system of equations is consistent and has an infinite number of solutions.

Let $z = k$

Now the first two equations of the system can be written as

$$x - 2y = 1 - k$$
$$-2x + y = 1 - k$$

Now, these equations in the matrix form can be written as

$$\begin{bmatrix} 1 & -2 \\ -2 & 1 \end{bmatrix} \begin{bmatrix} x \\ y \end{bmatrix} = \begin{bmatrix} 1-k \\ 1-k \end{bmatrix}$$

$$AX = B$$

where $A = \begin{bmatrix} 1 & -2 \\ -2 & 1 \end{bmatrix}, X = \begin{bmatrix} x \\ y \end{bmatrix}$ and $B = \begin{bmatrix} 1-k \\ 1-k \end{bmatrix}$

$$\text{Now,} \qquad |A| = \begin{vmatrix} 1 & -2 \\ -2 & 1 \end{vmatrix} = 1 - 4 = -3 \neq 0$$

$$\text{The co-factor matrix of } A = \begin{bmatrix} 1 & 2 \\ 2 & 1 \end{bmatrix}$$

$$\therefore \qquad adj\ A = \begin{bmatrix} 1 & 2 \\ 2 & 1 \end{bmatrix}$$

$$\text{Now,} \qquad A^{-1} = \frac{adj\ A}{|A|} = -\frac{1}{3}\begin{bmatrix} 1 & 2 \\ 2 & 1 \end{bmatrix}$$

$$\therefore \qquad X = A^{-1}B = -\frac{1}{3}\begin{bmatrix} 1 & 2 \\ 2 & 1 \end{bmatrix} \begin{bmatrix} 1-k \\ 1-k \end{bmatrix}$$

$$\begin{bmatrix} x \\ y \end{bmatrix} = \begin{bmatrix} k-1 \\ k-1 \end{bmatrix}$$

or $\qquad x = k - 1, y = k - 1, z = k$ **Ans.**

Note that these values satisfies the third equation.

Example 13: Solve the following system of equations by Gauss elimination method:

$$x+2y+z=3$$
$$2x+3y+3z=10$$
$$3x-y+2z=13$$

Solution: The above equations in the matrix form can be written as

$$\begin{bmatrix} 1 & 2 & 1 \\ 2 & 3 & 3 \\ 3 & -1 & 2 \end{bmatrix} \begin{bmatrix} x \\ y \\ z \end{bmatrix} = \begin{bmatrix} 3 \\ 10 \\ 13 \end{bmatrix}$$

$$AX=B$$

So, $$[A \vdots B] = \begin{bmatrix} 1 & 2 & 1 & \vdots & 3 \\ 2 & 3 & 3 & \vdots & 10 \\ 3 & -1 & 2 & \vdots & 13 \end{bmatrix}$$

Now, we have to make the matrix A as upper triangular using elementary row transformations.

$$[A \vdots B] = \begin{bmatrix} 1 & 2 & 1 & \vdots & 3 \\ 2 & 3 & 3 & \vdots & 10 \\ 3 & -1 & 2 & \vdots & 13 \end{bmatrix}$$

$$\sim \begin{bmatrix} 1 & 2 & 1 & \vdots & 3 \\ 0 & -1 & 1 & \vdots & 4 \\ 0 & -7 & -1 & \vdots & 4 \end{bmatrix} \text{(by } R_2 \to R_2 - 2R_1, R_3 \to R_3 - 3R_1\text{)}$$

$$\sim \begin{bmatrix} 1 & 2 & 1 & \vdots & 3 \\ 0 & -1 & 1 & \vdots & 4 \\ 0 & 0 & -8 & \vdots & -24 \end{bmatrix} \text{(by } R_3 \to R_3 - 7R_2\text{)}$$

From this, we have

$$x+2y+z=3$$
$$-y+z=4$$
$$-8y=-24$$

By back substitution, we have

$$z=3, y=-1, x=2$$

Hence, $$x=2, y=-1, z=3$$ **Ans.**

Example 14. Solve the following system of equations by Gauss elimination method:

$$x+3y+2z=9$$
$$2x+4y+3z=13$$
$$3x+6y+z=16$$

Solution: The above equations in the matrix form can be written as

$$\begin{bmatrix} 1 & 3 & 2 \\ 2 & 4 & 3 \\ 3 & 6 & 1 \end{bmatrix}\begin{bmatrix} x \\ y \\ z \end{bmatrix} = \begin{bmatrix} 9 \\ 13 \\ 16 \end{bmatrix}$$

$$AX = B$$

So, $$[A \vdots B] = \begin{bmatrix} 1 & 3 & 2 & \vdots & 9 \\ 2 & 4 & 3 & \vdots & 13 \\ 3 & 6 & 1 & \vdots & 16 \end{bmatrix}$$

Now, we have to make the matrix A as upper triangular using elementary row transformations.

$$[A \vdots B] = \begin{bmatrix} 1 & 3 & 2 & \vdots & 9 \\ 2 & 4 & 3 & \vdots & 13 \\ 3 & 6 & 1 & \vdots & 16 \end{bmatrix}$$

$$\sim \begin{bmatrix} 1 & 3 & 2 & \vdots & 9 \\ 0 & -2 & -1 & \vdots & -5 \\ 0 & -3 & -5 & \vdots & -11 \end{bmatrix} \text{(by } R_2 \to R_2 - 2R_1, R_3 \to R_3 - R_1\text{)}$$

$$\sim \begin{bmatrix} 1 & 3 & 2 & \vdots & 9 \\ 0 & 1 & \frac{1}{2} & \vdots & \frac{5}{2} \\ 0 & -3 & -5 & \vdots & -11 \end{bmatrix} \left(\text{by } R_2 \to -\frac{1}{2}R_2\right)$$

$$\sim \begin{bmatrix} 1 & 3 & 2 & \vdots & 9 \\ 0 & 1 & \frac{1}{2} & \vdots & \frac{5}{2} \\ 0 & 0 & -\frac{7}{2} & \vdots & -\frac{7}{2} \end{bmatrix} \text{(by } R_3 \to R_3 + 3R_2\text{)}$$

From this, we have

$$x + 3y + 2z = 9$$

$$y + \frac{1}{2}z = \frac{5}{2}$$

$$-\frac{7}{2}z = -\frac{7}{2}$$

By back substitution, we have

$$z = 1, y = 2, x = 1$$

Hence, $$x = 1, y = 2, z = 1$$ **Ans.**

Example 15. Solve the following system of equations by Gauss elimination method:

$$2x - y + 3z = 8$$
$$-x + 2y + z = 4$$
$$3x + y - 4z = 0$$

Solution: The above equations in the matrix form can be written as

$$\begin{bmatrix} 2 & -1 & 3 \\ -1 & 2 & 1 \\ 3 & 1 & -4 \end{bmatrix} \begin{bmatrix} x \\ y \\ z \end{bmatrix} = \begin{bmatrix} 8 \\ 4 \\ 0 \end{bmatrix}$$

$$AX = B$$

So,

$$[A \vdots B] = \begin{bmatrix} 2 & -1 & 3 & : & 8 \\ -1 & 2 & 1 & : & 4 \\ 3 & 1 & -4 & : & 0 \end{bmatrix}$$

Now, we have to make the matrix A as upper triangular using elementary row transformations.

$$[A \vdots B] = \begin{bmatrix} 2 & -1 & 3 & : & 8 \\ -1 & 2 & 1 & : & 4 \\ 3 & 1 & -4 & : & 0 \end{bmatrix}$$

$$\sim \begin{bmatrix} -1 & 2 & 1 & : & 4 \\ 2 & -1 & 3 & : & 8 \\ 3 & 1 & -4 & : & 0 \end{bmatrix} \text{(by } R_1 \leftrightarrow R_2\text{)}$$

$$\sim \begin{bmatrix} 1 & -2 & -1 & : & -4 \\ 2 & -1 & 3 & : & 8 \\ 3 & 1 & -4 & : & 0 \end{bmatrix} \text{(by } R_1 \to -R_1\text{)}$$

$$\sim \begin{bmatrix} 1 & -2 & -1 & : & -4 \\ 0 & 3 & 5 & : & 16 \\ 0 & 7 & -1 & : & 12 \end{bmatrix} \text{(by } R_2 \to R_2 - 2R_1, R_3 \to R_3 - 3R_1\text{)}$$

$$\sim \begin{bmatrix} 1 & -2 & -1 & : & -4 \\ 0 & 1 & \frac{5}{3} & : & \frac{16}{3} \\ 0 & 7 & -1 & : & 12 \end{bmatrix} \left(by\ R_2 \to \frac{1}{3} R2 \right)$$

$$\sim \begin{bmatrix} 1 & -2 & -1 & : & -4 \\ 0 & 1 & \frac{5}{3} & : & \frac{16}{3} \\ 0 & 0 & -\frac{38}{3} & : & -\frac{76}{3} \end{bmatrix} \text{(by } R_3 \to R_3 - 7R_2\text{)}$$

From this, we have

$$\begin{aligned} x - 2y - z &= -4 \\ y + \frac{5}{3}z &= \frac{16}{3} \\ -\frac{38}{3}z &= -\frac{76}{3} \end{aligned}$$

By back substitution, we have

$$z = 2, y = 2, x = 2$$

Hence, $x = 2, y = 2, z = 2$ **Ans.**

Example 16. Solve the following system of equations by Gauss elimination method:

$$\begin{aligned} x - 2y + z &= 1 \\ -2x + y + z &= 1 \\ x + y - 2z &= -2 \end{aligned}$$

Solution: The above equations in the matrix form can be written as

$$\begin{bmatrix} 1 & -2 & 1 \\ -2 & 1 & 1 \\ 1 & 1 & -2 \end{bmatrix} \begin{bmatrix} x \\ y \\ z \end{bmatrix} = \begin{bmatrix} 1 \\ 1 \\ -2 \end{bmatrix}$$

$$AX = B$$

So,

$$[A \vdots B] = \begin{bmatrix} 1 & -2 & 1 & : & 1 \\ -2 & 1 & 1 & : & 1 \\ 1 & 1 & -2 & : & -2 \end{bmatrix}$$

Now, we have to make the matrix A as upper triangular using elementary row transformations.

$$[A \vdots B] = \begin{bmatrix} 1 & -2 & 1 & : & 1 \\ -2 & 1 & 1 & : & 1 \\ 1 & 1 & -2 & : & -2 \end{bmatrix}$$

$$\sim \begin{bmatrix} 1 & -2 & 1 & : & 1 \\ 0 & -3 & 3 & : & 3 \\ 0 & 3 & -3 & : & -3 \end{bmatrix} \text{(by } R_2 \to R_2 + 2R_1, R_3 \to R_3 - R_1\text{)}$$

$$\sim \begin{bmatrix} 1 & -2 & 1 & : & 1 \\ 0 & -3 & 3 & : & 3 \\ 0 & 0 & 0 & : & 0 \end{bmatrix} \text{(by } R_3 \to R_3 + R_2\text{)}$$

Form this, we have

$$\begin{aligned} x - 2y + z &= 1 \\ -3y + 3z &= 3 \end{aligned}$$

Let $z = k.$

By back substitution, we have

$$z = k, y = k - 1, x = k - 1$$

Hence, $x = k - 1, y = k - 1, z = k$ **Ans.**

Example 17. Solve the following system of equations by Gauss elimination method:

$$x + 2y - z = 1$$
$$3x - 2y + 2z = 2$$
$$7x - 2y + 3z = 5$$

Solution: The above equations in the matrix form can be written as

$$\begin{bmatrix} 1 & 2 & -1 \\ 3 & -2 & 2 \\ 7 & -2 & 3 \end{bmatrix} \begin{bmatrix} x \\ y \\ z \end{bmatrix} = \begin{bmatrix} 1 \\ 2 \\ 5 \end{bmatrix}$$

$$AX = B$$

So, $$[A \vdots B] = \begin{bmatrix} 1 & 2 & -1 & : & 1 \\ 3 & -2 & 2 & : & 2 \\ 7 & -2 & 3 & : & 5 \end{bmatrix}$$

Now, we have to make the matrix A as upper triangular using elementary row transformations.

$$[A \vdots B] = \begin{bmatrix} 1 & 2 & -1 & : & 1 \\ 3 & -2 & 2 & : & 2 \\ 7 & -2 & 3 & : & 5 \end{bmatrix}$$

$$\sim \begin{bmatrix} 1 & 2 & -1 & : & 1 \\ 0 & -8 & 5 & : & -1 \\ 0 & -16 & 10 & : & -2 \end{bmatrix} \text{(by } R_2 \to R_2 - 3R_1, R_3 \to R_3 - 7R_1\text{)}$$

$$\sim \begin{bmatrix} 1 & 2 & -1 & : & 1 \\ 0 & -8 & 5 & : & -1 \\ 0 & 0 & 0 & : & 0 \end{bmatrix} \text{(by } R_2 \to R_3 - 2R_2\text{)}$$

From this, we have

$$x + 2y - z = 1$$
$$-8y + 5z = -1$$

Let $z = k$

By back substitution, we have

$$z = k, y = \frac{5k + 1}{8}, x = \frac{-k + 3}{4}$$

Hence, $x = \frac{-k + 3}{4}, y = \frac{5k + 1}{8}, z = k$ **Ans.**

Example 18: Solve the following system of equations by Gauss elimination method:

$$\begin{aligned} x+y+z &= 6 \\ x+2y+3z &= 14 \\ x+4y+7z &= 30 \end{aligned}$$

Solution: The above equations in the matrix form can be written as

$$\begin{bmatrix} 1 & 1 & 1 \\ 1 & 2 & 3 \\ 1 & 4 & 7 \end{bmatrix} \begin{bmatrix} x \\ y \\ z \end{bmatrix} = \begin{bmatrix} 6 \\ 14 \\ 30 \end{bmatrix}$$

$$AX = B$$

So, $$[A \vdots B] = \begin{bmatrix} 1 & 1 & 1 & \vdots & 6 \\ 1 & 2 & 3 & \vdots & 14 \\ 1 & 4 & 7 & \vdots & 30 \end{bmatrix}$$

Now, we have to make the matrix A as upper triangular using elementary row transformations.

$$[A \vdots B] = \begin{bmatrix} 1 & 1 & 1 & : & 6 \\ 1 & 2 & 3 & : & 14 \\ 1 & 4 & 7 & : & 30 \end{bmatrix}$$

$$\sim \begin{bmatrix} 1 & 1 & 1 & : & 6 \\ 0 & 1 & 2 & : & 8 \\ 0 & 3 & 6 & : & 24 \end{bmatrix} \text{(by } R_2 \to R_2 - R_1, R_3 \to R_3 - R_1\text{)}$$

$$\sim \begin{bmatrix} 1 & 1 & 1 & : & 6 \\ 0 & 1 & 2 & : & 8 \\ 0 & 0 & 0 & : & 0 \end{bmatrix} \text{(by } R_3 \to R_3 - 3R_2\text{)}$$

From this, we have

$$\begin{aligned} x+y+z &= 6 \\ y+2z &= 8 \end{aligned}$$

Let $z = k$

By back substitution, we have

$$z = k, y = 8-2k, x = -2+k$$ **Ans.**

Hence, $$x = -2+k, y = 8-2k, z = k$$

EXERCISE 2.6

1. Solve the following system of equations by Cramer's rule:

$$\begin{aligned} x-2y &= 4 \\ -3x+5y &= -7 \end{aligned}$$

2. Solve the following system of equations by Cramer's rule:

$$\begin{aligned} 7x+2y &= 25 \\ 2x-y &= 4 \end{aligned}$$

3. Show that the following system of equations is inconsistent:

$$\begin{aligned} 2x - y &= 5 \\ 6x - 3y &= 11 \end{aligned}$$

4. Solve the following system of equations by Cramer's rule:

$$\begin{aligned} 3x - 5y &= 6 \\ 6x - 10y &= 12 \end{aligned}$$

5. Solve the following system of equations by Cramer's rule:

$$\begin{aligned} x + 2y + 3z &= 6 \\ 2x + 4y + z &= 7 \\ 3x + 2y + 9z &= 14 \end{aligned}$$

6. Show that the following system of equations is inconsistent:

$$\begin{aligned} 4x + y + z &= 1 \\ x + 2y - z &= 1 \\ 3x - y + 2z &= 6 \end{aligned}$$

7. Solve the following system of equations by Cramer's rule:

$$\begin{aligned} 2x + y - 2z &= 4 \\ x - 2y + z &= -2 \\ 5x - 5y + z &= -2 \end{aligned}$$

8. Solve the following system of equations by matrix method:

$$\begin{aligned} 3x + 7y &= 24 \\ 2x + 5y &= 17 \end{aligned}$$

9. Solve the following system of equations by matrix method:

$$\begin{aligned} x + y + z &= 4 \\ 2x - y + 2z &= 5 \\ x - 2y - z &= -3 \end{aligned}$$

10. Solve the following system of equations by matrix method:

$$\begin{aligned} x + y + z &= 3 \\ x + 2y + z &= 4 \\ x + y + 2z &= 5 \end{aligned}$$

11. Solve the following system of equations by matrix method:

$$\begin{aligned} 2x - y + 3z &= 1 \\ x + 2y - z &= 2 \\ 5y - 5z &= 3 \end{aligned}$$

12. Solve the following system of equations by matrix method:

$$\begin{aligned} x - y + 2z &= 4 \\ 3x + y + 4z &= 6 \\ x + y + z &= 1 \end{aligned}$$

13. Solve the following system of equations by Gauss elimination method:

$$2x + 3y + 4z = 11$$

$$x + 5y + 7z = 15$$
$$3x + 11y + 13z = 25$$

14. Solve the following system of equations by Gauss elimination method:

$$x - y + 2z = 3$$
$$x + 2y + 3z = 5$$
$$3x - 4y - 5z = -13$$

15. Solve the following system of equations by Gauss elimination method:

$$2x - y + 3z = 9$$
$$x + y + z = 6$$
$$x - y + z = 2$$

16. Solve the following system of equations by Gauss elimination method:

$$x - 2y - z = 5$$
$$x + 8y - 3z = -1$$
$$2x + y - 3z = 7$$

17. Solve the following system of equations by Gauss elimination method:

$$5x + 3y + 7z = 4$$
$$3x + 26y + 2z = 9$$
$$7x + 2y + 10z = 5$$

ANSWERS

1. $x = -6, y = 5$

2. $x = 3, y = 2$

4. $x = \dfrac{6 + 5k}{3}, y = k$

5. $x = 1, z = 1$

7. $x = \dfrac{6 + 5k}{5}, y = \dfrac{8 + 4k}{5}, z = k$

8. $x = 1, y = 3$

9. $x = 1, y = 1, z = 2$

10. $x = 0, y = 1, z = 2$

11. $x = \dfrac{4 + 5k}{5}, y = \dfrac{3 + 5k}{5}, z = k$

12. $x = \dfrac{5 - 3k}{2}, y = \dfrac{-3 + k}{2}, z = k$

13. $x = 2, y = -3, z = 4$

14. $x = -1, y = 0, z = 2$

15. $x = 1, y = 2, z = 3$

16. $x = \dfrac{19 + 7k}{5}, y = \dfrac{-3 + k}{5}, z = k$

17. $x = \dfrac{7 - 16k}{11}, y = \dfrac{3 + k}{11}, z = k$

2.30 EIGEN VALUES AND EIGEN VECTORS

Characteristic Matrix: Let $A = [a_{ij}]$ be an square matrix of the type $n \times n$. The matrix $[A - \lambda I]$ is called the characteristic matrix, where λ is a scalar and I is the identity (unit) matrix.

e.g.,
$$A = \begin{bmatrix} 2 & 1 & 1 \\ 0 & 1 & 0 \\ 1 & 1 & 2 \end{bmatrix}$$

$$[A - \lambda I] = \begin{bmatrix} 2 & 1 & 1 \\ 0 & 1 & 0 \\ 1 & 1 & 2 \end{bmatrix} - \lambda \begin{bmatrix} 1 & 0 & 0 \\ 0 & 1 & 0 \\ 0 & 0 & 1 \end{bmatrix} = \begin{bmatrix} 2-\lambda & 1 & 1 \\ 0 & 1-\lambda & 0 \\ 1 & 1 & 2-\lambda \end{bmatrix}$$

Characteristic Polynomial: The determinant $|A - \lambda I|$ when expanded will give a polynomial, is called a characteristic polynomial of matrix A.

e.g.,
$$\begin{vmatrix} 2-\lambda & 1 & 1 \\ 0 & 1-\lambda & 0 \\ 1 & 1 & 2-\lambda \end{vmatrix} = \lambda^3 - 5\lambda^2 + 7\lambda - 3$$

This is a characteristic polynomial of matrix A.

Characteristic Equation: The equation $|\mathrm{A} - \lambda I| = 0$ is called the characteristic equation of the matrix A.

e.g., $\lambda^3 - 5\lambda^2 + 7\lambda - 3 = 0$

Characteristic Roots or Eigen Values: The roots of the characteristic equation $|A - \lambda I| = 0$ are called characteristic roots or eigen values or latent roots of the matrix A.

e.g.,
$$\begin{aligned} \lambda^3 - 5\lambda^2 + 7\lambda - 3 &= 0 \\ (\lambda - 1)(\lambda^2 - 4\lambda + 3) &= 0 \\ (\lambda - 1)(\lambda - 1)(\lambda - 3) &= 0 \\ \lambda &= 1, 1, 3 \end{aligned}$$

The eigen values of the matrix A are 1, 1 and 3.

Spectrum of A: The set of all eigen values of the matrix A is called the spectrum of A.

Trace of A: The sum of the elements of the principal diagonal of a matrix is called the trace of A.

Properties of Eigen Values:

(1) The sum of the eigen values of a matrix is equal to the trace of the matrix.

(2) The product of the eigen values of a matrix is equal to the value of determinant of that matrix.

(3) Any matrix and its transpose both have the same eigen values.

Characteristic Vectors or Eigen Vectors: If λ is a characteristic root or an eigen value of a square matrix A of $n \times n$, then a non-zero vector X satisfying the equation

$$[a - \lambda I]X = 0,$$

is called a characteristic vector or eigen vector of A corresponding to the characteristic root or an eigen value λ.

Note: The eigen vector corresponding to an eigen value is not unique.

ILLUSTRATIVE EXAMPLES

Example 1. Find the eigen values of the matrix $\begin{bmatrix} 5 & 4 \\ 1 & 2 \end{bmatrix}$.

Solution: Let $A = \begin{bmatrix} 5 & 4 \\ 1 & 2 \end{bmatrix}$

The characteristic equation of the matrix A is

$$|A - \lambda I| = 0$$

$$\begin{vmatrix} 5-\lambda & 4 \\ 1 & 2-\lambda \end{vmatrix} = 0$$

$$(5-\lambda)(2-\lambda) - 4 = 0$$

$$\lambda^2 - 7\lambda + 6 = 0$$

$$(\lambda - 1)(\lambda - 6) = 0$$

$$\lambda = 1, 6$$

Hence, the eigen values of the given matrix are 1 and 6. **Ans.**

Example 2. Find the eigen values of the matrix $\begin{bmatrix} 1 & 2 & 3 \\ 0 & -4 & 2 \\ 0 & 0 & 7 \end{bmatrix}$.

Solution: Let $A = \begin{bmatrix} 1 & 2 & 3 \\ 0 & -4 & 2 \\ 0 & 0 & 7 \end{bmatrix}$

The characteristic equation of the matrix A is

$$|A - \lambda I| = 0$$

$$\begin{vmatrix} 1-\lambda & 2 & 3 \\ 0 & -4-\lambda & 2 \\ 0 & 0 & 7-\lambda \end{vmatrix} = 0$$

$$(1-\lambda)(-4-\lambda)(7-\lambda) = 0$$

$$\lambda = -4, 1, 7$$

Hence, the eigen values of the given matrix are –4, 1, and 7. **Ans.**

Example 3. Find the eigen values and eigen vectors of the matrix $\begin{bmatrix} 4 & -5 \\ 1 & -2 \end{bmatrix}$.

Solution: Let $A = \begin{bmatrix} 4 & -5 \\ 1 & -2 \end{bmatrix}$

The characteristic equation of the matrix A is

$$|A - \lambda I| = 0$$

$$\begin{vmatrix} 4-\lambda & -5 \\ 1 & -2-\lambda \end{vmatrix} = 0$$

$$(4-\lambda)(-2-\lambda) + 5 = 0$$

$$\lambda^2 - 2\lambda - 3 = 0$$

$$(\lambda + 1)(\lambda - 3) = 0$$

$$\lambda = -1, 3$$

Hence, the eigen values of the given matrix are –1 and 3.

Now the equation $[A - \lambda I]X = 0$, for the matrix A is

$$\begin{bmatrix} 4-\lambda & -5 \\ 1 & -2-\lambda \end{bmatrix} \begin{bmatrix} x \\ y \end{bmatrix} = \begin{bmatrix} 0 \\ 0 \end{bmatrix}$$

For $\lambda = -1$, we have

$$\begin{bmatrix} 5 & -5 \\ 1 & -1 \end{bmatrix} \begin{bmatrix} x \\ y \end{bmatrix} = \begin{bmatrix} 0 \\ 0 \end{bmatrix}$$

$$5x - 5y = 0$$

$$x - y = 0$$

or $$\frac{x}{1} = \frac{y}{1} = k. \text{ (say)}$$

Hence, the eigen vector corresponding to $\lambda = -1$ is given by

$$X_1 = \begin{bmatrix} k \\ k \end{bmatrix} \text{ or } k\begin{bmatrix} 1 \\ 1 \end{bmatrix} \text{ or } \begin{bmatrix} 1 \\ 1 \end{bmatrix}$$

For $\lambda = 3$, we have

$$\begin{bmatrix} 1 & -5 \\ 1 & -5 \end{bmatrix} \begin{vmatrix} x \\ y \end{vmatrix} = \begin{vmatrix} 0 \\ 0 \end{vmatrix}$$

$$x - 5y = 0$$

$$x - 5y = 0$$

or $$\frac{x}{5} = \frac{y}{1} = k \text{ (say)}$$

Hence, the eigen vector corresponding to $\lambda = 3$ is given by

$$X_2 = \begin{bmatrix} 5k \\ k \end{bmatrix} \text{ or } k\begin{bmatrix} 5 \\ 1 \end{bmatrix} \text{ or } \begin{bmatrix} 5 \\ 1 \end{bmatrix}$$ **Ans.**

Example 4. Find the eigen values and eigen vectors of the matrix $\begin{bmatrix} 1 & 2 & -2 \\ 1 & 2 & 1 \\ -1 & -1 & 0 \end{bmatrix}$.

Solution: Let $A = \begin{bmatrix} 1 & 2 & -2 \\ 1 & 2 & 1 \\ -1 & -1 & 0 \end{bmatrix}$

The characteristic equation of the matrix A is

$$|A - \lambda I| = 0$$

$$\begin{vmatrix} 1-\lambda & 2 & -2 \\ 1 & 2-\lambda & 1 \\ -1 & -1 & 0-\lambda \end{vmatrix} = 0$$

$$(1-\lambda)[(2-\lambda)(-\lambda)+1] - 2(-\lambda+1) - 2(-1+2-\lambda) = 0$$

$$(1-\lambda)(\lambda^2 - 2\lambda - 3) = 0$$

$$(1-\lambda)(\lambda+1)(\lambda-3) = 0$$

$$\lambda = -1, 1, 3$$

Hence, the eigen values of the given matrix are –1, 1, and 3.

Now the equation $[A - \lambda I]X = 0$, for the matrix A is

$$\begin{bmatrix} 1-\lambda & 2 & -2 \\ 1 & 2-\lambda & 1 \\ -1 & -1 & -\lambda \end{bmatrix} \begin{bmatrix} x \\ y \\ z \end{bmatrix} = \begin{bmatrix} 0 \\ 0 \\ 0 \end{bmatrix}$$

For $\lambda = -1$, we have

$$\begin{bmatrix} 2 & 2 & -2 \\ 1 & 3 & 1 \\ -1 & -1 & 1 \end{bmatrix} \begin{bmatrix} x \\ y \\ z \end{bmatrix} = \begin{bmatrix} 0 \\ 0 \\ 0 \end{bmatrix}$$

$$2x + 2y - 2z = 0$$

$$x + 3y + z = 0$$

$$-x - y + z = 0$$

On solving first two equations

$$\frac{x}{2+6} = \frac{y}{-2-2} = \frac{z}{6-2}$$

$$\frac{x}{8} = \frac{y}{-4} = \frac{z}{4}$$

$$\frac{x}{2} = \frac{y}{-1} = \frac{z}{1} = k \text{ (say)}$$

Hence, the eigen vector corresponding to $\lambda = -1$ is given by

$$X_1 = \begin{bmatrix} 2k \\ -k \\ k \end{bmatrix} \text{ or } k\begin{bmatrix} 2 \\ -1 \\ 1 \end{bmatrix} \text{ or } \begin{bmatrix} 2 \\ -1 \\ 1 \end{bmatrix}$$

For $\lambda = 1$, we have

$$\begin{bmatrix} 0 & 2 & -2 \\ 1 & 1 & 1 \\ -1 & -1 & -1 \end{bmatrix} \begin{bmatrix} x \\ y \\ z \end{bmatrix} = \begin{bmatrix} 0 \\ 0 \\ 0 \end{bmatrix}$$

$$0 \cdot x + 2y - 2z = 0$$

$$x + y + z = 0$$

On solving these equations we have

$$\frac{x}{2+2} = \frac{y}{-2-0} = \frac{z}{0-2}$$

$$\frac{x}{4} = \frac{y}{-2} = \frac{z}{-2}$$

$$\frac{x}{2} = \frac{y}{-1} = \frac{z}{-1} = k \text{ (say)}$$

Hence, the eigen vector corresponding to $\lambda = 1$ is given by

$$X_2 = \begin{bmatrix} -2k \\ -k \\ -k \end{bmatrix} \text{ or } k \begin{bmatrix} 2 \\ -1 \\ -1 \end{bmatrix} \text{ or } \begin{bmatrix} 2 \\ -1 \\ -1 \end{bmatrix}$$

For $\lambda = 3$, we have

$$\begin{bmatrix} -2 & 2 & -2 \\ 1 & -1 & 1 \\ -1 & -1 & -3 \end{bmatrix} \begin{bmatrix} x \\ y \\ z \end{bmatrix} = \begin{bmatrix} 0 \\ 0 \\ 0 \end{bmatrix}$$

$$-2x + 2y - 2z = 0$$

$$-x - y - 3z = 0$$

or

$$2x - 2y + 2z = 0$$

$$x + y + 3z = 0$$

On solving these equations we have

$$\frac{x}{-6-2} = \frac{y}{2-6} = \frac{z}{2+2}$$

$$\frac{x}{-8} = \frac{y}{-4} = \frac{z}{4}$$

$$\frac{x}{-2} = \frac{y}{-1} = \frac{z}{1} = k \text{ (say)}$$

Hence, the eigen vector corresponding to $\lambda = 3$ is given by

$$X_3 = \begin{bmatrix} -2k \\ -k \\ k \end{bmatrix} \text{ or } k \begin{bmatrix} -2 \\ -1 \\ 1 \end{bmatrix} \text{ or } \begin{bmatrix} -2 \\ -1 \\ 1 \end{bmatrix} \text{ or } \begin{bmatrix} 2 \\ 1 \\ -1 \end{bmatrix}$$ **Ans.**

Example 5. Find the eigen values and eigen vectors or the matrix $\begin{bmatrix} 3 & 1 & 4 \\ 0 & 2 & 6 \\ 0 & 0 & 5 \end{bmatrix}$.

Solution: Let $A = \begin{bmatrix} 3 & 1 & 4 \\ 0 & 2 & 6 \\ 0 & 0 & 5 \end{bmatrix}$

The characteristic equation of the matrix A is

$$|A - \lambda I| = 0$$

$$\begin{vmatrix} 3-\lambda & 1 & 4 \\ 0 & 2-\lambda & 6 \\ 0 & 0 & 5-\lambda \end{vmatrix} \begin{bmatrix} x \\ y \\ z \end{bmatrix} = 0$$

$$(3-\lambda)(2-\lambda)(5-\lambda) = 0$$

$$\lambda = 2, 3, 5$$

Hence, the eigen values of the given matrix are 2, 3 and 5. Now the equation $[A - \lambda I]X = 0$, for the matrix A is

$$\begin{bmatrix} 3-\lambda & 1 & 4 \\ 0 & 2-\lambda & 6 \\ 0 & 0 & 5-\lambda \end{bmatrix} \begin{bmatrix} x \\ y \\ z \end{bmatrix} = \begin{bmatrix} 0 \\ 0 \\ 0 \end{bmatrix}$$

For $\lambda = 2$, we have

$$\begin{bmatrix} 1 & 1 & 4 \\ 0 & 0 & 6 \\ 0 & 0 & 3 \end{bmatrix} \begin{bmatrix} x \\ y \\ z \end{bmatrix} = \begin{bmatrix} 0 \\ 0 \\ 0 \end{bmatrix}$$

$$x + y + 4z = 0$$

$$0x + 0y + z = 0$$

On solving these equations we have

$$\frac{x}{1-0} = \frac{y}{0-1} = \frac{z}{0-0}$$

$$\frac{x}{1} = \frac{y}{-1} = \frac{z}{0} = k \text{ (say)}$$

Hence, the eigen vector corresponding $\lambda = 2$, is given by

$$X_1 = \begin{bmatrix} k \\ -k \\ 0 \end{bmatrix} \text{ or } k \begin{bmatrix} 1 \\ -1 \\ 0 \end{bmatrix} \text{ or } \begin{bmatrix} 1 \\ -1 \\ 0 \end{bmatrix}$$

For $\lambda = 3$, we have

$$\begin{bmatrix} 0 & 1 & 4 \\ 0 & -1 & 6 \\ 0 & 0 & 2 \end{bmatrix} \begin{bmatrix} x \\ y \\ z \end{bmatrix} = \begin{bmatrix} 0 \\ 0 \\ 0 \end{bmatrix}$$

$$0{\cdot}x + y + 4z = 0$$
$$0{\cdot}x - y + 6z = 0$$
$$0{\cdot}x + 0{\cdot}y + 2z = 0$$

On solving first two equations we have

$$\frac{x}{6+4} = \frac{y}{0-0} = \frac{z}{0-0}$$

$$\frac{x}{10} = \frac{y}{0} = \frac{z}{0} = k \text{ (say)}$$

Hence, the eigen vector corresponding to $\lambda = 3$ is given by

$$X_2 = \begin{bmatrix} k \\ 0 \\ 0 \end{bmatrix} \text{ or } k\begin{bmatrix} 1 \\ 0 \\ 0 \end{bmatrix} \text{ or } \begin{bmatrix} 1 \\ 0 \\ 0 \end{bmatrix}$$

For $\lambda = 5$, we have

$$\begin{bmatrix} -2 & 1 & 4 \\ 0 & -3 & 6 \\ 0 & 0 & 0 \end{bmatrix} \begin{bmatrix} x \\ y \\ z \end{bmatrix} = \begin{bmatrix} 0 \\ 0 \\ 0 \end{bmatrix}$$

$$-2x + y + 4z = 0$$
$$0{\cdot}x - 3y + 6z = 0$$

On solving these equations we have

$$\frac{x}{6+12} = \frac{y}{0+12} = \frac{z}{6-0}$$

$$\frac{x}{18} = \frac{y}{12} = \frac{z}{6}$$

$$\frac{x}{3} = \frac{y}{2} = \frac{z}{1} = k \text{ (say)}$$

Hence, the eigen vector corresponding to $\lambda = 5$ is given by

$$X_3 = \begin{bmatrix} 3k \\ 2k \\ k \end{bmatrix} \text{ or } k\begin{bmatrix} 3 \\ 2 \\ 1 \end{bmatrix} \text{ or } \begin{bmatrix} 3 \\ 2 \\ 1 \end{bmatrix}$$

Ans.

Example 6: Find the eigen values and the eigen vectors of the matrix $\begin{bmatrix} 6 & -2 & 2 \\ -2 & 3 & -1 \\ 2 & -1 & 3 \end{bmatrix}$.

Solution: Let $A = \begin{bmatrix} 6 & -2 & 2 \\ -2 & 3 & -1 \\ 2 & -1 & 3 \end{bmatrix}$

The characteristic equation of the matrix A is

$$|A-\lambda I| = 0$$

$$\begin{vmatrix} 6-\lambda & -2 & 2 \\ -2 & 3-\lambda & -1 \\ 2 & -1 & 3-\lambda \end{vmatrix} = 0$$

$$\begin{vmatrix} 6-\lambda & -2 & 2 \\ -2 & 3-\lambda & -1 \\ 0 & 2-\lambda & 2-\lambda \end{vmatrix} = 0 \text{ (by } R_3 \to R_3 + R_2)$$

$$(2-\lambda)\begin{vmatrix} 6-\lambda & -2 & 2 \\ -2 & 3-\lambda & -1 \\ 0 & 1 & 1 \end{vmatrix} = 0$$

$$(2-\lambda)\begin{vmatrix} 6-\lambda & -2 & 0 \\ -2 & 3-\lambda & 2-\lambda \\ 0 & 1 & 2 \end{vmatrix} = 0 \text{ (by } C_3 \to C_3 + C_2)$$

$$(2-\lambda)\,[(6-\lambda)(6-2\lambda)+2(-4)] = 0$$

$$(2-\lambda)(\lambda^2+10\lambda+16) = 0$$

$$(2-\lambda)(\lambda-2)(\lambda-8) = 0$$

$$\lambda = 2, 2, 8$$

Hence, the eigen values of the given matrix are 2, 2 and 8.

Now the equation $[A-\lambda I]X = 0$, for the matrix A is

$$\begin{bmatrix} 6-\lambda & -2 & 2 \\ -2 & 3-\lambda & -1 \\ 2 & -1 & 3-\lambda \end{bmatrix}\begin{bmatrix} x \\ y \\ z \end{bmatrix} = \begin{bmatrix} 0 \\ 0 \\ 0 \end{bmatrix}$$

For $\lambda = 2$, we have

$$\begin{bmatrix} 4 & -2 & 2 \\ -2 & 1 & -1 \\ 2 & -1 & 1 \end{bmatrix}\begin{bmatrix} x \\ y \\ z \end{bmatrix} = \begin{bmatrix} 0 \\ 0 \\ 0 \end{bmatrix}$$

$$4x-2y+2z = 0$$

$$-2x+y-z = 0$$

$$2x-y+z = 0$$

Now these equations are same. So we have only one equation

$$2x-y+z = 0$$

Let $z = k_1$, $y = k_2$ then $x = \dfrac{k_2-k_1}{2}$

Hence, the eigen vector corresponding to $\lambda = 2$ is given by

$$X_1 = \begin{bmatrix} \dfrac{k_2 - k_1}{2} \\ k_2 \\ k_1 \end{bmatrix}$$

Let $x = k_1, y = k_2$ then $z = k_2 - 2k_1$

Hence, the second eigen vector corresponding to $\lambda = 2$ is given by

$$X_2 = \begin{bmatrix} k_1 \\ k_2 \\ \dfrac{k_2 - k_1}{2} \end{bmatrix}$$

For $\lambda = 8$, we have

$$\begin{bmatrix} -2 & -2 & 2 \\ -2 & -5 & -1 \\ 2 & -1 & -5 \end{bmatrix} \begin{bmatrix} x \\ y \\ z \end{bmatrix} = \begin{bmatrix} 0 \\ 0 \\ 0 \end{bmatrix}$$

$$-2x - 2y + 2z = 0$$

$$-2x - 5y - z = 0$$

$$2x - y - 5z = 0$$

On solving first two equations we have

$$\frac{x}{2 + 10} = \frac{y}{-4 - 2} = \frac{z}{10 - 4}$$

$$\frac{x}{12} = \frac{y}{-6} = \frac{z}{6}$$

$$\frac{x}{2} = \frac{y}{-1} = \frac{z}{1} = k \text{ (say)}$$

Hence, the eigen vector corresponding to $\lambda = 8$ is given by

$$X_3 = \begin{bmatrix} 2k \\ -k \\ k \end{bmatrix} \text{ or } k\begin{bmatrix} 2 \\ -1 \\ 1 \end{bmatrix} \text{ or } \begin{bmatrix} 2 \\ -1 \\ 1 \end{bmatrix}$$

Ans.

Example 7. Find the eigen values and the eigen vectors of the matrix $\begin{bmatrix} 8 & -6 & 2 \\ -6 & 7 & -4 \\ 2 & -4 & 3 \end{bmatrix}$.

Solution: Let $A = \begin{bmatrix} 8 & -6 & 2 \\ -6 & 7 & -4 \\ 2 & -4 & 3 \end{bmatrix}$

The characteristic equation of the matrix A is

$$|A-\lambda I| = 0$$

$$\begin{vmatrix} 8-\lambda & -6 & 2 \\ -6 & 7-\lambda & -4 \\ 2 & -4 & 3-\lambda \end{vmatrix} = 0$$

$$\lambda^3 - 18\lambda^2 + 45\lambda = 0$$

$$\lambda(\lambda^2 - 18\lambda + 45) = 0$$

$$\lambda(\lambda-3)(\lambda-15) = 0$$

$$\lambda = 0, 3, 15$$

Hence, the eigen values of the given matrix are 0, 3 and 15.

Now the equation $[A-\lambda I]X = 0$, for the matrix A is

$$\begin{bmatrix} 8-\lambda & -6 & 2 \\ -6 & 7-\lambda & -4 \\ 2 & -4 & 3-\lambda \end{bmatrix}\begin{bmatrix} x \\ y \\ z \end{bmatrix} = \begin{bmatrix} 0 \\ 0 \\ 0 \end{bmatrix}$$

For $\lambda = 0$, we have

$$\begin{bmatrix} 8 & -6 & 2 \\ -6 & 7 & -4 \\ 2 & -4 & 3 \end{bmatrix}\begin{bmatrix} x \\ y \\ z \end{bmatrix} = \begin{bmatrix} 0 \\ 0 \\ 0 \end{bmatrix}$$

$$8x - 6y + 2z = 0$$

$$-6x + 7y - 4z = 0$$

$$2x - 4y + 3z = 0$$

On solving first two equations we have

$$\frac{x}{24-14} = \frac{y}{-12+32} = \frac{z}{56-36}$$

$$\frac{x}{10} = \frac{y}{20} = \frac{z}{20}$$

$$\frac{x}{1} = \frac{y}{2} = \frac{z}{2} = k \text{ (say)}$$

Hence, the eigen vector corresponding to $\lambda = 0$ is given by

$$X_1 = \begin{bmatrix} k \\ 2k \\ 2k \end{bmatrix} \text{ or } k\begin{bmatrix} 1 \\ 2 \\ 2 \end{bmatrix} \text{ or } \begin{bmatrix} 1 \\ 2 \\ 2 \end{bmatrix}$$

For $\lambda = 3$, we have

$$\begin{bmatrix} 5 & -6 & 2 \\ -6 & 4 & -4 \\ 2 & -4 & 0 \end{bmatrix}\begin{bmatrix} x \\ y \\ z \end{bmatrix} = \begin{bmatrix} 0 \\ 0 \\ 0 \end{bmatrix}$$

$$5x - 6y + 2z = 0$$
$$-6x + 4y - 4z = 0$$
$$2x - 4y + 0z = 0$$

On solving first two equations we have

$$\frac{x}{24-8} = \frac{y}{-12+20} = \frac{z}{20-36}$$

$$\frac{x}{16} = \frac{y}{8} = \frac{z}{-16}$$

$$\frac{x}{2} = \frac{y}{1} = \frac{z}{-2} = k \text{ (say)}$$

Hence, the eigen vector corresponding to $\lambda = 3$ is given by

$$X_2 = \begin{bmatrix} 2k \\ k \\ -2k \end{bmatrix} \text{ or } k\begin{bmatrix} 2 \\ 1 \\ -2 \end{bmatrix} \text{ or } \begin{bmatrix} 2 \\ 1 \\ -2 \end{bmatrix}$$

For $\lambda = 15$, we have

$$\begin{bmatrix} -7 & -6 & 2 \\ -6 & -8 & -4 \\ 2 & -4 & -12 \end{bmatrix} \begin{bmatrix} x \\ y \\ z \end{bmatrix} = \begin{bmatrix} 0 \\ 0 \\ 0 \end{bmatrix}$$

$$-7x - 6y + 2z = 0$$
$$-6x - 8y - 4z = 0$$
$$2x - 4y - 12z = 0$$

On solving first two equations we have

$$\frac{x}{24+16} = \frac{y}{-12-28} = \frac{z}{56-36}$$

$$\frac{x}{40} = \frac{y}{-40} = \frac{z}{20}$$

$$\frac{x}{2} = \frac{y}{-2} = \frac{z}{1} = k \text{ (say)}$$

Hence, the eigen vector corresponding to $\lambda = 15$ is given by

$$X_3 = \begin{bmatrix} 2k \\ -2k \\ k \end{bmatrix} \text{ or } k\begin{bmatrix} 2 \\ -2 \\ 1 \end{bmatrix} \text{ or } \begin{bmatrix} 2 \\ -2 \\ 1 \end{bmatrix}$$ **Ans.**

EXERCISE 2.7

1. Find the eigen values and the eigen vectors of the matrix

$$\begin{bmatrix} 1 & -2 \\ -5 & 4 \end{bmatrix}.$$

2. Find the eigen values and the eigen vectors of the matrix

$$\begin{bmatrix} 8 & -8 & -2 \\ 4 & -3 & -2 \\ 3 & -4 & 1 \end{bmatrix}.$$

3. Find the eigen values and the eigen vectors of the matrix

$$\begin{bmatrix} -1 & 2 & -2 \\ 1 & 2 & 1 \\ -1 & -1 & 0 \end{bmatrix}.$$

4. `Find the eigen values and the eigen vectors of the matrix

$$\begin{bmatrix} 1 & 0 & -1 \\ 1 & 2 & 1 \\ 2 & 2 & 3 \end{bmatrix}.$$

5. Find the eigen values and the eigen vectors of the matrix

$$\begin{bmatrix} 4 & 2 & -2 \\ -5 & 3 & 2 \\ -2 & 4 & 1 \end{bmatrix}.$$

6. Find the eigen values and the eigen vectors of the matrix

$$\begin{bmatrix} -2 & 2 & -3 \\ 2 & 1 & -6 \\ -1 & -2 & 0 \end{bmatrix}.$$

7. Find the eigen values and the eigen vectors of the matrix

$$\begin{bmatrix} 1 & -6 & -4 \\ 0 & 4 & 2 \\ 0 & -6 & -3 \end{bmatrix}.$$

8. Find the eigen values and the eigen vectors of the matrix

$$\begin{bmatrix} -2 & 5 & 4 \\ 5 & 7 & 5 \\ 4 & 5 & -2 \end{bmatrix}.$$

9. Find the eigen values and the eigen vectors of the matrix

$$\begin{bmatrix} 2 & 2 & 0 \\ 0 & 1 & 1 \\ -7 & 2 & -3 \end{bmatrix}.$$

10. Find the eigen values and the eigen vectors of the matrix

$$\begin{bmatrix} 2 & -2 & 3 \\ 1 & 1 & 1 \\ 1 & 3 & -1 \end{bmatrix}.$$

11. Find the eigen values and the eigen vectors of the matrix

$$\begin{bmatrix} -9 & 2 & 6 \\ 5 & 0 & -3 \\ -16 & 4 & 11 \end{bmatrix}.$$

ANSWERS

1. $-1, 6; (1, 1), (2, -5)$

2. $1, 2, 3; (4, 3, 2), (3, 2, 1), (2, 1, 1)$

3. $-\sqrt{5}, 1, \sqrt{5}; (1+\sqrt{5}, -1, 1), (-1, 0, 1), (1-\sqrt{5}, -1, 1)$

4. $1, 2, 3; (1, -1, 0), (2, -1, -2), (1, -1, -2)$

5. $1, 2, 5; (2, 1, 4), (1, 1, 2), (0, 1, 1)$

6. $-3, -3, 5; (0, 3, 2), (3, 0, 1), (1, 2, -1)$

7. $0, 1, 1; (2, -1, 2), (1, -2, 3), (0, 2, -3)$

8. $-6, -3, 12; (1, 0, -1), (1, -1, 1), (1, 2, 1)$

9. $-4, 1, 3; (1, -1, 13), (-2, 1, 4), (2, 1, -2)$

10. $-2, 1, 3; (11, 1, -14), (-1, 1, 1), (1, 1, 1)$

11. $-1, 1, 2; (2, -1, 3), (1, -1, 2), (2, -1, 4)$

3

Differential Calculus

3.1 LIMIT

Introduction: The word limit derived from the Latin word limes which means boundary and this word was used by many Greek Mathematicians namely Euclid, Archimedes, Anaxagoras, while Archimedes used the basic ideas of limit while approximating the length of the circumference of a circle as the limiting value of the perimeter of an inscribed or circumscribed regular polygon of n sides as n is increased indefinitely.

3.2 CONCEPT OF LIMIT

Consider the function $f(x) = \dfrac{x^2 - 9}{x - 3}$

Clearly, $f(3) = \dfrac{0}{0}$, which is indeterminate form

Hence $f(x)$ is not defined at $x = 3$

Again, $f(x) = \dfrac{x^2 - 9}{(x-3)} = \dfrac{(x-3)(x+3)}{(x-3)} = (x + 3)$ only when $x \neq 3$

If we given to x, a value not exactly 3 but slightly more than 3 then the value of $f(x)$ is slightly more than 6. Now, if we go on decreasing this value and take it nearer to 3 then the value of $f(x)$ will come nearer to 6 as shown below.

If $x = 3.1$, then $f(x) = 6.1$

$x = 3.01$, then $f(x) = 6.01$

$x = 3.001$, then $f(x) = 6.001$

$x = 3.0001$, then $f(x) = 6.0001$

If $x = 3.000001$, then $f(x) = 6.000001$

Thus, the value of x tends to 3, the value of $f(x)$ tends to 6, and we write as $x \to 3$, $f(x) \to 6$

(The symbol $\to$ stands for 'tends to')

Similarly, if we given to x, a value slightly less than 3 then the value of $f(x)$ is slightly less than 6. Now, if we go on increasing this value and take it closer to 3, the value of $f(x)$ will come nearer to 6, as shown below.

If $x = 2.9$, then $f(x) = 5.9$
$x = 2.99$, then $f(x) = 5.99$
$x = 2.999$, then $f(x) = 5.999$
$x = 2.9999$, then $f(x) = 5.9999$
$x = 2.999999$, then $f(x) = 5.999999$

Thus, in this case also as $x \to 3,\ f(x) \to 6$

On the basis of the above discussion, we express this as $\lim_{x \to 3}\left(\frac{x^2-9}{x-3}\right) = 6$.

3.3 DEFINITION OF LIMIT

The function $f(x)$ is said to have limit l as x tends to a i.e. $\lim_{x \to a} f(x) = l$, if given any positive number $\in$ (however small), we can find a number δ such that $|f(x) - l| < \in$, whenever $|x - a| < \delta$.

In short we can say that $\lim_{x \to a} f(x) = l$ if whenever $x \to a,\ f(x) \to l.$

Alternatively, we have the following definition
A function $f(x)$ is said to have the limit l, if

(i) $\lim_{x \to a^+} f(x)$ or $\lim_{h \to 0} f(a+h)$ exists,

(ii) $\lim_{x \to a^-} f(x)$ or $\lim_{h \to 0} f(a-h)$ exists

(iii) $\lim_{x \to a^+} f(x) = \lim_{x \to a^-} f(x) = l$

However, if $\lim_{x \to a^+} f(x) \neq \lim_{x \to a^-} f(x)$, we say that $\lim_{x \to a} f(x)$ does not exist.

Case I. If a function $f(x)$ is defined at $x = a$ then $\lim_{x \to a} f(x)$ can be obtained directly by putting $x = a$ in $f(x)$ i.e. $\lim_{x \to a} f(x) = f(a)$

Example 1. Evaluate $\lim_{x \to 2} (x^3 + x^2 - 5x + 3)$

Solution: $\lim_{x \to 2} (x^3 + x^2 - 5x + 3) = 2^3 + 2^2 - 5 \times 2 + 3 = 5$

Example 2. (i) $\lim_{x \to 3} (x + 3)$ (ii) $\lim_{x \to 2} x(x + 1)$

Solution: (i) $\lim_{x \to 3} (x + 3) = 3 + 3 = 6$

(ii) $\lim_{x \to 2} x(x + 1) = 2(2 + 1) = 6$

3.4 PROPERTIES OF LIMIT

1. $\lim_{x \to a} \{f(x) + g(x)\} = \left\{\lim_{x \to a} f(x)\right\} + \left\{\lim_{x \to a} g(x)\right\}$

2. $\lim_{x \to a} \{f(x) - g(x)\} = \left\{\lim_{x \to a} f(x)\right\} - \left\{\lim_{x \to a} g(x)\right\}$

3. $\lim_{x\to a}\{\alpha f(x)\} = \alpha.\left\{\lim_{x\to a} f(x)\right\}$, where α is a constant

4. $\lim_{x\to a}\{f(x)\cdot g(x)\} = \left\{\lim_{x\to a} f(x)\right\}.\left\{\lim_{x\to a} g(x)\right\}$

5. $\lim_{x\to a}\left\{\frac{f(x)}{g(x)}\right\} = \frac{\left\{\lim_{x\to a} f(x)\right\}}{\left\{\lim_{x\to a} g(x)\right\}}$ provided $\lim_{x\to a} g(x) \neq 0$.

6. $\lim_{x\to a}\{f(x)\}^2 = \left\{\lim_{x\to a} f(x)\right\}^2$

7. $\lim_{x\to a}\sqrt{f(x)} = \sqrt{\lim_{x\to a} f(x)}$

8. If $f(x) \le g(x)$ for all x, then $\lim_{x\to a} f(x) \le \lim_{x\to a} g(x)$.

9. $\lim_{x\to a}\{f(x)\}^{p/q} = \left\{\lim_{x\to a} f(x)\right\}^{p/q}$

3.5 SOME IMPORTANT EXPANSION

1. $(1+x)^n = \left\{1+nx+\frac{n(n-1)}{2!}x^2+\frac{n(n-1)(n-2)}{3!}x^3+...\right\}$, if $|x|<1$.

2. $e^x = 1+x+\frac{x^2}{2!}+\frac{x^3}{3!}+...+\frac{x^4}{4!}+...$

3. $\frac{x^n-a^n}{x-a} = \left(x^{n-1}+ax^{n-2}+a^2x^{n-3}+...+a^{n-1}\right)$

4. $\log(1+x) = x-\frac{x^2}{2}+\frac{x^3}{3}-\frac{x^4}{4}+...$ if $|x|<1$

5. $a^x = \left\{1+x(\log a)+\frac{x^2}{2!}(\log a)^2+...\right\}$

6. $\sin x = \left\{x-\frac{x^3}{3!}+\frac{x^5}{5!}-\frac{x^7}{7!}+...\right\}$

7. $\cos x = \left\{1-\frac{x^2}{2!}+\frac{x^4}{4!}-\frac{x^6}{6!}+...\right\}$

8. $\tan x = \left\{x+\frac{x^3}{3}+\frac{2}{15}x^5+...\right\}$

3.6 SOME IMPORTANT THEOREMS ON LIMITS

Theorem 1. Prove that $\lim_{x\to a}\frac{x^n-a^n}{x-a} = na^{n-1}, a>0$

Proof: We know that

$$\left(\frac{x^n - a^n}{x-a}\right) = \left(x^{n-1} + ax^{n-2} + a^2x^{n-3} + \ldots + a^{n-1}\right)$$

$$\therefore \quad \lim_{x\to a}\left(\frac{x^n - a^n}{x-a}\right) = \lim_{x\to a}\,(x^{n-1} + ax^{n-2} + a^2x^{n-3} + \ldots + a^{n-1}) = na^{n-1}$$

(by putting $x = a$)

Theorem 2. prove that

$$\lim_{x\to 0}\left(\frac{a^x - 1}{x}\right) = \log a$$

Proof:

$$\lim_{x\to 0}\left(\frac{a^x - 1}{x}\right) = \lim_{x\to 0}\left[\frac{\left\{1 + x(\log a) + \frac{x^2}{2!}(\log a)^2 + \ldots\right\} - 1}{x}\right]$$

$$= \lim_{x\to 0}\frac{x\left\{\log a + \frac{x}{2!}(\log a)^2 + \ldots\right\}}{x}$$

$$= \lim_{x\to 0}\left[\log a + \frac{x}{2!}(\log a)^2 + \frac{x^2}{3!}(\log a)^3 + \ldots\right]$$

$$= \log a$$

Theorem 3. Prove that $\lim_{x\to 0}\frac{\log(1+x)}{x} = 1$

Proof:

$$\lim_{x\to 0}\left[\frac{\log(1+x)}{x}\right] = \lim_{x\to 0}\frac{\left[x - \frac{x^2}{2} + \frac{x^3}{3}\ldots\right]}{x}$$

$$= \lim_{x\to 0}\left[1 - \frac{x}{2} + \frac{x^2}{3} - \ldots\right] = 1$$

Theorem 4. Prove that $\lim_{x\to 0}\left(\frac{e^x - 1}{x}\right) = 1$.

Proof: We have

$$\lim_{x\to 0}\left(\frac{e^x - 1}{x}\right) = \lim_{x\to 0}\left[\frac{\left(1 + x + \frac{x^2}{2!} + \frac{x^3}{3!} + \ldots\right) - 1}{x}\right]$$

$$= \lim_{x\to 0}\left[\frac{x + \frac{x^2}{2!} + \frac{x^3}{3!} + \ldots}{x}\right] = \lim_{x\to 0}\left[1 + \frac{x}{2!} + \frac{x^2}{3!} + \ldots\right] = 1.$$

Case II. If $f(x)$ is undefined at $x = a$ and takes any of the indeterminate forms like $\frac{0}{0}$, $\frac{\infty}{\infty}$, then we adopt the following technique. If $f(x)$ is a rational function whose limiting value is of the form $\frac{0}{0}$, then we factorise the numerator and denominator of the given function cancel out the common factors and then put $x = a$.

ILLUSTRATIVE EXAMPLES

Example 1. Evaluate

$$(i)\ \lim_{x\to 2}\frac{x^2-4}{x-2}\quad (ii)\ \lim_{x\to -2}\frac{x^3+8}{x+2}\quad (iii)\ \lim_{x\to 1}\frac{x^2-3x+2}{x^2-1}\quad (iv)\ \lim_{x\to 1}\frac{x^2-4x+3}{(x-1)}$$

Solution:

$$\text{(i)}\quad \lim_{x\to 2}\frac{x^2-4}{(x-2)} = \lim_{x\to 2}\frac{(x-2)(x+2)}{(x-2)} = \lim_{x\to 2}(x+2) = 4$$

$$(ii)\quad \lim_{x\to -2}\frac{x^3+8}{x+2} = \lim_{x\to -2}\frac{(x+2)(x^2-2x+4)}{(x+2)} = \underset{x\to -2}{dt}\left(x^2-2x+4\right) = 12$$

$$(iii)\quad \lim_{x\to 1}\frac{x^2-3x+2}{x^2-1} = \lim_{x\to 1}\frac{(x-1)(x-2)}{(x-1)(x+1)} = \lim_{x\to 1}\frac{x-2}{x+1} = -\frac{1}{2}$$

$$(iv)\quad \lim_{x\to 1}\frac{x^2-4x+3}{(x-1)} = \lim_{x\to 1}\frac{(x-1)(x-3)}{(x-1)} = \lim_{x\to 1}(x-3) = -2.$$

Case III. If the given function contains a surd of the form $\sqrt{a}\pm\sqrt{b}$, then simplify the given function by using conjugate surds. After simplification put $x = a$.

Example 2: Evaluate

$$(i)\quad \lim_{x\to 0}\frac{\sqrt{1+x}-\sqrt{1-x}}{\sqrt{2+3x}-\sqrt{2-3x}}$$

$$(ii)\quad \lim_{x\to 0}\frac{\sqrt{1+x}-\sqrt{1-x}}{x}$$

$$(iii)\quad \lim_{x\to 1}\left[\frac{(2x-3)\left(\sqrt{x}-1\right)}{2x^2+x-3}\right]$$

Solution:

$$(i)\quad \lim_{x\to 0}\frac{\sqrt{1+x}-\sqrt{1-x}}{\sqrt{2+3x}-\sqrt{2-3x}}$$

$$= \lim_{x\to 0}\frac{\left(\sqrt{1+x}-\sqrt{1-x}\right)}{\left(\sqrt{2+3x}-\sqrt{2-3x}\right)}\times\frac{\left(\sqrt{1+x}+\sqrt{1-x}\right)\left(\sqrt{2+3x}+\sqrt{2-3x}\right)}{\left(\sqrt{1+x}+\sqrt{1-x}\right)\left(\sqrt{2+3x}+\sqrt{2-3x}\right)}$$

$$= \lim_{x\to 0}\frac{\left[(1+x)-(1-x)\right]\left[\sqrt{2+3x}+\sqrt{2-3x}\right]}{\left[(2+3x)-(2-3x)\right]\left[\sqrt{1+x}+\sqrt{1-x}\right]}$$

$$= \lim_{x\to 0}\frac{(2x)\left(\sqrt{2+3x}+\sqrt{2-3x}\right)}{(6x)\left(\sqrt{1+x}+\sqrt{1-x}\right)}$$

$$= \frac{1}{3}\lim_{x\to 0}\frac{\left(\sqrt{2+3x}+\sqrt{2-3x}\right)}{\left(\sqrt{1+x}+\sqrt{1-x}\right)}$$

$$= \frac{1}{3}\frac{\left(\sqrt{2}+\sqrt{2}\right)}{\left(\sqrt{1}+\sqrt{1}\right)} = \frac{2\sqrt{2}}{3\times 2} = \frac{\sqrt{2}}{3}$$

(*ii*) $$\lim_{x\to 0}\left[\frac{\sqrt{1+x}-\sqrt{1-x}}{x}\right] = \lim_{x\to 0}\frac{\left(\left[\sqrt{1+x}-\sqrt{1-x}\right]\right)}{x}\frac{\left(\sqrt{1+x}+\sqrt{1+x}\right)}{\left(\sqrt{1+x}+\sqrt{1-x}\right)}$$

$$= \lim_{x\to 0}\frac{\left[(1+x)-(1-x)\right]}{x\left(\sqrt{1+x}+\sqrt{1-x}\right)}$$

$$= \lim_{x\to 0}\frac{2x}{x\left(\sqrt{1+x}+\sqrt{1-x}\right)}$$

$$= 2\lim_{x\to 0}\frac{1}{\left(\sqrt{1+x}+\sqrt{1-x}\right)} = 2\times\frac{1}{(1+1)} = 1.$$

(*iii*) $$\lim_{x\to 1}\left[\frac{(2x-3)\sqrt{x}-1}{2x^2+x-3}\right] = \lim_{x\to 1}\frac{(2x-3)\left(\sqrt{x}-1\right)}{(2x+3)(x-1)}$$

$$= \lim_{x\to 1}\frac{(2x-3)\left(\sqrt{x}-1\right)}{(2x+3)\left(\sqrt{x}-1\right)\left(\sqrt{x}+1\right)}$$

$$= \lim_{x\to 1}\left[\frac{2x-3}{(2x+3)\sqrt{x}+1}\right] = \frac{-1}{5\times 2} = \frac{-1}{10}$$

Example 3: Evaluate

(*i*) $\lim_{x\to 3}\frac{x^{12}-3^{12}}{x^{10}-3^{10}}$ (*ii*) $\lim_{x\to a}\frac{(x+3)^{3/2}-(a+3)^{3/2}}{x-a}$

(*iii*) $\lim_{x\to 0}\left(\frac{3^x-2^x}{x}\right)$ (*iv*) $\lim_{x\to 0}\frac{(1+x)^n-1}{x}$

Solution: (*i*) $$\lim_{x\to 3}\frac{x^{12}-3^{12}}{x^{10}-3^{10}} = \lim_{x\to 3}\left[\frac{x^{12}-3^{12}}{x-3}.\frac{x-3}{x^{10}-3^{10}}\right]$$

$$= \lim_{x\to 3}\left[\frac{x^{12}-3^{12}}{x-3}\div\frac{x^{10}-3^{10}}{x-3}\right]$$

$$= \lim_{x\to 3}\left[\frac{x^{12}-3^{12}}{x-3}\right]\div\lim_{x\to 3}\left[\frac{x^{10}-3^{10}}{x-3}\right]$$

$$= 12(3)^{11}\div 10(3)^{9}$$

$$= \frac{12\times(3)^{11}}{10\times(3)^{9}} = \frac{6}{5}\times 3^2 = \frac{54}{5}$$

(*ii*) $$\lim_{x\to a}\frac{(x+3)^{3/2}-(a+3)^{3/2}}{x-a} = \lim_{x+3\to a+3}\frac{(x+3)^{3/2}-(a+3)3/2}{(x+3)-(a+3)}$$

$$= \frac{3}{2}(a+3)^{\frac{3}{2}-1} = \frac{3}{2}(a+3)^{\frac{1}{2}}$$

(*iii*) $$\lim_{x\to 0}\left(\frac{3^x-2^x}{x}\right) = \lim_{x\to 0}\left[\frac{(3^x-1)(2^x-1)}{x}\right]$$

$$= \lim_{x\to 0}\left(\frac{3^x-1}{x}\right)-\lim_{x\to 0}\left(\frac{2^x-1}{x}\right)$$

$$= (\log 3-\log 2) = \log\frac{3}{2}$$

(*iv*) $\lim_{x\to 0}\frac{(1+x)^n-1}{x}$ Put $1+x=t$, so when $x\to 0$ then $t\to 1$.

$$\therefore \lim_{x\to 0}\frac{(1+x)^n-1}{x} = \lim_{t\to 1}\left[\frac{t^n-1}{t-1}\right]$$

$$= \lim_{t\to 1}\left[\frac{t^n-1^n}{t-1}\right] = n\times(1)^{n-1} = n$$

EXERCISE 3.1

Evaluate the following limits, if exists.

1. $\lim_{x\to 3}\frac{x^2+3x}{x+1}$

2. $\lim_{x\to 2}(x^2-3x+4)$

3. $\lim_{x\to 1}\frac{(x^3-1)}{(x-1)}$

4. $\lim_{x\to 4}\left(\frac{e^x-e^4}{x-4}\right)$

5. $\lim_{x\to 0}\frac{(1+x)^4-1}{x}$

6. $\lim_{a\to 0}\frac{\sqrt{x+a}-\sqrt{x}}{a}, x\neq 0$

7. $\lim_{x\to 2}\dfrac{x^2-4}{\sqrt{x+2}-\sqrt{3x-2}}$

8. $\lim_{x\to a}\dfrac{\sqrt{a+x}-\sqrt{a}}{x\sqrt{a(a+x)}}$

9. $\lim_{x\to 0}\left(\dfrac{3^{2+x}-9}{x}\right)$

10. $\lim_{x\to 0}\dfrac{\sqrt{1+x+x^2}-\sqrt{1-x+x^2}}{x^2-x}$

11. If $\lim_{x\to a}\dfrac{x^9+a^9}{x+a}=9$, find all possible values of a

12. $\lim_{x\to 0}\left(\dfrac{8^x-4^x-2^x+1}{x^2}\right)$

13. $\lim_{x\to 0}\left(\dfrac{2^x-1}{\sqrt{1+x}-1}\right)$

14. $\lim_{x\to 0}\dfrac{\log(1+2x)}{(3^x-1)}$

15. $\lim_{x\to 2}\dfrac{x^2-4}{\sqrt{3x-2}-\sqrt{x+2}}$

16. If $\lim_{x\to 2}\dfrac{x^n-2^n}{x-2}=80$ and $n\in N$, find n

17. $\lim_{x\to a}\dfrac{(x+2)^{5/2}-(a+2)^{5/2}}{x-a}$

ANSWERS

1. $\dfrac{9}{2}$	**2.** 2	**3.** 3	**4.** e^4
5. 4	**6.** $\dfrac{1}{2\sqrt{x}}, x\neq 0$	**7.** –8	**8.** $\dfrac{1}{2}a^{-3/2}, a\neq 0$
9. $9\log 3$	**10.** –1	**11.** ±1	**12.** (log 2) (log4)
13. $\log 4$	**14.** $\dfrac{2}{\log 3}$	**15.** 8	**16.** $n=5$
17. $\dfrac{5}{2}(a+2)^{3/2}$			

3.7 LIMIT OF THE TRIGONOMETRIC FUNCTIONS

For finding the limits of trigonometric functions. we use trigonometric transformation and simplify.

Theorem. Prove that (*i*) $\lim_{\theta\to 0}\sin\theta=0$ (*ii*) $\lim_{\theta\to 0}\cos\theta=1$

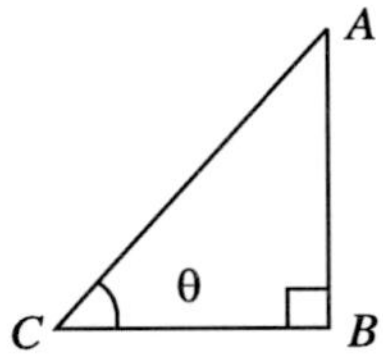

Proof: In ΔABC let $\angle B$ is a right angle and $\angle ACB=\theta$ radians.

$$\sin\theta=\frac{AB}{AC}\text{ and }\cos\theta=\frac{BC}{AC}$$

If we fixed BC and go on decreasing θ, we find that A coming nearer and nearer to B.

$\therefore$ When $\theta\to 0$ then $A\to B$.

$\therefore$ $\theta\to 0\Rightarrow AB\to 0$ and $CA\to CB$

$\Rightarrow \quad \dfrac{AB}{AC} \to 0$ and $\dfrac{BC}{AC} \to 1$

$\Rightarrow \quad \sin\theta \to 0$ and $\cos\theta \to 1$.

Hence, We have

(*i*) $\lim\limits_{\theta\to 0} \sin\theta = 0$ and (*ii*) $\lim\limits_{\theta\to 0} \cos\theta = 1$

ILLUSTRATIVE EXAMPLES

Example 1. Prove that $\lim\limits_{\theta\to 0} \dfrac{\sin\theta}{\theta} = 1$

Solution: $$\lim_{\theta\to 0} \frac{\sin\theta}{\theta} = \lim_{\theta\to 0} \frac{\left(\theta - \dfrac{\theta^3}{3!} + \dfrac{\theta^5}{5!}\ldots\right)}{\theta} = \lim_{\theta\to 0}\left(1 - \frac{\theta^2}{3!} + \frac{\theta^4}{5!}\ldots\right) = 1$$

$$\therefore \quad \lim_{\theta\to 0} \frac{\sin\theta}{\theta} = 1$$

Example 2. Evaluate $\lim\limits_{x\to 0} \dfrac{\sin 5x}{3x}$

Solution: $$\lim_{x\to 0} \frac{\sin 5x}{3x} = \lim_{x\to 0} \frac{5}{3}.\left(\frac{\sin 5x}{5x}\right) = \frac{5}{3}\lim_{x\to 0}\left(\frac{\sin 5x}{5x}\right) = \frac{5}{3}\times 1 = \frac{5}{3}$$

Example 3. Evaluate $\lim\limits_{x\to 0} \dfrac{\tan x}{x}$

Solution: $$\lim_{x\to 0}\left(\frac{\tan x}{x}\right) = \lim_{x\to 0}\left[\frac{\sin x}{x}.\frac{1}{\cos x}\right]$$

$$= \lim_{x\to 0}\left(\frac{\sin x}{x}\right).\lim_{x\to 0}\left(\frac{1}{\cos x}\right) = 1 \times 1 = 1$$

Example 4. Evaluate $\lim\limits_{x\to 0} \dfrac{\sin 3x}{\tan 5x}$

Solution: $$\lim_{x\to 0} \frac{\sin 3x}{\tan 5x} = \lim_{x\to 0}\left[\frac{3x\left(\dfrac{\sin 3x}{3x}\right)}{5x\left(\dfrac{\tan 5x}{5x}\right)}\right] = \frac{3}{5}\lim_{x\to 0}\left[\frac{\dfrac{\sin 3x}{3x}}{\dfrac{\tan 5x}{5x}}\right]$$

$$= \frac{3}{5}\frac{\lim\limits_{3x\to 0}\left(\dfrac{\sin 3x}{3x}\right)}{\lim\limits_{5x\to 0}\left(\dfrac{\tan 5x}{5x}\right)} = \frac{3}{5}\times\frac{1}{1} = \frac{3}{5}$$

Example 5. Evaluate $\lim\limits_{x\to \pi/2} (\sec x - \tan x)$

Proof: $$\lim_{x\to \pi/2} (\sec x - \tan x) = \lim_{x\to \pi/2}\left[\frac{1-\sin x}{\cos x}\right]$$

$$= \lim_{x\to\pi/2} \frac{[\cos x/2 - \sin x/2]^2}{\cos^2 x/2 - \sin^2 x/2}$$

$$= \lim_{x\to\pi/2} \frac{\cos x/2 - \sin x/2}{\cos x/2 + \sin x/2} = 0$$

Example 6. Evaluate $\lim_{x\to 0} \frac{\sin\frac{x}{\pi}}{x}$

Solution: $\lim_{x\to 0} \frac{\sin\frac{x}{\pi}}{x} = \lim_{x\to 0} \frac{\sin\frac{x}{\pi}}{\frac{x}{\pi}.\pi} = \frac{1}{\pi}\lim_{x\to 0}\left(\frac{\sin\frac{x}{\pi}}{\frac{x}{\pi}}\right) = \frac{1}{\pi}\times 1 = \frac{1}{\pi}$

Example 7. Evaluate $\lim_{x\to 0} \frac{\sin(3+x) - \sin(3-x)}{x}$

Solution :

$$\lim_{x\to 0} \frac{\sin(3+x) - \sin(3-x)}{x} = \lim_{x\to 0} \frac{2\cos\left(\frac{3+x+3-x}{2}\right)\sin\left(\frac{3+x-3+x}{2}\right)}{x}$$

$$= \lim_{x\to 0} \frac{2\cos 3 \sin x}{x} = 2\cos 3 \lim_{x\to 0} \frac{\sin x}{x}$$

$$= 2\cos 3 \times 1 = 2\cos 3$$

Example 8. Evaluate $\lim_{x\to 0}\left(\frac{\sin 3x + \sin 7x}{\sin 5x - \sin 3x}\right)$

Solution :

$$\lim_{x\to 0}\left(\frac{\sin 3x + \sin 7x}{\sin 5x - \sin 3x}\right) = \lim_{x\to 0} \frac{2\sin 5x \cos 2x}{2\cos 4x \sin x}$$

$$= \lim_{x\to 0}\left(\frac{\sin 5x \cos 2x}{\cos 4x \sin x}\right)$$

$$= \lim_{x\to 0}\left\{\frac{\sin 5x}{5x}\times\frac{x}{\sin x}\times\cos 2x\times\frac{1}{\cos 4x}\times 5\right\}$$

$$= 5\lim_{x\to 0}\left(\frac{\sin 5x}{5x}\right)\times\lim_{x\to 0}\left(\frac{x}{\sin x}\right)\lim_{2x\to 0}\cos 2x$$

$$\times\frac{1}{\lim_{4x\to 0}\cos 4x}$$

$$= 5\times 1\times 1\times 1\times\frac{1}{1} = 5$$

Example 9. Evaluate $\lim\limits_{x\to\pi}\left(\dfrac{1+\cos x}{\tan^2 x}\right)$

Solution:

$$\lim_{x\to\pi}\left(\frac{1+\cos x}{\tan^2 x}\right) = \lim_{x\to\pi}\frac{\cos^2 x(1+\cos x)}{\sin^2 x}$$

$$= \lim_{x\to\pi}\frac{\cos^2 x(1+\cos x)}{\left(1-\cos^2 x\right)} = \lim_{x\to\pi}\frac{\cos^2 x}{(1-\cos x)} = \frac{1}{2}$$

Example 10. Evaluate $\lim\limits_{x\to 0}\dfrac{\log\left(1+x^3\right)}{\sin^3 x}$

Solution:

$$\lim_{x\to 0}\frac{\log\left(1+x^3\right)}{\sin^3 x} = \lim_{x\to 0}\left[\frac{1}{x^3}\log(1+x^3)\times\frac{x^3}{\sin^3 x}\right]$$

$$= \lim_{x\to 0}\left[\frac{1}{x^3}\log(1+x^3)\right]\times\lim_{x\to 0}\left(\frac{x}{\sin x}\right)^3$$

$$= \lim_{x\to 0}\left[\log(1+x^3)^{\frac{1}{x^3}}\right]\times(1)^3$$

$$= \log e\times 1 = 1\times 1 = 1.$$

$$\left[\because \lim_{t\to 0}\log(1+t)^{\frac{1}{t}} = \log e\right]$$

Example 11. Evaluate $\lim\limits_{x\to 0}\left(\dfrac{e^x - e^{\sin x}}{x-\sin x}\right)$

Solution :

$$\lim_{x\to 0}\left(\frac{e^x - e^{\sin x}}{x-\sin x}\right) = \lim_{x\to 0}\left[\frac{e^{\sin x}\left(e^{x-\sin x}-1\right)}{(x-\sin x)}\right]$$

$$= \lim_{x\to 0}e^{\sin x}\times\lim_{x\to 0}\left[\frac{e^{x-\sin x}-1}{x-\sin x}\right]$$

$$= 1\times\lim_{t\to 0}\left(\frac{e^t-1}{t}\right) \text{ where } x-\sin x = t$$

$$= 1\times 1 = 1$$

Example 12. Evaluate (*i*) $\lim\limits_{x\to 0}\dfrac{\sin^{-1}x}{x}$ (*ii*) $\lim\limits_{x\to 0}\dfrac{\tan^{-1}x}{x}$

Solution:

(*i*) $\lim\limits_{x\to 0}\dfrac{\sin^{-1}x}{x}$ Put $x=\sin\theta$, As $x\to 0\Rightarrow \theta\to 0$

$$\therefore \lim_{x\to 0}\frac{\sin^{-1}x}{x} = \lim_{\theta\to 0}\frac{\sin^{-1}(\sin\theta)}{\sin\theta}$$

$$= \lim_{\theta\to 0}\frac{\theta}{\sin\theta}=1$$

(*ii*) Put $x=\tan\theta$, As $x\to 0\Rightarrow \theta=0$

$$\therefore \lim_{x\to 0}\frac{\tan^{-1}x}{x} = \lim_{x\to 0}\frac{\tan^{-1}(\tan\theta)}{\tan\theta}$$

$$= \lim_{x\to 0}\frac{\theta}{\tan\theta}=1$$

Example 13. Evaluate $\lim\limits_{x\to 0}\dfrac{x\left(1-\sqrt{1-x^2}\right)}{\left(\sqrt{1-x^2}\right)\left(\sin^{-1}x\right)^3}$

Solution: Put $x=\sin\theta$, As $x\to 0\Rightarrow \theta\to 0$

$$\therefore \lim_{x\to 0}\frac{x\left(1-\sqrt{1-x^2}\right)}{\left(\sqrt{1-x^2}\right)\left(\sin^{-1}x\right)^3} = \lim_{\theta\to 0}\frac{\sin\theta\left(1-\sqrt{1-\sin^2\theta}\right)}{\sqrt{1-\sin^2\theta}\quad(\theta)^3}$$

$$= \lim_{\theta\to 0}\frac{\sin\theta(1-\cos\theta)}{\cos\theta\times\theta^3}$$

$$= \lim_{\theta\to 0}\frac{\sin\theta(1-\cos\theta)}{\cos\theta\times\theta^3}\times\frac{1+\cos\theta}{1+\cos\theta}$$

$$= \lim_{\theta\to 0}\frac{\sin\theta}{\theta}\times\frac{1}{\cos\theta}\times\frac{1-\cos^2\theta}{\theta^2}\times\frac{1}{(1+\cos\theta)}$$

$$= \lim_{\theta\to 0}\frac{\sin\theta}{\theta}\lim_{\theta\to 0}\left(\frac{1}{\cos\theta}\right)\lim_{\theta\to 0}\left(\frac{\sin\theta}{\theta}\right)^2 \lim_{\theta\to 0}\left(\frac{1}{1+\cos\theta}\right)$$

$$= 1\times 1\times 1\times\frac{1}{1+1}=\frac{1}{2}$$

Example 14: Evaluate $\lim\limits_{x\to 0}\dfrac{a^{3x}-a^{2x}-a^{x}+1}{x^2}$

Solution:

$$\lim_{x\to 0}\frac{a^{3x}-a^{2x}-a^{x}+1}{x^2} = \lim_{x\to 0}\frac{a^{2x}(a^x-1)-(a^x-1)}{x^2}$$

$$= \lim_{x \to 0} \frac{(a^x - 1)(a^{2x} - 1)}{x^2}$$

$$= \lim_{x \to 0}\left(\frac{a^x - 1}{x}\right) \lim_{x \to 0}\left(\frac{a^{2x} - 1}{2x} \times 2\right)$$

$$= (\log a) \,.\, (\log a \times 2) = 2\,(\log a)^2$$

EXERCISE 3.2

Evaluate the following limits

1. $\lim_{x \to 0} \dfrac{\sin ax}{\tan bx}$

2. $\lim_{x \to 0}\left(\dfrac{\sin 2x + 3x}{2x + \sin 3x}\right)$

3. $\lim_{x \to 0} (x \operatorname{cosec} x)$

4. $\lim_{x \to 0} \dfrac{\sin(x/5)}{x}$

5. $\lim_{x \to 0} \dfrac{4^x - 1}{\sqrt{1+x} - 1}$

6. $\lim_{x \to 0} \dfrac{a^x - 1}{\sin x}$

7. $\lim_{x \to 0} \dfrac{\log(1+x)}{5^x - 1}$

8. $\lim_{x \to 3} \dfrac{x-3}{\sqrt{x-3}}$

9. $\lim_{x \to 1}\left(\dfrac{1}{x-1} - \dfrac{2}{x^2 - 1}\right)$

10. $\lim_{x \to 2} \dfrac{x^2 - 4}{\sqrt{3x-2} - \sqrt{x+2}}$

11. $\lim_{x \to 0} \dfrac{\sin 3x \cos 2x}{\sin^2 x}$

12. $\lim_{x \to 0} \dfrac{1 - \sqrt{\cos x}}{x^2}$

13. $\lim_{x \to a} \dfrac{\sin x - \sin a}{\sqrt{x} - \sqrt{a}}$

14. $\lim_{x \to 0} \dfrac{\sqrt{1+2x} - \sqrt{1-2x}}{\sin x}$

15. $\lim_{x \to 0}\left(\dfrac{a^{\sin x} - 1}{\sin x}\right)$

16. $\lim_{x \to 1} \dfrac{(1-x)}{(\pi - 2\sin^{-1} x)}$

17. $\lim_{x \to \pi} \dfrac{\sin 3x - 3\sin x}{(\pi - x)3}$

18. $\lim_{x \to a} \dfrac{\sqrt[m]{x} - \sqrt[m]{a}}{x - a}$

19. $\lim_{x \to 0} \dfrac{\cos^{-1}(1-x)}{\sqrt{x}}$

20. $\lim_{x \to 4} \dfrac{\log_e(x-3)}{4-x}$

21. $\lim_{x \to 0} \dfrac{x(e^{2+x} - e^2)}{1 - \cos x}$

22. $\lim_{x \to 0} \dfrac{3^{2+x} - 9}{x}$

23. $\lim_{x \to 0} \dfrac{3^x - 2^x}{x}$

24. $\lim_{x \to 0} \dfrac{e^{3+x} - \sin x - e^3}{x}$

ANSWERS

1. $\dfrac{a}{b}$ **2.** 1 **3.** 1 **4.** $\dfrac{1}{5}$

5. $2\log 4$	**6.** $\log a$	**7.** $\dfrac{1}{\log 5}$	**8.** 0
9. $\dfrac{1}{2}$	**10.** 8	**11.** $\dfrac{3}{2}$	**12.** $\dfrac{1}{4}$
13. $2\sqrt{a}\cos a$	**14.** 2	**15.** $\log a$	
16. 0	**17.** -1	**18.** $\dfrac{\sqrt[m]{a}}{ma}$	**19.** $\sqrt{2}$
20. -1	**21.** $2e^2$	**22.** $9\log_e 3$	**23.** $\log\dfrac{3}{2}$
24. e^3-1			

3.8 LIMITS AT INFINITY AND INFINITE LIMITS

Meaning of $x \to +\infty$

If x is always positive and becomes larger and larger without limit and in this case x can become larger than any positive number, we say that x tends to plus infinity and we write $x \to +\infty$.

Rule for finding: $\lim\limits_{x\to\infty} f(x) \to$ Replace x by $\dfrac{1}{y}$ in the given function and take the limit as $y \to 0$.

ILLUSTRATIVE EXAMPLES

Example 1. Show that $\lim\limits_{x\to\infty} \dfrac{1}{x^3} = 0$

Solution: Let $x = \dfrac{1}{y}$ so that as $x \to \infty$, $y \to 0$

$$\therefore \qquad \lim_{x\to\infty} \frac{1}{x^3} = \lim_{y\to 0} y^3 = 0$$

Note: In case of a rational function, divide the numerator and denominator by the highest power of x and then replace x by $\dfrac{1}{y}$ and take the limit as $y \to 0$.

Example 2. Evaluate $\lim\limits_{x\to\infty} \dfrac{(3x+1)(4x+2)}{(x+8)(x+1)}$

Solution:

$$\lim_{x\to\infty} \frac{(3x+1)(4x+2)}{(x+8)(x+1)} = \lim_{x\to\infty} \frac{\left(3+\frac{1}{x}\right)\left(4+\frac{2}{x}\right)}{\left(1+\frac{8}{x}\right)\left(1+\frac{1}{x}\right)}$$

(Dividing numerator and denominator by x^2)

$$= \frac{(3+0)(4+0)}{(1+0)(1+0)} = \frac{12}{2} = 6$$

Example 3. Evaluate $\lim\limits_{x\to\infty} \frac{3x^2+5x+2}{2x^3+3x+1}$

Solution:

$$\lim_{x\to\infty} \frac{3x^2+5x+2}{2x^3+3x+1} = \lim_{x\to\infty} \frac{3+\frac{5}{x^2}+\frac{2}{x^3}}{2+\frac{3}{x^2}+\frac{1}{x^3}}$$

$$= \frac{3+0+0}{2+0+0} = \frac{3}{2}$$

Example 4. Evaluate $\lim\limits_{x\to\infty} \frac{1}{(1+x)^2}$

Solution:

$$\lim_{x\to\infty} \frac{1}{(1+x)^2} = \lim_{x\to\infty} \frac{1}{(1+2x+x^2)}$$

$$= \lim_{x\to\infty} \frac{\left(\frac{1}{x^2}\right)}{\left(\frac{1}{x^2}+\frac{2}{x}+1\right)}$$

$$= \lim_{y\to 0} \left(\frac{y^2}{y^2+2y+1}\right) = \frac{0}{1} = 0$$

Example 5. Evaluate $\lim\limits_{n\to\infty} \frac{1+2+3+...+n}{n^2}$

Solution:

$$\lim_{n\to\infty} \frac{1+2+3+...+n}{n^2} = \lim_{n\to\infty} \frac{n(n+1)}{2n^2}$$

$$= \lim_{n\to\infty} \frac{n^2+n}{2n^2} = \lim_{n\to\infty} \frac{1+\frac{1}{n}}{2} = \frac{1}{2}$$

Example 6. Evaluate $\lim\limits_{n\to\infty} \frac{1^2+2^2+3^2+...+n^2}{n^3}$

Solution: We know that $(1^2+2^2+3^2+...+n^2) = \frac{n(n+1)(2n+1)}{6}$

$$\therefore \lim_{n\to\infty} \left(\frac{1^2+2^2+3^2+....+n^2}{n^3}\right) = \lim_{n\to\infty} \frac{n(n+1)(2n+1)}{6n^3}$$

$$= \lim_{n\to\infty} \frac{n^3\left(1+\frac{1}{n}\right)\left(2+\frac{1}{n}\right)}{6n^3}$$

$$= \lim_{n\to\infty} \frac{\left(1+\frac{1}{n}\right)\left(2+\frac{1}{n}\right)}{6} = \frac{(1+0)(2+0)}{6} = \frac{1}{3}$$

Example 7. Evaluate $\lim_{x\to\infty} \frac{\sin x}{x}$

Solution: $\lim_{x\to\infty} \frac{\sin x}{x}$

Put $x = \frac{1}{y}$ so that as $x \to \infty, y \to 0$

$$\therefore \quad \lim_{x\to\infty} \frac{\sin x}{x} = \lim_{y\to 0} \frac{\sin\left(\frac{1}{y}\right)}{\left(\frac{1}{y}\right)}$$

$$= \lim_{y\to 0}\left(y\sin\frac{1}{y}\right)$$

$$\left[\therefore \sin\left(\frac{1}{y}\right) \text{lies between } -1 \,\&\, 1\right]$$

$$= (0 \times a \text{ finite quantity}) = 0$$

Example 8. Evaluate $\lim_{x\to\infty} \frac{4}{\pi}\tan^{-1} bx; b > 0.$

Solution: Let $bx = \tan\theta$, so that as $x \to \infty$, $\theta = \frac{\pi}{2}$ as $b > 0$

$$\therefore \quad \lim_{x\to\infty} \frac{4}{\pi}\tan^{-1} bx = \frac{4}{\pi}\lim_{\theta\to\pi/2} \tan^{-1}(\tan\theta)$$

$$= \frac{4}{\pi}\lim_{\theta\to\pi/2} \theta = \frac{4}{\pi}\times\frac{\pi}{2} = 2$$

Example 9. Evaluate $\lim_{x\to\infty}\left(\sqrt{x^2+1}-x\right)$

Solution:

$$\lim_{x\to\infty}\left(\sqrt{x^2+1}-x\right) = \lim_{x\to\infty}\left[\left(\sqrt{x^2+1}-x\right)\times\frac{\sqrt{x^2+1}+x}{\sqrt{x^2+1}+x}\right]$$

$$= \lim_{x\to\infty}\left(\frac{x^2+1-x^2}{\sqrt{x^2+1}+x}\right) = \lim_{x\to\infty}\left(\frac{1}{\sqrt{x^2+1}+x}\right)$$

$$= \lim_{x\to\infty}\frac{\left(\frac{1}{x}\right)}{\sqrt{1+\frac{1}{x^2}}+1} = \lim_{y\to 0}\frac{y}{\left(\sqrt{1+y^2}+1\right)}$$

$$= \frac{0}{2} = 0$$

EXERCISE 3.3

Evaluate the following limits

1. $\lim\limits_{x\to\infty}\dfrac{x-1}{x+1}$

2. $\lim\limits_{x\to 1}\dfrac{1}{|1-x|}$

3. $\lim\limits_{x\to\infty} x\sin\dfrac{1}{x}$

4. $\lim\limits_{x\to\infty} 2x\left(\sqrt{x^2+1}-x\right)$

5. $\lim\limits_{x\to\infty}\dfrac{3^x-1}{3^x+1}$

6. $\lim\limits_{x\to\infty}\dfrac{1^3+2^3+3^3+\ldots+n^3}{n^4}$

7. $\lim\limits_{x\to\infty}\left(1+\dfrac{1}{2}+\dfrac{1}{2^2}+\ldots+\dfrac{1}{2^{n-1}}\right)$

8. $\lim\limits_{x\to\infty}\left(\sqrt{x^2-8x}+x\right)$

9. $\lim\limits_{x\to\infty}\left(\sqrt{x^2+x}-\sqrt{x^2-10}\right)$

10. $\lim\limits_{n\to\infty}\left(\dfrac{1}{3}+\dfrac{1}{3^2}+\dfrac{1}{3^3}+\ldots+\dfrac{1}{3^n}\right)$

11. $\lim\limits_{x\to\infty}\dfrac{4x^2+2x+5}{6x^2+2x+1}$

12. $\lim\limits_{x\to\infty}\dfrac{2x^3+5x^2+7}{5x^3+3x^2+8}$

13. $\lim\limits_{x\to\infty} x^2$

14. $\lim\limits_{x\to\infty}\dfrac{x}{\left(\sqrt{4x^2+1}-1\right)}$

ANSWERS

1. 1	**2.** ∞	**3.** 1	**4.** 1
5. 1	**6.** 1	**7.** 2	**8.** 4
9. $\frac{1}{2}$	**10.** $\frac{1}{2}$	**11.** $\frac{2}{3}$	**12.** $\frac{2}{5}$
13. ∞	**14.** $\frac{1}{2}$		

3.9 MEANING OF $x \to -\infty$

If x is always negative and becomes larger and larger without limit in magnitude, we say that x tends to minus infinity and we write $x \to -\infty$.

ILLUSTRATIVE EXAMPLES

Example 1. Evaluate $\lim_{x \to -\infty} e^x$

Solution: Put $x = -y$, we get

$$\lim_{x \to -\infty} e^x = \lim_{y \to \infty} e^{-y} = \lim_{y \to \infty} \frac{1}{\left(1 + y + \frac{y^2}{2!} + ...\right)}$$

Since y is very large so $\left(1 + y + \frac{y^2}{2!} + ...\right)$ becomes very larger and therefore

$\frac{1}{\left(1 + y + \frac{y^2}{2!} + ...\right)}$ becomes very small.

Hence, $\lim_{y \to \infty} \frac{1}{\left(1 + y + \frac{y^2}{2!} + ...\right)} = 0.$

Example 2. Evaluate $\lim_{x \to -\infty} \sqrt{x^2 - 8x} + x$

Solution:

$$\lim_{x \to -\infty} \sqrt{x^2 - 8x} + x = \lim_{x \to -\infty} \left[\sqrt{x^2 - 8x} + x \times \frac{\sqrt{x^2 - 8x} - x}{\sqrt{x^2 - 8x} - x}\right]$$

$$= \lim_{x \to -\infty} \left[\frac{x^2 - 8x - x^2}{\sqrt{x^2 - 8x} - x}\right]$$

$$= \lim_{x \to -\infty} \left[\frac{-8x}{\sqrt{x^2 - 8x} - x}\right]$$

Put $x = -y$ so that $x \to -\infty \Rightarrow y \to +\infty$

$$\therefore \quad \lim_{y \to \infty} \left[\frac{8y}{\sqrt{y^2 + 8y} + y}\right] = \lim_{y \to \infty} \left[\frac{8}{\sqrt{1 + \frac{8}{y}} + 1}\right] = \frac{8}{2} = 4$$

EXERCISE 3.4

Evaluate the following limits

1. $\lim_{x \to -\infty} \sqrt{4x^2 + 7x} + 2x$

2. $\lim_{x \to -\infty} \left(\sqrt{x^2 - 8x} + x\right)$

3. $\lim_{x \to -\infty} 2^x$

4. $\lim_{x \to +\infty} 5^x \sin\left(\frac{a}{5^x}\right)$

5. $\lim_{x \to \infty} [\tan x + \sec x \ (5 \cos x - \sin x)]$

ANSWERS

1. $\frac{-7}{4}$ **2.** 4 **3.** 0 **4.** a

5. 5

3.10 ONE SIDED LIMITS

Left hand limit

$\lim_{x \to a^-} f(x)$ is called the left hand limit (L.H.L) of $f(x)$ at $x = a$.

Right hand limit

$\lim_{x \to a^+} f(x)$ is called the right hand limit (R.H.L) of $f(x)$ at $x = a$.

Both of these are called one sided limit of $f(x)$ at $x = a$.

Note 1. If the left hand limit and the right hand limit exists at a point and are not equal, then we can say that the limit at that point does not exist.

2. If the left hand limit and the right hand limit exists and are equal, then we conclude that $\lim_{x \to a} f(x)$ exists and

$$\lim_{x \to a} f(x) = \lim_{x \to a^+} f(x) = \lim_{x \to a^-} f(x)$$

Rule of finding $\lim_{x \to a^-} f(x)$ *i.e.*, L.H.L

Replace x by $(a - h)$ and take the limit as $h \to 0$

i.e., $$\lim_{x \to a^-} f(x) = \lim_{h \to 0} f(a - h)$$

Rule for finding $\lim_{x \to a^+} f(x)$ i.e. R.H.L

Replace x by $(a + h)$ and take the limit as $h \to 0$

i.e., $$\lim_{x \to a^+} f(x) = \lim_{h \to 0} f(a + h)$$

ILLUSTRATIVE EXAMPLES

Example 1. Show that $\lim_{x\to 0} \log(1+x) = 0$

Solution:

$$\lim_{x\to 0^+} \log(1+x) = \lim_{h\to 0} \log(1+0+h)$$

$$= \lim_{h\to 0} \log(1+h) = \lim_{h\to 0}\left[h-\frac{h^2}{2}+\frac{h^3}{3}-...\right]=0$$

$$\lim_{x\to 0^-} \log(1+x) = \lim_{h\to 0} \log(1+0-h)$$

$$= \lim_{h\to 0} \log(1-h) = \lim_{h\to 0}\left[h+\frac{h^2}{2}+\frac{h^3}{3}+...\right]=0$$

$$\lim_{x\to 0^+} \log(1+x) = \lim_{x\to 0^-} \log(1+x) = 0$$

$$\therefore \lim_{x\to 0^-} \log(1+x) = 0$$

Example 2. Show that $\lim_{x\to 1} [x]$ does not exist

Solution: We have

$$\lim_{x\to 1^+} [x] = 1 \text{ [Since for } 1 \le x < 2, [x]=1]$$

and $$\lim_{x\to 1^-} [x] = 0 \text{ [Since for } 0 \le x < 1, [x]=0]$$

Thus $$\lim_{x\to 1^+} [x] \ne \lim_{x\to 1^-} [x]$$

Hence $\lim_{x\to 1} [x]$ does not exist.

Example 3. Show that $\lim_{x\to 0} e^x = 1$

Solution:

$$\lim_{x\to 0^+} e^x = \lim_{h\to 0} e^{0+h}$$

$$= \lim_{h\to 0} e^h = \lim_{h\to 0}\left[1+h+\frac{h^2}{2!}+\frac{h^3}{3!}+...\right]=1$$

$$\lim_{x\to 0^-} e^x = \lim_{h\to 0} e^{(0-h)}$$

$$= \lim_{h\to 0} e^{-h} = \lim_{h\to 0}\left[1-h+\frac{h^2}{2!}-\frac{h^3}{3!}+...\right]=1$$

Since $$\lim_{x\to 0^+} e^x = \lim_{x\to 0^-} e^x = 1, \ \lim_{x\to 0} e^x = 1$$

Example 4. Show that $\lim_{x \to 0} \frac{x}{|x|}$ does not exist.

Solution:

$$\lim_{x \to 0^+} \frac{x}{|x|} = \lim_{h \to 0} \frac{(0+h)}{|0+h|} = \lim_{h \to 0} \frac{h}{|h|} = \lim_{h \to 0} \frac{h}{h} = 1$$

$$\lim_{x \to 0^-} \frac{x}{|x|} = \lim_{h \to 0} \frac{0-h}{|0-h|} = \lim_{h \to 0} \frac{-h}{|-h|} = \lim_{h \to 0} \frac{-h}{h} = -1$$

Since $\lim_{x \to 0^+} \frac{x}{|x|} \neq \lim_{x \to 0^-} \frac{x}{|x|}$

Hence $\lim_{x \to 0} \frac{x}{|x|}$ does not exist.

Example 5. Show that $\lim_{x \to 0} |x| = 0$

Solution: Let

$$f(x) = |x| = \begin{cases} x, & \text{when } x \geq 0 \\ -x, & \text{when } x < 0 \end{cases}$$

Now, $\lim_{x \to 0^+} |x| = \lim_{h \to 0} |0+h| = \lim_{h \to 0} |h| = \lim_{h \to 0} h = 0$

$$\lim_{x \to 0^-} |x| = \lim_{h \to 0} |0-h| = \lim_{h \to 0} |-h| = \lim_{h \to 0} h = 0$$

$\therefore$ $\lim_{x \to 0^+} |x| = \lim_{x \to 0^-} |x| = 0$

Hence $\lim_{x \to 0} |x| = 0$

Example 6. Find $\lim_{x \to 4} f(x)$, where $f(x) = \begin{cases} 5, & \text{if } x \geq 4 \\ x+1, & \text{if } x < 4 \end{cases}$

Solution: We have $\lim_{x \to 4^+} f(x) = \lim_{x \to 4^+} 5 = 5$

$$\lim_{x \to 4^-} f(x) = \lim_{x \to 4^-} (x+1) = 4+1 = 5$$

$$\lim_{x \to 4^+} f(x) = 5 = \lim_{x \to 4^-} f(x)$$

Hence $\lim_{x \to 4} f(x)$ exists and is equal to 5.

EXERCISE 3.5

1. Evaluate $\lim_{x \to 2} (3x+4)$

2. Evaluate $\lim_{x \to 0} \frac{1}{3x+6}$

3. Show that $\lim_{x \to 0} 2^{1/x}$ does not exist.

4. Show that $\lim_{x \to 0} \left(\frac{e^{1/x} - 1}{e^{1/x} + 1} \right)$ does not exist.

5. Prove that $\lim_{x \to 0} \frac{1}{x}$ does not exist.

6. Show that $\lim_{x \to 0} e^{-1/x}$ does not exist.

7. Show that $\lim_{x \to 0} \cos \frac{1}{x}$ does not exist.

ANSWERS

1. 10 2. $\frac{1}{6}$

3.11 CONTINUITY

In this section, we define the concept of a continuous function and prove the continuity of some well known functions.

Continuity at a point: A real valued function $f(x)$ is said to be continuous at a point $x = a$ of its domain if $\lim_{x \to a} f(x)$ exists and equal to $f(a)$.

i.e., $f(x)$ is continuous at $x = a$, if

$$\lim_{x \to a^+} f(x) = \lim_{x \to a^-} f(x) = f(a)$$

Note: If a function $f(x)$ is not continuous at a point, we say that $f(x)$ is discontinuous at that point.

ILLUSTRATIVE EXAMPLES

Example 1: Show that the function $f(x) = x^2$ is continuous at $x = 1$.

Solution: We have $f(1) = (1)^2 = 1$

$$\lim_{x \to 1^+} f(x) = \lim_{x \to 1^+} x^2 = \lim_{h \to 0} (1+h)^2 = \lim_{h \to 0} (1 + 2h + h^2) = 1$$

$$\lim_{x \to 1^-} f(x) = \lim_{x \to 1^-} x^2 = \lim_{h \to 0} (1-h)^2 = \lim_{h \to 0} (1 - 2h + h^2) = 1$$

$$\therefore \quad \lim_{x \to 1^+} f(x) = \lim_{x \to 1^-} f(x) = f(1)$$

Hence $f(x)$ is continuous at $x = 1$.

Example 2. Show that the function

$$f(x) = \begin{cases} x \sin \frac{1}{x}, & \text{when } x \neq 0 \\ 0, & \text{when } x = 0 \end{cases}$$

is continuous at $x = 0$.

Solution: Since $f(0) = 0$

$$\lim_{x\to 0} f(x) = \lim_{x\to 0} x\sin\frac{1}{x} = 0 \times (\text{a finite quantity}) = 0$$

$$\therefore \quad \lim_{x\to 0} f(x) = f(0) = 0$$

Hence $f(x)$ is continuous at $x = 0$.

Example 3. Determine the value of k for which the function.

$$f(x) = \begin{cases} \dfrac{\sin 3x}{7x}, & \text{if } x \neq 0 \\ k, & \text{if } x = 0 \end{cases}$$

is continuous at $x = 0$

Solution: Since $f(0) = k$

$$\lim_{x\to 0} f(x) = \lim_{x\to 0} \frac{\sin 3x}{7x} = \lim_{x\to 0}\left(\frac{\sin 3x}{3x} \times \frac{3}{7}\right)$$

$$= \frac{3}{7}\lim_{x\to 0}\left(\frac{\sin 3x}{3x}\right) = \frac{3}{7}\times 1 = \frac{3}{7}$$

For continuity of $f(x)$ at $x = 0$, we must have $f(0) = \lim_{x\to 0} f(x)$ *i.e.* $k = \dfrac{3}{7}$

Example 4. Show that the function

$$f(x) = \begin{cases} \dfrac{x}{|x|}, & \text{when } x \neq 0 \\ 1, & \text{when } x = 0 \end{cases}$$

is discontinuous at $x = 0$.

Solution: Since $f(0) = 1$

$$\lim_{x\to 0^+} f(x) = \lim_{h\to 0} f(0+h) = \lim_{h\to 0}\frac{h}{|h|} = \lim_{h\to 0}\frac{h}{h} = 1$$

$$\lim_{x\to 0^-} f(x) = \lim_{h\to 0} f(0-h) = \lim_{h\to 0}\frac{-h}{|-h|} = \lim_{h\to 0}\frac{-h}{h} = -1$$

Since $\lim_{x\to 0^+} f(x) \neq \lim_{x\to 0^-} f(x)$

So, $\lim_{x\to 0} f(x)$ does not exist.

Hence $f(x)$ is discontinuous at $x = 0$.

Example 5. Discuss the continuity of the function $f(x) = \begin{cases} 3x+2, & \text{when } x \leq 0 \\ x+2, & \text{when } x > 0 \end{cases}$ at $x = 0$.

Solution: Since $f(0) = 3 \times 0 + 2 = 2$

$$\lim_{x\to 0^+} f(x) = \lim_{h\to 0} f(0+h) = \lim_{h\to 0} f(h) = \lim_{h\to 0}(h+2) = 2$$

$$\lim_{x\to 0^-} f(x) = \lim_{h\to 0} f(0-h) = \lim_{h\to 0} f(-h)$$

$$= \lim_{h\to 0} f[3(-h)+2] = \lim_{h\to 0} [-3h+2] = 2$$

Thus $\lim_{x\to 0^+} f(x) = \lim_{x\to 0^-} f(x) = f(0)$

Hence, the given function is continuous at $x = 0$.

3.12 REMOVABLE DISCONTINUITY

If the limit of a function exist at point a but is not equal to the value of the function at a i.e. if

$$\lim_{x\to a} f(x) \text{ exists and} \neq f(a)$$

then by changing the value of the function at a, we can make the function continuous at a. Such type of discontinuity is called removable discontinuity.

Example 6. Show that the function $f(x) = \begin{cases} 2x, \text{ if } x<2 \\ 2, \text{ if } x=2 \\ x^2, \text{ if } x>2 \end{cases}$ has a removable discontinuity at $x = 2$.

Solution: We have to show that $\lim_{x\to 2} f(x)$ exists but it is not equal to $f(2)$.

Clearly $f(2)=2$

Now $\lim_{x\to 2^+} f(x) = \lim_{h\to 0} f(2+h) = \lim_{h\to 0} (2+h)^2$

$$= \lim_{h\to 0} (4+h^2+4h) = 4$$

$$\lim_{x\to 2^-} f(x) = \lim_{h\to 0} f(2-h) = \lim_{h\to 0} 2(2-h)$$

$$= \lim_{h\to 0} (4-2h) = 4$$

$\therefore$ $\lim_{x\to 2^+} f(x) = \lim_{x\to 2^-} f(x) = 4$

Thus $\lim_{x\to 2} f(x) \neq f(2)$

Hence, $f(x)$ has a removable discontinuity at $x = 2$.

EXERCISE 3.6

1. Prove that $f(x) = \sin^2 x + x^2 - 2x$ is continuous at $x = 0$

2. Show that $f(x) = \begin{cases} \dfrac{\sin x}{x}, \text{ if } x<0 \\ x+1, \text{ if } x\geq 0 \end{cases}$ is continuous at $x = 0$.

3. Show that $f(x) = \begin{cases} \dfrac{x^2-1}{x+1}, \text{ if } x \neq -1 \\ -2, \text{ if } x = -1 \end{cases}$ is continuous at $x = -1$.

4. Find the value of k so that the function

$$f(x) = \begin{cases} kx^2, \text{ if } x \leq 2 \\ 3, \text{ if } x > 2 \end{cases}$$ is continuous at $x = 2$.

5. Prove that $f(x) = \begin{cases} \sin\dfrac{1}{x}, \text{ if } x \neq 0 \\ 0, \text{ if } x = 0 \end{cases}$ is discontinuous at $x = 0$.

6. Prove that $f(x) = \begin{cases} \dfrac{|x-a|}{x-a}, \text{ if } x \neq a \\ 1, \text{ if } x = a \end{cases}$ is discontinuous at $x = a$.

7. Prove that $f(x) = \begin{cases} \dfrac{\sin 3x}{x}, \text{ if } x \neq 0 \\ 1, \text{ if } x = 0 \end{cases}$ is discontinuous at $x = 0$.

8. Show that the function $f(x) = \begin{cases} \dfrac{x^2-1}{x-1}, \text{ if } x \neq 1 \\ 2, \text{ if } x = 1 \end{cases}$ is continuous at $x = 1$.

9. Find the value of λ for which

$$f(x) = \begin{cases} \dfrac{x^2-2x-3}{x+1}, \text{ when } x \neq -1 \\ \lambda, \text{ when } x = -1 \end{cases}$$ is continuous at $x = -1$.

ANSWERS

4. $k = \dfrac{3}{4}$ 9. $\lambda = -4$

3.13 CONTINUITY IN AN INTERVAL

A real valued function is said to be continuous in an interval (open or closed) if it is continuous at each point of the interval. When a function f is considered on a closed interval $[a, b]$ then f is continuous on $[a, b]$ means that it is continuous at each point of $[a, b]$ including the end points a and b.

If $f(x)$ is defined on a closed internal $[a, b]$, we say that

(*i*) f is continuous at a if $\lim\limits_{x \to a^+} f(x) = f(a)$;

(*ii*) f is continuous at b if $\lim\limits_{x \to b^-} f(x) = f(b)$;

(*iii*) f is continuous in $[a, b]$ if it is continuous at a, at b and at each point of (a, b).

Note: When we say that a real function f is continuous (with out mentioning where) we mean that it is continuous at each point of its domain.

3.14 PROPERTIES OF CONTINUOUS FUNCTIONS

Theorem 1: Let f and g be two real functions continuous at $x = a$ and let α be a real number, then

(*i*) $f + g$ is continuous at a

(*ii*) $f - g$ is continuous at a

(*iii*) αf is continuous at a

(*iv*) fg is continuous at a

(*v*) $\dfrac{1}{f}$ is continuous at a, provided $f(a) \neq 0$

(*vi*) $\dfrac{f}{g}$ is continuous at a, provided $g(a) \neq 0$

Proof: (*i*) Since f and g are continuous at a,

$$\therefore \quad \lim_{x \to a} f(x) = f(a) \text{ and } \lim_{x \to a} g(x) = g(a)$$

$$\text{Then} \quad \lim_{x \to a} f(x) + \lim_{x \to a} g(x) = f(a) + g(a) = (f + g)(a)$$

$$\text{Implies that} \quad \lim_{x \to a} (f + g)(x) = \lim_{x \to a} [f(x) + g(x)]$$

$$= \lim_{x \to a} f(x) + \lim_{x \to a} g(x) = (f + g)(a)$$

Thus $f + g$ is continuous at a.

The proofs of the remaining parts is left to the reader.

Theorem 2. Every constant function is continuous.

Proof: Let $f(x) = C$, where C is a constant.

Since the domain of a constant function is R. Let a be an arbitrary real number, then $f(a) = C$ and $\lim_{x \to a} f(x) = \lim_{x \to a} C = C$ $\quad [\because \; f(x) = C]$

$$\therefore \quad \lim_{x \to a} f(x) = f(a) = C$$

Thus $f(x)$ is continuous at $x = a$ for all $a \in R$

Hence $f(x)$ is continuous.

Theorem 3. Show that the identity function is continuous.

Proof: Let $f(x) = x$, for all $x \in R$

Let a be an arbitrary real number, then $f(a) = a$

Now, $\lim_{x \to a} f(x) = \lim_{x \to a} x = a$ $[\because \; f(x) = x]$

$\therefore$ $\lim_{x \to a} f(x) = f(a) = a$

This implies that $f(x)$ is continuous at $x = a$ for all $a \in R$.

Hence, the identity function is continuous.

Theorem 4. Every polynomial function is continuous.

Proof: Let $f(x) = a_0 + a_1x + a_2x^2 + \ldots\ldots + a_nx^n$ be a polynomial. We prove it by mathematical induction.

when $n = 0$, then $f(x) = a_0$, which being a constant function is continuous.

when $n = 1$, then $f(x) = a_0 + a_1x$

Since $f(x)$ is the sum of a constant function and a multiple of the identity function. It, being the sum of two continuous function is continuous. Let polynomial of degree n be continuous. Consider a general polynomial of degree $(n + 1)$ *i.e.*,

$$a_0 + a_1x + a_2x^2 + \ldots + a_nx^n + a_{n+1}\, x^{n+1}$$

This can be written as $a_0 + x\,(\,a_1 + a_2x + \ldots. + a_nx^{n+1} + a_{n+1}\,x^n)$

This is the sum of a constant function a_0 and the product of the identity function x and the polynomial function $a_1 + a_2x + \ldots.\; a_{n+|}\,x^n$ of degree at most n (which we assumed to be continuous).

Therefore, it is continuous.

Thus, the continuity of a polynomial of degree n implies that the continuity of a polynomial of degree $(n + 1)$.

Hence, by principle of mathematical induction, we can say that every polynomial function is continuous.

Theorem 5. The exponential function is continuous.

Proof: Let $f(x) = e^x$, then dom $(f) = R$

Let a be arbitrary real number, then,

$$\lim_{x \to a} e^x = \lim_{h \to 0} e^{a+h} = \lim_{h \to 0} e^a . e^h$$

$$= e^a \lim_{h \to 0} e^h = e^a \times 1 = e^a$$

$\therefore$ $f(x) = e^x$ is continuous at $x = a$ for all $a \in R$

Hence, the exponential function is continuous.

In the same way, we can prove the following theorems.

Theorem 6. Every rational function is continuous.

Theorem 7. The sine function is continuous

Theorem 8. The cosine function is continuous.

Theorem 9. The tangent function is continuous.

Theorem 10. The logarithmic function is continuous.

ILLUSTRATIVE EXAMPLES

Example 1. Show that the function defined by $f(x) = |x|, x \in R$ is continuous at every point of R

Solution: We have

$$f(x) = \begin{cases} x, & \text{if } x > 0 \\ 0, & x = 0 \\ -x, & \text{if } x < 0 \end{cases}$$

Let a be any real number. If $a = 0$, then

$$\lim_{x\to 0^+} f(x) = 0 = \lim_{x\to 0^-} f(x)$$

$$\therefore \quad \lim_{x\to 0} f(x) = 0 = f(0)$$

$\Rightarrow f$ is continuous at $a = 0$.

If $a > 0$, then

$$\lim_{x\to a^+} f(x) = a = \lim_{x\to a^-} f(x)$$

$$\Rightarrow \quad \lim_{x\to a} f(x) = a = f(a). \qquad (\text{as } a > 0)$$

$\Rightarrow f$ is continuous at all $a > 0$.

If $a < 0$, then

$$\lim_{x\to a^+} f(x) = -a = \lim_{x\to a^-} f(x)$$

$$\Rightarrow \quad \lim_{x\to a} f(x) = -a = f(a) \qquad (\text{as } a < 0)$$

$\Rightarrow f$ is continuous at all $a < 0$.

Hence, f is continuous at any real number a.

Example 2. Prove that the function defined by

$$f(x) = x - |x|, \quad x \in R$$

is continuous at $x = 0$.

Solution: The given function may be written as

$$f(x) = \begin{cases} 0, \text{ if } x \geq 0 \text{ as } |x| = x \text{ for } x \geq 0 \\ 2x, \text{ if } x < 0, \text{ as } |x| = -x \text{ for } x < 0 \end{cases}$$

$$\therefore \quad \lim_{x\to 0^+} f(x) = \lim_{x\to 0^+} 0 = 0$$

$$\lim_{x \to 0^-} f(x) = \lim_{x \to 0^-} (2x) = 0$$

Also $f(0) = 0$.

Thus $\lim_{x \to 0} f(x) = 0 = f(0)$

Hence $f(x)$ is continuous at $x = 0$.

Example 3. Discuss the continuity of the function $f(x) = \begin{cases} \dfrac{x}{|x|}, \text{ if } x \neq 0 \\ 0, \text{ if } x = 0 \end{cases}$

Solution: We know that identity function x is continuous and the modulus function $|x|$ is continuous. So the quotient function $\dfrac{x}{|x|}$ is continuous at each $x \neq 0$.

Since we know that $f(x)$ is discontinuous at $x = 0$.

Hence the given function is continuous at each point, except at $x = 0$.

Example 4. Show that the function f given by

$$f(x) = \begin{cases} x, \text{ if } x \geq 1 \\ x^2, \text{ if } x < 1 \end{cases}$$

is continuous everywhere on R.

Solution: Let a be any real number.

If $a > 1$, then

$$\lim_{x \to a} f(x) = \lim_{x \to a} x = a = f(a) \qquad [\therefore f(x) = x]$$

$\Rightarrow f$ is continuous at $a > 1$.

If $a < 1$, then

$$\lim_{x \to a} f(x) = \lim_{x \to a} x^2 = a^2 = f(a) \qquad (\because f(x) = x^2)$$

$\Rightarrow f$ is continuous at $a < 1$.

If $a = 1$, then

$$\lim_{x \to 1^+} f(x) = \lim_{x \to 1^+} x = 1.$$

and $$\lim_{x \to 1^-} f(x) = \lim_{x \to 1^-} x^2 = 1.$$

$$\Rightarrow \quad \lim_{x \to 1} f(x) = 1 = f(1)$$

$\Rightarrow f$ is continuous at $a = 1$.

Example 5. Determine the value of k, So that the function

$$f(x) = \begin{cases} kx^2, \text{ if } x \leq 2 \\ 5, \text{ if } x > 2 \end{cases} \text{ is continuous.}$$

Solution: Since a polynomial function is continuous and a constant function is continuous, the given function is continuous for all $x < 2$ and for all $x > 2$.

So, consider the point $x = 2$. We have $f(2) = 4k$

Also, $\lim_{x \to 2^+} f(x) = \lim_{h \to 0} f(2+h) = \lim_{h \to 0} 5 = 5$

and, $\lim_{x \to 2^-} f(x) = \lim_{h \to 0} f(2-h) = \lim_{h \to 0} k(2-h)^2 = 4k$

Therefore for continuity, we must have $4k = 5$

$$\therefore \quad k = \frac{5}{4}$$

Example 6. Discuss the continuity of the function

$$f(x) = \begin{cases} \dfrac{\sin x}{x}, \text{ if } x < 0 \\ x+1, \text{ if } x \geq 0 \end{cases}$$

Solution: We know that sin x as well as identity function x is continuous, so the quotient function $\dfrac{\sin x}{x}$ is continuous at each $x < 0$.

Also, when $x > 0$, we have $f(x) = x + 1$, which being a polynomial function, is continuous.

Let us consider the point $x = 0$.

$$f(0) = 0+1 = 1$$

$$f(0) = 0+1 = 1$$

$$\lim_{x \to 0^+} f(x) = \lim_{h \to 0} f(0+h) = \lim_{h \to 0} (h+1) = 1$$

$$\lim_{x \to 0^-} f(x) = \lim_{h \to 0} f(0-h) = \lim_{h \to 0} f(-h)$$

$$= \lim_{h \to 0} \frac{\sin(-h)}{-h} = \lim_{h \to 0} \left(\frac{-\sinh}{-h} \right) = \lim_{h \to 0} \frac{\sinh}{h} = 1$$

$$\therefore \quad \lim_{x \to 0^+} f(x) = \lim_{x \to 0^-} f(x) = 1$$

$$\lim_{x \to 0} f(x) = 1 = f(0)$$

$\Rightarrow f(x)$ is continuous at $x = 0$ also

Hence $f(x)$ is continuous at all points.

Example 7. Discuss the continuity of the function $f(x) = \begin{cases} 7x+5, \text{ when } x \geq 0 \\ 5+2x, \text{ when } x < 0 \end{cases}$

Solution: When $x < 0$. We have $f(x) = 2x + 5$, which being a polynomial function is continuous at each point where $x < 0$.

Also, when $x > 0$, we have $f(x) = 7x + 5$, which being a polynomial function, is continuous at each points where $x > o$.

Consider at point $x = 0$.

$$f(0) = 7 \times 0 + 5 = 5$$

$$\lim_{x \to 0^+} f(x) = \lim_{h \to 0} f(0+h) = \lim_{h \to 0} f(h) = \lim_{h \to 0} [7h+5] = 5$$

and, $$\lim_{x \to 0^-} f(x) = \lim_{h \to 0} f(0-h) = \lim_{h \to 0} f(-h) = \lim_{h \to 0} [5-2h] = 5$$

$\therefore$ $$\lim_{x \to 0^+} f(x) = \lim_{x \to 0^-} f(x) = 5.$$

Hence $f(x)$ is continuous at every point.

EXERCISE 3.7

1. Show that $f(x) = \begin{cases} x^2, \text{ if } x < 1 \\ x, \text{ if } x \geq 1 \end{cases}$ is continuous for all $x \in R$.

2. Let $f(x) = \begin{cases} 1, \text{ if } x \leq 3 \\ ax+b, \text{ if } 3 < x < 5 \\ 7, \text{ if } 5 \leq x \end{cases}$, find the values of a and b so that $f(x)$ is continuous.

3. Show that the function $f(x) = |\sin x + \cos x|$ is continuous at $x = \pi$.

4. Discuss the continuity of $f(x) = [x]$.

5. Show that $f(x) = \begin{cases} x, \text{ if } x \neq 0 \\ 1, \text{ if } x = 0 \end{cases}$ is continuous at each point except 0.

6. Discuss the continuity of the function $f(x) = |x| + |x-1|$ in the interval $[-1, 2]$

7. Prove that $f(x) = |\sin x|$ is continuous at all point of its domain.

8. Show that $f(x) = \begin{cases} x^2 \sin\dfrac{1}{x}, \text{ if } x \neq 0 \\ 0, \text{ if } x = 0 \end{cases}$ is continuous at $x = 0$.

9. Show that $f(x) = \begin{cases} 2x-1, \text{ if } x < 2 \\ \dfrac{3x}{2}, \text{ if } x \geq 2 \end{cases}$ is continuous at $x = 2$.

10. Find the value of k so that the function

$$f(x) = \begin{cases} k(x^2 - 2x), \text{ if } x < 0 \\ \cos x, \text{ if } x \geq 0 \end{cases}$$ is continuous at $x = 0$.

ANSWERS

2. $a = 3, b = -8$ **3.** continuous

4. discontinuous at $x = n$, where n is an integer.

6. continuous **7.** continuous

8. continuous **9.** continuous

10. No value of k will make f continuous at 0.

3.15 DIFFERENTIABILITY

Let $f(x)$ be a real function and a be any real number. Then we define:

1. Right hand derivative

The right hand derivative of a function f at a point a is denoted by $Rf'(a)$ and defined as $Rf'(a) = \lim_{h \to 0} \dfrac{f(a+h) - f(a)}{h}$, provided the limit exists.

2. Left hand derivative

The left hand derivative of a function f at a point a is denoted by $Lf'(a)$ and defined as $Lf'(a) = \lim_{h \to 0} \dfrac{f(a-h) - f(a)}{-h}$, provided the limit exists.

3. Differentiability at a point

A function $f(x)$ is said to be differentiable at $x = a$, if

(i) Both $Rf'(a)$ and $Lf'(a)$ exist and are finite. (ii) $Rf'(a) = Lf'(a)$

Note: $R\,f'(a) \neq L\,f'(a)$, we say that $f(x)$ is not differentiable at $x = a$.

3.16 RELATION BETWEEN CONTINUITY AND DIFFERENTIABILITY

Theorem. Every differentiable function is continuous, but every continuous function need not be differentiable.

Proof: Let $f(x)$ be a differentiable function and let a be any real number, then

$$\lim_{h \to 0} \frac{f(a+h) - f(a)}{h} = f'(a) \qquad ...(1)$$

$$\lim_{h \to 0} [f(a+h) - f(a)] = \lim_{h \to 0} \left[\frac{f(a+h) - f(a)}{h}\right] \times h$$

$$= \lim_{h \to 0} \frac{f(a+h) - f(a)}{h} \times \lim_{h \to 0} h = f'(a) \times 0 = 0$$

[Using (1)]

Thus, $\lim_{h \to 0} [f(a+h) - f(a)] = 0$ or $\lim_{h \to 0} f(a+h) = f(a)$

This shows that $f(x)$ is continuous at a for all a.

Hence, every differentiable function is continuous.

In order to show that a continuous function need not be differentiable, it is sufficient to give an example of a function which is continuous but not differentiable.

For example, the function given by $f(x) = |x|$ is continuous at $x = 0$, but not differentiable at $x = 0$.

Since $\quad f(0) = 0$

$$\lim_{x \to 0^+} f(x) = \lim_{h \to 0} f(0 + h) = \lim_{h \to 0} f(h) = \lim_{h \to 0} |h| = \lim_{h \to 0} h = 0$$

Also $$\lim_{x \to 0^-} f(x) = \lim_{h \to 0} f(0 - h) = \lim_{h \to 0} f(-h) = \lim_{h \to 0} |-h| = \lim_{h \to 0} h = 0$$

Thus, $$\lim_{x \to 0^+} f(x) = \lim_{x \to 0^-} f(x) = f(0)$$

So, $\quad f(x) = |x|$ is continuous at $x = 0$.

But $$Lf'(0) = \lim_{h \to 0} \frac{f(0 - h) - f(0)}{-h}$$

$$= \lim_{h \to 0} \frac{f(-h) - f(0)}{-h} = \lim_{h \to 0} \frac{|-h| - 0}{-h}$$

$$= \lim_{h \to 0} \frac{h}{-h} = -1.$$

Also, $$R f'(0) = \lim_{h \to 0} \frac{f(0 + h) - f(0)}{h}$$

$$= \lim_{h \to 0} \frac{f(h) - f(0)}{h} = \lim_{h \to 0} \frac{|h| - 0}{h} = \lim_{h \to 0} \frac{h}{h} = 1.$$

Thus $\quad L f'(0) \neq R f'(0)$

This shows that $f(x) = |x|$ is not differentiable at $x = 0$. Hence a continuous function need not be differentiable.

ILLUSTRATIVE EXAMPLES

Example 1. Show that $f(x) = x^{5/3}$ is differentiable at $x = 0$.

Solution: $$Lf'(0) = \lim_{h \to 0} \frac{f(0 - h) - f(0)}{-h} = \lim_{h \to 0} \frac{f(-h) - 0}{-h}$$

$$= \lim_{h \to 0} \frac{(-h)^{5/3} - 0}{-h}$$

$$= \lim_{h \to 0} \frac{(-h)^{5/3}}{(-h)} = \lim_{h \to 0} (-h)^{2/3} = 0$$

$$Rf'(0) = \lim_{h \to 0} \frac{f(0 + h) - f(0)}{h} = \lim_{h \to 0} \frac{f(h) - f(0)}{h}$$

$$= \lim_{h \to 0} \frac{h^{5/3} - 0}{h} = \lim_{h \to 0} \frac{h^{5/3}}{h} = \lim_{h \to 0} (h)^{2/3} = 0$$

Thus, $\quad L\,f'(0) = R\,f'(0)$

Hence $\quad f(x) = x^{5/3}$ is differentiable at $x = 0$.

Example 2. Show that $f(x) = x^2$ is differentiable at $x = 2$.

Solution:
$$Lf'(2) = \lim_{h\to 0}\frac{f(2-h)-f(2)}{-h} = \lim_{h\to 0}\frac{(2-h)^2-(2)^2}{-h}$$
$$= \lim_{h\to 0}\left[\frac{4+h^2-4h-4}{-h}\right] = \lim_{h\to 0}(-h+4) = 4$$
$$Rf'(2) = \lim_{h\to 0}\frac{f(2+h)-f(2)}{h} = \lim_{h\to 0}\left[\frac{(2+h)^2-(2)^2}{h}\right]$$
$$= \lim_{h\to 0}\left[\frac{4+h^2+4h-4}{h}\right] = \lim_{h\to 0}(h+4) = 4$$

Since $\quad L\,f'(2) = R\,f'(2) = 4$

Hence $f(x)$ is differentiable at $x = 2$.

Example 3. Show that $f(x) = [x]$ is not differentiable at $x = 2$.

Solution:
$$L\,f'(2) = \lim_{h\to 0}\frac{f(2-h)-f(2)}{-h} = \lim_{h\to 0}\frac{[2-h]-[2]}{-h} = \infty$$
$$[\because [2-h] = 1 \text{ and } [2] = 2\,]$$

Also,
$$R\,f'(2) = \lim_{h\to 0}\frac{f(2+h)-f(2)}{h} = \lim_{h\to 0}\frac{[2+h]-[2]}{h} = 0$$
$$[\because [2+h] = 2 \text{ and } [2] = 2]$$

Thus, $\quad L\,f'(2) \neq R\,f'(2)$

Hence $f(x) = [x]$ is not differentiable at $x = 2$.

Example 4. Let $f(x) = \begin{cases} 1+\sin x, & \text{if } 0 \le x < \pi/2 \\ 1 & \text{, if } x < 0 \end{cases}$

Show that $f'(0)$ does not exists.

Solution: $Lf'(0) = \lim\limits_{h\to 0}\dfrac{f(0-h)-f(0)}{-h} = \lim\limits_{h\to 0}\dfrac{f(h)-f(0)}{-h} = \lim\limits_{h\to 0}\dfrac{1-1}{-h} = 0$

and
$$Rf'(0) = \lim_{h\to 0}\frac{f(0+h)-f(0)}{h} = \lim_{h\to 0}\frac{f(h)-f(0)}{h}$$
$$= \lim_{h\to 0}\frac{(1+\sinh)-1}{h} = \lim_{h\to 0}\frac{\sinh}{h} = 1$$

Since $\quad L\,f'(0) \neq R\,f'(0)$

Hence $f'(0)$ does not exist.

Example 5. Show that $f(x) = |x-1|$ is continuous but not differentiable at $x = 1$.

Solution: We have $f(1) = |1-1| = 0$

$$\lim_{x\to 1^-} f(x) = \lim_{h\to 0} f(1-h)$$

$$= \lim_{h\to 0} |1-h-1| = \lim_{h\to 0} |-h| = \lim_{h\to 0} h = 0$$

and $$\lim_{x\to 1^+} f(x) = \lim_{h\to 0} f(1+h) = \lim_{h\to 0} |1+h-1| = \lim_{h\to 0} |h| = \lim_{h\to 0} h = 0$$

$\therefore$ $$\lim_{x\to 1^-} f(x) = \lim_{x\to 1^+} f(x) = f(1) = 0$$

So, $f(x)$ is continuous at $x = 1$.

But $$L f'(1) = \lim_{h\to 0} \frac{f(1-h)-f(1)}{-h} = \lim_{h\to 0} \frac{|1-h-1|-0}{-h}$$

$$= \lim_{h\to 0} \frac{|-h|}{-h} = \lim_{h\to 0} \frac{h}{-h} = -1$$

and $$R f'(1) = \lim_{h\to 0} \frac{f(1+h)-f(1)}{h} = \lim_{h\to 0} \frac{|1+h-1|-0}{h}$$

$$= \lim_{h\to 0} \frac{|h|}{h} = \lim_{h\to 0} \frac{h}{h} = 1$$

Thus $L f'(1) \neq R' f'(1)$

This show that $f(x)$ is not differentiable at $x = 1$.

Example 6. Show that the function $f(x) = \begin{cases} x \sin\frac{1}{x}, & \text{when } x \neq 0 \\ 0 & \text{, when } x = 0 \end{cases}$

is continuous but not differentiable at $x = 0$.

Solution: We have already discussed that the above function is continuous at $x = 0$.

Now $$R f'(0) = \lim_{h\to 0} \frac{f(0+h)-f(0)}{h} = \lim_{h\to 0} \frac{h \sin\frac{1}{h} - 0}{h}$$

$$= \lim_{h\to 0} \sin\frac{1}{h}, \text{ which does not exist.}$$

Hence $f(x)$ is not differentiable at $x = 0$.

EXERCISE 3.8

1. show that $f(x) = [x]$ is not differentiable at $x = 1$.
2. show that $f(x) = x^2$ is differentiable at $x = 1$.
3. show that the function $f(x) = \begin{cases} 1+x, \text{ if } x \le 2 \\ 5-x, \text{ if } x > 2 \end{cases}$ is not differentiable at $x = 2$.
4. if $f(x) = |x|$, show that $f'(2) = 1$.

5. show that the function $f(x) = |x-4|$ is continuous but not differentiable at $x = 4$.
6. show that $f(x) = x^{3/2}$ is not differentiable at $x = 0$.
7. Find the values of m and n so that the function

$$f(x) = \begin{cases} (x^2 + 3x + m), & \text{when } x \le 1; \\ (nx + 2), & \text{when } x > 1 \end{cases}$$ is differentiable at each $x \in R$.

ANSWERS

7. $m = 3, n = 5$.

3.17 DIFFERENTIATION

The first and the most important application of the limit concept will be to the problems of finding the instantaneous rate of change of a function with respect to the independent variable. Geometrically this problem will be seen to be equivalent to that of finding the slope of the tangent line at a point of the graph of the function. These problems are solved by finding the derivative of the given function.

Before we define the derivative of a function, let us consider the behavior of the function as the value of the independent variable changes. The change in the independent variable will definitely result in the change in the dependent variable. These changes in the variable are called increments.

3.18 DERIVATIVE OF A FUNCTION AT A POINT

Let $y = f(x)$ be a given continuous function. Let δy be an increment in y corresponding to an increment δx in x. Then,

$$y = f(x) \text{ and } y + \delta y = f(x + \delta x)$$

$$\therefore \quad \delta y = f(x + \delta(x) - f(x)$$

so
$$\frac{\delta y}{\delta x} = \frac{f(x+\delta x) - f(x)}{\delta x}$$

$$\lim_{\delta x \to 0} \frac{\delta y}{\delta x} = \lim_{\delta x \to 0} \frac{f(x+\delta x) - f(x)}{\delta x}$$

The value of above limit if it exists finitely is called the derivative or differential coefficient of f at any point x and is denoted by $\frac{dy}{dx}$, *i.e.*,

$$\frac{dy}{dx} = \lim_{\delta x \to 0} \frac{f(x+\delta x) - f(x)}{\delta x}$$

The derivative of $y = f(x)$ is also denoted as $\frac{df(x)}{dx}, \frac{df}{dx}, f'(x), f_1(x), y'(x), y_1(x)$.

The process of finding the derivative is known as differentiation. Note $\frac{dy}{dx}$ is called differential operator.

3.19 GEOMETRICAL INTERPRETATION OF DERIVATIVE AT A POINT

Let $y = f(x)$ be a continuous function. Let a be a point in the domain of the given function. Let $P[a, f(a)]$ be a point on the graph of this function. Let $Q[a+h, f(a+h)]$ be another point close to P.

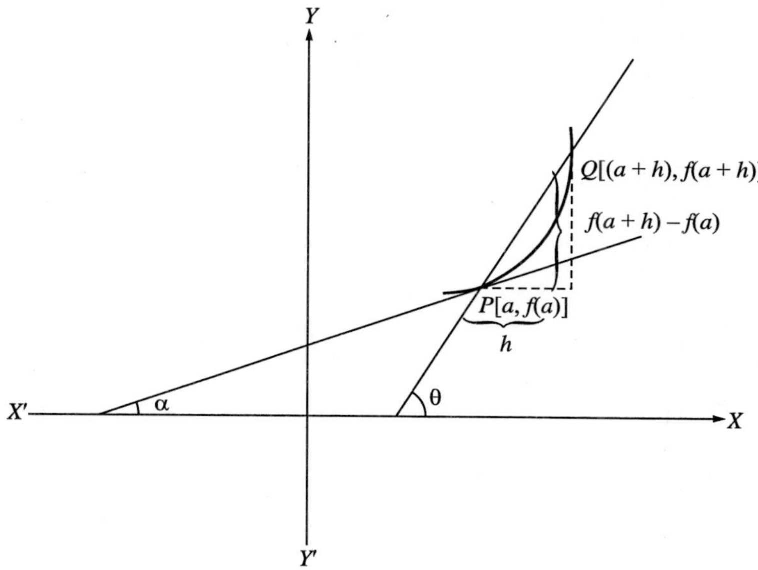

$$\text{Slope of chord } PQ = \frac{f(a+h)-f(a)}{a+h-a}$$

$$= \frac{f(a+h)-f(a)}{h}$$

now as h tends to zero, Q tend to P. The chord PQ will tend to the tangent at P.

$$\therefore \quad \lim_{h\to 0} \frac{f(a+h)-f(a)}{h} = \lim_{Q\to P}$$

or $\quad f'(a) =$ the slope of the tangent at P.

Hence, the derivative of $f(x)$ at $x =$ a is the slope of the tangent to the curve $y = f(x)$ at the point $[a, f(a)]$.

3.20 PHYSICAL INTERPRETATION OF A DERIVATIVE

Let $s = f(t)$ be a function representing the distance of a moving particle at time t.

Let $(s + \delta s)$ be the distance travelled in time $(t + \delta t)$.

$$\therefore \quad s + \delta s = f(t + \delta t)$$

$$\therefore \quad \delta s = f(t + \delta t) - f(t)$$

$$\therefore \quad \text{average velocity} = \frac{\delta s}{\delta t} = \frac{f(t+\delta(t)-f(t)}{\delta t}$$

When $\delta t \to 0$, the average velocity becomes the instantaneous velocity

$$\therefore \qquad \text{velocity at time } t = \lim_{\delta t \to 0} \frac{\delta s}{\delta t} = \lim_{\delta t \to 0} \frac{f(t + \delta(t) - f(t)}{\delta t} = f'(t)$$

3.21 SOME STANDARD FORMULAE

Some standard formulae of differential calculus are given below:

1. $\frac{d}{dx}(x^n) = nx^{n-1}$ **2.** $\frac{d}{dx}(\log_e x) = \frac{1}{x}$ **3.** $\frac{d}{dx}(\sin x) = \cos x$

4. $\frac{d}{dx}(\cos x) = -\sin x$ **5.** $\frac{d}{dx}(\tan x) = \sec^2 x$

6. $\frac{d}{dx}(\cot x) = -\text{cosec}^2 x$

7. $\frac{d}{dx}(\sec x) = \sec x \tan x$ **8.** $\frac{d}{dx}(\text{cosec } x) = -\text{cosec } x \cot x$

9. $\frac{d}{dx}(ax) = ax \log_e a$, a is constant

10. $\frac{d}{dx}(e^x) = e^x$ **11.** $\frac{d}{dx}(\sin^{-1} x) = \frac{1}{\sqrt{1-x^2}}$

12. $\frac{d}{dx}(\cos^{-1} x) = -\frac{1}{\sqrt{1-x^2}}$

13. $\frac{d}{dx}(\tan^{-1} x) = \frac{1}{1+x^2}$ **14.** $\frac{d}{dx}(\cot^{-1} x) = -\frac{1}{1+x^2}$

15. $\frac{d}{dx}(\sec^{-1} x) = \frac{1}{\sqrt[x]{x^2-1}}$

16. $\frac{d}{dx}(\text{cosec}^{-1} x) = -\frac{1}{\sqrt[x]{x^2-1}}$

17. $\frac{d}{dx}(\sinh x) = \cosh x$

18. $\frac{d}{dx}(\cosh x) = \sinh x$ **19.** $\frac{d}{dx}(\tanh x) = \text{sech}^2 x$

20. $\frac{d}{dx}(\coth x) = -\text{cosech}^2 x$

21. $\frac{d}{dx}(\text{sech } x) = -\text{sech } x \tanh x$

22. $\frac{d}{dx}(\text{cosech x}) = -\text{cosech } x \coth x$

23. $\frac{d}{dx}(\sinh^{-1} x) = \frac{1}{\sqrt{1+x^2}}$

24. $\dfrac{d}{dx}(\cosh^{-1}x) = \dfrac{1}{\sqrt{x^2-1}}$

25. $\dfrac{d}{dx}(\tanh^{-1}x) = \dfrac{1}{1-x^2}$ 26. $\dfrac{d}{dx}(\coth^{-1}x) = \dfrac{1}{x^2-1}$

27. $\dfrac{d}{dx}(\text{sech}^{-1}x) = -\dfrac{1}{x\sqrt{1-x^2}}$

28. $\dfrac{d}{dx}(\text{cosech}^{-1}x) = -\dfrac{1}{x\sqrt{x^2+1}}$

29. $\dfrac{d}{dx}(u.v) = u.\dfrac{dv}{dx} + v.\dfrac{du}{dx}$

30. $\dfrac{d}{dx}\left(\dfrac{u}{v}\right) = \dfrac{v\dfrac{du}{dx} - u.\dfrac{dv}{dx}}{v^2}$

31. $\dfrac{dy}{dx} = \dfrac{dy}{du}\cdot\dfrac{du}{dx}$ **32.** $\dfrac{d}{dx}(\log_a x) = \dfrac{1}{x}\log_a e$

SOME OTHER FORMULAE

1. $\sin x = \dfrac{e^{ix} - e^{-ix}}{2i}$ **2.** $\cos x = \dfrac{e^{ix} + e^{-ix}}{2}$

3. $\sinh x = \dfrac{e^{x} - e^{-x}}{2}$ **4.** $\cosh x = \dfrac{e^{x} + e^{-x}}{2}$

5. $\cosh^2 x - \sinh^2 x = 1$ **6.** $\text{sech}^2 x + \tanh^2 x = 1$

7. $\text{cosech}^2 x = \coth^2 x - 1$

3.22 DIFFERENTIATION FROM THE FIRST PRINCIPLE

The derivative obtained by using the definition is called differentiation from the first principle or *ab*-initio method.

Derivative of x^n

Let $y = x^n$

$\therefore$ $y + \delta y = (x + \delta x)^n$

$$\delta y = (x + \delta x)^n - x^n$$

$$\therefore \quad \lim_{\delta x \to 0} \frac{\delta y}{\delta x} = \lim_{\delta x \to 0} \frac{(x+\delta x)^n - x^n}{\delta x}$$

$$= \lim_{\delta x \to 0} \frac{x^n\left[1 + \dfrac{\delta x}{x}\right]^n - x^n}{\delta x}$$

$$= \lim_{\delta x \to 0} \frac{x^n\left[1 + n\frac{\delta x}{x} + \frac{n(n-1)}{2!}\left(\frac{\delta x}{x}\right)^2 + \dots\right] - x^n}{\delta x}$$

$$= \lim_{\delta x \to 0} \frac{n\,x^n\,\frac{\delta x}{x} + \frac{n(n-1)}{2!}x^n\left(\frac{\delta x}{x}\right)^2 + \dots}{\delta x}$$

$$= \lim_{\delta x \to 0}\left[\frac{nx^n}{x} + \frac{n(n-1)}{2}x^n\left(\frac{\delta x}{x^2}\right) + \dots\right]$$

$$\frac{dy}{dx} = \frac{nx^n}{x} = nx^{n-1}$$

i.e., $$\frac{d}{dx}(x^n) = nx^{n-1}$$

ILLUSTRATIVE EXAMPLES

Example 1. Find $\frac{dy}{dx}$ if

(*i*) $y = 2x^2 - 3x^6 + x^{15}$ (*ii*) $y = \frac{5}{x^{3/2}}$ (*iii*) $y = x^2 + \frac{3}{x}$

Solution:

(*i*) $$y = 2x^2 - 3x^6 + x^{15}$$

$$\frac{dy}{dx} = 2.2x^1 - 3.6x^5 + 15x^{14}$$

$$\frac{dy}{dx} = 4x - 18x^5 + 15x^{14}$$

(*ii*) $$y = \frac{5}{x^{3/2}} = 5x^{-3/2}$$

$$\frac{dy}{dx} = 5 \times -\frac{3}{2}\,x^{-\frac{3}{2}-1}$$

$$= -\frac{15}{2}x^{-5/2} = \frac{-15}{2x^{5/2}}$$

(*iii*) $$y = x^2 + \frac{3}{x} = x^2 + 3x^{-1}$$

$$\frac{dy}{dx} = 2x + 3(-1)x^{-1-1} = 2x - 3x^{-2} = 2x - \frac{3}{x^2}$$

Example 2. Find the derivative of the following functions from the first principle.

(*i*) $\sqrt{x}$ (*ii*) $2x^2 + 5x + 7$ (*iii*) $\frac{x^2+1}{2x}, x \neq 0$

Solution:

(*i*) Let
$$y = \sqrt{x}$$
$$y + \delta y = \sqrt{x + \delta x}$$
$\therefore$
$$\delta y = \sqrt{x+\delta x} - \sqrt{x}$$

By first principle

$$\frac{dy}{dx} = \lim_{\delta x \to 0} \frac{\delta y}{\delta x} = \lim_{\delta x \to 0} \frac{\sqrt{x+\delta x} - \sqrt{x}}{\delta x}$$
$$= \lim_{\delta x \to 0} \frac{[\sqrt{x+\delta x} - \sqrt{x}][\sqrt{x+\delta x} + \sqrt{x}]}{\delta x \cdot [\sqrt{x+\delta x} + \sqrt{x}]}$$
$$= \lim_{\delta x \to 0} \frac{x + \delta x - x}{\delta x[\sqrt{x+\delta x} + \sqrt{x}}$$
$$= \lim_{\delta x \to 0} \frac{1}{\sqrt{x+\delta x} + \sqrt{x}} = \frac{1}{\sqrt{x} + \sqrt{x}}$$
$$= \frac{1}{2\sqrt{x}}$$
$$\frac{d}{dx}(\sqrt{x}) = \frac{1}{2\sqrt{x}} = \frac{1}{2}x^{-\frac{1}{2}}$$

(*ii*) Let
$$y = 2x^2 + 5x + 7$$
$\therefore$
$$y + \delta y = 2(x+\delta x)^2 + 5(x+\delta x) + 7$$
$\therefore$
$$\delta y = [2(x+\delta x)^2 + 5(x+\delta x) + 7] - [2x^2 + 5x + 7]$$
$$= 2[(x+\delta x)^2 - x^2] + 5[x + \delta x - x]$$
$$\delta y = 2[2x\delta x + \delta x^2] + 5\delta x$$
$$\frac{\delta y}{\delta x} = \frac{2[2x\delta x + \delta x^2] + 5\delta x}{\delta x} = (4x + \delta x + 5)$$
$\therefore$
$$\frac{dy}{dx} = \lim_{\delta x \to 0} \frac{\delta y}{\delta x} = \lim_{\delta x \to 0} (4x + \delta x + 5) = (4x + 5)$$
$\therefore$
$$\frac{dy}{dx} = 4x + 5$$

(*iii*) Let
$$y = \frac{x^2 + 1}{2x} = \frac{x}{2} + \frac{1}{2x}, x \neq 0$$
$\therefore$
$$y + \delta y = \frac{(x+\delta x)}{2} + \frac{1}{2(x+\delta x)}$$
$\therefore$
$$\delta y = \left[\frac{(x+\delta x)}{2} + \frac{1}{2(x+\delta x)}\right] - \left[\frac{x}{2} + \frac{1}{2x}\right]$$
$$= \left(\frac{x + \delta x - x}{2}\right) + \frac{1}{2}\left(\frac{1}{x+\delta x} - \frac{1}{x}\right)$$

$$\delta y = \frac{\delta x}{2} + \frac{x - x - \delta x}{2(x+\delta x)x} = \frac{\delta x}{2} - \frac{\delta x}{2x(x+\delta x)}$$

$\therefore$
$$\frac{\delta y}{\delta x} = \frac{1}{2} - \frac{1}{2x(x+\delta x)}$$

$$\frac{dy}{dx} = \lim_{\delta x \to 0} \frac{\delta y}{\delta x} = \lim_{\delta x \to 0} \left[\frac{1}{2} - \frac{1}{2x(x+\delta x)}\right]$$

$$= \frac{1}{2} - \frac{1}{2x^2}$$

$\therefore$
$$\frac{dy}{dx} = \frac{1}{2} - \frac{1}{2x^2}$$

Derivatives of trigonometric functions: In this section we shall prove the derivatives of trigonometric functions from first principle.

1. Derivative of sin *x*

Let
$$y = \sin x$$
$$y + \delta y = \sin(x + \delta x)$$
$$\delta y = \sin(x+\delta x) - \sin x$$

$$\lim_{\delta x \to 0} \frac{\delta y}{\delta x} = \lim_{\delta x \to 0} \frac{\sin(x+\delta x) - \sin x}{\delta x}$$

$$= \lim_{\delta x \to 0} \frac{2\sin\frac{x+\delta x - x}{2}\cos\frac{x+\delta x + x}{2}}{\delta x}$$

$$= \lim_{\delta x \to 0} \frac{2\sin\left(\frac{\delta x}{2}\right)\cos\left(x + \frac{\delta x}{2}\right)}{\delta x}$$

$$= \lim_{\delta x \to 0} \frac{2\cos\left(x+\frac{\delta x}{2}\right)\sin\left(\frac{\delta x}{2}\right)}{2.\left(\frac{\delta x}{2}\right)}$$

$$= \lim_{\delta x \to 0} \frac{2\cos\left(x+\frac{\delta x}{2}\right)}{2} . \lim_{\frac{\delta x}{2} \to 0} \frac{\sin\left(\frac{\delta x}{2}\right)}{\left(\frac{\delta x}{2}\right)}$$

$$\left[\lim_{x \to 0} \frac{\sin x}{x} = 1 \text{ and } \delta x \to 0 \Rightarrow \frac{\delta x}{2} \to 0\right]$$

$\therefore$
$$\frac{dy}{dx} = \cos x.1$$

$$\frac{dy}{dx} = \cos x$$

2. Derivative of cos x

Similarly, we can prove that

$$\frac{d}{dx}(\cos x) = -\sin x$$

3. Derivative of tan x

Let

$$y = \tan x$$

$$y + \delta y = \tan(x + \delta x)$$

$$\delta y = \tan(x + \delta x) - \tan x$$

$\therefore$

$$\lim_{\delta x \to 0} \frac{\delta y}{\delta x} = \lim_{\delta x \to 0} \frac{\tan(x+\delta x) - \tan x}{\delta x}$$

$$= \lim_{\delta x \to 0} \frac{\dfrac{\sin(x+\delta x)}{\cos(x+\delta x)} - \dfrac{\sin x}{\cos x}}{\delta x}$$

$$\frac{dy}{dx} = \lim_{\delta x \to 0} \frac{\sin(x+\delta x)\cos x - \cos(x+\delta x)\sin x}{\delta x \cos(x+\delta x)\cos x}$$

$$= \lim_{\delta x \to 0} \frac{\sin(x+\delta x - x)}{\delta x \cos(x+\delta x).\cos x}$$

$$= \lim_{\delta x \to 0} \frac{\sin \delta x}{\delta x} \lim_{\delta x \to 0} \frac{1}{\cos(x+\delta x)\cos x}$$

$$\frac{dy}{dx} = 1.\frac{1}{\cos x.\cos x} = \frac{1}{\cos^2 x} = \sec^2 x$$

$$\frac{d}{dx}(\tan x) = \sec^2 x.$$

4. Derivative of cot x

Similarly, we can prove that

$$\frac{d}{dx}(\cot x) = -\text{cosec}^2 x$$

5. Derivative of sec x

Let

$$y = \sec x$$

$$y + \delta y = \sec(x + \delta x).$$

$\therefore$

$$\delta y = \sec(x + \delta x) - \sec x$$

$$\lim_{\delta x \to 0} \frac{\delta y}{\delta x} = \lim_{\delta x \to 0} \frac{\sec(x+\delta x) - \sec x}{\delta x}$$

$$= \lim_{\delta x \to 0} \frac{1}{\delta x}\left[\frac{1}{\cos(x+\delta x)} - \frac{1}{\cos x}\right]$$

$$= \lim_{\delta x \to 0} \frac{1}{\delta x}\left[\frac{\cos x - \cos(x+\delta x)}{\cos x \cos(x+\delta x)}\right]$$

$$= \lim_{\delta x \to 0} \frac{1}{\delta x} \frac{\left[2\sin\left(\frac{x+x+\delta x}{2}\right)\sin\left(\frac{x+\delta x-x}{2}\right)\right]}{\cos x\cos(x+\delta x)}$$

$$\left[\because\ \cos C - \cos D = 2\sin\frac{C+D}{2}\sin\frac{D-C}{2}\right]$$

$$= \lim_{\delta x \to 0} \frac{2\sin\left(x+\frac{\delta x}{2}\right)\sin\left(\frac{\delta x}{2}\right)}{\delta x.\cos x.\cos(x+\delta x)}$$

$$= \lim_{\delta x \to 0} \frac{\sin\left(\frac{\delta x}{2}\right)\sin\left(x+\frac{\delta x}{2}\right)}{\left(\frac{\delta x}{2}\right)\cos x\cos(x+\delta x)}$$

$$= \lim_{\frac{\delta x}{2} \to 0} \frac{\sin\left(\frac{\delta x}{2}\right)}{\left(\frac{\delta x}{2}\right)} \cdot \lim_{\delta x \to 0} \frac{\sin\left(x+\frac{\delta x}{2}\right)}{\cos x\cos(x+\delta x)}$$

$$\frac{dy}{dx} = \frac{1.\sin x}{\cos x.\cos x} = \tan x.\sec x$$

i.e. $$\frac{d}{dx}(\sec x) = \sec x \tan x$$

6. Derivative of cosec *x*

Similarly, we can prove that

$$\frac{d}{dx}(\text{cosec } x) = -\text{cosec } x \cot x.$$

7. Derivative of a^x, where a is a positive constant

Let $$y = a^x \qquad (a > 0, x \in R \text{ but } x \neq 1)$$

$$y + \delta y = a^{(x+\delta x)}$$

$$\delta y = a^{(x+\delta x)} - a^x$$

$$\therefore \quad \lim_{\delta x \to 0} \frac{\delta y}{\delta x} = \frac{dy}{dx} = \lim_{\delta x \to 0} \frac{a^{(x+\delta x)} - a^x}{\delta x}$$

$$= \lim_{\delta x \to 0} \frac{a^x(a^{\delta x} - 1)}{\delta x}$$

$$\frac{dy}{dx} = \lim_{\delta x \to 0} \frac{a^x\left[1 + \frac{\delta x \log_e a}{1!} + \frac{(\delta x)^2}{2!}(\log_e a)^2 + \ldots - 1\right]}{\delta x}$$

$$= \lim_{\delta x \to 0} a^x\left[\frac{\log_e a}{1!} + \frac{\delta x}{2!}(\log_e a)^2 + \ldots\right]$$

$$\frac{dy}{dx} = a^x \log_e a + 0 + 0 + ... = a^x \log_e a$$

i.e $$\frac{d}{dx}(a^x) = a^x(\log_e a)$$

Note: If we put $a = e$, then

$$\frac{d}{dx}(e^x) = e^x \log_e e = e^x \qquad [\because \log_e e = 1]$$

$$\frac{d}{dx}(e^x) = e^x.$$

8. Derivative of $\log_e x$

Let $$y = \log_e x$$

$$y + \delta y = \log_e(x + \delta x)$$

$\therefore$ $$\delta y = \log_e(x + \delta x) - \log_e x$$

$\therefore$ $$\lim_{\delta x \to 0} \frac{\delta y}{\delta x} = \frac{dy}{dx} = \lim_{\delta x \to 0} \frac{\log_e(x + \delta x) - \log_e x}{\delta x}$$

$$= \lim_{\delta x \to 0} \frac{1}{\delta x}\left[\log_e\left(\frac{x + \delta x}{2}\right)\right]$$

$$= \lim_{\delta x \to 0} \frac{1}{\delta x}\left[\log_e\left(1 + \frac{\delta x}{x}\right)\right]$$

$$= \lim_{\delta x \to 0} \frac{1}{\delta x}\left[\frac{\delta x}{x} - \frac{1}{2}\left(\frac{\delta x}{x}\right)^2 + \frac{1}{3}\left(\frac{\delta x}{x}\right)^3 - ...\right]$$

$$\frac{dy}{dx} = \lim_{\delta x \to 0}\left[\frac{1}{x} - \frac{1}{2}\frac{\delta x}{x^2} + \frac{1}{3}\frac{(\delta x)^2}{3x^3} - ...\right] = \frac{1}{x} - 0 + 0$$

$$\frac{dy}{dx} = \frac{1}{x}$$

i.e., $$\frac{d}{dx}(\log_e x) = \frac{1}{x}.$$

9. Derivative of a constant

Let $$y = c, \text{ where } c \text{ is a constant}$$

$$y + \delta y = c \quad \Rightarrow \quad \delta y = c - y = c - c = 0$$

$$\delta y = 0$$

$\therefore$ $$\frac{\delta y}{\delta x} = \frac{0}{\delta x}$$

$$\lim_{\delta x \to 0} \frac{\delta y}{\delta x} = \frac{dy}{dx} = \lim_{\delta x \to 0} \frac{0}{\delta x} = 0$$

$$\frac{dy}{dx} = 0$$

i.e. $$\frac{d}{dx}(c) = 0$$

3.23 DERIVATIVE OF THE PRODUCT OF A CONSTANT AND A FUNCTION

Let $y = cf(x)$, where c is a constant

$$y + \delta y = cf(x + \delta x)$$

$$\delta y = cf(x + \delta x) - cf(x)$$

$$\lim_{\delta x \to 0} \frac{\delta y}{\delta x} = \lim_{\delta x \to 0} c\left[\frac{f(x+\delta x) - f(x)}{\delta x}\right]$$

$$= c \lim_{\delta x \to 0}\left[\frac{f(x+\delta x) - f(x)}{\delta x}\right]$$

i.e. $$\frac{d}{dx}(cf(x) = c\frac{d}{dx}(f(x)$$

$$\left[\because \lim_{\delta x \to 0} \frac{f(x+\delta x) - f(x)}{\delta x} = \frac{df(x)}{dx}\right]$$

3.24 DERIVATIVE OF THE SUM (OR DIFFERENCE) OF FUNCTIONS

Let $y = f(x) + g(x)$

$$y + \delta y = f(x + \delta x) + g(x + \delta x)$$

$$\delta y = f(x + \delta x) + g(x + \delta x) - f(x) - g(x)$$

$$\delta y = f(x + \delta x) - f(x) + g(x + \delta x) - g(x)$$

$\therefore$ $$\lim_{\delta x \to 0} \frac{\delta y}{\delta x} = \lim_{\delta x \to 0} \frac{f(x+\delta x) - f(x) + g(x+\delta x) - g(x)}{\delta x}$$

$$\frac{dy}{dx} = \lim_{\delta x \to 0} \frac{f(x+\delta x) - f(x)}{\delta x} + \lim_{\delta x \to 0} \frac{g(x+\delta x) - g(x)}{\delta x}$$

i.e. $$\frac{d}{dx}[f(x) + g(x)] = \frac{d}{dx}[f(x)] + \frac{d}{dx}[g(x)]$$

$$\frac{d}{dx}[f(x) + g(x)] = \frac{d}{dx}[f(x)] + \frac{d}{dx}[g(x)]$$

Similarly, we can prove that

$$\frac{d}{dx}[f(x) - g(x)] = \frac{d}{dx}[f(x)] - \frac{d}{dx}[g(x)].$$

ILLUSTRATIVE EXAMPLE

Example 1. Find the derivative of 2^x.

Solution: $\frac{d}{dx}(2^x) = 2^x \log 2$

Example 2. Find the derivative of the following function by first principle.

(*i*) $\sin 2x$ (*ii*) $\sec 2x$ (*iii*) $\cot(3x+1)$

Solution:

(*i*) Let

$$y = \sin 2x$$
$$y+\delta y = \sin 2(x+\delta x) = \sin(2x+2\delta x)$$
$$\therefore \quad \delta y = \sin(2x+2\delta x) - \sin 2x$$
$$\lim_{\delta x\to 0}\frac{\delta y}{\delta x} = \frac{dy}{dx} = \lim_{\delta x\to 0}\frac{\sin(2x+2\delta x)-\sin 2x}{\delta x}$$
$$= \lim_{\delta x\to 0}\frac{2\cos(2x+\delta x)\sin\delta x}{\delta x}$$
$$= \lim_{\delta x\to 0} 2\cos(2x+\delta x).\frac{\sin\delta x}{\delta x}$$
$$= 2\lim_{\delta x\to 0}\cos(2x+\delta x).\lim_{\delta x\to 0}\frac{\sin\delta x}{\delta x}$$
$$= 2\cos 2x \times 1$$
$$\frac{dy}{dx} = 2\cos 2x$$

(*ii*) Let

$$y = \sec 2x$$
$$y+\delta y = \sec 2(x+\delta x) = \sec(2x+2\delta x).$$
$$\therefore \quad \delta y = \sec(2x+2\delta x) - \sec 2x$$
$$\lim_{\delta x\to 0}\frac{\delta y}{\delta x} = \frac{dy}{dx} = \lim_{\delta x\to 0}\left[\frac{\sec(2x+2\delta x)-\sec 2x}{\delta x}\right]$$
$$= \lim_{\delta x\to 0}\frac{1}{\delta x}\left[\frac{1}{\cos(2x+2\delta x)}-\frac{1}{\cos 2x}\right]$$
$$= \lim_{\delta x\to 0}\frac{1}{\delta x}\left[\frac{\cos 2x-\cos(2x+2\delta x)}{\cos(2x+2\delta x)\cos 2x}\right]$$
$$= \lim_{\delta x\to 0}\frac{1}{\delta x}\left[\frac{2\sin(2x+\delta x)\sin(\delta x)}{\cos(2x+2\delta x)\cos 2x}\right]$$
$$= \lim_{\delta x\to 0}\left[\frac{2\sin(2x+\delta x)}{\cos(2x+2\delta x).(\cos 2x)}\right]\cdot\lim_{\delta x\to 0}\frac{\sin\delta x}{\delta x}$$
$$= \frac{2.\sin 2x}{\cos 2x\cos 2x}.1 = 2\sec 2x\tan 2x.$$
$$\frac{dy}{dx} = 2\sec 2x\tan 2x.$$

(*iii*) Let

$$y = \cot(3x+1)$$
$$y+\delta y = \cot(3x+3\delta x+1)$$
$$\delta y = \cot(3x+3\delta x+1) - \cot(3x+1)$$
$$\lim_{\delta x\to 0}\frac{\delta y}{\delta x} = \frac{dy}{dx} = \lim_{\delta x\to 0}\left[\frac{\cot(3x+3\delta x+1)-\cot(3x+1)}{\delta x}\right]$$

$$\frac{dy}{dx} = \lim_{\delta x \to 0} \left[\frac{\dfrac{\cos(3x+1+3\delta x)}{\sin(3x+1+3\delta x)} - \dfrac{\cos(3x+1)}{\sin(3x+1)}}{\delta x} \right]$$

$$= \lim_{\delta x \to 0} \frac{\sin(3x+1-3x-1-3\delta x)}{\delta x \sin(3x+1)\cdot\sin(3x+1+3\delta x)}$$

$$= \lim_{\delta x \to 0} \frac{\sin(-3\delta x)}{\delta x \sin(3x+1)\sin(3x+1+\delta x)}$$

$$= \frac{-3}{\sin^2(3x+1)} \delta x = -3 \operatorname{cosec}^2(3x+1)$$

Example 3. Find the derivative of the following w.r.t. x:

(*i*) $x^3 + e^x + 3^x + \cot x$

(*ii*) $x^2 + \dfrac{2}{x} + 4 \sin x$

(*iii*) $\dfrac{1}{\sqrt[7]{x5}}$

(*iv*) $\left(\sqrt{x} + \dfrac{1}{\sqrt{x}}\right)^2$

(*v*) $\dfrac{3x^2 + 2x + 5}{\sqrt{x}}$

(*vi*) $(x^2 - 5x + 6)(x - 3)$

Solution: (*i*) $\dfrac{d}{dx}(x^3 + e^x + 3^x + \cot x) = \dfrac{d}{dx}(x^3) + \dfrac{d}{dx}(e^x) + \dfrac{d}{dx}(3^x) + \dfrac{d}{dx}(\cot x)$

$$= 3x^2 + e^x + 3^x \log 3 - \operatorname{cosec}^2 x$$

(*ii*) $$\frac{d}{dx}\left(x^2 + \frac{2}{x} + 4 \sin x\right) = \frac{d}{dx}(x^2) + \frac{d}{dx}\left(\frac{2}{x}\right) + 4\frac{d}{dx}(\sin x)$$

$$= \frac{d}{dx}(x^2) + 2\frac{d}{dx}(x^{-1}) + 4\frac{d}{dx}(\sin x)$$

$$= 2x + 2(-1)x^{-2} + 4 \cos x = 2x - 2x^{-2} + 4 \cos x$$

$$= 2x - \frac{2}{x^2} + 4 \cos x$$

(*iii*) Let $$y = \frac{1}{\sqrt[7]{x5}} = \frac{1}{(x5)^{1/7}} = \frac{1}{x^{5/7}} = x^{-5/7}$$

$$\frac{d}{dx}(y) = \frac{dy}{dx} = \frac{d}{dx}(x^{-5/7}) = -\frac{5}{7}x^{-\frac{5}{7}-1} = -\frac{5}{7}x^{-\frac{12}{7}}$$

$$\frac{d}{dx}(x^{-5/7}) = -\frac{5}{7}x^{-12/7}$$

(*iv*) $$\frac{d}{dx}\left\{\left(\sqrt{x} + \frac{1}{\sqrt{x}}\right)^2\right\} = \frac{d}{dx}\left\{\left(x + \frac{1}{x} + 2\right)\right\}$$

$$= \frac{d}{dx}(x) + \frac{d}{dx}(x^{-1}) + \frac{d}{dx}(2)$$

$$= 1 + (-1)x^{-2} + 0 = \left(1 - \frac{1}{x^2}\right)$$

(*v*) $$\frac{dx}{dx}\left(\frac{3x^2 + 2x + 5}{\sqrt{x}}\right) = \frac{d}{dx}(3x^{3/2} + 2x^{1/2} + 5x^{-1/2})$$

$$= 3\frac{d}{dx}(x^{3/2}) + 2\frac{d}{dx}(x^{1/2}) + 5\frac{d}{dx}(x^{-1/2})$$

$$= 3.\frac{3}{2}x^{1/2} + 2.\frac{1}{2}x^{-1/2} + 5\left(-\frac{1}{2}\right)x^{-3/2}$$

$$= \frac{9}{2}\sqrt{x} + \frac{1}{\sqrt{x}} - \frac{5}{2}x^{-3/2}$$

(*vi*) $$\frac{d}{dx}[(x^2 - 5x + 6)(x - 3)] = \frac{d}{dx}(x^3 - 8x^2 + 21x - 18)$$

$$= \frac{d}{dx}(x^3) - 8\frac{d}{dx}(x^2) + 21\frac{d}{dx}(x) - \frac{d}{dx}(18)$$

$$= 3x^2 - 8 \times 2x + 21 \times 1 - 0 = 3x^2 - 16x + 21$$

Example 4. Differentiate the following w.r.t. x

(*i*) $\log x^2 - 3e^x$ (*ii*) $x^4 - 2\cos x + 3\sin x$ (*iii*) $\log_2 x + 2\log_e x + 3\cot x$

(*iv*) $\log\left(\frac{1}{\sqrt{x}}\right) + 5x^a - 3a^x + \sqrt[3]{x^2} + 6\sqrt[4]{x^{-5}}$

Solution:

(*i*) $$\frac{d}{dx}(\log x^2 - 3e^x) = \frac{d}{dx}(\log x^2) - 3\frac{d}{dx}(e^x)$$

$$= \frac{d}{dx}(2\log x) - 3\frac{d}{dx}(e^x) \quad [\because \log x^2 = 2\log x]$$

$$= 2\frac{d}{dx}(\log x) - 3\frac{d}{dx}(e^x)$$

$$= 2 \times \frac{1}{x} - 3 \times e^x = \frac{2}{x} - 3e^x$$

(*ii*) $$\frac{d}{dx}(x^4 - 2\cos x + 3\sin x) = \frac{d}{dx}(x^4) - 2\frac{d}{dx}(\cos x) + 3\frac{d}{dx}(\sin x)$$

$$= 4x^3 - 2(-\sin x) + 3\cos x$$

$$= 4x^3 + 2\sin x + 3\cos x$$

(*iii*) Let $$y = (\log_2 x + 2\log_e x + 3\cot x)$$

$$y = \frac{\log_e x}{\log_e 2} + 2\log_e x + 3\cot x$$

$$y = \log_e x \,.\, \log_2 e + 2\log_e x + 3\cot x$$

$\therefore$
$$\frac{dy}{dx} = \frac{d}{dx}(\log_e x)\log_2 e + 2\frac{d}{dx}(\log_e x) + 3\frac{d}{dx}(\cot x)$$
$$= \frac{1}{x}.\log_2 e + 2.\frac{1}{x} - 3\text{cosec}^2 x$$
$$= \frac{1}{x}\log_2 e + \frac{2}{x} - 3\text{cosec}^2 x$$

(iv) Let
$$y = \log_e\left(\frac{1}{\sqrt{x}}\right) + 5x^a - 3a^x + \sqrt[3]{x^2} + 6\sqrt[4]{X^{-5}}$$
$$y = \log_e(x^{-1/2}) + 5x^a - 3a^x + x^{2/3} + 6x^{-5/4}$$
$$y = -\frac{1}{2}(\log_e x) + 5x^a - 3a^x + x^{2/3} + 6x^{-5/4}$$

$\therefore$
$$\frac{dy}{dx} = \frac{d}{dx}\left(-\frac{1}{2}\log_e x\right) + 5\frac{d}{dx}(x^a) - 3\frac{d}{dx}(a^x) + \frac{d}{dx}(x^{2/3}) + 6\frac{d}{dx}(x^{-5/4})$$
$$= -\frac{1}{2}.\frac{1}{x} + 5ax^{a-1} - 3a^x \log_e a + \frac{2}{3}x^{2/3-1} + 6\left(-\frac{5}{4}\right)x^{-\frac{5}{4}-1}$$
$$= -\frac{1}{2x} + 5ax^{a-1} - 3a^x \log_e a + \frac{2}{3}x^{-1/3} - \frac{15}{2}x^{-9/4}$$
$$\frac{dy}{dx} = -\frac{1}{2x} + 5ax^{a-1} - 3a^x \log_e a + \frac{2}{3.\sqrt[3]{x}} - \frac{15}{2}x^{9/4}$$

Example 5. Find the differential coefficient of the following w.r.t. x:

(i) $y = \sqrt{a + 2a^2e^x + a^3e^{2x}}$ (ii) $y = \dfrac{a^n}{x^n} + a^x$ (iii) $y = a\log_e x + x^2 + x^{7/2}$

(iv) $y = \dfrac{x^3 + 8x + 6}{x}$ (v) $4\log_{10} x + 2e^x + 3a^x + 1$

Solution:

(i)
$$y = \sqrt{a + 2a^2e^x + a^3e^{2x}} = \sqrt{a}\sqrt{1 + 2ae^x + a^2e^{2x}}$$
$$y = \sqrt{a}\sqrt{(1 + ae^x)^2} = \sqrt{a}(1 + ae^x) = \sqrt{a} + \sqrt{a}ae^x$$
$$y = a^{1/2} + a^{3/2}e^x$$

$\therefore$
$$\frac{dy}{dx} = \frac{d}{dx}(a^{1/2} + a^{3/2}e^x) = \frac{d}{dx}(a^{1/2}) + a^{3/2}\frac{d}{dx}(e^x)$$
$$= 0 + a^{3/2}e^x$$
$$\frac{dy}{dx} = a^{3/2}e^x$$

(ii)
$$y = \frac{a^n}{x^n} + a^x$$

$$y = a^n x^{-n} + a^x$$

$$\frac{dy}{dx} = \frac{d}{dx}(a^n x^{-n}) + \frac{d}{dx}(a^x) = a^n \frac{d}{dx}(x^{-n}) + \frac{d}{dx}(a^x)$$

$$= a^n(-n)x^{-n-1} + a^x \log_e a$$

$$= -na^n x^{-n-1} + a^x \log_e a$$

$$\frac{dy}{dx} = -\frac{na^n}{x^{n+1}} + a^x \log_e a$$

(*iii*)
$$y = a \log_e x + x^2 + x^{7/2}$$

∴
$$\frac{dy}{dx} = \frac{d}{dx}(a \log_e x) + \frac{d}{dx}(x^2) + \frac{d}{dx}(x^{7/2})$$

$$= a\frac{d}{dx}(\log_e x) + \frac{d}{dx}(x^2) + \frac{d}{dx}(x^{7/2})$$

$$= a\frac{1}{x} + 2x + \frac{7}{2}x^{\frac{7}{2}-1}$$

$$\frac{dy}{dx} = \frac{a}{x} + 2x + \frac{7}{2}x^{5/2}$$

(*iv*)
$$y = \frac{x^3 + 8x + 6}{x} = x^2 + 8 + \frac{6}{x}$$

∴
$$\frac{dy}{dx} = \frac{d}{dx}(x^2) + \frac{d}{dx}(8) + \frac{d}{dx}\left(\frac{6}{x}\right)$$

$$\frac{dy}{dx} = 2x + 0 - \frac{6}{x^2} = 2x - \frac{6}{x^2}$$

$$\frac{dy}{dx} = 2x - \frac{6}{x^2}$$

(*v*) Let
$$y = 4 \log_e x + 2e^x + 3a^x + 1$$

$$\frac{dy}{dx} = 4\frac{d}{dx}(\log_e x) + 2\frac{d}{dx}({}_e x) + 3\frac{d}{dx}({}_a x) + \frac{d}{dx}(1)$$

$$= 4.\frac{1}{x} + 2e^x + 3a^x \log_e a + 0$$

$$\frac{dy}{dx} = \frac{4}{x} + 2e^x + 3a^x \log_e a$$

Example 6. If y = $\sqrt{\dfrac{1-\cos 2x}{1+\cos 2x}}$, find $\dfrac{dy}{dx}$

Solution: $y = \sqrt{\dfrac{1-\cos 2x}{1+\cos 2x}} = \sqrt{\dfrac{1-(1-2\sin^2 x)}{1+2\cos^2 x - 1}} = \sqrt{\dfrac{2\sin^2 x}{2\cos^2 x}} = \tan x$

∴
$$\frac{dy}{dx} = \frac{d}{dx}(\tan x) = \sec^2 x$$

Example 7. Find the differential coefficient of the following w.r.t. x:

(*i*) $4 \operatorname{cosec} x \tan x - 6 \sec x \cot x$

(*ii*) $5 \sec x \sin x + 3 \operatorname{cosec} x \cos x + 5 \tan x \cot x$

Solution: Let

$$y = 4 \operatorname{cosec} x \tan x - 6 \sec x \cot x$$

$$y = 4 \times \frac{1}{\sin x} \times \frac{\sin x}{\cos x} - 6 \times \frac{1}{\cos x} \times \frac{\cos x}{\sin x}$$

$$y = \frac{4}{\cos x} - \frac{6}{\sin x}$$

$$y = 4 \sec x - 6 \operatorname{cosec} x$$

$\therefore$

$$\frac{dy}{dx} = 4\frac{d}{dx}(\sec x) - 6\frac{d}{dx}(\operatorname{cosec} x)$$

$$= 4 \sec x \tan x - 6(-\operatorname{cosec} x \cot x)$$

$$= 4 \sec x \tan x + 6 \operatorname{cosec} x \cot x$$

(*ii*) Let

$$y = 5 \sec x \sin x + 3 \operatorname{cosec} x \cos x + 5 \tan x \cot x$$

$$y = 5 \times \frac{1}{\cos x} \times \sin x + 3 \times \frac{1}{\sin x} \times \cos x + 5 \times \frac{\sin x}{\cos x} \times \frac{\cos x}{\sin x}$$

$$y = 5 \tan x + 3 \cot x + 5$$

$$\frac{dy}{dx} = \frac{d}{dx}(5 \tan x) + \frac{d}{dx}(3 \cot x) + \frac{d}{dx}(5)$$

$$= 5\frac{d}{dx}(\tan x) + 3\frac{d}{dx}(\cot x) + 0$$

$$= 5 \sec^2 x + 3(-\operatorname{cosec}^2 x)$$

$$= 5 \sec^2 x - 3 \operatorname{cosec}^2 x$$

EXERCISE 3.9

Find the differential coefficient of the following w.r.t. x:

1. $\sqrt[3]{x^2}$

2. $(x+2)(x+3)$

3. $\tan x + \frac{1}{\tan x}$

4. $x(1+x)^2$

5. $\sin x - \cos x + \tan x$

6. $5 \log_e x - 4x^{1/3} + e^x$

7. $6x^5 + 4x^3 - 3x^2 + 2x + 5$

8. $\frac{1}{\sqrt[3]{x}}$

9. $5x^a + 6a^x + 2 \log_5 x + 7e - 14$

10. $\left(x^2 + \frac{1}{x^2}\right)^3$

11. $\dfrac{a+b\cos x}{\sin x}$ 12. $\dfrac{3x^2+4x-5}{x}$

13. $5x^{-3/2}+\dfrac{4}{\sqrt{x}}+\sqrt{x}-\dfrac{7}{x}$

ANSWERS

1. $\dfrac{2}{3}x^{-1/3}$ 2. $(2x+5)$ 3. $\sec^2 x$ – cosce$^2 x$

4. $1+4x+3x^2$ 5. $\cos x+\sin x+(\sec^2 x)$

6. $\dfrac{5}{x}-\dfrac{4}{3}x^{-2/3}+e^x$ 7. $30x^4+12x^2-6x+2$

8. $-\dfrac{1}{3}x^{-4/3}$ 9. $5ax^{a-1}+(6\log a)a^x+\dfrac{2(\log_5 e)}{x}$

10. $6x^5-\dfrac{6}{x^7}+6x-\dfrac{6}{x^3}$

11. $-\operatorname{cosec} x\,(a\cot x+b\operatorname{cosec} x)$ 12. $\left(3+\dfrac{5}{x^2}\right)$

13. $\dfrac{-15}{2}x^{-5/2}-2x^{-3/2}+\dfrac{1}{2\sqrt{x}}+\dfrac{7}{x^2}$

3.25 DERIVATIVE OF THE PRODUCT OF TWO FUNCTIONS

Product Rule: Let $f(x)$ and $g(x)$ are two functions of x, then

$$\frac{dy}{dx}=\frac{d}{dx}[f(x).g(x)]=f(x)\frac{d}{dx}[g(x)]+g(x)\frac{d}{dx}[f(x)]$$

Note: The above result may be expressed as

$$\frac{d}{dx}[u.v]=u.\frac{dv}{dx}+v.\frac{du}{dx}$$

i.e., the derivative of the product of two functions

= first function × derivative of second

+ second function × derivative of first

Generalization of the Product Rule

If u, v, w are differentiable functions of x, then

$$\frac{d}{dx}(uvw)=(uv)\frac{dw}{dx}+(wu)\frac{dv}{dx}+(wv)\frac{du}{dx}$$

ILLUSTRATIVE EXAMPLES

Example 1. Find the derivative of $x^3\cot x$

Solution: Let $y=x^3.\cot x$

$$\frac{dy}{dx}=\frac{d}{dx}(x^3\cot x)=x^3\frac{d}{dx}(\cot x)+\cot x\frac{d}{dx}(x^3)$$

$$= x^3(-\text{cosec}^2 x) + \cot x(3x^2)$$
$$= x^2(3 \cot x - x \,\text{cosec}^2 x)$$

Example 2. Differentiate $x^2 \tan x - x \log x$

Solution:
$$\frac{dy}{dx} = \frac{d}{dx}(x^2 \tan x - x \log x)$$
$$= \frac{d}{dx}(x^2 \tan x) - \frac{d}{dx}(x \log x)$$
$$= \left[x^2 \frac{d}{dx}(\tan x) + \tan x \frac{d}{dx}(x^2)\right] - \left[x \frac{d}{dx}(\log x) + \log x . \frac{d}{dx}(x)\right]$$
$$\frac{dy}{dx} = (x^2 \sec^2 x + 2x \tan x) - \left(x . \frac{1}{x} + \log x . 1\right)$$
$$= (x^2 \sec^2 x + 2x \tan x - 1 - \log x)$$

Example 3. Differentiate $x.\log.x \tan x$

Solution:
$$\frac{dy}{dx} = \frac{d}{dx}(x \log x \tan x)$$
$$= \left[\frac{d}{dx}(x)\right] \log x \tan x + x . \frac{d}{dx}(\log x) \tan x + x \log x \frac{d}{dx}(\tan x)$$
$$= \log x \tan x + x . \frac{1}{x} \tan x + x \log x . \sec^2 x$$
$$= \log x . \tan x + \tan x + x \log x . \sec^2 x$$

Example 4. Differentiate the following functions w.r.t. x:

(*i*) $(x^3 + 2x + 2) \tan x$ (*ii*) $x^4 \sin x(\cot x + \text{cosec}\, x)$

(*iii*) $x^5 e^x + x^2 \log x$ (*iv*) $e^x(1 + \log x)$ (*v*) $e^x(x^3 + \sqrt{x})$

Solution: (*i*) $(x^3 + 2x + 2) \tan x$

Let $y = (x^3 + 2x + 2) \tan x$

$$\therefore \quad \frac{dy}{dx} = (x^3 + 2x + 2) \frac{d}{dx}(\tan x) + \tan x \frac{d}{dx}(x^3 + 2x + 2)$$
$$\frac{dy}{dx} = (x^3 + 2x + 2)\sec^2 x + \tan x(3x^2 + 2)$$

(*ii*) Let $y = x^4 \sin x(\cot x + \text{cosec}\, x)$

$$y = x^4 \sin x \cot x + x^4 \sin x . \text{cosec}\, x$$
$$y = x^4 \cos x + x^4 = x^4(1 + \cos x)$$

$$\therefore \qquad \frac{dy}{dx} = x^4 \frac{d}{dx}(1+\cos x) + (1+\cos x)\frac{d}{dx}(x^4)$$

$$\frac{dy}{dx} = -x^4 \sin x + (1+\cos x)(4x^3)$$

(*iii*) Let $\qquad y = x^5 e^x + x^2 \log x$

$$\frac{dy}{dx} = \frac{d}{dx}(x^5 e^x) + \frac{d}{dx}(x^2 \log x)$$

$$= x^5 \frac{d}{dx}(e^x) + e^x \frac{d}{dx}(x^5) + x^2 \frac{d}{dx}(\log x) + \log x \frac{d}{dx}(x^2)$$

$$= x^5 e^x + 5x^4 e^x + x^2 . \frac{1}{x} + \log x . 2x$$

$$\frac{dy}{dx} = x^4 e^x (x+5) + x + 2x \log x$$

(*iv*) Let $\qquad y = e^4 (1 + \log x)$

$$\frac{dy}{dx} = e^x \frac{d}{dx}(1+\log x) + (1+\log x)\frac{d}{dx}(e^x)$$

$$= e^x . \frac{1}{x} + (1+\log x).e^x$$

$$= \frac{1}{x} e^x + e^x + e^x \log x = e^x \left(\frac{1}{x} + 1 + \log x \right)$$

(*v*) Let $\qquad y = e^x (x^3 + \sqrt{x})$

$$\frac{dy}{dx} = e^x \frac{d}{dx}(x^3 + \sqrt{x}) + (x^3 + \sqrt{x})\frac{d}{dx}(e^x)$$

$$= e^x \left(3x^2 + \frac{1}{2} x^{-1/2} \right) + (x^3 + \sqrt{x}) e^x$$

$$\frac{dy}{dx} = e^x \left[x^3 + 3x^2 + \frac{1}{2\sqrt{x}} + \sqrt{x} \right]$$

Example 5. Differentiat the following functions w.r.t. x:

(*i*) $a^x \log_a x - \sin x \cos x$ (*ii*) $(x-2)(x-3) \sin x$ (*iii*) $\log x^x$

Solution: (*i*) Let $\qquad y = a^x \log_a x - \sin x \cos x$

$$\frac{dy}{dx} = \frac{d}{dx}(a^x \log_a x) - \frac{d}{dx}(\sin x \cos x)$$

$$= \left[a^x \frac{d}{dx}(\log_a x) + \log_a x \frac{d}{dx}(a^x) \right]$$

$$- \left[\sin x \frac{d}{dx}(\cos x) + \cos x \frac{d}{dx}(\sin x) \right]$$

$$= a^x . \frac{1}{x} \log_a e + \log_a x . a^x \log_e a$$

$$- [\sin x(-\sin x) + \cos x(\cos x)$$

$$= a^x \left[\frac{1}{x} \log_a e + (\log_a x)(\log_e a) \right] - [-\sin^2 x + \cos^2 x]$$

$$= a^x \left[\frac{1}{x} \log_a e + (\log_a x).(\log_e a) \right] - \cos 2x$$

(*ii*) Let $y = (x-2)(x-3) \sin x$

$$y = (x^2 - 5x + 6) \sin x$$

$$\therefore \quad \frac{dy}{dx} = (x^2 - 5x + 6) \frac{d}{dx} (\sin x) + \sin x \frac{d}{dx} (x^2 - 5x + 6)$$

$$= (x^2 - 5x + 6) \cos x + \sin x (2x - 5)$$

$$= x^2 \cos x + x(2 \sin x - 5 \cos x) - 5 \sin x + 6 \cos x$$

(*iii*) Let $y = \log x^x = x \log x$

$$\frac{dy}{dx} = \frac{d}{dx} (x \log x) = x \frac{d}{dx} (\log x) + \log x \frac{d}{dx} (x)$$

$$= x . \frac{1}{x} + \log x \times 1 = 1 + \log x$$

EXERCISE 3.10

Find the derivatives of the following functions w.r.t x:

1. $e^x \log_e x \cos x$
2. $e^x \sin x + x^k \cos x$
3. $4\sqrt{x} \log_a x$
4. $\sin x . \log x$
5. $7x\, a^x \sin x$
6. $x^n e^x + a^x \sec x$
7. $(x^2 + 3x + 1) \sin x$
8. $a^x \log x$
9. $x^3 \cos x - 2^x \tan x$
10. $a^x \sin x \log x$
11. $(x^5 \log x + x^2 . 3^x)$
12. $(x^2 - 4x + 5)(x^3 - 2)$
13. $(1 + 2 \tan x)(5 + 4 \cos x)$
14. $(x^2 + 11x + 7)(x + 5)$ 15. $\sec x \tan x$

ANSWERS

1. $e^x \log x \cos x + \frac{1}{x} e^x \cos x - e^x \sin x \log_e x$
2. $e^x(\cos x + \sin x) - x^{k-1}(x \sin x - k \cos x)$
3. $\frac{4}{\sqrt{x}} \left(\log_a e + \frac{1}{2} \log_a x \right)$
4. $\frac{\sin x}{x} + \cos x \log x$
5. $7\, [x a^x \cos x + x a^x \log_e a \sin x + a^x \sin x]$

6. $e^x x^{n-1}(x+n) + a^x \sec x(\tan x + \log_e a)$

7. $(x^2+3x+1)\cos x + (2^x+3)\sin x$

8. $a^x\left[(\log a)(\log x) + \frac{1}{x}\right]$

9. $x^2(3\cos x - x\sin x) - 2^x[(\log 2)\tan x + \sec^2 x]$

10. $a^x\left[\frac{\sin x}{x} + (\log x)(\cos x + \sin x \log a)\right]$

11. $x^4 + 5x^4 \log x + [x^2(\log 3) + 2x]3^x$ 12. $5x^4 - 16x^3 + 15x^2 - 4x + 8$

13. $-4\sin x + 10\sec^2 x + 8\cos x$ 14. $3x^2 + 32x + 62$

15. $\sec x(1 + 2\tan^2 x)$

3.26 DERIVATIVE OF THE QUOTIENT OF TWO FUNCTIONS

Quotient Rule: If $f(x)$ and $g(x)$ are two differentiable functions and $g(x) \neq 0$, then $\frac{f(x)}{g(x)}$ is also differentiable and

$$\frac{d}{dx}\left[\frac{f(x)}{g(x)}\right] = \frac{g(x)\frac{d}{dx}[f(x)] - f(x)\cdot\frac{d}{dx}[g(x)]}{[g(x)]^2}$$

Note: The above result may be expressed as

$$\frac{d}{dx}\left(\frac{u}{v}\right) = \left[\frac{v\frac{du}{dx} - u\frac{dv}{dx}}{v^2}\right]$$

i.e., Derivative of the quotient of two functions

$$= \frac{(\text{Denominator}) \times (\text{derivative of num.}) - (\text{num.} \times \text{derivative of denom.})}{(\text{denominator})^2}$$

ILLUSTRATIVE EXAMPLES

Example 1. Find the derivative of the function $\frac{x-1}{\sqrt{x+1}}, x \neq -1$

Solution: Let $y = \frac{x-1}{\sqrt{x+1}}$

$$\frac{dy}{dx} = \frac{(x+1)^{\frac{1}{2}}\frac{d}{dx}(x-1) - (x-1)\frac{d}{dx}(x+1)^{\frac{1}{2}}}{(x+1)}$$

$$= \frac{(x+1)^{\frac{1}{2}} - (x-1)\left(\frac{1}{2}\right)(x+1)^{-\frac{1}{2}}}{(x+1)}$$

$$= \frac{2(x+1)-(x-1)}{2(x+1)^{3/2}}$$

$$\frac{dy}{dx} = \frac{x+3}{2(x+1)3/2}$$

$$\frac{d}{dx}\left(\frac{x-1}{\sqrt{x+1}}\right) = \frac{x+3}{2(x+1)3/2}$$

Example 2. Differentiate the following w. r.t. x:

$(i)\ \frac{x+e^x}{1+\log x}$ $(ii)\ \frac{\sqrt{a}+\sqrt{x}}{\sqrt{a}-\sqrt{x}}$ $(iii)\ \frac{x\sin x}{1+\cos x}$ $(iv)\ \frac{x-\cos x}{x+\sin x}$

$(v)\ \frac{1+\log x}{1-\log x}$ $(vi)\ \frac{2x+3}{3x+4}$

Solution: (i) Let $y = \frac{x+e^x}{1+\log x}$

$$\frac{dy}{dx} = \frac{(1+\log x)\frac{d}{dx}(x+e^x)-(x+e^x)\frac{d}{dx}(1+\log x)}{(1+\log x)^2}$$

$$= \frac{(1+\log x)(1+e^x)-(x+e^x)\left(\frac{1}{x}\right)}{(1+\log x)^2}$$

$$\frac{dy}{dx} = \frac{\log x(1+e^x)+e^x\left(1-\frac{1}{x}\right)}{(1+\log x)^2}$$

(ii) Let $y = \frac{\sqrt{a}+\sqrt{x}}{\sqrt{a}-\sqrt{x}}$

$$\frac{dy}{dx} = \frac{\left(\sqrt{a}-\sqrt{x}\right)\frac{d}{dx}\left(\sqrt{a}+\sqrt{x}\right)-\left(\sqrt{a}+\sqrt{x}\right)\frac{d}{dx}\left(\sqrt{a}-\sqrt{x}\right)}{\left(\sqrt{a}-\sqrt{x}\right)^2}$$

$$= \frac{\left(\sqrt{a}-\sqrt{x}\right)\left(\frac{1}{2\sqrt{x}}\right)-\left(\sqrt{a}+\sqrt{x}\right)\left(-\frac{1}{2\sqrt{x}}\right)}{\left(\sqrt{a}-\sqrt{x}\right)^2}$$

$$\frac{dy}{dx} = \frac{\frac{1}{2}\sqrt{\frac{a}{x}}-\frac{1}{2}+\frac{1}{2}\sqrt{\frac{a}{x}}+\frac{1}{2}}{\left(\sqrt{a}-\sqrt{x}\right)^2} = \frac{\sqrt{a}}{\sqrt{x}(\sqrt{a}-\sqrt{x})^2}$$

(*iii*) Let $y = \dfrac{x \sin x}{1 + \cos x}$

$$\frac{dy}{dx} = \frac{(1+\cos x)\dfrac{d}{dx}(x\sin x) - x\sin x\dfrac{d}{dx}(1+\cos x)}{(1+\cos x)^2}$$

$$= \frac{(1+\cos x)(x\cos x + \sin x) - x\sin x(-\sin x)}{(1+\cos x)^2}$$

$$= \frac{x\cos x + \sin x + x\cos^2 x + \sin x\cos x + x\sin^2 x}{(1+\cos x)^2}$$

$$= \frac{x\cos x + x + \sin x + \sin x\cos x}{(1+\cos x)^2}$$

$$\frac{dy}{dx} = \frac{x(1+\cos x) + \sin x(1+\cos x)}{(1+\cos x)^2} = \frac{x + \sin x}{(1+\cos x)}$$

(*iv*) Let $y = \dfrac{x - \cos x}{x + \sin x}$

$$\frac{dy}{dx} = \frac{(x+\sin x)\dfrac{d}{dx}(x-\cos x) - (x-\cos x)\dfrac{d}{dx}(x+\sin x)}{(x+\sin x)^2}$$

$$= \frac{(x+\sin x)(1+\sin x) - (x-\cos x)(1+\cos x)}{(x+\sin x)^2}$$

$$= \frac{\sin x + \sin^2 x + x + x\sin x - x\cos x - x + \cos^2 x + \cos x}{(\sin x + x)^2}$$

$$= \frac{1 + \sin x + \cos x + x(\sin x - \cos x)}{(\sin x + x)^2}$$

(*v*) Let $y = \dfrac{1 + \log x}{1 - \log x}$

$$\frac{dy}{dx} = \frac{(1-\log x)\dfrac{d}{dx}(1+\log x) - (1+\log x)\dfrac{d}{dx}(1-\log x)}{(1-\log x)^2}$$

$$= \frac{(1-\log x)\left(\dfrac{1}{x}\right) - (1+\log x)\left(-\dfrac{1}{x}\right)}{(1-\log x)^2}$$

$$= \frac{\dfrac{1}{x} - \dfrac{\log x}{x} + \dfrac{1}{x} + \dfrac{\log x}{x}}{(1-\log x)^2} = \frac{2}{x(1-\log x)^2}$$

(*vi*) Let $y = \dfrac{2x+3}{3x+4}$

$$\frac{dy}{dx} = \frac{(3x+4)\dfrac{d}{dx}(2x+3) - (2x+3)\dfrac{d}{dx}(3x+4)}{(3x+4)^2}$$

$$= \frac{(3x+4)\cdot 2 - (2x-3)\cdot 3}{(3x+4)^2}$$

$$= \frac{6x+8-6x-9}{(3x+4)^2} = -\frac{1}{(3x+4)^2}$$

EXERCISE 3.11

Differentiate the following function w.r.t. x:

1. $\dfrac{x}{1+\tan x}$ **2.** $\dfrac{e^x - \tan x}{\cot x - x^n}$

3. $\dfrac{x^2 - x + 1}{x^2 + x + 1}$ **4.** $\dfrac{3^x}{x + \tan x}$

5. $\dfrac{2^x \cot x}{\sqrt{x}}$ **6.** $\dfrac{\sec x + \tan x}{\sec x - \tan x}$

7. $\dfrac{e^x (x-1)}{(x+1)}$ **8.** $\dfrac{2^x}{x}$

9. $\dfrac{x \tan x}{\sec x + \tan x}$ **10.** $\dfrac{\sin x - x\cos x}{x \sin x + \cos x}$

11. $\dfrac{x - \cos 2x}{x + \cos 2x}$ **12.** $\dfrac{1 - \tan x}{1 + \tan x}$

13. $\dfrac{x^2 + 3x - 1}{x + 2}$ **14.** $\dfrac{e^x}{(1 + x^2)}$

ANSWERS

1. $\dfrac{1 + \tan x - x\sec^2 x}{(1+\tan x)^2}$

2. $\dfrac{(\cot x - x^n)(e^x - \sec^2 x) + (e^x - \tan x)(\cos ec^2 x + nx^{n-1})}{(\cot x - x^n)^2}$

3. $\dfrac{2(x^2-1)}{(x^2+x+1)^2}$ **4.** $\dfrac{3^x[(x+\tan x)\log 3 - (1+\sec^2 x)]}{(x+\tan x)^2}$

5. $\dfrac{2^x\,[-x\operatorname{cosec}^2 x + x(\log 2)\cot x - \frac{1}{2}\log x]}{x^{3/2}}$

6. $\dfrac{2\cos x}{(1-\sin x)^2}$ 7. $\dfrac{e^x\,(x^2+1)}{(x+1)^2}$ 8. $\dfrac{2^x\,(x\log 2 - 1)}{x^2}$

9. $\dfrac{x\sec x(\sec x - \tan x) + \tan x}{(\sec x + \tan x)}$ 10. $\dfrac{x^2}{(x\sin x + \cos x)^2}$

11. $\dfrac{2(2x\sin 2x + \cos 2x)}{(x+\cos 2x)^2}$, when $x + \cos 2x \neq 0$

12. $\dfrac{-2\sec^2 x}{(1+\tan x)^2}$ 13. $\dfrac{x^2+4x+7}{(x+2)^2}$ 14. $\dfrac{e^x\,(1-x)^2}{(1+x^2)^2}$

3.27 DERIVATIVE OF A FUNCTION OF A FUNCTION

Chain Rule: If $y = f(t)$ and $t = g(x)$ then $\dfrac{dy}{dx} = \dfrac{dy}{dt} \times \dfrac{dt}{dx}$

Generalization of Chain Rule:

If $y = f(t)$, $t = g(u)$ and $u = h(x)$ then

$$\frac{dy}{dx} = \frac{dy}{dt} \times \frac{dt}{du} \times \frac{du}{dx}$$

ILLUSTRATIVE EXAMPLES

Example 1. Differentiate (i) $e^{\sin x}$ (ii) $\sqrt{ax^2 + 2bx} + c$

Solution: (*i*) Let $y = e^{\sin x}$

Put $\sin x = t$, so that $y = e^t$ and $t = \sin x$

$$\therefore \quad \frac{dy}{dt} = e^t \quad \text{and} \quad \frac{dt}{dx} = \cos x$$

$$\frac{dy}{dx} = \frac{dy}{dt} \times \frac{dt}{dx} = e^t \cos x = e^{\sin x} \cdot \cos x$$

(*ii*) $y = \sqrt{ax^2 + 2bx + c}$

put $ax^2 + 2bx + c = t$ so that $y = \sqrt{t}$ and $t = ax^2 + 2bx + c$

$$\therefore \quad \frac{dy}{dt} = \frac{1}{2}t^{-\frac{1}{2}} = \frac{1}{2\sqrt{t}} \quad \text{and} \quad \frac{dt}{dx} = 2ax + 2b$$

$$\frac{dy}{dx} = \frac{dy}{dt} \times \frac{dt}{dx} = \frac{1}{2\sqrt{t}} \times (2ax + 2b)$$

$$\frac{dy}{dx} = \frac{(ax+b)}{\sqrt{t}} = \frac{(ax+b)}{\sqrt{ax^2 + 2bx + c}}$$

Example 2. If $y = \log\ \tan\frac{x}{2}$, find $\frac{dy}{dx}$

Solution: Put $\frac{x}{2} = t,\ \tan\frac{x}{2} = \tan t = u$, so that

$$y = \log u,\ u = \text{tant } t \text{ and } t = \frac{x}{2}$$

$$\therefore \quad \frac{dy}{du} = \frac{1}{u},\ \frac{du}{dt} = \sec^2 t \text{ and } \frac{dt}{dx} = \frac{1}{2}$$

$$\therefore \quad \frac{dy}{dx} = \left(\frac{dy}{du} \times \frac{du}{dt} \times \frac{dt}{dx}\right) = \frac{1}{2u}\sec^2 t = \frac{\sec^2 t}{2\tan t}$$

$$\frac{dy}{dx} = \frac{1}{\sin 2t} = \frac{1}{\sin x} = \text{cosec } x \quad \left(\because t = \frac{x}{2}\right)$$

Example 3. Find the differential coefficient of cos (sin x)

Solution: Let $y = \cos(\sin x)$

Put $\sin x = t$, so that $y = \cos t$

$$\therefore \quad \frac{dy}{dt} = -\sin t$$

$$\frac{dx}{dt} = \cos x$$

$$\therefore \quad \frac{dy}{dx} = \frac{dy}{dt} \times \frac{dt}{dx} = -\sin t \cos x = -\cos x \sin(\sin x)$$

$$\frac{dy}{dx} = -\cos x \sin(\sin x)$$

Example 4. If $y = \sin\sqrt{(\sin x + \cos x)}$, find $\frac{dy}{dx}$

Solution: Put $\sin x + \cos x = t \Rightarrow \sqrt{(\sin x + \cos x)} = \sqrt{t} = u$, we get

$y = \sin u,\ u = \sqrt{t}$ and $t = (\sin x + \cos x)$

$$\therefore \quad \frac{dy}{du} = \cos u,\ \frac{du}{dt} = \frac{1}{2\sqrt{t}} \text{ and } \frac{dt}{dx} = (\cos x - \sin x)$$

$$\frac{dy}{dx} = \left(\frac{dy}{du} \times \frac{du}{dt} \times \frac{dt}{dx}\right) = \frac{\cos u}{2\sqrt{t}}(\cos x - \sin x)$$

$$= \frac{\cos\sqrt{t}}{2\sqrt{t}}(\cos x - \sin x)$$

$$\frac{dy}{dx} = \frac{\cos\sqrt{(\sin x + \cos x)}(\cos x - \sin x)}{2\sqrt{(\sin x + \cos x)}} \qquad [\because t = (\sin x + \cos x)]$$

Example 5. Differentiate $\sqrt{\dfrac{1+x}{1-x}}$

Solution: Let $y = \sqrt{\dfrac{1+x}{1-x}}$

Put $\dfrac{1+x}{1-x} = t$, so that $y = \sqrt{t}$ and $t = \dfrac{1+x}{1-x}$

$$\therefore \qquad \frac{dy}{dt} = \frac{1}{2}t^{-\frac{1}{2}} = \frac{1}{2\sqrt{t}}$$

$$\frac{dt}{dx} = \frac{(1-x)\dfrac{d}{dx}(1+x) - (1+x)\dfrac{d}{dx}(1-x)}{(1-x)^2}$$

$$\frac{dy}{dx} = \frac{(1-x)\cdot 1 - (1+x)(-1)}{(1-x)^2} = \frac{2}{(1-x)^2}$$

$$\frac{dy}{dx} = \left(\frac{dy}{dt} \times \frac{dt}{dx}\right) = \frac{1}{2\sqrt{t}} \times \frac{2}{(1-x)^2}$$

$$= \frac{1}{(1-x)^2\sqrt{\dfrac{1+x}{1-x}}} = \frac{\sqrt{1-x}}{(1-x)^2\sqrt{1+x}}$$

$$\frac{dy}{dx} = \frac{1}{(1-x)^{3/2}(1+x)^{1/2}}$$

$$\frac{dy}{dx} = \frac{1}{(1-x)^{3/2}(1+x)^{1/2}}$$

Example 6. Differentiate $\dfrac{1}{\log(\cos x)}$

Solution: Let $y = \dfrac{1}{\log(\cos x)} = [\log(\cos x)]^{-1}$

Put $\cos x = t \Rightarrow \log(\cos x) = \log t = u$, then $y = u^{-1}$

where $u = \log t$ and $t = \cos x$

$$\therefore \qquad \frac{dy}{du} = -\frac{1}{u^2}, \ \frac{du}{dt} = \frac{1}{t} \text{ and } \frac{dt}{dx} = -\sin x$$

$$\therefore \qquad \frac{dy}{dx} = \left(\frac{dy}{du} \times \frac{du}{dt} \times \frac{dt}{dx}\right)$$

$$= \frac{\sin x}{u^2 t} = \frac{\sin x}{(\log t)^2 t} \qquad [\because u = \log t]$$

$$= \frac{\sin x}{[\log(\cos x)]^2 \cdot \cos x}$$

$$= \frac{\tan x}{[\log(\cos x)]^2} \quad [\because t = \cos x]$$

Example 7. Differentiate $e^{\sqrt{\cot x}}$

Solution: Let $y = e^{\sqrt{\cot x}}$

Put $\cot x = t$ and $\sqrt{\cot x} = \sqrt{t} = u$

so that $y = e^u$, $u = \sqrt{t}$ and $t = \cot x$

$$\therefore \quad \frac{dy}{du} = e^u, \; \frac{du}{dt} = \frac{1}{2}t^{-\frac{1}{2}} = \frac{1}{2\sqrt{t}} \text{ and } \frac{dt}{dx} = -\text{cosec}^2 \text{ x}$$

$$\frac{dy}{dx} = \left(\frac{dy}{du} \times \frac{du}{dt} \times \frac{dt}{dx}\right) = -\frac{1}{2}\frac{\cos ec^2 x}{\sqrt{t}} e^u$$

$$= -\frac{\cos ec^2 x}{2\sqrt{t}} e^{\sqrt{t}} \quad \left[\because u = \sqrt{t}\right]$$

$$\frac{dy}{dx} = -\frac{\cos ec^2 x}{2\sqrt{\cot x}} e^{\sqrt{\cot x}} \quad [\because t = \cot x]$$

EXERCISE 3.12

1. If $y = \cos^2 x^2$, find $\frac{dy}{dx}$

2. Differentiate (*i*) $\sqrt{\frac{1 - \tan x}{1 + \tan x}}$ (*ii*) $\frac{x}{\sqrt{1 - x^2}}$

Differentiate the following w. r. t. x:

3. $\sqrt{\sin x}$ **4.** $\log(\log x)$ **5.** $\log \sin x/2$ **6.** $\log\sqrt{\tan x}$

7. $\sqrt{\frac{1 + \sin x}{1 - \sin x}}$ **8.** $\sqrt{\frac{1 - x^2}{1 + x^2}}$ **9.** $\sin x^3$ **10.** $(ax + b)^m$

ANSWERS

1. $-2x \sin(2x^2)$ **2.** (*i*) $\frac{-\sec^2 x}{(1 + \tan x)^{3/2}(1 - \tan x)^{1/2}}$ (*ii*) $\frac{1}{(1 - x^2)^{3/2}}$

3. $\frac{\cos x}{2\sqrt{\sin x}}$ **4.** $\frac{1}{x \log x}$ **5.** $\frac{1}{2}\cot\frac{x}{2}$ **6.** cosec 2x

7. $\dfrac{1}{(1-\sin x)}$ or $\sec x(\sec x + \tan x)$ **8.** $\dfrac{-2x}{(1+x^2)^{3/2}(1-x^2)^{1/2}}$

9. $3x^2 \cos x^3$ **10.** $am(ax+b)^{m-1}$

3.28 DERIVATIVE OF INVERSE TRIGONOMETRIC FUNCTIONS

Before we find the derivative of inverse trigonometric functions, let us recall that

$$(i)\ \sin^{-1} x + \cos^{-1} x = \frac{\pi}{2} \quad (ii)\ \tan^{-1} x + \cot^{-1} x = \frac{\pi}{2}$$

$$(iii)\ \sec^{-1} x + \text{cosec}^{-1} x = \frac{\pi}{2} \quad (iv)\ \tan^{-1} x \pm \tan^{-1} y = \tan^{-1} \frac{x \pm y}{1 \mp xy}$$

$$(v)\ 2\tan^{-1} x = \sin^{-1} \frac{2x}{1+x^2} = \cos^{-1} \frac{1-x^2}{1+x^2} = \tan^{-1} \frac{2x}{1-x^2}$$

$$(vi)\ \sin^{-1} x \pm \sin^{-1} y = \sin^{-1}\left[x\sqrt{1-y^2} \pm y\sqrt{1-x^2}\right]$$

$$(vii)\ \cos^{-1} x \pm \cos^{-1} y = \cos^{-1}\left[xy \mp \sqrt{1-x^2}\,\sqrt{1-y^2}\right]$$

$$(viii)\ \sin^{-1}(-x) = -\sin^{-1} x \quad (ix)\ \cos^{-1}(-x) = \pi - \cos^{-1} x$$

$$(x)\ \tan^{-1}(-x) = -\tan^{-1} x \quad (xi)\ \cot^{-1} x = \tan^{-1} \frac{1}{x}$$

$$(xii)\ \text{cosec}^{-1} x = \sin^{-1} \frac{1}{x} \quad (xiii)\ \sec^{-1} x = \cos^{-1} \frac{1}{x}$$

$$(xiv)\ \lim_{x \to 0} \frac{\sin^{-1} x}{x} = 1 \quad (xv)\ \lim_{x \to 0} \frac{\tan^{-1} x}{x} = 1$$

1. Derivative of $\sin^{-1} x$, where $x \in (-1, 1)$

Let $y = \sin^{-1} x \Rightarrow x = \sin y$

$y + \delta y = \sin^{-1}(x + \delta x) \Rightarrow x + \delta x = \sin(y + \delta y)$

$\delta x = \sin(y + \delta y) - \sin y$

$$\therefore \quad \frac{\delta y}{\delta x} = \frac{\delta y}{\sin(y+\delta y) - \sin y}$$

$$\lim_{\delta x \to 0} \frac{\delta y}{\delta x} = \lim_{\delta y \to 0} \frac{\delta y}{\sin(y+\delta y) - \sin y} \quad [\delta x \to 0 \Rightarrow \delta y \to 0]$$

$$= \lim_{\delta y \to 0} \frac{\delta y}{2\cos\left(\dfrac{y+\delta y+y}{2}\right)\sin\left(\dfrac{y+\delta y-y}{2}\right)}$$

$$\left[\begin{aligned} &\because \sin C - \sin D \\ &= 2\cos\frac{C+D}{2}\sin\frac{C-D}{2} \end{aligned}\right]$$

$$= \lim_{\delta y \to 0} \frac{\delta y}{2\cos\left(y + \frac{\delta y}{2}\right)\sin\left(\frac{\delta y}{2}\right)}$$

$$= \lim_{\delta y \to 0} \frac{1}{\cos\left(y + \frac{\delta y}{2}\right)} \lim_{\frac{\delta y}{2} \to 0} \frac{(\delta y/2)}{\sin(\delta y/2)}$$

$$\left[\because \delta y \to 0 \Rightarrow \frac{\delta y}{2} \to 0\right]$$

$$\frac{dy}{dx} = \frac{1}{\cos y}\cdot 1 = \frac{1}{\cos y} = \frac{1}{\sqrt{1-\sin^2 y}}$$

$$\frac{dy}{dx} = \frac{1}{\sqrt{1-\sin^2 y}} = \frac{1}{\sqrt{1-x^2}}$$

i.e., $\frac{d}{dx}(\sin^{-1} x) = \frac{1}{\sqrt{1-x^2}}$

2. Derivative of $\cos^{-1} x$, where $x \in (-1, 1)$

Similarly, we can prove that

$$\frac{d}{dx}(\cos^{-1} x) = -\frac{1}{\sqrt{1-x^2}}$$

3. Derivative of $\tan^{-1} x$, where $x \in R$

Let $y = \tan^{-1} x \Rightarrow x = \tan y$

$y + \delta y = \tan^{-1}(x + \delta x) \Rightarrow x + \delta x = \tan(y + \delta y)$

$\therefore\ \delta x = \tan(y + \delta y) - \tan y$

$$\frac{\delta y}{\delta x} = \frac{\delta y}{\tan(y + \delta y) - \tan y}$$

$$\lim_{\delta x \to 0} \frac{\delta y}{\delta x} = \lim_{\delta y \to 0} \frac{\delta y}{\tan(y + \delta y) - \tan y}$$

$$[\because \delta x \to 0 \Rightarrow \delta y \to 0]$$

$$= \lim_{\delta y \to 0}\left[\frac{\delta y}{\frac{\sin(y+\delta y)}{\cos(y+\delta y)} - \frac{\sin y}{\cos y}}\right]$$

$$= \lim_{\delta y \to 0} \frac{\delta y \cos(y+\delta y)\cos y}{\sin(y+\delta y)\cos y - \sin y \cos(y+\delta y)}$$

$$= \lim_{\delta y \to 0} \frac{\delta y \cos(y+\delta y)\cos y}{\sin(y+\delta y - y)}$$

$$= \lim_{\delta y \to 0} \frac{\delta y \cos(y+\delta y)\cos y}{\sin \delta y}$$

$$= \lim_{\delta y \to 0} \frac{\delta y}{\sin \delta y} \cdot \lim_{\delta y \to 0} \cos(y+\delta y)\cos y$$

$$= 1 \cdot \cos y \cdot \cos y = \cos^2 y$$

$$\therefore \quad \frac{dy}{dx} = \frac{1}{\sec^2 y} = \frac{1}{1+\tan^2 y}$$

i.e., $\dfrac{d}{dx}(\tan^{-1} x) = \dfrac{1}{1+x^2}$

4. Derivative of $\cot^{-1} x$

Similarly, we can prove that

$$\frac{d}{dx}(\cot^{-1} x) = -\frac{1}{1+x^2}$$

5. Derivative of $\sec^{-1} x$

Let $y = \sec^{-1} x \Rightarrow \sec y = x$

$\sec(y + \delta y) = x + \delta x$

$\delta x = \sec(y + \delta y) - \sec y$

$$\therefore \quad \lim_{\delta x \to 0} \frac{\delta y}{\delta x} = \lim_{\delta y \to 0} \frac{\delta y}{\sec(y+\delta y) - \sec y} \quad [\text{as } \delta x \to 0 \Rightarrow \delta y \to 0]$$

$$= \lim_{\delta y \to 0} \frac{\delta y}{\dfrac{1}{\cos(y+\delta y)} - \dfrac{1}{\cos y}}$$

$$= \lim_{\delta y \to 0} \frac{\delta y \cdot \cos(y+\delta y)\cos y}{\cos y - \cos(y+\delta y)}$$

$$= \lim_{\delta y \to 0} \frac{\delta y \cdot \cos y \cos(y+\delta y)}{2\sin\left(\dfrac{y+y+\delta y}{2}\right)\sin\left(\dfrac{y+\delta y - y}{2}\right)}$$

$$\left[\because \cos C - \cos D = 2\sin\frac{C+D}{2}\sin\frac{D-C}{2}\right]$$

$$= \lim_{\delta y \to 0} \frac{\delta y \cdot \cos y \cos(y+\delta y)}{2\sin\left(y+\dfrac{\delta y}{2}\right)\sin\dfrac{\delta y}{2}}$$

$$= \lim_{\frac{\delta y}{2}\to 0} \frac{(\delta y/2)}{\sin\left(\frac{\delta y}{2}\right)} \cdot \lim_{\delta y \to 0} \frac{\cos y \cdot \cos(y+\delta y)}{\sin\left(y+\frac{\delta y}{2}\right)}$$

$$= 1 \cdot \frac{\cos y \cos y}{\sin y} = \frac{1}{\sec y \tan y}$$

$$\frac{dy}{dx} = \frac{1}{\sec y\sqrt{\sec^2 y - 1}} \quad \frac{1}{x\sqrt{x^2-1}}$$

6. Derivative of $\text{cosec}^{-1} x$, where $x \in R-(-1, 1)$

Similarly, we can prove that

$$\frac{d}{dx}(\text{cosec}^{-1}x) = -\frac{1}{x\sqrt{x^2-1}}$$

ILLUSTRATIVE EXAMPLES

Example 1. Find the derivative of

(*i*) $\cos^{-1}(\cot x)$ (*ii*) $\sin^{-1} 2x$ (iii) $\cos^{-1}\sqrt{\frac{1+\cos x}{2}}$

Solution: (*i*) Let $y = \cos^{-1}(\cot x)$

Put $\cot x = t \Rightarrow y = \cos^{-1} t$ and $t = \cot x$

$$\therefore \quad \frac{dy}{dt} = -\frac{1}{\sqrt{1-t^2}} \text{ and } \frac{dt}{dx} = -\text{cosec}^2 x$$

$$\therefore \quad \frac{dy}{dx} = \left(\frac{dy}{dt}\times\frac{dt}{dx}\right) = \frac{\cos ec^2 x}{\sqrt{1-t^2}}$$

$$= \frac{\cos ec^2 x}{\sqrt{1-\cot^2 x}} \quad [\because t = \cot x]$$

$$\frac{dy}{dx} = \frac{\cos ec^2 x}{\sqrt{1-\cot^2 x}}$$

(*ii*) Let $y = \sin^{-1} 2x$

Put $2x = t \Rightarrow y = \sin^{-1} t$ and $t = 2x$

$$\therefore \quad \frac{dy}{dt} = \frac{1}{\sqrt{1-t^2}} \text{ and } \frac{dt}{dx} = 2$$

$$\therefore \quad \frac{dy}{dt} = \left(\frac{dy}{dt}\times\frac{dt}{dx}\right) = \frac{2}{\sqrt{1-t^2}}$$

$$\frac{dy}{dx} = \frac{2}{\sqrt{1-4x^2}} \quad [\because t = 2x]$$

(*iii*) Let $y = \cos^{-1}\sqrt{\frac{1+\cos x}{2}} = \cos^{-1}\sqrt{\frac{2\cos^2 x/2}{2}}$

$$y = \cos^{-1}(\cos x/2) = x/2$$

$$\frac{dy}{dx} = \frac{d}{dx}\left(\frac{x}{2}\right) = \frac{1}{2}$$

$$\frac{dy}{dx} = \frac{1}{2}$$

Example 2. Find the derivative of

(*i*) sin ($\tan^{-1} x$) (*ii*) sec ($\tan^{-1} x$)

Solution: (*i*) Let $y = \sin(\tan^{-1}x)$

Put $\tan^{-1}x = t \Rightarrow y = \sin t$ and $t = \tan^{-1} x$

$$\therefore \quad \frac{dy}{dt} = \cos t \text{ and } \frac{dt}{dx} = \frac{1}{1+x^2}$$

$$\frac{dy}{dx} = \left(\frac{dy}{dt} \times \frac{dt}{dx}\right) = \frac{\cos t}{(1+x^2)}$$

$$= \frac{1}{\sqrt{1+x^2}(1+x^2)} = \frac{1}{(1+x^2)3/2}$$

$$\left[\because \tan t = x \Rightarrow \cos t = \frac{1}{\sqrt{1+x^2}}\right]$$

$$\therefore \quad \frac{dy}{dx} = \frac{1}{(1+x^2)3/2}$$

(*ii*) Let $y = \sec(\tan^{-1} x)$

Put $\tan^{-1} x = t \Rightarrow y = \sec t$ and $t = \tan^{-1} x$

$$\therefore \quad \frac{dy}{dt} = \sec t \text{ tant and } \frac{dt}{dx} = \frac{1}{1+x^2}$$

$$\therefore \quad \frac{dy}{dx} = \left(\frac{dy}{dt} \times \frac{dt}{dx}\right) = \frac{\sec t \tan t}{(1+x^2)}$$

$$= \frac{\sqrt{1+\tan^2 t}\,(\tan t)}{(1+x^2)}$$

$$\therefore \quad \frac{dy}{dx} = \frac{\left(\sqrt{1+x^2}\right)x}{(1+x^2)} = \frac{x}{\sqrt{1+x^2}}$$

$$[\because t = \tan^{-1} x \Rightarrow \tan t = x]$$

$$\frac{dy}{dx} = \frac{x}{\sqrt{1+x^2}}$$

Example 3. If $y = \dfrac{x\sin^{-1} x}{\sqrt{1-x^2}}$, find $\dfrac{dy}{dx}$

Solution: By the quotient rule, we have

$$\frac{dy}{dx} = \frac{1}{(1-x^2)}\left[\sqrt{1-x^2}\,\frac{d}{dx}(x\sin^{-1} x) - (x\sin^{-1} x)\frac{d}{dx}\sqrt{1-x^2}\right]$$

$$= \frac{\sqrt{1-x^2}\left\{x\cdot\dfrac{1}{\sqrt{1-x^2}} + \sin^{-1} x\cdot 1\right\} - (x\sin^{-1} x)\cdot\dfrac{1}{2}(1-x^2)^{-\frac{1}{2}}(-2x)}{(1-x^2)}$$

$$= \frac{x + \sin^{-1} x\sqrt{1-x^2} + \dfrac{x^2\sin^{-1} x}{\sqrt{1-x^2}}}{(1-x^2)}$$

$$= \frac{x\sqrt{1-x^2} + \sin^{-1} x\cdot(1-x^2) + x^2\sin^{-1} x}{(1-x^2)^{3/2}}$$

$$\frac{dy}{dx} = \frac{x\sqrt{1-x^2} + \sin^{-1} x}{(1-x^2)^{3/2}}$$

Example 4. Differentiate $e^{\cot^{-1} x^2}$ w. r. t. x

Solution: Let $y = e^{\cot^{-1} x^2}$

Put $e^{\cot^{-1} x^2} = u$, then

$y = e^u$ and $u = e^{\cot^{-1} x^2}$

$$\frac{dy}{du} = e^u \text{ and } \frac{du}{dx} = \frac{-1}{1+x^4}\,\frac{d}{dx}(x^2) = \frac{-2x}{(1+x^4)}$$

$$\Rightarrow \qquad \frac{dy}{du} = e^{\cot^{-1} x^2} \text{ and } \frac{du}{dx} = \frac{-2x}{(1+x^4)}$$

$$\therefore \qquad \frac{dy}{dx} = \left(\frac{dy}{du}\times\frac{du}{dt}\right) = \frac{-2x\,e^{\cot^{-1} x^2}}{(1+x^4)}$$

$$\frac{dy}{dx} = \frac{-2x\,e^{\cot^{-1} x^2}}{(1+x^4)}$$

Example 5. Find $\dfrac{d}{dx}\left[\sqrt{1-x^2}\sin^{-1} x - x\right]$

Solution: $\dfrac{d}{dx}\left[\sqrt{1-x^2}\sin^{-1} x - x\right] = \dfrac{d}{dx}\left[\sqrt{1-x^2}\sin^{-1} x\right] - \dfrac{d}{dx}(x)$

$$= \sqrt{1-x^2}\,\frac{d}{dx}(\sin^{-1}x) + \sin^{-1}x\,\frac{d}{dx}\sqrt{1-x^2} - 1$$

$$= \sqrt{1-x^2}\cdot\frac{1}{\sqrt{1-x^2}} + \sin^{-1}x\cdot\frac{1}{2}(1-x^2)^{-\frac{1}{2}}\times(-2x) - 1$$

$$= \left[1 - \frac{x\sin^{-1}x}{\sqrt{1-x^2}} - 1\right] = \frac{-x\sin^{-1}x}{\sqrt{1-x^2}}$$

$$= \frac{-x\sin^{-1}x}{\sqrt{1-x^2}}$$

EXERCISE 3.13

Differentiate the following w. r. t. x:

1. $\sin^{-1}(\cos x)$

2. $\tan^{-1}(\cot x)$

3 $\sin^{-1} x/a$

4. $\cos^{-1} 2x$

5. $\tan^{-1} x^2$

6. $\tan^{-1}\left(\cos\sqrt{x}\right)$

7. $\log(\tan^{-1}x)$

8. $\tan^{-1}(\log_e x)$

9. If $y = \dfrac{x\sin^{-1}x}{\sqrt{1-x^2}} + \log\sqrt{1-x^2}$, prove that $\dfrac{dy}{dx} = \dfrac{\sin^{-1}x}{(1-x^2)^{3/2}}$

10. If $y = \sin(\tan^{-1}2x)$, prove that $\dfrac{dy}{dx} = \dfrac{2}{(1+4x^2)^{3/2}}$

ANSWERS

1. -1 **2.** -1 **3.** $\dfrac{1}{\sqrt{a^2-x^2}}$ **4.** $\dfrac{-2}{\sqrt{1-4x^2}}$

5. $\dfrac{2x}{(1+x^4)}$ **6.** $\dfrac{-\sin\sqrt{x}}{\left(2\sqrt{x}\,(1+\cos^2\sqrt{x}\right)}$

7. $\dfrac{1}{(1+x^2)\tan^{-1}x}$ **8.** $\dfrac{1}{x[1+(\log x)^2]}$

3.29 DIFFERENTIATIONBY USING TRIGONOMETRICAL TRANSFORMATIONS

Some functions can be differentiated easily by using trigonometrical transformations. Some of the important trigonometrical results are given below:

(*i*) $1 + \cos x = 2\cos^2 x/2$ (*ii*) $1 - \cos x = 2\sin^2 x/2$

(*iii*) $\sin x = \dfrac{2\tan x/2}{1+\tan^2 x/2}$ (*iv*) $\cos x = \dfrac{1-\tan^2 x/2}{1+\tan^2 x/2}$

(*v*) $\sin 3x = 3\sin x - 4\sin^3 x$ (*vi*) $\cos 3x = 4\cos^3 x - 3\cos x$

The following substitutions are generally useful in finding the derivatives:

1. If $a^2 + x^2$ or $\sqrt{a^2 + x^2}$ then substitute $x = a \tan \theta$ or $a \cot \theta$
2. If $a^2 - x^2$ or $\sqrt{a^2 - x^2}$ then substitute $x = a \sin \theta$ or $a \cos \theta$
3. If $x^2 - a^2$ or $\sqrt{x^2 - a^2}$ then substitute $x = a \sec \theta$ or $a \operatorname{cosec} \theta$
4. If $\sqrt{\dfrac{x-a}{x+a}}$ then substitute $x = a \cos \theta$ or $x = a \cos 2\theta$

ILLUSTRATIVE EXAMPLES

Example 1. Differentiate the following w. r. t. x:

(*i*) $\tan^{-1}\left(\sqrt{\dfrac{1+\sin x}{1-\sin x}}\right)$ (*ii*) $\tan^{-1}\sqrt{\dfrac{1-\cos x}{1+\cos x}}$

(*iii*) $\tan^{-1}\left(\dfrac{\cos x - \sin x}{\cos x + \sin x}\right)$ (*iv*) $\tan^{-1}\left(\sqrt{\dfrac{1+\cos x}{1-\cos x}}\right)$

(*v*) $\tan^{-1}\left(\dfrac{\cos x}{1+\sin x}\right)$ (*vi*) $\cos^{-1}\sqrt{\dfrac{1+\cos x}{2}}$

Solution: (*i*) $y = \tan^{-1}\sqrt{\dfrac{1+\sin x}{1-\sin x}}$

$$= \tan^{-1}\sqrt{\frac{\sin^2 x/2 + \cos^2 x/2 + 2\sin x/2\,\cos x/2}{\sin^2 x/2 + \cos^2 x/2 - 2\sin x/2\,\cos x/2}}$$

$$y = \tan^{-1}\sqrt{\frac{(\sin x/2 + \cos x/2)^2}{(\sin x/2 - \cos x/2)^2}}$$

$$= \tan^{-1}\left(\frac{1+\tan x/2}{1-\tan x/2}\right)$$

$$y = \tan^{-1}\left(\frac{\tan\pi/4 + \tan x/2}{1-\tan\pi/4\tan x/2}\right) \quad \left[\because \tan\frac{\pi}{4} = 1\right]$$

$$y = \tan^{-1}\left[\tan\left(\frac{\pi}{4} + \frac{x}{2}\right)\right]$$

$$y = \left(\frac{\pi}{4} + \frac{x}{2}\right)$$

$$\therefore \qquad \frac{dy}{dx} = \frac{d}{dx}\left[\tan^{-1}\sqrt{\frac{1+\sin x}{1-\sin x}}\right] = \frac{d}{dx}\left(\frac{x}{2}+\frac{\pi}{4}\right) = \frac{1}{2}$$

$$\frac{dy}{dx} = \frac{1}{2}$$

(*ii*) let $y = \tan^{-1}\sqrt{\dfrac{1-\cos x}{1+\cos x}} = \tan^{-1}\sqrt{\dfrac{1-(1-2\sin^2 x/2}{1+2\cos^2 x/2-1}}$

$$y = \tan^{-1}\sqrt{\frac{2\sin^2 x/2}{2\cos^2 x/2}} = \tan^{-1}(\tan x/2) = \frac{x}{2}$$

$$\therefore \qquad \frac{dy}{dx} = \frac{1}{2}$$

(*iii*) Let $y = \tan^{-1}\left(\dfrac{\cos x-\sin x}{\cos x+\sin x}\right)$

$$y = \tan^{-1}\left(\frac{1-\tan x}{1+\tan x}\right) = \tan^{-1}\left[\tan\left(\frac{\pi}{4}-x\right)\right]$$

$$y = \left(\frac{\pi}{4}-x\right)$$

$$\therefore \qquad \frac{dy}{dx} = -1$$

(*iv*) Let $y = \tan^{-1}\sqrt{\dfrac{1+\cos x}{1-\cos x}} = \tan^{-1}\sqrt{\dfrac{1+2\cos^2\dfrac{x}{2}-1}{1-(1-2\sin^2 x/2)}}$

$$y = \tan^{-1}(\cot x/2) = \tan^{-1}\left[\tan\left(\frac{\pi}{2}-\frac{x}{2}\right)\right]$$

$$y = \left(\frac{\pi}{2}-\frac{x}{2}\right)$$

$$\therefore \qquad \frac{dy}{dx} = -\frac{1}{2}$$

(*v*) Let $y = \tan^{-1}\left(\dfrac{\cos x}{1+\sin x}\right) = \tan^{-1}\left[\dfrac{\sin\left(\dfrac{\pi}{2}-x\right)}{1+\cos\left(\dfrac{\pi}{2}-x\right)}\right]$

$$y = \tan^{-1}\left[\frac{2\sin\left(\frac{\pi}{4}-\frac{x}{2}\right)\cos\left(\frac{\pi}{4}-\frac{x}{2}\right)}{2\cos^2\left(\frac{\pi}{4}-\frac{x}{2}\right)}\right]$$

$$y = \tan^{-1}\left[\tan\left(\frac{\pi}{4} - \frac{x}{2}\right)\right] = \left(\frac{\pi}{4} - \frac{x}{2}\right)$$

$$\therefore \quad \frac{dy}{dx} = -\frac{1}{2}$$

(vi) Let
$$y = \cos^{-1}\sqrt{\frac{1+\cos x}{2}} = \cos^{-1}\sqrt{\frac{2\cos^2 x/2}{2}}$$

$$y = \cos^{-1}[\cos x/2] = \frac{x}{2}$$

$$\therefore \quad \frac{dy}{dx} = \frac{1}{2}$$

Example 2. Differentiate the following w. r. t. x:

(*i*) $\tan^{-1}\left(\frac{2x}{1-x^2}\right)$ (*ii*) $\sin^{-1}\left(\frac{2x}{1+x^2}\right)$

(*iii*) $\sin^{-1}\left(\frac{1-x^2}{1+x^2}\right)$ (*iv*) $\cos^{-1}\left(\frac{1-x^2}{1+x^2}\right)$

(*v*) $\cot^{-1}\left(\frac{1-x}{1+x}\right)$ (*vi*) $\sec^{-1}\left(\frac{2x}{x^2-1}\right)$

(*vii*) $\tan^{-1}\sqrt{\frac{1-x^2}{1+x^2}}$ (*viii*) $\tan^{-1}\sqrt{\frac{1-x}{1+x}}$

(*ix*) $\tan^{-1}\left[\frac{\sqrt{1+\sin x}+\sqrt{1-\sin x}}{\sqrt{1+\sin x}-\sqrt{1-\sin x}}\right]$

(*x*) $\sin^{-1}\left(\frac{4\sin x + 3\cos x}{5}\right)$

Solution: (*i*) Let $y = \tan^{-1}\left(\frac{2x}{1-x^2}\right)$

Put $x = \tan\theta$

$$y = \tan^{-1}\left(\frac{2\tan\theta}{1-\tan^2\theta}\right) = \tan^{-1}(\tan 2\theta) = 2\theta = 2\tan^{-1}x$$

$$\therefore \quad \frac{dy}{dx} = \frac{2}{1+x^2}$$

Hence $\frac{d}{dx}\left[\tan^{-1}\left(\frac{2x}{1-x^2}\right)\right] = \frac{2}{1+x^2}$

(*ii*) Let $y = \sin^{-1}\left(\frac{2x}{1+x^2}\right)$

Put $x = \tan\theta$

$$y = \sin^{-1}\left(\frac{2\tan\theta}{1+\tan^2\theta}\right) = \sin^{-1}(\sin 2\theta)$$

$$= 2\theta = 2\tan^{-1} x$$

$$\therefore \qquad \frac{dy}{dx} = \frac{2}{1+x^2}$$

Hence $\quad \dfrac{d}{dx}\left[\sin^{-1}\left(\dfrac{2x}{1+x^2}\right)\right] = \dfrac{2}{1+x^2}$

(*iii*) Let $y = \sin^{-1}\left(\dfrac{1-x^2}{1+x^2}\right)$

Put $x = \tan\theta$

$$y = \sin^{-1}\left(\frac{1-\tan^2\theta}{1+\tan^2\theta}\right) = \sin^{-1}(\cos 2\theta)$$

$$= \sin^{-1}\left[\sin\left(\frac{\pi}{2} - 2\theta\right)\right]$$

$$y = \left(\frac{\pi}{2} - 2\theta\right) = \left(\frac{\pi}{2} - 2\tan^{-1} x\right)$$

$$\therefore \qquad \frac{dy}{dx} = \frac{-2}{(1+x^2)}$$

(*iv*) Let $y = \cos^{-1}\dfrac{1-x^2}{1+x^2}$

Put $x = \tan\theta$

$$y = \cos^{-1}\left[\frac{1-\tan^2\theta}{1+\tan^2\theta}\right] = \cos^{-1}(\cos 2\theta).$$

$$= 2\theta = 2\tan^{-1} x$$

$$\therefore \qquad \frac{dy}{dx} = \frac{2}{1+x^2}$$

Hence $\quad \dfrac{d}{dx}\left[\cos^{-1}\left(\dfrac{1-x^2}{1+x^2}\right)\right] = \dfrac{2}{1+x^2}$

(*v*) Let $y = \cot^{-1}\left(\dfrac{1-x}{1+x}\right)$

Put $x = \tan\theta$

$$y = \cot^{-1}\left(\frac{1-\tan\theta}{1+\tan\theta}\right) = \cot^{-1}\left[\tan\left(\frac{\pi}{4}-\theta\right)\right]$$

$$y = \cot^{-1}\left[\cot\left\{\frac{\pi}{2}-\left(\frac{\pi}{4}-\theta\right)\right\}\right]$$

$$y = \cot^{-1}\left[\cot\left(\frac{\pi}{4}+\theta\right)\right] = \frac{\pi}{4}+\theta = \frac{\pi}{4}+\tan^{-1}x$$

$$\therefore \qquad \frac{dy}{dx} = \frac{1}{1+x^2}$$

Hence $\dfrac{d}{dx}\left[\cot^{-1}\left(\dfrac{1-x}{1+x}\right)\right] = \dfrac{1}{1+x^2}$

(*vi*) Let $y = \sec^{-1}\dfrac{x^2+1}{x^2-1}$

Put $x = \cot\theta$

$$y = \sec^{-1}\left[\frac{\cot^2\theta+1}{\cot^2\theta-1}\right] = \sec^{-1}\left[\frac{1+\tan^2\theta}{1-\tan^2\theta}\right]$$

$$y = \sec^{-1}(\sec 2\theta) = 2\theta = 2\cot^{-1}x$$

$$\therefore \qquad \frac{dy}{dx} = \frac{-2}{(1+x^2)}$$

Hence $\dfrac{d}{dx}\left[\sec^{-1}\left(\dfrac{x^2+1}{x^2-1}\right)\right] = \dfrac{-2}{(1+x^2)}$

(*vii*) Let $y = \tan^{-1}\sqrt{\dfrac{1-x^2}{1+x^2}}$

Put $x^2 = \cos\theta$

$$y = \tan^{-1}\sqrt{\frac{1-\cos\theta}{1+\cos\theta}} = \tan^{-1}\sqrt{\frac{2\sin^2\theta/2}{2\cos^2\theta/2}}$$

$$= \tan^{-1}[\tan\theta/2]$$

$$y = \frac{\theta}{2} = \frac{1}{2}\cos^{-1}x^2$$

$$\therefore \qquad \frac{dy}{dx} = -\frac{1}{2}\frac{2x}{\sqrt{1-x^4}} = \frac{-x}{\sqrt{1-x^4}}$$

Hence $\dfrac{d}{dx}\left[\tan^{-1}\sqrt{\dfrac{1-x^2}{1+x^2}}\right] = \dfrac{-x}{\sqrt{1-x^4}}$

(*viii*) Let $y = \tan^{-1}\sqrt{\dfrac{1-x}{1+x}}$

Put $x = \cos 2\theta$

$$y = \tan^{-1}\sqrt{\frac{1-\cos 2\theta}{1+\cos 2\theta}} = \tan^{-1}\sqrt{\frac{2\sin^2\theta}{2\cos^2\theta}}$$

$$= \tan^{-1}(\tan\theta)$$

$$y = \theta = \frac{1}{2}\cos^{-1}x$$

$$\therefore \qquad \frac{dy}{dx} = -\frac{1}{2\sqrt{1-x^2}}$$

Hence $\dfrac{d}{dx}\left[\tan^{-1}\sqrt{\dfrac{1-x}{1+x}}\right] = -\dfrac{1}{2\sqrt{1-x^2}}$

(*ix*) Let $y = \tan^{-1}\left[\dfrac{\sqrt{1+\sin x}+\sqrt{1-\sin x}}{\sqrt{1+\sin x}-\sqrt{1-\sin x}}\right]$

$$= \tan^{-1}\left[\frac{\cos\frac{x}{2}+\sin\frac{x}{2}+\cos\frac{x}{2}-\sin\frac{x}{2}}{\cos\frac{x}{2}+\sin\frac{x}{2}-\cos\frac{x}{2}+\sin\frac{x}{2}}\right]$$

$$y = \tan^{-1}\left[\cot\frac{x}{2}\right] = \tan^{-1}\left[\tan\left(\frac{\pi}{2}-\frac{x}{2}\right)\right]$$

$$= \left(\frac{\pi}{2}-\frac{x}{2}\right)$$

$$\therefore \qquad \frac{dy}{dx} = -\frac{1}{2} \quad \left[\text{Assuming } 0 < x < \frac{\pi}{2}\right]$$

(*x*) Let $y = \sin^{-1}\left(\dfrac{4\sin x + 3\cos x}{5}\right)$

$$= \sin^{-1}\left[\frac{4}{5}\sin x + \frac{3}{5}\cos x\right] = \sin^{-1}[\sin x \sin\alpha + \cos x \cos\alpha] \quad \left(\text{Let } \frac{4}{5} = \sin\alpha\right)$$

$$= \sin^{-1}[\sin(x+\alpha)] = x + \alpha$$

$$y = x + \sin^{-1}\left(\frac{4}{5}\right)$$

$$\therefore \qquad \frac{dy}{dx} = 1$$

Example 3. If $y = \tan^{-1}\left[\dfrac{\sqrt{1+x^2}+\sqrt{1+x^2}}{\sqrt{1+x^2}-\sqrt{1-x^2}}\right]$ find $\dfrac{dy}{dx}$

Solution: Put $x^2 = \cos 2\theta$

$$y = \tan^{-1}\left[\frac{\sqrt{1+\cos 2\theta}+\sqrt{1-\cos 2\theta}}{\sqrt{1+\cos 2\theta}-\sqrt{1-\cos 2\theta}}\right]$$

$$= \tan^{-1}\left[\frac{\sqrt{2\cos^2\theta}+\sqrt{2\sin^2\theta}}{\sqrt{2\cos^2\theta}-\sqrt{2\sin^2\theta}}\right]$$

$$y = \tan^{-1}\left[\frac{\cos\theta+\sin\theta}{\cos\theta-\sin\theta}\right] = \tan^{-1}\left[\frac{1+\tan\theta}{1-\tan\theta}\right]$$

$$y = \tan^{-1}\left[\tan\left(\frac{\pi}{4}+\theta\right)\right] = \left(\frac{\pi}{4}+\theta\right)$$

$$= \frac{\pi}{4}+\frac{1}{2}\cos^{-1}x^2$$

$$\frac{dy}{dx} = \frac{1}{2}\ \frac{-1}{\sqrt{1-x^4}}(2x) = \frac{-x}{\sqrt{1-x^4}}$$

Example 4. If $y = \sin^{-1}\left(\dfrac{x+\sqrt{1-x^2}}{\sqrt{2}}\right)$, find $\dfrac{dy}{dx}$

Solution: $y = \sin^{-1}\left(\dfrac{x+\sqrt{1-x^2}}{\sqrt{2}}\right)$

Put $x = \sin\theta$

$$y = \sin^{-1}\left(\frac{\sin\theta+\cos\theta}{\sqrt{2}}\right)$$

$$= \sin^{-1}\left(\sin\theta\cos\frac{\pi}{4}+\cos\theta\sin\frac{\pi}{4}\right)$$

$$y = \sin^{-1}\left[\sin\left(\theta+\frac{\pi}{4}\right)\right] = \left(\theta+\frac{\pi}{4}\right)$$

$$y = \sin^{-1}x+\frac{\pi}{4}$$

$$\therefore \qquad \frac{dy}{dx} = \frac{1}{\sqrt{1-x^2}}$$

Example 5. If $y = \sec^{-1}\left(\dfrac{1}{2x^2-1}\right)$, find $\dfrac{dy}{dx}$

Solution: $y = \sec^{-1}\left(\dfrac{1}{2x^2-1}\right)$

Put $x = \cos\theta$

$$y = \sec^{-1}\left(\frac{1}{2\cos^2\theta - 1}\right) = \sec^{-1}\left(\frac{1}{\cos 2\theta}\right)$$

$$y = \sec^{-1}(\sec 2\theta) = 2\theta = 2\cos^{-1} x$$

$$\therefore \quad \frac{dy}{dx} = \frac{-2}{\sqrt{1-x^2}}$$

EXERCISE 3.14

Differentiate the following w. r. t. x:

1. $\tan^{-1}\sqrt{\dfrac{\sin x}{1+\cos x}}$
2. $\tan^{-1}(\cot x) + \cot^{-1}(\tan x)$
3. $\sin^{-1}\left(\sqrt{1-x^2}\right)$
4. $\sin^{-1}\left(2x\sqrt{1-x^2}\right)$
5. $\sin^{-1}\sqrt{\dfrac{1-x}{2}}$
6. $\sin^{-1}\left(\dfrac{1}{\sqrt{1+x^2}}\right)$
7. $\sec^{-1}\left(\dfrac{1}{\sqrt{1-x^2}}\right)$
8. $\tan^{-1}\left(\dfrac{x}{\sqrt{a^2-x^2}}\right)$
9. $\sin^{-1}\dfrac{x^2}{\sqrt{x^4+a^4}}$
10. $\tan^{-1}\left(\dfrac{\sqrt{1+a^2x^2}-1}{ax}\right)$
11. $\tan^{-1}\left(\dfrac{3-2x}{1+6x}\right)$
12. $\tan^{-1}\left(\dfrac{2x}{1+15x^2}\right)$
13. $\sec^{-1}\left(\dfrac{1+\tan^2 x}{1-\tan^2 x}\right)$
14. $\sin^{-1}\left(\dfrac{\sqrt{1+x}-\sqrt{1-x}}{2}\right)$
15. $\cos^{-1}(2x) + 2\cos^{-1}\sqrt{1-4x^2}$

ANSWERS

1. $\dfrac{1}{2}$ **2.** -2 **3.** $-\dfrac{1}{\sqrt{1-x^2}}$ **4.** $\dfrac{2}{\sqrt{1-x^2}}$ **5.** $\dfrac{-1}{2\sqrt{1-x^2}}$

6. $\dfrac{-1}{(1+x^2)}$ **7.** $\dfrac{1}{\sqrt{1-x^2}}$ **8.** $\dfrac{1}{\sqrt{a^2-x^2}}$ **9.** $\dfrac{2a^2x}{(a^4+x^4)}$

10. $\dfrac{a}{2(1+a^2x^2)}$ **11.** $\dfrac{-2}{(1+4x^2)}$ **12.** $\dfrac{5}{(1+25x^2} - \dfrac{3}{(1+9x^2)}$

13. 2 **14.** $\dfrac{1}{2\sqrt{1-x^2}}$ **15.** $\dfrac{2}{\sqrt{1-4x^2}}$

3.30 THE IMPLICIT FUNCTION AND ITS DERIVATIVE

So for we have considered the derivative of explicit function i.e., the function of the form $y = f(x)$.

An equation of the form $f(x, y) = 0$ in which y is not expressible directly in terms of x, is known as an implicit function of x and y.

ILLUSTRATIVE EXAMPLES

Example 1. If $x^3 + y^3 + 3x^2y = a^3$, find $\dfrac{dy}{dx}$

Solution: Given $x^3 + y^3 + 3x^2y = a^3$

Differentiating both sides w. r. t. x, we get

$$\frac{d}{dx}(x^3 + y^3 + 3x^2y) = \frac{d}{dx}(a^3)$$

$$\frac{d}{dx}(x^3) + \frac{d}{dx}(y^3) + \frac{d}{dx}(3x^2y) = \frac{d}{dx}(a^3)$$

$$3x^2 + 3y^2\frac{dy}{dx} + 6xy + 3x^2\frac{dy}{dx} = 0$$

$$(y^2 + x^2)\frac{dy}{dx} = -(x^2 + 2xy)$$

$$\frac{dy}{dx} = -\frac{x(x+2y)}{(x^2+y^2)}$$

Example 2. If $x^2 + y^2 - 6x + 8y - 12 = 0$, find $\dfrac{dy}{dx}$

Solution: Given $x^2 + y^2 - 6x + 8y - 12 = 0$

Differentiating w. r. t. x, we get

$$\frac{d}{dx}(x^2) + \frac{d}{dx}(y^2) - 6\frac{d}{dx}(x) + 8\frac{d}{dx}(y) + \frac{d}{dx}(-12) = 0$$

$$2x + 2y\frac{dy}{dx} - 6 + 8\frac{dy}{dx} + 0 = 0$$

$$(y+4)\frac{dy}{dx} = 3 - x$$

$$\frac{dy}{dx} = \frac{3-x}{y+4}, y \neq -4$$

Example 3. If $\log(x^2+y^2) = 2\tan^{-1}\dfrac{y}{x}$, show that $\dfrac{dy}{dx} = \dfrac{x+y}{x-y}$

Solution: $\log(x^2+y^2) = 2\tan^{-1}\dfrac{y}{x}$

Differentiating w. r. t. x, we get

$$\frac{1}{x^2+y^2}\left(2x+2y\frac{dy}{dx}\right) = 2\frac{1}{1+y^2/x^2}\;\frac{x\frac{dy}{dx}-y\cdot 1}{x^2}$$

$$\frac{dy}{dx}\left(\frac{y}{x^2+y^2}-\frac{x}{x^2+y^2}\right) = -\frac{y}{x^2+y^2}-\frac{x}{x^2+y^2}$$

$$\Rightarrow \quad \frac{dy}{dx}(y-x) = -(y+x)$$

$$\Rightarrow \quad \frac{dy}{dx} = \frac{x+y}{x-y}$$

Example 4. If $y\sqrt{1-x^2}+x\sqrt{1-y^2} = 1$, prove that $\dfrac{dy}{dx} = \sqrt{\dfrac{1-y^2}{1-x^2}}$

Put $x = \sin\alpha$ and $y = \sin\beta$ we get $\sin\beta\cos\alpha + \sin\alpha\cos\beta = 1$

$$\Rightarrow \quad \sin(\alpha+\beta) = 1$$

$$\alpha+\beta = \sin^{-1} 1 = \frac{\pi}{2}$$

$$\Rightarrow \quad \sin^{-1}x + \sin^{-1}y = \frac{\pi}{2}$$

Differentiating w. r. t. x, we get

$$\frac{1}{\sqrt{1-x^2}}+\frac{1}{\sqrt{1-y^2}}\;\frac{dy}{dx} = 0$$

$$\frac{dy}{dx} = -\frac{\sqrt{1-y^2}}{\sqrt{1-x^2}} = -\sqrt{\frac{1-y^2}{1-x^2}}$$

Example 5. If $\dfrac{x^2}{a^2}+\dfrac{y^2}{b^2} = 1$, find $\dfrac{dy}{dx}$

Solution: $\dfrac{x^2}{a^2}+\dfrac{y^2}{b^2} = 1$

Differentiating w. r. t. x, we get

$$\frac{2x}{a^2}+\frac{2y}{b^2}\;\frac{dy}{dx} = 0$$

$$\Rightarrow \quad \frac{dy}{dx} = -\frac{x}{a^2}\cdot\frac{b^2}{y} = -\frac{b^2x}{a^2y}$$

Example 6. If $ax^2 + 2hxy + by^2 + 2gx + 2fy + c = 0$, find $\frac{dy}{dx}$

Solution: $ax^2 + 2hxy + by^2 + 2gx + 2fy + c = 0$

Differentiating w .r. t. x, we get

$$2ax + 2h\left(x \cdot \frac{dy}{dx} + y \cdot 1\right) + 2by\frac{dy}{dx} + 2g + 2f \cdot \frac{dy}{dx} = 0$$

$$(2ax + 2hy + 2g) + (2hx + 2by + 2f)\,\frac{dy}{dx} = 0$$

$$\therefore \qquad \frac{dy}{dx} = -\left(\frac{ax + hy + g}{hx + by + f}\right)$$

EXERCISE 3.15

Find $\frac{dy}{dx}$, when

1. $\sqrt{x} + \sqrt{y} = \sqrt{a}$
2. $x^2 + y^2 + 5xy = 0$
3. $\sin(x + y) = x^2 + y^2$
4. $y \cdot \log x = x - y$
5. $x^{2/3} + y^{2/3} = a^{2/3}$
6. $x^2 + y^2 = \log(xy)$
7. $xy \log(x + y) = 1$
8. $x \sin^2 y = y \cos 2x$
9. If $y = a^{x+y}$, prove that $\frac{dy}{dx} = \frac{y \log a}{1 - y \log a}$
10. If $x^y = y^x$, prove that $\frac{dy}{dx} = \frac{(\log y)^2}{\log y - 1}$
11. If $\sqrt{y + x} + \sqrt{y - x} = a$ show that $\frac{dy}{dx} = \frac{y}{x} - \sqrt{\frac{y^2}{x^2} - 1}$
12. If $\sin y = x \sin(a + y)$, prove that $\frac{dy}{dx} = \frac{\sin^2(a + y)}{\sin a}$

ANSWERS

1. $-\sqrt{\frac{y}{x}}$
2. $-\frac{2x + 5y}{2y + 5x}$
3. $\frac{2x - \cos(x + y)}{\cos(x + y) - 2y}$
4. $\frac{x - y}{x(1 + \log x)}$
5. $\frac{-y^{\frac{1}{3}}}{x^{\frac{1}{3}}}$
6. $\frac{y(1 - 2x^2)}{x(2y^2 - 1)}$
7. $\frac{-y}{x}\left[\frac{x + (x + y)\log(x + y)}{y + (x + y)\log(x + y)}\right]$
8. $\frac{2y \sin 2x + \sin 2y}{\cos 2x - 2x \cos 2y}$

3.31 DERIVATIVE OF LOGARITHMIC FUNCTIONS

When the given function is a power of some expression or a product of expressions, we take the logarithmic on both sides differentiate the implicit function so obtained.

ILLUSTRATIVE EXAMPLES

Example 1. If $y = x^{(x^x)}$, find $\frac{dy}{dx}$

Solution: Let $y = x^{(x^x)}$, then $\log y = x^x(\log x)$

Differentiating both sides w. r. t. x, we get

$$\frac{1}{y}\cdot\frac{dy}{dx} = x^x\cdot\frac{1}{x} + (\log x)\cdot\frac{d}{dx}(x^x)$$

$$\frac{dy}{dx} = y[x^{x-1} + \log x)\, x^x(1 + \log x)]$$

$$\left[\begin{array}{l} \because u = x^x \Rightarrow \log u = x\log x \Rightarrow \frac{1}{u}\cdot\frac{du}{dx} = \left(x\cdot\frac{1}{x} + \log x\cdot 1\right) \\ \Rightarrow \frac{du}{dx} = u(1+\log x) = x^x(1+\log x) \end{array}\right]$$

$$\therefore \qquad \frac{dy}{dx} = x^{(x^x)}[x^{x-1} + (\log x)\, x^x(1 + \log x)]$$

Example 2. If $x^y = e^{x-y}$, prove that $\frac{dy}{dx} = \frac{\log x}{(1+\log x)^2}$

Solution: $x^y = e^{x-y}$

Taking log on both sides, we get

$y\log x = (x - y)$

or $(1 + \log x)y = x$

$$\therefore \qquad y = \frac{x}{(1+\log x)}$$

$$\therefore \qquad \frac{dy}{dx} = \frac{(1+\log x)\cdot 1 - x\frac{1}{x}}{(1+\log x)^2} = \frac{\log x}{(1+\log x)^2}$$

Example 3. If $y = \frac{5^x}{x^5}$, find $\frac{dy}{dx}$

Solution: $y = \frac{5^x}{x^5}$

Taking log on both sides, we get

$\log y = x\log 5 - 5\log x$

Differentiating both sides w. r. t. x, we get

$$\frac{1}{y}\cdot\frac{dy}{dx} = (\log 5).1 - \frac{5}{x}$$

$$\therefore \qquad \frac{dy}{dx} = y\left[\log 5 - \frac{5}{x}\right] = \frac{5^x}{x^5}\left[\log 5 - \frac{5}{x}\right]$$

Example 4. Differentiate $\tan x \tan 2x \tan 3x \tan 4x$

Solution: Let $y = \tan x \tan 2x \tan 3x \tan 4x$

Taking log on both sides, we get

$\log y = \log(\tan x \tan 2x \tan 3x \tan 4x)$

$\log y = \log(\tan x) + \log(\tan 2x) + \log(\tan 3x) + \log(\tan 4x)$

Differentiating both sides w. r. t. x, we get

$$\frac{1}{y}\cdot\frac{dy}{dx} = \left[\frac{\sec^2 x}{\tan x} + \frac{2\sec^2 2x}{\tan 2x} + \frac{3\sec^2 3x}{\tan 3x} + \frac{4\sec^2 4x}{\tan 4x}\right]$$

$$\therefore \quad \frac{dy}{dx} = y\left[\frac{1}{\sin x \cos x} + \frac{2}{\sin 2x \cos 2x} + \frac{3}{\sin 3x \cos 3x} + \frac{4}{\sin 4x \cos 4x}\right]$$

$$\frac{dy}{dx} = y\left[\frac{2}{\sin 2x} + \frac{4}{\sin 4x} + \frac{6}{\sin 6x} + \frac{8}{\sin 8x}\right]$$

$$\frac{dy}{dx} = [2\tan x \tan 2x \tan 3x \tan 4x] \times [\text{cosec } 2x + 2\text{cosec } 4x + 3\text{cosec } 6x + 4\text{cosec } 8x]$$

Example 5. If $y = (\sin x)^{\tan x} + (\cos x)^{\sec x}$, find $\frac{dy}{dx}$

Solution: Let $y = u + v$, where $u = (\sin x)^{\tan x}$ and $v = (\cos x)^{\sec x}$

then
$$\frac{dy}{dx} = \frac{du}{dx} + \frac{dv}{dx}$$

Now $u = (\sin x)^{\tan x}$

$$\Rightarrow \qquad \log u = \tan x \cdot \log(\sin x)$$

Differentiating both sides w. r. t. x, we get

$$\frac{1}{u}\frac{du}{dx} = \tan x \cdot \frac{1}{\sin x} \cdot \cos x + \log(\sin x)\sec^2 x$$

$$\frac{du}{dx} = u[1 + \log(\sin x) \cdot \sec^2 x]$$

$$= (\sin x)^{\tan x}[1 + \log(\sin x)\sec^2 x]$$

$$v = (\cos x)\sec x$$

$$\Rightarrow \qquad \log v = \sec x \cdot \log(\cos x)$$

Differentiating both sides w. r. t. x, we get

$$\frac{1}{v}\frac{dv}{dx} = \left[\sec x \cdot \frac{1}{\cos x}(-\sin x) + \log(\cos x) \cdot \sec x \cdot \tan x\right]$$

$$\Rightarrow \qquad \frac{dv}{dx} = v[-\sec x \tan x + \log(\cos x)\cdot \sec x \tan x]$$

$$\frac{dv}{dx} = (\cos x)^{\sec x}\cdot \sec x \tan x\,[\log(\cos x) - 1]$$

$$\therefore \qquad \frac{dy}{dx} = \frac{du}{dx} + \frac{dv}{dx} = (\sin x)^{\tan x}[1 + \log(\sin x)\sec^2 x]$$

$$+ (\cos x)^{\sec x}\cdot \sec x \tan x\,[\log(\cos x) - 1]$$

EXERCISE 3.16

Find the derivatives of the following:

1. $\sin(x^x)$ **2.** $(\log x)^x$

3. $x^{\sin x}$ **4.** $x^{\sin 2x}$

5. $(\cos x)^{(\log x)}$ **6.** $\dfrac{9^x}{x^9}$ **7.** $(\cos x)^{\cos x}$

8. $x^y = y^x$ **9.** $\dfrac{e^{x^2}\tan^{-1} x}{\sqrt{1+x^2}}$

10. $e^x \cos^3 x \sin^2 x$

ANSWERS

1. $\cos(x^x)\, x^x (1 + \log x)$ **2.** $(\log x)^x \left[\dfrac{1}{\log x} + \log(\log x)\right]$

3. $x^{\sin x}\left[\dfrac{\sin x}{x} + (\cos x)\log x\right]$ **4.** $x^{\sin 2x}\left[(2\cos 2x)\log x + \dfrac{\sin 2x}{x}\right]$

5. $(\cos x)^{\log x}\left[\dfrac{\log(\cos x)}{x} - \tan x \cdot \log x\right]$

6. $\dfrac{9^x}{x^9}\left[\log 9 - \dfrac{9}{x}\right]$ **7.** $(\cos x)^{\cos x}\cdot[(-\sin x)(1 + \log\cos x)]$

8. $\dfrac{y(x\log y - y)}{x(y\log x - x)}$ **9.** $\dfrac{e^{x^2}\tan^{-1} x}{\sqrt{1+x^2}}\left[2x + \dfrac{1}{(1+x^2)\tan^{-1} x} - \dfrac{x}{(1+x^2)}\right]$

10. $(e^x \cos^3 x \sin^2 x)(1 - 3\tan x + 2\cot x)$

3.32 DERIVATIVE OF AN INFINITE SERIES

If we take out a single term from an infinite series, it remains unaffected. We utilize this result in finding the derivative of an infinite series.

ILLUSTRATIVE EXAMPLES

Example 1. If $y = \sqrt{\sin x + \sqrt{\sin x + \sqrt{\sin x + \ldots \text{to} \infty}}}$, prove that $\frac{dy}{dx} = \frac{\cos x}{(2y-1)}$

Solution: $y = \sqrt{\sin x + \sqrt{\sin x + \sqrt{\sin x + \ldots \text{to} \infty}}}$

$$\text{or } y = \sqrt{\sin x + y}$$

Squaring both sides, we get

$$y^2 = \sin x + y$$

Differentiating both sides w. r. t.x, we get

$$2y \cdot \frac{dy}{dx} = \cos x + \frac{dy}{dx}$$

$$\Rightarrow \quad \frac{dy}{dx} = \frac{\cos x}{(2y-1)}$$

Example 2. If $y = \left(\sqrt{x}\right)^{\left(\sqrt{x}\right)^{\left(\sqrt{x}\right)^{\ldots\infty}}}$, prove that $\frac{dy}{dx} = \frac{y^2}{x(2 - y\log x)}$

Solution: $y = \left(\sqrt{x}\right)^{\left(\sqrt{x}\right)^{\left(\sqrt{x}\right)^{\ldots\infty}}}$

$$\text{or} \quad y = \left(\sqrt{x}\right)^y = x^{y/2}$$

Taking log both sides, we get

$$\log y = \frac{y}{2} - \log x$$

Differentiating both sides w. r. t. x, we get

$$\frac{1}{y} \cdot \frac{dy}{dx} = \frac{y}{2} \cdot \frac{1}{x} + \frac{1}{2} \log x \cdot \frac{dy}{dx}$$

$$\left(\frac{1}{y} - \frac{1}{2}\log x\right)\frac{dy}{dx} = \frac{y}{2x}$$

$$\left(\frac{2 - y\log x}{2y}\right)\frac{dy}{dx} = \frac{y}{2x}$$

$$\therefore \quad \frac{dy}{dx} = \frac{y^2}{x(2 - y\log x)}$$

Example 3. If $y = x^{x^{x\ldots\infty}}$, prove that $\frac{dy}{dx} = \frac{y^2}{x(1 - y\log x)}$

Solution: $y = x^{x^{x\ldots\infty}}$

$$\text{or } y = x^y$$

Taking log both sides, we get

$$\log y = y \log x$$

Differentiating both sides, w. r. t. x, we get

$$\frac{1}{y}\cdot\frac{dy}{dx} = y\cdot\frac{1}{x} + \log x\cdot\frac{dy}{dx}$$

$$\left(\frac{1}{y} - \log x\right)\frac{dy}{dx} = \frac{y}{x}$$

$$\left(\frac{1 - y\log x}{y}\right)\frac{dy}{dx} = \frac{y}{x}$$

$$\therefore \qquad \frac{dy}{dx} = \frac{y^2}{x(1 - y\log x)}$$

Example 4. If $y = e^{x+e^{x+e^{x}+\ldots\text{to}\infty}}$, prove that $\dfrac{dy}{dx} = \dfrac{y}{1-y}$

Solution: $y = e^{x+e^{x+e^{x}+\ldots\text{to}\infty}}$

$$\text{or } y = e^{x+y}$$

Taking log both sides, we get

$$\log y = (x + y) \log_e e = (x + y)$$

$$\log y = x + y$$

Differentiating both sides w. r. t. x, we get

$$\frac{1}{y}\cdot\frac{dy}{dx} = 1 + \frac{dy}{dx}$$

$$\left(\frac{1}{y} - 1\right)\frac{dy}{dx} = 1$$

$$\left(\frac{1-y}{y}\right)\frac{dy}{dx} = 1$$

$$\therefore \qquad \frac{dy}{dx} = \frac{y}{1-y}$$

EXERCISE 3.17

1. If $y = \sqrt{x + \sqrt{x + \sqrt{x + \ldots\infty}}}$, prove that $\dfrac{dy}{dx} = \dfrac{1}{2y-1}$

2. If $y = \sqrt{\log x + \sqrt{\log x + \sqrt{\log x + \ldots\infty}}}$, prove that $(2y-1)\dfrac{dy}{dx} = \dfrac{1}{x}$

3. If $y = (\cos x)^{(\cos x)^{(\cos x)\ldots\infty}}$, prove that $\dfrac{dy}{dx} = \dfrac{-y^2\tan x}{(1 - y\log\cos x)}$

4. If $y = a^{x^{a^{x\ldots\infty}}}$, prove that $\dfrac{dy}{dx} = \dfrac{y^2 \log y}{x[1 - y(\log x)(\log y)]}$

5. If $y = (\sin x)^{(\sin x)^{(\sin x)\ldots\infty}}$., prove that $\dfrac{dy}{dx} = \dfrac{y^2 \cot x}{(1 - y\log \sin x)}$

3.33 DIFFERENTIATION OF PARAMETRIC FUNCTIONS

Some times x and y are given as functions of another variable t. Then t is called a parameter.

The equation $x = f(t)$ and $y = g(t)$ expressing the coordinates of any point (x, y) on the curve as the functions of the parameter 't' are called the parametric representation of the curve.

In order to find differentiation in such form, we have by chain rule.

$$\frac{dy}{dx} = \frac{dy}{dt}\cdot\frac{dt}{dx}$$

or
$$\frac{dy}{dx} = \frac{dy/dt}{dx/dt}, \text{ provided } \frac{dx}{dt} \neq 0$$

ILLUSTRATIVE EXAMPLES

Example 1. If $x = a\cos\theta$, $y = a\sin\theta$, find $\dfrac{dy}{dx}$

Solution: $x = a\cos\theta$, $y = a\sin\theta$

$$\frac{dx}{d\theta} = -a\sin\theta, \quad \frac{dy}{d\theta} = a\cos\theta$$

$$\therefore \quad \frac{dy}{dx} = \frac{dy/d\theta}{dx/d\theta} = \frac{a\cos\theta}{-a\sin\theta} = -\cot\theta$$

$$\frac{dy}{dx} = -\cot\theta$$

Example 2. If $x = a(\theta - \sin\theta)$ and $y = a(1 - \cos\theta)$, find $\dfrac{dy}{dx}$

Solution: $x = a(\theta - \sin\theta)$, $y = a(1 - \cos\theta)$

$$\frac{dx}{d\theta} = a(1 - \cos\theta), \quad \frac{dy}{d\theta} = a(\sin\theta)$$

$$\therefore \quad \frac{dy}{dx} = \frac{a\sin\theta}{a(1 - \cos\theta)} = \frac{2\sin\theta/2\,\cos\theta/2}{2\sin^2\theta/2} = \cot\frac{\theta}{2}$$

$$\frac{dy}{dx} = \cot\frac{\theta}{2}$$

Example 3. If $x = a\cos^3 t$ and $y = a\sin^3 t$, find $\frac{dy}{dx}$

Solution: $x = a\cos^3 t$, $y = a\sin^3 t$

$$\frac{dx}{dt} = -3a\cos^2 t \sin t, \quad \frac{dy}{dt} = 3a\sin^2 t \cos t$$

$$\frac{dy}{dx} = \frac{3a\sin^2 t\cos t}{-3a\cos^2 t\sin t} = -\tan t$$

$$\frac{dy}{dx} = -\tan t$$

Example 4. If $x = e^{\cos 2t}$ and $y = e^{\sin 2t}$, prove that $\frac{dy}{dx} = -\frac{y\log x}{x\log y}$

Solution: If $x = e^{\cos 2t}$ and $y = e^{\sin 2t}$

$$\Rightarrow \quad \log x = \cos 2t \text{ and } \log y = \sin 2t$$

$$\frac{1}{x}\cdot\frac{dx}{dt} = -2\sin 2t \text{ and } \frac{1}{y}\cdot\frac{dy}{dt} = 2\cos 2t$$

$$\frac{dx}{dt} = -2x\sin 2t \text{ and } \frac{dy}{dt} = 2y\cos 2t$$

$$\therefore \quad \frac{dy}{dx} = \frac{2y\cos 2t}{-2x\sin 2t} = -\frac{y\cdot\log x}{x\cdot\log y}$$

Example 5. If $x^{2/3} + y^{2/3} = a^{2/3}$, find $\frac{dy}{dx}$

Solution: $x^{2/3} + y^{2/3} = a^{2/3}$

Put $x = a\cos^3\theta$, $y = a\sin^3\theta$

then $$\frac{dx}{d\theta} = -3a\cos^2\theta\sin\theta \text{ and } \frac{dy}{d\theta} = 3a\sin^2\theta\cos\theta$$

$$\therefore \quad \frac{dy}{dx} = \frac{3a\sin^2\theta\cos\theta}{-3a\cos^2\theta\sin\theta} = -\tan\theta$$

$$\frac{dy}{dx} = -\sqrt[3]{y/x}$$

EXERCISE 3.18

Find $\frac{dy}{dx}$, if

1. $x = a\log t$, $y = b\sin t$
2. $x = at^2$, $y = 2at$
3. $x = \sqrt{\sin 2\theta}$, $y = \sqrt{\cos 2\theta}$
4. $x = \frac{3at}{1+t^2}$, $y = \frac{3at^2}{1+t^2}$

5. $x = \cos^{-1}\dfrac{1}{\sqrt{1+t^2}}, y = \sin^{-1}\dfrac{1}{\sqrt{1+t^2}}$

6. $x = e^{\theta}(\sin\theta + \cos\theta), y = e^{\theta}(\sin\theta - \cos\theta)$

7. $x = a\sin 2\theta(1+\cos 2\theta), y = b\cos 2\theta(1-\cos 2\theta)$

8. $x = a\left(t + \dfrac{1}{t}\right), y = a\left(t - \dfrac{1}{t}\right)$

9. If $x = \dfrac{1+\log t}{t^2}, y = \dfrac{3+2\log t}{t}$, show that $\dfrac{dy}{dx} = t.$

10. If $x = a\sec^2\theta, y = b\tan^2\theta$, find $\dfrac{dy}{dx}$

ANSWERS

1. $\dfrac{bt\cos t}{a}$ **2.** $\dfrac{1}{t}$ **3.** $-(\tan 2\theta)^{3/2}$ **4.** $\dfrac{2t}{(1-t^2)}$

5. 1 **6.** $\tan\theta$ **7.** $\tan\theta$ **8.** $\dfrac{x}{y}$

10. $\dfrac{b}{a}$

3.34 DERIVATIVES OF HIGHER ORDER

Let $y = f(x)$ be a differentiable function of x whose second and higher order derivatives exist.

The Ist, IInd, IIIrd, ... and the nth derivatives of this function are denoted by $\dfrac{dy}{dx}, \dfrac{d^2y}{dx^2}, \dfrac{d^3y}{dx^3}, \ldots \dfrac{d^ny}{dx^n}$ or $y_1, y_2, y_3, \ldots y_n$ respectively.

ILLUSTRATIVE EXAMPLES

Example 1. If $y = \dfrac{\log x}{x}$, prove that $\dfrac{d^2y}{dx^2} = \dfrac{2\log x - 3}{x^3}$

Solution: $y = \dfrac{\log x}{x}$

Differentiating both sides w. r. t. x, we get

$$\frac{dy}{dx} = \frac{x\cdot\frac{1}{x} - \log x\cdot 1}{x^2} = \frac{1-\log x}{x^2}$$

Again, Differentiating w. r. to x, we get

$$\frac{d^2y}{dx^2} = \frac{x^2 \times -\frac{1}{x} - (1-\log x)2x}{x^4}$$

$$= \frac{-x - 2x + 2x \log x}{x^4}$$

$$\frac{d^2y}{dx^2} = \frac{2\log x - 3}{x^3}$$

Example 2. If $y = \tan^{-1} x$, show that $(1 + x^2)\dfrac{d^2y}{dx^2} + 2x\dfrac{dy}{dx} = 0$

Solution: $y = \tan^{-1} x$

Differentiating both sides w. r. to x, we get

$$\frac{dy}{dx} = \frac{1}{1+x^2}$$

$$\Rightarrow \quad (1+x^2)\frac{dy}{dx} = 1$$

Differentiating again w. r. t. x, we get

$$(1+x^2)\frac{d^2y}{dx^2} + 2x\frac{dy}{dx} = 0$$

Example 3. If $y = (\sin^{-1} x)^2$, prove that $(1-x^2)\dfrac{d^2y}{dx^2} - x\dfrac{dy}{dx} - 2 = 0$

Solution: $y = (\sin^{-1} x)^2$

Differentiating both sides w. r. t. x, we get

$$\frac{dy}{dx} = \frac{2\sin^{-1} x}{\sqrt{1-x^2}}$$

$$\Rightarrow \quad \sqrt{1-x^2}\,\frac{dy}{dx} = 2\sin^{-1} x$$

Squaring both sides, we get

$$(1-x^2)\left(\frac{dy}{dx}\right)^2 = 4(\sin^{-1} x)^2 = 4y$$

$$(1-x^2)\left(\frac{dy}{dx}\right)^2 - 4y = 0$$

Differentiating again w. r. t. x, we get

$$(1-x^2)2\left(\frac{dy}{dx}\right)\left(\frac{d^2y}{dx^2}\right) - 2x\left(\frac{dy}{dx}\right)^2 - 4\frac{dy}{dx} = 0$$

$$\Rightarrow \quad (1-x^2)\frac{d^2y}{dx^2} - x\frac{dy}{dx} - 2 = 0$$

Example 4. If $y = a\cos(\log x) + b\sin(\log x)$, prove that $x^2\dfrac{d^2y}{dx^2} + x\dfrac{dy}{dx} + y = 0$

Solution: $y = a\cos(\log x) + b\sin(\log x)$

Differentiating both sides w. r. t. x, we get

$$\frac{dy}{dx} = -a\sin(\log x)\cdot\frac{1}{x} + b\cos(\log x)\cdot\frac{1}{x}$$

$$\Rightarrow \quad x\frac{dy}{dx} = -a\sin(\log x) + b\cos(\log x)$$

Differentiating again w. r. to x, we get

$$x\frac{d^2y}{dx^2} + \frac{dy}{dx}\cdot 1 = -a\cos(\log x)\cdot\frac{1}{x} - b\sin(\log x)\cdot\frac{1}{x}$$

$$\Rightarrow \quad x^2\frac{d^2y}{dx^2} + x\frac{dy}{dx} = -[a\cos(\log x) + b\sin(\log x) = -y$$

$$\Rightarrow \quad x^2\frac{d^2y}{dx^2} + x\frac{dy}{dx} + y = 0$$

Example 5. If $y = e^{4x}\sin 3x$, find $\dfrac{d^2y}{dx^2}$

Solution: $y = e^{4x}\sin 3x$

Differentiating both sides w. r. t. x, we get

$$\frac{dy}{dx} = 3e^{4x}\cos 3x + 4e^{4x}\sin 3x = (3\cos 3x + 4\sin 3x)e^{4x}$$

Again, Differentiating both sides w. r. t. x, we get

$$\frac{d^2y}{dx^2} = e^{4x}(-9\sin 3x + 12\cos 3x) + 4e^{4x}(3\cos 3x + 4\sin 3x)$$

$$\frac{d^2y}{dx^2} = e^{4x}(7\sin 3x + 24\cos 3x)$$

Example 6. If $x = a(t + \sin t)$ and $y = a(1 - \cos t)$

$$\therefore \quad \frac{dx}{dt} = a(1 + \cos t) \text{ and } \frac{dy}{dt} = a\sin t$$

$$\Rightarrow \quad \frac{dy}{dx} = \frac{\sin t}{1 + \cos t}$$

Differentiating again w. r. to x, we get

$$\frac{d^2y}{dx^2} = \frac{(1 + \cos t)\cos t - \sin t(-\sin t)}{(1 + \cos t)^2}\left(\frac{dt}{dx}\right)$$

$$\therefore \qquad \frac{d^2y}{dx^2} = \frac{1+\cos t}{(1+\cos t)^2} \cdot \frac{1}{a(1+\cos t)} = \frac{1}{a(1+\cos t)^2}$$

$$\frac{d^2y}{dx^2} = \frac{1}{a(1+\cos t)^2}$$

EXERCISE 3.19

1. If $y = e^{-x} \cos x$, show that $\dfrac{d^2y}{dx^2} = 2e^{-x} \sin x$

2. If $y = e^{\tan x}$, prove that $\cos^2 x \dfrac{d^2y}{dx^2} - (1 + \sin 2x) \dfrac{dy}{dx} = 0$

3. If $y = \sin(\sin x)$, prove that $\dfrac{d^2y}{dx^2} + \tan x \dfrac{dy}{dx} + y \cos^2 x = 0$

4. If $y = (\tan^{-1} x)^2$, prove that $(x^2 + 1)^2 \dfrac{d^2y}{dx^2} + 2x(x^2 + 1)\dfrac{dy}{dx} - 2 = 0$

5. If $y = e^x(\sin x + \cos x)$, show that $\dfrac{d^2y}{dx^2} - 2\dfrac{dy}{dx} + 2y = 0$

6. If $x = \tan\left(\dfrac{1}{a}\log y\right)$, show that $(1 + x^2)\dfrac{d^2y}{dx^2} + (2x - a)\dfrac{dy}{dx} = 0$

7. Find the second derivative of the following functions

(*i*) $x^3 \log x$ (*ii*) $x\cos x$ (*iii*) $\log x - x$ (*iv*) $\log(2x + 5)$

(*v*) $\log(\log x)$ (*vi*) $\cos^{-1} x$ (*vii*) $e^x + \tan x$

8. If $y = x + \cot x$, show that $\sin^2 x \dfrac{d^2y}{dx^2} - 2y + 2x = 0$

9. If $y = x^x$, prove that $\dfrac{d^2y}{dx^2} - \dfrac{1}{y}\left(\dfrac{dy}{dx}\right)^2 - \dfrac{y}{x} = 0$

10. If $y = \log\left(x + \sqrt{x^2 + a^2}\right)$, prove that $(x^2 + a^2)\dfrac{d^2y}{dx^2} + x\dfrac{dy}{dx} = 0$

ANSWERS

7. (i) $x(5 + 6\log x)$ (*ii*) $-(2\sin x + x\cos x)$

(*iii*) $-\dfrac{1}{x^2}$ (*iv*) $-\dfrac{4}{(2x+5)^2}$ (*v*) $-\dfrac{1+\log x}{(x\log x)^2}$ (*vi*) $\dfrac{-x}{(1-x^2)^{3/2}}$

(*vii*) $e^x + 2\sec^2 x \cdot \tan x$

3.35 TANGENTS AND NORMALS

Equation of the tangent

The equation of a tangent to a curve $y = f(x)$ at a point $p(x_1, y_1)$ is given by

$$y - y_1 = \left(\frac{dy}{dx}\right)_{(x_1,\, y_1)} (x - x_1)$$

or $\boxed{y - y_1 = m(x - x_1)}$

where $m = \left(\frac{dy}{dx}\right)_{(x_1,\, y_1)}$ denotes the value of $\frac{dy}{dx}$ at $x = x_1$ and $y = y_1$.

Equation of the Normal

The equation of the normal to a curve $y = f(x)$ at a point $P(x_1, y_1)$ is given by

$$y - y_1 = -\frac{1}{\left(\frac{dy}{dx}\right)_{(x_1,\, y_1)}} (x - x_1)$$

or $\boxed{y - y_1 = -\frac{1}{m}(x - x_1)}$

where $$m = \left(\frac{dy}{dx}\right)_{(x_1, x_1)}$$

Note 1. The tangent to a curve $y = f(x)$ at a point $P(x_1, y_1)$ is parallel to the x-axis iff $\left(\frac{dy}{dx}\right)_{(x_1, y_1)} = 0$

2. The tangent to a curve $y = f(x)$ at a point $P(x_1, y_1)$ is parallel to the y-axis iff $\left(\frac{dy}{dx}\right)_{(x_1, y_1)} = 0.$

ILLUSTRATIVE EXAMPLES

Example 1. Show that the equation of the tangent to the ellipse $\frac{x^2}{a^2} + \frac{y^2}{b^2} = 1$ at (x_1, y_1) is $\frac{xx_1}{a^2} + \frac{yy_1}{b^2} = 1$

Solution: $\frac{x^2}{a^2} + \frac{y^2}{b^2} = 1$

Differentiating w. r. to x, we get

$$\frac{2x}{a^2} + \frac{2y}{b^2}\frac{dy}{dx} = 0$$

$$\Rightarrow \qquad \frac{dy}{dx} = -\frac{b^2 x}{a^2 y}$$

$$\left(\frac{dy}{dx}\right)_{(x_1, y_1)} = -\frac{b^2 x_1}{a^2 y_1}$$

$\therefore$ The equation of the tangent at (x_1, y_1) is

$$y - y_1 = -\frac{b^2 x_1}{a^2 y_1}\,(x - x_1)$$

$$\Rightarrow \qquad b^2 xx_1 + a^2 yy_1 = b^2 x_1^{\,2} + a^2 y_1^{\,2}$$

Dividing throughout by $a^2 b^2$, we get

$$\frac{xx_1}{a^2} + \frac{yy_1}{b^2} = \frac{x_1^{\,2}}{a^2} + \frac{y_1^{\,2}}{b^2}$$

$$\text{or } \frac{xx_1}{a^2} + \frac{yy_1}{b^2} = 1 \left[\because (x_1, y_1) \text{ lies on } \frac{x^2}{a^2} + \frac{y^2}{b^2} = 1 \Rightarrow \frac{x_1^{\,2}}{a^2} + \frac{y_1^{\,2}}{b^2} = 1\right]$$

Example 2. Find the equations of the tangent and the normal to the curve $y = x^4 - 5x^3 + 11x^2 - 5x + 2$ at the point (2, 3).

Solution: The equation of the curve is $y = x^4 - 5x^3 + 11x^2 - 5x + 2$

$$\therefore \qquad \frac{dy}{dx} = 4x^3 - 15x^2 + 22x - 5$$

$$\left(\frac{dy}{dx}\right)_{(2,3)} = 4 \times (2)^3 - 15(2)^2 + 22 \times 2 - 5 = 11$$

$\therefore$ The equation of the tangent at (2, 3) is

$$y - 3 = 11(x - 2) \Rightarrow 11x - y - 19 = 0$$

and equation of the normal at (2, 3) is

$$y - 3 = -\frac{1}{11}(x - 2)$$

$$11y - 33 = -x + 2$$

$$x + 11y - 35 = 0$$

Example 3. Find the equation of the tangent to the curve $y = \sqrt{5x - 3} - 2$ which is parallel to the line $4x - 2y + 3 = 0$.

Solution: The given line is $4x - 2y + 3 = 0$

$$\text{or} \qquad y = 2x + \frac{3}{2}$$

$\therefore$ slope of the given line = 2

So, the slope of the tangent = 2

Let the point of contact be (x_1, y_1)

$$y = \sqrt{5x - 3} - 2$$

$$\Rightarrow \qquad \frac{dy}{dx} = \frac{5}{2\sqrt{5x - 3}}$$

$$\Rightarrow \qquad \left(\frac{dy}{dx}\right)_{(x_1, y_1)} = \frac{5}{2\sqrt{5x_1 - 3}}$$

$$\therefore \qquad \frac{5}{2\sqrt{5x_1 - 3}} = 2 \Rightarrow \frac{25}{4(5x_1 - 3)} = 4$$

$$\Rightarrow \qquad x_1 = \frac{73}{80}$$

$$\therefore \qquad y_1 = \sqrt{5x_1 - 3} - 2 = \sqrt{5 \times \frac{73}{80} - 3} - 2$$

$$= \frac{-3}{4}$$

So, the point of contact is $\left(\frac{73}{80}, \frac{-3}{4}\right)$

Hence, the required equation of the tangent at $\left(\frac{73}{80}, \frac{-3}{4}\right)$ is

$$\left(y + \frac{3}{4}\right) = 2\left(x - \frac{73}{80}\right)$$

$$\Rightarrow \qquad 80x - 40y - 103 = 0$$

Example 4. Prove that the curve $\left(\frac{x}{a}\right)^n + \left(\frac{y}{b}\right)^n = 2$ touches the straight line $\frac{x}{a} + \frac{y}{b}$ $= 2$ at the point (a, b) what ever be the value of n.

Solution: The equation of the curve is $\left(\frac{x}{a}\right)^n + \left(\frac{y}{b}\right)^n = 2$

Differentiating w. r. t. x, we get

$$n \cdot \left(\frac{x}{a}\right)^{n-1} \frac{1}{a} + n \cdot \left(\frac{y}{b}\right)^{n-1} \frac{1}{b} \cdot \frac{dy}{dx} = 0$$

$$\text{or} \quad \frac{nx^{n-1}}{a^n} + \frac{ny^{n-1}}{b^n} \frac{dy}{dx} = 0$$

$$\text{or} \qquad \frac{dy}{dx} = -\frac{b^n a^{n-1}}{a^n y^{n-1}}$$

$$\therefore \qquad \left(\frac{dy}{dx}\right)_{(a,b)} = -\frac{b^n a^{n-1}}{a^n b^{n-1}} = -b/a$$

So, the equation of the tangent at (a, b) is

$$y - b = -\frac{b}{a}(x - a)$$

$$\text{or} \qquad ay + bx = 2ab$$

$$\therefore \qquad \frac{x}{a} + \frac{y}{b} = 2, \text{ which is independent of } n.$$

Example 5. Prove that all points of the curve $y^2 = 4a[x + a\sin x/a]$ at which the tangent is parallel to the axis of x, lie on a parabola.

Solution: Let the required point be (x_1, y_1).

Then, we must have $\left(\frac{dy}{dx}\right)_{(x_1,y_1)} = 0$

$$y^2 = 4a[x + a\sin x/a] \Rightarrow 2y\cdot\frac{dy}{dx} = 4a[1+\cos x/a]$$

$$\Rightarrow \qquad \frac{dy}{dx} = \frac{2a\left[1+\cos\frac{x}{a}\right]}{y}$$

$$\Rightarrow \qquad \left(\frac{dy}{dx}\right)_{(x_1,y_1)} = \frac{2a\left[1+\cos\frac{x_1}{a}\right]}{y_1}$$

$$\therefore \qquad \left(\frac{dy}{dx}\right)_{(x_1,y_1)} = 0 \Rightarrow \frac{2a\left[1+\cos\frac{x_1}{a}\right]}{y_1} = 0$$

$$\Rightarrow \qquad \left[1+\cos\frac{x_1}{a}\right] = 0$$

$$\Rightarrow \qquad \cos\frac{x_1}{a} = -1$$

$$\Rightarrow \qquad \sin\frac{x_1}{a} = \sqrt{1-\cos^2\frac{x_1}{a}} = 0$$

But (x_1, y_1) lies on the given curve

$$\therefore \qquad y_1^2 = 4a\left[x_1 + a\sin\frac{x_1}{a}\right]$$

$$\Rightarrow \qquad y_1^2 = 4ax_1 \quad \left[\because \sin\frac{x_1}{a} = 0\right]$$

This shows that (x_1, y_1) lies on the parabola $y^2 = 4ax$.

Example 6. Find the points on the curve $4x^2 + 9y^2 = 1$, where the tangents are perpendicular to the line $2y + x = 0$.

Solution: The equation of the given line is $y = -\frac{1}{2}x$.

$$\therefore \qquad \text{Slope of this line} = -\frac{1}{2} \qquad ...(1)$$

Let the required point be (x_1, y_1)

Now, $\qquad 4x^2 + 9y^2 = 1$

Differentiating w. r. t. x, we get

$$8x + 18y\frac{dy}{dx} = 0$$

$$\Rightarrow \qquad \frac{dy}{dx} = -\frac{4x}{9y}$$

$$\Rightarrow \qquad \left(\frac{dy}{dx}\right)_{(x_1, y_1)} = -\frac{4x_1}{9y_1}$$

$\therefore$ Slope of the tangent at $(x_1, y_1) = -\dfrac{4x_1}{9y_1}$ ...(2)

Form (1) and (2), we have

$$-\frac{4x_1}{9y_1} \times \left(-\frac{1}{2}\right) = -1$$

$$\Rightarrow \qquad y_1 = -\frac{2}{9}x_1 \qquad \text{...(3)}$$

Since (x_1, y_1)lies on the curve $4x^2 + 9y^2 = 1$,

$$\therefore \qquad 4{x_1}^2 + 9{y_1}^2 = 1$$

or $\qquad 4x_1^2 + 9 \times \dfrac{4}{81}x_1^2 = 1$ [using (3)]

$$\therefore \qquad {x_1}^2 = \frac{9}{40} \text{ or } x_1 = \pm\frac{3}{2\sqrt{10}}$$

So, $\qquad y_1 = \pm\dfrac{2}{9} \times \dfrac{3}{2\sqrt{10}} = \pm\dfrac{1}{3\sqrt{10}}$

Hence, the required points are

$$\left(\frac{3}{2\sqrt{10}}, \frac{1}{3\sqrt{10}}\right) \text{ and } \left(\frac{-3}{2\sqrt{10}}, \frac{-1}{3\sqrt{10}}\right)$$

Example 7. If the straight line $x\cos\alpha + y\sin\alpha = p$ touches the ellipse $\dfrac{x^2}{a^2} + \dfrac{y^2}{b^2}$ $= 1$, prove that $p^2 = a^2\cos^2\alpha + b^2\sin^2\alpha$.

Solution: Let the point of contact be (x_1, y_1)

Then, the equation of the tangent at (x_1, y_1) is

$$\frac{xx_1}{a^2} + \frac{yy_1}{b^2} = 1 \qquad \text{...(1)}$$

This equation is identical with $x\cos\alpha + y\sin\alpha = p$ ...(2)

comparing (1) & (2), we get

$$\frac{(x_1/a^2)}{\cos\alpha} = \frac{(y_1/b^2)}{\sin\alpha} = \frac{1}{p} \Rightarrow \frac{x_1/a}{a\cos\alpha} = \frac{y_1/b}{b\sin\alpha} = \frac{1}{p}$$

$$\therefore \qquad \frac{x_1}{a} = \frac{a\cos\alpha}{p} \qquad ...(3)$$

and $$\frac{y_1}{b} = \frac{b\sin\alpha}{p} \qquad ...(4)$$

Squaring (3) & (4) and then adding, we get

$$\frac{x_1^2}{a^2} + \frac{y_1^2}{b^2} = \frac{a^2\cos^2\alpha + b^2\sin^2\alpha}{p^2}$$

or $$\frac{a^2\cos^2\alpha + b^2\sin^2\alpha}{p^2} = 1 \left[\because \frac{x_1^2}{a^2} + \frac{y_1^2}{b^2} = 1\right]$$

or $$a^2\cos^2\alpha + b^2\sin^2\alpha = p^2.$$

Example 8. Show that $\frac{x}{a} + \frac{y}{b} = 1$ touches the curve $y = be^{-x/a}$ at the point where the curve crosses the axis of y.

Solution: The equation of the curve is $y = be^{-x/a}$

It crosses the y-axis at the point where $x = 0$.

Putting $x = 0$ in the equation of the curve, we get $y = b$.

$\therefore$ point of contact is $(0, b)$.

$$y = be^{-x/a} \Rightarrow \frac{dy}{dx} = \frac{-be^{-x/a}}{a}$$

$$\therefore \qquad \left(\frac{dy}{dx}\right)_{(0,b)} = -b/a$$

So, the equation of the tangent is $y - b = \frac{-b}{a}(x - 0)$

or $$bx + ay = ab$$

or $$\frac{x}{a} + \frac{y}{b} = 1$$

Hence $\frac{x}{a} + \frac{y}{b} = 1$ touches the curve $y = b_e^{-x/a}$ at $(0, b)$.

Example 9. Find the equation of the normal to the curve $y = 2\sin^2 3x$ at $x = \pi/6$.

Solution: When $x = \pi/6$, then $y = 2\sin^2\left(\frac{3\pi}{6}\right) = 2$

$\therefore$ the point of contact is $(\pi/6, 2)$

Now $$y = 2\sin^2 3x$$

$$\Rightarrow \qquad \frac{dy}{dx} = 4\sin 3x \times 3 \times \cos 3x$$

$$\Rightarrow \qquad \frac{dy}{dx} = 12\sin 3x \cos 3x = 6\sin 6x.$$

$$\therefore \qquad \left(\frac{dy}{dx}\right)_{(x=\pi/6)} = 6\sin(6\times\pi/6) = 6\sin\pi = 0$$

So, the tangent is parallel to the x-axis. Thus the normal is parallel to the y-axis and passes through the point $(\pi/6, 2)$.

So, the equation of the normal is $x = \dfrac{\pi}{6}$.

EXERCISE 3.20

1. Find the equation of the tangent to the curve $\sqrt{x}+\sqrt{y} = a$ at $\left(\dfrac{a^2}{4}, \dfrac{a^2}{4}\right)$.

2. Find the equation of the tangent to the curve $y = (\sec^4 x - \tan^4 x)$ at $x = \pi/3$.

3. Find the equation of the tangent to the curve $16x^2 + 9y^2 = 144$ at $(2, y_1)$, where $y_1 > 0$.

4. Show that the equation of the tangent to the hyperbola $\dfrac{x^2}{a^2} - \dfrac{y^2}{b^2} = 1$ at (x_1, y_1) is $\dfrac{xx_1}{a^2} - \dfrac{yy_1}{b^2} = 1$.

5. Find the equation of the normal to the curve $\dfrac{x^2}{a^2} - \dfrac{y^2}{b^2} = 1$ at $(a\sec\theta, b\tan\theta)$.

6. Find the equation of the normal to the curve $y^2 = 4ax$ at $\left(\dfrac{a}{m^2}, \dfrac{2a}{m}\right)$.

7. Find the equations of the tangent and the normal to the curve $y = x^4 - 6x^3 + 13x^2 - 10x + 5$ at the point where $x = 1$.

8. Find the equation of the tangent at $t = \dfrac{\pi}{4}$ for the curve $x = \sin 3t$, $y = \cos 2t$.

9. Show that the curves $xy = a^2$ and $x^2 + y^2 = 2a^2$ touch each other.

10. Show that the curves $2x = y^2$ and $2xy = k$ cut at right angles if $k^2 = 8$.

11. Find the equation of the tangent to the curve $x^2 + 2y = 8$, which is perpendicular to the line $x - 2y + 1 = 0$

12. Find the equations of the tangent and the normal to the curve $y(x-2)(x-3) - x + 7 = 0$ at the point where it cuts the x-axis.

13. If $\mathrm{x}\cos\alpha + \mathrm{y}\sin\alpha = \mathrm{p}$ touches curve $\mathrm{x^m y^n = a^{m+n}}$, prove that $\mathrm{p^{m+n} m^m . n^n} = (m+n)^{m+n} \cos^m\alpha \sin^n\alpha$.

14. Find the coordinates of the points on the curve $y = x^2 + 3x + 4$, the tangents at which pass through the origin.

ANSWERS

1. $x + y = \frac{a^2}{2}$ 2. $\sqrt{3}y - 48x - \left(7\sqrt{3} - 16a\right) = 0$

3. $8x + 3\sqrt{5}y - 36 = 0$

5. $ax\cos\theta + by\cot\theta = a^2 + b^2$

6. $m^2x + m^3y - 2am^2 - a = 0$

7. $y - 2x - 1 = 0; 2y + x - 7 = 0$ 8. $3y - 2\sqrt{2}x + 2 = 0$

11. $y + 2x - 8 = 0$ 12. $x - 20y - 7 = 0; 20x + y - 140 = 0$

14. $(-2, 2)$ and $(2, 14)$

3.36 MAXIMA AND MINIMA

In this section, we apply differentiation to calculate the maximum or minimum values of functions.

Definition: Let f be a function defined on an interval I. Then

(*i*) f is said to have a maximum value in I if there exists a point a in I such that $f(a) \geq f(x), \forall x \in$ I.

The number $f(a)$ is called the maximum value of f in I and the point a is called a point of maximum value of f in I.

(*ii*) f is said to have a minimum value in I if there exists a point a in I such that $f(a) \leq f(x), \forall x \in$ I.

The number $f(a)$ is called the minimum value of f in I and the point a is called a point of minimum value of f in I.

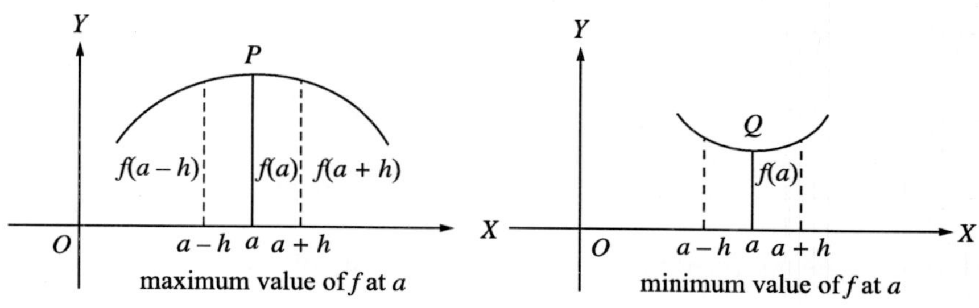

Fig. 1

(*iii*) f is said to have an extreme value in I if there exists a point a in I such that $f(a)$ is either a maximum value or a minimum value of f in I.

The number $f(a)$ is called an extreme value of f in I and the point a is called an extreme point. The points P and Q in Fig.1 are extremums.

3.37 A NECESSARY CONDITION FOR EXTREME VALUES

A necessary condition for $f(a)$ to be an extreme value of f is that $f'(a) = 0$. i.e.,

$$f(a) \text{ is an extreme value} \Rightarrow \boxed{f'(a) = 0 \text{ i.e.,} \left(\frac{dy}{dx}\right)_{x=a} = 0}$$

Note: 1. The vanishing of $f'(a)$ is only a necessary but not a sufficient for $f(a)$ to be an extreme value as we now show.

Consider $f(x) = x^3$ *for* $x = 0$

Now

$$x > 0 \Rightarrow f(x) > 0 = f(0);$$

$$x < 0 \Rightarrow f(x) < 0 = f(0).$$

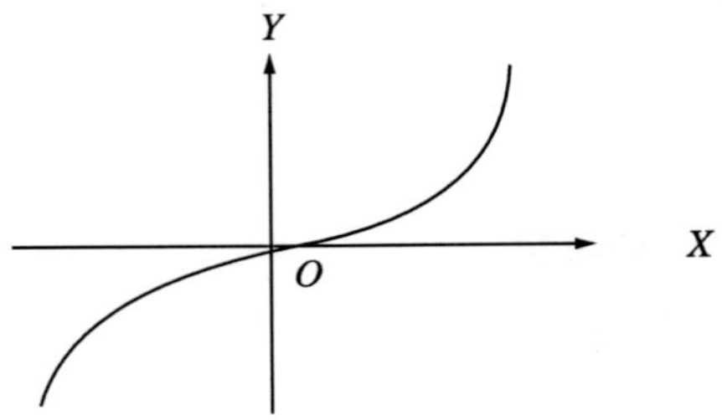

Thus, $f(0)$ is not an extreme value even though $f'(0) = 0$.

2. $f(0) = 0$ is a minimum value of $f(x) = |x|$ ever though f is not derivable at 0.

STATIONARY VALUES

A function f is said to be stationary for a and $f(a)$ a stationary value of f if $f'(a) = 0$. The rate of change of a function is zero at a stationary point. While a maximum or a minimum value is a stationary value, a stationary value may neither be a maximum nor a minimum value.

Sufficient condition for extreme values

(*i*) If $\frac{dy}{dx}$ changes sign from +ve to –ve, then $f(x)$ is maximum at $x = a$.

(*ii*) If $\frac{dy}{dx}$ changes sign from –ve to +ve, then $f(x)$ is minimum at $x = a$.

(*iii*) If $\frac{dy}{dx}$ does not change sign, then $x = a$ is a point of neither maximum nor minimum

3.38 METHODS FOR FINDING MAXIMA AND MINIMA

First Method

1. Find $\frac{dy}{dx}$ and equate it to zero. Solve this equation for real values of x. Let these values be $a, b, c, \ldots$

2. Consider the value $x = a$, study the sign of $\frac{dy}{dx}$ for value slightly less than $ä$ and slightly greater than $ä$.

3.(*i*) If $\frac{dy}{dx}$ changes sign from + ve to –ve, then $f(x)$ is maximum at $x = a$.

(*ii*) If $\frac{dy}{dx}$ changes sign from –ve to +ve, then $f(x)$ is minimum at $x = a$.

(*iii*) If $\frac{dy}{dx}$ does not change sign, then $x = a$ is a point of neither maximum nor minimum.

Second Method (Second derivative test)

1. Find $\frac{dy}{dx}$ and equate it to zero. Solve this equation for real values of x. Let these values be $a, b, c, \ldots$

2. Find $\frac{d^2y}{dx^2}$, put $a, b, c, \ldots$ one by one. Consider the value $x = a$.

(*i*) If $\frac{d^2y}{dx^2}$ is –ve, $f(x)$ is maximum at $x = a$.

(*ii*) if $\frac{d^2y}{dx^2}$ is +ve, $f(x)$ is minimum at $x = a$.

(*iii*) If $\frac{d^2y}{dx^2} = 0$ at $x = a$ but $\frac{d^3y}{dx^3} \neq 0$ at $x = a$, then there is neither maximum nor minimum at $x = a$.

But if $\frac{d^3y}{dx^3} = 0$ then find $\frac{d^4y}{dx^4}$ and if $\frac{d^4y}{dx^4}$ is –ve we have a maximum, if it is +ve we have a minimum at $x = a$.

Similarly for $x = b, x = c$, etc.

ILLUSTRATIVE EXAMPLES

Example 1. Find the maximum value of $x^3 - 6x^2 + 11x - 6$

Solution: Let $y = f(x) = x^3 - 6x^2 + 11x - 6$, then

$$\frac{dy}{dx} = 3x^2 - 12x + 11$$

For a Maximum or Minimum $\frac{dy}{dx} = 0$

$$\therefore \quad 3x^2 - 12x + 11 = 0$$

$$x = \frac{12 \pm \sqrt{144 - 4 \times 3 \times 11}}{2 \times 3}$$

$$x = 2 \pm \frac{1}{\sqrt{3}}$$

Now $$\frac{d^2y}{dx^2} = 6x - 12 = 6(x - 2)$$

When $x = 2 + \frac{1}{\sqrt{3}}$

$$\frac{d^2y}{dx^2} = 6\left[2 + \frac{1}{\sqrt{3}} - 2\right] = \frac{6}{\sqrt{3}} = +\text{ve}$$

$\therefore$ The function is minimum when $x = 2 + \frac{1}{\sqrt{3}}$

When $x = 2 - \frac{1}{\sqrt{3}}$

$$\frac{d^2y}{dx^2} = 6\left[2 - \frac{1}{\sqrt{3}} - 2\right] = -\frac{6}{\sqrt{3}} = -\text{ve}$$

$\therefore$ The function is maximum when $x = 2 - \frac{1}{\sqrt{3}}$

The maximum value of the function

$$= x^3 - 6x^2 + 11x - 6$$

$$= (x-1)(x-2)(x-3)$$

$$= \left(2 - \frac{1}{\sqrt{3}} - 1\right)\left(2 - \frac{1}{\sqrt{3}} - 2\right)\left(2 - \frac{1}{\sqrt{3}} - 3\right)$$

$$= \left(1 - \frac{1}{\sqrt{3}}\right)\left(-\frac{1}{\sqrt{3}}\right)\left(-1 - \frac{1}{\sqrt{3}}\right)$$

$$= \left(1 - \frac{1}{\sqrt{3}}\right)\left(1 + \frac{1}{\sqrt{3}}\right)\frac{1}{\sqrt{3}}$$

$$= \left(1 - \frac{1}{3}\right)\frac{1}{\sqrt{3}} = \frac{2}{3\sqrt{3}}$$

Example 2. Prove that $\sin x(1 + \cos x)$ is maximum when $x = \frac{\pi}{3}$.

Solution: Let $y = \sin x(1 + \cos x)$

$$\frac{dy}{dx} = \cos x(1 + \cos x) + \sin x(-\sin x)$$

$$= \cos x + \cos^2 x - \sin^2 x = \cos x + \cos 2x$$

For maximum or minimum $\frac{dy}{dx} = 0$

$$\cos x + \cos 2x = 0$$

$$\cos 2x = -\cos x = \cos(\pi - x)$$

$$2x = \pi - x$$

$$3x = \pi$$

$$x = \pi/3$$

Hence the function is maximum or minimum at $x = \pi/3$

Now $\dfrac{d^2y}{dx^2} = \dfrac{d}{dx}(\cos x + \cos 2x) = -\sin x - 2\sin 2x$

At $x = \pi/3$, we have

$$\frac{d^2y}{dx^2} = -\left[2\sin\frac{2\pi}{3} + \sin\frac{\pi}{3}\right]$$

$$= -\left[2\cdot\frac{\sqrt{3}}{2} + \frac{\sqrt{3}}{2}\right] \text{ which is –ve}$$

Hence the function is maximum at $x = \pi/3$.

Example 3. Find the maximum value of $\dfrac{\log x}{x}$, when $x > 0$.

Solution: Let $y = \dfrac{\log x}{x}$

$$\therefore \quad \frac{dy}{dx} = \frac{\frac{1}{x}\cdot x - (\log x)\cdot 1}{x^2} = \frac{1 - \log x}{x^2}$$

For maximum or minimum value of y, we have

$$\frac{dy}{dx} = 0$$

$$\Rightarrow \quad \frac{1 - \log x}{x^2} = 0 \Rightarrow \log x = 1$$

$$\Rightarrow \quad \log_e x = \log_e e$$

$$\Rightarrow \quad x = e$$

Now
$$\frac{d^2y}{dx^2} = \frac{\left(0 - \frac{1}{x}\right)x^2 - (1 - \log x)\cdot 2x}{x^4}$$

$$= \frac{2\log x - 3}{x^3}$$

$$\therefore \quad \text{At } x = e, \ \frac{d^2y}{dx^2} = \frac{2\log e - 3}{e^3} = \frac{2-3}{e^3} \quad (\because \log e = 1)$$

$$= -\frac{1}{e^3} \text{ which is –ve}$$

$\therefore$ At $x = e$ the function is maximum

The maximum value of the function

$$= \frac{\log e}{e} = \frac{1}{e} = e^{-1}$$

Example 4. Find the maximum and minimum values of $x^3 - 2x^2 + x + 6$

Solution: Let $y = x^3 - 2x^2 + x + 6$

$$\frac{dy}{dx} = 3x^2 - 4x + 1$$

For maximum and minimum value of y, we have

$$\frac{dy}{dx} = 0$$

$$\Rightarrow \quad 3x^2 - 4x + 1 = 0$$

$$x = 1, \frac{1}{3}$$

Now $\quad \frac{d^2y}{dx^2} = 6x - 4$. At $x = 1$, $\frac{d^2y}{dx^2} = 6 \times 1 - 4 = 2 = +\text{ve}$

$\therefore$ At $x = 1$, $\frac{d^2y}{dx^2}$ is +ve, therefore the function is minimum at $x = 1$.

At $x = \frac{1}{3}$, $\frac{d^2y}{dx^2} = 6 \times \frac{1}{3} - 4 = -2 = -\text{ ve}$

$\therefore$ At $x = \frac{1}{3}$, $\frac{d^2y}{dx^2}$ is –ve therefore the function is maximum at $x = \frac{1}{3}$.

The maximum value of the function $y = x^3 - 2x^2 + x + 6$

$$= \left(\frac{1}{3}\right)^3 - 2\left(\frac{1}{3}\right)^2 + \frac{1}{3} + 6$$

$$= \frac{1}{27} - \frac{2}{9} + \frac{1}{3} + 6$$

$$= \frac{1 - 6 + 9}{27} + 6$$

$$= \frac{4}{27} + 6$$

$$= 6\frac{4}{27}$$

The minimum value of the function $= x^3 - 2x^2 + x + 6$

$$= (1)^3 - 2(1)^2 + 1 + 6$$

$$= 1 - 2 + 1 + 6 = 6$$

Example 5. Find the maximum and minimum values of the function $3x^4 - 10x^3 + 6x^2 + 5$

Solution: Let $y = 3x^4 - 10x^3 + 6x^2 + 5$

$$\frac{dy}{dx} = 12x^3 - 30x^2 + 12x$$

For maximum and minimum value of y, we have

$$\frac{dy}{dx} = 0$$

$$\Rightarrow \quad 12x^3 - 30x^2 + 12x = 0$$

$$x(12x^2 - 30x + 12) = 0$$

$$\Rightarrow \quad 6x(2x^2 - 5x + 2) = 0$$

$$6x(x-2)(2x-1) = 0$$

$$\therefore \quad x = 0, 2, \frac{1}{2}$$

Now $$\frac{d^2y}{dx^2} = 36x^2 - 60x + 12$$

$$\text{At } x = 0, \frac{d^2y}{dx^2} = 12 = +\text{ve}$$

$$\text{At } x = 2, \frac{d^2y}{dx^2} = 36 = +\text{ve}$$

$$\text{At } x = \frac{1}{2}, \frac{d^2y}{dx^2} = -9 = -\text{ve}$$

At $x = 0$ and $x = 2$, the function is minimum

Minimum value of $x = 0$ is 5

Minimum value of $x = 2$ is –3

$$\text{At } x = \frac{1}{2}, \text{ the function is maximum}$$

Maximum value at $x = \frac{1}{2}$ is $\frac{87}{16}$

EXERCISE 3.21

1. For what value of x are the following functions maximum or minimum?
(*i*) $2x^2 - 6x + 3$ (*ii*) $x^5 - 5x^4 + 5x^3 + 10$

2. Find the maximum or minimum values of the functions
(*i*) $(x+1)(x-2)^2$ (*ii*) $2x^3 - 15x^2 + 36x + 10$

3. Show that the function $x^5 - 5x^4 + 5x^3 - 1$ is maximum at $x = 1$ minimum at $x = 3$.

4. Show that the maximum and minimum values of $\frac{(x+1)(x+4)}{(x-1)(x-4)}$ are –9 and $-\frac{1}{9}$.

5. Prove that $\sin^p \theta \cos^q \theta$ is maximum when $\theta = \tan^{-1} \sqrt{\frac{p}{q}}$.

6. Prove that the maximum value of $\left(\frac{1}{x}\right)^x$ is $e^{\frac{1}{e}}$.

7. Prove that the maximum value of the function $\sin x + \cos x$ is $\sqrt{2}$.

8. Find the maxima and minima of the function $x^5 - 5x^4 + 5x^3 - 1$.

9. Show that the function $(1-x)^2 e^x$ is maximum at $x = -1$ and minimum at $x = 1$.

10. Show that the maximum and minimum values of $8x^5 - 15x^4 + 10x^2$ are $\frac{21}{16}$ and 0 respectively.

ANSWERS

1. (*i*) Min at $x = \frac{3}{2}$ (*ii*) Max. at $x = 1$, Min. at $x = 3$

2. (*i*) max at $x = 0, 4$; min at $x = 2, 0$

   (*ii*) max 38 at $x = 2$, min 37 at $x = 3$

8. At $x = 0$, neither maximum nor minimum;

   At $x = 1$, maximum value is zero,

   At $x = 3$, minimum value is -28

3.39 ASYMPTOTES

Asymptote A straight line is said to be an asymptotes of a curve $y = f(x)$, if the perpendicular distance of the point $p(x, y)$ on the curve from the line tends to zero when x or y or both tends to infinity.

We shall consider separately the cases which arise when an asymptote is (*i*) parallel to either coordinate axis or (*ii*) parallel to neither.

An asymptote parallel to y-axis may be referred to as a vertical asymptote and that parallel to x-axis as a horizontal asymptote. An asymptote which is not parallel to either axis may be described as an oblique asymptote.

Asymptote parallel to y-axis for the curve $f(x, y) = 0$

Working Rule: To find asymptote parallel to y-axis, equate to zero the coefficient of the highest power of y, present in the given equation of the curve. Resolve it now into real linear factors.

If the coefficient of the highest power of y is either a constant or not resolvable in to real linear factors, then there are no asmptote parallel to y-axis.

Asymptote parallel to x-axis for the curve $f(x, y) = 0$

Working Rule: To find asymptote parallel to x-axis, equate to zero the coefficient of the highest power of x, present in the given equation of the curve. Resolve it now into real linear factors.

If the coefficient of the highest power of x is either a constant or not resolvable in to real linear factors, there are no asymptotes parallel to x-axis.

ILLUSTRATIVE EXAMPLES

Example 1. Find the asymptotes, parallel to axes, of the following curves:

$$(i)\ \frac{a^2}{x^2}-\frac{b^2}{y^2}=1\ (ii)\ \frac{a^2}{x^2}+\frac{b^2}{y^2}=1\ (iii)\ x^2y^2-y^2-2=0$$

Solution: (*i*) $\frac{a^2}{x^2}-\frac{b^2}{y^2}=1$

$\Rightarrow$ $$a^2y^2-b^2x^2=x^2y^2$$

or $$x^2y^2+b^2x^2-a^2y^2=0$$

Equating to zero the coefficient of the highest power of x (i.e., x^2), we get

$$y^2+b^2=0 \Rightarrow y=\pm ib,$$

which gives imaginary values of y, and therefore there is no asymptote parallel to x-axis.

Again, equating to zero the coefficient of the highest power of y (i.e., y^2), we get

$$x^2-a^2=0 \Rightarrow x=\pm a,$$

which are the asymptotes parallel to y-axis.

(*ii*) $$\frac{a^2}{x^2}+\frac{b^2}{y^2}=1$$

$\Rightarrow$ $$a^2y^2+b^2x^2=x^2y^2$$

or $$x^2y^2-b^2x^2-a^2y^2=0$$

Equating to zero the coefficient of the highest power of x (i.e., x^2), we get

$$y^2-b^2=0 \Rightarrow y=\pm b,$$

which are the asymptote parallel to x-axis.

Again, equating to zero the coefficient of the highest power of y (i.e., y^2), we get

$$x^2-a^2=0 \Rightarrow x=\pm a,$$

which are the asymptote parallel to y-axis.

(*iii*) $$x^2y^2-y^2-2=0$$

Equating to zero the coefficient of the highest power of x (i.e., x^2), we get

$$y^2=0 \Rightarrow y=0.$$

Again, equating to zero the coefficient of the highest power of y (i.e., y^2), we get

$$x^2-1=0 \Rightarrow x=\pm 1,$$

which are the asymptote parallel to y-axis.

Example 2. Find the asymptotes of the curve $xy^3-x^3=a(x^2+y^2)$ which are parallel to the axes.

Solution: $xy^3-x^3-a(x^2+y^2)=0$

Equating to zero the coefficient of the highest power x(i.e., x^3), we get $-1 = 0$, which is meaningless

Hence, asymptote parallel to x-axis does not exist.

Again equating to zero the coefficient of the highest power of y(i.e., y^3), we get

$$x = 0,$$

which are the asymptote parallel to y-axis.

Example 3. Find the asymptotes parallel to the x-axis of the curve $x^2y - 2x^2 - 3xy + 4y + 1 = 0$

Solution: Equating to zero the coefficient of the highest power of x (i.e., coefficient of x^2), we get

$$y - 2 = 0 \Rightarrow y = 2,$$

which is the required asymptote.

Example 4. Find the asymptotes, parallel to y-axis, of the curve $\dfrac{a^2}{x} + \dfrac{b^2}{y} = 1.$

Solution: $\dfrac{a^2}{x} + \dfrac{b^2}{y} = 1$

$$\Rightarrow \qquad a^2y + b^2x = xy$$

$$\text{or} \qquad a^2y + b^2x - xy = 0$$

Equating to zero the coefficient of highest power of y, we get

$$a^2 - x = 0 \Rightarrow x = a^2$$

which is the required asymptote parallel to y-axis of the given curve.

Example 5. Find the asymptote parallel to axes for the curve $x^2y^2 = b^2(x^2 + y^2)$

Solution: $x^2y^2 - b^2(x^2 + y^2) = 0$

Equating to zero the coefficient of the highest power of x(i.e., coeff. of x^2), we get

$$y^2 - b^2 = 0 \Rightarrow y = \pm b,$$

which is the asymptote parallel to x-axis.

Again, equating to zero the coefficient of the highest power of y(i.e., coeff. of y^2), we get

$$x^2 - b^2 = 0 \Rightarrow x = \pm b,$$

which is the asymptote parallel to y-axis.

EXERCISE 3.21

Find the asymptotes parallel to axes of the following curves:

1. $(x^2 + y^2)x - ay^2 = 0$

2. $x^2y^2 - a^2(x^2 + y^2) - a^3(x + y) + a^4 = 0$

3. $x^3 - y^3 = 3ax^2$

4. $\dfrac{a^2}{x^2} + \dfrac{b^2}{y^2} = 1$

5. $y^2(x^2 - 2a) = x^3 - a^3$

ANSWERS

1. $x = a$ 2. $y = \pm a, x = \pm a$ 3. $y - x + a = 0$
4. $x = 3, y = 1$ 5. $x = \pm\sqrt{2a}$

3.40 OBLIQUE ASYMPTOTES

If $y = mx + c$ is an oblique asymptote to any curve $f(x, y) = 0$, then

$$m = \lim_{x\to\infty} \frac{y}{x} \text{ and } c = \lim_{x\to\infty} (y - mx)$$

Note: The above method requires m and c to be finite. In particular m may be zero and therefore the asymptotes parallel to x-axis can also be found by using this method. Thus by this method, all the asymptotes of a given curve can be determined, excepting those which are parallel to y-axis

ILLUSTRATIVE EXAMPLES

Example 1. Find the asymptotes of $y = \dfrac{x^2 + 2x - 1}{x}$.

Solution: $y = \dfrac{x^2 + 2x - 1}{x}$

$$\lim_{x\to\infty} \frac{y}{x} = \lim_{x\to\infty}\left(\frac{x^2 + 2x - 1}{x}\right) = \lim_{x\to\infty}\left(\frac{1 + \frac{2}{x} - \frac{1}{x^2}}{1}\right) = 1$$

$\therefore$ $m = 1$

Now, $$c = \lim_{x\to\infty} (y - x) = \lim_{x\to\infty}\left(\frac{x^2 + 2x - 1}{x} - x\right)$$

$$c = \lim_{x\to\infty}\left(\frac{x^2 + 2x - 1 - x^2}{x}\right)$$

$$= \lim_{x\to\infty}\left(\frac{2x - 1}{x}\right) = \lim_{x\to\infty}\left(2 - \frac{1}{x}\right) \Rightarrow c = 2$$

Hence the asymptotes are $y = x + 2$

Example 2. Find the asymptotes of $\dfrac{x^2}{a^2} - \dfrac{y^2}{b^2} = 1$

Solution: $\dfrac{x^2}{a^2} - \dfrac{y^2}{b^2} = 1$

$$\Rightarrow \quad \frac{1}{a^2} - \frac{1}{b^2}\frac{y^2}{x^2} = \frac{1}{x^2} \Rightarrow \frac{1}{b^2}\frac{y^2}{x^2} = \frac{1}{x^2} + \frac{1}{a^2}$$

$$\therefore \qquad \lim_{x\to\infty}\left(\frac{y}{x}\right)^2 = \lim_{x\to\infty}\left(\frac{b^2}{x^2}+\frac{b^2}{a^2}\right) = \frac{b^2}{a^2}$$

$$\lim_{x\to\infty}\frac{y}{x} = \pm\frac{b}{a}$$

Now,
$$c = \lim_{x\to\infty}(y-mx) = \lim_{x\to\infty}\left(y - x\frac{dy}{dx}\right)$$

$$= \lim_{x\to\infty}\left(y - x\cdot\frac{b^2x}{a^2y}\right) = \lim_{x\to\infty}\frac{b^2}{y}\left(\frac{y^2}{b^2}-\frac{x^2}{a^2}\right)$$

$$= \lim_{x\to\infty}\left(\frac{-b^2}{y}\right) = 0 \left[\because \lim_{x\to\infty} y = \pm\frac{b}{a}x\right]$$

Hence the asymptotes are $y = \pm\frac{b}{a}x$

3.41 ASYMPTOTES OF ALGEBRAIC CURVES

Working rule: Suppose $y = mx + c$ is an asymptote of the curve.

Step 1. Put $y = mx + c$ in the equation of the curve and arrange it in descending powers of x.

Step 2. Equate the coefficients of two highest degree terms to zero.

Step 3. Solve these equations to find the values of m and c.

Step 4. For every pair of values of m and c, there is one asymptote $y = mx + c$ of the curve.

ILLUSTRATIVE EXAMPLES

Example 1. Find the asymptotes of the curve

$$x^3 + 2x^2y - xy^2 - 2y^3 + 4y^2 + 2xy + y - 1 = 0$$

Solution: Put $y = mx + c$ is the given equation, we get

$$x^3 + 2x^2(mx+c) - x(mx+c)^2 - 2(mx+c)^3 + 4(mx+c)^2 + 2x(mx+c) + mx + c - 1 = 0$$

$$\Rightarrow (1 + 2m - m^2 - 2m^3)x^3 + 2(c - mc - 3m^2c + 2m^2 + m)x^2 + \ldots = 0$$

Equating the coefficient of x^2 and x^3 to zero, we get

$$1 + 2m - m^2 - 2m^3 = 0 \qquad \ldots(1)$$

$$c - mc - 3m^2c + 2m^2 + m = 0 \qquad \ldots(2)$$

From equation (1), we have

$$2m^3 + m^2 - 2m - 1 = 0$$

$$\therefore \qquad m = -1, 1, -\frac{1}{2}$$

From equation (2)

$$c = -\left(\frac{2m^2 + m}{1 - m - 3m^2}\right)$$

If $m = -1$, we get $c = 1$

If $m = 1$, we get $c = 1$

If $m = -\dfrac{1}{2}$, we get $c = 0$

Hence the asymptotes are $y = -x + 1$, $y = x + 1$, $y = -\dfrac{1}{2}x$

Example 2. Find the asymptotes of the curve

$y^3 - x^2y + 2y^2 + 4y + 1 = 0$

Solution: Put $y = mx + c$ in the given equation of the curve, we get

$(mx + c)^3 - x^2(mx + c) + 2(mx + c)^2 + 4(mx + c) + 1 = 0$

$\Rightarrow (m^3 - m)x^3 + (3m^2c - c + 2m^2)x^2 + \ldots = 0$

Equating to zero the coefficients of x^3 and x^2, we get

$$m^3 - m = 0 \quad \ldots (1)$$

$$3m^2c - c + 2m^2 = 0 \quad \ldots (2)$$

From equation (1), we get $m = 0, 1, -1$

From equation (2), we get

$$c = \frac{2m^2}{1 - 3m^2}$$

If $m = 0$, we get $c = 0$

If $m = 1$, we get $c = -1$

If $m = -1$, we get $c = -1$

$\therefore$ Asymptotes are $y = 0$, $y = x - 1$ and $y = -x - 1$.

3.42 THE ASYMPTOTES OF THE GENERAL RATIONAL ALGEBRAIC CURVE OF nth DEGREE

Working rule 1. Find the polynomial $\phi_n(m)$ by putting $x = 1$, $y = m$ in the highest degree terms of the given equation of the curve.

2. Put $\phi_n(m)$ equal to zero solve for m, we get the values of m. If some values of m are imaginary we reject them.

3. Find the polynomial $\phi_{n-1}(m)$ by putting $x = 1$, $y = m$ in $(n - 1)$th degree terms of the given equation. Similarly find polynomial $\phi_{n-2}(m)$ by putting $x = 1$, $y = m$ in the $(n - 2)^{th}$ degree terms of the given equation and so on.

4. Find the values $c_1, c_2, c_3, \ldots$ corresponding to values of $m_1, m_2, m_3, \ldots$ (the values of m) by using the relation $c\,\phi'_n(m) + \phi_{n-1}(m) = 0$.

i.e., $$c = -\frac{\phi_{n-1}(m)}{\phi'_n(m)} \quad (\text{provided } \phi'_n(m) \neq 0)$$

5. The required asymptotes are $y = m_1 x + c_1, y = m_2 x + c_2, y = m_3 x + c_3, \ldots$

6. If $\phi'_n(m) = 0$ for some values of m, but $\phi_{n-1}(m) \neq 0$, then corresponding to that value of m there is no asymptotes.

7. If $\phi'_n(m) = 0 = \phi_{n-1}(m)$, for some value of m(i.e., two roots of $\phi_n(m) = 0$ are equal), then the values of c are given by equation

$$\frac{c^2}{2}\phi''_n(m) + \frac{c}{1!}\phi'_{n-1}(m) + \phi_{n-2}(m) = 0.$$

8. Similarly, if three roots of $\phi_n(m) = 0$ are equal [i.e., $\phi''_n(m) = 0 = \phi'_{n-1}(m) = \phi_{n-2}(m)$], then the corresponding values of c are given by the equation

$$\frac{c^3}{3!}\phi'''_n(m) + \frac{c^2}{2!}\phi''_{n-1}(m) + \frac{c}{1!}\phi'_{n-2}(m) + \phi_{n-3}(m) = 0.$$

Note 1. The number of asymptotes, real or imaginary, of an algebraic curve of the nth degree cannot exceed n.

Hence, a curve of degree n can not have more than n asymptotes. But the number of real asymptotes can be less than n. Some roots of the equation $\phi_n(m) = 0$ may come out to be imaginary.

2. When it is required to find the asymptotes of a curve it is advisable to first find the asymptotes parallel to axes, if there be any, and then search for the oblique asymptotes.

ILLUSTRATIVE EXAMPLES

Example 1. Find all the asymptotes of the curve

$$y^2(a^2 - x^2) = x^4$$

Solution: The equation of the curve is

$$y^2x^2 + x^4 - a^2y^2 = 0$$

Since the curve is of degree four so it cannot have more than four asymptotes.

Equating to zero the coefficient of the highest power of x(i.e., x^4), it is constant. Hence there is no asymptote parallel to x-axis.

Equating to zero the coefficient of the highest power of y(i.e., y^2), we get

$$x^2 - a^2 = 0$$

$$\Rightarrow \quad x = \pm a$$

Therefore asymptote parallel to y-axis is $x = +a$.

To find the remaining oblique asymptotes, we put $x = 1$, $y = m$ in the highest degree terms (i.e., four degree terms), we get $\phi_4(m) = m^2 + 1$

$$\therefore \quad \phi_4(m) = 0 \Rightarrow m^2 + 1 = 0$$

$$m = \pm i$$

Since the roots of the equation $\phi_4(m) = 0$ are imaginary and therefore the corresponding asymptotes are imaginary.

Hence the only real asymptotes of the curve are $x = \pm a$.

Example 2. Find all the asymptotes of the curve

$$x^3 + 2x^2y - xy^2 - 2y^3 + 2xy + 4y^2 + y - 1 = 0$$

Solution: $x^3 + 2x^2y - xy^2 - 2y^3 + 2xy + 4y^2 + y - 1 = 0$

Since the curve is of degree three, so it can not have more than three asymptotes. Since the coefficient of x^3 and y^3, the highest degree terms is x and y are constant, therefore these are no asymptotes parallel to x-axis or y-axis.

To find the remaining oblique asymptotes, we put $x = 1$, $y = m$ in the highest degree terms, we get

$$\phi_3(m) = 1 + 2m - m^2 - 2m^3$$

Now again put $x = 1$, $y = m$ in the next highest degree terms (i.e., degree 2), we get

$$\phi_2(m) = 4m^2 + 2m$$

Now $\qquad \phi_3(m) = 0 \Rightarrow 1 + 2m - m^2 - 2m^3 = 0$

$\Rightarrow \qquad 2m^3 + m^2 - 2m - 1 = 0$

$\Rightarrow \qquad m = -1, 1, -\frac{1}{2}$

Now c is given by

$$c = -\frac{\phi_2(m)}{\phi'_3(m)} = \frac{2m + 4m^2}{6m^2 + 2m - 2}$$

$$\text{if } m = -1, c = \frac{-2+4}{6-2-2} = \frac{2}{2} = 1$$

$$\text{if } m = 1, c = \frac{2+4}{6+2-2} = \frac{6}{6} = 1$$

$$\text{if } m = -\frac{1}{2}, c = \frac{2\times-\frac{1}{2}+4\times\frac{1}{4}}{\frac{6}{4}+2\times-\frac{1}{2}-2} = \frac{0}{\frac{3}{2}-3} = 0$$

Putting these values of m and c in $y = mx + c$, the asymptotes are

$$y = -x + 1, y = x + 1, y = -\frac{1}{2}x$$

Example 3. Find all the asymptotes of the curve

$$x^3 + 3x^2y - 4y^3 - x + y + 3 = 0$$

Solution: Since the curve is of degree three, so it can not have more than three asymptotes.

Since the coefficients of x^3 and y^3, the highest degree terms in x and y are constant, therefore there are no asymptotes parallel to x-axis or y-axis.

To find the remaining oblique asymptotes, we put $x = 1$, $y = m$ in the highest degree terms (i.e., degree 3), we get $\phi_3(m) = 1 + 3m - 4m^3$

Now again put $x = 1, y = m$ in the next highest degree terms (i.e., degree 2), we get

$\phi_2(m) = 0$ [because there is no second degree term]

$$\therefore \qquad \phi'_3(m) = 3 - 12m^2$$

Now $\phi_3(m) = 0$ gives $1 + 3m - 4m^3 = 0$

$$(1 - m)(1 + 2m)^2 = 0$$

$$m = 1, -\frac{1}{2}, -\frac{1}{2}$$

c is given by the equation

$$c = -\frac{\phi_2(m)}{\phi_3'(m)} = -\frac{0}{3 - 12m^2} \qquad \text{...(1)}$$

when $m = 1, c = 0$ and the corresponding asymptote is $y = x + 0$ i.e., $y = x$

when $m = -\frac{1}{2}, -\frac{1}{2}$ [equal roots] the equation (1) fails to give c. so in this case c is given by the equation

$$\frac{c^2}{2!}\phi''_3(m) + \frac{c}{1!}\phi'_2(m) + \phi_1(m) = 0$$

putting $x = 1, y = m$ in the first degree terms of the equation of the curve, we get $\phi_1(m) = -1 + m$.

Also $\phi''_3(m) = -24m$ and $\phi'_2(m) = 0$

Hence for $m = -\frac{1}{2}$, c is to be given by

$$\frac{1}{2}c^2(-24m) + 0 + m - 1 = 0$$

putting $\qquad m = -\frac{1}{2}$, we get

$$6c^2 - \frac{3}{2} = 0 \Rightarrow c^2 = \frac{1}{4} \text{ or } c = \pm\frac{1}{2}$$

Hence $y = -\frac{1}{2}x + \frac{1}{2}$ and $y = -\frac{1}{2}x - \frac{1}{2}$ are two asymptotes corresponding to the slope $m = -\frac{1}{2}$

Hence the required asymptotes are $y = x$ and $x + 2y = \pm 1$.

Example 4. Find the asymptotes of the curve

$$x^2(x^2 + y^2) = a^2(y^2 - x^2)$$

Solution: The equation of the curve

$$x^2(x^2 + y^2) - a^2(y^2 - x^2) = 0$$

Since the curve is of degree 4, therefore it cannot have more than four asymptotes.

Equating to zero the coefficient of the highest power of y(i.e., y^2), the asymptotes parallel to y-axis are given by

$$x^2 - a^2 = 0$$

$$x = \pm a$$

The coefficient of the highest power of x(i.e., x^4) is a constant. Hence there is no asymptote parallel to x-axis.

To find the remaining oblique asymptotes, we put $x = 1$, $y = m$ in the highest degree terms (i.e., 4 th degree terms), we get

$$\phi_4(m) = 1 + m^2$$

The roots of the equation $\phi_4(m) = 0$

i.e. $$1 + m^2 = 0$$

are imaginary and consequently the corresponding asymptotes are imaginary.

Hence the only real asymptotes of the curve are $x = \pm a$.

Example 5. Find the asymptotes of

$$(x-y)^2(x^2+y^2) - 10(x-y)x^2 + 12y^2 + 2x + y = 0$$

Solution: The equation of the curve is

$$(x-y)^2(x^2+y^2) - 10(x-y)x^2 + 12y^2 + 2x + y = 0 \quad \text{...(1)}$$

Since coefficient of x^4 and y^4, the highest degree terms in x and y are constants, therefore there are no asymptotes parallel to x-axis or y-axis.

To find oblique asymptotes putting $x = 1$, $y = m$ in the fourth, third and second degree terms in (1), we get

$$\phi_4(m) = (1-m)^2(1+m^2) = m^4 - 2m^3 + 2m^2 - 2m + 1$$

$$\phi'_4(m) = 4m^3 - 6m^2 + 4m - 2$$

$$\phi''_4(m) = 12m^2 - 12m + 4$$

$$\phi_3(m) = -10(1-m)\ \phi'_3(m) = 10$$

$$\phi_2(m) = 12m^2$$

Now $$\phi_4(m) = 0 \text{ gives}$$

$$(1-m)^2(1+m^2) = 0$$

$m = 1, 1$ [because $1 + m^2 = 0$ gives only imaginary values of m]

For $m = 1, 1$ [equal roots] the value of c is given by

$$\frac{c^2}{2!}\phi''_4(m) + \frac{c}{1!}\phi'_3(m) + \phi_2(m) = 0$$

or $$c^2(6m^2 - 6m + 2) + 10c + 12m^2 = 0$$

$$\text{for } m = 1 \Rightarrow 2c^2 + 10c + 12 = 0$$

or $$c^2 + 5c + 6 = 0 \Rightarrow c = -2, -3$$

$\therefore$ Asymptotes are $y = mx + c$

i.e., $$y = x - 2 \text{ and } y = x - 3$$

Example 6. Find all the asymptotes of the curve

$$(x+y)^2(x+y+2) = x + 9y - 2$$

Solution: The equation of the curve is

$$(x+y)^3 + 2(x+y)^2 - x - 9y + 2 = 0 \qquad \text{... (1)}$$

Since the coefficient of x^3 and y^3, the highest degree terms of x and y are constants, therefore there are no asymptotes parallel to x-axis or y-axis.

To find oblique asymptotes, putting $x = 1$, $y = m$ in the third, second and first degree terms and constant term in (1), we have

$$\phi_3(m) = (1+m)^3 \qquad \phi'_3(m) = 3(1+m)^2$$
$$\phi_2(m) = 2(1+m)^2 \qquad \phi''_3(m) = 6(1+m)$$
$$\phi_1(m) = -1 - 9m \qquad \phi'''(m) = 6$$
$$\phi_0(m) = 2 \qquad \phi'_2(m) = 4(1+m)$$
$$\phi''_2(m) = 4$$
$$\phi'_1(m) = -9$$

Now $\phi_3(m) = 0$ gives

$$(1+m)^3 = 0 \Rightarrow m = -1, -1, -1 \text{ [Three equal roots]}$$

For these three equal values of $m = -1$, the value of c is given by

$$\frac{c^3}{3!}\phi'''_3(m) + \frac{c^2}{2!}\phi''_2(m) + \frac{c}{1!}\phi'_1(m) + \phi_0(m) = 0$$

$$\frac{c^3}{6}\cdot 6 + \frac{c^2}{2}(4) + c \times -9 + 2 = 0$$

$$\Rightarrow \qquad c^3 + 2c^2 - 9c + 2 = 0$$

$$\Rightarrow \qquad (c-2)(c^2 + 4c - 1) = 0$$

Solving for c, we get

$$c = -2 \pm \sqrt{5}$$

$$\therefore \qquad c = 2,\ -2 + \sqrt{5} \text{ and } -2 - \sqrt{5}$$

Hence the asymptotes are $y = mx + c$

i.e., $y = -x + 2$, $y = -x - 2 + \sqrt{5}$ and $y = -x - 2 - \sqrt{5}$.

EXERCISE 3.22

Find all the asymptotes of the following curves:

1. $xy^2 + y^3 - x - y = 1$

2. $x^3 - 6x^2y + 11xy^2 - 6y^3 + x + y + 1 = 0$

3. $x^3 + 2x^2y + xy^2 - x^2 - xy + 2 = 0$

4. $(x-1)(x-2)(x+y) + x^2 + x + 1 = 0$

5. $x^3 + y^3 - 3axy = 0$

6. $3x^3 + 2x^2y - 7xy^2 + 2y^3 - 14xy + 7y^2 + 4x + 5y = 0$

7. $x^2y^3 + x^3y^2 = x^3 + y^3$

8. $\dfrac{a^2}{x^2} - \dfrac{b^2}{y^2} = 1$

9. $4x^3 + 2x^2 - 3xy^2 - y^3 - 1 - xy - y^2 = 0.$

ANSWERS

1. $y = \pm 1, y + x = 0$
2. $x - y = 0, x - 2y = 0, x - 3y = 0$
3. $x = 0, x + y = 0, x + y = 1$
4. $x = 1, x = 2, x + y + 1 = 0$
5. $x + y + a = 0$
6. $6y - 6x + 7 = 0; 2y - 6x + 3 = 0$ and $2y + x + 5/3 = 0$
7. $x = \pm 1, y = \pm 1, y = -x$
8. $x = \pm a$
9. $y = x, y = -2x, y = -2x - 1$

3.43 ASYMPTOTES TO NON-ALGEBRAIC CURVES

The asymptotes of non-algebraic curves is clear from the following example:

Example Find the asymptotes of the curve $y = \sec x$.

Solution: $y = \sec x$

$$\therefore \qquad \frac{dy}{dx} = \sec x \tan \text{x}$$

Therefore, the tangent to given curve at (x, y)

$$Y - \sec x = \sec x \tan x(X - x)$$

i.e.,
$$y \cos^2 x - \cos x = (X - x) \sin x \qquad \text{...(1)}$$

Now as $x \to \frac{\pi}{2}$, $y \to \infty$ and the distance of (x, y) from the origin tends to infinity, therefore taking limit of (1) as $x \to \frac{\pi}{2}$, we get

$$Y{\cdot}0 - 0 = \left(X - \frac{\pi}{2}\right){\cdot}1$$

i.e.,
$$X = \frac{\pi}{2}$$

This is one asymptote. The other asymptotes are

$$X = -\frac{\pi}{2}, \pm\frac{3}{2}\pi, \ldots$$
$$r = f(\theta)$$

3.44 ASYMPTOTES OF POLAR CURVE $r = f(\theta)$ WORKING RULE

1. Put $r = \frac{1}{u}$ in the given equation. Also change all trigonometric ratios, if any, in to $\sin \theta$ and $\cos \theta$.

2. Find the limit of θ as $u \to 0$. Let θ_1 be this limit

3. Determine $p = \lim_{\substack{\theta \to \theta_1 \\ u \to 0}} \left(-\frac{d\theta}{du}\right)$ for value or values of θ obtained in step 2.

4. Putting the values of p and θ_1 in the equation $p = r\sin(\theta_1 - \theta)$ we get the corresponding asymptote.

Note: If $\lim\limits_{\theta \to \theta_1} \left(\dfrac{-d\theta}{du} \right)$ does not tend to a finite limit then there is no asymptote corresponding to the value $\theta = \theta_1$.

ILLUSTRATIVE EXAMPLES

Example 1. Find the asymptotes of the curve

$$r\cos\theta = a\sin\theta$$

Solution: $r\cos\theta = a\sin\theta$

put $r = \dfrac{1}{u}$, we get

$$\frac{1}{u}\cos\theta = a\sin\theta$$

or $$u = \frac{\cos\theta}{a\sin\theta}$$

As $u \to 0$, $\dfrac{\cos\theta}{a\sin\theta} \to 0$ i.e., $\cos\theta \to 0$

$$\therefore \quad \theta \to (2n+1)\frac{\pi}{2} \text{ i.e., } \theta_1 = (2n+1)\frac{\pi}{2}$$

Since $$u = \frac{\cos\theta}{a\sin\theta}$$

$$\therefore \quad \frac{du}{d\theta} = \frac{1}{a}\left[\frac{\sin\theta \times -\sin\theta - \cos\theta \times \cos\theta}{\sin^2\theta}\right]$$

$$\frac{du}{d\theta} = \frac{1}{a}\left[\frac{-\sin^2\theta - \cos^2\theta}{\sin^2\theta}\right] = \frac{-1}{a\sin^2\theta}$$

$$\therefore \quad p = \lim_{\theta \to \theta_1}\left(\frac{-d\theta}{du}\right) = \lim_{\theta \to (2n+1)\frac{\pi}{2}} (a\sin^2\theta)$$

$$= a \lim_{\theta \to (2n+1)\frac{\pi}{2}} \sin^2\theta$$

$$p = a[(-1)^n]^2 = a(-1)^{2n} \qquad \left[\because \sin(2n+1)\frac{\pi}{2} = (-1)^n\right]$$

$\therefore$ Equation of the asymptote is

$$p = r\sin(\theta_1 - \theta)$$

or $$a(-1)^{2n} = r\sin\left[(2n+1)\frac{\pi}{2} - \theta\right]$$

$$a(-1)^{2n} = r\sin\left[n\pi + \frac{\pi}{2} - \theta\right]$$

$$a(-1)^{2n} = (-1)^n\ r\sin\left(\frac{\pi}{2} - \theta\right)$$

or $$a(-1)^{2n} = (-1)^n\, r\cos\theta$$

$$a(-1)^n = r\cos\theta$$

or $$r\cos\theta = \pm a$$

Example 2. Find the asymptotes of the curve

$$r\sin\theta = 2\cos 2\theta$$

Solution: $r\sin\theta = 2\cos 2\theta$ ...(1)

put $$r = \frac{1}{u}, \text{ we get}$$

$$\frac{1}{u}\sin\theta = 2\cos 2\theta \Rightarrow u = \frac{\sin\theta}{2\cos 2\theta}$$

As $u \to 0$, $\frac{\sin\theta}{2\cos 2\theta} \to 0$ i.e., $\sin\theta \to 0$

$\therefore$ $$\theta \to n\pi \text{ i.e., } \theta_1 = n\pi$$

Since $$u = \frac{\sin\theta}{2\cos 2\theta}$$

$\therefore$ $$\frac{du}{d\theta} = \frac{1}{2}\left[\frac{\cos 2\theta \times \cos\theta - \sin\theta \times -2\sin 2\theta}{(\cos 2\theta)^2}\right]$$

$$\frac{du}{d\theta} = \left[\frac{\cos 2\theta\cos\theta + 2\sin\theta\sin 2\theta}{2\cos^2 2\theta}\right]$$

$\therefore$ $$p = \lim_{\theta\to\theta_1}\left(-\frac{d\theta}{du}\right)$$

$$p = \lim_{\theta\to n\pi}\left[\frac{-2\cos^2 2\theta}{\cos 2\theta\cos\theta + 2\sin\theta\sin 2\theta}\right]$$

$$p = -2\lim_{\theta\to n\pi}\left[\frac{\cos^2 2\theta}{\cos 2\theta\cos\theta + 2\sin\theta\sin 2\theta}\right]$$

$$p = -2\left[\frac{\cos^2(2n\pi)}{\cos 2n\pi\cos n\pi + 2\sin n\pi\sin 2n\pi}\right]$$

$$p = -2\left[\frac{1}{1\cdot\cos n\pi}\right]$$

$$\left[\begin{array}{l}\because \cos 2n\pi = 1 \\ \sin n\pi = 0 \\ \sin 2n\pi = 0\end{array}\right]$$

$$p = \frac{-2}{\cos n\pi} = \frac{-2}{(-1)^n}$$

$[\because \cos n\pi = (-1)^n]$

$\therefore$ Asymptotes are given by

$$p = r\sin(\theta_1 - \theta)$$

$$\frac{-2}{(-1)^n} = r\sin(n\pi - \theta)$$

or $$\frac{2}{(-1)^n} = -r\sin(n\pi - \theta) = -r\{(-1)^{n-1}\sin\theta\}$$

$\Rightarrow$ $$r\sin\theta = 2$$

Example 3. Prove that the curve $r = \dfrac{a}{1-\cos\theta}$ has no asymptotes.

Solution: $r = \dfrac{a}{1-\cos\theta}$

put $$r = \frac{1}{u}$$

$\therefore$ $$\frac{1}{u} = \frac{a}{1-\cos\theta} \Rightarrow u = \frac{1-\cos\theta}{a}$$

A s $u \to 0$, $\dfrac{1-\cos\theta}{a} \to 0$ i.e., $\cos\theta \to 1$ or $\cos\theta \to \cos 0$

$\therefore$ $$\theta \to 2n\pi \pm 0 = 2n\pi$$

$\therefore$ $\theta_1 = 2n\pi$, where n is any integer.

Since $$u = \frac{1-\cos\theta}{a}$$

$\therefore$ $$\frac{du}{d\theta} = \frac{\sin\theta}{a}$$

$\therefore$ $$p = \lim_{\theta\to\theta_1}\left(-\frac{d\theta}{du}\right) = \lim_{\theta\to 2n\pi}\left(\frac{-a}{\sin\theta}\right)$$

$$= \frac{-a}{\sin 2n\pi} \to -\infty$$

$\therefore$ $\displaystyle\lim_{\theta\to\theta_1}\left(-\frac{d\theta}{du}\right)$ does not exist. $[\because \sin 2n\pi = 0]$

Hence the curve $r = \dfrac{a}{1-\cos\theta}$ has no asymptote.

EXERCISE 3.23

Find the asymptotes of the following curves:

1. $r\cos\theta = u\sin^2\theta$ **2.** $r\sin 2\theta = a$

3. $r\sin\theta = 2\cos\theta$ **4.** $r\theta = a$ **5.** $r = 4(\sec\theta + \tan\theta)$

6. $r = a\sec\theta + b\tan\theta$ **7.** $r\cos\theta = a\sin^2\theta$

8. $r = \dfrac{2\theta}{\sin\theta}$ **9.** $r = a\operatorname{cosec}\theta + b$

ANSWERS

1. $r\cos\theta = 4$ 2. $r\sin\theta = \pm\frac{1}{2}a$, $r\cos\theta = \pm\frac{1}{2}a$
3. $r\sin\theta = \pm 2$ 4. $r\sin\theta = a$
5. $r\cos\theta = 8$ 6. $r\cos\theta = (a-b)$ **7.** $a = r\cos\theta$
8. $r\sin\theta = 2n\pi, n = \pm 1, \pm 2, \ldots$ 9. $r\sin\theta = a$.

3.45 CIRCULAR ASYMPTOTES

Let the equation of a curve be $r = f(\theta)$. If $\lim\limits_{\theta\to\infty} f(\theta) = a$, then the circle $r = a$ is called the circular asymptote of the curve $r = f(\theta)$.

ILLUSTRATIVE EXAMPLES

Example 1. Find the circular asymptote of the curve

$$r = \frac{3\theta^2 + 2\theta + 1}{2\theta^2 + \theta + 1}$$

Solution: The given equation can be written as

$$r = \frac{3 + \frac{2}{\theta} + \frac{1}{\theta^2}}{2 + \frac{1}{\theta} + \frac{1}{\theta^2}} = f(\theta)$$

$$\lim_{\theta\to\infty} f(\theta) = \lim_{\theta\to\infty} r = r = \frac{3}{2}$$

$$\therefore \qquad r = \frac{3}{2}$$

Hence circular asymptote is $r = \frac{3}{2}$.

Example 2. Find the circular asymptote of the curve

$$r = \frac{\theta^2 - 1}{\theta^2 + 1}$$

Solution: $f(\theta) = \dfrac{\theta^2 - 1}{\theta^2 + 1} = \dfrac{1 - \frac{1}{\theta^2}}{1 + \frac{1}{\theta^2}}$

$$\lim_{\theta \to \infty} f(\theta) = 1$$

Hence the circular asymptote is $r = 1$.

EXERCISE 3.24

Find the circular asymptotes of the following curves:

1. $r = \dfrac{a\theta}{\theta - 1}$
2. $r(e^{\theta} - 1) = a(e^{\theta} + 1)$

ANSWERS

1. $r = a$
2. $r = a$

4

Integral Calculs

4.1 INTRODUCTION

Integration is the 'reverse' or 'opposite' process of differentiation.

If $\frac{d}{dx}[F(x)] = f(x)$, then

$$\int f(x)\,dx = F(x) + C$$

If $f(x)$ is the derivative of the function $F(x)$, then $F(x)$ is called the anti-derivative or primitive or integral of the function $f(x)$ and C is called constant of integration or arbitrary constant.

The process of finding function whose derivative is given is called anti-differentiation or Integration.

Here the symbol $\int$ stands for integral and dx after $f(x)$ indicates that x is the variable of integration. The function $f(x)$ which is to be integrated is called the integrand.

In the formula,

$$\int f(x)\,dx = F(x) + C,$$

the constant of integration C is not definitely known to us.For this reason,these integrals are called indefinite integrals.

Remarks : If $f(x)$ be a function such that $f'(x)$ is identically zero,then $f(x)$ is a constant function *i.e.* if $\frac{df(x)}{dx} = f'(x) = 0$

then $$\int f'(x)\,dx = f(x) = \text{constant.}$$

4.2 SOME STANDARD FORMULA

	Derivative	*Primitive or Indefinite Integral*
1.	$\because \frac{d}{dx}\left[\frac{x^{n+1}}{n+1}\right] = x^n$	$\Rightarrow \int x^n dx = \frac{x^{n+1}}{n+1} + C, \quad n \neq -1$

2. $\because \frac{d}{dx}(\log x) = \frac{1}{x}$ $\Rightarrow \int \frac{1}{x} dx = \log x + C.$

3. $\because \frac{d}{dx}(e^x) = e^x$ $\Rightarrow \int e^x dx = e^x + C$

4. $\because \frac{d}{dx}(\sin x) = \cos x$ $\Rightarrow \int \cos x \, dx = \sin x + C$

5. $\because \frac{d}{dx}(-\cos x) = \sin x$ $\Rightarrow \int \sin x dx = -\cos x + C$

6. $\because \frac{d}{dx}(\tan x) = \sec^2 x$ $\Rightarrow \int \sec^2 x \, dx = \tan x + C$

7. $\because \frac{d}{dx}(-\cot x) = \text{cosec}^2 x$ $\Rightarrow \int \text{cosec}^2 x \, dx = -\cot x + C$

8. $\because \frac{d}{dx}(\sec x) = \sec x \tan x$ $\Rightarrow \int \sec x \tan x \, dx = \sec x + C$

9. $\because \frac{d}{dx}(-\text{cosec}\, x) = \text{cosec}\, x \cot x$ $\Rightarrow \int \text{cosec}\, x \cot x \, dx = -\text{cosec}\, x + C$

10. $\because \frac{d}{dx}(\sin^{-1} x) = \frac{1}{\sqrt{1-x^2}}$ $\Rightarrow \int \frac{1}{\sqrt{1-x^2}} dx = \sin^{-1} x + C$

11. $\because \frac{d}{dx}(-\cos^{-1} x) = \frac{1}{\sqrt{1-x^2}}$ $\Rightarrow \int \frac{dx}{\sqrt{1-x^2}} = -\cos^{-1} x + C$

12. $\because \int \frac{d}{dx}(\tan^{-1} x) = \frac{1}{1+x^2}$ $\Rightarrow \int \frac{dx}{1+x^2} = \tan^{-1} x + C$

13. $\because \frac{d}{dx}(-\cot^{-1} x) = \frac{1}{1+x^2}$ $\Rightarrow \int \frac{dx}{1+x^2} = -\cot^{-1} x + C$

14. $\because \frac{d}{dx}(\sec^{-1} x) = \frac{1}{x\sqrt{x^2-1}}$ $\Rightarrow \int \frac{dx}{x\sqrt{x^2-1}} = \sec^{-1} x + C$

15. $\because \frac{d}{dx}(-\text{cosec}^{-1} x) = \frac{1}{x\sqrt{x^2-1}}$ $\Rightarrow \int \frac{dx}{x\sqrt{x^2-1}} = -\text{cosec}^{-1} x + C$

16. $\because \frac{d}{dx}\left(\frac{a^x}{\log_e^a}\right) = a^x$ $\Rightarrow \int a^x dx = \frac{a^x}{\log_e^a} + C$

17. $\because \frac{d}{dx}(x) = 1$ $\Rightarrow \int 1 dx = x + C$

4.3 INTEGRAL OF THE PRODUCT OF A CONSTANT AND A FUNCTION

Let k be a constant then,

$$\int kf(x)dx = k\int f(x)dx$$

i.e. the integral of the product of a function by a constant is equal to the product of the constant by the integral of the function.

Example 1. $\int 5x^3\, dx = 5\int x^3\, dx = 5\left[\frac{x^4}{4}\right] + C = \frac{5}{4}x^4 + C$

2. $\int 2\cos x\, dx = 2\int \cos x\, dx = 2\sin x + C$

4.4 INTEGRAL OF THE SUM OR DIFFERENCE OF TWO FUNCTIONS

Let $f(x)$ and $g(x)$ be two functions of x, then

$$\int [f(x) \pm g(x)]\, dx = \int f(x)\, dx \pm \int g(x)\, dx$$

i.e. the integral of the sum or difference of two functions is equal to the sum or difference of the integrals of the two functions taken one at a time.

COROLLARY: If $f_1(x), f_2(x), f_3(x), ..., f_n(x)$ be n functions of x, then

$$\int [f_1(x) \pm f_2(x) \pm ... \pm f_n(x)]\, dx = \int f_1(x)\, dx \pm \int f_2(x)\, dx \pm ... \pm \int f_n(x)\, dx$$

Theorem: The differential of the integral of a function is equal to the function itself,

i.e. $$\frac{d}{dx}\left(\int f(x)\, dx\right) = f(x)$$

ILLUSTRATIVE EXAMPLE

Example 1. Evaluate the following integrals:

(i) $\int \left(x - \frac{1}{x}\right)^3 dx$ (ii) $\int (4x^3 - 3\sin x + 2\sec^2 x + e^x)\, dx$

(iii) $\int \left(\cos x - \frac{2}{x} + a^x\right) dx$ (iv) $\int \frac{(x+2)(4x^2-5)}{x}\, dx$

Solution: (i) $\int \left(x - \frac{1}{x}\right)^3 dx = \int \left(x^3 - \frac{1}{x^3} - 3x + \frac{3}{x}\right) dx$

$$= \int \left(x^3 - x^{-3} - 3x + 3\cdot\frac{1}{x}\right) dx$$

$$= \int x^3 dx - \int x^{-3} dx - 3\int x\, dx + 3\int \frac{1}{x}\, dx$$

$$= \frac{x^4}{4} - \frac{x^{-2}}{-2} - 3\frac{x^2}{2} + 3\log x + C$$

$$= \frac{x^4}{4} + \frac{1}{2x^2} - \frac{3}{2}x^2 + 3\log x + C$$

(ii) $\int (4x^3 - 3\sin x + 2\sec^2 x + e^x)\, dx$

$$= 4\int x^3 dx - 3\int \sin x\, dx + 2\int \sec^2 x\, dx + \int e^x dx$$

$$= 4\frac{x^4}{4} - 3(-\cos x) + 2\tan x + e^x + C$$

$$= x^4 + 3\cos x + 2\tan x + e^x + C$$

(iii) $\int (\cos x - \frac{2}{x} + a^x)\, dx = \int \cos x\, dx - 2\int \frac{1}{x}\, dx + \int a^x dx$

$$= \sin x - 2\log x + \frac{a^x}{\log a} + C$$

(iv) $\int \frac{(x+2)(4x^2-5)}{x} dx = \int \left(\frac{4x^3 + 8x^2 - 5x - 10}{x} \right) dx$

$$= \int \left(4x^2 + 8x - 5 - \frac{10}{x} \right) dx$$

$$= 4\int x^2 dx + 8\int x dx - 5\int 1 dx - 10\int \frac{1}{x} dx$$

$$= 4\frac{x^3}{3} + 8\frac{x^2}{2} - 5x - 10\log x + C$$

$$= \frac{4}{3}x^3 + 4x^2 - 5x - 10\log x + C$$

Example 2. Find the integrals of the following functions:

(i) $\frac{\left(a^x + b^x\right)^2}{a^x b^x}$ (ii) $\tan^2 x$ (iii) $\sqrt{1 - \sin 2x}$

(iv) $(\tan x + \cot x)^2$ (v) $\sin^2 \frac{x}{2}$ (vi) $\frac{1}{1 + \sec x}$

Solution: (i) $\int \frac{(a^x + b^x)^2}{a^x b^x} dx = \int \frac{a^{2x} + b^{2x} + 2a^x b^x}{a^x b^x} dx$

$$= \int \left(\frac{a^x}{b^x} + \frac{b^x}{a^x} + 2 \right) dx$$

$$= \int \left(\frac{a}{b} \right)^x dx + \int \left(\frac{b}{a} \right)^x dx + 2\int 1 dx$$

$$= \frac{\left(\frac{a}{b}\right)^x}{\log\frac{a}{b}} + \frac{\left(\frac{b}{a}\right)^x}{\log\frac{b}{a}} + 2x + C, a \neq b$$

In particular, if $a = b$ then

$$\int \frac{(a^x + b^x)^2}{a^x b^x} = \int \frac{(2a^x)^2}{a^x a^x} = \int 4\,dx = 4x + C$$

(ii) $$\int \tan^2 x\,dx = \int (\sec^2 x - 1)dx$$

$$= \int \sec^2 x\,dx - \int 1 dx$$

$$= \tan x - x + C$$

(iii) $$\int \sqrt{1 - \sin 2x}\,dx = \int \sqrt{\cos^2 x + \sin^2 x - 2\sin x \cos x}\,dx$$

$$= \int (\cos x - \sin x)\,dx$$

$$= \int \cos x\,dx - \int \sin x\,dx$$

$$= \sin x + \cos x + C$$

(iv) $$\int (\tan x + \cot x)^2 dx = \int (\tan^2 x + \cot^2 x + 2\tan x \cot x)\,dx$$

$$= \int (\sec^2 x - 1 + \text{cosec}^2 x - 1 + 2)\,dx$$

$$= \int (\sec^2 x + \text{cosec}^2 x)dx$$

$$= \int \sec^2 x\,dx + \int \text{cosec}^2 x\,dx = \tan x - \cot x + C$$

(v) $$\int \sin^2 \frac{x}{2} dx = \int \frac{(1 - \cos x)}{2} dx$$

$$= \frac{1}{2}\int 1\,dx - \frac{1}{2}\int \cos x\,dx$$

$$= \frac{1}{2}x - \frac{1}{2}\sin x = \frac{1}{2}(x - \sin x)$$

(vi) $$\int \frac{1}{1 + \sec x} dx = \int \frac{1}{\sec x + 1} \times \frac{\sec x - 1}{\sec - 1} dx$$

$$= \int \frac{\sec x - 1}{\sec^2 x - 1} dx$$

$$= \int \frac{\sec x - 1}{\tan^2 x} dx$$

$$= \int \frac{\sec x}{\tan^2 x} dx - \int \frac{1}{\tan^2 x} dx$$

$$= \int \text{cosec}\, x \cot x\,dx - \int \cot^2 x\,dx$$

$$= \int \text{cosec}\, x \cot x\, dx - \int (\text{cosec}^2 x - 1)\, dx$$

$$= -\text{cosec}\, x - (-\cot x - x) + C$$

$$= -\text{cosec}\, x + \cot x + x + C$$

Example 3. Evaluate the following integrals:

(i) $\int \frac{\sec x + \tan x}{\sec x - \tan x} dx$ (ii) $\int \frac{\sin x}{1 + \sin x} dx$

(iii) $\int \frac{x^4}{1 + x^2} dx$ (iv) $\int \frac{1 - \cos 2x}{1 + \cos 2x} dx$

(v) $\int \frac{\cos 2x - \cos 2\alpha}{\cos x - \cos \alpha} dx$ (vi) $\int \frac{\sec x}{\sec x + \tan x} dx$

Solution: (i) $\int \frac{\sec x + \tan x}{\sec x - \tan x} dx = \int \frac{\sec x + \tan x}{\sec x - \tan x} \times \frac{\sec x + \tan x}{\sec x + \tan x} dx$

$$= \int \frac{\sec^2 x + \tan^2 x + 2 \sec x \tan x}{\sec^2 x - \tan^2 x} dx$$

$$= \int \frac{\sec^2 x + \sec^2 x - 1 + 2 \sec x \tan x}{1} dx$$

$$= 2\int \sec^2 x\, dx - \int 1 dx + 2\int \sec x \tan x\, dx$$

$$= 2 \tan x - x + 2 \sec x + C$$

(ii) $\int \frac{\sin x}{1 + \sin x} dx = \int \frac{1 + \sin x - 1}{1 + \sin x} dx$

$$= \int \left(1 - \frac{1}{1 + \sin x}\right) dx$$

$$= \int 1 dx - \int \frac{1}{1 + \sin x} \times \frac{1 - \sin x}{1 - \sin x} dx$$

$$= x - \int \frac{1 - \sin x}{\cos^2 x} dx$$

$$= x - \int \sec^2 x\, dx + \int \sec x \tan x\, dx$$

$$= x - \tan x + \sec x + C$$

(iii) $\int \frac{x^4}{1 + x^2} dx = \int \frac{x^4 + 1 - 1}{1 + x^2} dx$

$$= \int \frac{(x^2 - 1)(x^2 + 1) + 1}{1 + x^2} dx$$

$$= \int \left(x^2 - 1 + \frac{1}{1 + x^2}\right) dx$$

$$= \int x^2 dx - \int 1 dx + \int \frac{1}{1+x^2} dx$$

$$= \frac{x^3}{3} - x + \tan^{-1} x + C$$

(iv) $$\int \frac{1-\cos 2x}{1+\cos 2x} dx = \int \frac{1-(1-2\sin^2 x)}{1+(2\cos^2 x - 1)} dx$$

$$= \int \frac{\sin^2 x}{\cos^2 x} dx$$

$$= \int \tan^2 x \, dx$$

$$= \int (\sec^2 x - 1) \, dx$$

$$= \int \sec^2 x \, dx - \int 1 \, dx$$

$$= \tan x - x + C$$

(v) $$\int \frac{\cos 2x - \cos 2\alpha}{\cos x - \cos \alpha} dx = \int \frac{(2\cos^2 x - 1) - (2\cos^2 \alpha - 1)}{\cos x - \cos \alpha} dx$$

$$= \int \frac{2(\cos^2 x - \cos^2 \alpha)}{\cos x - \cos \alpha} dx$$

$$= 2\int (\cos x + \cos \alpha) \, dx$$

$$= 2\left[\int (\cos x \, dx + \int \cos \alpha \, dx\right]$$

$$= 2\left[\int \cos x \, dx + \cos \alpha \int 1 \, dx\right]$$

$$= 2(\sin x + x \cos \alpha) + C$$

(vi) $$\int \frac{\sec x}{\sec x + \tan x} dx = \int \frac{\sec x + \tan x - \tan x}{\sec x + \tan x} dx$$

$$= \int \left(1 - \frac{\tan x}{\sec x + \tan x}\right) dx$$

$$= \int 1 \, dx - \int \frac{\tan x}{\sec x + \tan x} dx$$

$$= x - \int \frac{\sin x}{1 + \sin x} dx$$

$$= x - \int \frac{\sin x(1 - \sin x)}{1 - \sin^2 x} dx$$

$$= x - \int \frac{\sin x - \sin^2 x}{\cos^2 x}\, dx$$

$$= x - \int (\sec x \tan x - \tan^2 x)\, dx$$

$$= x - \int (\sec x \tan x - \sec^2 x + 1)\, dx$$

$$= x - \sec x + \tan x - x + C$$

$$= \tan x - \sec x + C$$

Example 4: Evalute the following integrals :

(i) $\int \frac{5\cos^3 x + 6\sin^3 x}{\sin^2 x \cos^2 x}\, dx$ (ii) $\int \cos^{-1}(\sin x)\, dx$

(iii) $\int \tan^{-1}(\sec x + \tan x)\, dx$ (iv) $\int \cot^{-1}\left(\frac{\sin 2x}{1 - \cos 2x}\right) dx$

Solution: (i) $\int \frac{5\cos^3 x + 6\sin^3 x}{\sin^2 x \cos^2 x}\, dx = \int \left(\frac{5 \cdot \cos x}{\sin^2 x} + \frac{6 \cdot \sin x}{\cos^2 x}\right) dx$

$$= \int (5 \operatorname{cosec} x \cot x + 6 \sec x \tan x)\, dx$$

$$= 5\int \operatorname{cosec} x \cot x\, dx + 6\int \sec x \tan x\, dx$$

$$= -5 \operatorname{cosec} x + 6 \sec x + C$$

(ii) $\int \cos^{-1}(\sin x)\, dx = \int \cos^{-1}\left(\cos\left(\frac{\pi}{2} - x\right)\right) dx$

$$= \int \left(\frac{\pi}{2} - x\right) dx$$

$$= \frac{\pi}{2}\int 1\, dx - \int x\, dx$$

$$= \frac{\pi}{2} \cdot x - \frac{x^2}{2} + C$$

(iii) $\int \tan^{-1}(\sec x + \tan x)\, dx = \int \tan^{-1}\left(\frac{1 + \sin x}{\cos x}\right) dx$

$$= \int \tan^{-1}\left(\frac{\cos^2 \frac{x}{2} + \sin^2 \frac{x}{2} + 2\sin\frac{x}{2}\cos\frac{x}{2}}{\cos^2\frac{x}{2} - \sin^2\frac{x}{2}}\right) dx$$

$$= \int \tan^{-1}\left(\frac{(\cos\frac{x}{2}+\sin\frac{x}{2})^2}{\cos^2\frac{x}{2}-\sin^2\frac{x}{2}}\right)dx$$

$$= \int \tan^{-1}\left(\frac{\cos\frac{x}{2}+\sin\frac{x}{2}}{\cos\frac{x}{2}-\sin\frac{x}{2}}\right)dx$$

$$= \int \tan^{-1}\left(\frac{1+\tan\frac{x}{2}}{1-\tan\frac{x}{2}}\right)dx$$

$$= \int \tan^{-1}\left[\tan\left(\frac{\pi}{4}+\frac{x}{2}\right)\right]dx = \int\left(\frac{\pi}{4}+\frac{x}{2}\right)dx$$

$$= \frac{\pi}{4}\cdot x+\frac{1}{2}\cdot\frac{x^2}{2}+C$$

$$= \frac{x}{4}(\pi+x)+C$$

(iv) $\int \cot^{-1}\left(\frac{\sin 2x}{1-\cos 2x}\right)dx = \int \cot^{-1}\left(\frac{2\sin x\cos x}{2\sin^2 x}\right)dx$

$$= \int \cot^{-1}(\cot x)\,dx$$

$$= \int x\,dx$$

$$= \frac{x^2}{2}+C$$

Example 5: If $f'(x) = (2x^3+3)^2$ and $f(1) = 0$, find $f(x)$

Solution: given, $f'(x) = (2x^3+3)^2 = 4x^6+9+12x^3$

$$\therefore f(x) = \int f'(x)\,dx$$

$$= \int (4x^6+12x^3+9)\,dx$$

$$= 4\cdot\frac{x^7}{7}+12\cdot\frac{x^4}{4}+9x+C$$

$$\Rightarrow f(x) = \frac{4}{7}x^7 + 3x^4 + 9x + C$$

But $$f(1) = 0 \Rightarrow \frac{4}{7} + 3 + 9 + C = 0$$

$$\Rightarrow C = -\frac{88}{7}$$

$$\therefore f(x) = \frac{4}{7}x^7 + 3x^4 + 9x - \frac{88}{7}$$

Example 6 : Find the relation connecting distance and time when the relation between velocity and time is given by $v = \dfrac{t^2-1}{t^2}$, and it is given that $S = 6$ when $t = 1$

Solution: Given, $$V = \frac{ds}{dt} = \frac{t^2-1}{t^2} = 1 - t^{-2}$$

$$\therefore S = \int\left(\frac{ds}{dt}\right)dt = \int(1-t^{-2})\,dt$$

$$= t + t^{-1} + C = t + \frac{1}{t} + C$$

$$\Rightarrow S = t + \frac{1}{t} + C$$

But $S = 6$ when $t = 1$

$$\therefore 1 + \frac{1}{1} + C = 6$$

$$\Rightarrow C = 4$$

$$\therefore S = t + \frac{1}{t} + C$$

Example 7: Find the value of the following integrals:

(i) $\displaystyle\int\left(\frac{3}{\sqrt{1-x^2}} + \frac{7}{x\sqrt{x^2-1}} + \cot^2 x\right)dx$ (ii) $\displaystyle\int\frac{e^{\log x}}{x}\,dx$

Solution: (i) $\displaystyle\int\left(\frac{3}{\sqrt{1-x^2}} + \frac{7}{x\sqrt{x^2-1}} + \cot^2 x\right)dx$

$$= 3\int \frac{dx}{\sqrt{1-x^2}} + 7\int \frac{dx}{x\sqrt{x^2-1}} + \int (\text{cosec}^2 x - 1)\, dx$$

$$= 3 \sin^{-1} x + 7 \sec^{-1} x - \cot x - x + C$$

(ii) $\int \frac{e^{\log x}}{x}\, dx = \int \frac{x}{x}\, dx = \int 1\, dx = x + c \left[\because e^{\log x} = x\right]$

Example 8: Integrate the following functions *w.r.t.x* :

(i) $\tan^{-1}\left[\sqrt{\frac{1-\sin x}{1+\sin x}}\right]$ (ii) $\frac{1}{\sin^2 x \cos^2 x}$ (iii) $(e^{3a\log x} + e^{x \log a})$

Solution: (i) $\int tan^{-1}\left[\sqrt{\frac{1-\sin x}{1+\sin x}}\right] dx = \int tan^{-1}\sqrt{\frac{1-\cos\left(\frac{\pi}{2}-x\right)}{1+\cos\left(\frac{\pi}{2}-x\right)}}\, dx$

$$= \int tan^{-1}\sqrt{\frac{2\sin^2\left(\frac{\pi}{4}-\frac{x}{2}\right)}{2\cos^2\left(\frac{\pi}{4}-\frac{x}{2}\right)}}\, dx$$

$$= \int \tan^{-1}\left[\tan\left(\frac{\pi}{4}-\frac{x}{2}\right)\right] dx$$

$$= \int \left(\frac{\pi}{4}-\frac{x}{2}\right) dx$$

$$= \frac{\pi}{4}\cdot x - \frac{x^2}{4} + C$$

$$= \frac{x}{4}(\pi - x) + C$$

(ii) $\int \frac{1}{\sin^2 x \cos^2 x}\, dx = \int \frac{\sin^2 x + \cos^2 x}{\sin^2 x \cos^2 x}\, dx = \int (\sec^2 x + \text{cosec}^2 x)\, dx$

$$= \tan x - \cot x + C$$

(iii) $\int (e^{3a \log x} + e^{x \log a})\, dx = \int \left(e^{\log x^{3a}} + e^{\log a^x}\right) dx$

$$= \int (x^{3a} + a^x)\, dx \qquad \left[\because e^{\log x} = x\right]$$

$$= \frac{x^{3a+1}}{3a+1} + \frac{a^x}{\log_e^a} + C$$

Example 9. Integrate the following functions $w.r.t.x$:

(i) $\int \frac{1-\cos 2x}{1+\cos 2x}\, dx$ (ii) $\int \frac{\sin 2x}{\sin x}\, dx$ **(I.U. 2007–08)**

Solution: (i)
$$\int \frac{1-\cos 2x}{1+\cos 2x}\, dx = \int \frac{1-(1-2\sin^2 x)}{1+(2\cos^2 x-1)}\, dx$$
$$= \int \frac{\sin^2 x}{\cos^2 x}\, dx$$
$$= \int \tan^2 x\, dx$$
$$= \int (\sec^2 x - 1)\, dx$$
$$= \int \sec^2 x\, dx - \int 1\, dx$$
$$= \tan x - x + C$$

(ii)
$$\int \frac{\sin 2x}{\sin x}\, dx = \int \frac{2\sin x \cos x}{\sin x}\, dx$$
$$= 2\int \cos x\, dx$$
$$= 2\sin x + C$$

EXERCISE 4.1

1. Evaluate:

(i) $\int \left(\sqrt{x} - \frac{1}{\sqrt{x}}\right)^2 dx$ (ii) $\int (4x^3 + 3x^2 + 2x + 1)\, dx$

(iii) $\int \frac{(x^2+3)(x^2-4)}{x^2}\, dx$ (iv) $\int \left(\tan^2 x - 7x^2 + \frac{1}{x}\right) dx$

(v) $\int \frac{(x+a)^2}{\sqrt{x}}\, dx$

2. Evaluate:

(i) $\int \frac{2\sin x + 5}{\cos^2 x}\, dx$ (ii) $\int \left(\sqrt{\frac{a}{x}} + \sqrt{\frac{x}{a}}\right) dx$

(iii) $\int (\sec^2 x - 5\operatorname{cosec}^2 x + 7)\, dx$ (iv) $\int \cot^2 x\, dx$

(v) $\int (10^x + 10)\, dx$

3. Evaluate the indefinite integrals:

(i) $\int \frac{\sin 2x}{\sin x}\, dx$ (ii) $\int \sqrt{1+\sin 2x}\, dx$

(iii) $\int \frac{\sin^2 x}{1+\cos x}\, dx$ (iv) $\int \frac{1}{1+\cos 2x}\, dx$

(v) $\int \frac{\cos 2x}{\cos^2 x \sin^2 x}\, dx$

4. Evaluate:

(i) $\int \frac{\sin x+\cos x}{\sqrt{1+\sin 2x}}\, dx$ (ii) $\int \frac{\cos x-\cos 2x}{1-\cos x}\, dx$

(iii) $\int (e^x + a^x)dx$ (iv) $\int (2\tan x - 3\cot x)^2\, dx$

(v) $\frac{dx}{1+\sin x}$

5. Evaluate :

(i) $\int \cos x \operatorname{cosec}^2 x\, dx$ (ii) $\int \frac{\sin^3 x-\cos^3 x}{\sin^2 x \cos^2 x}\, dx$

(iii) $\int \tan^{-1}\left(\sqrt{\frac{1-\cos 2x}{1+\cos 2x}}\right) dx$ (iv) $\int \frac{\sin^6 x+\cos^6 x}{\sin^2 x \cos^2 x}\, dx$

(v) $\int \left(e^{2\log x} + e^{\log \cos x}\right) dx$

6. If $f'(x) = x^2 - \frac{1}{x^2}$ and $f(1) = \frac{1}{3}$, find $f(x)$

7. Find the relation between distance and time , when the relation between velocity and time is given by $v = t^2 - 4t + 3$ and distance = 5 when time, $t = 0$

8. Find the equation of the curve whose slope at any point $P(x,y)$ of the curve is $\cos x$ and the curve passes through the point $\left(\frac{\pi}{2}, 1\right)$

ANSWERS

1. (i) $\frac{x^2}{2} + \log x - 2x + C$ (ii) $x^4 + x^3 + x^2 + x + C$

(iii) $\frac{x^3}{3} - x + \frac{12}{x} + C$ (iv) $\tan x - x - \frac{7}{3}x^3 + \log x + C$

(v) $\frac{2}{5}x^{5/2} + \frac{4}{3}a\, x^{3/2} + 2a^2 x^{1/2} + C$

2. (i) $2\sec x + 5\tan x + C$ (ii) $2\sqrt{ax} + \frac{2}{3\sqrt{a}}x^{3/2} + C$

(iii) $\tan x + 5\cot x + 7x + C$ (iv) $-\cot x - x + C$

(v) $\frac{10^x}{\log_e 10} + 10x + C$

3. (i) $2\sin x + C$ (ii) $\sin x - \cos x + C$

(iii) $x - \sin x + C$ (iv) $\frac{1}{2}\tan x + C$

(v) $-\cot x - \tan x + C$

4. (i) x+C (ii) $2\sin x + x + C$

(iii) $e^x + \frac{a^x}{\log_e a} + C$ (iv) $4\tan x - 9\cot x - 25x + C$

(v) $\tan x - \sec x + C$

5. (i) $-\operatorname{cosec} x + C$ (ii) $\sec x + \operatorname{cosec} x + C$

(iii) $\frac{x^2}{2} + C$ (iv) $\tan x - \cot x - 3x + C$

(v) $\frac{x^3}{3} + \sin x + C$

6. $\frac{1}{3}x^3 + \frac{1}{x} - 1$ **7.** $S = \frac{t^3}{3} - 2t^2 + 3t + 5$

8. $y = \sin x$

4.5 METHODS OF INTEGRATION

There are two main methods of integration:

(a) Integration by Substitution

(b) Integration by Parts.

4.6 INTEGRATION BY SUBSTITUTION OR CHANGE OF VARIABLE

Some function can be integrated directly by the use of standard integrals as studied earlier while there exist some integrals which cannot be integrated directly but can be reduced to the standard integrals by proper substitution i.e. by the introduction of new variable. The method of evaluating an integral by reducing it to standard form by a substitution is called integration by substitution.

Rule I: $\int f(x)\,dx = F(x) + C$

then, $\int f(ax+b)\,dx = \frac{F(ax+b)}{a} + C, \quad a \neq 0$

Proof: Put $ax + b = t \Rightarrow a\,dx = dt \Rightarrow dx = \frac{dt}{a}$

$$\therefore \quad \int f(ax+b)\,dx = \int f(t)\cdot\frac{1}{a}\,dt$$

$$= \frac{1}{a}\int f(t)\,dt$$

$$= \frac{1}{a}F(t) + C$$

$$= \frac{F(ax+b)}{a} + C, \quad a \neq 0$$

Hence, we have the following important rule, "if the integral of a function of x is known,then if x is multiplied by a constant and to the product is added another constant,the integral is of the same function but it is divided by the coefficient of *x*."

4.7 SOME RESULT OF SUBSTITUTION

1. $\int (ax+b)^n\,dx = \frac{1}{a}\cdot\frac{(ax+b)^{n+1}}{n+1} + C$

2. $\int \frac{1}{ax+b}\,dx = \frac{1}{a}\log(ax+b) + C$

3. $\int \sin(ax+b)\,dx = \frac{-\cos(ax+b)}{a} + C$

4. $\int \cos(ax+b)\,dx = \frac{\sin(ax+b)}{a} + C$

5. $\int \sec^2(ax+b)\,dx = \frac{\tan(ax+b)}{a} + C$

6. $\int e^{ax}\,dx = \frac{e^{ax}}{a} + C$

7. $\int \sec(ax+b)\tan(ax+b)\,dx = \frac{\sec(ax+b)}{a} + C$

8. $\int \operatorname{cosec}(ax+b)\cot(ax+b)\,dx = \frac{-\operatorname{cosec}(ax+b)}{a} + C$

9. $\int \operatorname{cosec}^2(ax+b)\,dx = \frac{-\cot(ax+b)}{a} + C$

10. $\int a^{mx+b}\;C = \frac{a^{mx+b}}{m\log_e a} + C$

ILLUSTRATIVE EXAMPLE

Example 1. Evaluate:

(i) $\int (5x-2)^3\, dx$ (ii) $\int \frac{dx}{2+3x}$

(iii) $\int \sqrt{3x-2}\, dx$ (iv) $\int \frac{dx}{\sqrt[3]{4x+5}}$

(v) $\int 5^{-2x} e^{-2x} dx$

Solution: (i) $\int (5x-2)^3 dx$, Let $5x-2=t \Rightarrow 5\,dx = dt \Rightarrow dx = \frac{dt}{5}$

$$\therefore \quad \int (5x-2)^3\, dx = \int \frac{t^3}{5}\, dt = \frac{1}{5}\int t^3\, dt$$

$$= \frac{1}{5}\frac{t^4}{4} + C = \frac{(5x-2)^4}{20} + C$$

Alternatively:

$$\int (5x-2)^3\, dx = \frac{(5x-2)^{3+1}}{5\cdot(3+1)} + C$$

$$= \frac{(5x-2)^4}{20} + C \qquad \text{[By Rule I]}$$

(ii) $\int \frac{dx}{2+3x}$, Let $2+3x=t \Rightarrow 3dx = dt \Rightarrow dx = \frac{dt}{3}$

$$\therefore \quad \int \frac{dx}{2+3x} = \int \frac{1}{3}\frac{dt}{t} = \frac{1}{3}\log t + C = \frac{1}{3}\log(2+3x) + C$$

Alternatively: $\int \frac{dx}{2+3x} = \frac{1}{3}\log(2+3x) + C$

(iii) $$\int \sqrt{3x-2}\, dx = \int (3x-2)^{\frac{1}{2}}\, dx$$

Let $3x-2 = t \Rightarrow dx = \frac{dt}{3}$

$$\therefore \quad \int \sqrt{3x-2}\, dx = \int t^{\frac{1}{2}} \cdot \frac{dt}{2}$$

$$= \frac{1}{2}\int t^{\frac{1}{2}}\, dt = \frac{1}{2}\frac{t^{\frac{3}{2}}}{\frac{3}{2}} + C$$

$$= \frac{(3x-2)^{\frac{3}{2}}}{3} + C$$

(iv) $$\int \frac{dx}{\sqrt[3]{4x+5}} = \int (4x+5)^{\frac{-1}{3}}\, dx$$

Let $$4x+5 = t \;\Rightarrow\; 4\,dx = dt \;\Rightarrow\; dx = \frac{dt}{4}$$

$$\therefore \quad \int \frac{dx}{3\sqrt{4x+5}} = \int \frac{t^{\frac{-1}{3}}}{4}\, dt$$

$$= \frac{1}{4}\frac{t^{\frac{2}{3}}}{\frac{2}{3}} + C = \frac{3}{8}(4x+5)^{\frac{2}{3}} + C$$

(v) $$\int 5^{-2x} e^{-2x}\, dx = \int (5e)^{-2x}\, dx$$

$$= \frac{(5e)^{-2x}}{\log 5e \cdot (-2)} + C = -\frac{(5e)^{-2x}}{2\log 5e} + C$$

Example 2. Evaluate:

(i) $\int \frac{1}{\sqrt{2x+3}+\sqrt{2x-3}}\, dx$ (ii) $\int \frac{2x}{(2x+1)^2}\, dx$

(iii) $\int \frac{x+2}{\sqrt{3x+1}}\, dx$ (iv) $\int \frac{x^3}{2x+1}\, dx$

(v) $\int \sqrt{e^x}\, dx$ (vi) $\int 10^{2x+1}\, dx$

Solution : (i)

$$\int \frac{1}{\sqrt{2x+3}+\sqrt{2x-3}}\, dx = \int \frac{1}{\sqrt{2x+3}+\sqrt{2x-3}} \times \frac{\sqrt{2x+3}-\sqrt{2x-3}}{\sqrt{2x+3}-\sqrt{2x-3}}\, dx$$

$$= \int \frac{\sqrt{2x+3}-\sqrt{2x-3}}{(2x+3)-(2x-3)}\, dx = \int \frac{\left(\sqrt{2x+3}-\sqrt{2x-3}\right)}{6}\, dx$$

$$= \frac{1}{6}\left[\int (2x+3)^{\frac{1}{2}}\, dx - \int (2x-3)^{\frac{1}{2}}\, dx\right]$$

$$= \frac{1}{6}\left[\frac{(2x+3)^{\frac{3}{2}}}{\frac{3}{2}\cdot 2} - \frac{(2x-3)^{\frac{3}{2}}}{\frac{3}{2}\cdot 2}\right] + C$$

$$= \frac{1}{18}\left[(2x+3)^{\frac{3}{2}} - (2x-2)^{\frac{3}{2}}\right] + C$$

(ii) $$\int \frac{2x\,dx}{(2x+1)^2} = \int \frac{(2x+1)-1}{(2x+1)^2}\,dx$$

$$= \int \left[\frac{1}{(2x+1)} - \frac{1}{(2x+1)^2}\right] dx$$

$$= \frac{\log(2x+1)}{2} - \frac{(2x+1)^{-1}}{-1\cdot 2} + C$$

$$= \frac{\log(2x+1)}{2} + \frac{1}{2(2x+1)} + C$$

(iii) $$\int \frac{(x+2)\,dx}{\sqrt{3x+1}} = \int \frac{\frac{1}{3}(3x+1) + \frac{5}{3}}{(3x+1)^{\frac{1}{2}}}\,dx$$

$$= \int \left[\frac{(3x+1)^{1/2}}{3} + \frac{5}{3}(3x+1)^{-1/2}\right] dx$$

$$= \frac{1}{3}\,\frac{(3x+1)^{\frac{3}{2}}}{\frac{3}{2}\cdot 3} + \frac{5}{3}\,\frac{(3x+1)^{\frac{1}{2}}}{3\cdot\frac{1}{2}} + C$$

$$= \frac{2}{27}(3x+1)^{\frac{3}{2}} + \frac{10}{9}\sqrt{3x+1} + C$$

(iv) $$\int \frac{x^3}{2x+1}\,dx = \int \left[\frac{x^2}{2} - \frac{x}{4} + \frac{1}{8} - \frac{1}{8(2x+1)}\right] dx$$

$$= \frac{1}{2}\int x^2 dx - \frac{1}{4}\int x\,dx + \frac{1}{8}\int 1\,dx - \frac{1}{8}\int \frac{1}{2x+1}\,dx$$

$$= \frac{1}{2}\cdot\frac{x^3}{3} - \frac{1}{4}\frac{x^2}{2} + \frac{1}{8}\cdot x - \frac{1}{8}\frac{\log(2x+1)}{2} + C$$

$$= \frac{x^3}{6} - \frac{x^2}{8} + \frac{x}{8} - \frac{1}{16}\cdot\log(2x+1) + C$$

(v) $$\int \sqrt{e^x}\;dx = \int (e^x)^{\frac{1}{2}}\,dx = \int e^{\frac{x}{2}}\,dx$$

Let $\dfrac{x}{2} = t \Rightarrow dx = 2\,dt$

$$\therefore \int \sqrt{e^x}\ dx = 2\int e^t dt = 2\ e^t + C = 2\ e^{\frac{x}{2}} + C = 2\sqrt{e^x} + C$$

(vi) $\int 10^{2x+1} dx,$

Let $2x + 1 = t \Rightarrow 2dx = dt \Rightarrow dx = \dfrac{dt}{2}$

$$\therefore \int 10^{2x+1} dx = \frac{1}{2}\int 10^t\, dt = \frac{1}{2}\frac{10^t}{\log 10} + C$$

$$= \frac{10^{2x+1}}{2\log 10} + C$$

Example 3. Evaluate :

(i) $\int \sin(3x+5)\ dx$

(ii) $\int \sin 3x \cos 5x\ dx$

(iii) $\int \operatorname{cosec}^2(9x-7)\ dx$

(iv) $\int \sqrt{1-\sin x}\ dx$

Solution: (i) $\int \sin(3x+5)\ dx$, Let $3x + 5 = t \Rightarrow dx = \dfrac{dt}{3}$

$$\therefore \int \sin(3x+5)\ dx\ \frac{1}{3}\int \sin t \cdot dt = -\frac{1}{3}\cos t + C$$

$$= -\frac{1}{3}\cos(3x+5) + C$$

(ii) $\int \sin 3x \cos 5x\ dx = \dfrac{1}{2}\int 2\sin 3x \cos 5x\ dx$

$$= \frac{1}{2}\int \left(\sin(3x+5x) + \sin(3x-5x)\right) dx$$

$$= \frac{1}{2}\int \left(\sin 8x - \sin 2x\right) dx$$

$$= \frac{1}{2}\left[\frac{-\cos 8x}{5} + \frac{\cos 2x}{2}\right] + C$$

(iii) $\int \operatorname{cosec}^2(9x-7)\ dx = \dfrac{-\cot(9x-7)}{9} + C$ [By Rule I]

(iv) $\int \sqrt{1-\sin x}\ dx = \int \sqrt{\sin^2\dfrac{x}{2} + \cos^2\dfrac{x}{2} - 2\sin\dfrac{x}{2}\cos\dfrac{x}{2}}\ dx$

$$= \int \sqrt{\left(-\sin\frac{x}{2}+\cos\frac{x}{2}\right)^2}\, dx$$

$$= \int \left(-\sin\frac{x}{2}+\cos\frac{x}{2}\right) dx$$

$$= \frac{\cos\frac{x}{2}}{\frac{1}{2}} - \frac{\sin\frac{x}{2}}{\frac{1}{2}} + C$$

$$= 2\left(\cos\frac{x}{2}+\sin\frac{x}{2}\right)+C$$

Example 4. Evalute:

(i) $\int \sin^3 x\, dx$ (ii) $\int \cos^2(2x-5)\, dx$

(iii) $\int \cos^4 2x\ dx$ (iv) $\int \frac{\sin 4x}{\sin x}\, dx$

Solution: (i) $\int \sin^3 x\, dx = \int \frac{3\sin x - \sin 3x}{4}\, dx \qquad [\because \sin 3x = 3\sin x - 4\sin^3 x]$

$$= \frac{3}{4}\int \sin x\, dx - \frac{1}{4}\int \sin 3x\, dx$$

$$= \frac{3}{4}\frac{(-\cos x)}{1} - \frac{1}{4}\frac{(-\cos 3x)}{3} + C$$

$$= \frac{1}{12}(\cos 3x - 9\cos x) + C$$

(ii) $\int \cos^2(2x-5)\, dx = \int \frac{[1+\cos(4x-10)]}{2}\, dx \quad [\because \cos 2x = 2\cos^2 x - 1]$

$$= \frac{1}{2}\left[x + \frac{\sin(4x-10)}{4}\right] + C$$

(iii) $\int \cos^4 2x\ dx = \int (\cos^2 2x)^2\, dx \qquad [\because \cos 4x = 2\cos^2 2x - 1]$

$$= \int \left(\frac{1+\cos 4x}{2}\right)^2 dx$$

$$= \int \left(\frac{1+\cos^2 4x + 2\cos 4x}{4} \right) dx$$

$$= \frac{1}{4}\int \left(1 + \frac{1+\cos 8x}{2} + 2\cos 4x \right) dx$$

$$= \frac{1}{4}\int \left(\frac{3}{2} + \frac{1}{2}\cdot \cos 8x + 2\cos 4x \right) dx$$

$$= \frac{1}{4}\left[\frac{3}{2}x + \frac{1}{2}\frac{\sin 8x}{8} + \frac{2\sin 4x}{4} \right] + C$$

$$= \frac{1}{64}[24x + \sin 8x + 8\sin 4x] + C$$

(iv) $$\int \frac{\sin 4x}{\sin x}\, dx = \int \frac{2\sin 2x \cos 2x}{\sin x}\, dx$$

$$= \int \frac{2.2\sin x \cos x \cos 2x}{\sin x}\, dx$$

$$= 2\int 2\cos x \cos 2x\, dx$$

$$= 2\int [\cos(x+2x) + \cos(x-2x)\ ]\, dx$$

$$= 2\int (\cos 3x + \cos x)\, dx$$

$$= 2\left(\frac{\sin 3x}{3} + \frac{\sin x}{1} \right) + C$$

$$= \frac{2}{3}(\sin 3x + 3\sin x) + C$$

Example 5. Find $\int \sin mx \cos nx\, dx$ where m, n are positive integers and $m \neq n$. What happens if $m = n$?

Solution: $$\int \sin mx \sin nx\, dx = \frac{1}{2}\int 2\sin mx \sin nx\, dx$$

$$= \frac{1}{2}\int [\cos(mx - nx) - \cos(mx + nx)]\, dx$$

$$= \frac{1}{2}\int [\cos(m-n)x - \cos(m+n)x]\, dx$$

$$= \frac{1}{2}\left[\frac{\sin(m-n)x}{m-n} - \frac{\sin(m+n)x}{m+n}\right] + C, m \neq n$$

In particular, if $m = n$, then

$$\int \sin mx \sin nx \, dx = \int \sin^2 mx \, dx = \int \frac{1-\cos 2mx}{2} dx$$

$$= \frac{1}{2}\left[x - \frac{\sin 2mx}{2m}\right] + C$$

Rule II : $\int [f(x)]^n f'(x)\, dx = \frac{[f(x)]^{n+1}}{n+1} + C, n \neq -1$

Proof: Let $f(x) = t \Rightarrow f'(x)\, dx = dt$

$$\therefore \quad \int [f(x)]^n f'(x)\, dx = \int t^n dt = \frac{t^{n+1}}{n+1} + C = \frac{[f(x)]^{n+1}}{n+1} + C, n \neq -1$$

Hence, "if the integrand is the product of a power of a function $f(x)$ and its derivative $f'(x)$, then the integral is obtaind by increasing the index of power of $f(x)$ by 1and dividing the result by new index, and adding constant of integration C".

ILLUSTRATIVE EXAMPLE

Example 6. Evaluate the following integrals:

(i) $\int x^3(1+x^4)^3\, dx$ (ii) $\int (2x+4)\sqrt{x^2+4x+3}\, dx$

(iii) $\int \frac{4x+1}{\sqrt{2x^2+x-7}} dx$

Solution: (i) $\int x^3(1+x^4)^3\, dx = \frac{1}{4}\int 4x^3 \cdot (1+x^4)^3\, dx$

Let $1 + x^4 = t \Rightarrow 4x^3 dx = dt$

$$\therefore \quad \int x^3(1+x^4)^3\, dx = \frac{1}{4}\int t^3\, dx = \frac{1}{4}\frac{t^4}{4} + C$$

$$= \frac{(1+x^4)^4}{16} + C$$

Alternatively,

$$\int x^3(1+x^4)^3\, dx = \frac{1}{4}\int (1+x^4)^3 \cdot 4x^3 dx \quad [\text{use} \int [f(x)]^n f'(x)\, dx\,]$$

$$= \frac{1}{4}\frac{(1+x^4)^4}{4}+C$$

$$= \frac{(1+x^4)^4}{16}+C$$

(ii) $\int (2x+4)\sqrt{x^2+4x+3}\,dx = \int (x^2+4x+3)^{1\backslash 2}\cdot(2x+4)\,dx$

$$\left[\because \frac{d}{dx}(x^2+4x+3)=(2x+4)\right]$$

$$= \frac{(x^2+4x+3)^{3/2}}{\frac{3}{2}}+C$$

$$= \frac{2}{3}\cdot(x^2+4x+3)^{3/2}+C$$

(iii) $\int \frac{4x+1}{\sqrt{2x^2+x-7}}\,dx = \int (2x^2+x-7)^{-1/2}(4x+1)\,dx$

$$= \frac{(2x^2+x-7)^{-1/2}}{\frac{1}{2}}+C$$

$$\left[\because \frac{d}{dx}(2x^2+x-7)=4x+1\right]$$

$$= 2\sqrt{2x^2+x-7}+C$$

Example 7. Evaluate:

(i) $\int \sec^3 x \tan x\,dx$ (ii) $\int \frac{\cot x}{\sqrt{\sin x}}\,dx$

(iii) $\int \tan^4 x\,dx$ (iv) $\int \frac{e^{4x}}{(1+e^{4x})^2}\,dx$

Solution: (i) $\int \sec^3 x \tan x\,dx = \int (\sec x)^2\cdot(\sec x \tan x)\,dx$

$$= \frac{(\sec x)^3}{3}+C$$

$$= \frac{\sec^3 x}{3}+C$$

(ii)
$$\int \frac{\cot x}{\sqrt{\sin x}}\,dx = \int \frac{\cos x}{\sin x\sqrt{\sin x}}\,dx$$

$$= \int (\sin x)^{-3/2} \cdot \cos x\, dx$$

$$= \frac{(\sin x)^{\frac{-3}{2}+1}}{\frac{-3}{2}+1} + C$$

$$= \frac{(\sin x)^{-1/2}}{-1/2} + C$$

$$= \frac{-2}{\sqrt{\sin x}} + C$$

(iii)
$$\int \tan^4 x\, dx = \int \tan^2 x \cdot \tan^2 x\, dx$$

$$= \int \tan^2 x(\sec^2 x - 1)\, dx$$

$$= \int (\tan^2 x \sec x^2 - \tan^2 x)\, dx$$

$$= \int (\tan^2 x \sec^2 x - \sec^2 x + 1)\, dx$$

$$= \int (\tan x)^2 \cdot \sec^2 x\, dx - \int \sec^2 x\, dx + \int 1 \cdot dx$$

$$= \frac{(\tan x)^3}{3} - \tan x + x + C$$

$$= \frac{\tan^3 x}{3} - \tan x + x + C$$

(iv)
$$\int \frac{e^{4x}}{(1+e^{4x})^2}\,dx = \frac{1}{4}\int (1+e^{4x})^{-2} \cdot (4e^{4x})\, dx$$

$$\left[\because \frac{d}{dx}(1+e^{4x}) = 4e^{4x}\right]$$

$$= \frac{1}{4}\,\frac{(1+e^{4x})^{-1}}{-1} + C$$

$$= -\frac{1}{4(1+e^{4x})} + C$$

Example 8. Evaluate:

(i) $\int \sec x \cdot \log(\sec x + \tan x)\, dx$ (ii) $\int \frac{(x+1)(x+\log x)}{2x}\, dx$

(iii) $\int \frac{(a+b\sin^{-1} x)^n}{\sqrt{1-x^2}}\, dx$ (iv) $\int \frac{(3\tan^2 x + 2)\sec^2 x}{(\tan^3 x + 2\tan x + 5)^2}\, dx$

Solution: (i) $\int \log(\sec x + \tan x) \cdot \sec x\, dx = \frac{[\log(\sec x + \tan x)]^2}{2} + C$

$$\left[\because \frac{d}{dx}\log(\sec x + \tan x) \frac{\sec x \tan x + \sec^2 x}{\sec x + \tan x} = \sec x\right]$$

(ii) $$\int \frac{(x+1)(x+\log x)}{2x}\, dx = \frac{1}{2}\int \left(1 + \frac{1}{x}\right) \cdot (x + \log x)\, dx$$

$$= \frac{1}{2}\int (x+\log x)\left(1+\frac{1}{x}\right) dx$$

$$= \frac{1}{2}\frac{(x+\log x)^2}{2} + C$$

$$= \frac{(x+\log x)^2}{4} + C \qquad \left[\because \frac{d}{dx}(x+\log x) = 1 + \frac{1}{x}\right]$$

(iii) $$\int \frac{(a+b\sin^{-1} x)^n}{\sqrt{1-x^2}}\, dx = \frac{1}{b}\int (a+b\sin^{-1} x)^n \cdot \frac{b}{\sqrt{1-x^2}}\, dx$$

$$= \frac{(a+b\sin^{-1} x)^{n+1}}{b(n+1)} + C$$

$$\left[\because \frac{d}{dx}(a+b\sin^{-1} x) = \frac{b}{\sqrt{1-x^2}}\right]$$

(iv) $$\int \frac{(3\tan^2 x + 2)\sec^2 x}{(\tan^3 x + 2\tan x + 5)^2}\, dx$$

$$= \int (\tan^3 x + 2\tan x + 5)^{-2}(3\tan^2 x + 2)\sec^2 x\, dx$$

$$= \frac{(\tan^3 x + 2\tan x + 5)^{-1}}{-1} + C$$

$$= -\frac{1}{(\tan^3 x + 2\tan x + 5)} + C$$

$$\left[\because \frac{d}{dx}(\tan^3 x + 2\tan x + 5) = 3\tan^2 x \cdot \sec^2 x + 2\sec^2 x\right]$$

Rule III: $\int \frac{f'(x)}{f(x)}\,dx = \log|f(x)| + C$

Proof: Let $f(x) = t \Rightarrow f'(x)\,dx = dt$

$\therefore$ $\int \frac{f'(x)}{f(x)}\,dx = \int \frac{dt}{t} = \log|t| + C = \log|f(x)| + C$

Remark: Rule III is special case of Rule II. By Rule II we know that

$$\int [f(x)]^n f'(x)\,dx = \frac{[f(x)]^{n+1}}{n+1} + C$$

It fails if $n = -1$ and hence it becomes Rule III.

This rule states that,

"If the numerator is the derivative of the denominator, then the integral is log |denominator| + C.

Some Important Integrals Based on Rule III

1. $\int \tan x\,dx = \log|\sec x| + C = -\log|\cos x| + C$

2. $\int \cot x\,dx = \log|\sin x| + C$

3. $\int \sec x\,dx = \log|\sec x + \tan x| + C = \log\left|\tan\left(\frac{\pi}{4} + \frac{x}{2}\right) + C\right|$

4. $\int \operatorname{cosec} x\,dx = \log|\operatorname{cosec} x - \cot x| + C = \log\left|\tan\frac{x}{2}\right| + C$.

ILLUSTRATIVE EXAMPLE

Example 9. Evaluate:

(i) $\int \frac{1 + \cot x}{x + \log\sin x}\,dx$ (ii) $\int \frac{x^5}{1 + x^6}\,dx$ (iii) $\int \frac{e^x}{(1 + e^x)}\,dx$

(iv) $\int \frac{dx}{a + be^{cx}}$ (v) $\int \frac{\sin 2x}{a^2\cos^2 x + b^2\sin^2 x}\,dx$

Solution: (i) $\int \frac{1+\cot x}{x+\log\sin x}\,dx = \log|x + \log\sin x| + C$

$$\left[\because \frac{d}{dx}(x+\log\sin x) = 1 + \frac{\cos x}{\sin x} = 1 + \cot x\right]$$

(ii) $\int \frac{x^5}{1+x^6}\,dx$,

$$\because \quad \frac{d}{dx}(1+x^6) = 6x^5$$

$$\therefore \quad \int \frac{x^5}{1+x^6}\,dx = \frac{1}{6}\int \frac{6x^5}{1+x^6}\,dx = \frac{1}{6}\log|1+x^6| + C$$

(iii) $\int \frac{e^x}{(1+e^x)}\,dx$,

$$\because \quad \frac{d}{dx}(1+e^x) = e^x$$

$$\therefore \quad \int \frac{e^x}{1+e^x}\,dx = \log|1+e^x| + C$$

(iv) $\int \frac{dx}{a+be^{cx}} = \int \frac{e^{-cx}}{ae^{-cx}+b}\,dx$

Let $\quad ae^{-cx} + b = t \Rightarrow -ac\,e^{-cx}\,dx = dt$

$$\Rightarrow \quad e^{-cx}\,dx = \frac{-dt}{ac}$$

$$\therefore \quad \int \frac{dx}{a+be^{cx}} = -\frac{1}{ac}\int \frac{1}{t}\,dt$$

$$= -\frac{1}{ac}\log|t| + C$$

$$= -\frac{1}{ac}\log|ae^{-cx}+b| + C$$

(v) $\int \frac{\sin 2x}{a^2\cos^2 x + b^2\sin^2 x}\,dx = \int \frac{2\sin x\cos x}{a^2\cos^2 x + b^2\sin^2 x}\,dx$

$$= \frac{1}{b^2-a^2}\int \frac{2(b^2-a^2)\sin x\cos x}{a^2\cos^2 x + b^2\sin^2 x}\,dx$$

$$\because \quad \frac{d}{dx}(a^2\cos^2 x + b^2\sin^2 x) = 2a^2\cos x(-\sin x) + 2b^2\sin x\cos x$$

$$= 2(b^2 - a^2)\sin x \cos x$$

$$\therefore \quad \int \frac{\sin 2x\, dx}{a^2\cos^2 x + b^2 \sin^2 x} = \frac{1}{b^2 - a^2}\log|a^2\cos^2 x + b^2\sin^2 x| + C$$

The above formula fails if $a = \pm b$. If $a = \pm b$, then

$$\int \frac{\sin 2x\, dx}{a^2\cos^2 x + a^2 \sin^2 x} = \int \frac{\sin 2x}{a^2(\cos^2 x + \sin^2 x)}$$

$$= \frac{1}{a^2}\int \sin 2x\, dx$$

$$= \frac{1}{a^2}\left(-\frac{\cos 2x}{2}\right) + C = -\frac{1}{2a^2}\cdot\cos 2x + C$$

Example 10. Evaluate:

(i) $\int \frac{dx}{\sqrt{1-x^2}\cos^{-1} x}$ (ii) $\int \frac{dx}{\sqrt{x}\left(1+\sqrt{x}\right)}$ (iii) $\int \frac{dx}{a\sin x + b\cos x}$

(iv) $\int \frac{dx}{x\log x \cdot \log(\log x)}$ (v) $\int \frac{1+\tan x}{1-\tan x}\, dx$

Solution: (i) $\int \frac{dx}{\sqrt{1-x^2}\cos^{-1} x} = -\int \frac{\frac{-1}{\sqrt{1-x^2}}}{\cos^{-1} x}\, dx = -\log|\cos^{-1} x| + C$

$$\left[\because \frac{d}{dx}(\cos^{-1} x) = \frac{-1}{\sqrt{1-x^2}}\right]$$

(ii) since $\frac{d}{dx}\left(1+\sqrt{x}\right) = \frac{1}{2}x^{-1/2} = \frac{1}{2\sqrt{x}}$

$$\therefore \quad \int \frac{dx}{\sqrt{x}\left(1+\sqrt{x}\right)} = 2\int \frac{dx}{2\sqrt{x}\left(1+\sqrt{x}\right)}$$

$$= 2\int \frac{\frac{1}{2\sqrt{x}}}{1+\sqrt{x}}\, dx$$

$$= 2\log\left|1+\sqrt{x}\right| + C$$

$$= \log\left(1+\sqrt{x}\right)^2 + C$$

(iii) $\int \frac{dx}{a\sin x + b\cos x}$,

Let $\qquad a = r\cos\theta$ and $b = r\sin\theta$

$$\Rightarrow \qquad r = \sqrt{a^2+b^2} \text{ and } \theta = \tan^{-1}\frac{b}{a}$$

$$\therefore \int \frac{dx}{r\cos\theta\sin x + r\sin\theta\cos x} = \frac{1}{r}\int \frac{dx}{\sin(\theta+x)} = \frac{1}{r}\int \operatorname{cosec}(\theta+x)\,dx$$

$$= \frac{1}{r}\log\left(\tan\frac{(\theta+x)}{2}\right)+C$$

$$\left[\because \int \operatorname{cosec} x\,dx = \log\tan\frac{x}{2}+C\right]$$

$$= \frac{1}{\sqrt{a^2+b^2}}\log\left(\tan\frac{(x+\tan^{-1}b/a)}{2}\right)+C$$

(iv) $$\int \frac{dx}{x\log x\cdot\log(\log x)} = \int \frac{\frac{1}{x\log x}}{\log(\log x)}\,dx = \log[\log(\log x)]+C$$

$$\left[\because \frac{d}{dx}(\log\log x) = \frac{1}{\log x}\cdot\frac{1}{x} = \frac{1}{x\log x}\right]$$

(v) $$\int \frac{1+\tan x}{1-\tan x}\,dx = \int \frac{1+\frac{\sin x}{\cos x}}{1-\frac{\sin x}{\cos x}}\,dx = \int \frac{\cos x+\sin x}{\cos x-\sin x}\,dx$$

$$= -\int \frac{-(\cos x+\sin x)}{\cos x-\sin x}\,dx$$

$$= -\log|\cos x-\sin x|+C$$

$$\left[\because \frac{d}{dx}(\cos x-\sin x) = -\sin x-\cos x = -(\cos x+\sin x)\right]$$

Example 11. Evaluate:

(i) $\displaystyle\int \frac{dx}{1+\sqrt{x+1}}$ (ii) $\displaystyle\int \frac{e^x(1+x)}{\cos^2(xe^x)}\,dx$

(iii) $\displaystyle\int \frac{\sin x\,dx}{\sin(x-\alpha)}$ (iv) $\displaystyle\int \frac{x\,dx}{\sqrt{1-x^2}\cos^2\sqrt{1-x^2}}$

Solution: (i) Let $\sqrt{x+1} = t \Rightarrow x+1 = t^2 \therefore dx = 2t\,dt$

$$\therefore \qquad \int \frac{dx}{1+\sqrt{x+1}} = \int \frac{2t\,dt}{1+t} = 2\int\left[1-\frac{1}{1+t}\right]dt$$

$$= 2[t-\log|1+t|]+C$$

$$= 2\sqrt{x+1}-2\log\left|1+\sqrt{x+1}\right|+C$$

(ii) Let $xe^x = t \Rightarrow (e^x \cdot 1 + xe^x)dx = dt$

$$\Rightarrow \quad e^x(1+x)dx = dt$$

$$\therefore \quad \int \frac{e^x(1+x)}{\cos^2(xe^x)}dx = \int \frac{dt}{\cos^2 t}$$

$$= \int \sec^2 t\, dt$$

$$= \tan t + C$$

$$= \tan(xe^x) + C$$

(iii) Let $x - \alpha = t \Rightarrow dx = dt$ and $x = \alpha + t$

$$\therefore \quad \int \frac{\sin x}{\sin(x-a)}dt = \int \frac{\sin(\alpha + t)}{\sin t}dt$$

$$= \int \frac{\sin\alpha\cos t + \cos\alpha\sin t}{\sin t}dt$$

$$= \sin\alpha \int \cot t\, dt + \cos\alpha \int dt$$

$$= \sin\alpha \log|\sin t| + t\cos\alpha + C$$

$$= \sin\alpha \log|\sin(x-\alpha)| + (x-\alpha)\cos\alpha + C$$

(iv) $\displaystyle\int \frac{x\,dx}{\sqrt{1-x^2}\cos^2\sqrt{1-x^2}} = \int \frac{x\sec^2\sqrt{1-x^2}}{\sqrt{1-x^2}}dx$

Let $\sqrt{1-x^2} = t \Rightarrow \dfrac{-1}{2\sqrt{1-x^2}} \cdot 2x\,dx = dt$

$$\Rightarrow \quad \frac{x}{\sqrt{1-x^2}}dx = -dt$$

$$\therefore \int \frac{x\,dx}{\sqrt{1-x^2}\cos^2\sqrt{1-x^2}} = -\int \sec^2 t\, dt$$

$$= -\tan t + C$$

$$= -\tan\sqrt{1-x^2} + C$$

Example 12. Evaluate:

(i) $\int \tan(2-3x)\,dx$ (ii) $\int \cot^3 x\,dx$

(iii) $\displaystyle\int \frac{1+\tan^2 x}{1-\tan^2 x}dx$ (iv) $\displaystyle\int \frac{\sec x}{\sec 2x}dx$

Solution: $\displaystyle\int \tan(2-3x)\,dx = \frac{-\log|\cos(2-3x)|}{-3} + C$

$$= \frac{1}{3}\log|\cos(2-3x)| + C$$

(ii) $\int \cot^3 x\, dx = \int \cot x \cdot \cot^2 x\, dx = \int \cot x(\operatorname{cosec}^2 x - 1)\, dx$

$$= -\int \cot x(-\operatorname{cosec}^2 x) - \int \cot x\, dx$$

$$= \frac{-(\cot x)^2}{2} - \log|\sin x| + C$$

$$= -\frac{1}{2}\cot^2 x - \log|\sin x| + C$$

(iii) $\int \frac{1+\tan^2 x}{1-\tan^2 x}\, dx = \int \frac{1}{\cos 2x}\, dx$

$$\left[\because \cos 2x = \frac{1-\tan^2 x}{1+\tan^2 x}\right]$$

$$= \int \sec 2x\, dx$$

$$= \frac{\log|\sec 2x + \tan 2x|}{2} + C$$

(iv) $\int \frac{\sec x}{\sec 2x}\, dx = \int \frac{\cos 2x}{\cos x}\, dx = \int \frac{2\cos^2 x - 1}{\cos x}\, dx = \int (2\cos x - \sec x)\, dx$

$$= 2\sin x - \log|\sec x + \tan x| + C$$

Example 13. Evaluate:

(i) $\int \frac{dx}{\sin x \cos^2 x}$ (ii) $\int \frac{\sin 2x}{\sin 5x \sin 3x}\, dx$

(iii) $\int \frac{1+\tan^2 x}{\tan x}\, dx$

Solution: (i) $\int \frac{dx}{\sin x \cos^2 x} = \int \frac{\sin^2 x + \cos^2 x}{\sin x \cos^2 x}\, dx$

$$= \int (\sec x \tan x + \operatorname{cos}ecx)\, dx$$

$$= \sec x + \log|\operatorname{cosec} x - \cot x| + C$$

(ii) $\int \frac{\sin 2x}{\sin 5x \sin 3x}\, dx = \int \frac{\sin(5x-3x)}{\sin 5x \sin 3x}\, dx$

$$= \int \frac{\sin 5x \cos 3x - \cos 5x \sin 3x}{\sin 5x \sin 3x}\, dx$$

$$= \int (\cot 3x - \cot 5x)\, dx$$

$$= \frac{\log|\sin 3x|}{3} - \frac{\log|\sin 5x|}{5} + C$$

$$= \frac{1}{3}\log|\sin 3x| - \frac{1}{5}\log|\sin 5x| + C$$

(iii) $\int \frac{1+\tan^2 x}{\tan x}\,dx = \int \frac{\sec^2 x}{\tan x}\,dx$

$$= \int \frac{1}{\sin x \cos x}\,dx = \int \frac{2}{\sin 2x}\,dx$$

$$= 2\int \operatorname{cosec} 2x\,dx$$

$$= 2\frac{\log|\operatorname{cosec} 2x - \cot 2x|}{2} + C$$

$$= \log|\operatorname{cosec} 2x - \cot 2x| + C$$

EXERCISE 4.2

1. Evaluate:

(i) $\int \frac{1}{\sqrt{3x-2}}\,dx$

(ii) $\int (3x-7)^{9/2}\,dx$

(iii) $\int \frac{x^2}{1-x}\,dx$

(iv) $\int e^{2x+3}\,dx$

(v) $\int \sin(3x+1)\,dx$

(vi) $\int \sin 5x \sin 7x\,dx$

(vii) $\int \sec^2(7x+11)\,dx$

(viii) $\int \sin x \sin 2x \sin 3x\,dx$

(ix) $\int \frac{dx}{\sqrt{2x+3}-\sqrt{2x}}$

(x) $\int \cos^2 x\,dx$

(xi) $\int \frac{\sin^2 x}{(1+\cos x)^2}\,dx$

(xii) $\int \cos mx \cos nx\,dx, m \neq n$

(xiii) $\int \frac{dx}{x + x\log x}$

(xiv) $\int \cot(5x-3)\,dx$

(xv) $\int \sqrt{1+\cos x}\,dx$

(xvi) $\int \sin x\sqrt{1-\cos 2x}\,dx$

(xvii) $\int 2^{3x+4}\,dx$

2. Evaluate:

(i) $\int x(5x^2-7)^6\,dx$

(ii) $\int (2x+1)\sqrt{x^2+x+1}\,dx$

(iii) $\int \frac{6x^2+1}{(4x^3+2x-5)^3}\,dx$

(iv) $\int \sec^4 x \tan x\,dx$

(v) $\int \cos^3 x \sin 2x \, dx$

(vi) $\int \frac{\sqrt{\tan x}}{\sin x \cos x} dx$

(vii) $\int \frac{e^{4x}}{(3+e^{4x})^3} dx$

(viii) $\int \frac{3(\cos^{-1} x)^2}{\sqrt{1-x^2}} dx$

(ix) $\int \frac{\cos x - \sin x}{1+\sin 2x} dx$

(x) $\int e^{-x} \operatorname{cosec}^2(2e^{-x}+5) \, dx$

(xi) $\int x^3 \tan^4 x^4 \sec^2 x^4 \, dx$

(xii) $\int \frac{e^{m\tan^{-1} x}}{1+x^2} dx$

(xiii) $\int \frac{(\log x)^2}{x} dx$

(xiv) $\int \frac{(\sin^{-1} 2x)^3}{\sqrt{1-4x^2}} dx$

(xv) $\int (1+\tan^2 x)\sqrt{\tan x} \, dx$

(xvi) $\int \frac{\tan x}{\sqrt{\cos x}} dx$

(xvii) $\int \frac{\sin x \, dx}{\sqrt{3+2\cos x}}$

(xviii) $\int (e^{3x}+1)^2 e^{3x} \, dx$

(xix) $\int \cot^3 x \operatorname{cosec}^2 x \, dx$

(xx) $\int \frac{1}{2\sqrt{x}\sqrt{1-x}} dx$

3. Evaluate:

(i) $\int \frac{1-\sin 2x}{x+\cos^2 x} dx$

(ii) $\int \frac{x-2}{2x^2-8x+5} dx$

(iii) $\int \frac{x-1}{x(x-\log x)} dx$

(iv) $\int \frac{1-\tan x}{1+\tan x} dx$

(v) $\int \frac{dx}{x \log |x|}$

(vi) $\int \frac{dx}{\sin(x-\alpha)\sin(\alpha-\beta)} dx$

(vii) $\int (\tan x + \cot x) \, dx$

(viii) $\int \frac{dx}{x\sqrt{x^2-1}\sec^{-1} x}$

(ix) $\int \frac{x+1}{x^2+2x-3} dx$

(x) $\int \frac{dx}{e^{ax}+1}$

(xi) $\int \frac{\sin 2x}{\sin^4 x + \cos^4 x} dx$

(xii) $\int \frac{e^{\sqrt{x}}}{\sqrt{x}} dx$

(xiii) $\int \frac{\log(x+1) - \log x}{x(x+1)^3} dx$

(xiv) $\int \operatorname{cosec} x \log(\operatorname{cosec} x - \cot x)\, dx$

(xv) $\int \frac{x^{e-1} + e^{x-1}}{x^e + e^x} dx$

(xvi) $\int \frac{\tan x}{a + b\tan^2 x} dx$

(xvii) $\int \frac{2\cos x - 3\sin x}{6\cos x + 4\sin x} dx$

(xviii) $\int \frac{x^2\, dx}{(a+bx)^2}$

(xix) $\int \frac{dx}{(ax+b) + \sqrt{ax+b}}$

(xx) $\int \frac{1+\tan x}{x + \log\sec x} dx$

[**Hint:** 3. (x) $\int \frac{dx}{e^{ax}+1} = \int \frac{e^{-ax}}{1+e^{-ax}} dx$

(xi) $\int \frac{\sin 2x\, dx}{\sin^4 x + \cos^4 x}$, Divide numerator and denominator by $\cos^4 x$.

$= \int \frac{2\tan x \sec^2 x}{1+\tan^4 x} dx$, Let $\tan^2 x = t$

(xviii) $\int \frac{x^2\, dx}{(a+bx)^2}$, Since powers of numerator and denominator are same, divide nume. by deno.

(xix) $\int \frac{dx}{(ax+b) + \sqrt{ax+b}} = \int \frac{1}{\sqrt{ax+b}\left[1+\sqrt{ax+b}\right]} dx$]

4. Evaluate:

(i) $\int \frac{dx}{\sin x + \sqrt{3}\cos x}$

(ii) $\int \frac{dx}{\cos 3x - \cos x}$

(iii) $\int \frac{dx}{\sqrt{1+\cos x}}$

(iv) $\int \frac{\cos x\, dx}{(1+\sin x)(2+\sin x)}$

(v) $\int \frac{\sin x\, dx}{\cos^2 x + \cos x - 2}$

(vi) $\int \sec^6 x\, dx$

(vii) $\int \sin^4 x \cos^2 x\, dx$

(viii) $\int \sin^4 x\, dx$

(ix) $\int \frac{2x^2 - 1}{x(x^2 - \log x)} dx$

(x) $\int \sqrt{\frac{1-\sqrt{x}}{1+\sqrt{x}}} dx$

ANSWERS

1. (i) $\frac{2}{3}\sqrt{3x-2} + C$

(ii) $\frac{2}{33}(3x-7)^{11/2} + C$

(iii) $-\left(\frac{x^2}{2}+x+\log|1-x|\right)+C$

(iv) $\frac{e^{2x+3}}{2}+C$

(v) $-\frac{1}{3}\cos(3x+1)+C$

(vi) $\frac{1}{24}(6\sin 2x-\sin 12x)+C$

(vii) $\frac{1}{7}\tan(7x+11)+C$

(viii) $\frac{1}{48}$ (2 cos 6x – 3 cos 4x – 6 cos 2x) + C

(ix) $\frac{1}{9}[(2x+3)]^{3/2}+(2x)^{3/2}]+C$

(x) $\frac{1}{4}(2x+\sin 2x)+C$

(xi) $2\tan\frac{x}{2}-x+C$

(xii) $\frac{1}{2}\left(\frac{\sin(m+n)x}{m+n}+\frac{\sin(m-n)x}{m-n}\right)+C$

(xiii) $\log(1+\log x)+C$

(xiv) $\frac{1}{5}\log|\sin(5x-3)|+C$

(xv) $2\sqrt{2}\sin\frac{x}{2}+C$

(xvi) $\frac{1}{2\sqrt{2}}(2x-\sin 2x)+C$

(xvii) $\frac{2^{3x+4}}{3\log e^2}$

2. (i) $\frac{1}{70}(5x^2-7)^7+C$

(ii) $\frac{2}{3}(x^2+x+1)^{3/2}+C$

(iii) $-\frac{1}{4(4x^3+2x-5)^2}+C$

(iv) $\frac{\sec^4 x}{4}+C$

(v) $-\frac{2}{5}\cos^5 x+C$

(vi) $2\sqrt{\tan x}+C$

(vii) $-\frac{1}{8(3+e^{4x})^2}+C$

(viii) $-(\cos^{-1}x)^3+C$

(ix) $\frac{-1}{\cos x+\sin x}+C$

(x) $\frac{1}{2}\cot(2e^{-x}+5)+C$

(xi) $\frac{1}{20}\tan^5 x^4+C$

(xii) $\frac{1}{m}e^{m\tan^{-1}x}+C$

(xiii) $\frac{1}{3}(\log x)^3+C$

(xiv) $\frac{1}{8}(\sin^{-1}2x)^4+C$

(xv) $\frac{2}{3}(\tan x)^{3/2}+C$

(xvi) $\frac{2}{\sqrt{\cos x}}+C$

(xvii) $-\sqrt{3+2\cos x}+C$

(xviii) $\frac{1}{9}(e^{3x}+1)^3+C$

(xix) $-\frac{1}{4}\cot^4 x + C$ (xx) $\sin^{-1}\left(\sqrt{x}\right) + C$

3. (i) $\log|x + \cos^2 x| + C$ (ii) $\frac{1}{4}\log|2x^2 - 8x + 5| + C$

(iii) $\log|x - \log x| + C$ (iv) $\log|\cos x + \sin x| + C$

(v) $\log(\log|x|) + C$ (vi) $\frac{1}{\sin(\beta - \alpha)}\log\left|\frac{\sin(x - \beta)}{\sin(x - \alpha)}\right| + C$

(vii) $\log|\tan x| + C$ (viii) $\log|\sec^{-1} x| + C$

(ix) $\frac{1}{2}\log|x^2 + 2x - 3| + C$ (x) $-\frac{1}{a}\log(1 + e^{-ax}) + C$

(xi) $\tan^{-1}(\tan^2 x) + C$ (xii) $2e^{\sqrt{x}} + C$

(xiii) $-\frac{1}{2}[\log(x + 1) - \log x]^2 + C$ (xiv) $\frac{1}{2}[\log|\operatorname{cosec} x - \cot x|]^2 + C$

(xv) $\frac{1}{e}\log|e^x + x^e| + C$ (xvi) $\frac{1}{2(b - a)}$, $\log|a\cos^2 x + b\sin^2 x| + C$ for $a \neq b$, for $a = b$, $\frac{1}{2a}\sin^2 x + C$

(xvii) $\frac{1}{2}|3\cos x + 2\sin x| + C$

(xviii) $\frac{1}{b^2}\left[(a + bx) - \frac{a^2}{(a + bx)} - 2a\log|a + bx|\right] + C$

(xix) $\frac{2}{a}\log\left[\sqrt{ax + b} + 1\right] + C$ (xx) $\log|x + \log\sec x| + C$

4. (i) $\frac{1}{2}\log\left|\operatorname{cosec}\left(x + \frac{\pi}{3}\right) - \cot\left(x + \frac{\pi}{3}\right)\right| + C$

(ii) $-\frac{1}{4}[\log|\sec x + \tan x| - \operatorname{cosec} x] + C$

(iii) $\sqrt{2}\log\left|\sec\frac{x}{2} + \tan\frac{x}{2}\right| + C$ (iv) $\log\left|\frac{\sin x + 1}{\sin x + 2}\right| + C$

(v) $\frac{1}{3}\log\left|\frac{2 + \cos x}{\cos x - 1}\right| + C$ (vi) $\frac{2}{3}\tan^3 x + \frac{1}{5}\tan^5 x + \tan x + C$

(vii) $\frac{1}{192}(12x - 3\sin 2x - 3\sin 4x + \sin 6x) + C$

(viii) $\frac{1}{32}(12x - 8\sin 2x + \sin 4x) + C$

(ix) $\log|x^2 - \log x| + C$ (x) $\left(\sqrt{x} - 2\right)\sqrt{1-x} + \cos^{-1} x + C$

4.8 SOME STANDARD INTEGRALS

1. $\int \frac{dx}{\sqrt{a^2 - x^2}} = \sin^{-1}\frac{x}{a} + C, |x| < a$

proof: Put $x = a \sin t \Rightarrow dx = a \cos t \, dt$ and $t = \sin^{-1}\frac{x}{a}$

$$\therefore \quad \int \frac{dx}{\sqrt{a^2 - x^2}} = \int \frac{a\cos t \, dt}{\sqrt{a^2 - a^2\sin^2 t}} = \int \frac{a\cos t}{a\cos t} dt = \int 1 \, dt$$

$$= t + C = \sin^{-1}\frac{x}{a} + C$$

2. $\int \frac{dx}{\sqrt{a^2 + x^2}} = \log\left| x + \sqrt{x^2 + a^2} \right| + C, a > 0$

Proof: Put $x = a \tan t \Rightarrow dx = a \sec^2 t \, dt$ and $t = \tan^{-1}\frac{x}{a}$

$$\therefore \quad \int \frac{dx}{\sqrt{a^2 + x^2}} = \int \frac{a\sec^2 t \, dt}{\sqrt{a^2 + a^2\tan^2 t}} = \int \sec t \, dt$$

$$= \log|\sec t + \tan t| + C_1$$

$$= \log\left|\tan t + \sqrt{1 + \tan^2 t}\right| + C_1$$

$$= \log\left|\frac{x}{a} + \sqrt{1 + \frac{x^2}{a^2}}\right| + C_1$$

$$= \log\frac{\left|x + \sqrt{a^2 + x^2}\right|}{a} + C_1$$

$$= \log\left|x + \sqrt{a^2 + x^2}\right| - \log a + C_1$$

$$= \log\left|x + \sqrt{a^2 + x^2}\right| + C$$

[$C_1 - \log a = \text{constant} = C$ (let)]

3. $\int \frac{dx}{\sqrt{x^2 - a^2}} = \log\left| x + \sqrt{x^2 - a^2} \right| + C, |x| > a > 0$

Proof: Put $x = a \sec t \Rightarrow dx = a \sec t \tan t\, dt$ and $t = \sec^{-1}\frac{x}{a}$

$$\therefore \quad \int \frac{dx}{\sqrt{x^2 - a^2}} = \int \frac{a \sec t \tan t\, dt}{\sqrt{a^2 \sec^2 t - a^2}} = \int \sec t\, dt$$

$$= \log|\sec t + \tan t| + C_1$$

$$= \log\left|\sec t + \sqrt{\sec^2 t - 1}\right| + C_1$$

$$= \log\left|\frac{x}{a} + \sqrt{\frac{x^2}{a^2} - 1}\right| + C_1$$

$$= \log\frac{\left|x + \sqrt{x^2 - a^2}\right|}{a} + C_1$$

$$= \log\left|x + \sqrt{x^2 - a^2}\right| + C$$

4. $\int \frac{dx}{a^2 + x^2} = \frac{1}{a}\tan^{-1}\frac{x}{a} + C$

Proof: Put $x = a \tan t \Rightarrow dx = a \sec^2 t\, dt$ and $t = \tan^{-1}\frac{x}{a}$

$$\therefore \quad \int \frac{dx}{a^2 + x^2} = \int \frac{a \sec^2 t\, dt}{a^2 + a^2 \tan^2 t} = \frac{1}{a}\int \frac{\sec^2 t\, dt}{\sec^2 t}$$

$$= \frac{1}{a}\int 1\, dt = \frac{1}{a}t + C$$

$$= \frac{1}{a}\tan^{-1}\frac{x}{a} + C$$

5. $\int \frac{dx}{a^2 - x^2} = \frac{1}{2a}\log\left|\frac{a + x}{a - x}\right| + C$

Proof: $\frac{1}{a^2 - x^2} = \frac{1}{(a - x)(a + x)} = \frac{1}{2a}\left(\frac{1}{a + x} + \frac{1}{a - x}\right)$

$$\therefore \quad \int \frac{dx}{a^2 - x^2} = \frac{1}{2a}\int\left(\frac{1}{a + x} + \frac{1}{a - x}\right)dx$$

$$= \frac{1}{2a}[\log|a + x| - \log|a - x|] + C$$

$$= \frac{1}{2a}\log\left|\frac{x + a}{x - a}\right| + C$$

6. $\int \frac{dx}{x^2 - a^2} = \frac{1}{2a} \log\left|\frac{x-a}{x+a}\right| + C$

Proof: same as (5)

7. $\int \frac{dx}{x\sqrt{x^2 - a^2}} = \frac{1}{a}\sec^{-1}\frac{x}{a} + C$

Proof: Put $x = a \sec t \Rightarrow dx = a \sec t \tan t\, dt$ and $t = \sec^{-1}\frac{x}{a}$.

$$\therefore \quad \int \frac{dx}{x\sqrt{x^2 - a^2}} = \int \frac{a \sec t \tan t\, dt}{a \sec t\sqrt{a^2 \sec^2 t - a^2}} = \int \frac{a \sec t \tan t\, dt}{a^2 \sec t \tan t}$$

$$= \frac{1}{a}\int 1\, dt = \frac{1}{a}\cdot t + C$$

$$= \frac{1}{a}\sec^{-1}\frac{x}{a} + C$$

ILLUSTRATIVE EXAMPLES

Example 1. Evaluate the following integrals:

(i) $\int \frac{dx}{\sqrt{9 - 25x^2}}$ (ii) $\int \frac{dx}{16x^2 + 25}$

(iii) $\int \frac{dx}{a^2 - b^2x^2}$ (iv) $\int \frac{dx}{\sqrt{1 - e^{2x}}}$

(v) $\int \frac{dx}{5x\sqrt{3x^2 - 12}}$ (vi) $\int \frac{\sin x}{1 + \cos^2 x}\, dx$

(vii) $\int \frac{dx}{2 + \cos x}$

Solution: (i) $\int \frac{dx}{\sqrt{9 - 25x^2}} = \frac{1}{5}\int \frac{dx}{\sqrt{\frac{9}{25} - x^2}} = \frac{1}{5}\int \frac{dx}{\sqrt{\left(\frac{3}{5}\right)^2 - x^2}}$

$$\left[\int \frac{dx}{\sqrt{a^2 - x^2}} \text{ form}\right]$$

$$= \frac{1}{5}\sin^{-1}\frac{x}{3/5} + C = \frac{1}{5}\sin^{-1}\frac{5x}{3} + C$$

(ii) $\int \frac{dx}{16x^2 + 25} = \frac{1}{16}\int \frac{dx}{x^2 + \frac{25}{16}} = \frac{1}{16}\int \frac{dx}{x^2 + \left(\frac{5}{4}\right)^2}$ $\left[\int \frac{dx}{x^2 + a^2} \text{ form}\right]$

$$= \frac{1}{16}\cdot\frac{1}{5/4}\tan^{-1}\frac{x}{5/4} + C = \frac{1}{20}\tan^{-1}\frac{4x}{5} + C$$

(iii) $\int \frac{dx}{a^2 - b^2x^2} = \frac{1}{b^2}\int \frac{dx}{\left(\frac{a}{b}\right)^2 - x^2}$ $\left[\int \frac{dx}{a^2 - x^2} \text{ form}\right]$

$$= \frac{1}{b^2}\cdot\frac{1}{2a/b}\log\left|\frac{a/b + x}{a/b - x}\right| + C$$

$$= \frac{1}{2ab}\log\left|\frac{a + bx}{a - bx}\right| + C$$

(iv) $\int \frac{dx}{\sqrt{1 - e^{2x}}} = \int \frac{e^{-x}\,dx}{\sqrt{e^{-2x - 1}}}$

Let $e^{-x} = t \Rightarrow e^{-x}\,dx = -dt$

$$\therefore \quad \int \frac{dx}{\sqrt{1 - e^{2x}}} = -\int \frac{dt}{\sqrt{t^2 - 1}}$$

$$= -\log\left|t + \sqrt{t^2 - 1}\right| + C \qquad \left[\int \frac{dx}{\sqrt{x^2 - a^2}} \text{ form}\right]$$

$$= -\log\left|e^{-x} + \sqrt{e^{-2x} - 1}\right| + C$$

(v) $\int \frac{dx}{5x\sqrt{3x^2 - 12}} = \frac{1}{5\times\sqrt{3}}\int \frac{dx}{x\sqrt{x^2 - 2^2}}$ $\left[\int \frac{dx}{x\sqrt{x^2 - a^2}} \text{ form}\right]$

$$= \frac{1}{5\sqrt{3}}\cdot\frac{1}{2}\sec^{-1}\frac{x}{2} + C = \frac{1}{10\sqrt{3}}\sec^{-1}\frac{x}{2} + C$$

(vi) $\int \frac{\sin x}{1 + \cos^2 x}\,dx$

Let $\cos x = t \Rightarrow -\sin x\,dx = dt \Rightarrow \sin x\,dx = -dt$

$$\therefore \quad \int \frac{\sin x\,dx}{1 + \cos^2 x} = -\int \frac{dt}{1 + t^2} = -\tan^{-1} t + C$$

$$= -\tan^{-1}(\cos x) + C$$

(vii) $\int \frac{dx}{2 + \cos x} = \int \frac{dx}{2\left(\cos^2\frac{x}{2} + \sin^2\frac{x}{2}\right) + \left(\cos^2\frac{x}{2} - \sin^2\frac{x}{2}\right)}$

$$= \int \frac{dx}{3\cos^2\frac{x}{2} + \sin^2\frac{x}{2}} = \int \frac{\sec^2\frac{x}{2}\,dx}{3 + \tan^2\frac{x}{2}}$$

Put $\tan\frac{x}{2} = t \Rightarrow \sec^2\frac{x}{2}\cdot\frac{1}{2}\,dx = dt$

$$\Rightarrow \sec^2\frac{x}{2}\,dx = 2\,dt$$

$$\therefore \quad \int\frac{dx}{2+\cos x} = \int\frac{2\,dt}{3+t^2} = 2\int\frac{dt}{\left(\sqrt{3}\right)^2+t^2}$$

$$= \frac{2}{\sqrt{3}}\tan^{-1}\frac{t}{\sqrt{3}} + C$$

$$= \frac{2}{\sqrt{3}}\tan^{-1}\left(\frac{1}{\sqrt{3}}\tan\frac{x}{2}\right) + C$$

Example 2. Evaluate the following integrals:

(i) $\int\frac{2x^2+1}{9x^2+5}\,dx$ (ii) $\int\frac{e^{2\log x}-1}{e^{2\log x}+1}\,dx$

(iii) $\int\frac{(x^2+\sin^2 x)}{1+x^2}\sec^2 x\,dx$ (iv) $\int\sqrt{\frac{1-x}{1+x}}\,dx$

(v) $\int\frac{dx}{\sqrt{x^2+2x}}$

Solution: (i) $\int\frac{2x^2+1}{9x^2+5}\,dx = \int\left(\frac{2}{9}-\frac{1/9}{9x^2+5}\right)dx$

$$= \frac{2}{9}\int 1\,dx - \frac{1}{9}\int\frac{dx}{9x^2+5}$$

$$= \frac{2}{9}x - \frac{1}{9}\cdot\frac{1}{9}\int\frac{dx}{x^2+\left(\frac{\sqrt{5}}{3}\right)^2}$$

$$= \frac{2}{9}x - \frac{1}{81}\cdot\frac{1}{\frac{\sqrt{5}}{3}}\tan^{-1}\left(\frac{x}{\frac{\sqrt{5}}{3}}\right) + C$$

$$= \frac{2}{9}x - \frac{1}{27\sqrt{5}}\tan^{-1}\left(\frac{3x}{\sqrt{5}}\right) + C$$

(ii) $\int\frac{e^{2\log x}-1}{e^{2\log x}+1}\,dx = \int\frac{e^{\log x^2}-1}{e^{\log x^2}+1}\,dx = \int\frac{x^2-1}{x^2+1}\,dx$

$$= \int\frac{x^2+1-2}{x^2+1}\,dx$$

$$= \int \left(1 - \frac{2}{1+x^2}\right) dx = \int 1\, dx - 2\int \frac{1}{1+x^2}\, dx$$

$$= x - 2\tan^{-1} x + C$$

(iii) $\displaystyle\int \frac{(x^2 + \sin^2 x)\sec^2 x}{1+x^2}\, dx = \int \frac{(x^2 + 1 - \cos^2 x)}{1+x^2} \sec^2 x\, dx$

$$= \int \left(\sec^2 x - \frac{1}{1+x^2}\right) dx$$

$$= \int \sec^2 x\, dx - \int \frac{dx}{1+x^2}$$

$$= \tan x - \tan^{-1} x + C$$

(iv) $\displaystyle\int \sqrt{\frac{1-x}{1+x}}\, dx = \int \sqrt{\frac{1-x}{1+x}} \times \sqrt{\frac{1-x}{1-x}}\, dx = \int \frac{1-x}{\sqrt{1-x^2}}\, dx$

$$= \int \frac{1}{\sqrt{1-x^2}}\, dx - \int \frac{x}{\sqrt{1-x^2}}\, dx$$

$$= \int \frac{1}{\sqrt{1^2 - x^2}}\, dx + \frac{1}{2}\int (1-x^2)^{-1/2}(-2x)\, dx$$

$$= \sin^{-1}\left(\frac{x}{1}\right) + \frac{1}{2}\,\frac{(1-x^2)^{1/2}}{\frac{1}{2}} + C$$

$$= \sin^{-1} x + \sqrt{1-x^2} + C$$

(v) $\displaystyle\int \frac{dx}{\sqrt{x^2+2x}} = \int \frac{dx}{\sqrt{x^2+2x+1-1}} = \int \frac{dx}{\sqrt{(x+1)^2 - 1}}$

Let $x + 1 = t \Rightarrow dx = dt$

$\therefore$ $\displaystyle\int \frac{dx}{\sqrt{x^2+2x}} = \int \frac{dt}{\sqrt{t^2-1}} = \log\left|t + \sqrt{t^2-1}\right| + C$

$$= \log\left|x + 1 + \sqrt{(x+1)^2 - 1}\right| + C$$

$$= \log\left|x + 1 + \sqrt{x^2 + 2x}\right| + C$$

Example 3. Evaluate the following integrals:

(i) $\displaystyle\int \frac{x\, dx}{\sqrt{9x^4 + 25}}$ (ii) $\displaystyle\int \frac{\cos x\, dx}{\sqrt{\frac{1}{4} + \cos^2 x}}$

(iii) $\displaystyle\int \frac{\tan x}{3 + 2\tan^2 x}\, dx$

Solution: (i) Let $x^2 = t \Rightarrow xdx = \dfrac{dt}{2}$

$$\therefore \quad \int \frac{x\,dx}{\sqrt{9x^4+25}} = \int \frac{\frac{dt}{2}}{\sqrt{9t^2+25}} = \frac{1}{2}\cdot\frac{1}{3}\int \frac{dt}{\sqrt{t^2+\frac{25}{9}}}$$

$$= \frac{1}{6}\int \frac{dt}{\sqrt{t^2+\left(\frac{5}{3}\right)^2}}$$

$$= \frac{1}{6}\log\left|t+\sqrt{t^2+\frac{25}{9}}\right| + C$$

$$= \frac{1}{6}\log\left|x^2+\sqrt{x^2+\frac{25}{9}}\right| + C$$

(ii) $\displaystyle\int \frac{\cos x\,dx}{\sqrt{\frac{1}{4}+\cos^2 x}} = \int \frac{\cos x\,dx}{\sqrt{\frac{1}{4}+1-\sin^2 x}} = \int \frac{\cos x\,dx}{\sqrt{\frac{5}{4}-\sin^2 x}}$

Let $\quad \sin x = t$

$\Rightarrow \quad \cos x\,dx = dt$

$$\therefore \quad \int \frac{\cos x\,dx}{\sqrt{\frac{1}{4}+\cos^2 x}} = \int \frac{dt}{\sqrt{\frac{5}{4}-t^2}} = \int \frac{dt}{\sqrt{\left(\frac{\sqrt{5}}{2}\right)^2 - t^2}}$$

$$= \sin^{-1}\left(\frac{t}{\frac{\sqrt{5}}{2}}\right) + C$$

$$= \sin^{-1}\left(\frac{2t}{\sqrt{5}}\right) + C = \sin^{-1}\left(\frac{2\sin x}{\sqrt{5}}\right) + C$$

(iii) $\displaystyle\int \frac{\tan x\,dx}{3+2\tan^2 x} = \int \frac{\frac{\sin x}{\cos x}}{3+2\frac{\sin^2 x}{\cos^2 x}}\,dx = \int \frac{\sin x\cos x}{3\cos^2 x+2\sin^2 x}\,dx$

$$= \int \frac{\sin x\cos x}{3\cos^2 x+2-2\cos^2 x}\,dx = \int \frac{\sin x\cos x}{2+\cos^2 x}\,dx$$

Let $2 + \cos^2 x = t \Rightarrow -2\cos x\sin x\,dx = dt$

$\Rightarrow \quad \cos x\sin x\,dx = -\dfrac{dt}{2}$

$$\therefore \quad \int \frac{\tan x\, dx}{3+2\tan^2 x} = -\frac{1}{2}\int \frac{dt}{t} = -\frac{1}{2}\log t + C$$

$$= -\frac{1}{2}\log(2+\cos^2 x) + C$$

EXERCISE 4.3

Evaluate the following integrals:

1. (i) $\int \frac{dx}{\sqrt{36x^2-25}}$ (ii) $\int \frac{dx}{x^2+25}$

(iii) $\int \frac{dx}{4x^2+9}$ (iv) $\int \frac{dx}{\sqrt{25-x^2}}$

(v) $\int \frac{dx}{\sqrt{a^2+b^2x^2}}$ (vi) $\int \frac{dx}{1-4x^2}$

2. (i) $\int \frac{x+6}{\sqrt{4x^2+1}}\, dx$ (ii) $\int \frac{2x^3\, dx}{4+x^8}$

(iii) $\int \frac{e^x}{\sqrt{4+e^{2x}}}\, dx$ (iv) $\int \frac{x^2-1}{x^2+4}\, dx$

(v) $\int \frac{x^2\, dx}{x^2-a^2}$ (vi) $\int \frac{x^2}{9+4x^2}\, dx$

3. (i) $\int \frac{x^2}{a^6-x^6}\, dx$ (ii) $\int \frac{dx}{\sin^2 x+4\cos^2 x}$

(iii) $\int \frac{dx}{4-5\sin^2 x}$ (iv) $\int \frac{dx}{x\sqrt{3x^2-1}}$

(v) $\int \frac{\sin x\, dx}{\sqrt{9-\cos^2 x}}$ (vi) $\int \frac{dx}{\sin x+\cos x}$

4. (i) $\int \frac{1}{x}\sqrt{\frac{x+a}{x-a}}\, dx$ [**Hint.** Multiply and divide by $\sqrt{x+a}$]

(ii) $\int \frac{x-1}{\sqrt{x^2+1}}\, dx$ (iii) $\int \frac{\sqrt{x^2-a^2}}{x}\, dx$

(iv) $\int \sqrt{\frac{a-x}{a+x}}\, dx$ (v) $\int \frac{dx}{5+4\cos x}$

(vi) $\int \frac{dx}{2\sqrt{x(1-x)}}$

ANSWERS

1. (i) $\frac{1}{6}\log\left|6x+\sqrt{36x^2-25}\right|+C$ (ii) $\frac{1}{5}\tan^{-1}\frac{x}{5}+C$

(iii) $\frac{1}{6}\tan^{-1}\frac{2x}{3}+C$ (iv) $\frac{1}{2}\sin^{-1}\left(\frac{2x}{5}\right)+C$

(v) $\frac{1}{b}\log\left|bx+\sqrt{a^2+b^2x^2}\right|+C$ (vi) $\frac{1}{4}\log\left|\frac{1+2x}{1-2x}\right|+C$

2. (i) $\frac{1}{4}\sqrt{4x^2+1}+3\log\left|2x+\sqrt{4x^2+1}\right|+C$

(ii) $\frac{1}{4}\tan^{-1}\left(\frac{x^4}{2}\right)+C$ (iii) $\log\left|e^x+\sqrt{4+e^{2x}}\right|+C$

(iv) $x-\frac{5}{2}\tan^{-1}\frac{x}{2}+C$ (v) $x+\frac{a}{2}\log\left|\frac{x-a}{x+a}\right|+C$

(vi) $\frac{x}{4}-\frac{3}{8}\tan^{-1}\left(\frac{2x}{3}\right)+C$

3. (i) $\frac{1}{6a^3}\log\left|\frac{a^3+x^3}{a^3-x^3}\right|+C$ (ii) $\frac{1}{2}\tan^{-1}\left(\frac{\tan x}{2}\right)+C$

(iii) $\frac{1}{4}\log\left|\frac{2+\tan x}{2-\tan x}\right|+C$ (iv) $\sec^{-1}\left(\sqrt{3}x\right)+C$

(v) $-\sin^{-1}\left(\frac{\cos x}{3}\right)+C$ (vi) $\frac{1}{\sqrt{2}}\log\left[\sec\left(x-\frac{\pi}{4}\right)-\tan\left(x-\frac{\pi}{4}\right)\right]+C$

4. (i) $\log\left|x+\sqrt{x^2-a^2}\right|+\sec^{-1}\frac{x}{a}+C$

(ii) $\sqrt{x^2+1}-\log\left|x+\sqrt{x^2+1}\right|+C$

(iii) $\sqrt{x^2-a^2}-a\sec^{-1}\frac{x}{a}+C$ (iv) $a\sin^{-1}\frac{x}{a}+\sqrt{a^2-x^2}+C$

(v) $\frac{2}{3}\tan^{-1}\left[\frac{1}{3}\tan\frac{x}{2}\right]+C$ (vi) $\sin^{-1}\sqrt{x}+C$

4.9 INTEGRATION BY PARTS:

Theorem: If $f(x)$ and $g(x)$ are two differentiable functions then

$$\int f(x)\cdot g(x)\,dx = f(x)\int g(x)\,dx-\int\left[\left\{\frac{d}{dx}f(x)\right\}\int g(x)\,dx\right]dx$$

i.e., Integral of the product of two functions = First function × Integral of second – Integral of (Derivative of first × Integral of second).

The above rule of integration is known as the rule of integration by parts.

Note: For the selection of the first function, we consider the function which comes first in the word ILATE as the first function.

I: inverse trigonometric functions

L: logarithmic function

A: algebraic function

T: trigonometric functions

E: exponential functions

COROLLARY: $\int e^x [f(x) + f'(x)]\, dx = f(x)\, e^x + C$

Proof: $\text{L.H.S.} = \int f(x)\, e^x + \int e^x f'(x)\, dx$

$$= e^x f(x) - \int f'(x) e^x\, dx + \int e^x\, f'(x)\, dx$$

[integration by parts]

$$= e^x f(x) = \text{R.H.S.}$$

ILLUSTRATIVE EXAMPLES

Example 1. Prove that

(i) $\int e^{ax} \sin(bx + c)\, dx = \dfrac{e^{ax}}{\sqrt{a^2 + b^2}} \sin\left(bx + c - \tan^{-1}\dfrac{b}{a}\right) + C$

(ii) $\int e^{ax} \cos(bx + c)\, dx = \dfrac{e^{ax}}{\sqrt{a^2 + b^2}} \cos\left(bx + c - \tan^{-1}\dfrac{b}{a}\right) + C$

Solution: (i) $\text{L.H.S.} = \int e^{ax} \sin(bx + c)\, dx$

Integrating by parts taking e^{ax} as second function

$$= \sin(bx + c) \cdot \frac{e^{ax}}{a} - \int \cos(bx + c)\, b \cdot \frac{e^{ax}}{a}\, dx + C_1$$

$$= \frac{e^{ax}}{a} \sin(bx + c) - \frac{b}{a} \int e^{ax} \cos(bx + c)\, dx + C_1$$

Again integrating by parts

$$= \frac{e^{ax}}{a} \sin(bx + c) - \frac{b}{a}\left[\frac{e^{ax}}{a} \cos(bx + c) + \int e^{ax} \sin(bx + c) \cdot \frac{b}{a}\, dx\right] + C$$

$$= \frac{e^{ax}}{a} \sin(bx + c) - \frac{b}{a^2} e^{ax} \cos(bx + c) - \frac{b^2}{a^2} \int e^{ax} \sin(bx + c)\, dx + C$$

$$\Rightarrow \left(1+\frac{b^2}{a^2}\right)\int e^{ax}\sin(bx+c)\,dx = \frac{e^{ax}}{a^2}[a\sin(bx+c)-b\cos(bx+c)]+C$$

$$\Rightarrow \int e^{ax}\sin(bx+c)\,dx = \frac{e^{ax}}{a^2+b^2}[a\sin(bx+c)-b\cos(bx+c)]+C$$

Put $a = r\cos\alpha$ and $b = r\sin\alpha$

$$\Rightarrow \qquad \tan\alpha = \frac{b}{a} \text{ and } r = \sqrt{a^2+b^2}$$

$$\Rightarrow \quad \int e^{ax}\sin(bx+c)\,dx$$

$$= \frac{e^{ax}}{a^2+b^2}[r\cos\alpha\sin(bx+c)-r\sin\alpha\cos(bx+c)]+C$$

$$= \frac{re^{ax}}{a^2+b^2}[\sin(bx+c)\cos\alpha-\sin\alpha\cos(bx+c)]+C$$

$$= \frac{re^{ax}}{r^2}\cdot\sin(bx+c-\alpha)+C$$

$$= \frac{e^{ax}}{r}\cdot\sin(bx+c-\alpha)+C$$

$$\Rightarrow \int e^{ax}\sin(bx+c)\,dx = \frac{e^{ax}}{\sqrt{a^2+b^2}}\sin\left(bx+c-\tan^{-1}\frac{b}{a}\right)+C$$

(ii) Proof same as part (i)

Example 2. Evaluate the following integrals:

(i) $\int xe^{3x}\,dx$ (ii) $\int (2+7x)\cos 6x\,dx$

(iii) $\int \log x\,dx$ (iv) $\int x^2\sin 2x\,dx$

Solution: (i) $\int xe^{3x}\,dx$, Taking x as Ist function, integrating by parts, we get

$$= x\cdot\frac{e^{3x}}{3}-\int 1\cdot\frac{e^{3x}}{3}\,dx$$

$$= \frac{xe^{3x}}{3}-\frac{e^{3x}}{3\cdot 3}+C$$

$$= \frac{xe^{3x}}{3}-\frac{e^{3x}}{9}+C$$

$$= \frac{1}{9}(3x-1)e^{3x}+C$$

(ii) $\int (2+7x)\cos 6x\, dx = (2+7x)\dfrac{(\sin 6x)}{6} - \int (0+7)\dfrac{(\sin 6x)}{6}\, dx$

$$= \frac{(2+7x)}{6}\sin 6x - \frac{7}{6}\int \sin 6x\, dx$$

$$= \frac{(2+7x)}{6}\sin 6x + \frac{7}{6}\cdot\frac{1}{6}\cos 6x + C$$

$$= \frac{(2+7x)}{6}\sin 6x + \frac{7}{36}\cos 6x + C$$

(iii) $\int \log x\, dx$, Taking 1 as IInd function

$$= \int 1\cdot\log x\, dx$$

$$= \log x \times x - \int \frac{1}{x}\cdot x\, dx$$

$$= x\log x - \int 1\, dx$$

$$= x\log x - x + C$$

$$= x(\log x - 1) + C$$

(iv) $\int x^2 \sin 2x\, dx = x^2\cdot\dfrac{(-\cos 2x)}{2} - \int 2x\,\dfrac{(-\cos 2x)}{2}\, dx$

$$= -\frac{x^2}{2}\cos 2x + \int x\cos 2x\, dx$$

$$= -\frac{x^2}{2}\cos 2x + \left[x\cdot\frac{\sin 2x}{2} - \int 1\cdot\frac{\sin 2x}{2}\, dx\right]$$

$$= -\frac{x^2}{2}\cos 2x + \frac{x}{2}\sin 2x - \frac{1}{2}\int \sin 2x\, dx$$

$$= -\frac{x^2}{2}\cos 2x + \frac{x}{2}\sin 2x + \frac{1}{4}\cdot\cos 2x + C$$

Example 3. Evaluate:

(i) $\int x^3 \log x\, dx$ (ii) $\int \log(2+x^2)\, dx$

(iii) $\int x^3\sqrt{x^2+3}\, dx$ (iv) $\int \sin 2x\cdot\log(\cos 2x)\, dx$

Solution: $\int x^3 \log x\, dx = \log x\cdot\dfrac{x^4}{4} - \int \dfrac{1}{x}\cdot\dfrac{x^4}{4}\, dx$

$$= \frac{1}{4}\cdot x^4 \log x - \frac{1}{4}\int x^3\, dx$$

$$= \frac{1}{4}x^4 \log x - \frac{1}{4}\cdot\frac{1}{4}x^4 + C$$

$$= \frac{x^4}{16}(4\log x - 1) + C$$

(ii) $\int \log(2+x^2)\,dx = \int \log(2+x^2)\cdot 1\,dx$

$$= \log(2+x^2)\cdot x - \int \frac{1}{2+x^2}\cdot 2x\cdot x\,dx$$

$$= x\log(2+x^2) - 2\int \frac{x^2}{2+x^2}\,dx$$

$$= x\log(2+x^2) - 2\int \frac{x^2+2-2}{2+x^2}\,dx$$

$$= x\log(2+x^2) - 2\int \left[1 - \frac{2}{2+x^2}\right]dx$$

$$= x\log(2+x^2) - 2\int 1\,dx + 4\int \frac{dx}{2+x^2}$$

$$= x\log(2+x^2) - 2x + 4\int \frac{dx}{\left(\sqrt{2}\right)^2 + x^2}$$

$$= x\log(2+x^2) - 2x + \frac{4}{\sqrt{2}}\tan^{-1}\left(\frac{x}{\sqrt{2}}\right) + C$$

$$= x\log(2+x^2) - 2x + 2\sqrt{2}\tan^{-1}\left(\frac{x}{\sqrt{2}}\right) + C$$

(iii) $\int x^3\sqrt{x^2+3}\,dx$

Put $\sqrt{x^2+3} = t \Rightarrow x^2 + 3 = t^2$

$\Rightarrow x\,dx = t\,dt$

$\therefore \quad \int x^3\sqrt{x^2+3}\,dx = \int x\cdot x^2\cdot\sqrt{x^2+3}\,dx$

$$= \int (t^2-3)\cdot t\cdot t\,dt$$

$$= \int (t^4 - 3t^2)\,dt$$

$$= \frac{t^5}{5} - \frac{3t^3}{3} + C$$

$$= \frac{(x^2+3)^{5/2}}{5} - (x^2+3)^{3/2} + C$$

(iv) $\int \sin 2x \cdot \log(\cos 2x)\, dx$,

Let $\cos 2x = t \Rightarrow -\sin 2x \cdot 2\, dx = dt$

$\Rightarrow \quad \sin 2x\, dx = -\dfrac{dt}{2}$

$$\therefore \int \sin 2x \cdot \log(\cos 2x)\, dx = -\frac{1}{2}\int \log t \cdot dt$$

$$= -\frac{1}{2}\int \log t \cdot 1\, dt$$

$$= -\frac{1}{2}\left[t \log t - \int t \cdot \frac{1}{t}\, dt\right]$$

$$= -\frac{1}{2}[t \log t - t] + C$$

$$= -\frac{t}{2}[\log t - 1] + C$$

$$= -\frac{\cos 2x}{2}[\log(\cos 2x) - 1] + C$$

$$= \frac{\cos 2x}{2}[1 - \log(\cos 2x)] + C$$

Example 4. Evaluate:

(i) $\int \sin\sqrt{x}\, dx$ (ii) $\int \cos^{-1}(2x^2 - 1)\, dx$

(iii) $\int x \tan^{-1}(x^2)\, dx$ (iv) $\int \dfrac{e^{\sqrt{x}} \cos\sqrt{x}}{\sqrt{x}}\, dx$

Solution: (i) Let $\sqrt{x} = t \Rightarrow \dfrac{1}{2\sqrt{x}}\, dx = dt \Rightarrow dx = 2\sqrt{x}\, dt$

$\Rightarrow \quad dx = 2t\, dt$

$$\therefore \int \sin\sqrt{x}\, dx = \int \sin t \cdot 2t\, dt = 2\int t \cdot \sin t\, dt$$

$$= 2\left[-t \cos t + \int 1 \cdot \cos t\, dt\right]$$

$$= 2[-t \cos t + \sin t] + C$$

$$= 2\left[\sin\sqrt{x} - \sqrt{x}\cos\sqrt{x}\right] + C$$

(ii) Let $x = \cos t \Rightarrow dx = -\sin t\, dt$ and $t = \cos^{-1} x$

$$\therefore \int \cos^{-1}(2x^2 - 1)\, dx = \int \cos^{-1}(2\cos^2 t - 1)(-\sin t)\, dt$$

$$= \int \cos^{-1}(\cos 2t)(-\sin t)\, dt$$

$$= -\int 2t \sin t\, dt$$

$$= -2\int t\sin t\, dt$$

$$= -2[-t\cos t + \sin t] + C$$

$$= -2\left[-t\cos t + \sqrt{1-\cos^2 t}\right] + C$$

$$= -2\left[-x\cos^{-1} x + \sqrt{1-x^2}\right] + C$$

$$= 2\left[x\cos^{-1} x - \sqrt{1-x^2}\right] + C$$

(iii) $\int x\tan^{-1}(x^2)\, dx$,

Let $\quad x^2 = t \Rightarrow x\,dx = \dfrac{dt}{2}$

$$\therefore \quad \int x\tan^{-1} x^2\, dx = \int \tan^{-1} t \frac{dt}{2}$$

$$= \frac{1}{2}\int \tan^{-1} t\, dt$$

$$= \frac{1}{2}\int 1\cdot\tan^{-1} t\, dt$$

$$= \frac{1}{2}\left[t\tan^{-1} t - \int \frac{1}{1+t^2}\cdot t\, dt\right]$$

$$= \frac{1}{2}\left[t\tan^{-1} t - \frac{1}{2}\int (1+t^2)^{-1}(2t)\, dt\right]$$

$$= \frac{1}{2}\left[t\tan^{-1} t - \frac{1}{2}\log(1+t^2)\right] + C$$

$$= \frac{1}{2}\left[x^2\tan^{-1} x^2 - \frac{1}{2}\log(1+x^4)\right] + C$$

$$= \frac{1}{2}x^2\tan^{-1} x^2 - \frac{1}{4}\log(1+x^4) + C$$

(iv) Let $\sqrt{x} = t \Rightarrow \dfrac{1}{2\sqrt{x}}\, dx = dt \Rightarrow \dfrac{1}{\sqrt{x}}\, dx = 2dt$

$$\therefore \quad \int \frac{e^{\sqrt{x}}\cos\sqrt{x}}{\sqrt{x}}\, dx = 2\int e^t \cos t\, dt \qquad \text{...(i)}$$

Now, $\int e^t\cos t\, dt = e^t(\sin t) - \int e^t \sin t\, dt$

$$= e^t\sin t - \left[e^t(-\cos t) - \int e^t(-\cos t)\, dt\right]$$

$$= e^t\sin t + e^t\cos t - \int e^t\cos t\, dt + C$$

$$\therefore \qquad 2\int e^t \cos t\,dt = e^t(\sin t + \cos t) + C$$

$\therefore$ (i) $\Rightarrow$

$$\int \frac{e^{\sqrt{x}}\cos\sqrt{x}}{\sqrt{x}}\,dx = e^t(\sin t + \cos t) + C$$

$$= e^{\sqrt{x}}\left(\sin\sqrt{x} + \cos\sqrt{x}\right) + C$$

Example 5. Evaluate:

(i) $\int \left[\frac{1}{\log x} - \frac{1}{(\log x)^2}\right] dx$ (ii) $\int \frac{x\sin^{-1} x^2}{\sqrt{1-x^2}}\,dx$

(iii) $\int \cos^{-1}\left(\frac{1-x^2}{1+x^2}\right) dx$ (iv) $\int \sin^{-1}\left(\sqrt{\frac{x}{a+x}}\right) dx$

(v) $\int \frac{\cos^{-1} x}{x^2}\,dx$

Solution: (i) $\int \left[\frac{1}{\log x} - \frac{1}{(\log x)^2}\right] dx = \int \frac{dx}{\log x} - \int \frac{dx}{(\log x)^2}$

$$= x\cdot\frac{1}{\log x} - \int\left[-\frac{1}{(\log x)^2}\times\frac{1}{x}\right] x\,dx - \int \frac{dx}{(\log x)^2}$$

[Integrating I integral by parts]

$$= \frac{x}{\log x} + \int \frac{dx}{(\log x)^2} - \int \frac{dx}{(\log x)^2} = \frac{x}{\log x} + C$$

(ii) Put $x^2 = t \Rightarrow x\,dx = \frac{dt}{2}$

$$\therefore \qquad \int \frac{x\sin^{-1} x^2}{\sqrt{1-x^2}}\,dx = \frac{1}{2}\int \frac{\sin^{-1} t}{\sqrt{1-t}}\,dt = \frac{1}{2}\int \sin^{-1} t\cdot(1-t)^{-\frac{1}{2}}\,dt$$

$$= \frac{1}{2}\left[\sin^{-1}\frac{(1-t)^{1/2}}{\frac{1}{2}(-1)} - \int \frac{1}{\sqrt{1-t^2}}\cdot\frac{(1-t)^{1/2}}{\frac{1}{2}(-1)}\,dt\right]$$

$$= \frac{1}{2}\left[\frac{\sin^{-1} t\sqrt{1-t}}{-\frac{1}{2}} + 2\int \frac{\sqrt{1-t}}{\sqrt{1-t^2}}\,dt\right]$$

$$= -\sin^{-1} t \cdot \sqrt{1-t} + \int (1+t)^{1/2}\, dt$$

$$= -\sqrt{1-t}\sin^{-1} t + \frac{(1+t)^{1/2}}{\frac{1}{2}\cdot 1} + C$$

$$= -\sqrt{1-t}\sin^{-1} t + 2\sqrt{1+t} + C$$

$$= -\sqrt{1-x^2}\sin^{-1} x^2 + 2\sqrt{1+x^2} + C$$

(iii) Put $x = \tan\theta \Rightarrow dx = \sec^2\theta\, d\theta$ and $\theta = \tan^{-1}x$

$$\therefore \int \cos^{-1}\left(\frac{1-x^2}{1+x^2}\right) dx = \int \cos^{-1}\left(\frac{1-\tan^2\theta}{1+\tan^2\theta}\right)\sec^2\theta\, d\theta$$

$$= \int \cos^{-1}(\cos 2\theta)\sec^2\theta\, d\theta$$

$$= \int 2\theta\sec^2\theta\, d\theta$$

$$= 2\int \theta\sec^2\theta\, d\theta$$

$$= 2\left[\theta\cdot\tan\theta - \int 1\cdot\tan\theta\, d\theta\right]$$

$$= 2[\theta\tan\theta - \log\sec\theta] + C$$

$$= 2\left[x\tan^{-1} x - \log\sqrt{1+\tan^2\theta}\right] + C$$

$$= 2\left[x\tan^{-1} x - \log\sqrt{1+x^2}\right] + C$$

(iv) Put $x = a\tan^2 t \Rightarrow dx = 2a\tan t\sec^2 t\, dt$

and $\qquad t = \tan^{-1}\sqrt{\dfrac{x}{a}}$

$$\therefore \int \sin^{-1}\sqrt{\frac{x}{a+x}}\, dx = \int \sin^{-1}\sqrt{\frac{a\tan^2 t}{a+a\tan^2 t}}\cdot 2a\tan t\sec^2 t\, dt$$

$$= 2a\int \sin^{-1}\left(\frac{\tan t}{\sec t}\right)\cdot\tan t\cdot\sec^2 t\, dt$$

$$= 2a\int \sin^{-1}(\sin t)\cdot\tan t\cdot\sec^2 t\, dt$$

$$= 2a\int t\cdot\tan t\cdot\sec^2 t\, dt$$

$$= a\int t\cdot(2\tan t\sec^2 t)\, dt$$

$$= a\left[t\cdot\tan^2 t - \int 1\cdot\tan^2 t\, dt\right]$$

$$= at\tan^2 t - a\int (\sec^2 t - 1)\, dt$$

$$= at\tan^2 t - a(\tan t - t) + C$$

$$= a\cdot\frac{x}{a}\tan^{-1}\sqrt{\frac{x}{a}} - a\left(\sqrt{\frac{x}{a}} - \tan^{-1}\sqrt{\frac{x}{a}}\right) + C$$

$$= x\tan^{-1}\sqrt{\frac{x}{a}} - \sqrt{ax} + a\tan^{-1}\sqrt{\frac{x}{a}} + C$$

$$= (a+x)\tan^{-1}\left(\sqrt{\frac{x}{a}}\right) - \sqrt{ax} + C$$

(v) Put $\cos^{-1} x = t \Rightarrow x = \cos t \Rightarrow dx = -\sin t\, dt$

$$\therefore \quad \int \frac{\cos^{-1} x}{x^2}\, dx = \int \frac{t}{\cos^2 t}(-\sin t)\, dt$$

$$= -\int t(\sec t \tan t)\, dt$$

$$= -\left[t\cdot\sec t - \int 1\cdot\sec t\, dt\right]$$

$$= -t\sec t + \log|\sec t + \tan t| + C$$

$$= -\frac{t}{\cos t} + \log\left|\frac{1+\sin t}{\cos t}\right| + C$$

$$= -\frac{\cos^{-1} x}{x} + \log\left|\frac{1+\sqrt{1-x^2}}{x}\right| + C$$

$$\left[\because \sin t = \sqrt{1-\cos^2 t} = \sqrt{1-x^2}\right]$$

Example 6. Evaluate:

(i) $\int x^2\cdot(\log x)^2\, dx$ (ii) $\int \sec^3 x\, dx$

(iii) $\int e^x \sin x \cos 2x\, dx$ (iv) $\int \sin(\log x)\, dx$

Solution: (i) $\int x^2\cdot(\log x)^2\, dx = \int (\log x)^2\cdot x^2\, dx$

$$= (\log x)^2\cdot\frac{x^3}{3} - \int 2\log x\cdot\frac{1}{x}\cdot\frac{x^3}{3}\, dx$$

$$= \frac{x^3}{3}\cdot(\log x)^2 - \frac{2}{3}\int x^2\log x\, dx$$

$$= \frac{x^3}{3}(\log x)^2 - \frac{2}{3}\left[\log x\cdot\frac{x^3}{3} - \int\frac{1}{x}\cdot\frac{x^3}{3}\, dx\right]$$

$$= \frac{1}{3}\cdot x^3(\log x)^2 - \frac{2}{9}x^3\log x + \frac{2}{3}\int\frac{x^2}{3}\, dx$$

$$= \frac{1}{3}\cdot x^3\cdot(\log x)^2 - \frac{2}{9}x^3\log x + \frac{2}{9}\cdot\frac{x^3}{3} + C$$

$$= \frac{1}{3} \cdot x^3 \cdot (\log x)^2 - \frac{2}{9} x^3 \log x + \frac{2}{27} x^3 + C$$

$$= \frac{1}{27} x^3 [9(\log x)^2 - 6 \log x + 2] + C$$

(ii) $\int \sec^3 x \, dx = \int \sec x \cdot \sec^{2x} \, dx$

$$= \sec x \cdot \tan x - \int \sec x \cdot \tan x \cdot \tan x \, dx$$

$$= \sec x \tan x - \int \sec x (\sec^2 x - 1) \, dx$$

$$= \sec x \tan x - \int \sec^3 x \, dx + \int \sec x \, dx$$

$$\Rightarrow \quad 2 \int \sec^3 x \, dx = \sec x \tan x + \log |\sec x + \tan x| + 2C$$

$$\Rightarrow \quad \int \sec^3 x \, dx = \frac{1}{2} (\sec x \tan x + \log |\sec x + \tan x|) + C$$

(iii) $\int e^x \sin x \cos 2x \, dx = \frac{1}{2} \int e^x (2 \sin x \cos 2x) \, dx$

$$= \frac{1}{2} \int e^x (\sin 3x - \sin x) \, dx$$

$$= \frac{1}{2} \int e^x \sin 3x - \frac{1}{2} \int e^x \sin x \, dx$$

$$= \frac{1}{2} \left[\frac{e^x}{\sqrt{1+3^2}} \sin(3x - \tan^{-1} 3) \right] - \frac{1}{2} \left[\frac{e^x}{\sqrt{1+1^2}} \sin(x - \tan^{-1}(1)) \right] + C$$

$$= \frac{e^x}{2} \left[\frac{1}{\sqrt{10}} \sin(3x - \tan^{-1} 3) - \frac{1}{\sqrt{2}} \sin\left(x - \frac{\pi}{4} \right) \right] + C$$

[using Ex.(1)]

(iv) $\int \sin(\log x) \, dx = \int \sin(\log x) \cdot 1 \, dx$

$$= \sin(\log x) \cdot x - \int \cos(\log x) \cdot \frac{1}{x} \cdot x \, dx$$

$$= x \sin(\log x) - \int \cos(\log x) \cdot 1 \, dx$$

$$= x \sin(\log x) - \left[\cos(\log x) \cdot x - \int -\sin(\log x) \cdot \frac{1}{x} \cdot x \, dx \right]$$

$$= x \sin(\log x) - x \cos(\log x) - \int \sin(\log x) \, dx + 2C$$

$$\Rightarrow \quad 2 \int \sin(\log x) \, dx = x \sin(\log x) - x \cos(\log x) + 2C$$

$$\Rightarrow \quad \int \sin(\log x) \, dx = \frac{1}{2} \cdot x \cdot [\sin(\log x) - \cos(\log x)] + C$$

Example 7. Evaluate:

(*i*) $\int e^x\left(\dfrac{1}{x^2}-\dfrac{2}{x^3}\right)dx$ (ii) $\int \dfrac{x-1}{(x+1)^3}e^x\,dx$

(iii) $\int e^x\cdot\dfrac{1-\sin x}{1-\cos x}\,dx$ (iv) $\int e^x[\sec x+\log(\sec x+\tan x)]$

Solution: (i) $\int e^x\left(\dfrac{1}{x^2}-\dfrac{2}{x^3}\right)dx = \int e^x\left(\dfrac{1}{x^2}+\dfrac{d}{dx}\left(\dfrac{1}{x^2}\right)\right)dx$

$$= e^x\cdot\frac{1}{x^2}+C$$

$$\left[\because \int e^x\left(f(x)+f'(x)\right)dx = e^x f(x)+C\right]$$

(ii) $\int \dfrac{x-1}{(x+1)^3}e^x\,dx = \int \dfrac{x+1-2}{(x+1)^3}e^x\,dx$

$$= \int\left[\frac{1}{(1+x)^2}-\frac{2}{(1+x)^3}\right]e^x\,dx$$

$$= \int e^x\left[\frac{1}{(1+x)^2}+\frac{d}{dx}\left(\frac{1}{(1+x)^2}\right)\right]dx$$

$$= e^x\frac{1}{(1+x)^2}+C$$

(iii) $\int e^x\cdot\dfrac{1-\sin x}{1-\cos x}\,dx = \int e^x\cdot\dfrac{1-2\sin x\dfrac{x}{2}\cos\dfrac{x}{2}}{2\sin^2\dfrac{x}{2}}\,dx$

$$= \int e^x\left[\frac{1}{2}\text{cosec}^2\frac{x}{2}-\cot\frac{x}{2}\right]dx$$

$$= -\int e^x\left[\cot\frac{x}{2}-\frac{1}{2}\text{cosec}^2\frac{x}{2}\right]dx$$

$$= -\int e^x\left[\cot\frac{x}{2}+\frac{d}{dx}\left(\cot\frac{x}{2}\right)\right]dx$$

$$= -e^x\cot\frac{x}{2}+C$$

$$\left[\text{Using} \int e^x\left(f(x)+f'(x)\right)dx = e^x f(x)+C\right]$$

(iv) $\int e^x[\sec x+\log(\sec x+\tan x)]\,dx$

$$= \int e^x\left[\log(\sec x+\tan x)+\frac{d}{dx}(\log(\sec x+\tan x)\right]dx$$

$$\left[\because \frac{d}{dx}\big(\log(\sec x + \tan x)\big) = \frac{\sec x \tan x + \sec^2 x}{\sec x + \tan x} = \sec x\right]$$

$$= e^x \log(\sec x + \tan x) + C$$

Example 8. Evaluate:

(i) $\int \left[\log(\log x) + \frac{1}{(\log x)^2}\right] dx$ (ii) $\int \frac{\log x}{(1 + \log x)^2}\, dx$

(iii) $\int (-\sin x + 2\cos x)e^{2x}\, dx$

Solution: (i) $\int \left[\log(\log x) + \frac{1}{(\log x)^2}\right] dx$

$$= \int \log(\log x)\cdot 1\, dx + \int \frac{1}{(\log x)^2}\, dx$$

$$= \log(\log x)\cdot x - \int \frac{1}{\log x}\cdot\frac{1}{x}\cdot x\, dx + \int \frac{1}{(\log x)^2}\, dx$$

$$= x\log(\log x) - \int \frac{1}{\log x}\, dx + \int \frac{dx}{(\log x)^2}$$

$$= x\log(\log x) - \int (\log x)^{-1}\cdot 1\, dx + \int \frac{dx}{(\log x)^2}$$

$$= x\log(\log x) - \left[(\log x)^{-1}\cdot x - \int (-1)(\log x)^{-2}\cdot\frac{1}{x}\cdot x\, dx\right] + \int \frac{dx}{(\log x)^2}$$

$$= x\log(\log x) - \frac{x}{\log x} - \int \frac{dx}{(\log x)^2} + \int \frac{dx}{(\log x)^2} + C$$

$$= x\left[\log(\log x) - \frac{1}{\log x}\right] + C$$

(ii) $\int \frac{\log x}{(1 + \log x)^2}\, dx = \int \frac{1 + \log x - 1}{(1 + \log x)^2}\, dx$

$$= \int \left[\frac{1}{1 + \log x} - \frac{1}{(1 + \log x)^2}\right] dx$$

$$= \int (1 + \log x)^{-1}\cdot 1\, dx - \int \frac{dx}{(1 + \log x)^2}$$

$$= (1 + \log x)^{-1}\cdot x - \int (-1)\,(1 + \log x)^{-2}\left(0 + \frac{1}{x}\right)\cdot x\, dx - \int \frac{dx}{(1 + \log x)^2}$$

$$= \frac{x}{(1+\log x)} + \int \frac{dx}{(1+\log x)^2} - \int \frac{dx}{(1+\log x)^2} + C$$

$$= \frac{x}{1+\log x} + C$$

(iii) $\int (-\sin x + 2\cos x)e^x \, dx = -\int e^{2x} \sin x \, dx + \int e^{2x} \cdot 2\cos x \, dx$

$$= -e^{2x}(-\cos x) - \int e^{2x} \cdot 2(-\cos) \, dx + \int 2e^{2x} \cos x \, dx$$

$$= e^{2x} \cos x + \int 2e^{2x} \cos x + \int 2e^{2x} \cos x \, dx + C$$

$$= e^{2x} \cos x + C$$

EXERCISE 4.4

Evaluate the following integrals:

1. (i) $\int x\sec^2 x \, dx$ (ii) $\int x\sin x \, dx$

(iii) $\int (2x+5)\sin 2x \, dx$ (iv) $\int (2x-3)\sin^2 x \, dx$

(v) $\int x\tan^2 x \, dx$ (vi) $\int x\sin 3x \sin 5x \, dx$

2. (i) $\int x\log x \, dx$ (ii) $\int x\log(1+x) \, dx$

(iii) $\int \frac{\log(\log x)}{x} dx$ (iv) $\int \frac{\log x}{x^2} dx$

(v) $\int \frac{\log(x+2)}{(x+2)^2} dx$ (vi) $\int (e^{\log x} + \sin x)\cos x \, dx$

3. (i) $\int \sin^{-1} x \, dx$ (ii) $\int \frac{x\sin^{-1} x}{\sqrt{1-x^2}} dx$

(iii) $\int \tan^{-1} x \, dx$ (iv) $\int \sin^{-1}(3x-4x^3) \, dx$

(v) $\int \tan^{-1}\left(\frac{3x-x^3}{1-3x^2}\right) dx$ (vi) $\int \sec^{-1} x \, dx$

(vii) $\int \tan^{-1}\sqrt{\frac{1-x}{1+x}} \, dx$ (viii) $\int \frac{e^{m\tan^{-1} x}}{(1+x^2)^{3/2}} dx$

4. (i) $\int e^{3x} \sin 4x \, dx$ (ii) $\int e^{ax} \cos bx \, dx$

(iii) $\int e^{3x} \sin x \sin 3x \, dx$ (iv) $\int x^2 e^{x^3} \cos x^3 \, dx$

(v) $\int e^x x^3 \, dx$

5. (i) $\int e^x(\cot x - \operatorname{cosec}^2 x)\,dx$ (ii) $\int e^x\left(\sin^{-1}x + \frac{1}{\sqrt{1-x^2}}\right)dx$

(iii) $\int e^x\left(\frac{1+\sin x}{1+\cos x}\right)dx$ (iv) $\int \frac{x+\sin x}{1+\cos x}\,dx$

(v) $\int \frac{x-1}{x^2}e^x$ (vi) $\int e^x \sec x(1+\tan x)\,dx$

(vii) $\int \frac{\sin x\cos x - 1}{\sin^2 x}\cdot e^x\,dx$ (viii) $\int e^x\cdot\frac{1+x^2}{(1+x)^2}\,dx$

(ix) $\int \frac{1}{x}(x(\log x)^2 + 2\log x)e^x\,dx$ (x) $\int \frac{2+\sin x}{1+\cos x}e^{x/2}\,dx$

6. (i) $\int \cos(\log x)\,dx$ (ii) $\int \operatorname{cosec}^3 x\,dx$

(iii) $\int \cos^{-1}\left(\frac{1}{x}\right)dx$ $\left[\text{Hint. } \sec^{-1}x = \cos^{-1}\frac{1}{x}\right]$

(iv) $\int e^{\sin x}\cdot\sin 2x\,dx$ (v) $\int x^3\tan^2(x^2)\,dx$

ANSWERS

1. (i) $x\tan x + \log|\cos x| + C$ (ii) $\sin x - x\cos x + C$

(iii) $-\frac{1}{2}(2x+5)\cos 2x + \frac{\sin 2x}{2} + C$

(iv) $\frac{1}{8}(2x-3)^2 - \frac{1}{4}(2x-3)\sin 2x - \frac{1}{4}\cos 2x + C$

(v) $x\tan x - \log\sec x - \frac{x^2}{2} + C$

(vi) $\frac{1}{128}(32x\sin 2x + 16\cos 2x - 8x\sin 8x - \cos 8x) + C$

2. (i) $\frac{1}{4}x^2(2\log x - 1) + C$ (ii) $\frac{1}{2}(x^2-1)\log(1+x) - \frac{1}{4}x^2 + \frac{1}{2}x + C$

(iii) $(\log(\log x) - 1)\log x + C$ (iv) $-\frac{1}{x}(\log x + 1) + C$

(v) $-\frac{1}{x+2}(\log(x+2)+1) + C$ (vi) $x\sin x + \cos x + \frac{1}{2}\sin^2 x + C$

3. (i) $x\sin^{-1}x + \sqrt{1-x^2} + C$ (ii) $-\sqrt{1-x^2}\sin^{-1}x + x + C$

(iii) $x\tan^{-1}x - \frac{1}{2}\log(1+x^2) + C$ (iv) $3\left(x\sin^{-1}x + \sqrt{1-x^2}\right) + C$

(v) $3x\tan^{-1}x - \log|(1+x^2)^{3/2}| + C$ (vi) $x\sec^{-1}x - \log\left|x+\sqrt{x^2-1}\right| + C$

(vii) $\frac{x}{2}\cos^{-1}x - \frac{1}{2}\sqrt{1-x^2} + C$ (viii) $\frac{e^{m\tan^{-1}x}}{1+m^2}\cdot\frac{x+m}{\sqrt{1+x^2}} + C$

4. (i) $\frac{1}{25}e^{3x}(3\sin 4x - 4\cos 4x) + C$

(ii) $\frac{e^{ax}}{a^2+b^2}(a\cos bx + b\sin bx) + C$

(iii) $\frac{1}{2}\left[\frac{e^{3x}}{\sqrt{13}}\cos\left(2x - \tan^{-1}\frac{2}{3}\right) - \frac{1}{5}e^{3x}\cos\left(4x - \tan^{-1}\frac{4}{3}\right)\right] + C$

(iv) $\frac{1}{6}e^{x^3}(\sin x^3 + \cos x^3) + C$

(v) $(x^3 - 3x^2 + 6x - 6)e^x + C$

5. (i) $e^x\cot x + C$ (ii) $e^x\sin^{-1}x + C$

(iii) $e^x\tan\frac{x}{2} + C$ (iv) $x\tan\frac{x}{2} + C$

(v) $\frac{e^x}{x} + C$ (vi) $e^x\sec x + C$

(vii) $e^x\cot x + C$ (viii) $\frac{x-1}{x+1}e^x + C$

(ix) $e^x(\log x)^2 + C$ (x) $2e^{x/2} + \tan\frac{x}{2} + C$

6. (i) $\frac{1}{2}x(\cos(\log x) + \sin(\log x) + C$

(ii) $-\frac{1}{2}(\text{cosec } x\cot x - \log|\text{cosec } x - \cot x|) + C$

(iii) $x\sin^{-1}x - \log\left|x+\sqrt{x^2-1}\right| + C$

(iv) $2(\sin x - 1)e^{\sin x} + C$

(v) $\frac{1}{4}(2x^2\tan x^2 + 2\log|\cos x^2| - x^4) + C$

4.10 SOME MORE STANDARD INTEGRALS

1. $\int\sqrt{a^2-x^2}\,dx = \frac{x}{2}\sqrt{a^2-x^2} + \frac{a^2}{2}\sin^{-1}\frac{x}{a} + C$

Proof: $\int\sqrt{a^2-x^2}\,dx = \int\sqrt{a^2-x^2}\cdot 1\,dx$

Integrating by parts, taking $\sqrt{a^2 - x^2}$ as the first function.

$$= \sqrt{a^2 - x^2} \cdot x - \int \frac{(a^2 - x^2)^{-1/2}}{2}(-2x) \cdot x\, dx$$

$$= x\sqrt{a^2 - x^2} - \int \frac{-x^2}{\sqrt{a^2 - x^2}}\, dx$$

$$= x\sqrt{a^2 - x^2} - \int \frac{(a^2 - x^2) - a^2}{\sqrt{a^2 - x^2}}\, dx$$

$$= x\sqrt{a^2 - x^2} - \int \left(\sqrt{a^2 - x^2} - \frac{a^2}{\sqrt{a^2 - x^2}} \right) dx$$

$$= x\sqrt{a^2 - x^2} - \int \sqrt{a^2 - x^2}\, dx + a^2 \int \frac{dx}{\sqrt{a^2 - x^2}}$$

$$\Rightarrow\ 2\int \sqrt{a^2 - x^2}\, dx = x\sqrt{a^2 - x^2} + a^2 \int \frac{dx}{\sqrt{a^2 - x^2}}$$

$$= x\sqrt{a^2 - x^2} + a^2 \cdot \sin^{-1}\frac{x}{a} + 2C$$

$$\Rightarrow\ \int \sqrt{a^2 - x^2}\, dx = \frac{x}{2}\sqrt{a^2 - x^2} + \frac{a^2}{2}\sin^{-1}\frac{x}{a} + C$$

2. $\int \sqrt{a^2 + x^2}\, dx = \frac{x}{2}\sqrt{a^2 + x^2} + \frac{a^2}{2}\log\left| x + \sqrt{x^2 + a^2} \right| + C$

Proof: $\int \sqrt{a^2 + x^2}\, dx = \int \sqrt{a^2 + x^2} \cdot 1\, dx$

Integrating by parts taking $\sqrt{a^2 + x^2}$ as the first function

$$= \sqrt{a^2 + x^2} \cdot x - \int \frac{1}{2}(a^2 + x^2)^{-1/2}(2x) \cdot x\, dx$$

$$= x\sqrt{a^2 + x^2} - \int \frac{x^2}{\sqrt{a^2 + x^2}}\, dx$$

$$= x\sqrt{a^2 + x^2} - \int \frac{(a^2 + x^2) - a^2}{\sqrt{a^2 + x^2}}\, dx$$

$$= x\sqrt{a^2 + x^2} - \int \sqrt{a^2 + x^2} + a^2 \int \frac{dx}{\sqrt{a^2 + x^2}}$$

$$= x\sqrt{a^2 + x^2} - \int \sqrt{a^2 + x^2}\, dx + a^2 \log\left| x + \sqrt{x^2 + a^2} \right| + 2C$$

$$\Rightarrow\quad 2\int \sqrt{a^2 + x^2}\, dx = x\sqrt{a^2 + x^2} + a^2 \log\left| x + \sqrt{a^2 + x^2} \right| + 2C$$

$$\Rightarrow \qquad \int \sqrt{a^2+x^2}\,dx = \frac{x}{2}\sqrt{a^2+x^2}+\frac{a^2}{2}\log\left|x+\sqrt{a^2+x^2}\right|+C$$

3. $\int \sqrt{x^2-a^2}\,dx = \frac{x}{2}\sqrt{x^2-a^2}-\frac{a^2}{2}\log\left|x+\sqrt{x^2-a^2}\right|+C$

Proof: Same as above.

ILLUSTRATIVE EXAMPLES

Example 1: Evaluate the following integrals:

(i) $\int \sqrt{25-36x^2}\,dx$ (ii) $\int \sqrt{9x^2+49}\,dx$

(iii) $\int \sqrt{4x^2-1}\,dx$ (iv) $\int \frac{x^2}{\sqrt{x^2+6}}\,dx$

Solution: (i) $\int \sqrt{25-36x^2}\,dx = 6\int \sqrt{\left(\frac{5}{6}\right)^2-x^2}\,dx$

$$= 6\left[\frac{x}{2}\sqrt{\left(\frac{5}{6}\right)^2-x^2}+\frac{\left(\frac{5}{6}\right)^2}{2}\sin^{-1}\left(\frac{x}{\frac{5}{6}}\right)\right]+C$$

$$= 3x\sqrt{\frac{25}{36}-x^2}+\frac{25}{12}\sin^{-1}\left(\frac{6x}{5}\right)+C$$

$$= \frac{x}{2}\sqrt{25-36x^2}+\frac{25}{12}\sin^{-1}\left(\frac{6x}{5}\right)+C$$

(ii) $\int \sqrt{9x^2+49}\,dx = 3\int \sqrt{x^2+\frac{49}{9}}\,dx$

$$= 3\int \sqrt{x^2+\left(\frac{7}{3}\right)^2}\,dx$$

$$= 3\left[\frac{x}{2}\sqrt{x^2+\frac{49}{9}}+\frac{\left(\frac{7}{3}\right)^2}{2}\log\left|x+\sqrt{x^2+\frac{49}{9}}\right|+C\right.$$

$$= \frac{x}{2}\sqrt{9x^2+49}+\frac{49}{6}\log\left|x+\sqrt{x^2+\frac{49}{9}}\right|+C$$

(iii) $\int \sqrt{4x^2-1}\,dx = 2\int \sqrt{x^2-\left(\frac{1}{2}\right)^2}\,dx$

$$= 2\left[\frac{x}{2}\sqrt{x^2-\frac{1}{4}} - \frac{\left(\frac{1}{2}\right)^2}{2}\log\left|x+\sqrt{x^2-\frac{1}{4}}\right|\right] + C$$

$$= \frac{x}{2}\sqrt{4x^2-1} - \frac{1}{4}\log\left|x+\sqrt{x^2-\frac{1}{4}}\right| + C$$

(iv) $\int \frac{x^2}{\sqrt{x^2+6}}\,dx = \int \frac{(x^2+6)-6}{\sqrt{x^2+6}}\,dx$

$$= \int \sqrt{x^2+6}\,dx - 6\int \frac{dx}{\sqrt{x^2+6}}$$

$$= \frac{x}{2}\sqrt{x^2+6} + \frac{(\sqrt{6})^2}{2}\log\left|x+\sqrt{x^2+6}\right| - 6\log\left|x+\sqrt{x^2+6}\right| + C$$

$$= \frac{x}{2}\sqrt{x^2+6} + 3\log\left|x+\sqrt{x^2+6}\right| - 6\log\left|x+\sqrt{x^2+6}\right| + C$$

$$= \frac{x}{2}\sqrt{x^2+6} - 3\log\left|x+\sqrt{x^2+6}\right| + C$$

Example 2. Evaluate the following integrals:

(i) $\int (x-2)\sqrt{\frac{x+3}{x-3}}\,dx$ (ii) $\int \sqrt{x^2+3x}\,dx$

(iii) $\int (x+1)\sqrt{1-x-x^2}\,dx$

Solution: (i) $\int (x-2)\sqrt{\frac{x+3}{x-3}}\,dx = \int \frac{(x-2)(x+3)}{\sqrt{x-3}\sqrt{x+3}}\,dx$

$$= \int \frac{x^2+x-6}{\sqrt{x^2-9}}\,dx$$

$$= \int \frac{(x^2-9)+(x+3)}{\sqrt{x^2-9}}\,dx$$

$$= \int \sqrt{x^2-9}\,dx + \int \frac{x\,dx}{\sqrt{x^2-9}} + 3\int \frac{dx}{\sqrt{x^2-9}}$$

$$= \frac{x}{2}\sqrt{x^2-9} - \frac{9}{2}\log\left|x+\sqrt{x^2-9}\right| + \frac{1}{2}\int (x^2-9)^{-1/2}\cdot(2x)\,dx$$

$$+ 3\log\left|x+\sqrt{x^2-9}\right| + C$$

$$= \frac{x}{2}\sqrt{x^2-9} - \frac{3}{2}\log\left|x+\sqrt{x^2-9}\right| + \frac{1}{2}\frac{(x^2-9)^{1/2}}{\frac{1}{2}} + C$$

$$= \frac{1}{2}(x+2)\sqrt{x^2-9} - \frac{3}{2}\log\left|x+\sqrt{x^2-9}\right| + C$$

(ii) $\int \sqrt{x^2+3x}\,dx = \int \sqrt{\left(x+\frac{3}{2}\right)^2 - \left(\frac{3}{2}\right)^2}\,dx$

(Put $x+\frac{3}{2} = t \Rightarrow dx = dt$)

$$= \int \sqrt{t^2 - \left(\frac{3}{2}\right)^2}\,dt$$

$$= \frac{t}{2}\sqrt{t^2-\frac{9}{4}} - \frac{\left(\frac{3}{2}\right)^2}{2}\log\left|t+\sqrt{t^2-\left(\frac{3}{2}\right)^2}\right| + C$$

$$= \frac{\left(x+\frac{3}{2}\right)}{2}\sqrt{\left(x+\frac{3}{2}\right)^2 - \frac{9}{4}} - \frac{9}{8}\log\left|x+\frac{3}{2}+\sqrt{\left(x+\frac{3}{2}\right)^2-\left(\frac{3}{2}\right)^2}\right| + C$$

$$= \frac{2x+3}{4}\sqrt{x^2+3x} - \frac{9}{8}\log\left|x+\frac{3}{2}+\sqrt{x^2+3x}\right| + C$$

(iii) $\int (x+1)\sqrt{1-x-x^2}\,dx = \int \left[-\frac{1}{2}(-1-2x)+\frac{1}{2}\right]\sqrt{1-x-x^2}\,dx$

$$= -\frac{1}{2}\int (1-x-x^2)^{1/2}(-1-2x)\,dx + \frac{1}{2}\int \sqrt{1-(x+x^2)}\,dx$$

$$= -\frac{1}{2}\frac{(1-x-x^2)^{3/2}}{\frac{3}{2}} + \frac{1}{2}\int \sqrt{1-\left(x^2+x+\frac{1}{4}\right)+\frac{1}{4}}\,dx$$

$$= -\frac{1}{3}(1-x-x^2)^{3/2} + \frac{1}{2}\int \sqrt{\left(\frac{\sqrt{5}}{2}\right)^2 - \left(x+\frac{1}{2}\right)^2}\,dx$$

(Put $x+\frac{1}{2} = t \Rightarrow dx = dt$)

$$= -\frac{1}{3}(1-x-x^2)^{3/2} + \frac{1}{2}\int \sqrt{\left(\frac{\sqrt{5}}{2}\right)^2 - t^2}\,dt$$

$$= -\frac{1}{3}(1-x-x^2)^{3/2} + \frac{1}{2}\left[\frac{t}{2}\sqrt{\left(\frac{\sqrt{5}}{2}\right)^2 - t^2} + \frac{\left(\frac{\sqrt{5}}{2}\right)^2}{2}\sin^{-1}\left(\frac{t}{\frac{\sqrt{5}}{2}}\right)\right] + C$$

$$= -\frac{1}{3}(1-x-x^2)^{3/2} + \frac{1}{2}\left[\frac{\left(x+\frac{1}{2}\right)}{2}\sqrt{\frac{5}{4} - \left(x+\frac{1}{2}\right)^2} + \frac{5}{8}\sin^{-1}\left(\frac{2\left(x+\frac{1}{2}\right)}{\sqrt{5}}\right)\right] + C$$

$$= -\frac{1}{3}(1-x-x^2)^{3/2} + \frac{2x+1}{8}\sqrt{1-x-x^2} + \frac{5}{16}\sin^{-1}\left(\frac{2x+1}{\sqrt{5}}\right) + C$$

Example 3. Evaluate the following integrals:

(i) $\int \cos x\sqrt{4-\sin^2 x}\,dx$ (ii) $\int \sin^{-1}\sqrt{x}\,dx$

(iii) $\int \frac{\sin^{-1}\sqrt{x} - \cos^{-1}\sqrt{x}}{\sin^{-1}\sqrt{x} + \cos^{-1}\sqrt{x}}\,dx$ (iv) $\int \sec x \tan x\sqrt{\tan^2 x - 4}\,dx$

Solution: (i) Put $\sin x = t \Rightarrow \cos x\,dx = dt$

$$\therefore \quad \int \cos x\sqrt{4-\sin^2 x}\,dx = \int \sqrt{4-t^2}\,dt$$

$$= \int \sqrt{(2)^2 - t^2}\,dt$$

$$= \frac{1}{2}\sqrt{(2)^2 - t^2} + \frac{(2)^2}{2}\sin^{-1}\left(\frac{t}{2}\right) + C$$

$$= \frac{\sin x}{2}\sqrt{4-\sin^2 x} + 2\sin^{-1}\left(\frac{\sin x}{2}\right) + C$$

(ii) Put $\sqrt{x} = t \Rightarrow x = t^2 \Rightarrow dx = 2t\,dt$

$$\therefore \quad \int \sin^{-1}\sqrt{x}\,dx = \int \sin^{-1} t \cdot 2t\,dt$$

$$= 2\int t\sin^{-1} t\,dt$$

$$= 2\left[\sin^{-1} t \cdot \frac{t^2}{2} - \int \frac{1}{\sqrt{1-t^2}} \cdot \frac{t^2}{2}\,dt\right]$$

$$= t^2 \sin^{-1} t - \int \frac{t^2}{\sqrt{1-t^2}}\,dt$$

$$= t^2 \sin^{-1} t - \int \frac{-(1-t^2)+1}{\sqrt{1-t^2}}\,dt$$

$$= t^2 \sin^{-1} t + \int \sqrt{1-t^2}\, dt - \int \frac{dt}{\sqrt{1-t^2}}$$

$$= t^2 \sin^{-1} t + \left[\frac{t}{2}\sqrt{1-t^2} + \frac{1^2}{2}\sin^{-1} t\right] - \sin^{-1} t + C$$

$$= t^2 \sin^{-1} t + \frac{t}{2}\sqrt{1-t^2} - \frac{1}{2}\sin^{-1} t + C$$

$$= \left(t^2 - \frac{1}{2}\right)\sin^{-1} t + \frac{t}{2}\sqrt{1-t^2} + C$$

$$= \left(x - \frac{1}{2}\right)\sin^{-1}\sqrt{x} + \frac{\sqrt{x}}{2}\sqrt{1-x} + C$$

(iii) We know that,

$$\sin^{-1}\sqrt{x} + \cos^{-1}\sqrt{x} = \frac{\pi}{2} \Rightarrow \cos^{-1}\sqrt{x} = \frac{\pi}{2} - \sin^{-1}\sqrt{x}$$

$$\therefore \quad \int \frac{\sin^{-1}\sqrt{x} - \cos^{-1}\sqrt{x}}{\sin^{-1}\sqrt{x} + \cos^{-1}\sqrt{x}}\, dx$$

$$= \int \frac{\sin^{-1}\sqrt{x} - \left(\frac{\pi}{2} - \sin^{-1}\sqrt{x}\right)}{\frac{\pi}{2}}\, dx = \int \frac{2\sin^{-1}\sqrt{x} - \frac{\pi}{2}}{\frac{\pi}{2}}\, dx$$

$$= \frac{4}{\pi}\int \sin^{-1}\sqrt{x} - \int 1\, dx$$

$$= \frac{4}{\pi}\left[\left(x - \frac{1}{2}\right)\sin^{-1}\sqrt{x} + \frac{\sqrt{x}}{2}\sqrt{1-x}\right] - x + C$$

(By part (ii)]

$$= \frac{2}{\pi}\left(x - \frac{1}{2}\right)\sin^{-1}\sqrt{x} + \frac{2\sqrt{x}}{\pi}\sqrt{1-x} - x + C$$

(iv) $\int \sec x \tan x \sqrt{\tan^2 x - 4}\, dx = \int \sec x \tan x \sqrt{\sec^2 x - 5}\, dx$

(Put $\sec x = t \Rightarrow \sec x \tan x\, dx = dt$)

$$= \int \sqrt{t^2 - 5}\, dt = \int \sqrt{t^2 - \left(\sqrt{5}\right)^2}\, dt$$

$$= \frac{t}{2}\sqrt{t^2 - \left(\sqrt{5}\right)^2} - \frac{\left(\sqrt{5}\right)^2}{2}\log\left|t + \sqrt{t^2 - \left(\sqrt{5}\right)^2}\right| + C$$

$$= \frac{\sec x}{2}\sqrt{\sec^2 x - 5} - \frac{5}{2}\log\left|\sec x + \sqrt{\sec^2 x - 5}\right| + C$$

EXERCISE 4.5

Evaluate the following integrals:

1. (i) $\int \sqrt{6-x^2}\,dx$ (ii) $\int \sqrt{x^2-7}\,dx$

(iii) $\int \sqrt{x^2+3}\,dx$ (iv) $\int \sqrt{4x^2+9}\,dx$

(v) $\int \sqrt{16-25x^2}\,dx$ (vi) $\int \sqrt{2-3x^2}\,dx$

2. (i) $\int \sqrt{x^2+8x-6}\,dx$ (ii) $\int \sqrt{\dfrac{5-x}{x-2}}\,dx$ [Hint. put $x-2=t$]

(iii) $\int \sqrt{ax-x^2}\,dx$ (iv) $\int x^2\sqrt{x^6-1}\,dx$ [Hint. put $x^3=t$]

(v) $\int e^x\sqrt{e^{2x}+1}\,dx$ (vi) $\int \dfrac{\sqrt{x+1}-\sqrt{x-1}}{\sqrt{x+1}+\sqrt{x-1}}\,dx$

3. (i) $\int \cos x\sqrt{4+\sin^2 x}\,dx$ (ii) $\int x\sin^{-1}x\,dx$

(iii) $\int x\cos^{-1}x\,dx$ (iv) $\int \sec x\tan x\sqrt{\sec^2 x+1}\,dx$

(v) $\int \dfrac{x^2+1}{\sqrt{x^2+3}}\,dx$ (vi) $\int x\sqrt{x^4+1}\,dx$

ANSWERS

1. (i) $\dfrac{x}{2}\sqrt{6-x^2}+3\sin^{-1}\dfrac{x}{\sqrt{6}}+C$

(ii) $\dfrac{x}{2}\sqrt{x^2-7}-\dfrac{7}{2}\log\left|x+\sqrt{x^2-7}\right|+C$

(iii) $\dfrac{x}{2}\sqrt{x^2+3}+\dfrac{3}{2}\log\left|x+\sqrt{x^2+3}\right|+C$

(iv) $\dfrac{x}{2}\sqrt{4x^2+9}+\dfrac{9}{4}\log\left|2x+\sqrt{4x^2+9}\right|+C$

(v) $\dfrac{x}{2}\sqrt{16-25x^2}+\dfrac{8}{3}\sin^{-1}\dfrac{5x}{4}+C$

(vi) $\dfrac{x}{2}\sqrt{2-3x^2}+\dfrac{1}{\sqrt{3}}\sin^{-1}\left(\dfrac{\sqrt{3}}{\sqrt{2}}x\right)+C$

2. (i) $\dfrac{(x+4)\sqrt{x^2+8x-6}}{2}-11\log\left|x+4+\sqrt{x^2+8x-6}\right|+C$

(ii) $\sqrt{7x^2-x^2-10}+3\sin^{-1}\sqrt{\dfrac{x-2}{3}}+C$

(iii) $\frac{2x-a}{4}\sqrt{ax-x^2}+\frac{a^2}{8}\sin^{-1}\left(\frac{2x-a}{a}\right)+C$

(iv) $\frac{1}{6}\left[x^3\sqrt{x^6-1}-\log\left|x^3+\sqrt{x^6-1}\right|\right]+C$

(v) $\frac{1}{2}\left[e^x\sqrt{e^{2x}+1}+\log\left|e^x+\sqrt{e^{2x}+1}\right|\right]+C$

(vi) $\frac{1}{2}\left[x^2-x\sqrt{x^2-1}+\log\left|x+\sqrt{x^2-1}\right|\right]+C$

3. (i) $\frac{\sin x}{2}\sqrt{4+\sin^2 x}+2\log\left|\sin x+\sqrt{4+\sin^2 x}\right|+C$

(ii) $\frac{1}{4}\left[(2x^2-1)\sin^{-1}x+x\sqrt{1-x^2}\right]+C$

(iii) $\frac{1}{4}\left[(2x^2-1)\cos^{-1}x-x\sqrt{1-x^2}\right]+C$

(iv) $\frac{1}{2}\sec x\sqrt{\sec^2 x+1}+\frac{1}{2}\log\left|\sec x+\sqrt{\sec^2 x+1}\right|+C$

(v) $\frac{1}{2}\left[x\sqrt{x^2+3}-\log\left|x+\sqrt{x^2+3}\right|\right]+C$

(vi) $\frac{1}{4}x^2\sqrt{x^4+1}+\frac{1}{4}\log\left|x^2+\sqrt{x^4+1}\right|+C$

4.11 INTEGRATION BY PARTIAL FRACTIONS

ILLUSTRATIVE EXAMPLES

Example 1. Evaluate:

(i) $\int\frac{2x+1}{(x+1)(x-2)}dx$ (ii) $\int\frac{x^2+x+3}{(x-2)(x+1)}dx$

(iii) $\int\frac{2x-3}{(x^2-1)(2x+3)}dx$ (iv) $\int\frac{3x-2}{(x+1)^2(x+3)}dx$

Solution: (i) Let $\frac{2x+1}{(x+1)(x-2)}=\frac{A}{(x+1)}+\frac{B}{(x-2)}$

$\Rightarrow \qquad 2x+1=A(x-2)+B(x+1)$

On putting $x=2$ and $x=-1$, we get, $A=\frac{1}{3}$ and $B=\frac{5}{3}$

$$\therefore \int\frac{2x+1}{(x+1)(x-2)}dx=\int\left[\frac{\frac{1}{3}}{(x+1)}+\frac{\frac{5}{3}}{(x-2)}\right]dx$$

$$= \frac{1}{3}\int \frac{dx}{x+1} + \frac{5}{3}\int \frac{dx}{(x-2)}$$

$$= \frac{1}{3}\log|x+1| + \frac{5}{3}\log|x-2| + C$$

(ii) Let $\dfrac{x^2+x+3}{(x-2)(x+1)} = 1 + \dfrac{2x+5}{(x-2)(x+1)}$ [Dividing N by D]

$\Rightarrow \quad \dfrac{x^2+x+3}{(x-2)(x+1)} = 1 + \dfrac{A}{x-2} + \dfrac{B}{x+1}$

$\Rightarrow \quad x^2+x+3 = (x-2)(x+1) + A(x+1) + B(x-2)$

On putting $x = -1$ and $x = 2$, we get

$A = 3$ and $B = -1$

$\therefore \quad \dfrac{x^2+x+3}{(x-2)(x+1)} = 1 + \dfrac{3}{x-2} - \dfrac{1}{x+1}$

$$\Rightarrow \int \frac{x^2+x+3}{(x-2)(x+1)}dx = \int \left[1 + \frac{3}{x-2} - \frac{1}{x+1}\right]dx$$

$$= \int 1\,dx + 3\int \frac{dx}{x-2} - \int \frac{dx}{x+1}$$

$$= x + 3\log|x-2| - \log|x+1| + C$$

(iii) Let $\dfrac{2x-3}{(x^2-1)(2x+3)} = \dfrac{2x-3}{(x-1)(x+1)(2x+3)}$

$$= \frac{A}{x-1} + \frac{B}{x+1} + \frac{C}{2x+3}$$

$\Rightarrow \quad 2x-3 = A(x+1)(2x+3) + B(x-1)(2x+3) + C(x-1)(x+1)$

On putting $x = 1, x = -1$ and $x = -\dfrac{3}{2}$, we get

$$A = -\frac{1}{10}, B = \frac{5}{2}, C = -\frac{24}{5}$$

$\therefore \quad \dfrac{2x-3}{(x^2-1)(2x+3)} = -\dfrac{1}{10(x-1)} + \dfrac{5}{2(x+1)} - \dfrac{24}{5(2x+3)}$

$$\Rightarrow \int \frac{2x-3}{(x^2-1)(2x+3)}dx = -\frac{1}{10}\int \frac{dx}{x-1} + \frac{5}{2}\int \frac{dx}{x+1} - -\frac{24}{5}\int \frac{dx}{2x+3}$$

$$= -\frac{1}{10}\log|x-1| + \frac{5}{2}\log|x+1| - \frac{24}{5}\frac{\log|2x+3|}{2} + C$$

$$= -\frac{1}{10}\log|x-1| + \frac{5}{2}\log|x+1| - \frac{12}{5}\log|2x+3| + C$$

(iv) Let $\dfrac{3x-2}{(x+1)^2(x+3)} = \dfrac{A}{x+1} + \dfrac{B}{(x+1)^2} + \dfrac{C}{(x+3)}$

$\Rightarrow \quad 3x-2 = A(x+1)(x+3) + B(x+3) + C(x+1)^2$

On putting $x = -1$ and $x = -3$, we get

$$B = -\frac{5}{2} \text{ and } C = -\frac{11}{4}$$

Equating coefficient of x^2 on both side, we get

$$A + C = 0 \quad \Rightarrow \quad A = -C = \frac{11}{4}$$

$$\therefore \quad \frac{3x-2}{(x+1)^2(x+3)} = \frac{\frac{11}{4}}{x+1} - \frac{\frac{5}{2}}{(x+1)^2} - \frac{\frac{11}{4}}{(x+3)}$$

$$\Rightarrow \int \frac{3x-2}{(x+1)^2(x+3)}\,dx = \frac{11}{4}\int \frac{dx}{(x+1)} - \frac{5}{2}\int \frac{dx}{(x+1)^2} - \frac{11}{4}\int \frac{dx}{(x+3)}$$

$$= \frac{11}{4}\log|x+1| - \frac{5}{2}\frac{(x+1)^{-1}}{-1} - \frac{11}{4}\log|x+3| + C$$

$$= \frac{11}{4}\log\left|\frac{x+1}{x+3}\right| + \frac{5}{2(x+1)} + C$$

Example 2. Evaluate:

(i) $\displaystyle\int \frac{x}{(x^2+1)(x^2+2)}\,dx$ (ii) $\displaystyle\int \frac{dx}{x(x^4+1)}$

(iii) $\displaystyle\int \frac{dx}{1-x^3}$

Solution: Let $\dfrac{x}{(x^2+1)(x^2+2)} = \dfrac{Ax+B}{x^2+1} + \dfrac{Cx+D}{x^2+2}$

$$\Rightarrow \quad x = (Ax+B)(x^2+2) + (Cx+D)(x^2+1)$$

Equating coefficients of various powers of x on both sides, we get

$$A + C = 0$$
$$B + D = 0$$
$$2A + C = 1$$
$$2B + D = 0$$

Solving the above equations, we get

$$A = 1, C = -1, B = 0, D = 0$$

$$\therefore \quad \frac{x}{(x^2+1)(x^2+2)} = \frac{x}{x^2+1} - \frac{x}{x^2+2}$$

$$\Rightarrow \int \frac{x}{(x^2+1)(x^2+2)}\,dx = \int \frac{x}{x^2+1}\,dx - \int \frac{x}{x^2+2}\,dx$$

$$= \frac{1}{2}\int \frac{2x}{x^2+1}\,dx - \frac{1}{2}\int \frac{2x}{x^2+2}\,dx$$

$$= \frac{1}{2}\log|x^2+1| - \frac{1}{2}\log|x^2+2| + C$$

$$= \frac{1}{2}\log\left|\frac{x^2+1}{x^2+2}\right| C$$

Alternatively, put $x^2 = t \Rightarrow x\,dx = \dfrac{dt}{2}$

$$\therefore \quad \int \frac{x\,dx}{(x^2+1)(x^2+2)} = \frac{1}{2}\int \frac{dt}{(t+1)(t+2)}$$

$$= \frac{1}{2}\int \left[\frac{1}{t+1} - \frac{1}{t+2}\right] dt \qquad \text{(by partial fractions)}$$

$$\frac{1}{2}\log|t+1| - \frac{1}{2}\log|t+2| + C$$

$$= \frac{1}{2}\log\left|\frac{t+1}{t+2}\right| + C$$

$$= \frac{1}{2}\log\left|\frac{x^2+1}{x^2+2}\right| + C$$

(ii) $\displaystyle\int \frac{dx}{x(x^4+1)} = \int \frac{x^3}{x^4(x^4+1)}\,dx$

Put $\qquad x^4 = t \Rightarrow 4x^3\,dx = dt \Rightarrow x^3\,dx = \dfrac{dt}{4}$

$$\therefore \quad \int \frac{dx}{x(x^4+1)} = \frac{1}{4}\int \frac{dt}{t(t+1)}$$

$$= \frac{1}{4}\int \left[\frac{1}{t} - \frac{1}{t+1}\right] dt \qquad \text{(by partial fractions)}$$

$$= \frac{1}{4}\log|t| - \frac{1}{4}\log|t+1| + C$$

$$= \frac{1}{4}\log\left|\frac{t}{t+1}\right| + C$$

$$= \frac{1}{4}\log\left|\frac{x^4}{x^4+1}\right| + C$$

(iii) $\displaystyle\frac{1}{1-x^3} = \frac{1}{(1-x)(1+x+x^2)}$

$$= \frac{A}{1-x} + \frac{Bx+C}{x^2+x+1}$$

$$\Rightarrow \qquad 1 = A(x^2 + x + 1) + (Bx + C)(1 - x)$$

On putting $x = 1$, we get, $A = \dfrac{1}{3}$

Equating the coefficients of x^2 and constant terms, we get, $A - B = 0$ and $A + C = 1$

Solving the above two equations, we get

$$B = \frac{1}{3} \text{ and } C = \frac{2}{3}$$

$$\therefore \qquad \frac{1}{1-x^3} = \frac{\frac{1}{3}}{1-x} + \frac{\frac{1}{3}x + \frac{2}{3}}{x^2+x+1}$$

$$\Rightarrow \qquad \int \frac{dx}{1-x^3} = \frac{1}{3}\int \frac{dx}{1-x} + \frac{1}{3}\int \frac{(x+2)}{x^2+x+1}\,dx$$

$$= -\frac{1}{3}\int \frac{dx}{x-1} + \frac{1}{2}\cdot\frac{1}{3}\int \frac{2x+4}{x^2+x+1}\,dx$$

$$= -\frac{1}{3}\log|x-1| + \frac{1}{6}\int \left[\frac{2x+1}{x^2+x+1} + \frac{3}{x^2+x+1}\right]dx$$

$$= -\frac{1}{3}\log|x-1| + \frac{1}{6}\log|x^2+x+1| + \frac{1}{2}\int \frac{dx}{\left(x+\frac{1}{2}\right)^2 + \left(\frac{\sqrt{3}}{2}\right)^2}$$

$$= -\frac{1}{3}\log|x-1| + \frac{1}{6}\log|x^2+x+1| + \frac{1}{2}\cdot\frac{2}{\sqrt{3}}\tan^{-1}\left(\frac{x+\frac{1}{2}}{\frac{\sqrt{3}}{2}}\right) + C$$

$$= -\frac{1}{3}\log|x-1| + \frac{1}{6}\log|x^2+x+1| + \frac{1}{\sqrt{3}}\tan^{-1}\left(\frac{2x+1}{\sqrt{3}}\right) + C$$

Example 3. Evaluate the following integrals:

(i) $\displaystyle\int \frac{x+1}{x(1+xe^x)}\,dx$ (ii) $\displaystyle\int \frac{\cos x\,dx}{(1+\sin x)(2+\sin x)}$

(iii) $\displaystyle\int \frac{\tan x + \tan^3 x}{1+\tan^3 x}\,dx$ (iv) $\displaystyle\int \frac{dx}{e^{2x}+2e^x-3}$

Solution: (i) $\displaystyle\int \frac{x+1}{x(1+xe^x)}\,dx = \int \frac{(x+1)e^x}{x(1+x\,e^x)e^x}\,dx$

Put $x\,e^x = t \Rightarrow (x+1)e^x\,dx = dt$

$$\therefore \quad \int \frac{x+1}{x(1+x\,e^x)}\,dx = \int \frac{dt}{t(1+t)}$$

$$= \int \left[\frac{1}{t} - \frac{1}{1+t}\right] dt \qquad \text{(by partial fractions)}$$

$$= \log|t| - \log|1+t| + C$$

$$= \log\left|\frac{t}{1+t}\right| + C$$

$$= \log\left|\frac{xe^x}{1+xe^x}\right| + C$$

(ii) Put $\sin x = t \Rightarrow \cos x\, dx = dt$

$$\therefore \quad \int \frac{\cos x\, dx}{(1+\sin x)(2+\sin x)} = \int \frac{dt}{(1+t)(2+t)}$$

$$= \int \left[\frac{1}{1+t} - \frac{1}{2+t}\right] dt \qquad \text{(by partial fractions)}$$

$$= \log|1+t| - \log|2+t| + C$$

$$= \log\left|\frac{1+t}{2+t}\right| + C$$

$$= \log\left|\frac{1+\sin x}{2+\sin x}\right| + C$$

(iii) $\displaystyle\int \frac{\tan x + \tan^3 x}{1+\tan^3 x} dx = \int \frac{\tan x(1+\tan^2 x)}{1+\tan^3 x} dx$

$$= \int \frac{\tan x \cdot \sec^2 x}{1+\tan^3 x} dx$$

Put $\tan x = t \Rightarrow \sec^2 x\, dx = dt$

$$\therefore \quad \int \frac{\tan x + \tan^3 x}{1+\tan^3 x} dx = \int \frac{t\, dt}{1+t^3}$$

$$= \frac{1}{3}\int \left[\frac{-1}{1+t} + \frac{1+t}{t^2-t+1}\right] dt$$

$$= \frac{1}{3}(-\log|1+t|) + \frac{1}{2}\cdot\frac{1}{3}\int \frac{2t+2}{t^2-t+1} dt$$

$$= -\frac{1}{3}\log|1+t| + \frac{1}{6}\int \frac{2t-1+3}{t^2-t+1} dt$$

$$= -\frac{1}{3}\log|1+t| + \frac{1}{6}\int \frac{2t-1}{t^2-t+1} dt + \frac{1}{2}\int \frac{dt}{t^2-t+1}$$

$$= -\frac{1}{3}\log|1+t| + \frac{1}{6}\log|t^2 - t + 1| + \frac{1}{2}\int \frac{dt}{\left(t-\frac{1}{2}\right)^2 + \left(\frac{\sqrt{3}}{2}\right)^2}$$

$$= -\frac{1}{3}\log|1+t| + \frac{1}{6}\log|t^2 - t + 1| + \frac{1}{2}\cdot\frac{1}{\frac{\sqrt{3}}{2}}\tan^{-1}\left|\frac{t-\frac{1}{2}}{\frac{\sqrt{3}}{2}}\right| + C$$

$$= -\frac{1}{3}\log|1+t| + \frac{1}{6}\log|t^2 - t + 1| + \frac{1}{\sqrt{3}}\tan^{-1}\left|\frac{2t-1}{\sqrt{3}}\right| + C$$

$$= -\frac{1}{3}\log + |1+\tan x| + \frac{1}{6}\log|\tan^2 x - \tan x + 1| + \frac{1}{\sqrt{3}}\tan^{-1}\left|\frac{2\tan x - 1}{\sqrt{3}}\right| + C$$

(iv) Let $e^x = t \Rightarrow e^x\, dx = dt \Rightarrow t\, dx = dt$

$$\Rightarrow \qquad dx = \frac{dt}{t}$$

$$\therefore \qquad \int \frac{dx}{e^{2x} + 2e^x - 3} = \int \frac{dt}{t(t^2 + 2t - 3)}$$

$$= \int \frac{dt}{t(t-1)(t+3)}$$

$$= \int \left[\frac{-\frac{1}{3}}{t} + \frac{\frac{1}{4}}{t-1} + \frac{\frac{1}{12}}{t+3}\right] dt$$

$$= -\frac{1}{3}\log|t| + \frac{1}{4}\log|t-1| + \frac{1}{12}\log|t+3| + C$$

$$= -\frac{1}{3}\log|e^x| + \frac{1}{4}\log|e^x - 1| + \frac{1}{12}\log|e^x + 3| + C$$

Example 4. Evaluate the following integrals:

(i) $\displaystyle\int \frac{\log x}{(x+1)^2}\, dx$ (ii) $\displaystyle\int \frac{2e^{2x} + e^x + 1}{e^{2x} + 1}\, dx$

(iii) $\displaystyle\int \frac{(3\sin x - 2)\cos x}{5 - \cos^2 x - 4\sin x}\, dx$

Solution: (i) $\displaystyle\int \frac{\log x}{(x+1)^2}\, dx = -\frac{\log x}{(x+1)} + \int \frac{1}{x+1}\cdot\frac{1}{x}\, dx$

$$\left[\text{integrating by parts} \because \int \frac{dx}{(x+1)^2} = -\frac{1}{x+1}\right]$$

$$= -\frac{\log x}{x+1} + \int \left[\frac{1}{x} - \frac{1}{x+1}\right] dx \qquad \text{(By partial fractions)}$$

$$= -\frac{\log x}{x+1} + \log x - \log(x+1) + C$$

$$= \left(\frac{x}{x+1}\right)\log x - \log(x+1) + C$$

(ii) Let $e^x = t \Rightarrow e^x\, dx = dt \Rightarrow dx = \dfrac{dt}{t}$

$$\therefore \int \frac{2e^{2x} + e^x + 1}{e^{2x} + 1}\, dx = \int \frac{2t^2 + t + 1}{t^2 + 1} \cdot \frac{dt}{t}$$

$$= \int \frac{2t^2 + t + 1}{t(t^2 + 1)}\, dt$$

$$= \int \left[\frac{1}{t} + \frac{1 + t}{1 + t^2}\right] dt \qquad \text{(By partial fractions)}$$

$$= \int \frac{1}{t}\, dt + \int \frac{dt}{1 + t^2} + \frac{1}{2}\int \frac{2t}{1 + t^2}\, dt$$

$$= \log t + \tan^{-1}(t) + \frac{1}{2}\log(1 + t^2) + C$$

$$= \log(e^x) + \tan^{-1}(e^x) + \frac{1}{2}\log(1 + e^{2x}) + C$$

$$= x + \tan^{-1}(e^x) + \frac{1}{2}\log(1 + e^{2x}) + C$$

$$(\because \log e^x = x \log e = x \cdot 1 = x)$$

(iii) $\displaystyle\int \frac{(3\sin x - 2)\cos x}{5 - (1 - \sin^2 x) - 4\sin x}\, dx = \int \frac{(3\sin x - 2)\cos x}{4 + \sin^2 x - 4\sin x}\, dx$

Put $\sin x = t \Rightarrow \cos x\, dx = dt$

$$\therefore \quad \int \frac{(3\sin x - 2)\cos x}{5 - \cos^2 x - 4\sin x}\, dx$$

$$= \int \frac{(3t - 2)\, dt}{4 + t^2 - 4t}$$

$$= \int \frac{(3t - 2)\, dt}{(t - 2)^2}$$

$$= \int \left[\frac{3}{t - 2} + \frac{4}{(t - 2)^2}\right] dt \qquad \text{(By partial fractions)}$$

$$= 3\log(t-2) + \frac{4(t-2)^{-1}}{-1} + C$$

$$= 3\log(t-2) - \frac{4}{t-2} + C$$

$$= 3\log(\sin x - 2) - \frac{4}{\sin x - 2} + C$$

EXERCISE 4.6

Evaluate the following integrals:

1. (i) $\int \frac{2x\,dx}{x^2 - 7x + 12}$ (ii) $\int \frac{3}{(x-4)(x-1)}\,dx$

(iii) $\int \frac{x^2\,dx}{x^2 + 7x + 10}$ (iv) $\int \frac{4x^4 - 15x^2 + 11x + 13}{2x^3 + 3x^2 - 3x - 2}\,dx$

(v) $\int \frac{x^2}{x^2 - 4x + 3}\,dx$ (vi) $\int \frac{dx}{1 + x + x^2 + x^3}$

(vii) $\int \frac{dx}{(x^2+1)(x^2+2)}$ (viii) $\int \frac{dx}{1 + x^3}$

(ix) $\int \frac{dx}{(x+1)^2(x^2+1)}$ (x) $\int \frac{dx}{x(1 + 4x^3 + 3x^6)}$

2. (i) $\int \frac{dx}{e^x - 1}$ (ii) $\int \frac{dx}{(e^x - 1)^2}$

(iii) $\int \frac{dx}{1 + 3e^x + 2e^{2x}}$ (iv) $\int \frac{dx}{x(x^n + 1)}$ [Hint. put $x^n = t$]

(v) $\int \frac{dx}{(1 + e^x)(1 + e^{-x})}$

3. (i) $\int \frac{\sec x}{1 + \operatorname{cosec} x}\,dx$ (ii) $\int \frac{dx}{\sin x(3 + 2\cos x)}$

(iii) $\int \frac{\sin x \cos x}{\cos^2 x + 3\cos x + 2}\,dx$ (iv) $\int \frac{\log|x|}{(x+1)^3}\,dx$

(v) $\int \frac{dx}{x[(6\log x)^2 + 7\log x + 2]}$ (vi) $\int \frac{dx}{\sin x + \sin 2x}$

(vii) $\int \frac{\sec x\,dx}{1 + \sin x}$ (viii) $\int \frac{\cos x\,dx}{(\sin x - 1)(\sin x - 3)}$

ANSWERS

1. (i) $8\log|x-4|-6\log|x-3|+C$

(ii) $\log\left|\dfrac{x-4}{x-1}\right|+C$

(iii) $x-\dfrac{25}{3}\log(x+5)+\dfrac{4}{3}\log(x+2)+C$

(iv) $\dfrac{13}{9}\log(x-1)-\dfrac{5}{9}\log(x+2)-\dfrac{8}{9}\log(2x+1)+x^2-3x+C$

(v) $x+\dfrac{9}{2}\log|x-3|-\dfrac{1}{2}\log|x-1|+C$

(vi) $\dfrac{1}{2}\log(1+x)-\dfrac{1}{4}\log(1+x^2)+\dfrac{1}{2}\tan^{-1}x+C$

(vii) $\tan^{-1}x-\dfrac{1}{\sqrt{2}}\tan^{-1}\dfrac{x}{\sqrt{2}}+C$

(viii) $\dfrac{1}{3}\log|x+1|-\dfrac{1}{6}\log|x^2-x+1|+\dfrac{1}{\sqrt{3}}\tan^{-1}\left(\dfrac{2x-1}{\sqrt{3}}\right)+C$

(ix) $\dfrac{1}{2}\log|x+1|-\dfrac{1}{2(x+1)}-\dfrac{1}{4}\log(x^2+1)+C$

(x) $\dfrac{1}{3}\log|x^3|+\dfrac{1}{6}\log|1+x^3|-\dfrac{1}{2}\log|1+3x^3|+C$

2. (i) $-x+\log|e^x-1|+C$

(ii) $\log\left|\dfrac{e^x}{e^x-1}\right|-\dfrac{1}{e^x-1}+C$

(iii) $x-2\log(1+2e^x)+\log(1+e^x)+C$

(iv) $\dfrac{1}{n}\log\left|\dfrac{x^n}{x^n+1}\right|+C$

(v) $-\dfrac{1}{e^x+1}+C$

3. (i) $\dfrac{1}{4}\log\left|\dfrac{1+\sin x}{1-\sin x}\right|+\dfrac{1}{2(1+\sin x)}+C$

(ii) $\dfrac{1}{10}\log(1-\cos x)+\dfrac{2}{5}\log(3+2\cos x)-\dfrac{1}{2}\log\left|\dfrac{1+\sin x}{1-\sin x}\right|+C$

(iii) $\log(1+\cos x)-2\log(2+\cos x)+C$

(iv) $-\dfrac{\log|x|}{2(x+1)^2}+\dfrac{1}{2}\log\left|\dfrac{x}{x+1}\right|+\dfrac{1}{2(x+1)}+C$

(v) $\log\left|\dfrac{1+2\log x}{2+3\log x}\right|+C$

(vi) $\dfrac{1}{6}\log(1-\cos x)+\dfrac{1}{2}\log(1+\cos x)-\dfrac{2}{3}\log|1+2\cos x|+C$

(vii) $\dfrac{1}{4}\log\left|\dfrac{1+\sin x}{1-\sin x}\right|-\dfrac{1}{2(1+\sin x)}+C$

(viii) $\dfrac{1}{2}\log(\sin x-3)-\dfrac{1}{2}\log(\sin x-1)+C$

4.12 INTEGRATION OF THE TYPE $\dfrac{1}{ax^2+bx+c}$:

$$ax^2+bx+c = a\left(x^2+\frac{b}{a}x+\frac{c}{a}\right)$$

$$= a\left[\left(x+\frac{b}{2a}\right)^2-\left(\frac{b^2}{4a^2}-\frac{c}{a}\right)\right]$$

$$= a\left[\left(x+\frac{b}{2a}\right)^2-\left(\frac{b^2-4ac}{4a^2}\right)\right]$$

If b^2-4ac is positive, then

$$\frac{1}{ax^2+bx+c} = \frac{1}{a\left[\left(x+\frac{b}{2a}\right)^2-\left(\frac{b^2-4ac}{4a^2}\right)\right]}$$

and the integrand is of the type $\dfrac{1}{a(x^2-k^2)}$ and we use the formula:

$$\int\frac{dx}{x^2-k^2} = \frac{1}{2k}\log\left|\frac{x-k}{x+k}\right|+C$$

If b^2-4ac is negative, the integrand is of the type $\dfrac{1}{a(x^2+k^2)}$ and we use the formula:

$$\int\frac{dx}{x^2+k^2} = \frac{1}{k}\tan^{-1}\frac{x}{k}+C$$

4.13 INTEGRATION OF THE TYPE $\dfrac{Ax+B}{ax^2+bx+c}$:

Break $(Ax+B)$ into two parts. One part being the differential coefficient of the denominator and the other part, a constant.

$$\therefore \qquad Ax + B = M\cdot\frac{d}{dx}(ax^2 + bx + c) + N$$

where M and N are both constants.

Thus, $\qquad Ax + B = M(2ax + b) + N$

$\Rightarrow \qquad Ax + B = x(2aM) + Mb + N$

comparing the coefficients, we have

$$A = 2aM \Rightarrow M = \frac{A}{2a}$$

and $$B = Mb + N \Rightarrow N = B - Mb = B - \frac{Ab}{2a}$$

$$\therefore \qquad \frac{Ax + B}{ax^2 + bx + c} = \frac{M\cdot\frac{d}{dx}(ax^2 + bx + c) + N}{ax^2 + bx + c}$$

$$\therefore \int \frac{Ax + B}{ax^2 + bx + c}\,dx = \int \frac{M\frac{d}{dx}(ax^2 + bx + c)}{ax^2 + bx + c}\,dx + \int \frac{N\,dx}{ax^2 + bx + c}$$

$$= M\log(ax^2 + bx + c) + \int \frac{N\,dx}{ax^2 + bx + c}$$

The second term on the right is integrated as the previous article.

ILLUSTRATIVE EXAMPLES

Example 1. Evaluate the following integrals:

(i) $\displaystyle\int \frac{dx}{x^2 - x + 1}$ (ii) $\displaystyle\int \frac{dx}{9x^2 + 6x + 10}$

(iii) $\displaystyle\int \frac{dx}{2x^2 + x - 1}$

Solution: (i) $$\int \frac{dx}{x^2 - x + 1} = \int \frac{dx}{\left(x^2 - x + \frac{1}{4}\right) + \frac{3}{4}}$$

$$= \int \frac{dx}{\left(x - \frac{1}{2}\right)^2 + \frac{3}{4}}$$

$$= \int \frac{dx}{\left(x - \frac{1}{2}\right)^2 + \left(\frac{\sqrt{3}}{2}\right)^2}$$

$$= \frac{1}{\frac{\sqrt{3}}{2}}\tan^{-1}\left(\frac{x - \frac{1}{2}}{\frac{\sqrt{3}}{2}}\right) + C$$

$$= \frac{2}{\sqrt{3}}\tan^{-1}\left(\frac{2x-1}{\sqrt{3}}\right)+C$$

(ii) $\displaystyle\int \frac{dx}{9x^2+6x+10} = \frac{1}{9}\int \frac{dx}{x^2+\frac{2x}{3}+\frac{10}{9}}$

$$= \frac{1}{9}\int \frac{dx}{\left(x^2+\frac{2x}{3}+\frac{1}{9}\right)+\left(\frac{10}{9}-\frac{1}{9}\right)}$$

$$= \frac{1}{9}\int \frac{dx}{\left(x+\frac{1}{3}\right)^2+1^2}$$

$$= \frac{1}{9}\cdot\frac{1}{1}\cdot\tan^{-1}\left(\frac{x+\frac{1}{3}}{1}\right)+C$$

$$= \frac{1}{9}\tan^{-1}\left(\frac{3x+1}{3}\right)+C$$

(iii) $\displaystyle\int \frac{dx}{2x^2+x-1} = \frac{1}{2}\int \frac{dx}{x^2+\frac{x}{2}-\frac{1}{2}}$

$$= \frac{1}{2}\int \frac{dx}{\left(x^2+\frac{x}{2}+\frac{1}{16}\right)-\frac{1}{16}-\frac{1}{2}}$$

$$= \frac{1}{2}\int \frac{dx}{\left(x+\frac{1}{4}\right)^2-\left(\frac{3}{4}\right)^2}$$

$$= \frac{1}{2}\cdot\frac{1}{2\cdot\frac{3}{4}}\log\left|\frac{x+\frac{1}{4}-\frac{3}{4}}{x+\frac{1}{4}+\frac{3}{4}}\right| \qquad \left[\because \int \frac{dx}{x^2-a^2}=\frac{1}{2a}\log\left|\frac{x-a}{x+a}\right|\right]$$

$$= \frac{1}{3}\log\left|\frac{4x-2}{4x+4}\right|+C$$

$$= \frac{1}{3}\log\left|\frac{2x-1}{2x+2}\right|+C$$

Example 2. Evaluate the following integrals:

(i) $\int \frac{3x+1}{2x^2-2x+3}\,dx$ (ii) $\int \frac{5x-2}{3x^2+2x+1}\,dx$

(iii) $\int \frac{x^2\,dx}{x^2+6x-3}$

Solution: Let $3x+1 = M\frac{d}{dx}(2x^2-2x+3)+N$. Then

$$3x+1 = M(4x-2)+N$$

Equating coefficients on both sides, we have

$$4M = 3 \Rightarrow M = \frac{3}{4}$$

and $$N-2M = 1 \Rightarrow N = \frac{5}{2}$$

$\therefore$ $$3x+1 = \frac{3}{4}(4x-2)+\frac{5}{2}$$

$\therefore$ $$\int \frac{3x+1}{2x^2-2x+3}\,dx = \int \frac{\frac{3}{4}(4x-2)+\frac{5}{2}}{2x^2-2x+3}\,dx$$

$$= \frac{3}{4}\int \frac{4x-2}{2x^2-2x+3}\,dx + \frac{5}{2}\int \frac{dx}{2x^2-2x+3}$$

$$= \frac{3}{4}\log(2x^2-2x+3) + \frac{5}{2}\int \frac{dx}{2x^2-2x+3}$$

$$= \frac{3}{4}\log(2x^2-2x+3) + \frac{5}{4}\int \frac{dx}{x^2-x+\frac{3}{2}}$$

$$= \frac{3}{4}\log(2x^2-2x+3) + \frac{5}{4}\int \frac{dx}{\left(x^2-x+\frac{1}{4}\right)+\frac{3}{2}-\frac{1}{4}}$$

$$= \frac{3}{4}\log(2x^2-2x+3) + \frac{5}{4}\int \frac{dx}{\left(x-\frac{1}{2}\right)^2+\left(\frac{\sqrt{5}}{2}\right)^2}$$

$$= \frac{3}{4}\log(2x^2-2x+3) + \frac{5}{4}\cdot\frac{1}{\frac{\sqrt{5}}{2}}\tan^{-1}\left(\frac{x-\frac{1}{2}}{\frac{\sqrt{5}}{2}}\right)+C$$

$$= \frac{3}{4}\log(2x^2-2x+3) + \frac{\sqrt{5}}{2}\tan^{-1}\left(\frac{2x-1}{\sqrt{5}}\right)+C$$

(ii) Let $5x-2 = M\frac{d}{dx}(3x^2+2x+1)+N$

$$5x-2 = M(6x+2)+N$$

$$\Rightarrow \quad 6M = 5 \text{ or } M = \frac{5}{6}$$

and $$2M+N = -2 \text{ or } N = -\frac{11}{3}$$

$$\therefore \int \frac{5x-2}{3x^2+2x+1}dx = \int \frac{\frac{5}{6}(6x+2)-\frac{11}{3}}{3x^2+2x+1}dx$$

$$= \frac{5}{6}\int \frac{6x+2}{3x^2+2x+1}dx - \frac{11}{3}\int \frac{dx}{3x^2+2x+1}$$

$$= \frac{5}{6}\log(3x^2+2x+1) - \frac{11}{3}\cdot\frac{1}{3}\int \frac{dx}{x^2+\frac{2}{3}x+\frac{1}{3}}$$

$$= \frac{5}{6}\log(3x^2+2x+1) - \frac{11}{9}\int \frac{dx}{\left(x^2+\frac{2x}{3}+\frac{1}{9}\right)+\frac{1}{3}-\frac{1}{9}}$$

$$= \frac{5}{6}\log(3x^2+2x+1) - \frac{11}{9}\int \frac{dx}{\left(x+\frac{1}{3}\right)^2+\left(\frac{\sqrt{2}}{3}\right)^2}$$

$$= \frac{5}{6}\log(3x^2+2x+1) - \frac{11}{9}\cdot\frac{1}{\frac{\sqrt{2}}{3}}\tan^{-1}\left(\frac{x+\frac{1}{3}}{\frac{\sqrt{2}}{3}}\right)+C$$

$$= \frac{5}{6}\log(3x^2+2x+1) - \frac{11}{3\sqrt{2}}\tan^{-1}\left(\frac{3x+1}{\sqrt{2}}\right)+C$$

(iii) $$\int \frac{x^2\,dx}{x^2+6x-3} = \int\left(1-\frac{6x-3}{x^2+6x-3}\right)dx$$

$$= x - \int \frac{6x-3}{x^2+6x-3}dx$$

$$= x - \int \frac{3(2x+6)-21}{x^2+6x-3}dx$$

$$= x - 3\int \frac{2x+6}{x^2+6x-3} + 21\int \frac{dx}{x^2+6x-3}$$

$$= x - 3\log(x^2+6x-3) + 21\int \frac{dx}{(x^2+6x+9)-9-3}$$

$$= x - 3\log(x^2 + 6x - 3) + 21\int \frac{dx}{(x+3)^2 - \left(2\sqrt{3}\right)^2}$$

$$= x - 3\log(x^2 + 6x - 3) + 21 \cdot \frac{1}{2 \cdot 2\sqrt{3}} \log\left|\frac{x + 3 - 2\sqrt{3}}{x + 3 + 2\sqrt{3}}\right| + C$$

$$= x - 3\log(x^2 + 6x - 3) + \frac{7\sqrt{3}}{4} \log\left|\frac{x + 3 - 2\sqrt{3}}{x + 3 + 2\sqrt{3}}\right| + C$$

Example 3. Evaluate the following integrals:

(i) $\displaystyle\int \frac{x}{x^4 - x^2 + 1}\, dx$ (ii) $\displaystyle\int \frac{2\sin 2x - \cos x}{6 - \cos^2 x - 4\sin x}\, dx$

Solution: (i) Put $x^2 = t \Rightarrow x\, dx = \dfrac{dt}{2}$

$$\therefore \int \frac{x\, dx}{x^4 - x^2 + 1} = \frac{1}{2}\int \frac{dt}{t^2 - t + 1}$$

$$= \frac{1}{2}\int \frac{dt}{\left(t^2 - t + \frac{1}{4}\right) + 1 - \frac{1}{4}}$$

$$= \frac{1}{2}\int \frac{dt}{\left(t - \frac{1}{2}\right)^2 + \left(\frac{\sqrt{3}}{2}\right)^2}$$

$$= \frac{1}{2} \cdot \frac{1}{\frac{\sqrt{3}}{2}} \tan^{-1}\left(\frac{t - \frac{1}{2}}{\frac{\sqrt{3}}{2}}\right) + C$$

$$= \frac{1}{\sqrt{3}} \tan^{-1}\left(\frac{2t - 1}{\sqrt{3}}\right) + C$$

$$= \frac{1}{\sqrt{3}} \tan^{-1}\left(\frac{2x^2 - 1}{\sqrt{3}}\right) + C$$

(ii) $$\int \frac{2\sin 2x - \cos x}{6 - \cos^2 x - 4\sin x}\, dx = \int \frac{2 \cdot 2\sin x \cos x - \cos x}{6 - (1 - \sin^2 x) - 4\sin x}\, dx$$

$$= \int \frac{(4\sin x - 1)\cos x}{5 + \sin^2 x - 4\sin x}\, dx$$

(Put $\sin x = t \Rightarrow \cos x\, dx = dt$)

$$= \int \frac{(4t - 1)\, dt}{5 + t^2 - 4t} = \int \frac{2(2t - 4) + 7}{t^2 - 4t + 5}\, dt$$

$$= 2\int \frac{2t-4}{t^2-4t+5}\,dt + 7\int \frac{dt}{(t-2)^2+1^2}$$

$$= 2\log(t^2-4t+5) + 7\int \frac{dt}{(t-2)^2+1^2}$$

$$= 2\log(t^2-4t+5) + 7\cdot\frac{1}{1}\tan^{-1}\left(\frac{t-2}{1}\right) + C$$

$$= 2\log(\sin^2 x - 4\sin x + 5) + 7\tan^{-1}(\sin x - 2) + C$$

EXERCISE 4.7

Evaluate the following integrals:

1. (i) $\int \frac{dx}{x^2-4x+8}$ (ii) $\int \frac{dx}{1+x-x^2}$

(iii) $\int \frac{dx}{x^2+6x+12}$ (iv) $\int \frac{dx}{2x^2-2x+1}$

(v) $\int \frac{dx}{2x^2+4x+3}$

2. (i) $\int \frac{x+7}{x^2+2x+8}\,dx$ (ii) $\int \frac{x}{x^2+4x+5}\,dx$

(iii) $\int \frac{x^2}{x^2+6x-12}\,dx$ (iv) $\int \frac{4x-3}{3x^2+2x-5}\,dx$

(v) $\int \frac{x^2+x+1}{x^2-x+1}\,dx$

3. (i) $\int \frac{\cos x}{6+4\sin x-\cos^2 x}\,dx$ (ii) $\int x\tan^{-1}(2x+3)\,dx$

ANSWERS

1. (i) $\frac{1}{2}\tan^{-1}\left(\frac{x-2}{2}\right)+C$ (ii) $\frac{1}{\sqrt{5}}\log\left|\frac{\sqrt{5}-1+2x}{\sqrt{5}+1-2x}\right|+C$

(iii) $\frac{1}{\sqrt{3}}\tan^{-1}\left(\frac{x+3}{\sqrt{3}}\right)+C$ (iv) $\tan^{-1}(2x-1)+C$

(v) $\frac{1}{\sqrt{2}}\tan^{-1}\left[\sqrt{2}\,(x+1)\right]+C$

2. (i) $\frac{1}{2}\log(x^2+2x+8)+\frac{6}{\sqrt{7}}\tan^{-1}\left(\frac{x+1}{\sqrt{7}}\right)+C$

(ii) $\frac{1}{2}\log(x^2+4x+5)-2\tan^{-1}(x+2)+C$

(iii) $x-3\log(x^2+6x-12)+\frac{15}{\sqrt{21}}\log\left|\frac{x+3-\sqrt{21}}{x+3+\sqrt{21}}\right|+C$

(iv) $\frac{2}{5}\log(3x^2+2x-5)-\frac{13}{24}\log\left|\frac{3x-3}{3x+3}\right|+C$

(v) $x+\log(x^2-x+1)+\frac{2}{\sqrt{3}}\tan^{-1}\left(\frac{2x-1}{\sqrt{3}}\right)+C$

3. (i) $\tan^{-1}(\sin x+2)+C$

(ii) $\frac{1}{2}(x^2-2)\tan^{-1}(2x+3)-\frac{1}{4}x+\frac{3}{8}\log(2x^2+6x+5)+C$

4.14 INTEGRATION OF RATIONAL FUNCTIONS OF sin *x* AND cos *x*

If the integrand is a rational function of sin x or cos x or both, it can be reduced to a rational function of t by the substitution $t = \tan\frac{x}{2}$

ILLUSTRATIVE EXAMPLES

Example 1. Evaluate the following integrals:

(i) $\int \frac{dx}{2+\cos x}$ (ii) $\int \frac{dx}{4+5\sin x}$

Solution: (i) Let $\tan\frac{x}{2} = t \Rightarrow \sec^2\frac{x}{2}\cdot\frac{1}{2}\,dx = dt$

$$\Rightarrow \qquad dx = \frac{2\,dt}{\sec^2\frac{x}{2}} = \frac{2\,dt}{1+\tan^2\frac{x}{2}} = \frac{2\,dt}{1+t^2}$$

and
$$\cos x = \frac{1-\tan^2\frac{x}{2}}{1+\tan^2\frac{x}{2}} = \frac{1-t^2}{1+t^2}$$

$$\therefore \qquad \int\frac{dx}{2+\cos x} = \int\frac{\frac{2dt}{(1+t^2)}}{2+\frac{1-t^2}{1+t^2}}$$

$$= \int\frac{2\,dt}{2+2t^2+1-t^2}$$

$$= \int\frac{2\,dt}{t^2+3} = \int\frac{2\,dt}{t^2+\left(\sqrt{3}\right)^2} = \frac{2}{\sqrt{3}}\tan^{-1}\left(\frac{t}{\sqrt{3}}\right)+C$$

$$= \frac{2}{\sqrt{3}} \tan^{-1}\left(\frac{1}{\sqrt{3}} \tan\frac{x}{2}\right) + C$$

(ii) Let $\tan\frac{x}{2} = t \Rightarrow dx = \frac{2\,dt}{1+t^2}$ [as part (i)]

and $$\sin x = \frac{2\tan\frac{x}{2}}{1+\tan^2\frac{x}{2}} = \frac{2t}{1+t^2}$$

$$\therefore \quad \int \frac{dx}{4+5\sin x} = \int \frac{\frac{2\,dt}{1+t^2}}{4+5\cdot\frac{2t}{1+t^2}}$$

$$= \int \frac{2\,dt}{4+4t^2+10t} = \frac{1}{2}\int \frac{dt}{t^2+\frac{5}{2}t+1}$$

$$= \frac{1}{2}\int \frac{dt}{t^2+\frac{5}{2}t+1+\frac{25}{16}-\frac{25}{16}} = \frac{1}{2}\int \frac{dt}{\left(t+\frac{5}{4}\right)^2-\left(\frac{3}{4}\right)^2}$$

$$= \frac{1}{2}\cdot\frac{1}{2\cdot\frac{3}{4}}\log\left|\frac{t+\frac{5}{4}-\frac{3}{4}}{t+\frac{5}{4}+\frac{3}{4}}\right| + C$$

$$= \frac{1}{3}\log\left|\frac{\tan\frac{x}{2}+\frac{1}{2}}{\tan\frac{x}{2}+2}\right| + C = \frac{1}{3}\log\left|\frac{2\tan\frac{x}{2}+1}{2\tan\frac{x}{2}+4}\right| + C$$

Example 2. Evaluate $\int \frac{1+\sin x}{\sin x(1+\cos x)}\,dx$

Solution: Let $\tan\frac{x}{2} = t \Rightarrow \sec^2\frac{x}{2}\cdot\frac{1}{2}\,dx = dt \Rightarrow dx = \frac{2\,dt}{1+t^2}$ and

$$\sin x = \frac{2\tan\frac{x}{2}}{1+\tan^2\frac{x}{2}} = \frac{2t}{1+t^2}$$

$$\cos x = \frac{1-\tan^2\frac{x}{2}}{1+\tan^2\frac{x}{2}} = \frac{1-t^2}{1+t^2}$$

$$\therefore \int \frac{1+\sin x}{\sin x(1+\cos x)}\,dx = \int \frac{1+\frac{2t}{1+t^2}}{\frac{2t}{1+t^2}\left(1+\frac{1-t^2}{1+t^2}\right)}\cdot\frac{2\,dt}{1+t^2}$$

$$= \int \frac{1+t^2+2t}{t(1+t^2+1-t^2)}\,dt = \frac{1}{2}\int \frac{t^2+2t+1}{t}\,dt$$

$$= \frac{1}{2}\int \left[t+2+\frac{1}{t}\right]dt$$

$$= \frac{1}{2}\int \left[\frac{t^2}{2}+2t+\log t\right]+C$$

$$= \frac{1}{4}\tan^2\frac{x}{2}+\tan\frac{x}{2}+\frac{1}{2}\log\tan\frac{x}{2}+C$$

4.15 EVALUATION OF $\int \frac{a\sin x+b\cos x}{c\sin x+d\cos x}\,dx$:

Let,

numerator = l(denominator) + m(derivative of denominator)

i.e., $a \sin x + b \cos x = l(c \sin x + d \cos x) + m(c \cos x - d \sin x)$

Equating the coefficients of $\sin x$ and $\cos x$ on both sides, we get.

$$a = lc - md \text{ and } b = ld + mc$$

i.e., $$l = \frac{ac+bd}{c^2+d^2} \text{ and } m = \frac{bc-ad}{c^2+d^2} \quad \text{... (i)}$$

$$\therefore \int \frac{a\sin x+b\cos x}{c\sin x+d\cos x}\,dx = \int \frac{l(c\sin x+d\cos x)+m(c\cos x-d\sin x)}{c\sin x+d\cos x}\,dx$$

$$= \int \left[l+m\cdot\frac{c\cos x-d\sin x}{c\sin x+d\cos x}\right]dx$$

$$= lx + m\log(c\sin x + d\cos x) + C$$

$$= \frac{ac+bd}{c^2+d^2}\cdot x+\frac{bc-ad}{c^2+d^2}\cdot\log(c\sin x+d\cos x)+C \qquad \text{(Using (i)]}$$

ILLUSTRATIVE EXAMPLES

Example 1. Evaluate $\int \frac{dx}{1+\cot x}\,dx$

Solution: $\int \frac{dx}{1+\cot x} = \int \frac{dx}{1+\frac{\cos x}{\sin x}} = \int \frac{\sin x\,dx}{\sin x+\cos x}$

Let, $\sin x = l(\sin x + \cos x) + m(\cos x - \sin x)$

Equating coefficients of $\sin x$ and $\cos x$ on both sides, we get

$$1 = l - m \text{ and } 0 = l + m$$

solving these, we get

$$l = \frac{1}{2} \text{ and } m = -\frac{1}{2}$$

$$\therefore \quad \int \frac{dx}{1+\cot x} = \int \frac{l(\sin x + \cos x) + m(\cos x - \sin x)}{\sin x + \cos x} dx$$

$$= \int l\, dx + m\int \frac{\cos x - \sin x}{\sin x + \cos x} dx$$

$$= lx + m \log(\sin x + \cos x) + C$$

$$= \frac{1}{2}\cdot x - \frac{1}{2}\log(\sin x + \cos x) + C$$

Example 2. Evaluate $\int \frac{dx}{2+3\tan x}$

Solution: $\int \frac{dx}{2+3\tan x} = \int \frac{dx}{2+3\frac{\sin x}{\cos x}} = \int \frac{\cos x\, dx}{2\cos x + 3\sin x}$

Let, $\cos x = l(2\cos x + 3\sin x) + m(-2\sin x + 3\cos x)$

$\Rightarrow \quad 2l + 3m = 1 \text{ and } 3l - 2m = 0$

solving these, we get.

$$m = \frac{3}{13} \text{ and } l = \frac{2}{13}$$

$$\therefore \quad \int \frac{dx}{2+3\tan x} = \int \frac{l(2\cos x + 3\sin x) + m(-2\sin x + 3\cos x)}{2\cos x + 3\sin x}\cdot dx$$

$$= \int l\, dx + m\int \frac{-2\sin x + 3\cos x}{2\cos x + 3\sin x} dx$$

$$= lx + m \log(2\cos x + 3\sin x) + C$$

$$= \frac{2}{13}x + \frac{3}{13}\log(2\cos x + 3\sin x) + C$$

EXERCISE 4.8

Evaluate the following integrals:

1. (i) $\int \frac{dx}{4+5\cos x}$ (ii) $\int \frac{dx}{1-2\sin x}$

2. (i) $\int \frac{dx}{5+4\cos x}$ (ii) $\int \frac{\cos x\, dx}{3\cos x - 5}$

$$\left[\text{Hint.(ii)} \frac{\cos x}{3\cos x - 5} = \frac{\frac{1}{3}(3\cos x - 5) + \frac{5}{3}}{3\cos x - 5} = \frac{1}{3} + \frac{5}{3} \cdot \frac{1}{3\cos x - 5}\right]$$

3. (i) $\int \frac{dx}{1 - \sin x + \cos x}$ (ii) $\int \frac{dx}{3 + 2\sin x + \cos x}$

4. (i) $\int \frac{dx}{3 + 5\cot x}$ (ii) $\int \frac{dx}{1 - \tan x}$

5. (i) $\int \frac{2\sin x + 3\cos x}{3\sin x + 4\cos x} dx$ (ii) $\int \frac{\sin x + 2\cos x}{2\sin x + \cos x} dx$

ANSWERS

1. (i) $\frac{1}{3}\log\left|\frac{3 + \tan\frac{x}{2}}{3 - \tan\frac{x}{2}}\right| + C$ (ii) $\frac{1}{\sqrt{3}}\log\left|\frac{\tan\frac{x}{2} - 2 - \sqrt{3}}{\tan\frac{x}{2} - 2 + \sqrt{3}}\right| + C$

2. (i) $\frac{2}{3}\tan^{-1}\left(\frac{1}{3}\tan\frac{x}{2}\right) + C$ (ii) $\frac{1}{3}x - \frac{5}{6}\tan^{-1}\left(2\tan\frac{x}{2}\right) + C$

3. (i) $-\log\left(1 - \tan\frac{x}{2}\right) + C$ (ii) $\tan^{-1}\left(1 + \tan\frac{x}{2}\right) + C$

4. (i) $\frac{3}{34}x - \frac{5}{34}\log(3\sin x + 5\cos x) + C$

(ii) $\frac{1}{2}x - \frac{1}{2}\log(\cos x - \sin x) + C$

5. (i) $\frac{18}{25}x + \frac{1}{25}\log(3\sin x + 4\cos x) + C$

(ii) $\frac{4}{5}x + \frac{3}{5}\log(2\sin x + \cos x) + C$

4.16 INTEGRATION OF THE FORM $\int \frac{dx}{\sqrt{ax^2 + bx + c}}$:

Express $\sqrt{ax^2 + bx + c}$ in the form $\sqrt{X^2 + A^2}$ or $\sqrt{X^2 - A^2}$ or $\sqrt{A^2 - X^2}$, where X is a linear function of X & A is a constant and then use the following formulas:

$$\int \frac{dx}{\sqrt{x^2 - a^2}} = \log\left| x + \sqrt{x^2 - a^2} \right| + C$$

$$\int \frac{dx}{\sqrt{a^2 - x^2}} = \sin^{-1}\frac{x}{a} + C$$

$$\int \frac{dx}{\sqrt{x^2 + a^2}} = \log\left| x + \sqrt{x^2 + a^2} \right| + C$$

4.17 INTEGRATION OF THE FORM $\int \frac{px + q}{\sqrt{ax^2 + bx + c}}\,dx$:

Let, $px + q = M\frac{d}{dx}(ax^2 + bx + c) + N$.

Then equating the coefficients of equal powers of x, find M and N. Now the integral breaks up into two parts which are integrable.

ILLUSTRATIVE EXAMPLES

Example 1. Evaluate the following integrals:

(i) $\int \frac{dx}{\sqrt{x^2 - 4x + 5}}$ (ii) $\int \frac{dx}{\sqrt{4 + 8x - 5x^2}}$

Solution: (i) $\int \frac{dx}{\sqrt{x^2 - 4x + 5}} = \int \frac{dx}{\sqrt{(x^2 - 4x + 4) - 4 + 1}} = \int \frac{dx}{\sqrt{(x-2)^2 + 1^2}}$

$$= \log\left| (x-2) + \sqrt{(x-2)^2 + 1^2} \right| + C$$

$$= \log\left| x - 2 + \sqrt{x^2 - 4x + 5} \right| + C$$

(ii) $\int \frac{dx}{\sqrt{4 + 8x - 5x^2}} = \int \frac{dx}{\sqrt{4 - (5x^2 - 8x)}} = \int \frac{dx}{\sqrt{4 - 5\left(x^2 - \frac{8}{5}x\right)}}$

$$= \int \frac{dx}{\sqrt{4 - 5\left(x^2 - \frac{8}{5}x + \frac{16}{25} - \frac{16}{25}\right)}}$$

$$= \int \frac{dx}{\sqrt{4 - 5\left(x - \frac{4}{5}\right)^2 + \frac{16}{5}}}$$

$$= \int \frac{dx}{\sqrt{\frac{36}{5} - 5\left(x - \frac{4}{5}\right)^2}} = \frac{1}{\sqrt{5}}\int \frac{dx}{\sqrt{\frac{36}{25} - \left(x - \frac{4}{5}\right)^2}}$$

$$= \frac{1}{\sqrt{5}} \int \frac{dx}{\sqrt{\left(\frac{6}{5}\right)^2 - \left(x - \frac{4}{5}\right)^2}}$$

$$= \frac{1}{\sqrt{5}} \sin^{-1}\left(\frac{x - \frac{4}{5}}{\frac{6}{5}}\right) + C$$

$$= \frac{1}{\sqrt{5}} \sin^{-1}\left(\frac{5x - 4}{6}\right) + C$$

Example 2. Evaluate the following integrals:

(i) $\int \frac{2x + 1}{\sqrt{x^2 + 4x + 3}}\, dx$ (ii) $\int \frac{3x + 1}{\sqrt{5 - 2x - x^2}}\, dx$

Solution: (i) Let $2x + 1 = M\frac{d}{dx}(x^2 + 4x + 3) + N$. Then

$$2x + 1 = M(2x + 4) + N$$

$$\Rightarrow \quad 2M = 2 \Rightarrow M = 1$$

$$\text{and} \quad 4M + N = 1 \Rightarrow N = 1 - 4 = -3$$

$$\therefore \int \frac{2x + 1}{\sqrt{x^2 + 4x + 3}}\, dx = \int \frac{(2x + 4) - 3}{\sqrt{x^2 + 4x + 3}}\, dx$$

$$= \int (x^2 + 4x + 3)^{-1/2}(2x + 4)\, dx - 3\int \frac{dx}{\sqrt{x^2 + 4x + 3}}$$

$$= \frac{(x^2 + 4x + 3)^{1/2}}{\frac{1}{2}} - 3\int \frac{dx}{\sqrt{(x + 2)^2 - 1^2}}$$

$$= 2\sqrt{x^2 + 4x + 3} - 3\log\left|x + 2 + \sqrt{(x + 2)^2 - 1^2}\right| + C$$

$$= 2\sqrt{x^2 + 4x + 3} - 3\log\left|x + 2 + \sqrt{x^2 + 4x + 3}\right| + C$$

(ii) Let $3x + 1 = M\frac{d}{dx}(5 - 2x - x^2) + N$

$$\Rightarrow \quad 3x + 1 = M((-2 - 2x) + N$$

$$\Rightarrow \quad -2M = 3 \Rightarrow M = -\frac{3}{2}$$

$$\text{and} \quad -2M + N = 1 \Rightarrow N = -2$$

$$\therefore \int \frac{3x + 1}{\sqrt{5 - 2x - x^2}}\, dx = \int \frac{-\frac{3}{2}(-2 - 2x)}{\sqrt{5 - 2x - x^2}}\, dx - 2\int \frac{dx}{\sqrt{5 - 2x - x^2}}$$

$$= -\frac{3}{2}\frac{(5 - 2x - x^2)^{1/2}}{\frac{1}{2}} - 2\int \frac{dx}{\sqrt{5 - (x^2 + 2x)}}$$

$$= -3\sqrt{5-2x-x^2} - 2\int \frac{dx}{\sqrt{6-(x+1)^2}}$$

$$= -3\sqrt{5-2x-x^2} - 2\int \frac{dx}{\left(\sqrt{6}\right)^2 - (x+1)^2}$$

$$= -3\sqrt{5-2x-x^2} - 2\sin^{-1}\left(\frac{x+1}{\sqrt{6}}\right) + C$$

Example 3. Evaluate: $\int \frac{x\,dx}{\sqrt{1-x^2+x^4}}$

Solution: Let $x^2 = t \Rightarrow 2x\,dx = dt \Rightarrow x\,dx = \frac{dt}{2}$

$$\therefore \quad \int \frac{x\,dx}{\sqrt{1-x^2+x^4}} = \frac{1}{2}\int \frac{dt}{\sqrt{1-t+t^2}}$$

$$= \frac{1}{2}\int \frac{dt}{\sqrt{1+\left(t-\frac{1}{2}\right)^2 - \frac{1}{4}}}$$

$$= \frac{1}{2}\int \frac{dt}{\sqrt{\left(\frac{\sqrt{3}}{2}\right)^2 + \left(t-\frac{1}{2}\right)^2}}$$

$$= \frac{1}{2}\cdot\log\left| t - \frac{1}{2} + \sqrt{\left(\frac{\sqrt{3}}{2}\right)^2 + \left(t-\frac{1}{2}\right)^2}\right| + C$$

$$= \frac{1}{2}\log\left| t - \frac{1}{2} + \sqrt{t^2 - t + 1}\right| + C$$

$$= \frac{1}{2}\log\left| x^2 - \frac{1}{2} + \sqrt{x^4 - x^2 + 1}\right| + C$$

$$= \frac{1}{2}\log\frac{\left|2x^2 - 1 + 2\sqrt{x^4 - x^2 + 1}\right|}{2} + C$$

$$= \frac{1}{2}\log\left|2x^2 - 1 + 2\sqrt{x^4 - x^2 + 1}\right| - \frac{1}{2}\log 2 + C$$

$$= \frac{1}{2}\log\left|2x^2 - 1 + 2\sqrt{x^4 - x^2 + 1}\right| + C'$$

where $\qquad C' = C - \frac{1}{2}\log 2$

EXERCISE 4.9

Evaluate the following integrals:

1. (i) $\int \frac{dx}{\sqrt{7-6x-x^2}}$ (ii) $\int \frac{dx}{\sqrt{(x-a)(x-b)}}$

(iii) $\int \frac{dx}{\sqrt{1-4x-2x^2}}$

2. (i) $\int \frac{dx}{\sqrt{1-x-x^2}}$ (ii) $\int \frac{dx}{\sqrt{x(1-2x)}}$

(iii) $\int \frac{dx}{\sqrt{2x^2-x+2}}$

3. (i) $\int \frac{x\,dx}{\sqrt{8+x-x^2}}$ (ii) $\int \frac{(2x+3)\,dx}{\sqrt{x^2+x+1}}$

(iii) $\int \frac{(6x+5)}{\sqrt{6+x-2x^2}}\,dx$

4. (i) $\int \frac{4x+5}{\sqrt{2x^2+x-3}}\,dx$ (ii) $\int \frac{x+1}{\sqrt{2x^2+x-3}}\,dx$

(iii) $\int \frac{2x}{\sqrt{1-x^2-x^4}}\,dx$

5. (i) $\int \frac{x^2+x+1}{\sqrt{x^2-x+1}}\,dx$ (ii) $\int \sqrt{\frac{a+x}{x}}\,dx$

(iii) $\int \frac{6x+7}{\sqrt{(x-5)(x-4)}}\,dx$

ANSWERS

1. (i) $\sin^{-1}\left(\frac{x+3}{4}\right)+C$

(ii) $\log\left|\frac{2x-a-b}{2}+\sqrt{(x-a)(x-b)}\right|+C$

(iii) $\frac{1}{\sqrt{2}}\sin^{-1}\left(\frac{(x+1)\sqrt{2}}{\sqrt{3}}\right)+C$

2. (i) $\sin^{-1}\left(\frac{2x+1}{\sqrt{5}}\right)+C$ (ii) $\frac{1}{\sqrt{2}}\sin^{-1}(4x-1)+C$

(iii) $\frac{1}{\sqrt{2}}\log\left|x-\frac{1}{4}+\sqrt{x^2-\frac{x}{2}+1}\right|+C$

3. (i) $-\sqrt{8+x-x^2}+\frac{1}{2}\sin^{-1}\left(\frac{2x-1}{\sqrt{33}}\right)+C$

(ii) $2\sqrt{x^2+x+1}+2\log\left|x+\frac{1}{2}+\sqrt{x^2+x+1}\right|+C$

(iii) $-3\sqrt{6+x-2x^2}+\frac{13}{2\sqrt{2}}\sin^{-1}\left(\frac{4x-1}{7}\right)+C$

4. (i) $2\sqrt{2x^2+x-3}+2\sqrt{2}\log\left|x+\frac{1}{4}+\sqrt{x^2+\frac{x}{2}-\frac{3}{2}}\right|+C$

(ii) $\frac{1}{2}\sqrt{2x^2+x-3}+\frac{3}{4\sqrt{2}}\log\left|4x+1+\sqrt{2x^2+x-3}\right|+C$

(iii) $\sin^{-1}\left(\frac{2x^2+1}{\sqrt{5}}\right)+C$

5. (i) $\frac{2x+7}{4}\sqrt{x^2-x+1}+\frac{11}{8}\log\left|x-\frac{1}{2}+\sqrt{x^2-x+1}\right|+C$

(ii) $\sqrt{x^2+ax}+\frac{a}{2}\log\left|x+\frac{a}{2}+\sqrt{x^2+ax}\right|+C$

(iii) $6\sqrt{x^2-9x+20}+34\log\left|x-\frac{9}{2}+\sqrt{x^2-9x+20}\right|+C$

4.18 DEFINITE INTEGRALS

Fundamental Theorem of Definite Integral:

Let $f(x)$ be a continuous functions on the closed interval $[a, b]$ and $\phi(x)$ be an anti-derivative of $f(x)$, then

$$\int_a^b f(x)\,dx = [\phi(x)]_a^b = \phi(b)-\phi(a)$$

= (value of anti-derivative at b, the upper limit)

– (value of the same anti-derivative at a, the lower limit)

The interval $[a, b]$ is called the range of integration. In definite integral there is no constant of integration and its value is unique.

ILLUSTRATIVE EXAMPLES

Example 1. Evaluate the following:

(i) $\int_1^2 x\,dx$ (ii) $\int_1^2 \frac{dx}{x}$

(iii) $\int_0^1 (2x^3 + 3)\, dx$ (iv) $\int_0^{\pi/2} \cos^2 x\, dx$

(v) $\int_0^{\pi/6} (\sin x + \sin 2x)\, dx$ (vi) $\int_0^{\pi/4} \cos\frac{3x}{2}\cos\frac{x}{2}\, dx$

(vii) $\int_{-2}^{1} \left(2x + \frac{3}{2}\right)^4 dx$ (viii) $\int_0^1 \frac{dx}{1+x^2}$

Solution: (i) $\int_1^2 x\, dx = \left[\frac{x^2}{2}\right]_1^2 = \frac{4}{2} - \frac{1}{2} = 2 - \frac{1}{2} = \frac{3}{2}$

(ii) $\int_1^2 \frac{dx}{x} = [\log x]_1^2 = \log 2 - \log 1 = \log 2 - 0 = \log 2$

(iii) $\int_0^1 (2x^3 + 3)\, dx = \left[\frac{2x^3}{3} + 3x\right]_0^1 = \left[\frac{2}{3} + 3 - 0 - 0\right] = \frac{11}{3}$

(iv) $\int_0^{\pi/2} \cos^2 x\, dx = \int_0^{\pi/2} \frac{1+\cos 2x}{2}\, dx = \frac{1}{2}\int_0^{\pi/2} (1 + \cos 2x)\, dx$

$$= \frac{1}{2}\left[x + \frac{\sin 2x}{2}\right]_0^{\pi/2}$$

$$= \frac{1}{2}\left[\left(\frac{\pi}{2} + \frac{1}{2}\sin\pi\right) - \left(0 + \frac{\sin 0}{2}\right)\right]$$

$$= \frac{1}{2}\left[\frac{\pi}{2} + \frac{1}{2}\cdot 0 - \frac{1}{2}\cdot 0\right] = \frac{\pi}{4}$$

(v) $\int_0^{\pi/6} (\sin x + \sin 2x)\, dx = \left[-\cos x - \frac{\cos 2x}{2}\right]_0^{\pi/6}$

$$= \left[\left(-\cos\frac{\pi}{6} - \frac{\cos\frac{\pi}{3}}{2}\right) - \left(-\cos 0 - \frac{\cos 0}{2}\right)\right]$$

$$= \left[\left(-\frac{\sqrt{3}}{2} - \frac{1}{2\cdot 2}\right) - \left(-1 - \frac{1}{2}\right)\right]$$

$$= -\frac{\sqrt{3}}{2} - \frac{1}{4} + 1 + \frac{1}{2}$$

$$= -\frac{\sqrt{3}}{2} + \frac{5}{4} = \frac{5 - 2\sqrt{3}}{4}$$

(vi) $\int_0^{\pi/4} \cos\frac{3x}{2}\cos\frac{x}{2}\,dx = \frac{1}{2}\int_0^{\pi/4} 2\cos\frac{3x}{2}\cos\frac{x}{2}\,dx$

$$= \frac{1}{2}\int_0^{\pi/4}\left[\cos\left(\frac{3x}{2}+\frac{x}{2}\right)+\cos\left(\frac{3x}{2}-\frac{x}{2}\right)\right]dx$$

$$= \frac{1}{2}\int_0^{\pi/4}[\cos 2x + \cos x]\,dx$$

$$= \frac{1}{2}\left[\frac{\sin 2x}{2}+\sin x\right]_0^{\pi/2}$$

$$= \frac{1}{2}\left[\left(\frac{\sin\frac{\pi}{2}}{2}+\sin\frac{\pi}{4}\right)-\left(\frac{\sin 0}{2}+\sin 0\right)\right]$$

$$= \frac{1}{2}\left[\left(\frac{1}{2}\cdot 0+\frac{1}{\sqrt{2}}\right)-(0+0)\right]$$

$$= \frac{1}{2}\left[\frac{1}{2}+\frac{1}{\sqrt{2}}\right] = \frac{\sqrt{2}+1}{4}$$

(vii) $\int_{-2}^{1}\left(2x+\frac{3}{2}\right)^4 dx = \left[\frac{\left(2x+\frac{3}{2}\right)^5}{5\cdot 2}\right]_{-2}^{1}$

$$= \frac{1}{10}\left[\left(2x+\frac{3}{2}\right)^5\right]_{-2}^{1}$$

$$= \frac{1}{10}\left[\left(2+\frac{3}{2}\right)^5-\left(-4+\frac{3}{2}\right)^5\right]$$

$$= \frac{1}{10}\left(\frac{16807}{32}+\frac{3125}{32}\right)$$

$$= \frac{1}{10}\cdot\frac{19932}{32}$$

$$= \frac{4983}{80} = 62\frac{23}{80}$$

(viii) $\int_0^1 \frac{dx}{1+x^2} = [\tan^{-1} x]_0^1 = \tan^{-1}(1) - \tan^{-1}(0)$

$$= \frac{\pi}{4} - 0 = \frac{\pi}{4}$$

Example 2. Evaluate the following:

(i) $\int_0^{\pi/2} \sqrt{1+\cos 2x}\, dx$ (ii) $\int_0^{\pi/2} \sqrt{1+\sin 2x}\, dx$

(iii) $\int_0^{2\pi} \sqrt{1+\sin\frac{x}{2}}\, dx$ (iv) $\int_0^{\pi/2} \sin^3 x\, dx$

(v) $\int_0^{\pi/4} \sec x\sqrt{\frac{1-\sin x}{1+\sin x}}\, dx$ (vi) $\int_{\pi/3}^{\pi/4} (\tan x + \cot x)^2\, dx$

(viii) $\int_0^{\pi/4} \sin 2x \sin 3x\, dx$

Solution: (i) $\int_0^{\pi/2} \sqrt{1+\cos 2x}\, dx = \int_0^{\pi/2} \sqrt{1+2\cos^2 x - 1}\, dx$

$$= \int_0^{\pi/2} \sqrt{2}\cos x\, dx$$

$$= \sqrt{2}[\sin x]_0^{\pi/2} = \sqrt{2}\left[\sin\frac{\pi}{2} - \sin 0\right]$$

$$= \sqrt{2}[1-0] = \sqrt{2}$$

(ii) $\int_0^{\pi/2} \sqrt{1+\sin 2x}\, dx = \int_0^{\pi/2} \sqrt{(\sin^2 x + \cos^2 x) + 2\sin x\cos x}\, dx$

$$= \int_0^{\pi/2} \sqrt{(\sin x + \cos x)^2}\, dx = \int_0^{\pi/2} (\sin x + \cos x)\, dx$$

$$= \int_0^{\pi/2} \sin x\, dx + \int_0^{\pi/2} \cos x\, dx$$

$$= [-\cos x]_0^{\pi/2} + [\sin x]_0^{\pi/2}$$

$$= \left[-\cos\frac{\pi}{2} + \cos 0\right] + \left[\sin\frac{\pi}{2} - \sin 0\right]$$

$$= [-0+1] + [1-0] = 2$$

(iii) $\int_0^{2\pi} \sqrt{1+\sin\frac{x}{2}}\, dx = \int_0^{2\pi} \sqrt{1+\cos\left(\frac{\pi}{2}-\frac{x}{2}\right)}\, dx \quad \left[\because \cos\left(\frac{\pi}{2}-x\right) = \sin x\right]$

$$= \int_0^{2\pi} \sqrt{2\cos^2\left(\frac{\pi}{4}-\frac{x}{4}\right)}\, dx$$

$$= \sqrt{2}\int_0^{2\pi} \cos\left(\frac{\pi}{4} - \frac{x}{4}\right)$$

$$= \sqrt{2}\left[\frac{\sin\left(\frac{\pi}{4} - \frac{x}{4}\right)}{-\frac{1}{4}}\right]_0^{2\pi} = \frac{\sqrt{2}}{-\frac{1}{4}}\left[\sin\left(-\frac{\pi}{4}\right) - \sin\frac{\pi}{4}\right]$$

$$= -4\sqrt{2}\left[-\frac{1}{\sqrt{2}} - \frac{1}{\sqrt{2}}\right] = 8$$

(iv) $$\int_0^{\pi/2} \sin^3 x\, dx = \int_0^{\pi/2} \frac{(3\sin x - \sin 3x)}{4}\, dx$$

$$= \frac{1}{4}\int_0^{\pi/2} (3\sin x - \sin 3x)\, dx$$

$$= \frac{1}{4}\left[-3\cos x + \frac{\cos 3x}{3}\right]_0^{\pi/2}$$

$$= \frac{1}{4}\left[\left(-3\cos\frac{\pi}{2} + \frac{\cos\frac{3\pi}{2}}{3}\right) - \left(-3\cos 0 + \frac{\cos 0}{3}\right)\right]$$

$$= \frac{1}{4}\left[0 - \left(-3 + \frac{1}{3}\right)\right] = \frac{2}{3}$$

(v) $$\int_0^{\pi/4} \sec x\sqrt{\frac{1-\sin x}{1+\sin x}}\, dx = \int_0^{\pi/4} \sec x\sqrt{\frac{1-\sin x}{1+\sin x}} \times \sqrt{\frac{1-\sin x}{1-\sin x}}\, dx$$

$$= \int_0^{\pi/4} \sec x\sqrt{\frac{(1-\sin x)^2}{1-\sin^2 x}}\, dx$$

$$= \int_0^{\pi/4} \sec x\frac{(1-\sin x)}{\cos x}\, dx$$

$$= \int_0^{\pi/4} (\sec^2 x - \sec x\tan x)\, dx$$

$$= [\tan x - \sec x]_0^{\pi/4}$$

$$= \left(\tan\frac{\pi}{4} - \sec\frac{\pi}{4}\right) - (\tan 0 - \sec 0)$$

$$= [1 - \sqrt{2} - (0 - 1)]$$

$$= 2 - \sqrt{2}$$

(vi) $$\int_{\pi/3}^{\pi/4} (\tan x + \cot x)^2 \, dx = \int_{\pi/3}^{\pi/4} \left(\frac{\sin x}{\cos x} + \frac{\cos x}{\sin x} \right)^2 dx$$

$$= \int_{\pi/3}^{\pi/4} \left(\frac{\sin^2 x + \cos^2 x}{\sin x \cos x} \right)^2 dx$$

$$= \int_{\pi/3}^{\pi/4} \left(\frac{1}{\sin x \cos x} \right)^2 dx$$

$$= 4 \int_{\pi/3}^{\pi/4} \left(\frac{1}{2 \sin x \cos x} \right)^2 dx$$

$$= 4 \int_{\pi/3}^{\pi/4} \operatorname{cosec}^2 2x \, dx$$

$$= 4 \left[-\frac{\cot 2x}{2} \right]_{\pi/3}^{\pi/4}$$

$$= \frac{4}{2} \left[-\cot \frac{\pi}{2} + \cot \frac{2\pi}{3} \right]$$

$$= 2 \left[0 + \left(-\frac{1}{\sqrt{3}} \right) \right] = -\frac{2}{\sqrt{3}}$$

(vii) $$\int_0^{\pi/4} \sin 2x \sin 3x \, dx = \frac{1}{2} \int_0^{\pi/4} 2 \sin 2x \sin 3x \, dx$$

$$= \frac{1}{2} \int_0^{\pi/4} [\cos(2x - 3x) - \cos(2x + 3x)] \, dx$$

$$= \frac{1}{2} \int_0^{\pi/4} [\cos x - \cos 5x] \, dx$$

$$= \frac{1}{2} \left[\sin x - \frac{\sin 5x}{5} \right]_0^{\pi/4}$$

$$= \frac{1}{2} \left[\left(\sin \frac{\pi}{4} - \frac{\sin \frac{5\pi}{4}}{5} \right) - \left(\sin 0 - \frac{\sin 0}{5} \right) \right]$$

$$= \frac{1}{2} \left[\frac{1}{\sqrt{2}} + \frac{1}{5\sqrt{2}} - 0 \right] = \frac{3}{5\sqrt{2}}$$

Example 3. Evaluate the following:

(i) $\int_0^4 \frac{dx}{\sqrt{x^2+2x+3}}$ (ii) $\int_0^1 \frac{(\tan^{-1} x)^2}{1+x^2}\,dx$

(iii) $\int_0^1 \frac{x^5}{1+x^6}\,dx$ (iv) $\int_{\pi/2}^{\pi} \frac{1-\sin x}{1-\cos x}\,dx$

(v) $\int_0^1 \frac{1}{\sqrt{1+x}+\sqrt{x}}\,dx$ (vi) $\int_0^2 \frac{dx}{4+x-x^2}$

(vii) $\int_0^1 \frac{4x^2+3}{8x^2+4x+5}\,dx$ (viii) $\int_0^1 \frac{x+1}{(x^2+2x+3)^2}\,dx$

Solution: (i) $\int_0^4 \frac{dx}{\sqrt{x^2+2x+3}} = \int_0^4 \frac{dx}{\sqrt{(x^2+2x+1)+3-1}}$

$$= \int_0^4 \frac{dx}{\sqrt{(x+1)^2+\left(\sqrt{2}\right)^2}}$$

$$= \left\{\log\left[(x+1)+\sqrt{x^2+2x+3}\right]\right\}_0^4$$

$$= \log\left(5+\sqrt{27}\right)-\log\left(1+\sqrt{3}\right)$$

$$= \log\left(\frac{5+3\sqrt{3}}{1+\sqrt{3}}\right)$$

(ii) $\int_0^1 \frac{(\tan^{-1} x)^2}{1+x^2}\,dx = \int_0^1 (\tan^{-1} x)^2 \cdot \frac{1}{1+x^2}\,dx$ $\left[\int f'(x)\cdot f''(x)\,dx \text{ form}\right]$

$$= \left[\frac{(\tan^{-1} x)^3}{3}\right]_0^1 = \frac{1}{3}[(\tan^{-1}(1))^3 - (\tan^{-1}(0))^3]$$

$$= \frac{1}{3}\left[\left(\frac{\pi}{4}\right)^3 - 0\right] = \frac{\pi^3}{192}$$

(iii) $\int_0^1 \frac{x^5\,dx}{1+x^6} = \frac{1}{6}\int_0^1 \frac{6x^5}{1+x^6}\,dx$ $\left[\int \frac{f'(x)}{f(x)} \text{ form}\right]$

$$= \frac{1}{6}[\log(1+x^6)]_0^1 = \frac{1}{6}[\log 2 - \log 1]$$

$$= \frac{1}{6}[\log 2 - 0] \ (\because \log 1 = 0)$$

$$= \frac{1}{6}\log 2$$

(iv) $$\int_{\pi/2}^{\pi} \frac{1-\sin x}{1-\cos x}\,dx = \int_{\pi/2}^{\pi} \frac{dx}{1-\cos x} - \int_{\pi/2}^{\pi} \frac{\sin x}{1-\cos x}\,dx$$

$$= \int_{\pi/2}^{\pi} \frac{dx}{2\sin^2 \frac{x}{2}} - \int_{\pi/2}^{\pi} \frac{\sin x}{1-\cos x}\,dx$$

$$= \frac{1}{2}\int_{\pi/2}^{\pi} \operatorname{cosec}^2 \frac{x}{2}\,dx - [\log(1-\cos x)]_{\pi/2}^{\pi}$$

$$= \frac{1}{2}\left[\frac{-\cot\frac{x}{2}}{\frac{1}{2}}\right]_{\pi/2}^{\pi} - \left[\log(1-\cos\pi) - \log\left(1-\cos\frac{\pi}{2}\right)\right]$$

$$= -\left(\cot\frac{\pi}{2} - \cot\frac{\pi}{4}\right) - [\log(1+1) - \log(1-0)]$$

$$= -(0-1) - (\log 2 - \log 1)$$

$$= 1 - (\log 2 - 0) = 1 - \log 2$$

(v) $$\int_0^1 \frac{dx}{\sqrt{1+x}+\sqrt{x}} = \int_0^1 \frac{1}{\sqrt{1+x}+\sqrt{x}} \times \frac{\sqrt{1+x}-\sqrt{x}}{\sqrt{1+x}-\sqrt{x}}\,dx$$

$$= \int_0^1 \frac{(1+x)^{1/2} - x^{1/2}}{1+x-x}\,dx$$

$$= \int_0^1 [(1+x)^{1/2} - x^{1/2}]\,dx$$

$$= \left[\frac{(1+x)^{3/2}}{\frac{3}{2}} - \frac{x^{3/2}}{3/2}\right]_0^1$$

$$= \frac{2}{3}[2^{3/2} - 1^{3/2}) - (1^{3/2} - 0)]$$

$$= \frac{2}{3}\left(2\sqrt{2} - 1 - 1\right) = \frac{2}{3}\left(2\sqrt{2} - 2\right)$$

$$= \frac{4}{3}\left(\sqrt{2} - 1\right)$$

(vi) $$\int_0^2 \frac{dx}{4+x-x^2} = \int_0^2 \frac{dx}{4-(x^2-x)}$$

$$= \int_0^2 \frac{dx}{4-\left(x^2 - x + \frac{1}{4}\right) + \frac{1}{4}}$$

$$= \int_0^2 \frac{dx}{\left(\frac{\sqrt{17}}{2}\right)^2 - \left(x - \frac{1}{2}\right)^2}$$

$$= \frac{1}{2\frac{\sqrt{17}}{2}} \left[\log \left| \frac{\frac{\sqrt{17}}{2} + x - \frac{1}{2}}{\frac{\sqrt{17}}{2} - x + \frac{1}{2}} \right| \right]_0^2$$

$$= \frac{1}{\sqrt{17}} \left[\log \left| \frac{2x + \sqrt{17} - 1}{\sqrt{17} - 2x + 1} \right| \right]_0^2$$

$$= \frac{1}{\sqrt{17}} \left[\log \left(\frac{3 + \sqrt{17}}{\sqrt{17} - 3} \right) - \log \left(\frac{\sqrt{17} - 1}{\sqrt{17} + 1} \right) \right]$$

$$= \frac{1}{\sqrt{17}} \log \left[\frac{\sqrt{17} + 3}{\sqrt{17} - 3} \times \frac{\sqrt{17} + 1}{\sqrt{17} - 1} \right]$$

$$= \frac{1}{\sqrt{17}} \log \left(\frac{17 + 4\sqrt{17} + 3}{17 + 3 - 4\sqrt{17}} \right)$$

$$= \frac{1}{\sqrt{17}} \log \left(\frac{5 + \sqrt{17}}{5 - \sqrt{17}} \right)$$

$$= \frac{1}{\sqrt{17}} \log \left(\frac{5 + \sqrt{17}}{5 - \sqrt{17}} \times \frac{5 + \sqrt{17}}{5 + \sqrt{17}} \right)$$

$$= \frac{1}{\sqrt{17}} \log \left(\frac{25 + 17 + 10\sqrt{17}}{25 - 17} \right)$$

$$= \frac{1}{\sqrt{17}} \log \left(\frac{42 + 10\sqrt{17}}{8} \right)$$

$$= \frac{1}{\sqrt{17}} \log \left(\frac{21 + 5\sqrt{17}}{4} \right)$$

(vii) $\displaystyle\int_0^1 \frac{4x^2 + 3}{8x^2 + 4x + 5} dx = \int_0^1 \left(\frac{1}{2} + \frac{\frac{1}{2} - 2x}{8x^2 + 4x + 5} \right) dx$

$$= \int_0^1 \frac{1}{2} dx + \int_0^1 \frac{\frac{1}{2} - 2x}{8x^2 + 4x + 5} dx$$

$$= \frac{1}{2}[x]_0^1 + \int_0^1 \frac{-\frac{1}{8}(16x+4)+1}{8x^2+4x+5}\,dx$$

$$= \frac{1}{2}(1-0) - \frac{1}{8}\int_0^1 \frac{16x+4}{8x^2+4x+5}\,dx + \int_0^1 \frac{dx}{8x^2+4x+5}$$

$$= \frac{1}{2} - \frac{1}{8}\left[\log(8x^2+4x+5)\right]_0^1 + \frac{1}{8}\int_0^1 \frac{dx}{x^2+\frac{x}{2}+\frac{5}{8}}$$

$$= \frac{1}{2} - \frac{1}{8}(\log 17 - \log 5) + \frac{1}{8}\int_0^1 \frac{dx}{\left(x+\frac{1}{4}\right)^2 + \left(\frac{3}{4}\right)^2}$$

$$= \frac{1}{2} - \frac{1}{8}\log\left(\frac{17}{5}\right) + \frac{1}{8}\cdot\frac{1}{\frac{3}{4}}\left[\tan^{-1}\left(\frac{x+\frac{1}{4}}{\frac{3}{4}}\right)\right]_0^1$$

$$= \frac{1}{2} - \frac{1}{8}\log\left(\frac{17}{5}\right) + \frac{1}{6}\left[\tan^{-1}\left(\frac{4x+1}{3}\right)\right]_0^1$$

$$= \frac{1}{2} - \frac{1}{8}\log\left(\frac{17}{5}\right) + \frac{1}{6}\left(\tan^{-1}\frac{5}{3} - \tan^{-1}\frac{1}{3}\right)$$

$$= \frac{1}{2} - \frac{1}{8}\log\left(\frac{17}{5}\right) + \frac{1}{6}\tan^{-1}\left(\frac{\frac{5}{3}-\frac{1}{3}}{1+\frac{5}{3}\cdot\frac{1}{3}}\right)$$

$$= \frac{1}{2} - \frac{1}{8}\log\left(\frac{17}{5}\right) + \frac{1}{6}\tan^{-1}\left(\frac{6}{7}\right)$$

(viii) $\displaystyle\int_0^1 \frac{x+1}{(x^2+2x+3)^2}\,dx = \frac{1}{2}\int_0^1 \frac{2x+2}{(x^2+2x+3)^2}\,dx$

$$= \frac{1}{2}\int_0^1 (x^2+2x+3)^{-2}(2x+2)\,dx$$

$$= \frac{1}{2}\left[\frac{(x^2+2x+3)^{-1}}{-1}\right]_0^1$$

$$= -\frac{1}{2}[(6)^{-1} - (3)^{-1}] = -\frac{1}{2}\left[\frac{1}{6} - \frac{1}{3}\right]$$

$$= -\frac{1}{2} \times -\frac{1}{6} = \frac{1}{12}$$

Example 4. Evaluate the following definite integrals:

(i) $\int_{\pi/4}^{\pi/2} \cos 2x \cdot \log \sin x \, dx$ (ii) $\int_0^1 3^{2x-2} \, dx$

(iii) $\int_0^1 \log(1+x) \, dx$ (iv) $\int_1^2 e^x \left(\frac{1}{x} - \frac{1}{x^2} \right) dx$

(v) $\int_a^b \frac{\log x}{x} \, dx$ (vi) $\int_0^1 \tan^{-1} x \, dx$

(viii) $\int_0^1 x^2 e^x \, dx$

Solution: (i) $\int_{\pi/4}^{\pi/2} \cos 2x \cdot \log \sin x \, dx = \int_{\pi/4}^{\pi/2} \log \sin x \cdot \cos 2x \, dx$

(Integrate by parts)

$$= \left[\log \sin x \cdot \frac{\sin 2x}{2} \right]_{\pi/4}^{\pi/2} - \int_{\pi/4}^{\pi/2} \frac{1}{\sin x} \cdot \cos x \cdot \frac{\sin 2x}{2} \, dx$$

$$= \frac{1}{2}\left[0 - \log \frac{1}{\sqrt{2}} \right] - \int_{\pi/4}^{\pi/2} \cos^2 x \, dx$$

$$= -\frac{1}{2}\left(\log 1 - \log \sqrt{2} \right) - \int_{\pi/4}^{\pi/2} \frac{1 + \cos 2x}{2} \, dx$$

$$= -\frac{1}{2}\left(0 - \frac{1}{2}\log 2 \right) - \frac{1}{2}\left[x + \frac{\sin 2x}{2} \right]_{\pi/4}^{\pi/2}$$

$$= \frac{1}{4}\log 2 - \frac{1}{2}\left[\left(\frac{\pi}{2} + \frac{\sin \pi}{2} \right) - \left(\frac{\pi}{4} + \frac{\sin \frac{\pi}{2}}{2} \right) \right]$$

$$= \frac{1}{4}\log 2 - \frac{1}{2}\left[\frac{\pi}{2} + 0 - \frac{\pi}{4} - \frac{1}{2} \right]$$

$$= \frac{1}{4}\log 2 - \frac{1}{2}\left(\frac{\pi}{4} - \frac{1}{2} \right)$$

$$= \frac{1}{4}\log 2 - \frac{\pi}{8} + \frac{1}{4}$$

(ii) $\int_0^1 3^{2x-2} \, dx = \left[\frac{3^{2x-2}}{\log_e^3} \cdot \frac{1}{2} \right]_0^1$

$$= \left[\frac{1}{\log_e^3}\cdot\frac{1}{2} - \frac{3^{-2}}{\log_e^3}\cdot\frac{1}{2}\right]$$

$$= \frac{1}{2\log 3} - \frac{1}{18\log 3}$$

$$= \frac{4}{9}\cdot\frac{1}{\log 3}$$

(iii) $\int_0^1 \log(1+x)\,dx = [\log(1+x)\cdot x]_0^1 - \int_0^1 \frac{1}{1+x}\cdot x\,dx$

(Integrating by parts)

$$= (\log 2 - 0) - \int_0^1 \frac{x+1-1}{1+x}\,dx$$

$$= \log 2 - \int_0^1 \left(1 - \frac{1}{1+x}\right)dx$$

$$= \log 2 - [x - \log(1+x)]_0^1$$
$$= \log 2 - [(1-\log 2) - (0 - \log 1)]$$
$$= \log 2 - [1 - \log 2 - 0 + 0]$$
$$= 2\log 2 - 1 = \log 2^2 - \log e \qquad [\because \log_e e = 1]$$

$$= \log 4 - \log e = \log\frac{4}{e}$$

(iv) $\int_1^2 e^x\left(\frac{1}{x} - \frac{1}{x^2}\right)dx = \int_1^2 \frac{e^x}{x} - \int_1^2 e^x\cdot\frac{1}{x^2}\,dx$

$$= \left[\frac{1}{x}e^x\right]_1^2 + \int_1^2 \frac{1}{x^2}e^x\,dx - \int_1^2 \frac{e^x}{x^2}\,dx$$

$$= \left[\frac{e^2}{2} - \frac{e}{1}\right] + \int_1^2 \frac{e^x}{x^2}\,dx - \int_1^2 \frac{e^x}{x^2}\,dx$$

$$= \frac{e(e-2)}{2}$$

(v) $\int_a^b \frac{\log x}{x}\,dx = \int_a^b (\log x)\cdot\frac{1}{x}\,dx \qquad \left[\int f(x)^n\, f'(x)\,dx\right]$

$$= \left[\frac{(\log x)^2}{2}\right]_a^b = \frac{1}{2}[(\log b)^2 - (\log a)^2]$$

$$= \frac{1}{2}(\log b + \log a)(\log b - \log a)$$

$$= \frac{1}{2}\log ab \cdot \log\left(\frac{a}{b}\right)$$

$$= \log\sqrt{ab} \cdot \log\left(\frac{a}{b}\right)$$

(vi) $\int_0^1 \tan^{-1} x\, dx = \int_0^1 \tan^{-1} \cdot 1\, dx$

$$= [\tan^{-1} x \cdot x]_0^1 - \int_0^1 \frac{1}{1+x^2} \cdot x\, dx$$

$$= [\tan^{-1} x - \tan^{-1} 0] - \frac{1}{2}\int_0^1 \frac{2x}{1+x^2}\, dx$$

$$= \left(\frac{\pi}{4} - 0\right) - \frac{1}{2}[\log(1+x^2)]_0^1$$

$$= \frac{\pi}{4} - \frac{1}{2}(\log 2 - \log 1)$$

$$= \frac{\pi}{4} - \frac{1}{2}\log 2 - \frac{1}{2}\log 1$$

$$= \frac{\pi}{4} - \log\sqrt{2} - 0$$

$$= \frac{\pi}{4} - \log\sqrt{2}$$

(vii) $\int_0^1 x^2 e^x\, dx = [x^2 e^x]_0^1 - \int_0^1 2x \cdot e^x\, dx$

$$= \left\{(e - 0) - 2 \cdot [x \cdot e^x]_0^1 - \int_0^1 1 \cdot e^x\, dx\right\}$$

$$= e - 2 \cdot \left\{(e - 0) - [e^x]_0^1\right\}$$

$$= e - 2[e - (e - 1)]$$

$$= e - 2(1) = e - 2$$

Example 5. Evaluate the following integrals:

(i) $\int_0^{\sqrt{2}} \sqrt{2 - x^2}\, dx$ (ii) $\int_{\pi/4}^{\pi/2} (\log(\sin x) + \cot x)\, dx$

(iii) $\int_2^3 \frac{x^3 + 1}{x(x-1)}\, dx$ (iv) $\int_1^2 \frac{2}{4x^2 - 1}\, dx$

Solution: (i) $\int_0^{\sqrt{2}} \sqrt{2-x^2}\, dx = \int_0^{\sqrt{2}} \sqrt{\left(\sqrt{2}\right)-x^2}\, dx$

$$= \left[\frac{x}{2}\sqrt{2-x^2} + \frac{(\sqrt{2})^2}{2}\sin^{-1}\frac{x}{\sqrt{2}}\right]_0^{\sqrt{2}}$$

$$= \left(\frac{\sqrt{2}}{2}\cdot 0 + \sin^{-1} 1\right) - \left(\frac{0\sqrt{2}}{2} + \sin^{-1} 0\right)$$

$$= \left(0 + \frac{\pi}{2}\right) - (0+0) = \frac{\pi}{2}$$

(ii) $\int_{\pi/4}^{\pi/2} e^x(\log(\sin x) + \cot x)\, dx = \int_{\pi/4}^{\pi/2} e^x \log(\sin x)\, dx + \int_{\pi/4}^{\pi/2} e^x \cot x\, dx$

$$= [\log(\sin x)\cdot e^x]_{\pi/4}^{\pi/2} - \int_{\pi/4}^{\pi/2} \frac{1}{\sin x}\cos x \cdot e^x\, dx + \int_{\pi/4}^{\pi/2} e^x \cot x\, dx$$

$$= \left[e^{\pi/2}\log\left(\sin\frac{\pi}{2}\right) - e^{\pi/4}\log\left(\sin\frac{\pi}{4}\right)\right]$$

$$- \int_{\pi/4}^{\pi/2} e^x \cot x\, dx + \int_{\pi/4}^{\pi/2} e^x \cot x\, dx$$

$$= e^{\pi/2}\log 1 - e^{\pi/4}\log\left(\frac{1}{\sqrt{2}}\right)$$

$$= 0 - e^{\pi/4}(\log 1 - \log\sqrt{2})$$

$$= -e^{\pi/4}\left(0 - \frac{1}{2}\log 2\right)$$

$$= \frac{1}{2}e^{\pi/4}\log 2$$

(iii) $\int_2^3 \frac{x^3+1}{x(x-1)}\, dx = \int_2^3 \left(x+1+\frac{x+1}{x(x-1)}\right) dx$ (By division)

$$= \int_2^3 \left(x+1+\frac{2}{x-1}-\frac{1}{x}\right) dx \quad \text{(By partial fractions)}$$

$$= \left[\frac{x^2}{2} + x + 2\log(x-1) - \log x\right]_2^3$$

$$= \left(\frac{9}{2} + 3 + 2\log 2 - \log 3\right) - \left(\frac{4}{2} + 2 + 2\log 1 - \log 2\right)$$

$$= \left(\frac{15}{2} + 2\log 2 - \log 3\right) - (2 + 2 + 0 - \log 2)$$

$$= \frac{15}{2} + 2\log 2 - \log 3 - 4 + \log 2$$

$$= \frac{7}{2} + 3\log 2 - \log 3$$

$$= \frac{7}{2} + \log 2^3 - \log 3$$

$$= \frac{7}{2} + \log 8 - \log 3 = \frac{7}{2} + \log\frac{8}{3}$$

(iv) $\displaystyle\int_1^2 \frac{2\,dx}{4x^2 - 1} = \frac{2}{4}\int_1^2 \frac{dx}{x^2 - \left(\frac{1}{2}\right)^2}$

$$= \frac{1}{2}\cdot\frac{1}{2\cdot\frac{1}{2}}\left[\log\left|\frac{x - \frac{1}{2}}{x + \frac{1}{2}}\right|\right]_1^2$$

$$= \frac{1}{2}\left[\log\left|\frac{2x - 1}{2x + 1}\right|\right]_1^2$$

$$= \frac{1}{2}\left(\log\frac{3}{5} - \log\frac{1}{3}\right)$$

$$= \frac{1}{2}\log\left(\frac{3}{5}\times\frac{3}{1}\right) = \frac{1}{2}\log\frac{9}{5}$$

Example 6. If $\int_0^a 3x^2\,dx = 8$, find the value of a.

Solution: Given $\int_0^a 3x^2\,dx = 8$

$$\Rightarrow \quad 3\left[\frac{x^3}{3}\right]_0^a = 8$$

$$\Rightarrow \quad [x^3]_0^a = 8$$

$$\Rightarrow \quad (a^3 - 0) = 2^3$$

$$\Rightarrow \quad a^3 = 2^3$$

$$\Rightarrow \quad a = 2$$

Example 7. If $\int_a^b x^3\,dx = 0$ and $\int_a^b x^2\,dx = \frac{2}{3}$ find the value of $\int_{a+1}^{b+1} 2x\,dx$

Solution: Given, $\int_a^b x^3\,dx = 0$

$$\Rightarrow \quad \left[\frac{x^4}{4}\right]_a^b = 0$$

$$\Rightarrow \quad \frac{b^4 - a^4}{4} = 0$$

$$\Rightarrow \quad b^4 - a^4 = 0 \qquad \text{... (i)}$$

Also, $\int_a^b x^2\,dx = \frac{2}{3}$

$$\Rightarrow \quad \left[\frac{x^3}{3}\right]_a^b = \frac{2}{3}$$

$$\Rightarrow \quad \frac{b^3 - a^3}{3} = \frac{2}{3}$$

$$\Rightarrow \quad b^3 - a^3 = 2 \qquad \text{... (ii)}$$

By (i), $b^4 - a^4 = 0$

$\Rightarrow (b - a)(b + a)(b^2 + a^2) = 0$

$\Rightarrow$ either $b = a$ or $b = -a$ or $b^2 = -a^2$

But $b = a$ and $b^2 = -a^2$ are not possible.

$\Rightarrow \quad b = -a$

$\therefore$ By (ii), $-a^3 - a^3 = 2$

$$-2a^3 = 2$$

$$a^3 = -1$$

$$\Rightarrow \quad a = -1$$

Here, $b = 1$

Now, $\int_{a+1}^{b+1} 2x \cdot dx = 2\int_{-1+1}^{2} x \cdot dx = 2\int_0^2 x \cdot dx = 2\left[\frac{x^2}{2}\right]_0^2$

$$= [x^2]_0^2 = 4 - 0 = 4$$

EXERCISE 4.10

Evaluate the following integrals:

1. (i) $\int_1^3 x^2\, dx$ (ii) $\int_1^2 (e^x + 3x^2)\, dx$

(iii) $\int_2^5 \left(\sqrt{x} + \frac{1}{\sqrt{x}}\right)^3 dx$ (iv) $\int_0^1 (2x^3 + 3)^2\, dx$

(v) $\int_3^5 \frac{dt}{1+3t}$ (vi) $\int_0^a \frac{dx}{\sqrt{a^2 - x^2}}$

(vii) $\int_1^2 \frac{dx}{x}$ (viii) $\int_0^1 \sqrt{5x+4}\, dx$

2. (i) $\int_0^{\pi/2} \cos^4 x\, dx$ (ii) $\int_0^{\pi} \frac{dx}{1+\sin x}$

(iii) $\int_0^{\pi/2} \sin x \sin 2x\, dx$ (iv) $\int_0^{\pi/4} \sqrt{1-\sin 2x}\, dx$

(v) $\int_{\pi}^{3\pi/2} \sqrt{1-\cos 2x}\, dx$ (vi) $\int_0^{\pi/2} \sin^2 x\, dx$

3. (i) $\int_{\pi/2}^{\pi} \frac{1+\cos x}{1-\cos x}\, dx$ (ii) $\int_0^{\pi/4} \tan^2 x\, dx$

(iii) $\int_{-\sqrt{3}}^{\sqrt{3}} \frac{dx}{1+x^2}$ (iv) $\int_1^3 (x^2+e^x)(x^3+3e^x+4)\, dx$

(v) $\int_0^{\pi/2} (a^2\cos^2 x + b^2 \sin^2 x)\, dx$ (vi) $\int_0^{\pi/2} \sin^2 x \cos x\, dx$

(vii) $\int_0^{\pi/4} (\tan x + \cot x)^{-1}\, dx$ (viii) $\int_0^{\pi/2} \frac{\sin^2 x}{(1+\cos x)^2}\, dx$

(ix) $\int_0^{\pi/2} \frac{\sin x}{\sqrt{1+\cos x}}\, dx$

4. (i) $\int_0^1 \frac{(\sin^{-1} x)^2}{\sqrt{1-x^2}}\, dx$ (ii) $\int_0^1 \frac{\sin(\tan^{-1} x)}{1+x^2}\, dx$

(ii) $\int_2^3 \frac{\cos(\log x)}{x}\, dx$ (iv) $\int_0^{\pi/3} \frac{\cos x\, dx}{3+4\sin x}$

(v) $\int_0^{\pi/4} \frac{2\cos 2x}{1+\sin 2x}\,dx$

5. (i) $\int_0^{\pi/2} x\cos 2x\,dx$ (ii) $\int_0^{\pi/2} x\sin^2 x\,dx$

(iii) $\int_0^1 x\tan^{-1} x\,dx$ (iv) $\int_1^5 \frac{\log x}{(x+1)^2}\,dx$

(v) $\int_0^1 xe^{2x}\,dx$ (vi) $\int_1^e \frac{e^x}{x}(1+x\log x)\,dx$

(vii) $\int_1^7 4^x\,dx$ (viii) $\int_0^{\pi/2} x\sin x\cos x\,dx$

6. (i) $\int_0^1 \frac{x^9\,dx}{5+x^{10}}$ (ii) $\int_0^1 \frac{xe^x}{(x+1)^2}\,dx$

(iii) $\int_{1/4}^{1/2} \frac{dx}{\sqrt{x-x^2}}$ (iv) $\int_1^2 \frac{x^3}{(x+1)(x^2-7x+12)}\,dx$

(v) $\int_0^2 \frac{dx}{\sqrt{3+2x-x^2}}$ (vi) $\int_1^3 \frac{dx}{x^3+x^2}$

(vii) $\int_{-a}^{a} \sqrt{\frac{a-x}{a+x}}\,dx$ (viii) $\int_0^a \frac{dx}{\sqrt{ax-x^2}}$

7. (i) $\int_{\pi/4}^{\pi/2} \cot^2 x\,dx$ (ii) $\int_0^{\sqrt{2}} \frac{dx}{x\sqrt{x^2-1}}$

(iii) $\int_0^1 (e^x - e^{-x})^2\,dx$

ANSWERS

1. (i) $\frac{26}{3}$ (ii) $e^2 - e + 7$

(iii) $\frac{1}{5}(128\sqrt{5} - 58\sqrt{2})$ (iv) $\frac{88}{7}$

(v) $\frac{1}{3}\log\frac{8}{5}$ (vi) $\frac{\pi}{2}$

(vii) $\log 2$ (viii) $\frac{38}{15}$

2. (i) $\frac{3\pi}{16}$ (ii) 2

(iii) $\frac{2}{3}$ (iv) $\sqrt{2}-1$

(v) $\sqrt{2}$ (vi) $\frac{\pi}{4}$

3. (i) $2-\frac{\pi}{2}$ (ii) $1-\frac{\pi}{4}$

(iii) $\frac{2\pi}{3}$ (iv) $\frac{1}{6}(31+3e^3)^2-(3e+5)^2]$

(v) $\frac{\pi}{4}(a^2+b^2)$ (vi) $\frac{1}{3}$

(vii) $\frac{1}{4}$ (viii) $2-\frac{\pi}{2}$

(ix) $2(\sqrt{2}-1)$

4. (i) $\frac{\pi^3}{24}$ (ii) $1-\frac{1}{\sqrt{2}}$

(iii) $\sin(\log 3)-\sin(\log 2)$ (iv) $\frac{1}{4}\log\left(1+\frac{2}{\sqrt{3}}\right)$

(v) $\log 2$

5. (i) $-\frac{1}{2}$ (ii) $\frac{\pi^2}{16}+\frac{1}{4}$

(iii) $\frac{\pi}{4}-\frac{1}{2}$ (iv) $\frac{5}{6}\log 5-\log 3$

(v) $\frac{e^2+1}{4}$ (vi) e^e

(vii) $16380\log_4 e$ (viii) $\frac{\pi}{8}$

6. (i) $\frac{1}{10}\log\frac{6}{5}$ (ii) $\frac{1}{2}e-1$

(iii) $\frac{\pi}{6}$ (iv) $1+\frac{98}{5}\log 2-\frac{257}{50}\log 3$

(v) $\frac{\pi}{3}$ (vi) $\frac{2}{3}+\log\frac{2}{3}$

(vii) $a\pi$ (viii) π

7. (i) $\frac{\pi}{4}+1$ (ii) $\frac{\pi}{4}$

(iii) $\frac{1}{2}(e^2-e^{-2}-4)$

4.19 DEFINITE INTEGRATION BY SUBSTITUTION

In definite integrals, when we have to make some substitutions, the limits for the new variable are also found, the integrand is written in the form of new variable and the differential (dx) is also changed.

ILLUSTRATIVE EXAMPLES

Example 1. Evaluate the following definite integrals:

(i) $\int_0^1 xe^{x^2}\,dx$ (ii) $\int_1^2 \frac{(1+\log x)^3}{x}\,dx$

(iii) $\int_0^1 \frac{e^{2x}}{1+e^{4x}}\,dx$ (iv) $\int_0^{\pi/2} \cos^3 x\,dx$

(v) $\int_0^1 \frac{\sin(\tan^{-1}x)}{1+x^2}\,dx$ (vi) $\int_0^{\pi/2} \frac{\sin x\,dx}{\sqrt{1+\cos x}}$

Solution: (i) Put $x^2 = t \Rightarrow 2x\,dx = dt \Rightarrow x\,dx = \frac{dt}{2}$

when $x = 0, t = 0$ and when $x = 1, t = 1^2 = 1$

$$\therefore \quad \int_0^1 xe^{x^2}\,dx = \int_0^1 e^t \frac{dt}{2} = \frac{1}{2}[e^t]_0^1 = \frac{e^1 - e^0}{2} = \frac{e^1 - e^0}{2} = \frac{e-1}{2}$$

(ii) Put $1 + \log x = t \Rightarrow 0 + \frac{1}{x}\,dx = dt \Rightarrow \frac{dx}{x} = dt$

when $x = 1, t = 1 + \log 1 = 1 + 0 = 1$

and, when $x = 2, t = 1 + \log 2$

$$\therefore \quad \int_1^2 \frac{(1+\log x)^3}{x}\,dx = \int_1^{1+\log 2} t^3\,dt = \left[\frac{t^4}{4}\right]_1^{1+\log 2}$$

$$= \frac{1}{4}[(1+\log 2)^4 - (1)^4] = \frac{1}{4}[(1+\log 2)^4 - 1]$$

(iii) Put $e^{2x} = t \Rightarrow 2e^{2x}\,dx = dt \Rightarrow e^{2x}\,dx = \frac{dt}{2}$

when $x = 0, \mathrm{t} = e^0 = 1$

and, when $x = 1, t = e^2$

$$\therefore \quad \int_0^1 \frac{e^{2x}}{1+e^{4x}}\,dx = \int_1^{e^2} \frac{\frac{dt}{2}}{1+t^2} = \frac{1}{2}\int_1^{e^2} \frac{dt}{1+t^2}$$

$$= \frac{1}{2}[\tan^{-1} t]_1^{e^2} = \frac{1}{2}[\tan^{-1} e^2 - \tan^{-1} 1]$$

$$= \frac{1}{2}\left[\tan^{-1} e^2 - \frac{\pi}{4}\right]$$

(iv) $\int_0^{\pi/2} \cos^3 x\, dx = \int_0^{\pi/2} \cos^2 x \cdot \cos x\, dx = \int_0^{\pi/2} (1 - \sin^2 x)\cos x\, dx$

Put $\sin x = t \Rightarrow \cos x\, dx = dt$

when $x = 0, t = \sin 0 = 0$

and, when $x = \frac{\pi}{2}, t = \sin\frac{\pi}{2} = 1$

$$\therefore \int_0^{\pi/2} \sin^3 x\, dx = \int_0^1 (1 - t^2)\, dt = \left[t - \frac{t^3}{3}\right]_0^1$$

$$= \left(1 - \frac{1}{3}\right) - (0 - 0) = \frac{2}{3}$$

(v) Put $\tan^{-1} x = t \Rightarrow \frac{1}{1+x^2} \cdot dx = dt$

when $x = 0, t = \tan^{-1} 0 = 0$ and,

when $x = 1, t = \tan^{-1} 1 = \frac{\pi}{4}$

$$\therefore \int_0^1 \frac{\sin(\tan^{-1} x)}{1+x^2}\, dx = \int_0^{\pi/4} \sin t\, dt = -[\cos t]_0^{\pi/4}$$

$$= -\left[\cos\frac{\pi}{4} - \cos 0\right] = -\left(\frac{1}{\sqrt{2}} - 1\right)$$

$$= \frac{\sqrt{2} - 1}{\sqrt{2}}$$

(vi) Put $1 + \cos x = t \Rightarrow -\sin x\, dx = dt \Rightarrow \sin x\, dx = -dt$

when $x = 0, t = 1 + \cos 0 = 1 + 1 = 2$

when $x = \frac{\pi}{2}, t = 1 + \cos\frac{\pi}{2} = 1 + 0 = 1$

$$\therefore \int_0^{\pi/2} \frac{\sin x}{\sqrt{1 + \cos x}}\, dx = \int_2^1 \frac{-dt}{\sqrt{t}} = -\int_2^1 t^{-1/2}\, dt$$

$$= -\left[\frac{t^{1/2}}{\frac{1}{2}}\right]_2^1 = -2[\sqrt{t}]_2^1$$

$$= -2[1 - \sqrt{2}] = 2[\sqrt{2} - 1]$$

Example 2. Evaluate the following:

(i) $\int_0^{\pi/6} \cos 3x\, dx$ (ii) $\int_0^{\pi/2} \frac{\cos x}{(1 + \sin x)(2 + \sin x)}\, dx$

(iii) $\int_2^3 \frac{\cos(\log x)}{x}\,dx$ (iv) $\int_0^{\pi/4} \frac{\sin x + \cos x}{9 + 16\sin 2x}\,dx$

Solution: (i) Put $3x = t \Rightarrow 3\,dx = dt \Rightarrow dx = \frac{dt}{3}$

when $x = 0, t = 0$ and $x = \frac{\pi}{6}, t = \frac{\pi}{2}$

$$\therefore \int_0^{\pi/6} \cos 3x\,dx = \int_0^{\pi/2} \cos t\,\frac{dt}{3} = \frac{1}{3}\int_0^{\pi/2} \cos t\,dt$$

$$= \frac{1}{3}[\sin t]_0^{\pi/2} = \frac{1}{3}\left[\sin\frac{\pi}{2} - \sin 0\right]$$

$$= \frac{1}{3}[1 - 0] = \frac{1}{3}$$

(ii) Put $\sin x = t \Rightarrow \cos x\,dx = dt$

when $x = 0, t = \sin 0 = 0$

and $x = \frac{\pi}{2}, t = \sin\frac{\pi}{2} = 1$

$$\therefore \int_0^{\pi/2} \frac{\cos x}{(1 + \sin x)(2 + \sin x)}\,dx = \int_0^1 \frac{dt}{(1 + t)(2 + t)}$$

$$= \int_0^1 \left[\frac{1}{1 + t} - \frac{1}{2 + t}\right] dt \quad \text{[Taking partial fractions]}$$

$$= [\log(1 + t) - \log(2 + t)]_0^1$$

$$= [\log 2 - \log 3 - \log 1 + \log 2]$$

$$= 2\log 2 - \log 3 - \log 1$$

$$= 2\log 2 - \log 3 - 0 \qquad [\because \log 1 = 0]$$

$$= \log 4 - \log 3 = \log\frac{4}{3}$$

(iii) Put $\log x = t \Rightarrow \frac{1}{x}\,dx = dt$

when $x = 2, t = \log 2$

and $x = 3, t = \log 3$

$$\therefore \int_2^3 \frac{\cos(\log x)}{x}\,dx = \int_{\log 2}^{\log 3} \cos t\,dt$$

$$= [\sin t]_{\log 2}^{\log 3}$$

$$= \sin(\log 3) - \sin(\log 2)$$

(iv) $$\int_0^{\pi/4} \frac{\sin x + \cos x}{9 + 16\sin 2x}\, dx = \int_0^{\pi/4} \frac{\sin x + \cos x}{9 + 16[1 - (1 - \sin 2x)]}\, dx$$

$$= \int_0^{\pi/4} \frac{\sin x + \cos x}{25 - 16(\sin x - \cos x)^2}\, dx$$

Put $\sin x - \cos x = t \Rightarrow (\cos x + \sin x)\, dx = dt$

when $x = 0, t = \sin 0 - \cos 0 = 0 - 1 = -1$

and $x = \frac{\pi}{4}, t = \sin\frac{\pi}{4} - \cos\frac{\pi}{4} = \frac{1}{\sqrt{2}} - \frac{1}{\sqrt{2}} = 0$

$$\therefore \int_0^{\pi/4} \frac{\sin x + \cos x}{9 + 16\sin 2x}\, dx = \int_{-1}^{0} \frac{dt}{25 - 16t^2}$$

$$= \frac{1}{16}\int_{-1}^{0} \frac{dt}{\frac{25}{16} - t^2} = \frac{1}{16}\int_{-1}^{0} \frac{dt}{\left(\frac{5}{4}\right)^2 - t^2}$$

$$= \frac{1}{16}\cdot\frac{1}{2\cdot\frac{5}{4}}\left[\log\left|\frac{\frac{5}{4} + t}{\frac{5}{4} - t}\right|\right]_{-1}^{0}$$

$$= \frac{1}{40}\left[\log\left|\frac{5 + 4t}{5 - 4t}\right|\right]_{-1}^{0}$$

$$= \frac{1}{40}\left[\log 1 - \log\frac{1}{9}\right]$$

$$= \frac{1}{40}\log\left[1 \times \frac{9}{1}\right]$$

$$= \frac{1}{40}\cdot\log 9 = \frac{1}{40}\cdot\log 3^2 = \frac{1}{20}\cdot\log 3$$

Example 3. Evaluate the following definite integrals:

(i) $\int_0^{\pi/2} \frac{2\cos x - \sin 2x}{2(1 + \sin x)}\, dx$ (ii) $\int_0^{\pi/2} \sqrt{\cos x - \cos^3 x}(\sec^2 x - 1)\cos^2 x\, dx$

(iii) $\int_0^{\pi} \frac{dx}{5 + 3\cos x}$ (iv) $\int_0^{1} \frac{x}{1 + \sqrt{x}}\, dx$

Solution: (i) $$\int_0^{\pi/2} \frac{2\cos x - \sin 2x}{2(1 + \sin x)}\, dx = \int_0^{\pi/2} \frac{2\cos x - 2\sin x\cos x}{2(1 + \sin x)}\, dx$$

$$= \int_0^{\pi/2} \frac{2\cos x(1 - \sin x)}{2(1 + \sin x)}\, dx$$

$$= \int_0^{\pi/2} \frac{\cos x(1-\sin x)}{1+\sin x}\,dx$$

Put $\quad 1+\sin x = t \Rightarrow \cos x\,dx = dt$ and $\sin x = t-1$

when $\quad x = 0, t = 1+\sin 0 = 1+0 = 1$

and $\quad x = \frac{\pi}{2}, t = 1+\sin\frac{\pi}{2} = 1+1 = 2$

$$\therefore \quad \int_0^{\pi/2} \frac{2\cos x - \sin 2x}{2(1+\sin x)}\,dx = \int_1^2 \frac{1-(t-1)}{t}\,dt$$

$$= \int_1^2 \frac{2-t}{t}\,dt = \int_1^2 \left(\frac{2}{t}-1\right)dt$$

$$= [2\log t - t]_1^2 = (2\log 2 - 2) - (2\log 1 - 1)$$

$$= (\log 4 - 2) - (0-1) = \log 4 - 2 = \log 4 - 1$$

(ii) $\displaystyle\int_0^{\pi/2} \sqrt{\cos x - \cos^3 x}(\sec^2 x - 1)\cos^2 x\,dx$

$$= \int_0^{\pi/2} \sqrt{\cos x}\sqrt{1-\cos^2 x}\cdot\tan^2 x\cdot\cos^2 x\,dx$$

$$= \int_0^{\pi/2} \sqrt{\cos x}\cdot\sin x\cdot\frac{\sin^2 x}{\cos^2 x}\cdot\cos^2 x\,dx$$

$$= \int_0^{\pi/2} \sqrt{\cos x}\sin^3 x\,dx = \int_0^{\pi/2} \sqrt{\cos x}\cdot\sin^2 x\cdot\sin x\cdot dx$$

$$= \int_0^{\pi/2} \sqrt{\cos x}\cdot(1-\cos^2 x)\cdot\sin x\,dx$$

Put $\quad \cos x = t \Rightarrow -\sin x\,dx = dt \Rightarrow \sin x\,dx = -dt$

when $\quad x = 0, t = \cos 0 = 1$

and $\quad x = \frac{\pi}{2}, t = \cos\frac{\pi}{2} = 0$

$$\therefore \int_0^{\pi/2} \sqrt{\cos x - \cos^3 x}(\sec^2 x - 1)\cos^2 x\,dx = -\int_1^0 \sqrt{t}(1-t^2)\,dt$$

$$= -\int_1^0 (t^{1/2} - t^{5/2})\,dt$$

$$= -\left[\frac{t^{3/2}}{\frac{3}{2}} - \frac{t^{7/2}}{\frac{7}{2}}\right]_1^0$$

$$= -\left[(0-0) - \left(\frac{1}{\frac{3}{2}} - \frac{1}{\frac{7}{2}}\right)\right] = -\left[0 - \frac{2}{3} + \frac{2}{7}\right] = \frac{8}{21}$$

(iii) $\int_0^{\pi} \frac{dx}{5+3\cos x} = \int_0^{\pi} \frac{dx}{5+3\left(\frac{1-\tan^2\frac{x}{2}}{1+\tan^2\frac{x}{2}}\right)}$

$$= \int_0^{\pi} \frac{\left(1+\tan^2\frac{x}{2}\right)}{5+5\tan^2\frac{x}{2}+3-3\tan^2\frac{x}{2}}\,dx$$

$$= \int_0^{\pi} \frac{\sec^2\frac{x}{2}}{8+2\tan^2\frac{x}{2}}\,dx$$

$$= \frac{1}{2}\int_0^{\pi} \frac{\sec^2\frac{x}{2}}{4+\tan^2\frac{x}{2}}\,dx$$

Put $t = \tan\frac{x}{2} \Rightarrow \sec^2\frac{x}{2}\cdot\frac{1}{2}\,dx = dt \Rightarrow \sec^2\frac{x}{2}\,dx = 2\,dt$

when $x = 0, t = \tan 0 = 0$

and $x = \pi, t = \tan\frac{\pi}{2} = \infty$

$\therefore$ $\int_0^{\pi} \frac{dx}{5+3\cos x} = \frac{1}{2}\int_0^{\infty} \frac{dt}{4+t^2} = \frac{1}{2}\left[\tan^{-1}\frac{t}{2}\right]_0^{\infty}$

$$= \frac{1}{2}[\tan^{-1}\infty - \tan^{-1}0]$$

$$= \frac{1}{2}\left[\frac{\pi}{2} - 0\right] = \frac{\pi}{4}$$

(iv) Put $\sqrt{x} = t \Rightarrow \frac{1}{2\sqrt{x}}\,dx = dt \Rightarrow dx = 2\sqrt{x}\,dt = 2t\,dt$

when $x = 0, t = 0$ and $x = 1, t = 1$

$\therefore$ $\int_0^1 \frac{x}{1+\sqrt{x}}\,dx = \int_0^1 \frac{t^2\cdot 2t\,dt}{1+t}$

$$= 2\int_0^1 \frac{t^3}{1+t}\,dt = 2\int_0^1 \left[t^2 - t + 1 - \frac{1}{1+t}\right]dt$$

$$= 2\left[\frac{t^3}{3} - \frac{t^2}{2} + t - \log(1+t)\right]_0^1$$

$$= 2\left[\left(\frac{1}{3} - \frac{1}{2} + 1 - \log 2\right) - (0 - 0 + 0 - \log 1)\right]$$

$$= 2\left[\frac{5}{6} - \log 2 - 0\right] = \frac{5}{3} - 2\log 2$$

$$= \frac{5}{3} - \log 4$$

Example 4. Evaluate the following definite integrals:

(i) $\displaystyle\int_0^{1/2} \frac{\sin^{-1} x}{(1-x^2)^{3/2}}\, dx$ (ii) $\displaystyle\int_0^{\pi/2} (\sqrt{\tan x} + \sqrt{\cot x})\, dx$

(iii) $\displaystyle\int_0^{\pi/2} \frac{dx}{2\cos x + 4\sin x}$ (iv) $\displaystyle\int_0^{\pi/2} \frac{\sin 2x}{\sin^4 x + \cos^4 x}\, dx$

Solution: (i) Put $\sin^{-1} x = t \Rightarrow x = \sin t \Rightarrow dx = \cos t\, dt$

when $x = 0, t = 0$ and $x = \dfrac{1}{2}, t = \dfrac{\pi}{6}$

$$\therefore \int_0^{1/2} \frac{\sin^{-1} x}{(1-x^2)^{3/2}}\, dx = \int_0^{\pi/6} \frac{t\cos t\, dt}{(1-\sin^2 t)^{3/2}}$$

$$= \int_0^{\pi/6} \frac{t\cos t\, dt}{\cos^3 t}$$

$$= \int_0^{\pi/6} t\cdot\sec^2 t\, dt$$

$$= [t\cdot\tan t]_0^{\pi/6} - \int_0^{\pi/6} 1\cdot\tan t\, dt \qquad \text{[By parts]}$$

$$= \left(\frac{\pi}{6}\tan\frac{\pi}{6} - 0\right) - [-\log\cos t]_0^{\pi/6}$$

$$= \frac{\pi}{6}\cdot\frac{1}{\sqrt{3}} + [\log\cos t]_0^{\pi/6}$$

$$= \frac{\pi}{6\sqrt{3}} + \left[\log\cos\frac{\pi}{6} - \log\cos 0\right]$$

$$= \frac{\pi}{6\sqrt{3}} + \left(\log\frac{\sqrt{3}}{2} - \log 1\right) = \frac{\pi}{6\sqrt{3}} + \log\frac{\sqrt{3}}{2} - 0$$

$$= \frac{\pi}{6\sqrt{3}} + \log\frac{\sqrt{3}}{2}$$

(ii) $\displaystyle\int_0^{\pi/2} (\sqrt{\tan x} + \sqrt{\cot x})\, dx = \int_0^{\pi/2} \left(\frac{\sqrt{\sin x}}{\sqrt{\cos x}} + \frac{\sqrt{\cos x}}{\sqrt{\sin x}} \right) dx$

$$= \int_0^{\pi/2} \frac{\sin x + \cos x}{\sqrt{\sin x \cdot \cos x}}\, dx$$

$$= \sqrt{2} \int_0^{\pi/2} \frac{\sin x + \cos x}{\sqrt{2 \sin x \cos x}}\, dx$$

$$= \sqrt{2} \int_0^{\pi/2} \frac{\sin x + \cos x}{\sqrt{1 - (1 - 2 \sin x(\cos x)}}\, dx$$

$$= \sqrt{2} \int_0^{\pi/2} \frac{\sin x + \cos x}{\sqrt{1 - (\sin x - \cos x)^2}}\, dx$$

Put $\sin x - \cos x = t \Rightarrow (\cos x + \sin x)\, dx = dt$

when $x = 0, t = \sin 0 - \cos 0 = 0 - 1 = -1$

and $x = \dfrac{\pi}{2}, t = \sin\dfrac{\pi}{2} - \cos\dfrac{\pi}{2} = 1 - 0 = 1$

$\therefore \displaystyle\int_0^{\pi/2} (\sqrt{\tan x} + \sqrt{\cot x})\, dx = \sqrt{2} \int_{-1}^{1} \frac{dt}{\sqrt{1 - t^2}}$

$$= \sqrt{2}[\sin^{-1} t]_{-1}^{1} = \sqrt{2}[\sin^{-1}(1) - \sin^{-1}(-1)]$$

$$= \sqrt{2}\left[\frac{\pi}{2} - \left(-\frac{\pi}{2}\right)\right] = \sqrt{2}\pi$$

(iii) $\displaystyle\int_0^{\pi/2} \frac{dx}{2 \cos x + 4 \sin x} = \int_0^{\pi/2} \frac{dx}{2\left(\dfrac{1 - \tan^2 \frac{x}{2}}{1 - \tan^2 \frac{x}{2}}\right) + 4 \cdot \left(\dfrac{2 \tan \frac{x}{2}}{1 + \tan^2 \frac{x}{2}}\right)}$

$$= \int_0^{\pi/2} \frac{\left(1 + \tan^2 \frac{x}{2}\right) dx}{2 - 2 \tan^2 \frac{x}{2} + 8 \tan \frac{x}{2}}$$

$$= \int_0^{\pi/2} \frac{\sec^2 \frac{x}{2}\, dx}{2 - 2 \tan^2 \frac{x}{2} + 8 \tan \frac{x}{2}}$$

Put $\tan\dfrac{x}{2} = t \Rightarrow \sec^2 \dfrac{x}{2} \cdot \dfrac{1}{2}\, dx = dt \Rightarrow \sec^2 \dfrac{x}{2}\, dx = 2\, dt$

when $x = 0, t = \tan 0 = 0$ and $x = \dfrac{\pi}{2}, t = \tan\dfrac{\pi}{4} = 1$

$$\therefore \quad \int_0^{\pi/2} \frac{dx}{2\cos x + 4\sin x} = \int_0^1 \frac{2\,dt}{2 - 2t^2 + 8t}$$

$$= \int_0^1 \frac{dt}{1 - t^2 + 4t} = \int_0^1 \frac{dt}{1 - (t^2 - 4t + 4) + 4}$$

$$= \int_0^1 \frac{dt}{5 - (t-2)^2} = \int_0^1 \frac{dt}{(\sqrt{5})^2 - (t-2)^2}$$

$$= \frac{1}{2\sqrt{5}} \left[\log \left| \frac{\sqrt{5} + (t-2)}{\sqrt{5} - (t-2)} \right| \right]_0^1$$

$$= \frac{1}{2\sqrt{5}} \left[\log \frac{\sqrt{5} - 1}{\sqrt{5} + 1} - \log \frac{\sqrt{5} - 2}{\sqrt{5} + 2} \right]$$

$$= \frac{1}{2\sqrt{5}} \log \frac{\sqrt{5} - 1}{\sqrt{5} + 1} \times \frac{\sqrt{5} + 2}{\sqrt{5} - 2}$$

$$= \frac{1}{2\sqrt{5}} \log \frac{3 + \sqrt{5}}{3 - \sqrt{5}}$$

$$= \frac{1}{2\sqrt{5}} \log \frac{3 + \sqrt{5}}{3 - \sqrt{5}} \times \frac{3 + \sqrt{5}}{3 + \sqrt{5}}$$

$$= \frac{1}{2\sqrt{5}} \log \frac{(3 + \sqrt{5})^2}{4} = \frac{1}{2\sqrt{5}} \log \left(\frac{(3 + \sqrt{5})}{2} \right)^2$$

$$= \frac{1}{\sqrt{5}} \log \left(\frac{3 + \sqrt{5}}{2} \right)$$

(iv) $\displaystyle\int_0^{\pi/2} \frac{\sin 2x}{\sin^4 x + \cos^4 x}\, dx = \int_0^{\pi/2} \frac{2\sin x \cos x}{\sin^4 x + \cos^4 x}\, dx$

Put $\sin^2 x = t \Rightarrow 2 \sin x \cos x \, dx = dt$

and $\cos^2 x = 1 - \sin^2 x = 1 - t$

when $x = 0, t = \sin^2 0 = 0$

and $x = \dfrac{\pi}{2}, t = \sin^2 \dfrac{\pi}{2} = 1^2 = 1$

$$\therefore \quad \int_0^{\pi/2} \frac{\sin 2x}{\sin^4 x + \cos^4 x}\, dx = \int_0^1 \frac{dt}{t^2 + (1-t)^2}$$

$$= \int_0^1 \frac{dt}{2t^2 - 2t + 1} = \frac{1}{2} \int_0^1 \frac{dt}{t^2 - t + \frac{1}{2}}$$

$$= \frac{1}{2}\int_0^1 \frac{dt}{\left(t-\frac{1}{2}\right)^2+\left(\frac{1}{2}\right)^2} = \frac{1}{2}\cdot\frac{1}{\frac{1}{2}}\left[\tan^{-1}\left(\frac{t-\frac{1}{2}}{\frac{1}{2}}\right)\right]_0^1$$

$$= [\tan^{-1}(2t-1)]_0^1 = \tan^{-1} 1 - \tan^{-1}(-1)$$

$$= \frac{\pi}{4}-\left(-\frac{\pi}{4}\right) = \frac{\pi}{2}$$

Example 5. Evaluate the following definite integrals:

(i) $\int_0^a \sin^{-1}\sqrt{\frac{x}{a+x}}\,dx$ (ii) $\int_0^1 \frac{x^3}{(1+x^2)^4}\,dx$

(iii) $\int_0^1 \tan^{-1}\left(\frac{2x}{1+x^2}\right)dx$ (iv) $\int_0^1 \frac{x^2}{1+x^6}\,dx$

Solution: (i) Put $x = a\tan^2\theta \Rightarrow dx = 2a\tan\theta\sec^2\theta\,d\theta$

when $x = 0, \theta = 0$ and $x = a, \theta = \frac{\pi}{4}$

$$\therefore \int_0^a \sin^{-1}\sqrt{\frac{x}{x+a}}\,dx = \int_0^{\pi/4} \sin^{-1}\left(\sqrt{\frac{a\tan^2\theta}{a\tan^2\theta+a}}\right)\cdot 2a\tan\theta\sec^2\theta\,d\theta$$

$$= \int_0^{\pi/4} \sin^{-1}\left(\sqrt{\frac{\tan^2\theta}{\sec^2\theta}}\right)\cdot 2a\tan\theta\cdot\sec^2\theta\,d\theta$$

$$= 2a\int_0^{\pi/4} \sin^{-1}(\sin\theta)\tan\theta\sec^2\theta\,d\theta$$

$$= 2a\int_0^{\pi/4} \theta\cdot(\tan\theta\sec^2\theta)\,d\theta$$

[Integrate by parts taking θ as first function]

$$= 2a\left[\theta\int(\tan\theta\sec^2\theta)^{d\theta} - \int 1\cdot\left\{\int(\tan\theta\sec^2\theta)\,d\theta\right\}d\theta\right]_0^{\pi/4}$$

$$= 2a\left[\theta\cdot\frac{\tan^2\theta}{2} - \int\frac{\tan^2\theta}{2}\,d\theta\right]_0^{\pi/4}$$

$$= a\left[\theta\tan^2\theta - \int(\sec^2\theta-1)\,d\theta\right]_0^{\pi/4}$$

$$= a[\theta\tan^2\theta - \tan\theta + \theta]_0^{\pi/4}$$

$$= a\left[\left(\frac{\pi}{4}\tan^2\frac{\pi}{4} - \tan\frac{\pi}{4} + \frac{\pi}{4}\right) - (0)\right]$$

$$= a\left[\frac{\pi}{4} - 1 + \frac{\pi}{4}\right] = a\left(\frac{\pi}{2} - 1\right)$$

(ii) Put $1 + x^2 = t \Rightarrow 2x\,dx = dt \Rightarrow x\,dx = \dfrac{dt}{2}$

and $\qquad x^2 = t - 1$

when $\qquad x = 0, t = 1$ and $x = 1, t = 2$

$$\therefore \qquad \int_0^1 \frac{x^3\,dx}{(1+x^2)^4} = \int_0^1 \frac{x^2 \cdot x\,dx}{(1+x^2)^4} = \int_1^2 \frac{(t-1)\dfrac{dt}{2}}{t^4}$$

$$= \frac{1}{2}\int_1^2 \left(\frac{1}{t^3} - \frac{1}{t^4}\right) dt = \frac{1}{2}\int_1^2 (t^{-3} - t^{-4})\,dt$$

$$= \frac{1}{2}\left[\frac{t^{-2}}{-2} - \frac{t^{-3}}{-3}\right]_1^2 = \frac{-1}{2}\left[\frac{1}{2t^2} - \frac{1}{3t^3}\right]_1^2$$

$$= \frac{-1}{2}\left[\left(\frac{1}{8} - \frac{1}{24}\right) - \left(\frac{1}{2} - \frac{1}{3}\right)\right] = \frac{1}{24}$$

(iii) Put $x = \tan\theta \Rightarrow dx = \sec^2\theta\,d\theta$ and $\theta = \tan^{-1} x$

when $\qquad x = 0, \theta = \tan^{-1} 0 = 0$

and $\qquad x = 1, \theta = \tan^{-1} 1 = \dfrac{\pi}{4}$

$$\therefore \qquad \int_0^1 \tan^{-1}\left(\frac{2x}{1+x^2}\right) dx = \int_0^{\pi/4} \tan^{-1}\left(\frac{2\tan\theta}{1+\tan^2\theta}\right)\sec^2\theta\,d\theta$$

$$= \int_0^{\pi/4} \tan^{-1}(\tan 2\theta)\sec^2\theta\,d\theta$$

$$= \int_0^{\pi/4} 2\theta\sec^2\theta\,d\theta = 2\int_0^{\pi/4} \theta\sec^2\theta\,d\theta$$

$$= 2\left[(\theta\cdot\tan\theta)_0^{\pi/4} - \int_0^{\pi/4} 1\cdot\tan\theta\,d\theta\right]$$

$$= 2\cdot\left\{\left(\frac{\pi}{4}\tan\frac{\pi}{4} - 0\right) - [-\log\cos t]_0^{\pi/4}\right\}$$

$$= 2\left[\frac{\pi}{4}\cdot 1 + \log\cos\frac{\pi}{4} - \log\cos 0\right]$$

$$= 2\left[\frac{\pi}{4} + \log\frac{1}{\sqrt{2}} - \log 1\right]$$

$$= 2\left[\frac{\pi}{4} + \frac{1}{2}\log\frac{1}{2}\right]$$

$$= \frac{\pi}{2} + \log\frac{1}{2} = \frac{\pi}{2} + \log 1 - \log 2 = \frac{\pi}{2} - \log 2$$

(iv) Put $x^3 = t \Rightarrow 3x^2\,dx = dt \Rightarrow x^2\,dx = \dfrac{dt}{3}$

If $\quad x = 0, t = 0$

and $\quad x = 1, t = 1$

$$\therefore \quad \int_0^1 \frac{x^2}{1+x^6}\,dx = \int_0^1 \frac{dt}{1+t^2} = [\tan^{-1} t]_0^1$$

$$= [\tan^{-1}(1) - \tan^{-1}(0)]$$

$$= \left[\frac{\pi}{4} - 0\right] = \frac{\pi}{4}$$

EXERCISE 4.11

Evaluate the following definite integrals:

1. (i) $\displaystyle\int_0^1 x^2 e^{x^3}\,dx$ (ii) $\displaystyle\int_0^1 \frac{x\,dx}{1+x^4}$

(iii) $\displaystyle\int_a^b \frac{\log x}{x}\,dx$ (iv) $\displaystyle\int_0^1 \frac{e^x}{1+e^{2x}}\,dx$

(v) $\displaystyle\int_0^{\pi/2} \frac{\sin x}{1+\cos^2 x}\,dx$

2. (i) $\displaystyle\int_0^{\pi/2} \sin^3 x\,dx$ (ii) $\displaystyle\int_0^{\pi/2} \cos^3 x(\sin x)^{1/4}\,dx$

(iii) $\displaystyle\int_0^{\pi} \frac{dx}{3+2\sin x+\cos x}$ (iv) $\displaystyle\int_0^{\pi} \frac{dx}{5+4\cos x}$

(v) $\displaystyle\int_0^{\pi/2} \frac{dx}{4\cos x+2\sin x}$ (vi) $\displaystyle\int_0^{\pi/2} \frac{dx}{5+4\sin x}$

(vii) $\displaystyle\int_{\pi/4}^{\pi/2} \frac{\cos x}{3\sin x-\sin 3x}\,dx$ (viii) $\displaystyle\int_0^{\pi/2} \sin 2x \tan^{-1}(\sin x)\,dx$

3. (i) $\displaystyle\int_0^{\pi/2} \frac{dx}{a^2\cos^2 x+b^2\sin^2 x}$ (ii) $\displaystyle\int_0^{\pi/2} \frac{dx}{(a^2\cos^2 x+b^2\sin^2 x)^2}$

(iii) $\displaystyle\int_0^{a} \sin^{-1}\left(\frac{2x}{1+x^2}\right)dx$ (iv) $\displaystyle\int_0^1 \frac{1-x^2}{(1+x^2)^2}\,dx$

(v) $\int_0^1 \frac{1-x^2}{1+x^2+x^4}\,dx$ (vi) $\int_0^1 \cos^{-1}\left(\frac{1-x^2}{1+x^2}\right)dx$

(vii) $\int_0^{\pi} \frac{\sin x}{\sin x+\cos x}\,dx$ (viii) $\int_0^{\pi/2} \frac{\sin^2 x}{(1+\cos x)^2}\,dx$

4. (i) $\int_0^1 \frac{dx}{2e^x-1}$ (ii) $\int_0^{\pi/2} \frac{dx}{2+\tan x}$

(iii) $\int_0^{\pi/2} \sqrt{\sin x}\cdot\cos^5 x\,dx$ (iv) $\int_0^{\pi/2} \frac{\sin x\,dx}{(1+\cos x)(2+\cos x)}$

(v) $\int_0^{\pi/2} \frac{\sin x\cos x\,dx}{(1+\sin x)(1+2\sin x)}$ (vi) $\int_a^b \frac{dx}{(x^2+a^2)(x^2+b^2)}$

5. (i) $\int_0^{\pi/2} \frac{\sin x\cos x}{1+\sin^4 x}\,dx$ (ii) $\int_0^4 \frac{dx}{\sqrt{x^2+2x+3}}$

(iii) $\int_1^2 \frac{dx}{(x+1)\sqrt{x^2-1}}$ (iv) $\int_0^{16} \frac{x^{1/4}}{1+\sqrt{x}}\,dx$

(v) $\int_0^{\sqrt{2}} \sqrt{2-x^2}\,dx$ (vi) $\int_{\pi/6}^{\pi/2} \frac{\operatorname{cosec} x\cot x}{1+\operatorname{cosec}^2 x}\,dx$

ANSWERS

1. (i) $\frac{1}{3}(e-1)$ (ii) $\frac{\pi}{8}$

(iii) $\frac{1}{2}\log\frac{b}{a}\cdot\log ab$ (iv) $\tan^{-1} e-\frac{\pi}{4}$

(v) $\frac{\pi}{4}$

2. (i) $\frac{2}{3}$ (ii) $\frac{32}{65}$

(iii) $\frac{\pi}{4}$ (iv) $\frac{\pi}{3}$

(v) $\frac{1}{\sqrt{5}}\log\left(\frac{\sqrt{5}+1}{\sqrt{5}-1}\right)$ (vi) $\frac{2}{3}\tan^{-1}\frac{1}{3}$

(vii) $\frac{1}{8}$ (viii) $\frac{\pi}{2}-1$

3. (i) $\dfrac{\pi}{2ab}$ (ii) $\dfrac{\pi}{2}\cdot\dfrac{a^2+b^2}{a^3b^3}$

(iii) $2a\tan^{-1}a-\log(1+a^2)$ (iv) $\dfrac{1}{2}$

(v) $\dfrac{1}{2}\log 3$ (vi) $\dfrac{\pi}{2}-\log 2$

(vii) $\dfrac{\pi}{2}$ (viii) $2-\dfrac{\pi}{2}$

4. (i) $\log\dfrac{2e-1}{e}$ (ii) $\dfrac{1}{5}[\pi-\log 2]$

(iii) $\dfrac{64}{231}$ (iv) $\log\dfrac{4}{3}$

(v) $\dfrac{1}{2}\log\dfrac{4}{3}$

(vi) $\dfrac{\pi}{4ab(a-b)}-\dfrac{1}{a^2-b^2}\left[\dfrac{1}{b}\tan^{-1}\dfrac{a}{b}+\dfrac{1}{a}\tan^{-1}\dfrac{b}{a}\right]$

5. (i) $\dfrac{\pi}{8}$ (ii) $\log(2+\sqrt{3})$

(iii) $\dfrac{1}{\sqrt{3}}$ (iv) $4\left(\dfrac{2}{3}+\tan^{-1}2\right)$

(v) $\dfrac{\pi}{2}$ (vi) $\tan^{-1}\left(\dfrac{1}{3}\right)$

4.20 PROPERTIES OF DEFINITE INTEGRAL

Let f be a continuous function on the closed interval $[a, b]$, then:

1. $\displaystyle\int_a^b f(x)\,dx = \int_a^b f(t)\,dt$

2. $\displaystyle\int_a^b f(x)\,dx = -\int_b^a f(x)\,dx$

3. $\displaystyle\int_a^b f(x)\,dx = \int_a^c f(x)\,dx + \int_c^b f(x)\,dx$, where $a < c < b$

4. $\displaystyle\int_0^a f(x)\,dx = \int_0^a f(a-x)\,dx$

5. $\displaystyle\int_a^b f(x)\,dx = \int_a^b f(a+b-x)\,dx$

6. $\int_{-a}^{a} f(x)\,dx = \begin{cases} 2\int_0^a f(x)\,dx, \text{ if f(x) is an even function} \\ 0 \qquad\qquad ,\text{if } f(x) \text{ is an odd function} \end{cases}$

[A function $f(x)$ is even if $f(-x) = f(x)$ and odd if $f(-x) = -f(x)$]

7. $\int_0^{2a} f(x)\,dx = \begin{cases} 2\int_0^a f(x)\,dx, \text{ if } f(2a-x) = f(x) \\ 0 \qquad\qquad , \text{ if } f(2a-x) = -f(x) \end{cases}$

The above properties are very helpful in evaluating definite integrals in a simple way.

ILLUSTRATIVE EXAMPLES

Example 1. Evaluate the following definite integrals:

(i) $\int_1^4 (|x|+|x-3|)\,dx$ (ii) $\int_1^4 [|x-1|+|x-2|+|x-3|]\,dx$

(iii) $\int_0^{2\pi} |\sin x|\,dx$ (iv) $\int_{-\pi/2}^{\pi/2} (2\sin|x|+\cos|x|)\,dx$

Solution: (i) Since for $1 \le x \le 3, x \ge 0 \Rightarrow |x| = x$

and $x-3 \le 0 \Rightarrow |x-3| = -(x-3)$

Also for $3 \le x \le 4, x \ge 0 \Rightarrow |x| = x$

and $x-3 \ge 0 \Rightarrow |x-3| = x-3$

$$\therefore \int_1^4 (|x|+|x-3|)\,dx = \int_1^3 (|x|+|x-3|)\,dx + \int_3^4 (|x|+|x-3|)\,dx$$

(Property 3)

$$= \int_1^3 (x-(x-3)\,dx + \int_3^4 (x+x-3)\,dx$$

$$= \int_1^3 3\,dx + \int_3^4 (2x-3)\,dx$$

$$= 3[x]_1^3 + \left[2\cdot\frac{x^2}{2} - 3x\right]_3^4$$

$$= 3(3-1) + (16-12) - (9-9)$$

$$= 6 + 4 - 0 = 10$$

(ii) When $1 \le x \le 2, f(x) = (x-1)-(x-2)-(x-3) = -x+4;$

when $2 \le x \le 3, f(x) = (x-1)+(x-2)-(x-3) = x;$

when $3 \le x \le 4, f(x) = (x-1)+(x-2)+(x+3) = 3x-6.$

$$\therefore \int_1^4 [|x-1|+|x-2|+|x-3|]\,dx = \int_1^2 (-x+4)\,dx + \int_2^3 x\,dx + \int_3^4 (3x-6)\,dx$$

(Property 3)

$$= \left[-\frac{x^2}{2} + 4x\right]_1^2 + \left[\frac{x^2}{2}\right]_2^3 + \left[\frac{3x^2}{2} - 6x\right]_3^4$$

$$= (-2+8) - \left(-\frac{1}{2} + 4\right) + \frac{1}{2}(9-4) + (24-24) - \left(\frac{27}{2} - 18\right)$$

$$= 6 - \frac{7}{2} + \frac{5}{2} + 0 + \frac{9}{2} = \frac{19}{2}$$

(iii) $\int_0^{2\pi} |\sin x|\, dx$

$$= \int_0^{\pi/2} |\sin x|\, dx + \int_{\pi/2}^{\pi} |\sin x|\, dx + \int_{\pi}^{3\pi/2} |\sin x|\, dx + \int_{3\pi/2}^{2\pi} |\sin x|\, dx$$

(Property 3)

$$= \int_0^{\pi/2} \sin x\, dx + \int_{\pi/2}^{\pi} \sin x\, dx + \int_{\pi}^{3\pi/2} (-\sin x)\, dx + \int_{3\pi/2}^{2\pi} (-\sin x)\, dx$$

= ($\because$ sine is + ve in I & II quadrant)

$$= [-\cos x]_0^{\pi/2} + [-\cos x]_{\pi/2}^{\pi} + [\cos x]_{\pi}^{3\pi/2} + [\cos x]_{3\pi/2}^{2\pi}$$

$$= \left(-\cos\frac{\pi}{2} + \cos 0\right) + \left(-\cos\pi + \cos\frac{\pi}{2}\right) + \left(\cos\frac{3\pi}{2} - \cos\pi\right) + \left(\cos 2\pi - \frac{\cos 3\pi}{2}\right)$$

$$= \cos 0 - 2\cos\pi + \cos 2\pi = 1 - 2(-1) + 1$$

$$= 4$$

(iv) $\int_{-\pi/2}^{\pi/2} (2\sin|x| + \cos|x|)dx$

$$= \int_{-\pi/2}^{0} (2\sin|x| + \cos|x|)\, dx + \int_0^{\pi/2} (2\sin|x| + \cos|x|)\, dx$$

(Property 3)

$$= \int_{-\pi/2}^{0} (2\sin(-x) + \cos(-x)\, dx + \int_0^{\pi/2} (2\sin x + \cos x)\, dx$$

$$= \int_{-\pi/2}^{0} (-2\sin x + \cos x)\, dx + \int_0^{\pi/2} (2\sin x + \cos x)\, dx$$

$$= [2\cos x + \sin x]_{-\pi/2}^{0} + [-2\cos x + \sin x]_0^{\pi/2}$$

$$= \left(2\cos 0 + \sin 0 - 2\cos\frac{\pi}{2} + \sin\frac{\pi}{2}\right)$$

$$+ \left(-2\cos\frac{\pi}{2} + \sin\frac{\pi}{2} + 2\cos 0 - \sin 0\right)$$

$$= (2 + 0 - 0 + 1) + (0 + 1 + 2 - 0)$$
$$= 6$$

Example 2. Evaluate the following definite integrals:

(i) $\int_{-1}^{1} \frac{|x|}{x} dx$

(ii) $\int_{-1}^{2} f(x)\, dx$, where $f(x) = \begin{cases} 2x + 1, & x \le 1 \\ x - 5, & x > 1 \end{cases}$

(iii) $\int_{1}^{3} f(x)\, dx$, where $f(x) = \begin{cases} x^2 - 8, & 1 \le x \le 2 \\ -2x, & 2 \le x \le 3 \end{cases}$

Solution: (i) Note that $\frac{|x|}{x}$ is discontinuous at $x = 0$,

$$\therefore \quad \int_{-1}^{1} \frac{|x|}{x} dx = \int_{-1}^{0} \frac{|x|}{x} dx + \int_{0}^{1} \frac{|x|}{x} dx$$

$$= \int_{-1}^{0} -\frac{x}{x} dx + \int_{0}^{1} \frac{x}{x} dx$$

$(\because -1 \le x \le 0 \Rightarrow |x| = -x \text{ and } 0 \le x \le 1 \Rightarrow |x| = x)$

$$= -\int_{-1}^{0} 1\, dx + \int_{0}^{1} 1\, dx$$

$$= -[x]_{-1}^{0} + [x]_{0}^{1} = -[0 + 1] + [1 - 0].$$
$$= -1 + 1 = 0$$

(ii) $$\int_{-1}^{2} f(x)\, dx = \int_{-1}^{1} f(x)\, dx + \int_{1}^{2} f(x)\, dx \quad \text{(Property 3)}$$

$$= \int_{-1}^{1} (2x + 1)\, dx + \int_{1}^{2} (x - 5)\, dx$$

$$= [x^2 + x]_{-1}^{1} + \left[\frac{x^2}{2} - 5x\right]_{1}^{2}$$

$$= (1 + 1) - (1 - 1) + (2 - 10) - \left(\frac{1}{2} - 5\right)$$

$$= 2 - 0 - 8 + \frac{9}{2} = -\frac{3}{2}$$

(iii) $\int_1^3 f(x)\,dx = \int_1^2 f(x)\,dx + \int_2^3 f(x)\,dx$ (Property 3)

$$= \int_1^2 (x^2 - 8)\,dx + \int_2^3 (-2x)\,dx$$

$$= \left[\frac{x^3}{3} - 8x\right]_1^2 - [x^2]_2^3$$

$$= \left[\left(\frac{8}{3} - 16\right) - \left(\frac{1}{3} - 8\right)\right] - [9 - 4]$$

$$= \frac{7}{3} - 8 - 5 = \frac{7}{3} - 13 = -\frac{32}{3}$$

Example 3. Evaluate the following definite integrals:

(i) $\int_{-\pi/2}^{\pi/2} \sin^4 x\,dx$ (ii) $\int_{-\pi/4}^{\pi/4} x^3 \sin^4 x\,dx$

(iii) $\int_{-\pi/2}^{\pi/2} \cos^4 x\,dx$ (iv) $\int_{-3}^{3} \frac{x^3 + 5x}{2x^2 + 7}\,dx$

Solution: (i) Let $f(x) = \sin^4 x \Rightarrow f(-x) = \sin^4(-x)$

$\Rightarrow (-\sin x)^4 = \sin^4 x = f(x)$

$\Rightarrow f(x)$ is an even function.

Therefore, by property 6

$$\int_{-\pi/2}^{\pi/2} \sin^4 x\,dx = 2\int_0^{\pi/2} \sin^4 x\,dx = 2\int_0^{\pi/2} \left(\frac{1 - \cos 2x}{2}\right)^2 dx$$

$$= \frac{1}{2}\int_0^{\pi/2} (1 - 2\cos 2x + \cos^2 2x)\,dx$$

$$= \frac{1}{2}\int_0^{\pi/2} \left(1 - 2\cos 2x + \frac{1 + \cos 4x}{2}\right) dx$$

$$= \frac{1}{4}\int_0^{\pi/2} (3 - 4\cos 2x + \cos 4x)\,dx$$

$$= \frac{1}{4}\left[3x - \frac{4\sin 2x}{2} + \frac{\sin 4x}{4}\right]_0^{\pi/2}$$

$$= \frac{1}{4}\left[\left(\frac{3\pi}{2} - 2\sin\pi + \frac{1}{4}\sin 2\pi\right) - \left(0 - 2\sin 0 + \frac{1}{4}\sin 0\right)\right]$$

$$= \frac{1}{4}\left[\left(\frac{3\pi}{2} - 0 + 0\right) - (0 - 0 + 0)\right]$$

$$= \frac{3\pi}{8}$$

(ii) Let $f(x) = x^3 \sin^4 x$

$\Rightarrow f(-x) = (-x)^3 (\sin(-x))^4 = -x^3(-\sin x)^4 = -x^3 \sin^4 x = -f(x)$

$\Rightarrow f(x)$ is an odd function.

therefore, by property 6

$$\int_{-\pi/4}^{\pi/4} x^3 \sin^4 x \, dx = 0$$

(iii) Let $f(x) = \cos^4 x$

$\Rightarrow f(-x) = (\cos(-x))^4 = \cos^4 x = f(x)$

$\Rightarrow f(x)$ is an even function.

Therefore, by property 6

$$\int_{-\pi/2}^{\pi/2} \cos^4 x \, dx = 2\int_0^{\pi/2} \cos^4 x \, dx = 2\int_0^{\pi/2} \left(\frac{1 + \cos 2x}{2}\right)^2 dx$$

$$= \frac{1}{2}\int_0^{\pi/2} (1 + \cos^2 2x + 2\cos 2x) \, dx$$

$$= \frac{1}{2}\int_0^{\pi/2} \left(1 + \frac{1 + \cos 4x}{2} + 2\cos 2x\right) dx$$

$$= \frac{1}{4}\int_0^{\pi/2} (3 + \cos 4x + 4\cos 2x) \, dx$$

$$= \frac{1}{4}\left[3x + \frac{\sin 4x}{4} + \frac{4\sin 2x}{2}\right]_0^{\pi/2}$$

$$= \frac{1}{4}\left[\left(\frac{3\pi}{2} + \frac{\sin 2\pi}{4} + 2\sin\pi\right) - (0 + 0 + 0)\right]$$

$$= \frac{1}{4}\left[\frac{3\pi}{2}\right] = \frac{3\pi}{8}$$

(iv) Let $f(x) = \dfrac{x^3 + 5x}{2x^2 + 7}$

$$\Rightarrow \quad f(-x) = \frac{(-x)^3 + 5(-x)}{2(-x)^2 + 7} = \frac{-x^3 - 5x}{2x^2 + 7} = -f(x)$$

$\Rightarrow f(x)$ is an odd function,

Therefore, by property 6

$$\int_{-3}^{3} \frac{x^3 + 5x}{2x^2 + 7}\, dx = 0$$

Example 4. Evaluate the following integrals:

(i) $\int_0^2 x^2 \sqrt{2-x}\, dx$ (ii) $\int_0^{\pi} \frac{x \tan x}{\sec x + \cos x}\, dx$

(iii) $\int_0^{\pi/2} \frac{\sqrt{\sin x}}{\sqrt{\sin x} + \sqrt{\cos x}}\, dx$ (iv) $\int_0^1 x(1-x)^5\, dx$

Solution: (i) $\int_0^2 x^2 \sqrt{2-x}\, dx = \int_0^2 (2-x)^2 \sqrt{2-(2-x)}\, dx$

$$\left[\text{Property 4, } \int_0^a f(x)\, dx = \int_0^a f(a-x)\, dx\right]$$

$$= \int_0^2 (4 - 4x + x^2)\sqrt{x}\, dx$$

$$= \int_0^2 (4x^{1/2} - 4x^{3/2} + x^{5/2})\, dx$$

$$= \left[4 \cdot \frac{x^{3/2}}{\frac{3}{2}} - 4 \cdot \frac{x^{5/2}}{\frac{5}{2}} + \frac{x^{7/2}}{\frac{7}{2}}\right]_0^2$$

$$= \left[\frac{8}{3} \cdot 2^{3/2} - \frac{8}{5} \cdot 2^{5/2} + \frac{2}{7} \cdot 2^{7/2} - (0)\right]$$

$$= \left[\frac{8}{3} \cdot 2\sqrt{2} - \frac{8}{5} \cdot 4\sqrt{2} + \frac{2}{7} \cdot 8\sqrt{2}\right]$$

$$= 16\sqrt{2}\left(\frac{1}{3} - \frac{2}{5} + \frac{1}{7}\right)$$

$$= 16\sqrt{2} \times \frac{8}{105}$$

$$= \frac{128\sqrt{2}}{105}$$

(ii) Let $I = \int_0^{\pi} \frac{x \tan x}{\sec x + \cos x}\, dx = \int_0^{\pi} \frac{(\pi - x)\tan(\pi - x)}{\sec(\pi - x) + \cos(\pi - x)}\, dx$ ...(1)

(property 4)

$$I = \int_0^{\pi} \frac{(\pi - x)(-\tan x)}{-\sec x - \cos x}\, dx = \int_0^{\pi} \frac{(\pi - x)\tan x}{\sec x + \cos x}\, dx \;...\,(2)$$

Adding (1) & (2), we get

$$2I = \int_0^{\pi} \frac{\pi \tan x}{\sec x + \cos x}\, dx = \pi \int_0^{\pi} \frac{\sin x}{1 + \cos^2 x}\, dx$$

$$\Rightarrow \qquad I = \frac{\pi}{2}\int_0^{\pi} \frac{\sin x}{1 + \cos^2 x}\, dx$$

Put $\cos x = t \Rightarrow \sin x\, dx = -dt$

when $x = 0, t = 1$ and $x = \pi, t = -1$

$$\therefore \qquad I = \frac{\pi}{2}\int_1^{-1} \frac{-dt}{1 + t^2} = -\frac{\pi}{2}[\tan^{-1} t]_1^{-1}$$

$$= -\frac{\pi}{2}[\tan^{-1}(-1) - \tan^{-1}(1)]$$

$$= -\frac{\pi}{2}\left[-\frac{\pi}{4} - \frac{\pi}{4}\right] = -\frac{\pi}{2} \times -\frac{\pi}{2} = \frac{\pi^2}{4}$$

(iii) Let $$I = \int_0^{\pi/2} \frac{\sqrt{\sin x}}{\sqrt{\sin x} + \sqrt{\cos x}}\, dx \;...\,(1)$$

$$= \int_0^{\pi/2} \frac{\sqrt{\sin\left(\frac{\pi}{2} - x\right)}}{\sqrt{\sin\left(\frac{\pi}{2} - x\right)} + \sqrt{\cos\left(\frac{\pi}{2} - x\right)}}\, dx \qquad \text{(Property 4)}$$

$$I = \int_0^{\pi/2} \frac{\sqrt{\cos x}}{\sqrt{\cos x} + \sqrt{\sin x}}\, dx \qquad ...\,(2)$$

Adding (1) & (2), we get

$$2I = \int_0^{\pi/2} 1\, dx = [x]_0^{\pi/2}$$

$$= \left(\frac{\pi}{2} - 0\right) = \frac{\pi}{2}$$

$$\Rightarrow \qquad I = \frac{\pi}{4}$$

(iv) $$\int_0^1 x(1 - x)^5\, dx = \int_0^1 (1 - x)[1 - (1 - x)]^5\, dx \qquad \text{(property 4)}$$

$$= \int_0^1 (1 - x)x^5\, dx = \int_0^1 (x^5 - x^6)\, dx$$

$$= \left[\frac{x^6}{6} - \frac{x^7}{7} \right]_0^1$$

$$= \left(\frac{1}{6} - \frac{1}{7} \right) - (0) = \frac{1}{42}$$

Example 5. Evaluate the following definite integrals:

(i) $\int_2^3 \frac{\sqrt{x}}{\sqrt{x} + \sqrt{5-x}}\, dx$ (ii) $\int_0^1 \log\left(\frac{1}{x} - 1 \right) dx$

(iii) $\int_0^{\pi/2} \frac{\sin x - \cos x}{1 + \sin x \cos x}\, dx$ (iv) $\int_{\pi/6}^{\pi/3} \frac{1}{1 + \tan x}\, dx$

Solution: (i) Let $I = \int_2^3 \frac{\sqrt{x}}{\sqrt{x} + \sqrt{5-x}}\, dx$ ...(1)

$$= \int_2^3 \frac{\sqrt{5-x}}{\sqrt{5-x} + \sqrt{5-(5-x)}}\, dx \qquad \text{(Property 5)}$$

$$I = \int_2^3 \frac{\sqrt{5-x}}{\sqrt{5-x} + \sqrt{x}}\, dx \qquad ...(2)$$

Adding (1) & (2), we get,

$$2I = \int_2^3 1\, dx = [x]_2^3 = 3 - 2 = 1$$

$$\Rightarrow \quad I = \frac{1}{2}$$

(ii) Let $I = \int_0^1 \log\left(\frac{1}{x} - 1 \right) dx = \int_0^1 \log\left(\frac{1-x}{x} \right) dx$ ...(1)

$$= \int_0^1 \log\left(\frac{1-(1-x)}{1-x} \right) dx \qquad \text{(Property 4)}$$

$$= \int_0^1 \log\left(\frac{x}{1-x} \right) dx$$

$$= \int_0^1 \log\left(\frac{1-x}{x} \right)^{-1} = \int_0^1 -1 \cdot \log\left(\frac{1-x}{x} \right) dx$$

$$= -\int_0^1 \log\left(\frac{1-x}{x} \right) dx = -I \qquad \text{(Using (1)}$$

$$\Rightarrow \quad 2I = 0 \Rightarrow I = 0$$

(iii) Let $\quad I = \int_0^{\pi/2} \frac{\sin x - \cos x}{1 + \sin x \cos x} dx$ ...(1)

$$= \int_0^{\pi/2} \frac{\sin\left(\frac{\pi}{2} - x\right) - \cos\left(\frac{\pi}{2} - x\right)}{1 + \sin\left(\frac{\pi}{2} - x\right)\cos\left(\frac{\pi}{2} - x\right)} dx \qquad \text{(Property 4)}$$

$$= \int_0^{\pi/2} \frac{\cos x - \sin x}{1 + \cos x \sin x} dx$$

$$= -\int_0^{\pi/2} \frac{\sin x - \cos x}{1 + \cos x \sin x} dx = -I \qquad \text{(Using 1)}$$

$\Rightarrow \quad 2I = 0 \Rightarrow I = 0$

(iv) $\int_{\pi/6}^{\pi/3} \frac{1}{1 + \tan x} dx = \int_{\pi/6}^{\pi/3} \frac{1}{1 + \frac{\sin x}{\cos x}} dx$

$$\Rightarrow \quad I = \int_{\pi/6}^{\pi/3} \frac{\cos x}{\cos x + \sin x} dx \qquad \text{...(1)}$$

$$= \int_{\pi/6}^{\pi/3} \frac{\cos\left(\frac{\pi}{2} - x\right)}{\cos\left(\frac{\pi}{2} - x\right) + \sin\left(\frac{\pi}{2} - x\right)} dx \quad \text{(Property 5)}$$

$$I = \int_{\pi/6}^{\pi/3} \frac{\sin x}{\sin x + \cos x} dx \qquad \text{...(2)}$$

Adding (1) & (2)

$$2I = \int_{\pi/6}^{\pi/3} 1\, dx = [x]_{\pi/6}^{\pi/3} = \frac{\pi}{3} - \frac{\pi}{6} = \frac{\pi}{6}$$

$$\Rightarrow \quad I = \frac{\pi}{12}$$

Example 6. Evaluate the following definite integrals:

(i) $\int_0^1 \frac{\log(1 + x)}{1 + x^2} dx$ (ii) $\int_0^a \frac{dx}{x + \sqrt{a^2 - x^2}}$

Solution: (i) Put $x = \tan\theta \Rightarrow dx = \sec^2\theta\, d\theta$

when $x = 1$, $\tan\theta = 1 \Rightarrow \theta = \frac{\pi}{4}$

and $x = 0$, $\tan\theta = 0 \Rightarrow \theta = 0$

$$\therefore I = \int_0^1 \frac{\log(1+x)}{1+x^2}\,dx = \int_0^{\pi/4} \frac{\log(1+\tan\theta)}{1+\tan^2\theta}\cdot\sec^2\theta\,d\theta$$

$$= \int_0^{\pi/4} \log(1+\tan\theta)\,d\theta$$

$$= \int_0^{\pi/4} \log\left[1+\tan\left(\frac{\pi}{4}-\theta\right)\right]d\theta \text{ (Property 4)}$$

$$= \int_0^{\pi/4} \log\left[1+\frac{1-\tan\theta}{1+\tan\theta}\right]d\theta$$

$$= \int_0^{\pi/4} \log\left(\frac{2}{1+\tan\theta}\right)d\theta$$

$$\Rightarrow I = \int_0^{\pi/4} \log 2\,d\theta - \int_0^{\pi/4} \log(1+\tan\theta)\,d\theta$$

$$\Rightarrow I = \log 2 \int_0^{\pi/4} d\theta - I$$

$$\Rightarrow 2I = \frac{\pi}{4}\log 2$$

$$\Rightarrow I = \frac{\pi}{8}\log 2$$

(ii) Put $x = a\sin\theta \Rightarrow dx = a\cos\theta\,d\theta$

when $x = 0, \theta = 0$ and $x = a, \theta = \frac{\pi}{2}$

$$\therefore \quad I = \int_0^a \frac{dx}{x+\sqrt{a^2-x^2}} = \int_0^{\pi/2} \frac{a\cos\theta\,d\theta}{a\sin\theta+\sqrt{a^2-a^2\sin^2\theta}}$$

$$= \int_0^{\pi/2} \frac{a\cos\theta\,d\theta}{a\sin\theta+a\cos\theta}$$

$$I = \int_0^{\pi/2} \frac{\cos\theta\,d\theta}{\sin\theta+\cos\theta} \quad \text{...(1)}$$

$$= \int_0^{\pi/2} \frac{\cos\left(\frac{\pi}{2}-\theta\right)}{\sin\left(\frac{\pi}{2}-\theta\right)+\cos\left(\frac{\pi}{2}-\theta\right)}\,d\theta$$

$$\Rightarrow \quad I = \int_0^{\pi/2} \frac{\sin\theta}{\cos\theta+\sin\theta}\,d\theta \quad \text{...(2)}$$

Adding (1) & (2), we get

$$2I = \int_0^{\pi/2} 1\, d\theta$$

$$= [\theta]_0^{\pi/2}$$

$$= \left(\frac{\pi}{2} - 0\right)$$

$$= \frac{\pi}{2}$$

$$\Rightarrow \quad I = \frac{\pi}{4}$$

Example 7. Evaluate the following integrals:

(i) $\displaystyle\int_0^{\pi} \frac{x\, dx}{1+\cos^2 x}$ (ii) $\displaystyle\int_0^{\pi/2} \log(\sin x)\, dx$

(iii) $\displaystyle\int_0^{\pi} \frac{x\, dx}{a^2\cos^2 x + b^2 \sin^2 x}$ (iv) $\displaystyle\int_0^{\pi} \frac{x \tan x}{\sec x \operatorname{cosec} x}\, dx$

Solution: (i) Let $I = \displaystyle\int_0^{\pi} \frac{x\, dx}{1+\cos^2 x}$ ...(1)

$$= \int_0^{\pi} \frac{(\pi - x)\, dx}{1+\cos^2(\pi - x)} \qquad \text{(Property 4)}$$

$$\Rightarrow \quad I = \int_0^{\pi} \frac{(\pi - x)\, dx}{1+\cos^2 x} \qquad \text{...(2)}$$

Adding (1) & (2), we get,

$$2I = \int_0^{\pi} \frac{\pi\, dx}{1+\cos^2 x} = \pi \int_0^{\pi} \frac{\sec^2 x\, dx}{1+\sec^2 x}$$

(Dividing num and deno. by $\cos^2 x$)

$$= 2\pi \int_0^{\pi/2} \frac{\sec^2 x}{1+1+\tan^2 x}\, dx$$

$$\Rightarrow \quad I = \pi \int_0^{\pi/2} \frac{\sec^2 x}{2+\tan^2 x}\, dx$$

Put $\tan x = t \Rightarrow \sec^2 x\, dx = dt$

when $x = 0, t = 0$ and $x = \dfrac{\pi}{2}, t = \infty$

$$\therefore \qquad I = \pi\int_0^{\infty}\frac{dt}{2+t^2} = \pi\left[\frac{1}{\sqrt{2}}\tan^{-1}\frac{t}{\sqrt{2}}\right]_0^{\infty}$$

$$= \frac{\pi}{\sqrt{2}}[\tan^{-1}\infty - \tan^{-1}0]$$

$$= \frac{\pi}{\sqrt{2}}\left(\frac{\pi}{2}-0\right)$$

$$\Rightarrow \qquad I = \frac{\pi^2}{2\sqrt{2}}$$

(ii) Let $I = \int_0^{\pi/2} \log(\sin x)\, dx$ ...(1)

$$= \int_0^{\pi/2} \log\left(\sin\left(\frac{\pi}{2}-x\right)\right)dx \qquad \text{(Property 4)}$$

$$\Rightarrow \quad I = \int_0^{\pi/2} \log(\cos x)\, dx \qquad ...(2)$$

Adding (1) and (2), we get

$$2I = \int_0^{\pi/2} [\log(\sin x) + \log(\cos x)]\, dx$$

$$= \int_0^{\pi/2} \log(\sin x \cdot \cos x)\, dx$$

$$= \int_0^{\pi/2} \log\left(\frac{\sin 2x}{2}\right) dx = \int_0^{\pi/2} \log \sin 2x\, dx - \int_0^{\pi/2} \log 2\, dx$$

$$= \int_0^{\pi/2} \log\sin 2x\, dx - \log 2\left(\frac{\pi}{2}\right)$$

[For I integral put $2x = t \Rightarrow dx = \frac{dt}{2}$

when $x = 0, t = 0$ and $x = \frac{\pi}{2}, t = \pi$]

$$= \int_0^{\pi}\frac{\log\sin t}{2}\, dt - \frac{\pi}{2}\log 2$$

$$= \frac{1}{2}\int_0^{\pi}\log\sin t\, dt - \frac{\pi}{2}\log 2$$

$$= \frac{1}{2} \cdot 2 \int_0^{\pi/2} \log \sin t \, dt - \frac{\pi}{2} \log 2 \qquad \left[\begin{array}{l} \text{Property 7} \\ \because f(\pi - x) = f(x) \\ \therefore \int_0^a f(x)\, dx = 2 \int_0^{a/2} f(x)\, dx \end{array} \right]$$

$$= \int_0^{\pi/2} \log \sin t \, dt - \frac{\pi}{2} \log 2$$

$$\Rightarrow \quad 2\text{I} = \text{I} - \frac{\pi}{2} \log 2 \qquad \text{(Property I)}$$

$$\Rightarrow \quad \text{I} = -\frac{\pi}{2} \log 2$$

(iii) Let $\text{I} = \int_0^{\pi} \frac{x \, dx}{a^2 \cos^2 x + b^2 \sin^2 x} = \int_0^{\pi} \frac{(\pi - x)\, dx}{a^2 \cos^2 (\pi - x) + b^2 \sin^2 (\pi - x)}$

$$= \int_0^{\pi} \frac{(\pi - x)\, dx}{a^2 \cos^2 x + b^2 \sin^2 x}$$

$$= \int_0^{\pi} \frac{\pi \, dx}{a^2 \cos^2 x + b^2 \sin^2 x} - \int_0^{\pi} \frac{x \, dx}{a^2 \cos^2 x + b^2 \sin^2 x}$$

$$= \pi \int_0^{\pi} \frac{dx}{a^2 \cos^2 x + b^2 \sin^2 x} - \text{I}$$

$$\Rightarrow \quad 2\text{I} = \pi \int_0^{\pi} \frac{dx}{a^2 \cos^2 x + b^2 \sin^2 x}$$

$$\Rightarrow \quad 2\text{I} = 2\pi \int_0^{\pi/2} \frac{dx}{a^2 \cos^2 x + b^2 \sin^2 x} \qquad \text{(Property 7)}$$

$$\Rightarrow \quad \text{I} = \pi \int_0^{\pi/2} \frac{\sec^2 x \, dx}{a^2 + b^2 \tan^2 x}$$

Put $\tan x = t \Rightarrow \sec^2 x \, dx = dt$

when $x = 0, t = 0$ and $x = \frac{\pi}{2}, t = \infty$

$$\therefore \quad \text{I} = \pi \int_0^{\infty} \frac{dt}{a^2 + b^2 t^2}$$

$$= \frac{\pi}{b^2}\int_0^{\infty} \frac{dt}{\left(\frac{a}{b}\right)^2 t^2}$$

$$= \frac{\pi}{b^2}\cdot\frac{1}{\frac{a}{b}}\left[\tan^{-1}\left(\frac{t}{\frac{a}{b}}\right)\right]_0^{\infty}$$

$$= \frac{\pi}{ab}\,[\tan^{-1}\infty - \tan^{-1}0]$$

$$= \frac{\pi}{ab}\left[\frac{\pi}{2} - 0\right]$$

$$\Rightarrow \qquad I = \frac{\pi^2}{2ab}$$

(iv) Let
$$I = \int_0^{\pi} \frac{x \tan x\, dx}{\sec x \operatorname{cosec} x}$$

$$= \int_0^{\pi} \frac{(\pi - x)\tan(\pi - x)}{\sec(\pi - x)\operatorname{cosec}(\pi - x)}\, dx \qquad \text{(Property 4)}$$

$$= \int_0^{\pi} \frac{(\pi - x)(-\tan x)}{(-\sec x)(\operatorname{cosec} x)}\, dx$$

$$= \int_0^{\pi} \frac{(\pi - x)\tan x}{\sec x \operatorname{cosec} x}\, dx$$

$$= \pi\int_0^{\pi} \frac{\tan x\, dx}{\sec x \operatorname{cosec} x} - \int_0^{\pi} \frac{x \tan x\, dx}{\sec x \operatorname{cosec} x}$$

$$= \pi\int_0^{\pi} \frac{\tan x\, dx}{\sec x \operatorname{cosec} x} - I$$

$$\Rightarrow \qquad 2I = \pi\int_0^{\pi} \frac{\tan x\, dx}{\sec x \operatorname{cosec} x}$$

$$\Rightarrow \qquad I = \frac{\pi}{2}\int_0^{\pi} \sin^2 x\, dx$$

$$= \frac{\pi}{2}\int_0^{\pi}\left(\frac{1 - \cos 2x}{2}\right) dx$$

$$= \frac{\pi}{4}\int_0^{\pi}(1 - \cos 2x)\, dx$$

$$= \frac{\pi}{4}\left[x - \frac{\sin 2x}{2}\right]_0^{\pi}$$

$$= \frac{\pi}{4}\left[\left(\pi - \frac{\sin 2\pi}{2}\right) - \left(0 - \frac{\sin 0}{2}\right)\right]$$

$$= \frac{\pi}{4}[(\pi - 0) - (0)]$$

$$\Rightarrow \qquad I = \frac{\pi^2}{4}$$

Example 8. Evaluate the following integrals:

(i) $\displaystyle\int_0^{\pi/2} \frac{x\,dx}{1 + \sin x + \cos x}$ (ii) $\displaystyle\int_0^{\pi} x \sin x \cos^4 x\,dx$

(iii) $\displaystyle\int_{-1}^{1} (x^4 \tan x^3 + x^2)\,dx$

Solution: (i) Let $I = \displaystyle\int_0^{\pi/2} \frac{x\,dx}{1 + \sin x + \cos x}$ ...(1)

$$= \int_0^{\pi/2} \frac{\left(\frac{\pi}{2} - x\right)dx}{1 + \sin\left(\frac{\pi}{2} - x\right) + \cos\left(\frac{\pi}{2} - x\right)} \qquad \text{(Property 4)}$$

$$\Rightarrow \qquad I = \int_0^{\pi/2} \frac{\left(\frac{\pi}{2} - x\right)dx}{1 + \cos x + \sin x} \qquad ...(2)$$

Adding (1) & (2), we get,

$$2I = \frac{\pi}{2}\int_0^{\pi/2} \frac{dx}{1 + \sin x + \cos x}$$

$$\Rightarrow \qquad I = \frac{\pi}{4}\int_0^{\pi/2} \frac{dx}{1 + \frac{2\tan\frac{x}{2}}{1 + \tan^2\frac{x}{2}} + \frac{1 - \tan^2\frac{x}{2}}{1 + \tan^2\frac{x}{2}}}$$

$$= \frac{\pi}{4}\int_0^{\pi/2} \frac{\left(1 + \tan^2\frac{x}{2}\right)dx}{2 + 2\tan\frac{x}{2}}$$

$$= \frac{\pi}{4}\int_0^{\pi/2} \frac{\sec^2\frac{x}{2}\,dx}{2\left(1 + \tan\frac{x}{2}\right)}$$

$$= \frac{\pi}{8}\int_0^{\pi/2} \frac{\sec^2 \frac{x}{2}\, dx}{1+\tan\frac{x}{2}}$$

Put $\tan\frac{x}{2} = t \Rightarrow \sec^2\frac{x}{2}\cdot\frac{1}{2}\, dx = dt \Rightarrow \sec^2\frac{x}{2}\, dx = 2\, dt$

when $x = 0, t = 0$ and $x = \frac{\pi}{2}, t = 1$

$$\therefore \qquad I = \frac{\pi}{8}\int_0^1 \frac{2\, dt}{1+t} = \frac{\pi}{4}\int_0^1 \frac{dt}{1+t}$$

$$= \frac{\pi}{4}[\log(1+t)]_0^1 = \frac{\pi}{4}[\log 2 - \log 1]$$

$$= \frac{\pi}{4}(\log 2 - 0)$$

$$\Rightarrow \qquad I = \frac{\pi}{4}\log 2$$

(ii) Let $I = \int_0^{\pi} x \sin x \cos^4 x\, dx$ ...(1)

$$= \int_0^{\pi} (\pi - x)\sin(\pi - x)\cos^4(\pi - x)\, dx \qquad \text{(Property 4)}$$

$$\Rightarrow \qquad I = \int_0^{\pi} (\pi - x)\sin x \cos^4 x\, dx \qquad ...(2)$$

Adding (1) & (2), we get

$$2I = \int_0^{\pi} \pi \sin x \cos^4 x\, dx$$

$$\Rightarrow \qquad I = \frac{\pi}{2}\int_0^{\pi} \sin x \cos^4 x\, dx$$

Put $\cos x = t \Rightarrow -\sin x\, dx = dt \Rightarrow \sin x\, dx = -dt$

when $x = 0, t = \cos 0 = 1$

and $x = \pi, t = \cos \pi = -1$

$$\therefore \qquad I = \frac{\pi}{2}\int_1^{-1} t^4(-dt)$$

$$= -\frac{\pi}{2}\int_1^{-1} t^4(dt)$$

$$= -\frac{\pi}{2}\int_{1}^{-1} t^4 \, dt$$

$$= -\frac{\pi}{2}\left[\frac{t^5}{5}\right]_1^{-1}$$

$$= -\frac{\pi}{2}\left[\frac{(-1)^5}{5} - \frac{(1)^5}{5}\right]$$

$$= -\frac{\pi}{2}\left[\frac{-1}{5} - \frac{1}{5}\right]$$

$$\Rightarrow \qquad I = -\frac{\pi}{2} \times -\frac{2}{5} = \frac{\pi}{5}.$$

(iii) $$\int_{-1}^{1} (x^4 \tan x^3 + x^2)\, dx = \int_{-1}^{1} x^4 \tan x^3 \, dx + \int_{-1}^{1} x^2 \, dx \qquad ...(1)$$

Now, let $f(x) = x^4 \tan x^3$

$\Rightarrow f(-x) = (-x)^4 \tan(-x)^3 = x^4 \tan(-x^3) = -x^4 \tan x^3$

$\Rightarrow f(-x) = -f(x)$

$\Rightarrow x^4 \tan x^3$ is an odd function.

$$\therefore \int_{-1}^{1} x^4 \tan x^3 \, dx = 0 \qquad \text{(Property 6)}$$

Also, let $f(x) = x^2$

$\Rightarrow f(-x) = (-x)^2 = x^2 = f(x)$

$\Rightarrow x^2$ is an even function

$$\therefore \int_{-1}^{1} x^2 \, dx = 2\int_{0}^{1} x^2 \, dx \qquad \text{(Property 6)}$$

Now, by equation (1)

$$\int_{-1}^{1} (x^4 \tan x^3 + x^2)\, dx = 0 + 2\int_{0}^{1} x^2 \, dx$$

$$= 2\left[\frac{x^3}{3}\right]_0^1$$

$$= 2\left(\frac{1}{3} - 0\right) = \frac{2}{3}$$

EXERCISE 4.12

1. Evaluate the following definite integrals:

(i) $\int_{-3}^{3} |x|\, dx$

(ii) $\int_{0}^{\pi} \cos^7 x\, dx$

(iii) $\int_{0}^{2} |x^2 + 2x - 3|\, dx$

(iv) $\int_{-\pi/4}^{\pi/4} |\sin 2x|\, dx$

(v) $\int_{1}^{3} |x^2 - 2x|\, dx$

(vi) $\int_{-1}^{1} \sqrt{|x| - x}\, dx$

(vii) $\int_{1}^{3} f(x)\, dx$, where $f(x) = \begin{cases} x^2 + x, & 1 \le x \le 2 \\ 3x, & 2 \le x \le 3 \end{cases}$

(viii) $\int_{0}^{2} f(x)\, dx$, where $f(x) = \begin{cases} x^2, & 0 \le x \le 1 \\ \sqrt{x}, & 1 \le x \le 2 \end{cases}$

(ix) $\int_{0}^{9} f(x)\, dx$, where $f(x) = \begin{cases} \sin x, & 0 \le x \le \dfrac{\pi}{2} \\ 1, & \dfrac{\pi}{2} \le x \le 3 \\ e^{x-3}, & 3 \le x \le 9 \end{cases}$

(x) $\int_{0}^{2} \dfrac{x-1}{|x-1|}\, dx$

(xi) $\int_{-5}^{5} |x+3|\, dx$

(xii) $\int_{-1}^{1} e^{|x|}\, dx$

2. Evaluate the following definite integrals:

(i) $\int_{0}^{\pi/2} \log \tan x\, dx$

(ii) $\int_{0}^{\pi/2} \dfrac{\sin x}{\sin x + \cos x}\, dx$

(iii) $\int_{0}^{\pi/2} \dfrac{\sqrt{\cot x}}{\sqrt{\cot x} + \sqrt{\tan x}}\, dx$

(iv) $\int_{0}^{3} \dfrac{\sqrt{x}}{\sqrt{x} + \sqrt{3-x}}\, dx$

(v) $\int_{0}^{2\pi} \dfrac{\sin 2x}{1 - 2\cos x}\, dx$

(vi) $\int_{0}^{\pi/2} \sin 2x \log \tan x\, dx$

(vii) $\int_{0}^{\pi/2} (2 \log \sin x - \log \sin 2x)\, dx$

(viii) $\int_{0}^{\pi/2} \dfrac{x}{\sin x + \cos x}\, dx$

3. Evaluate:

(i) $\int_0^2 x^2(2-x)^{3/2}\,dx$ (ii) $\int_0^7 \frac{\sqrt[3]{x}}{\sqrt[3]{x}+\sqrt[3]{7-x}}\,dx$

(iii) $\int_0^1 \frac{\log x}{\sqrt{1-x^2}}\,dx$ (iv) $\int_0^1 x(1-x)^{5/2}\,dx$

(v) $\int_2^5 \frac{\sqrt{x}}{\sqrt{7-x}+\sqrt{x}}\,dx$ (vi) $\int_2^8 \frac{\sqrt[3]{x+1}}{\sqrt[3]{x+1}+\sqrt[3]{11-x}}\,dx$

(vii) $\int_{\pi/6}^{\pi/3} \frac{dx}{1+\sqrt{\tan x}}$

4. Evaluate:

(i) $\int_{-\pi/2}^{\pi/2} \sin^2 x\,dx$ (ii) $\int_{-\pi}^{\pi} x^{10}\sin^7 x\,dx$

(iii) $\int_{-2}^{2} (|x|+|x-1|)\,dx$ (iv) $\int_{-1}^{1} \log\left(\frac{2-x}{2+x}\right)dx$

(v) $\int_{-1}^{1} \frac{x\sin^{-1}x}{\sqrt{1-x^2}}\,dx$

5. Evaluate the following definite integrals:

(i) $\int_0^{\pi/2} \frac{\cos x}{1+\sin x+\cos x}\,dx$ (ii) $\int_0^{\pi} \frac{x\sin x}{1+\cos^2 x}\,dx$

(iii) $\int_0^{\pi} x\log\sin x\,dx$ (iv) $\int_0^{\pi/2} \frac{x\sin x\cos x}{\sin^4 x+\cos^4 x}\,dx$

(v) $\int_0^{\pi/2} \frac{\sin x-\cos x}{\sqrt{1-\sin 2x}}\,dx$ (vi) $\int_0^1 \cot^{-1}(1-x+x^2)\,dx$

ANSWERS

1. (i) 9 (ii) 0

(iii) 4 (iv) 1

(v) 2 (vi) $\frac{2\sqrt{2}}{3}$

(vii) $\frac{34}{3}$ (viii) $\frac{1}{3}(4\sqrt{2}-1)$

(ix) $3-\frac{\pi}{2}+e^6$ (x) 0

(xi) 34 (xii) $2(e-1)$

2. (i) 0 (ii) $\frac{\pi}{4}$

(iii) $\frac{\pi}{4}$ (iv) $\frac{3}{2}$

(v) 0 (vi) 0

(vii) $-\frac{\pi}{2}\log 2$ (viii) $\frac{\pi}{2\sqrt{2}}\log(\sqrt{2}+1)$

3. (i) $\frac{256\sqrt{2}}{315}$ (ii) $\frac{5}{2}$

(iii) $-\frac{\pi}{2}\log 2$ (iv) $\frac{4}{63}$

(v) $\frac{3}{2}$ (vi) 3

(vii) $\frac{\pi}{12}$

4. (i) $\frac{\pi}{2}$ (ii) 0

(iii) 9 (iv) 0

(v) 2

5. (i) $\frac{\pi}{4}(1-\log 2)$ (ii) $\frac{\pi^2}{4}$

(iii) $-\frac{\pi^2}{2}\log 2$ (iv) $\frac{\pi^2}{16}$

(v) 0 (vi) $\frac{\pi}{2}-\log 2$

4.21 DOUBLE INTEGRALS:

The double integral of a function $f(x, y)$ over a region R is denoted by

$$\iint_R f(x, y)\,dx\,dy \text{ or } \iint_R f(x, y)\,dy\,dx$$

where one integral is for x and other for y.

Properties of Double Integrals:

If $f(x, y)$ and $g(x, y)$ are two functions over a region R, then

1. $\iint_R [f(x, y) \pm g(x, y)]\, dx\, dy = \iint_R f(x, y)\, dx\, dy \pm \iint_R g(x, y)\, dx\, dy$
2. $\iint_R k\, f(x, y)\, dx\, dy = k \iint_R f(x, y)\, dx\, dy$, where k is a constant
3. $\iint_R f(x, y)\, dx\, dy = \iint_{R_1} f(x, y)\, dx\, dy + \iint_{R_2} f(x, y)\, dx\, dy$, where R is partitioned into two parts R_1 and R_2.

4.22 EVALUATION OF DOUBLE INTEGRALS

Double integral over the region R can be evaluated by two successive integration. Let the region R be defined by the curves $x = x_1, x = x_2$ and $y = y_1, y = y_2$

Case I: When x_1, x_2 are functions of y and y_1, y_2 are constants.

Let PQ and RS be the curves $x_1 = \phi_1(y)$ and $x_2 = \phi_2(y)$. Take the horizontal strip MN. Here the double integral is first evaluated w.r.t. x (considering y as a constant). The resulting expression which is a function of y is integrated w.r.t. y between the limits $y = y_1$ and $y = y_2$. Thus.

$$\iint_R f(x, y)\, dx\, dy = \int_{y_1}^{y_2} \left[\int_{x_1=\phi_1(y)}^{x_2=\phi_2(y)} f(x, y)\, dx \right] dy$$

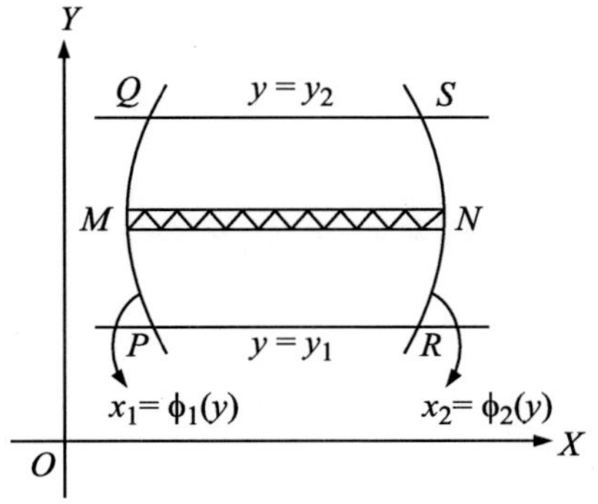

Fig.

Case II: When y_1, y_2 are functions of x and x_1, x_2 are constants.

Let PQ and RS are the curve $y_1 = \phi_1(x)$ and $y_2 = \phi_2(x)$. Take a vertical strip MN. Here the double integral is first evaluated w.r.t. y (considering x as a constant). The resulting expression which is a function of x is integrated w.r.t. y between the limits $y = y_1$ and $y = y_2$. Thus.

$$\iint_R f(x, y)\, dx\, dy = \int_{x_1}^{x_2} \left[\int_{y_1=\phi_1(x)}^{y_2=\phi_2(x)} f(x, y)\, dy \right] dx$$

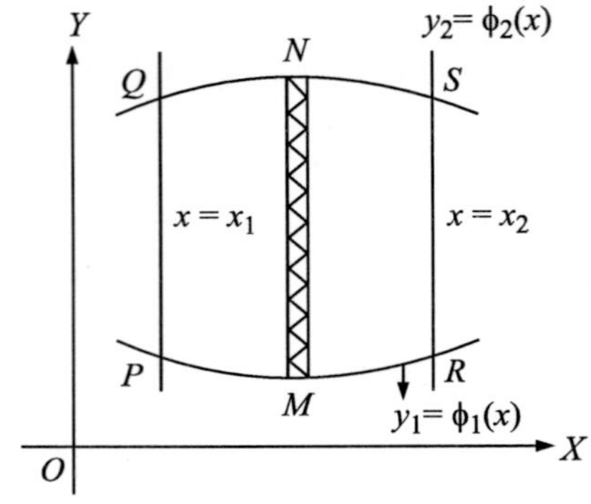

Fig.

Case III: When x_1, x_2, y_1, y_2 are constants.

Here the region of integration is the rectangle $PQRS$. Hence it does not matter whether we apply case I or case II. Thus.

$$\iint_R f(x, y)\, dx\, dy = \int_{y_1}^{y_2}\left[\int_{x_1}^{x_2} f(x, y)\, dx\right] dy = \int_{x_1}^{x_2}\left[\int_{y_1}^{y_2} f(x, y)\, dy\right] dx$$

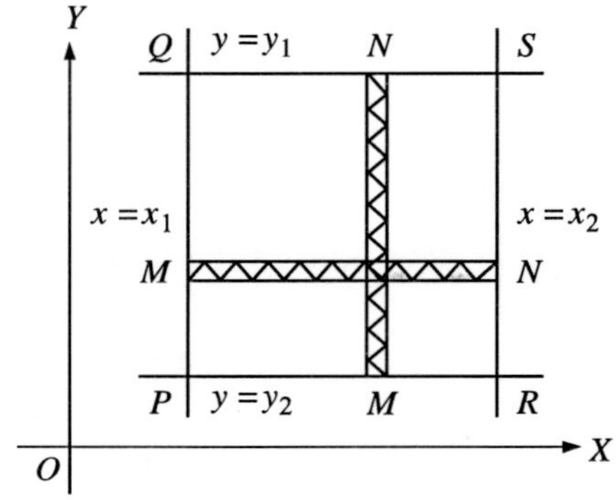

Fig.

ILLUSTRATIVE EXAMPLES

Example 1. Evaluate the following integrals:

(i) $\int_0^1\int_0^2 xy^2\, dx\, dy$ (ii) $\int_0^3\int_1^2 xy(1 + x + y)\, dy\, dx$

Solution: (i) $\int_0^1\int_0^2 xy^2\, dx\, dy = \int_0^1\left[\int_0^2 x\, dx\right] y^2\, dy$

$$= \int_0^1 \left[\frac{x^2}{2}\right]_0^2 y^2\, dy = \int_0^1 \left(\frac{4}{2} - 0\right) y^2\, dy$$

$$= 2\int_0^1 y^2\, dy = 2\left[\frac{y^3}{3}\right]_0^1 = 2\left(\frac{1}{3} - 0\right) = \frac{2}{3}$$

(ii) $$\int_0^3 \int_1^2 xy(1 + x + y)\, dy\, dx = \int_0^3 \int_1^2 (xy + x^2y + xy^2)\, dy\, dx$$

$$= \int_0^3 \left[\int_1^2 y\, dy\right] x\, dx + \int_0^3 \left[\int_1^2 y\, dy\right] x^2\, dx + \int_0^3 \left[\int_1^2 y^2\, dy\right] x\, dx$$

$$= \int_0^3 \left[\frac{y^2}{2}\right]_1^2 x\, dx + \int_0^3 \left[\frac{y^2}{2}\right]_1^2 x^2\, dx + \int_0^3 \left[\frac{y^3}{3}\right]_1^2 x\, dx$$

$$= \int_0^3 \left(\frac{4}{2} - \frac{1}{2}\right) x\, dx + \int_0^3 \left(\frac{4}{2} - \frac{1}{2}\right) x^2\, dx + \int_0^3 \left(\frac{8}{3} - \frac{1}{3}\right) x\, dx$$

$$= \frac{3}{2}\int_0^3 x\, dx + \frac{3}{2}\int_0^3 x^2\, dx + \frac{7}{3}\int_0^3 x\, dx$$

$$= \frac{3}{2}\left[\frac{x^2}{2}\right]_0^3 + \frac{3}{2}\left[\frac{x^3}{3}\right]_0^3 + \frac{7}{3}\left[\frac{x^2}{2}\right]_0^3$$

$$= \frac{3}{2}\left[\frac{9}{2} - 0\right] + \frac{3}{2}\left[\frac{27}{3} - 0\right] + \frac{7}{3}\left[\frac{9}{2} - 0\right]$$

$$= \frac{27}{4} + \frac{27}{2} + \frac{21}{2}$$

$$= \frac{123}{4} = 30\frac{3}{4}$$

Example 2. Evaluate $\displaystyle\int_0^1 \int_0^1 \frac{dx\, dy}{\sqrt{1 - x^2}\,\sqrt{1 - y^2}}$

Solution: $$\int_0^1 \int_0^1 \frac{dx\, dy}{\sqrt{1 - x^2}\,\sqrt{1 - y^2}} = \int_0^1 \left[\int_0^1 \frac{dx}{\sqrt{1 - x^2}}\right] \frac{1}{\sqrt{1 - y^2}}\, dy$$

$$= \int_0^1 [\sin^{-1} x]_0^1 \frac{1}{\sqrt{1 - y^2}}\, dy$$

$$= \int_0^1 \left[\frac{\pi}{2} - 0\right] \frac{1}{\sqrt{1 - y^2}}\, dy$$

$$= \frac{\pi}{2}\int_0^1 \frac{dy}{\sqrt{1-y^2}} = \frac{\pi}{2}[\sin^{-1} y]_0^1 = \frac{\pi}{2}\left[\frac{\pi}{2} - 0\right]$$

$$= \frac{\pi^2}{4}$$

Example 3. Evaluate $\int_0^1 \int_x^1 (x^2 + y^2)\, dx\, dy$

Solution: $\int_0^1 \int_x^1 (x^2 + y^2)\, dx\, dy = \int_0^1 \left[\int_x^1 (x^2 + y^2)\, dy\right] dx$

$$= \int_0^1 \left[x^2 y + \frac{y^3}{3}\right]_x^1 dx = \int_0^1 \left[\left(x^2 + \frac{1}{3}\right) - \left(x^3 + \frac{x^3}{3}\right)\right] dx$$

$$= \int_0^1 \left(x^2 + \frac{1}{3} - \frac{4x^3}{3}\right) dx = \left[\frac{x^3}{3} + \frac{x}{3} - \frac{4x^4}{4 \cdot 3}\right]_0^1$$

$$= \left[\left(\frac{1}{3} + \frac{1}{3} - \frac{1}{3}\right) - 0\right] = \frac{1}{3}$$

Example 4. Evaluate $\int_0^x \int_1^2 \frac{dx\, dy}{x^2 + y^2}$

Solution: $\int_0^x \int_1^2 \frac{dx\, dy}{x^2 + y^2} = \int_1^2 \left[\int_0^x \frac{dy}{x^2 + y^2}\right] dx$

$$= \int_1^2 \left[\frac{1}{x}\tan^{-1}\frac{y}{x}\right]_0^x dx = \int_1^2 \left[\frac{1}{x}\tan^{-1}(1) - \frac{1}{x}\tan^{-1}(0)\right] dx$$

$$= \int_1^2 \left(\frac{\pi}{4x} - 0\right) dx = \frac{\pi}{4}\int_1^2 \frac{dx}{x}$$

$$= \frac{\pi}{4}[\log x]_1^2 = \frac{\pi}{4}(\log 2 - \log 1) = \frac{\pi}{4}(\log 2 - 0)$$

$$= \frac{\pi}{4}\log 2$$

Example 5. Evaluate $\int_1^2 \int_1^y \frac{dx\, dy}{xy}$

Solution: $\int_1^2 \int_1^y \frac{dx\, dy}{xy} = \int_1^2 \left[\int_1^y \frac{dx}{x}\right] \frac{1}{y}\, dy$

$$= \int_1^2 [\log x]_1^y \frac{1}{y}\, dy = \int_1^2 (\log y - 0)\frac{1}{y}\, dy$$

$$= \int_1^2 \frac{\log y}{y}\, dy$$

Put $\log y = t \Rightarrow \frac{1}{y}\, dy = dt$

when $y = 1, t = \log 1 = 0$

and $y = 2, t = \log 2$

$$\therefore \quad \int_1^2 \int_1^y \frac{dx\, dy}{xy} = \int_0^{\log 2} t\, dt = \left[\frac{t^2}{2}\right]_0^{\log 2}$$

$$= \frac{(\log 2)^2}{2} - 0 = \frac{(\log 2)^2}{2}$$

Example 6. Evaluate $\int_0^1 \int_0^x e^{y/x}\, dx\, dy$

Solution: $\int_0^1 \int_0^x e^{y/x}\, dx\, dy = \int_0^1 \left[\int_0^x e^{y/x}\, dy\right] dx$

$$= \int_0^1 \left[\frac{e^{y/x}}{\frac{1}{x}}\right]_0^x dx$$

$$= \int_0^1 \left[\frac{e}{\frac{1}{x}} - \frac{1}{\frac{1}{x}}\right] dx = \int_0^1 [ex - x]\, dx$$

$$= (e-1)\int_0^1 x\, dx = (e-1)\left[\frac{x^2}{2}\right]_0^1$$

$$= \frac{e-1}{2} = \frac{1}{2}(e-1)$$

Example 7. Evaluate $\int_0^{\pi/2} \int_0^{\pi/2} \sin(x+2y)\, dx\, dy$

Solution: $\int_0^{\pi/2} \int_0^{\pi/2} \sin(x+2y)\, dx\, dy = \int_0^{\pi/2} [-\cos(x+2y)]_0^{\pi/2}\, dy$

$$= -\int_0^{\pi/2} \left[\cos\left(\frac{\pi}{2} + 2y\right) - \cos 2y\right] dy$$

$$= -\int_0^{\pi/2} (-\sin 2y - \cos 2y)\, dy$$

$$= \int_0^{\pi/2} (\sin 2y + \cos 2y)\, dy$$

$$= \left[-\frac{\cos 2y}{2} + \frac{\sin 2y}{2} \right]_0^{\pi} dy$$

$$= \left(-\frac{\cos 2\pi}{2} + \frac{\sin 2\pi}{2} \right) - \left(-\frac{\cos 0}{2} + \frac{\sin 0}{2} \right)$$

$$= \left(\frac{1}{2} + 0 \right) - \left(-\frac{1}{2} + 0 \right) = 1$$

Example 8. Evaluate $\int_0^{\pi} \int_0^{a(1+\cos\theta)} r\, dr\, d\theta$

Solution: $\int_0^{\pi} \int_0^{a(1+\cos\theta)} r\, dr\, d\theta = \int_0^{\pi} \left[\int_0^{a(1+\cos\theta)} r\, dr \right] d\theta$

$$= \int_0^{\pi} \left[\frac{r^2}{2} \right]_0^{a(1+\cos\theta)} d\theta$$

$$= \frac{1}{2} \int_0^{\pi} [a^2 (1 + \cos\theta)^2\, d\theta = \frac{1}{2} \int_0^{\pi} [a^2 (1+\cos\theta)^2 - 0]\, d\theta$$

$$= \frac{a^2}{2} \int_0^{\pi} (1 + \cos\theta)^2\, d\theta = \frac{a^2}{2} \int_0^{\pi} \left(2\cos^2 \frac{\theta}{2} \right)^2 d\theta$$

$$= \frac{a^2}{2} \int_0^{\pi} 4\cos^4 \frac{\theta}{2}\, d\theta = 2a^2 \int_0^{\pi} \cos^4 \frac{\theta}{2}\, d\theta$$

$$= \frac{a^2}{2} \int_0^{\pi/2} \cos^4 \phi\, d\phi = \frac{3\pi a^2}{4}$$

where $\theta = 2\phi$ i.e., $d\theta = 2\, d\phi$

Example 9. Evaluate $\int_0^{\pi/2} \int_0^{a\cos\theta} r \sin\theta\, dr\, d\theta$

Solution: $\int_0^{\pi/2} \int_0^{a\cos\theta} r \sin\theta\, dr\, d\theta = \int_0^{\pi/2} \left[\int_0^{a\cos\theta} r\, dr \right] \sin\theta\, d\theta$

$$= \int_0^{\pi/2} \left[\frac{r^2}{2} \right]_0^{a\cos\theta} \sin\theta\, d\theta$$

$$= \int_0^{\pi/2} \frac{a^2 \cos^2\theta}{2} \sin\theta\, d\theta = \frac{a^2}{2} \int_0^{\pi/2} \cos^2\theta \sin\theta\, d\theta$$

Put $\cos\theta = t \Rightarrow -\sin\theta\, d\theta = dt$ i.e., $\sin\theta\, d\theta = dt$

when $\theta = 0, t = \cos 0 = 1$ and $\theta = \frac{\pi}{2}, t = \cos\frac{\pi}{2} = 0.$

$$\therefore \int_0^{\pi/2}\int_0^{a\cos\theta} r\sin\theta\, dr\, d\theta = -\frac{a^2}{2}\int_1^0 t^2\, dt$$

$$= \frac{a^2}{2}\int_0^1 t^2\, dt = \frac{a^2}{2}\left[\frac{t^3}{3}\right]_0^1 dt$$

$$= \frac{a^2}{2}\left(\frac{1}{3} - 0\right) = \frac{a^2}{6}$$

Example 10. Evaluate $\int_0^1\int_{y^2}^{y}(1 + xy^2)\, dx\, dy$

Solution: $\int_0^1\int_{y^2}^{y}(1 + xy^2)\, dx\, dy = \int_0^1\left[\int_{y^2}^{y}(1 + xy^2)\, dx\right] dy$

$$= \int_0^1\left[x + \frac{x^2y^2}{2}\right]_{y^2}^{y} dy$$

$$= \int_0^1\left[\left(y + \frac{y^4}{2}\right) - \left(y^2 + \frac{y^6}{2}\right)\right] dy$$

$$= \left[\frac{y^2}{2} + \frac{y^5}{10} - \frac{y^3}{3} - \frac{y^7}{14}\right]_0^1$$

$$= \left(\frac{1}{2} + \frac{1}{10} - \frac{1}{3} - \frac{1}{14}\right) = \frac{41}{210}$$

Example 11. Evaluate $\int_0^{2a}\int_0^{\sqrt{2ax - x^2}} xy\, dy\, dx$

Solution: $\int_0^{2a}\int_0^{\sqrt{2ax - x^2}} xy\, dy\, dx = \int_0^{2a}\left[\frac{xy^2}{2}\right]_0^{\sqrt{2ax - x^2}} dx$

$$= \frac{1}{2}\int_0^{2a}[x(2ax - x^2)]\, dx$$

$$= \frac{1}{2}\int_0^{2a}(2ax^2 - x^3)\, dx$$

$$= \frac{1}{2}\left[\frac{2ax^3}{3} - \frac{x^4}{4}\right]_0^{2a}$$

$$= \frac{1}{2}\left(\frac{16a^4}{3} - \frac{16a^4}{4}\right)$$

$$= \frac{2}{3}a^4$$

Example 12. Evaluate $\int_0^1 \int_1^{e^x} \left(\frac{x}{y}\right) dy\, dx$

Solution: $\int_0^1 \int_1^{e^x} \left(\frac{x}{y}\right) dy\, dx = \int_0^1 x\left[\int_1^{e^x} \frac{1}{y}\, dy\right] dx$

$$= \int_0^1 x[\log y]_1^{e^x}\, dx = \int_0^1 x[\log e^x - \log 1]\, dx$$

$$= \int_0^1 x(x-0)\, dx = \int_0^1 x^2\, dx$$

$$= \left[\frac{x^3}{3}\right]_0^1 = \frac{1}{3}$$

Example 13. Evaluate $\int_0^a \int_0^{\sqrt{a^2-x^2}} \sqrt{a^2-x^2}\, dy\, dx$

Solution: $\int_0^a \left[\int_0^{\sqrt{a^2-x^2}} \cdot dy\right] \sqrt{a^2-x^2}\, dx = \int_0^a \int_0^{\sqrt{a^2-x^2}} \sqrt{a^2-x^2}\, dy\, dx$

$$= \int_0^a [y]_0^{\sqrt{a^2-x^2}} \cdot \sqrt{a^2-x^2}\, dx$$

$$= \int_0^a \left[\sqrt{a^2-x^2} - 0\right] \sqrt{a^2-x^2}\, dx$$

$$= \int_0^a (a^2-x^2)\, dx = \left[a^2x - \frac{x^3}{3}\right]_0^a$$

$$= \left(a^3 - \frac{a^3}{3}\right) = \frac{2a^3}{3}$$

Example 14. Evaluate $\iint xy\,dx\,dy$ over the region in the positive quadrant for which $x + y \le 1$.

Solution: The required region is OAB as shown in the figure.

Now draw MN parallel to y-axis the limits for x is O to A

i.e., $0 \to 1$ and for y is $0 \to (1 - x)$

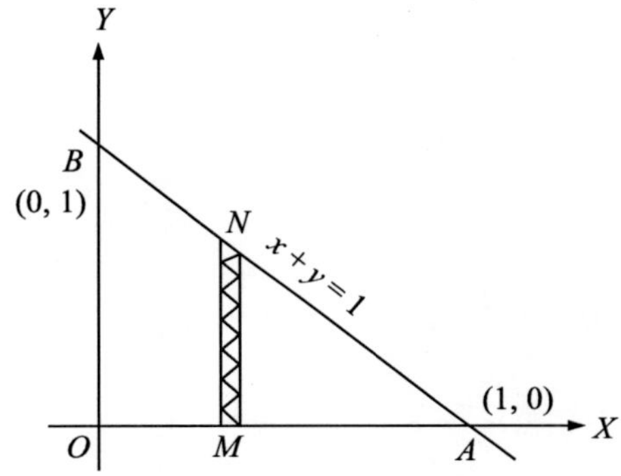

Fig.

$$\therefore \iint xy\,dx\,dy = \int_0^1 \int_0^{1-x} xy\,dx\,dy = \int_0^1 x\left[\int_0^{1-x} y\,dy\right] dx$$

$$= \int_0^1 x\left[\frac{y^2}{2}\right]_0^{1-x} dx = \int_0^1 x\left[\frac{(1-x)^2}{2} - 0\right] dx$$

$$= \frac{1}{2}\int_0^1 x(1 + x^2 - 2x)\,dx$$

$$= \frac{1}{2}\int_0^1 (x + x^3 - 2x^2)\,dx$$

$$= \frac{1}{2}\left[\frac{x^2}{2} + \frac{x^4}{4} - \frac{2x^3}{3}\right]_0^1 = \frac{1}{2}\left(\frac{1}{2} + \frac{1}{4} - \frac{2}{3}\right)$$

$$= \frac{1}{24}$$

Example 15. Evaluate $\iint_R y\,dx\,dy$ where R is the region bounded by the parabolas $y^2 = 4x$ and $x^2 = 4y$.

Solution: Solving $y^2 = 4x$ and $x^2 = 4y$, we have

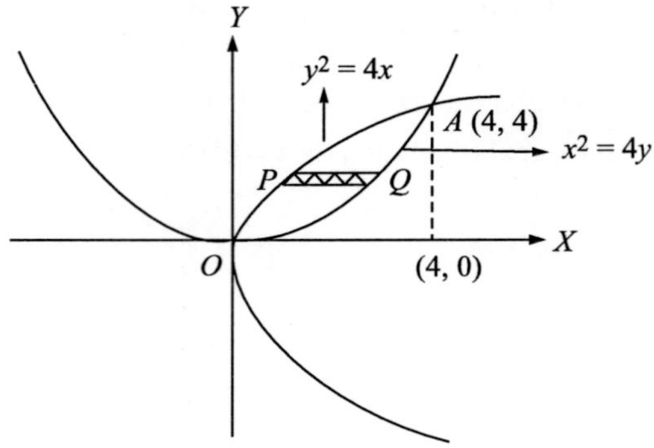

Fig.

$$\left(\frac{x^2}{4}\right)^2 = 4x \text{ or } x(x^3 - 64) = 0$$

$\therefore x = 0, 4$

when $x = 4, y = 4$

$\therefore$ Co-ordinates of A are $(4, 4)$

Taking horizontal strip PQ parallel to x-axis, the region R can be expressed as

$$\frac{y^2}{4} \le x \le 2\sqrt{y} \text{ and } 0 \le y \le 4$$

$$\therefore \iint_R y\, dx\, dy = \int_0^4 \int_{y^2/4}^{2\sqrt{y}} y\, dx\, dy = \int_0^4 y \left[\int_{y^2/4}^{2\sqrt{y}} dx\right] dy$$

$$= \int_0^4 y[x]_{y^2/4}^{2\sqrt{y}}\, dy = \int_0^4 y\left(2\sqrt{y} - \frac{y^2}{4}\right) dy$$

$$= \int_0^4 \left(2y^{3/2} - \frac{y^3}{4}\right) dy = \left[\frac{2y^{5/2}}{\frac{5}{2}} - \frac{y^4}{16}\right]_0^4$$

$$= 2(4)^{5/2} \cdot \frac{2}{5} - \frac{(4)^4}{16} = \frac{4}{5} \cdot (2)^5 - \frac{(4)^4}{16}$$

$$= \frac{128}{5} - 16 = \frac{48}{5}$$

Example 16. Evaluate $\iint_R xy(x + y)\, dx\, dy$ over the region R bounded the curve $y = x^2$ and $y = x$

Solution: Solving $y = x^2$ and $y = x$ we get, $x^2 = x \Rightarrow x = 0, 1$

$\therefore y = 0, 1$

coordinates of O is $(0, 0)$ and A is $(1, 1)$

Taking strip parallel to y-axis limits of x & y are.

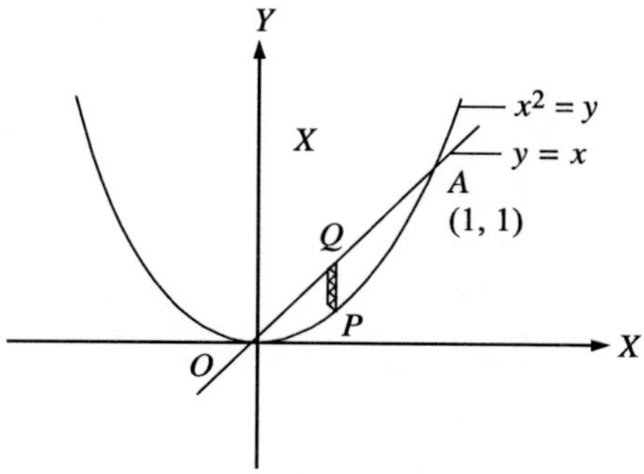

Fig.

$R \equiv 0 \le x \le 1$ and $x^2 \le y \le x$

$$\therefore \iint_R xy(x+y)\,dx\,dy = \int_0^1 \int_{x^2}^{x} xy(x+y)\,dx\,dy$$

$$= \int_0^1 \left[\int_{x^2}^{x} (x^2 y + xy^2)\,dy \right] dx$$

$$= \int_0^1 \left[\frac{x^2 y^2}{2} + \frac{xy^3}{3} \right]_{x^2}^{x} dx$$

$$= \int_0^1 \left[\left(\frac{x^4}{2} + \frac{x^4}{3} \right) - \left(\frac{x^6}{2} + \frac{x^7}{3} \right) \right] dx$$

$$= \int_0^1 \left(\frac{5x^4}{6} - \frac{x^6}{2} - \frac{x^7}{3} \right) dx$$

$$= \left[\frac{5x^5}{30} - \frac{x^7}{14} - \frac{x^8}{24} \right]_0^1$$

$$= \frac{1}{6} - \frac{1}{14} - \frac{1}{24} = \frac{3}{56}$$

Example 17. Evaluate $\iint_R \sqrt{xy - y^2}\,dx\,dy$, where R is the region bounded by the triangle whose vertices are (0, 0), (10, 1) and (1, 1).

Solution: Let OAB be the required triangle. The equation of OA is

$$y - 0 = \frac{1-0}{1-0}(x-0) \Rightarrow y = x$$

The equation of the line OB is

$$y - 0 = \frac{1-0}{1-0}(x-0) \Rightarrow x = 10y$$

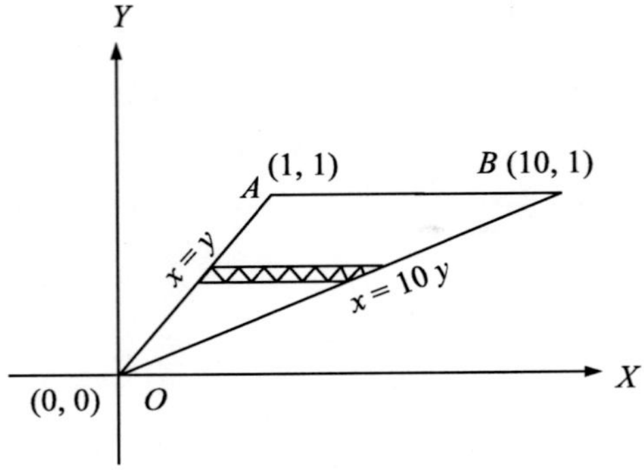

Fig.

Taking horizontal strip, the region R is expressed as,

$$y \le x \le 10y, 0 \le y \le 1$$

$$\therefore \iint_R \sqrt{xy - y^2}\, dx\, dy = \int_0^1 \int_y^{10y} \sqrt{xy - y^2}\, dx\, dy$$

$$= \int_0^1 \left[\int_y^{10y} (xy - y^2)^{1/2}\, dx \right] dy$$

$$= \int_0^1 \left[\frac{(xy - y^2)^{3/2}}{\frac{3}{2} \cdot y} \right]_y^{10y} dy$$

$$= \frac{2}{3} \int_0^1 \left[\frac{(xy - y^2)^{3/2}}{y} \right]_y^{10y} dy$$

$$= \frac{2}{3} \int_0^1 \left[\frac{(10y^2 - y^2)^{9/3}}{y} \right] dy$$

$$= \frac{2}{3} \int_0^1 \frac{(9y^2)^{3/2}}{y}\, dy$$

$$= \frac{2}{3} \int_0^1 27y^2\, dy = 18 \int_0^1 y^2\, dy$$

$$= 18\left[\frac{y^3}{3}\right]_0^1 = 18\cdot\frac{1}{3} = 6$$

Example 18. Evaluate $\iint_R \frac{xy}{\sqrt{1-y^2}}\, dx\, dy$ over the positive quadrant of the circle $x^2+y^2=1$

Solution: Taking strip MN parallel to y-axis, the region R is given as.

$$0 \le x \le 1 \quad \text{and} \quad 0 \le y \le \sqrt{1-x^2}$$

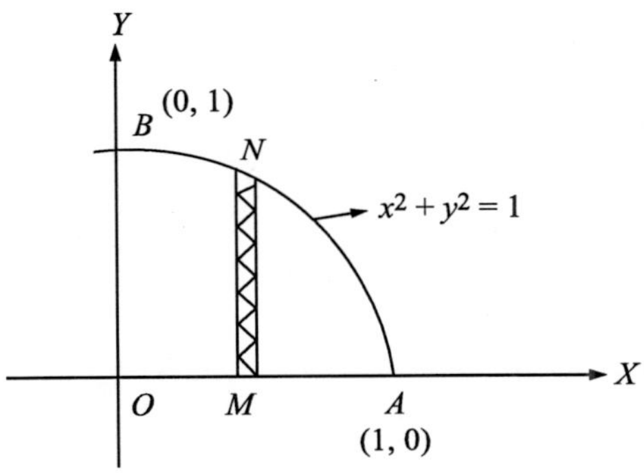

Fig.

$$\therefore \iint_R \frac{xy}{\sqrt{1-y^2}}\, dx\, dy = \int_0^1 \int_0^{\sqrt{1-x^2}} \frac{xy}{\sqrt{1-y^2}}\, dx\, dy$$

$$= \int_0^1 x\left[\int_0^{\sqrt{1-x^2}} \frac{y}{\sqrt{1-y^2}}\, dy\right] dx$$

$$= \int_0^1 x\left[-\sqrt{1-y^2}\right]_0^{\sqrt{1-x^2}} dx \qquad \left(\because \int \frac{y\, dy}{\sqrt{1-y^2}} = -\sqrt{1-y^2}\right)$$

$$= -\int_0^1 x\left[\sqrt{1-1+x^2} - 1\right] dx = -\int_0^1 x(x-1)\, dx$$

$$= -\int_0^1 (x^2 - x)\, dx = -\left[\frac{x^3}{3} - \frac{x^2}{2}\right]_0^1$$

$$= -\left(\frac{1}{3} - \frac{1}{2}\right) = \frac{1}{6}$$

Example 19. Evaluate $\iint_R e^{y/x} dx\, dy$, where R is the region bounded by the lines $y = x$, $y = 0$ and $x = 1$

Solution: The region R is the triangle OAB where $O(0, 0)$, $A(1, 0)$ and $B(1, 1)$ are the vertices. The region R is expressed as.

$$0 \le x \le 1 \quad \text{and} \quad 0 \le y \le x$$

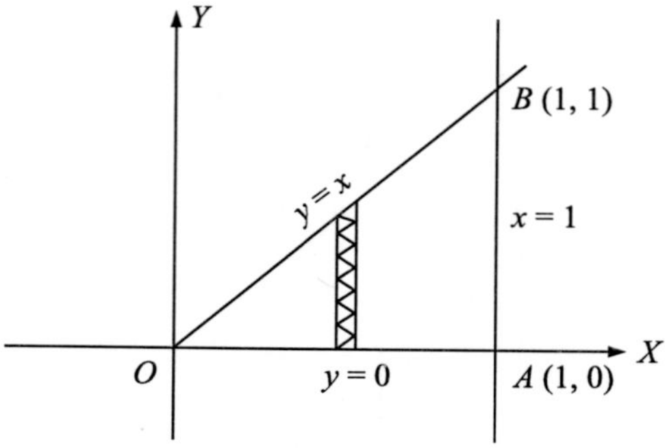

Fig.

$$\therefore \iint_R e^{y/x}\, dx\, dy = \int_0^1 \int_0^x e^{y/x} dx\, dy$$

$$= \frac{1}{2}(e-1) \qquad \text{[Solve as Example 6.]}$$

Example 20. Evaluate $\iint_R (x^2 + y^2)\, dx\, dy$, where R is the region bounded by $y = x^2$, $x = 2$ and $y = 1$

Solution: The region R is ABC as shown in the figure. The coordinates of A is (1, B), $B(2, 1)$ and $C(2, 4)$.

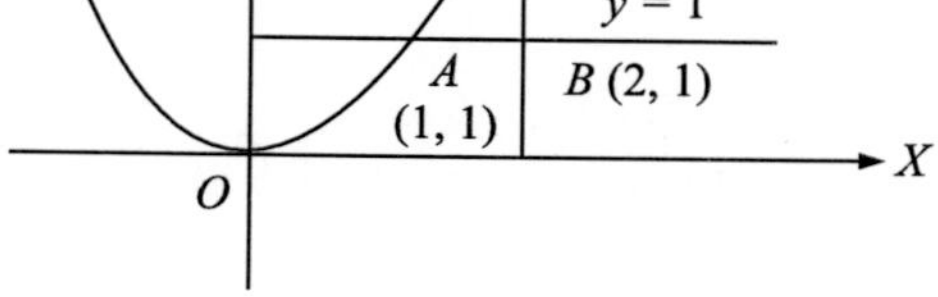

Fig.

In R, x varies from $1 \to 2$ and y from $1 \to x^2$ (taking vertical strip)

$$\therefore \iint_R (x^2 + y^2)\, dx\, dy = \int_1^2 \int_1^{x^2} (x^2 + y^2)\, dx\, dy$$

$$= \int_1^2 \left[\int_1^{x^2} (x^2 + y^2)\, dy \right] dx$$

$$= \int_1^2 \left[x^2 y + \frac{y^3}{3} \right]_1^{x^2} dx$$

$$= \int_1^2 \left[\left(x^4 + \frac{x^6}{3} \right) - \left(x^2 + \frac{1}{3} \right) \right] dx$$

$$= \left[\frac{x^5}{5} + \frac{x^7}{21} - \frac{x^3}{3} - \frac{x}{3} \right]_1^2$$

$$= \left(\frac{32}{5} + \frac{128}{21} - \frac{8}{3} - \frac{2}{3} \right) - \left(\frac{1}{5} + \frac{1}{21} - \frac{1}{3} - \frac{1}{3} \right) = \frac{1006}{105}$$

Example 21. Evaluate $\iint_R (4 - x^2 - y^2)\, dx\, dy$, where R is the region bounded by the lines $x = 0, x = 1, y = 0$ and $y = \frac{3}{2}$

Solution: The region R is given by $OABD$ and is expressed as,

$$0 \le x \le 1 \quad \text{and} \quad 0 \le y \le \frac{3}{2}$$

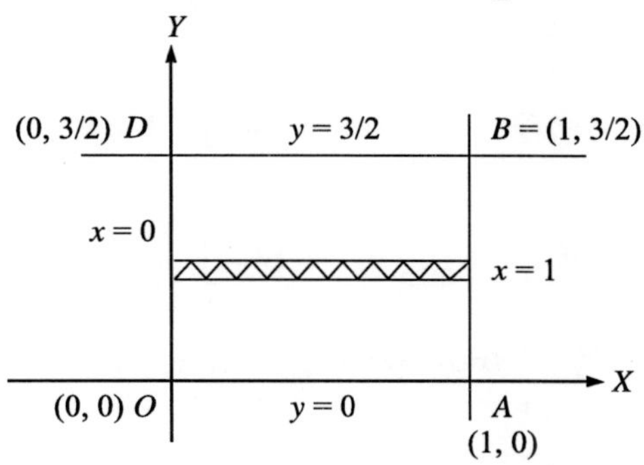

Fig.

$$\therefore \iint_R (4 - x^2 - y^2)\, dx\, dy = \int_0^1 \left[\int_0^{3/2} 4 - x^2 - y^2)\, dy \right] dx$$

$$= \int_0^1 \left[4y - x^2 y - \frac{y^3}{3} \right]_0^{3/2} dx$$

$$= \int_0^1 \left(6 - \frac{3}{2} x^2 - \frac{9}{8} \right) dx$$

$$= \left[6x - \frac{3}{2}\cdot\frac{x^3}{3} - \frac{9}{8}x\right]_0^1$$

$$= 6 - \frac{1}{2} - \frac{9}{8} = \frac{35}{8}$$

EXERCISE 4.13

Evaluate the following integrals:

1. $\int_1^2 \int_0^{y/2} y\, dy\, dx$

2. $\int_0^1 \int_0^{y^2+1} x^2 y\, dx\, dy$

3. $\int_3^4 \int_1^2 \frac{dx\, dy}{(x+y)^2}$

4. $\int_0^1 \int_y^{2y} e^{x-y}\, dx\, dy$

5. $\int_{-1}^3 \int_{x^2}^{x+2} dy\, dx$

6. $\int_1^a \int_1^b \frac{dy\, dx}{xy}$

7. $\int_0^1 \int_{x^2}^{x} (x^2 + 3y + 2)\, dy\, dx$

8. $\int_0^1 \int_0^{\sqrt{1+x^2}} \frac{dy\, dx}{1 + x^2 + y^2}$

9. $\int_0^{2\pi} \int_{a\sin\theta}^{a} r\, dr\, d\theta$

10. $\int_{b/2}^{b} \int_0^{\pi/2} r\, d\theta\, dr$

11. $\int_0^a \int_0^{\sqrt{ax}} x^2\, dy\, dx$

12. $\int_0^3 \int_1^{\sqrt{4-y}} (x+y)\, dx\, dy$

13. $\int_0^{\pi/2} \int_0^{a\cos\theta} r^4\, dr\, d\theta$

14. $\int_0^1 \int_x^{\sqrt{x}} (x^2 + y^2)\, dy\, dx$

15. Evaluate $\iint_R e^{2x+3y}\, dx\, dy$, where R is the region bounded by the lines $x = 0$, $y = 0$ and $x + y = 1$

16. Evaluate $\iint_R \sqrt{4x^2 - y^2}\, dx\, dy$, where R is the triangle given by $x = y$, $y = 0$, $x = 1$

17. Evaluate $\iint xy\, dx\, dy$, over the positive quadrant of the circle $x^2 + y^2 = 1$.

18. Evaluate $\iint_R xy\, dx\, dy$, where R is the rectangle bounded by $x = 2$, $x = 5$, $y = 1$ and $y = 2$

19. Evaluate $\iint (x^2 + y^2)\, dx\, dy$, over the region in the first quadrant for which $x + y \le 1$

20. Prove that $\int_1^2 \int_3^4 (xy + e^y)\, dy\, dx = \int_3^4 \int_1^2 (xy + e^y)\, dx\, dy$

ANSWERS

1. $\frac{7}{6}$
2. $\frac{67}{120}$
3. $\log\left(\frac{25}{24}\right)$
4. $(e-2)$
5. $\frac{8}{3}$
6. $\log a \log b$
7. $\frac{7}{12}$
8. $\frac{\pi}{4}\log(\sqrt{2}+1)$
9. $\frac{\pi a^2}{2}$
10. $\frac{3\pi b^2}{16}$
11. $\frac{2a^4}{7}$
12. $\frac{241}{60}$
13. $\frac{8a^5}{75}$
14. $\frac{3}{15}$
15. $\frac{1}{6}(e-1)^2(2e+1)$
16. $\frac{1}{3}\left(\frac{\pi}{3}+\frac{\sqrt{3}}{2}\right)$
17. $\frac{1}{8}$
18. $\frac{63}{4}$
19. $\frac{1}{6}$

4.23 TRIPLE INTEGRALS

The triple integral of a function $f(x, y, z)$ over a region V is denoted by

$$\iiint_V f(x, y, z)\, dx\, dy\, dz \text{ or } \iiint_V f(x, y, z)\, dV .$$

It can also be expressed as the repeated integral

$$\int_{z_1}^{z_2}\int_{y_1}^{y_2}\int_{x_1}^{x_2} f(x, y, z)\, dx\, dy\, dz ,$$

The order of integration depending upon the limits.

Let z_1, z_2 be function of x, y; y_1, y_2, are function of x and x_1, x_2 be constants, *i.e.*, let $z_1 = f_1(x, y)$, $z_2 = f_2(x, y)$, $y_1 = \phi_1(x)$, $y_2 = \phi_2(x)$ and $x_1 = a$, $x_2 = b$

Then, the above integral can be written as,

$$\int_{x_1=a}^{x_2=b}\left[\int_{y_1=\phi_1(x)}^{y_2=\phi_2(x)}\left[\int_{z_1=f_1(x,y)}^{z_2=f_2(x,y)} f(x, y, z)\, dz\right] dy\right] dx$$

Here, $f(x, y, z)$ is first integrated w.r.t. z (keeping y and z constant), between the limits z_1, and z_2. The resulting expression, which is a function of x and y is then integrated w.r.t. y (keeping x constant) between the limits y_1 and y_2. The resulting expression, which is a function of x is then integrated w.r.t. x between the limits x_1 and x_2.

Note: If x_1, x_2; y_1, y_2 and z_1, z_2 are all constants, then it does not matter what order of integration we take, *i.e.*,.

$$\int_{z_1}^{z_2}\int_{y_1}^{y_2}\int_{x_1}^{x_2} f(x, y, z)\, dx\, dy\, dz = \int_{x_1}^{x_2}\int_{z_1}^{z_2}\int_{y_1}^{y_2} f(x, y, z)\, dy\, dz\, dx$$

$$= \int_{x_1}^{x_2}\int_{y_1}^{y_2}\int_{z_1}^{z_2} f(x, y, z)\, dz\, dy\, dx$$

ILLUSTRATIVE EXAMPLES

Example 1. Evaluate the triple integral $\int_0^2\int_1^3\int_1^2 xy^2z\, dz\, dy\, dx$

Solution: $\int_{x=0}^{2}\int_{y=1}^{3}\int_{z=1}^{2} xy^2z\, dz\, dy\, dx = \int_{x=0}^{2}\int_{y=1}^{3} xy^2\left[\frac{3^2}{2}\right]_1^2 dy\, dx$

$$= \int_{x=0}^{2}\int_{y=1}^{3} xy^2\left(2-\frac{1}{2}\right) dy\, dx$$

$$= \frac{3}{2}\int_{x=0}^{2}\int_{y=1}^{3} xy^2\, dy\, dx = \frac{3}{2}\int_{x=0}^{2}\left[\frac{xy^3}{3}\right]_1^3 dx$$

$$= \frac{3}{2}\int_{x=0}^{2}\left(9x-\frac{x}{3}\right) dx = \frac{3}{2}\left[\frac{9}{2}x^2-\frac{x^2}{6}\right]_0^2$$

$$= \frac{3}{2}\left[18-\frac{2}{3}\right] = 26$$

Example 2. Evaluate $\int_0^1\int_{y^2}^1\int_0^{1-x} x\, dz\, dx\, dy$.

Solution: $\int_{y=0}^{1}\int_{x=y^2}^{1}\int_{z=0}^{1-x} x\, dz\, dx\, dy = \int_{y=0}^{1}\int_{x=y^2}^{1} x\left[\int_0^{1-x} dz\right] dx\, dy$

$$= \int_{y=0}^{1}\int_{x=y^2}^{1} x[z]_0^{1-x}\, dx\, dy$$

$$= \int_{y=0}^{1}\int_{x=y^2}^{1} x(1-x)\, dx\, dy$$

$$= \int_{y=0}^{1}\left[\int_{x=y^2}^{1} (x-x^2)\, dx\right] dy$$

$$= \int_{y=0}^{1}\left[\frac{x^2}{2}-\frac{x^3}{3}\right]_{y^2}^{1} dy$$

$$= \int_{y=0}^{1} \left[\left(\frac{1}{2} - \frac{1}{3} \right) - \left(\frac{y^4}{2} - \frac{y^6}{3} \right) \right] dy$$

$$= \int_{y=0}^{1} \left(\frac{1}{6} - \frac{y^4}{2} + \frac{y^6}{3} \right) dy = \left[\frac{y}{6} - \frac{y^5}{10} + \frac{y^7}{21} \right]_0^1$$

$$= \frac{1}{6} - \frac{1}{10} + \frac{1}{21} = \frac{4}{35}$$

Example 3. Evaluate $\int_0^1 \int_0^{\sqrt{1-x^2}} \int_0^{\sqrt{1-x^2-y^2}} \frac{dz\, dy\, dx}{\sqrt{1-x^2-y^2-z^2}}$

Solution: $\int_{x=0}^{1} \int_{y=0}^{\sqrt{1-x^2}} \int_{z=0}^{\sqrt{1-x^2-y^2}} \frac{dz\, dy\, dx}{\sqrt{(1-x^2-y^2)-z^2}}$

$$= \int_{x=0}^{1} \int_{y=0}^{\sqrt{1-x^2}} \left[\sin^{-1} \frac{z}{\sqrt{1-x^2-y^2}} \right]_0^{\sqrt{1-x^2-y^2}} dy\, dx$$

$$= \int_{x=0}^{1} \int_{y=0}^{\sqrt{1-x^2}} (\sin^{-1} 1 - \sin^{-1} 0)\, dy\, dx = \int_{x=0}^{1} \int_{y=0}^{\sqrt{1-x^2}} \frac{\pi}{2}\, dy\, dx$$

$$= \frac{\pi}{2} \int_{x=0}^{1} [y]_0^{\sqrt{1-x^2}}\, dx = \frac{\pi}{2} \int_{x=0}^{1} \sqrt{1-x^2}\, dx$$

$$= \frac{\pi}{2} \left[\frac{x\sqrt{1-x^2}}{2} + \frac{1}{2} \sin^{-1} x \right]_0^1$$

$$= \frac{\pi}{2} \left[0 + \frac{1}{2} \sin^{-1} 1 - 0 - 0 \right] = \frac{\pi}{2} \left[\frac{\pi}{4} \right]$$

$$= \frac{\pi^2}{8}$$

Example 4. Evaluate $\int_1^3 \int_{1/x}^{1} \int_0^{\sqrt{xy}} xyz\, dz\, dy\, dx$

Solution: $\int_{x=1}^{3} \int_{y=1/x}^{1} \int_{z=0}^{\sqrt{xy}} xyz\, dz\, dy\, dx = \int_{x=1}^{3} \int_{y=1/x}^{1} \left[xy \frac{z^2}{2} \right]_0^{\sqrt{xy}} dy\, dx$

$$= \int_{x=1}^{3} \int_{y=1/x}^{1} \frac{x^2 y^2}{2}\, dy\, dx$$

$$= \int_{x=1}^{3} \left[\frac{x^2 y^3}{6} \right]_{1/x}^{1} dx$$

$$= \int_{x=1}^{3} \left[\frac{x^2}{6} - \frac{1}{6x} \right]_{1}^{3}$$

$$= \left[\frac{x^3}{18} - \frac{1}{6} \log x \right]_{1}^{3}$$

$$= \left[\left(\frac{3}{2} - \frac{1}{6} \log 3 \right) - \left(\frac{1}{18} - \frac{1}{6} \log 1 \right) \right]$$

$$= \left[\frac{3}{2} - \frac{1}{6} \log 3 - \frac{1}{18} + 0 \right]$$

$$= \frac{13}{9} - \frac{1}{6} \log 3$$

Example 5. Evaluate $\int_0^a \int_0^x \int_0^{x+y} e^{x+y+z} \, dz \, dy \, dx$

Solution: $\int_{x=0}^{a} \int_{y=0}^{x} \int_{z=0}^{x+y} e^{x+y+z} \, dz \, dy \, dx = \int_{x=0}^{a} \int_{y=0}^{x} \left[\frac{e^{x+y+z}}{1} \right]_0^{x+y} dy \, dx$

$$= \int_{x=0}^{a} \int_{y=0}^{x} (e^{2x+2y} - e^{x+y}) \, dy \, dx$$

$$= \int_{x=0}^{a} \left[\frac{e^{2x+2y}}{2} - \frac{e^{x+y}}{1} \right]_0^{x} dx$$

$$= \int_{x=0}^{a} \left[\left(\frac{e^{4x}}{2} - e^{2x} \right) - \left(\frac{e^{2x}}{2} - e^{x} \right) \right] dx$$

$$= \int_{x=0}^{a} \left[\frac{e^{4x}}{2} - \frac{3e^{2x}}{2} + e^{x} \right] dx$$

$$= \left[\frac{e^{4x}}{8} - \frac{3}{2} \frac{e^{2x}}{2} + \frac{e^{x}}{1} \right]_0^{a}$$

$$= \left(\frac{e^{4a}}{8} - \frac{3}{4} e^{2a} + e^{a} \right) - \left(\frac{1}{8} - \frac{3}{4} + 1 \right)$$

$$= \frac{e^{4a}}{8} - \frac{3}{4} e^{2a} + e^{a} - \frac{3}{8}$$

Example 6. Evaluate $\int_0^1 \int_0^{1-x} \int_0^{1-x-y} \frac{dx\,dy\,dz}{(x+y+z+1)^3}$

Solution: $\int_{x=0}^{1} \int_{y=0}^{1-x} \int_{z=0}^{1-x-y} \frac{dz\,dy\,dx}{(x+y+z)^3}$

$$= \int_{x=0}^{1} \int_{y=0}^{1-x} \left[\frac{\frac{1}{(x+y+z+1)^2}}{-2} \right]_0^{1-x-y} dy\,dx$$

$$= \int_{x=0}^{1} \int_{y=0}^{1-x} \left[-\frac{1}{2} \cdot \frac{1}{(x+y+z+1)^2} \right]_0^{1-x-y} dy\,dx$$

$$= -\frac{1}{2} \int_{x=0}^{1} \int_{y=0}^{1-x} \left[\left(\frac{1}{2}\right)^2 - \frac{1}{(x+y+1)^2} \right] dy\,dx$$

$$= -\frac{1}{2} \int_{x=0}^{1} \int_{y=0}^{1-x} \left[\frac{1}{4} - \frac{1}{(x+y+1)^2} \right] dy\,dx$$

$$= -\frac{1}{2} \int_{x=0}^{1} \left[\frac{y}{4} + \frac{1}{(x+y+1)} \right]_0^{1-x} dx$$

$$= -\frac{1}{2} \int_{x=0}^{1} \left[\frac{1-x}{4} + \frac{1}{2} - \frac{1}{x+1} \right] dx$$

$$= -\frac{1}{2} \int_{x=0}^{1} \left[\frac{3}{4} - \frac{x}{4} - \frac{1}{x+1} \right] dx$$

$$= -\frac{1}{2} \left[\frac{3}{4}x - \frac{x^2}{8} - \log(x+1) \right]_0^1$$

$$= -\frac{1}{2} \left[\frac{3}{4} - \frac{1}{8} - \log 2 + \log 1 \right]$$

$$= -\frac{1}{2} \left[\frac{5}{8} - \log 2 \right] = \frac{1}{2} \left[\log 2 - \frac{5}{8} \right]$$

Example 7. Evaluate $\int_0^{\pi} \int_0^{\pi/2} \int_0^{1} r^2 \sin\theta\, dr\, d\theta\, d\phi$

Solution: $\int_{\phi=0}^{\pi} \int_{\theta=0}^{\pi/2} \int_{r=0}^{1} r^2 \sin\theta\, dr\, d\theta\, d\phi = \int_{\phi=0}^{\pi} \int_{\theta=0}^{\pi/2} \left[\frac{r^3}{3} \sin\theta \right]_0^1 d\theta\, d\phi$

$$= \int_{\phi=0}^{\pi}\int_{\theta=0}^{\pi/2}\left[\frac{1}{3}\sin\theta\right]d\theta\, d\phi$$

$$= \frac{1}{3}\int_{\phi=0}^{\pi}\int_{\theta=0}^{\pi/2}\sin\theta\, d\theta\, d\phi$$

$$= \frac{1}{3}\int_{\phi=0}^{\pi}[-\cos\theta]_0^{\pi/2}\, d\phi$$

$$= \frac{1}{3}\int_{\phi=0}^{\pi}\left[-\cos\frac{\pi}{2}+\cos 0\right]d\phi$$

$$= \frac{1}{3}\int_{\phi=0}^{\pi}[-0+1]\, d\phi = \frac{1}{3}[\phi]_0^{\pi} = \frac{\pi}{3}$$

Example 8. Evaluate $\displaystyle\int_0^{\pi/2}\int_0^{a\cos\theta}\int_0^{\sqrt{a^2-r^2}} r\, dz\, dr\, d\theta$

Solution: $\displaystyle\int_{\theta=0}^{\pi/2}\int_{r=0}^{a\cos\theta}\int_{z=0}^{\sqrt{a^2-r^2}} r\, dz\, dr\, d\theta = \int_{\theta=0}^{\pi/2}\int_{r=0}^{a\cos\theta} r[z]_0^{\sqrt{a^2-r^2}}\, dr\, d\theta$

$$= \int_{\theta=0}^{\pi/2}\int_{r=0}^{a\cos\theta} r\sqrt{a^2-r^2}\, dr\, d\theta$$

$$= \int_{\theta=0}^{\pi/2}\left[-\frac{1}{2}\frac{(a^2-r^2)^{3/2}}{\frac{3}{2}}\right]_0^{a\cos\theta} d\theta$$

$$= -\frac{1}{3}\int_{\theta=0}^{\pi/2}[(a^2-r^2)^{3/2}]_0^{a\cos\theta}\, d\theta$$

$$= -\frac{1}{3}\int_{\theta=0}^{\pi/2}[(a^2-a^2\cos^2\theta)^{3/2}-(a^2)^{3/2}]\, d\theta$$

$$= -\frac{1}{3}\int_{\theta=0}^{\pi/2}[(a^2\sin^2\theta)^{3/2}-a^3]\, d\theta$$

$$= -\frac{1}{3}\int_{\theta=0}^{\pi/2}[a^3\sin^3\theta-a^3]\, d\theta$$

$$= -\frac{a^3}{3}\int_{\theta=0}^{\pi/2}(\sin^3\theta-1)\, d\theta$$

$$= -\frac{a^3}{3}\int_{\theta=0}^{\pi/2}\left[\frac{3\sin\theta - \sin 3\theta}{4} - 1\right] d\theta$$

$(\because \sin 3x = 3 \sin x - 4 \sin^3 x)$

$$= -\frac{a^3}{12}\int_{\theta=0}^{\pi/2}(3\sin\theta - \sin 3\theta - 4)\, d\theta$$

$$= -\frac{a^3}{12}\left[-3\cos\theta + \frac{\cos 3\theta}{3} - 4\theta\right]_0^{\pi/2}$$

$$= -\frac{a^3}{12}\left[\left(-3\cos\frac{\pi}{2} + \frac{1}{3}\cos\frac{3\pi}{2} - 4\cdot\frac{\pi}{2}\right) - \left(-3\cos 0 + \frac{1}{3}\cos 0 - 0\right)\right]$$

$$= -\frac{a^3}{12}\left[(-0 + 0 - 2\pi) - \left(-3 + \frac{1}{3}\right)\right]$$

$$= -\frac{a^3}{12}\left(-2\pi + \frac{8}{3}\right) = \frac{a^3}{12}\left(2\pi - \frac{8}{3}\right)$$

$$= \frac{a^3}{3}\left(\frac{\pi}{2} - \frac{2}{3}\right)$$

Example 9. Evaluate $\int_0^1\int_0^1\int_{\sqrt{x^2+y^2}}^2 xyz\, dz\, dy\, dx$

Solution: $\int_{x=0}^1\int_{y=0}^1\int_{z=\sqrt{x^2+y^2}}^2 xyz\, dz\, dy\, dx = \int_{x=0}^1\int_{y=0}^1 xy\left[\frac{z^2}{2}\right]_{\sqrt{x^2+y^2}}^2 dy\, dx$

$$= \int_{x=0}^1\int_{y=0}^1 \frac{xy}{2}\left[4 - \left(\sqrt{x^2+y^2}\right)^2\right] dy\, dx$$

$$= \frac{1}{2}\int_{x=0}^1\int_{y=0}^1 [4xy - x^3y - xy^3]\, dy\, dx$$

$$= \frac{1}{2}\int_{x=0}^1\left[4x\frac{y^2}{2} - \frac{x^3y^2}{2} - \frac{xy^4}{4}\right]_0^1 dx$$

$$= \frac{1}{2}\int_{x=0}^1\left[2x - \frac{x^3}{2} - \frac{x}{4}\right] dx = \frac{1}{2}\int_{x=0}^1\left(\frac{7x}{4} - \frac{x^3}{2}\right) dx$$

$$= \frac{1}{2}\left[\frac{7x^2}{8} - \frac{x^4}{8}\right]_0^1 = \frac{1}{2}\left[\frac{7}{8} - \frac{1}{8}\right] = \frac{3}{8}$$

Example 10. Evaluate $\int_0^1 \int_0^1 \int_0^{x+y} e^z \, dx \, dy \, dz$

Solution: $\int_{x=0}^{1} \int_{y=0}^{1} \int_{z=0}^{x+y} e^z \, dz \, dy \, dx = \int_{x=0}^{1} \int_{y=0}^{1} [e^z]_0^{x+y} \, dy \, dx$

$$= \int_{x=0}^{1} \int_{y=0}^{1} [e^{x+y} - 1] \, dy \, dx$$

$$= \int_{x=0}^{1} [e^{x+y} - y]_0^1 \, dx$$

$$= \int_{x=0}^{1} [e^{x+1} - 1 - e^x] \, dx$$

$$= [e^{x+1} - x - e^x]_0^1$$

$$= (e^2 - 1 - e) - (e - 0 - 1)$$

$$= (e^2 - 2e) = e(e-2)$$

Example 11. Evaluate $\int_0^a \int_0^x \int_0^y xyz \, dz \, dy \, dx$

Solution: $\int_{x=0}^{a} \int_{y=0}^{x} \int_{z=0}^{y} xyz \, dz \, dy \, dx = \int_{x=0}^{a} \int_{y=0}^{x} xy \left[\frac{z^2}{2} \right]_0^y dy \, dx$

$$= \int_{x=0}^{a} \int_{y=0}^{x} xy \left(\frac{y^2}{2} \right) dy \, dx$$

$$= \frac{1}{2} \int_{x=0}^{a} \int_{y=0}^{x} xy^3 \, dy \, dx$$

$$= \frac{1}{2} \int_{x=0}^{a} \left[\frac{xy^4}{4} \right]_0^x dx = \frac{1}{2} \int_{x=0}^{a} \frac{1}{4}(x^5 - 0) \, dx$$

$$= \frac{1}{8} \int_{x=0}^{a} x^5 \, dx = \frac{1}{8} \left[\frac{x^6}{6} \right]_0^a = \frac{a^6}{48}$$

Example 12. Evaluate $\iiint_V xyz \, dx \, dy \, dz$, where V is the region bounded by the positive octant of the sphere $x^2 + y^2 + z^2 = a^2$

Solution: The projection of the given region in the $x - y$ plane $(z = 0)$ is the region bounded by the circle $x^2 + y^2 = a^2$ and lying in the first quadrant as shown in the figure.

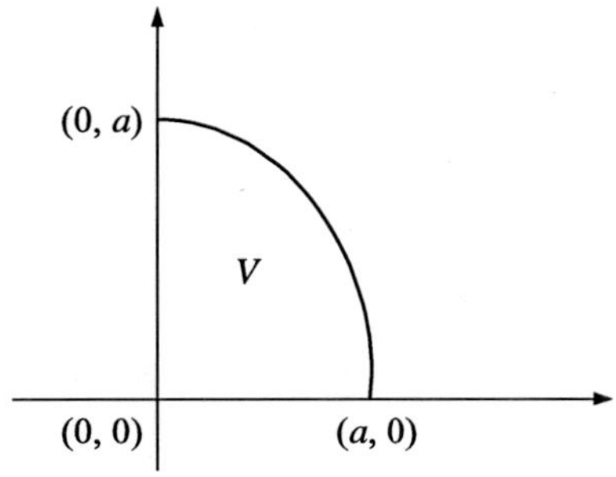

Fig.

In the given region x veries from $0 \to a$. For constant x, y varies from $0 \to \sqrt{a^2 - x^2}$. For a fixed x and y, z varies from $0 \to \sqrt{a^2 - x^2 - y^2}$.

$$\therefore \iiint_V xyz\, dx\, dy\, dz = \int_{x=0}^{a} \int_{y=0}^{\sqrt{a^2-x^2}} \int_{z=0}^{\sqrt{a^2-x^2-y^2}} xyz\, dz\, dy\, dx$$

$$= \int_{x=0}^{a} \int_{y=0}^{\sqrt{a^2-x^2}} xy\left[\frac{z^2}{2}\right]_0^{\sqrt{a^2-x^2-y^2}} dy\, dx$$

$$= \frac{1}{2}\int_{x=0}^{a} \int_{y=0}^{\sqrt{a^2-x^2}} xy(a^2 - x^2 - y^2)\, dy\, dx$$

$$= \frac{1}{2}\int_{x=0}^{a} \int_{y=0}^{\sqrt{a^2-x^2}} (a^2xy - x^3y - xy^3)\, dy\, dx$$

$$= \frac{1}{2}\int_{x=0}^{a} \left[a^2x\frac{y^2}{2} - \frac{x^3y^2}{2} - \frac{xy^4}{4}\right]_0^{\sqrt{a^2-x^2}} dx$$

$$= \frac{1}{8}\int_{x=0}^{a} [2a^2x(a^2 - x^2) - 2x^3(a^2 - x^2) - x(a^2 - x^2)^2]\, dx$$

$$= \frac{1}{8}\int_{x=0}^{a} [2a^4x - 2a^2x^3 - 2a^2x^3 + 2x^5 - a^4x - x^5 + 2a^2x^3]\, dx$$

$$= \frac{1}{8}\int_{x=0}^{a} (a^4x + x^5 - 2a^2x^3)\, dx$$

$$= \frac{1}{8}\left[a^4\frac{x^2}{2} + \frac{x^6}{6} - \frac{2a^2x^4}{4}\right]_0^a$$

$$= \frac{1}{8}\left[\frac{a^6}{2} + \frac{a^6}{6} - \frac{2a^6}{4}\right]$$

$$= \frac{a^6}{48}$$

Example 13. Evaluate $\iiint_V (x^2 + y^2 + z^2)\, dz\, dy\, dx$ where V is the region bounded by the coordinate planes and the planes $x = y = z = a$.

Solution: Here the column parallel to z-axis is bounded by the planes $z = 0$ and $z = a$ and the region S, on which the above region V stands is the region on $x - y$ plane bounded by the lines $x = 0, x = a, y = 0, y = a$.

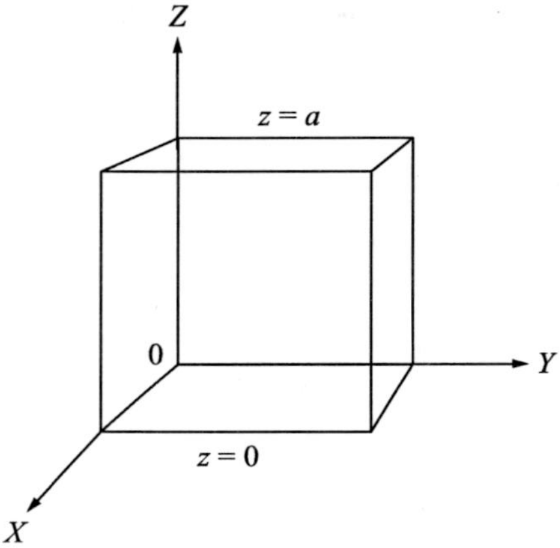

Fig.

$$\therefore \iiint_V (x^2 + y^2 + z^2)\, dz\, dy\, dx = \int_{x=0}^{a}\int_{y=0}^{a}\int_{z=0}^{a} (x^2 + y^2 + z^2)\, dz\, dy\, dx$$

$$= \int_{x=0}^{a}\int_{y=0}^{a}\left[x^2 z + y^2 z + \frac{z^3}{3}\right]_0^a dy\, dx$$

$$= \int_{x=0}^{a}\int_{y=0}^{a}\left[x^2 a + y^2 a + \frac{a^3}{3}\right] dy\, dx$$

$$= \int_{x=0}^{a}\left[x^2 ay + \frac{y^3}{3}a + \frac{a^3}{3}y\right]_0^a dx$$

$$= \int_{x=0}^{a}\left[x^2 a^2 + \frac{a^4}{3} + \frac{a^4}{3}\right] dx$$

$$= \int_{x=0}^{a}\left[x^2 a^2 + \frac{2a^4}{3}\right] dx = \left[\frac{x^3 a^2}{3} + \frac{2a^4}{3}x\right]_0^a$$

$$= \left[\frac{a^5}{3} + \frac{2a^5}{3}\right] = a^5$$

EXERCISE 4.14

Evaluate the following integrals:

1. $\int_0^a \int_0^b \int_0^c (x + y + z)\, dx\, dy\, dz$ **2.** $\int_0^1 \int_0^2 \int_1^2 x^2 y\, dz\, dy\, dx$

3. $\int_0^{\log 2} \int_0^x \int_0^{x+\log y} e^{x+y+z}\, dz\, dy\, dx$ **4.** $\int_{x=0}^{a} \int_{y=0}^{a-x} \int_{z=0}^{a-x-y} x^2\, dz\, dy\, dx$

5. $\int_{z=-1}^{1} \int_{x=0}^{z} \int_{y=x-z}^{x+z} (x + y + z)\, dy\, dx\, dz$ **6.** $\int_0^{\pi/2} d\theta \int_0^{a\sin\theta} dr \int_0^{(a^2-r^2)/a} r\, dz$

7. $\int_{x=0}^{2} \int_{y=0}^{x} \int_{z=0}^{x+y} e^x (y + 2z)\, dx\, dy\, dz$ **8.** $\int_0^{12} \int_0^{6-\frac{x}{2}} \int_0^{4-\frac{2y}{3}-\frac{x}{3}} x\, dz\, dy\, dx$

9. $\int_0^{\pi} \int_0^{\pi/2} \int_0^{k} r^2 \sin\theta\, dr\, d\theta\, d\phi$ **10.** $\int_0^4 \int_0^x \int_0^{\sqrt{x+y}} z\, dx\, dy\, dz$

11. Evaluate $\iiint_V (x^2 + y^2 + z^2)\, dx\, dy\, dz$ where V is the region bounded by the planes $x + y + z = a$, $x = 0$, $y = 0$ and $z = 0$

12. Evaluate $\iiint xyz\, dz\, dy\, dx$ over the region enclosed by three co-ordinate planes and the plane $x + y + z = 1$

13. Evaluate $\iiint dz\, dy\, dx$ over the hemisphere $x^2 + y^2 + z^2 = a^2$ above $x - y$ plane.

14. Evaluate $\iiint (x + y + z)\, dx\, dy\, dz$ over the tetrahedron $x = 0$, $y = 0$, $z = 0$, $x + y + z = 1$

ANSWERS

1. $\frac{abc(a+b+c)}{2}$ **2.** 1 **3.** $\frac{8}{3}\log 2 - \frac{19}{9}$

4. $\frac{a^5}{60}$ **5.** 0 **6.** $\frac{5a^3\pi}{64}$

7. $19\left(\frac{e^2}{3} + 1\right)$ **8.** 1444 **9.** $\frac{k^3\pi}{3}$

10. 16 **11.** $\frac{a^5}{20}$ **12.** $\frac{7}{21600}$

13. $\frac{2}{3}\pi a^3$ **14.** $\frac{1}{8}$

4.24 AREA

Definite integrals have a wide range of application we shall use definite integrals in computing the area of bounded regions and the volume of revolution.

4.25 AREA BY SINGLE INTEGRAL

The area of the region bounded by the curve $y = f(x)$. The x-axis and the ordinates $x = a$, $x = b$ is given by

$$\text{Area} = \int_a^b f(x)\,dx \text{ or } \int_a^b y\,dx$$

Similarly, the area of the region bounded by the curve $x = g(y)$, the y-axis and the abscissae $y = c$, $y = d$, is given by

$$\text{Area} = \int_c^d g(y)\,dy \text{ or } \int_c^d x\,dy$$

4.26 AREA BETWEEN CURVES

The area of the region between the curves $y = f(x)$, $y = g(x)$ and the ordinates $x = a$, $x = b$ is

$$\text{Area} = \int_a^b f(x)\,dx - \int_a^b g(x)\,dx = \int_a^b [f(x) - g(x)]\,dx$$

Similarly, the area of the region between the curves $x = f(y)$, $x = g(y)$ and the abscissae $y = c$, $y = d$ is

$$Area = \int_c^d (f(y) - g(y))\,dy$$

Remark: The above formula also holds when one or both of the curves $y = f(x)$ and $y = g(x)$ lie partially or completely below the x-axis.

4.27 AREA BY DOUBLE INTEGRAL

The area of the region bounded by the curve $y_1 = f_1(x)$, $y_2 = f_2(x)$ and the lines $x = a$, $x = b$ is given by

$$\text{Area} = \int_a^b \int_{y_1}^{y_2} dy\,dx$$

Similarly, the area of the region bounded by the curves $x_1 = f_1(y)$, $x_2 = f_2(y)$ and the lines $y = c$, $y = d$ is given by

$$\text{Area} = \int_c^d \int_{x_1}^{x_2} dx\,dy$$

ILLUSTRATIVE EXAMPLES

Example 1. Find the area enclosed between the curve $y = 4x - x^2$ and the x-axis.

Solution: The given curve is,

$$y = 4x - x^2$$

$$\Rightarrow \quad x^2 - 4x = -y$$

$$\Rightarrow \quad (x-2)^2 = -(y-4)$$

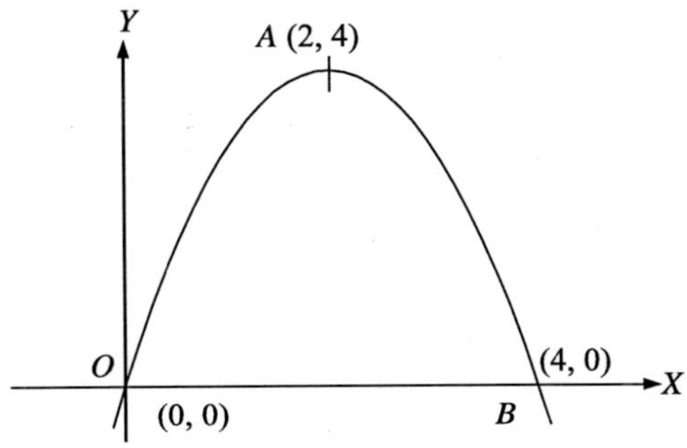

Fig.

which is a downward parabola with vertex at $(2, 4)$

The parabola meets the x-axis *i.e.*, $y = 0$ at

$4x - x^2 = 0$ *i.e.*, at $x = 0, x = 4$.

$$\therefore \text{Area} = \int_0^4 y\,dx = \int_0^4 (4x - x)^2\,dx = \left[2x^2 - \frac{x^3}{3}\right]_0^4$$

$$= \left(32 - \frac{64}{3}\right) = \frac{33}{3} \text{ sq. units.}$$

Example 2. Find the area enclosed by the curve $y = 2\sqrt{1 - x^2}$, and the x-axis

Solution: The given curve is

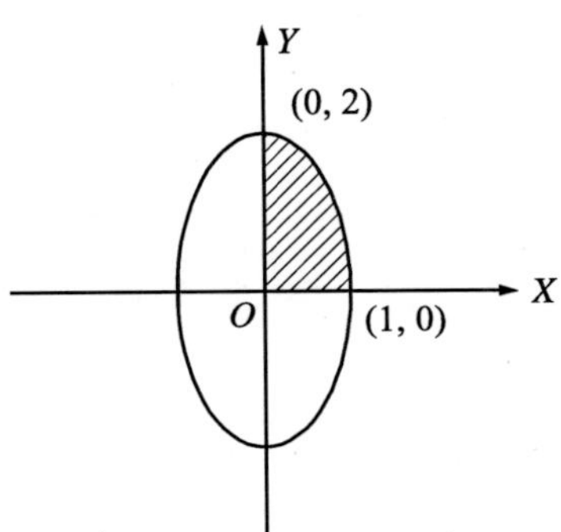

Fig.

$$y = 2\sqrt{1-x^2} \Rightarrow y^2 = 4(1-x^2)$$

$$\Rightarrow \frac{y^2}{4} = 1 - x^2$$

$\Rightarrow \dfrac{x^2}{1} + \dfrac{y^2}{4} = 1$, which is an ellipse. Hence, the curve $y = 2\sqrt{1-x^2}$ represents the portion of the ellipse lying in the first quadrant.

$\therefore$ The required area = The area of the shaded region

$$= \int_0^1 y\,dx = \int_0^1 2\sqrt{1-x^2}\,dx$$

$$= 2\left[\frac{x}{2}\sqrt{1-x^2} + \frac{1}{2}\sin^{-1}x\right]_0^1$$

$$= 2\left[\left(0 + \frac{1}{2}\sin^{-1}1\right) - \left(0 + \frac{1}{2}\sin^{-1}0\right)\right]$$

$$= 2\cdot\frac{1}{2}\cdot\frac{\pi}{2} = \frac{\pi}{2} \text{ sq. units}$$

Example 3. Find the area of the region bounded by the line $x = 2$ and the parabola $y^2 = 8x$.

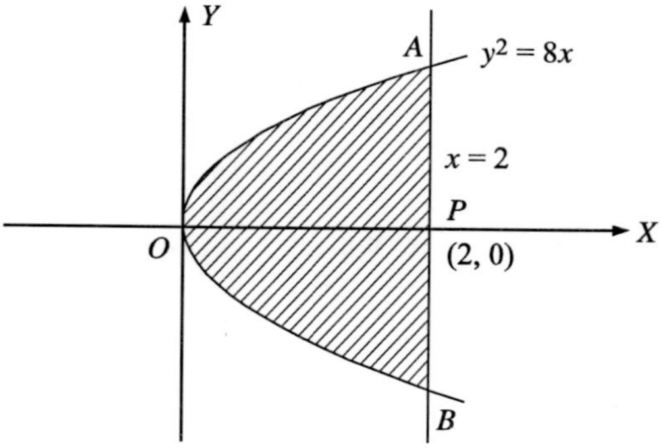

Fig.

Solution: Required area $AOBPA = 2(\text{Area of } AOPA)$

$$= 2\int_0^2 y\,dx = 2\int_0^2 \sqrt{8x}\,dx$$

$$= 2\cdot 2\sqrt{2}\int_0^2 x^{1/2}\,dx = 4\sqrt{2}\left[\frac{x^{3/2}}{\frac{3}{2}}\right]_0^2$$

$$= 4\sqrt{2}\cdot\frac{2}{3}[2^{3/2} - 0] = \frac{8\sqrt{2}}{3}\cdot 2\sqrt{2}$$

$$= \frac{32}{3} \text{ sq. units.}$$

Example 4. Find the area bounded by the curve $x = 8 + 2y - y^2$, the y-axis and the lines $y = -1, y = 3$

Solution: The given curve is $x = 8 + 2y - y^2$

It can be written as

$y^2 - 2y = -x + 8$

$\Rightarrow (y-1)^2 = -(x-9)$ which represents a parabola with vertex at (9, 1)

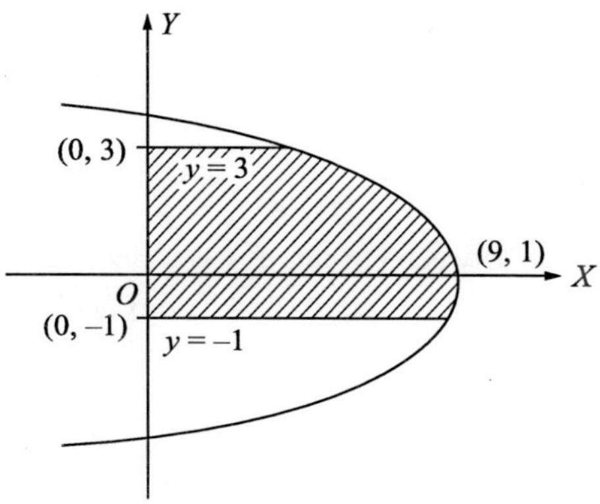

Fig.

$$\therefore \text{Required area} = \int_{-1}^{3} x\,dy = \int_{-1}^{3} (8 + 2y - y^2)\,dy$$

$$= \left[8y + y^2 - \frac{y^3}{3}\right]_{-1}^{3}$$

$$= (24 + 9 - 9) - \left(-8 + 1 + \frac{1}{3}\right)$$

$$= 24 + \frac{20}{3} = \frac{92}{3} \text{ sq. units.}$$

Example 5. Find the area of the region bounded by the line $x + 2y = 8$, x – axis and the line $x = 2$ and $x = 4$

Solution: Required area ***ABCD***

$$= \int_{2}^{4} y\,dx = \int_{2}^{4} \left(\frac{8-x}{2}\right) dx$$

$$= \frac{1}{2}\int_{2}^{4} (8-x)\,dx = \frac{1}{2}\left[8x - \frac{x^2}{2}\right]_{2}^{4}$$

$$= \frac{1}{2}[(32-8)-(16-2)] = \frac{1}{2}[24-14] = 5 \text{ sq. units.}$$

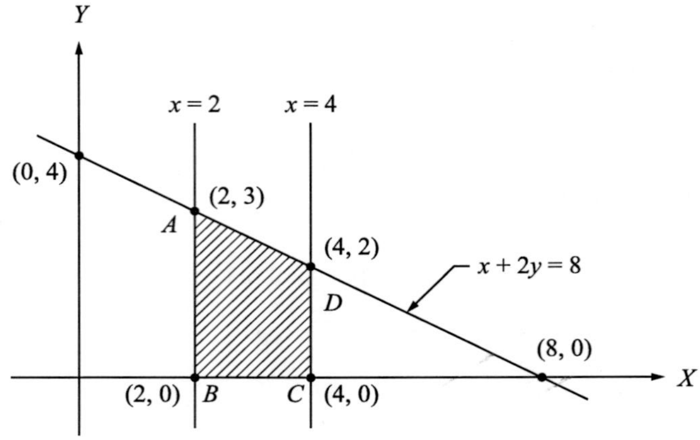

Fig.

Example 6. Find the area between the curves $y = x^2$ and $y = 2x - x^2$

Solution: The equation $x^2 = y$ is a parabola with vertex at (0, 0)

Aslo equation

$y = 2x - x^2$

$\Rightarrow x^2 - 2x = -y$

$\Rightarrow (x-1)^2 = -(y-1)$is a parabola with vertex (1, 1)

solving $y = x^2$ & $y = 2x - x^2$, we get

$x^2 = 2x - x^2$

$\Rightarrow 2x^2 - 2x = 0 \Rightarrow x = 0, 1$

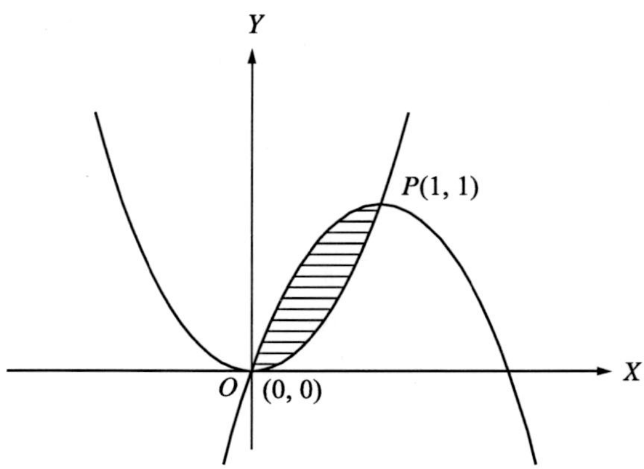

Fig.

$\therefore$ The point of intersection are $O(0, 0)$ and $P(1, 1)$

$\therefore$ The required area $=$ The area of the shaded part

$$= \int_0^1 [(2x - x^2) - x^2]\, dx$$

$$\left[\because \text{Area} = \int_a^b [f(x) - g(x)]\, dx\right]$$

$$= \int_0^1 (2x - 2x^2)\, dx = \left[x^2 - \frac{2x^3}{3}\right]_0^1$$

$$= \left(1 - \frac{2}{3}\right) - (0 - 0) = \frac{1}{3} \text{ sq. units.}$$

Example 7. Find the area of the region bounded by the curve $4y = 3x^2$ and the line $3x - 2y + 12 = 0$

Solution: The given curve is

$4y = 3x^2$

$\Rightarrow x^2 = \dfrac{4}{3}y$, which represents a parabola with vertex at (0, 0).

The given line is

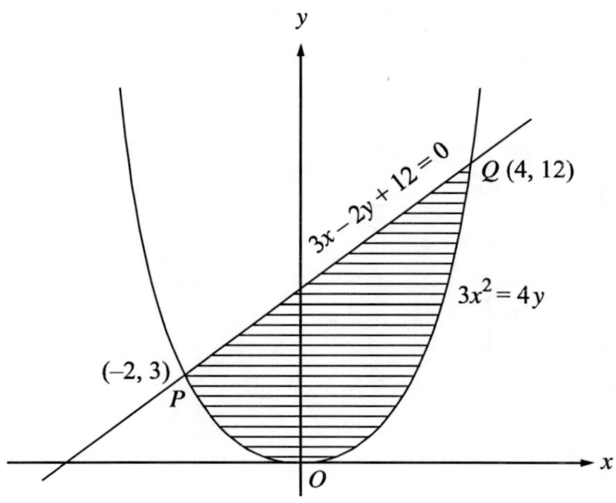

Fig.

$$3x - 2y + 12 = 0$$

$$\Rightarrow \quad y = \frac{3x + 12}{2}$$

$$\Rightarrow \quad \frac{3x^2}{4} = \frac{3x + 12}{2}$$

$$\Rightarrow \quad 3x^2 = 6x + 24$$

$$\Rightarrow \quad 3x^2 - 6x - 24 = 0$$

$$\Rightarrow \quad (x + 2)(x - 4) = 0$$

$$\Rightarrow \quad x = -2, 4$$

$\therefore$ The points of intersection are $P(-2, 3)$ and $Q(4, 12)$

$\therefore$ Required area = area of the shaded part.

$$= \int_{-2}^{4}\left[\frac{3x+12}{2} - \frac{3}{4}x^2\right] dx$$

$$= \left[\frac{3}{2}\cdot\frac{x^2}{2} + 6x - \frac{3}{4}\cdot\frac{x^3}{3}\right]_{-2}^{4}$$

$$= \left[\frac{3}{4}\cdot x^2 + 6x - \frac{1}{4}\cdot x^3\right]_{-2}^{4}$$

$$= (12 + 24 - 16) - (3 - 12 + 2)$$

$$= 20 + 7 = 27 \text{ sq. units.}$$

Example 8. Find the area of the region enclosed by the parabola $y^2 = 4ax$ and the chord $y = mx$.

Solution: The given curve,

$y^2 = 4ax$ is a parabola with vertex at $(0, 0)$

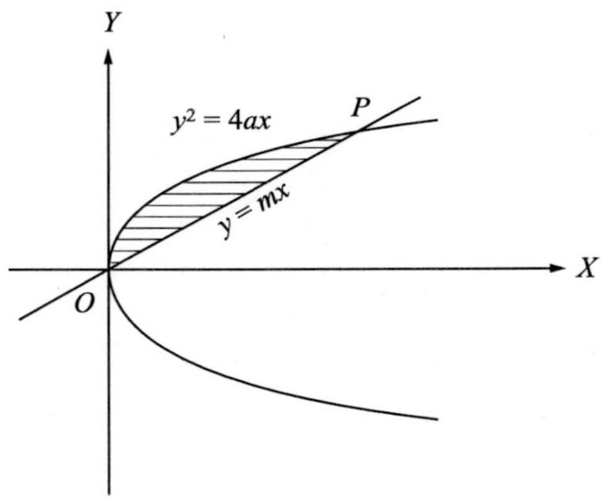

Fig.

Solving $y = mx$ & $y^2 = 4ax$, we get

$$\frac{y}{m} = \frac{y^2}{4a}$$

$\Rightarrow \quad 4ay = my^2$

$\Rightarrow \quad y(4a - my) = 0$

$\Rightarrow \quad y = 0, \frac{4a}{m}$ and hence $x = 0, \frac{4a}{m^2}$

$\therefore$ The point of intersection are $O(0, 0)$ and $P\left(\frac{4a}{m^2}, \frac{4a}{m}\right)$

$\therefore$ The required area = The area of shaded part

$$= \int_0^{\frac{4a}{m}} \left[\frac{y^2}{4a} - \frac{y}{m} \right] dx$$

$$= \left[\frac{y^3}{12a} - \frac{y^2}{2m} \right]_0^{4a/m}$$

$$= \left[\frac{64a^3}{12am^3} - \frac{16a^2}{2m^3} \right]$$

$$= \frac{16a^2}{3m^3} - \frac{8a^2}{m^3} = -\frac{8a^2}{3m^3} \text{ sq. units}$$

But area should be always positive

$$\therefore \text{ Required area} = \frac{8a^2}{3m^3} \text{ sq. units}$$

Example 9. Find the double integration, the area between the parabola $y = 4x - x^2$ and the line $y = x$.

Solution: The given curve

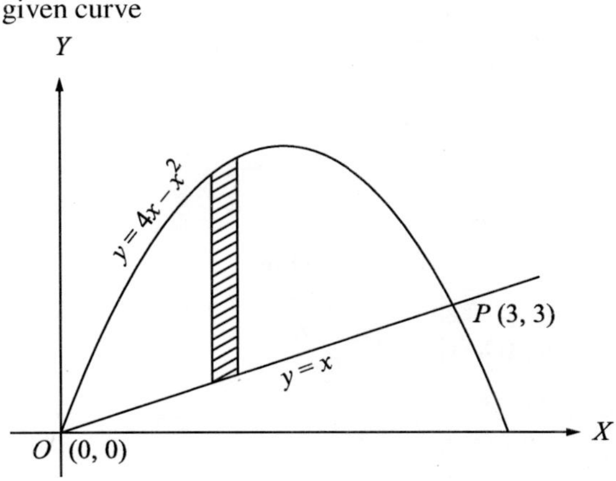

Fig.

$$y = 4x - x^2$$

$$\Rightarrow \quad x^2 - 4x = -y$$

$\Rightarrow \quad (x-2)^2 = -(y-4)$ represents a parabola with vertex at $(2, 4)$.

Solving $y = x$ and $y = 4x - x^2$, we get

$x = 4x - x^2 \Rightarrow x^2 - 3x = 0$

$\Rightarrow x(x-3) = 0 \Rightarrow x = 0, 3$

Hence, $y = 0, 3$

$\therefore$ The point of intersection of the parabola and the line are $O(0, 0)$ and $P(3, 3)$

$$\therefore \text{ Required area } = \int_0^3 \int_x^{4x-x^2} dy\, dx$$

$$= \int_0^3 [y]_x^{4x-x^2}\, dx$$

$$= \int_0^3 [4x - x^2 - x]\, dx = \int_0^3 (3x - x^2)\, dx$$

$$= \left[\frac{3x^2}{2} - \frac{x^3}{3}\right]_0^3 = \frac{27}{2} - 9 = \frac{9}{2} \text{ sq. units}$$

Example 10. Find by double integration, the area bounded by the parabola $y^2 = 4ax$ and its latus rectum

Solution: Required area $AO\,QPA$

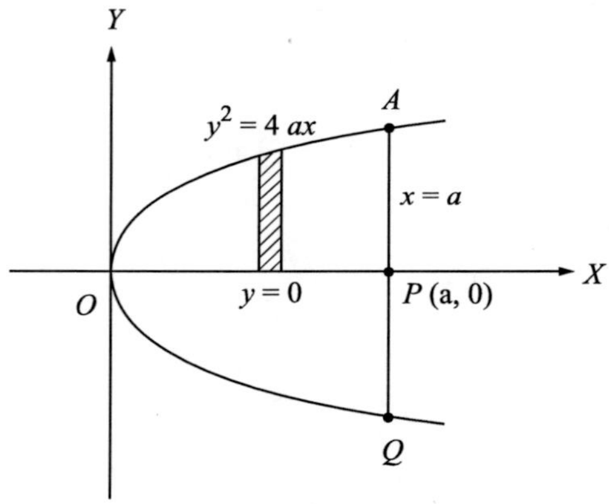

Fig.

$= 2(\text{area of } AOPA)$

$$= 2\int_0^a \int_0^{2\sqrt{ax}} dy\, dx$$

$$= 2\int_0^a [y]_0^{2\sqrt{ax}}\, dx$$

$$= 2\int_0^a 2\sqrt{ax}\, dx = 4\sqrt{a}\int_0^a x^{1/2}\, dx$$

$$= 4\sqrt{a}\left[\frac{x^{3/2}}{\frac{3}{2}}\right]_0^a = \frac{8}{3}\sqrt{a}\,[a^{3/2} - 0] = \frac{8a^2}{3} \text{ sq. units}$$

Example 11. Find the area of the region bounded by the curves $xy = 2$, $4y = x^2$, $y = 4$; using double integration.

Solution: Required area

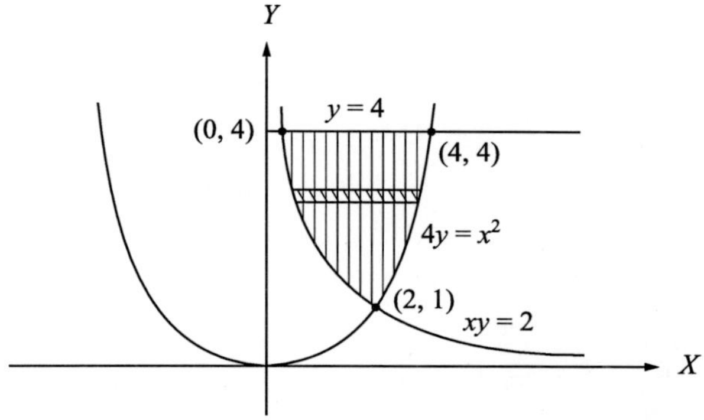

Fig.

$$= \int_1^4 \int_{2/y}^{2\sqrt{y}} dx\, dy = \int_1^4 [x]_{2/y}^{2\sqrt{y}}\, dy$$

$$= \int_1^4 \left(2\sqrt{y} - \frac{2}{y}\right) dy = 2\int_1^4 \left(y^{1/2} - \frac{1}{y}\right) dy = 2\left[\frac{y^{3/2}}{3/2} - \log y\right]_1^4$$

$$= 2\left[\left(\frac{2}{3}\cdot 4^{3/2} - \log 4\right) - \left(\frac{2}{3} - 0\right)\right]$$

$$= 2\left[\frac{16}{3} - \log 4 - \frac{2}{3}\right] = 2\left(\frac{14}{3} - \log 4\right) = \frac{28}{3} - 2\log 4$$

$$= \left(\frac{28}{3} - 4\log 2\right) \text{ sq. units}$$

Example 12. Find by double integration, the area enclosed by the ellipse $\frac{x^2}{a^2} + \frac{y^2}{b^2} = 1$

Solution: Required area of ellipse

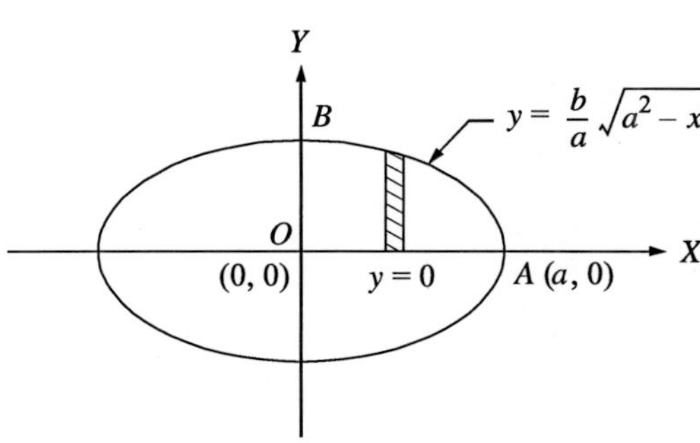

Fig.

$= 4(\text{area of } \boldsymbol{OABO})$

$$= 4\int_0^a \int_0^{\frac{b}{a}\sqrt{a^2-x^2}} dx\, dy = 4\int_0^a [x]_0^{\frac{b}{a}\sqrt{a^2-x^2}} dy = 4\int_0^a \frac{b}{a}\sqrt{a^2-x^2}\, dx$$

$$= \frac{4b}{a}\int_0^a \sqrt{a^2-x^2}\, dx$$

$$= \frac{4b}{a}\left[\frac{x\sqrt{a^2-x^2}}{2} + \frac{a^2}{2}\sin^{-1}\frac{x}{a}\right]_0^a$$

$$= \frac{4b}{a}\left[\left(0 + \frac{a^2}{2}\sin^{-1}1\right) - \left(0 + \frac{a^2}{2}\sin^{-1}0\right)\right]$$

$$= \frac{4b}{a}\left[\frac{\pi}{2}\cdot\frac{a^2}{2}\right] = \pi ab \text{ sq. units}$$

Example 13. Find by double integration, the area of the region between the parabolas $y^2 = 4ax$ and $x^2 = 4ay$

Solution: Solving $y^2 = 4ax$ and $x^2 = 4ay$ we get intersection points $O(0, 0)$ and $A(4a, 4a)$

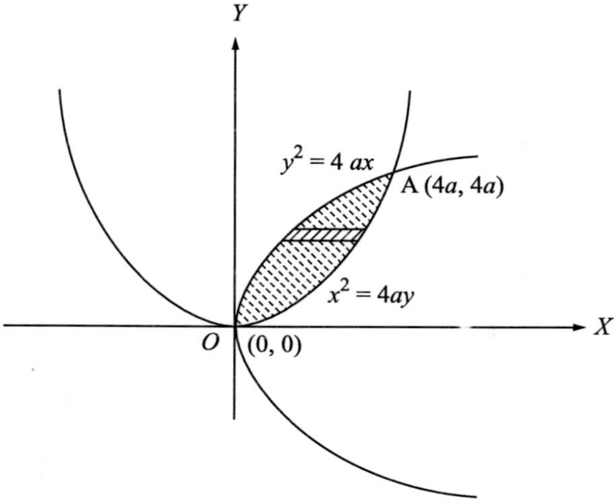

Fig.

$\therefore$ Required area

$$= \int_{y=0}^{4a} \int_{x=y^2/4a}^{2\sqrt{ay}} dx\, dy$$

$$= \int_{y=0}^{4a} [x]_{y^2/4a}^{2\sqrt{ay}}\, dy = \int_{y=0}^{4a} \left[2\sqrt{ay} - \frac{y^2}{4a}\right] dy$$

$$= \left[2\sqrt{a}\, y^{3/2} \cdot \frac{2}{3} - \frac{y^3}{12a}\right]_0^{4a}$$

$$= \left[\frac{4}{3}\sqrt{a}(4a)^{3/2} - \frac{64a^3}{12a}\right] = \left[\frac{32}{3}a^2 - \frac{16}{3}a^2\right]$$

$$= \frac{16}{3}a^2 \text{ sq. units}$$

Example 14. Find by double integration, the smaller of the area bounded by the circle $x^2 + y^2 = 9$ and the line $x + y = 3$

Solution: Solving $x^2 + y^2 = 9$ and $x = 3 - y$, we get

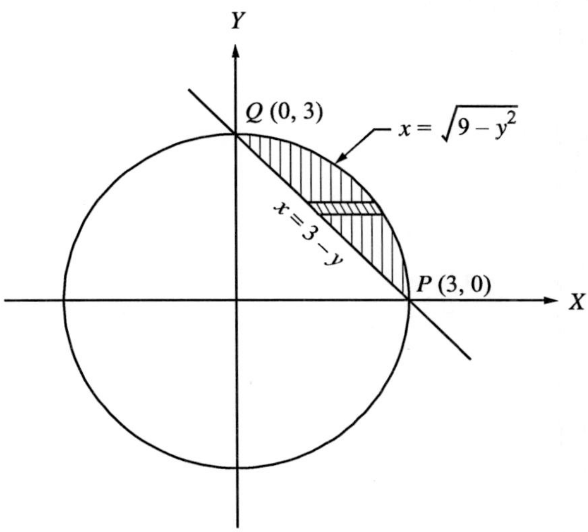

Fig.

$$(3 - y)^2 + y^2 = 9$$

$\Rightarrow \quad 9 + 2y^2 - 6y = 9$

$\Rightarrow \quad y^2 - 3y = 0$

$\Rightarrow \quad y = 0, 3$

Hence $x = 3, 0$

$\therefore$ The point of intersection are $P(0, 3)$ and $Q(3, 0)$.

$$\therefore \text{ The required area } = \int_0^3 \int_{3-y}^{\sqrt{9-y^2}} dx\, dy$$

$$= \int_0^3 [x]_{3-y}^{\sqrt{9-y^2}} \, dy$$

$$= \int_0^3 \left[\sqrt{9-y^2} - (3-y) \right] dy$$

$$= \left[\frac{y}{2}\sqrt{9-y^2} + \frac{9}{2}\sin^{-1}\frac{y}{3} - 3y + \frac{y^2}{2} \right]_0^3$$

$$= \left(0 + \frac{9}{2}\sin^{-1} 1 - 9 + \frac{9}{2} \right) - \left(0 + \frac{9}{2}\sin^{-1} 0 - 0 + 0 \right)$$

$$= \frac{9}{2}\cdot\frac{\pi}{2} - \frac{9}{2} = \frac{9}{2}\left(\frac{\pi}{2} - 1\right) = \frac{9}{4}(\pi - 2) \text{ sq. units.}$$

Example 15. Find the area of the triangle bounded by the lines $x = 0$, $y = 0$ and $x + y = 1$, using double integration.

Solution: The vertices of the triangle are $A(1, 0)$, $O(0, 0)$ and $B(0, 1)$

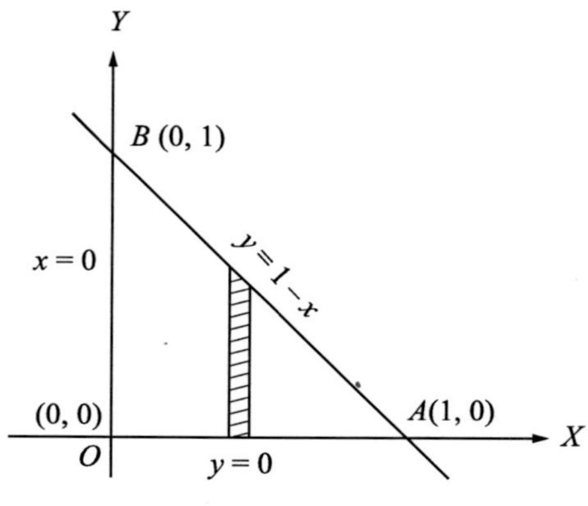

Fig.

$\therefore$ The required area of the triangle

$$= \int_{x=0}^{1} \int_{y=0}^{1-x} dy \, dx$$

$$= \int_{x=0}^{1} [y]_0^{1-x} \, dx = \int_{x=0}^{1} (1-x) \, dx$$

$$= \left[x - \frac{x^2}{2} \right]_0^1 = \left(1 - \frac{1}{2}\right) - (0 - 0)$$

$$= \frac{1}{2} = 0.5 \text{ sq. units.}$$

Example 16. Calculate the area bounded by the curve $y = x(2 - x)$ and the lines $x = 0, y = 0, x = 2$.

Solution: The given curve is

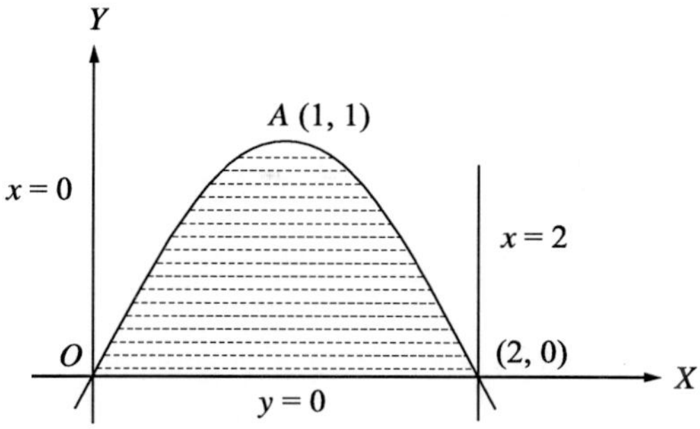

Fig.

$$y = 2x - x^2$$
$$\Rightarrow \quad x^2 - 2x + 1 = -y + 1$$
$$\Rightarrow \quad (x - 1)^2 = -(y - 1)$$

which is a parabola with vertex at $A(1, 1)$

$$\text{Required area } = \int_0^2 y\,dx = \int_0^2 (2x - x^2)\,dx$$

$$= \left[x^2 - \frac{x^3}{3}\right]_0^2 = 4 - \frac{8}{3} = \frac{4}{3} \text{ sq. units.}$$

Example 17. Find the area enclosedby the curve $y = -x^2$ and the straight line $x + y + 2 = 0$

Solution: The given curve is

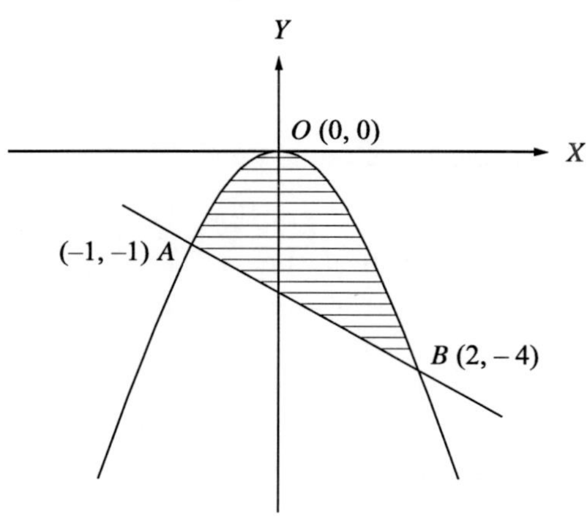

Fig.

$y = -x^2$ *i.e.*, $x^2 = -y$

which is a parabola with vertex at (0, 0)

The given line is $x + y + 2 = 0$,

Solving $x^2 = -y$ and $x + y + 2 = 0$, we get

$x^2 = x + 2 \Rightarrow x^2 - x - 2 = 0$

$\Rightarrow x = -1, 2$

$\therefore$ The point of intersection are $A(-1, -1)$ and $B(2, -4)$

$$\therefore \text{ The required area } = \int_{-1}^{2} [(-2 - x) - (-x^2)]\, dx$$

$$= \int_{-1}^{2} (-2 - x + x^2)\, dx$$

$$= \left[-2x - \frac{x^2}{2} + \frac{x^3}{3} \right]_{-1}^{2}$$

$$= \left(-4 - 2 + \frac{8}{3} \right) - \left(2 - \frac{1}{2} - \frac{1}{3} \right)$$

$$= \frac{-10}{3} - \frac{7}{6} = -\frac{9}{2} \text{ i.e., } \frac{9}{2} \text{ sq. units.}$$

EXERCISE 4.15

1. Find the area of the region bounded by the curve $y = x^2 - 6x + 5$, the x-axis and the lines $x = 2, x = 4$
2. Find the area bounded by the parabola $y^2 = 4x$ and the line $y = 2x - 4$.
3. Find the area bounded by the parabolas $y = 6x - x^2$ and $y = x^2 - 2x$.
4. Find area bounded by the ellipse $\frac{x^2}{a^2} + \frac{y^2}{b^2} = 1$ and the line $bx + ay = 1$ in the first quadrant.
5. Find the area bounded by the curve $y = x^2$, the x-axis and the lines $x = 1, x = 3$.
6. Find the area bounded by the curve $y = 9 - x^2$ and the axes.
7. Find the area between the curve $y = 4\sqrt{x - 1}$, x–axis and the line $x = 3$.
8. Compute the area bounded by the lines $x + 2y = 2$, $y = 1 + x$ and $2x + y = 7$.
9. Find the area of the triangle ABC whose vertices are $A(2, 5)$, $B(4, 7)$ and $C(6, 2)$.
10. Find the smaller of the areas bounded by the ellipse $4x^2 + 9y^2 = 36$ and the straight line $2x + 3y = 6$, using double integration.
11. Show by double integration, that the area between the parabolas $y^2 = 8x$ and $x^2 = 8y$ is $\frac{64}{3}$.

12. Find by double integration the area of the region bounded by the curves $y^2 = 4 - x$ and $y^2 = x$.

13. Using double integration, find the area bounded by the lines $x + y = 5, x - 2y + 4 = 0$ and x–axis.

14. Find the area enclosed between the curves $y = x^2 - 6x + 3$ and $y = 2x + 9$, using double integration.

15. Using double integration, find the area included between the curves $y^2 = 4a(x + a)$ and $y^2 = 4b(b - x)$

ANSWERS

1. $\frac{22}{3}$ sq. units	**2.** 9 sq. units	**3.** $\frac{64}{3}$ sq. units
4. $\frac{\pi ab}{4}$ sq. units	**5.** $\frac{26}{3}$ sq. units	**6.** 18 sq. units
7. $\frac{16\sqrt{2}}{3}$ sq. units	**8.** 6 sq. units	**9.** 7 sq. units
10. $\frac{3}{2}(\pi - 2)$ sq. units	**12.** $\frac{16\sqrt{2}}{3}$ sq. units	**13.** $\frac{27}{2}$ sq. units
14. $\frac{88\sqrt{22}}{3}$ sq. units	**15.** $\frac{8}{3}\sqrt{ab}$ sq. units	

5

Vector Calculus

5.1 INTRODUCTION

Vectors (quantities that are completely characterized by both magnitude and direction) are mathematical objects of great importance used in modelling either geometric problems or physical problems.

5.2 SCALAR QUANTITIES

A quantity which has only magnitude but no direction is called scalar. Mass, length, volume, density, area, time, work, temperature are some examples of scalar.

5.3 VECTOR QUANTITIES

A quantity that possesses both a magnitude and a specific direction and satisfies the laws of algebra of vectors is called a vector. Force, velocity, acceleration are some examples of vector.

Geometrically, a vectors is a directed line segment. A vector is represented as $\overrightarrow{AB}$ or $\vec{a}$. Its modulus is written as $|\overrightarrow{AB}|$ or $|\vec{a}|$ or a. Thus $|\overrightarrow{AB}| = |\vec{a}| = a$.

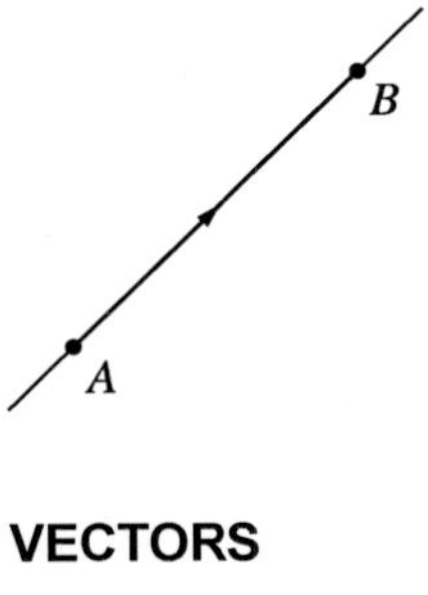

5.4 VARIOUS TYPES OF VECTORS

1. ***Zero vector.*** A vector whose magnitude is zero and direction is indeterminate is called zero (or null) vector.

   The null vector is the only vector whose direction is indeterminate. All null vectors are regarded as equal to each other without any consideration of direction. Vectors AA, PP etc. are null vectors.

2. ***Proper Vector.*** A vector with non zero magnitude is called a proper vector.

3. ***Unit vector.*** A vector with magnitude of unity is called a unit vector. Unit vector in the direction of $\vec{a}$ is $\hat{a} = \dfrac{\vec{a}}{|\vec{a}|}$.
4. ***Like vectors.*** Two (or more) vectors are called like vectors if their supports are same or parallel and are in the same sense.
5. ***Unlike vectors.*** Two vectors are called unlike vectors if their supports are same or parallel and are in the opposite sense.
6. ***Equal vectors.*** Two vectors are called equal (or equivalent) vectors if they have same length, same or parallel supports and same sense.

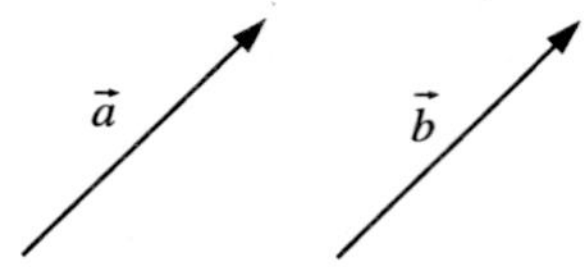

It the figure, vectors $\vec{a}$ and $\vec{b}$ are equal because they have the same length and the same direction on parallel lines of support.

7. ***Collinear vectors.*** Two (or more) vectors are called collinear vectors if they have same or parallel lines of support, what ever the magnitudes of the vectors may be. In short, two (or more) vectors are called collinear vectors if they have same or parallel supports.

The vectors $\vec{a}, \vec{b}, \vec{c}$ are collinear.

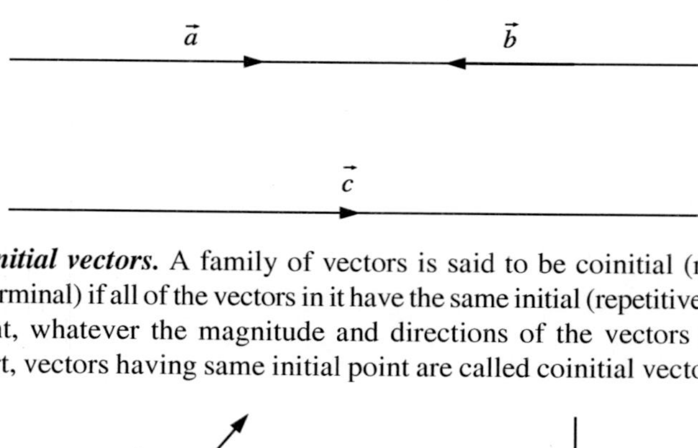

8. ***Coinitial vectors.*** A family of vectors is said to be coinitial (respectively coterminal) if all of the vectors in it have the same initial (repetitively terminal) point, whatever the magnitude and directions of the vectors may be. In short, vectors having same initial point are called coinitial vectors.

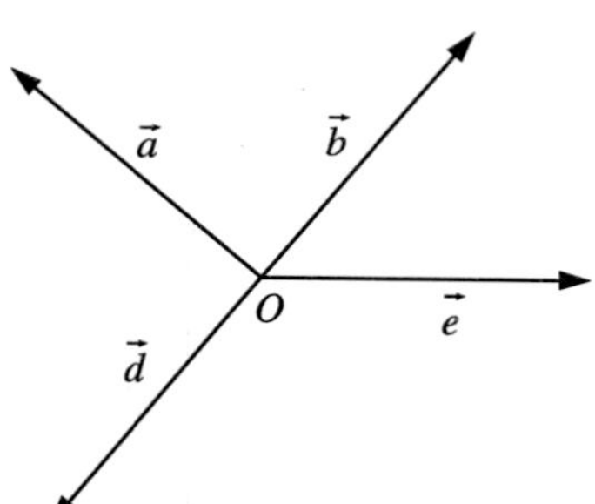

co-initial vectors

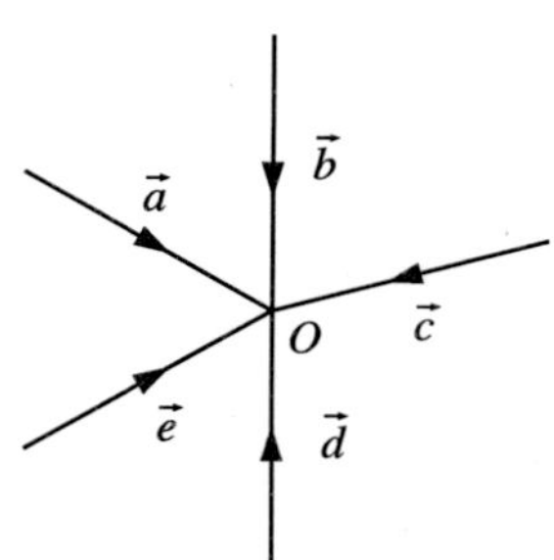

coterminal vectors

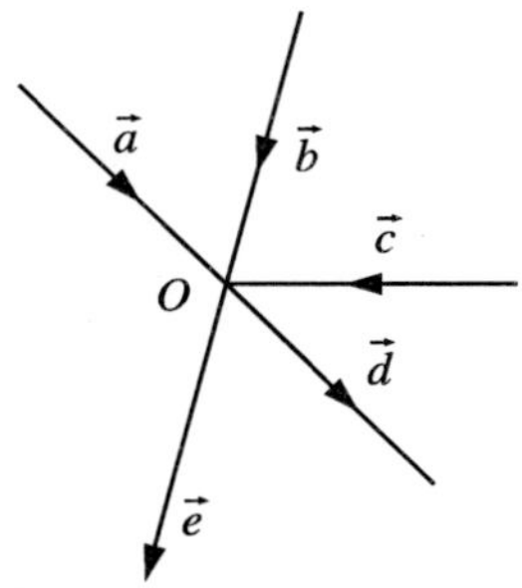

Neither coinitial nor coterminal vectors

9. ***Coplanar vector.*** Three (or more) vectors are called coplanar vectors if they lie in the same plane or parallel to the same plane. Two (free) vectors are always coplanar.

   The vectors $\vec{a}, \vec{b}, \vec{c}, \vec{d}$ and $\vec{e}$ are coplanar vectors.

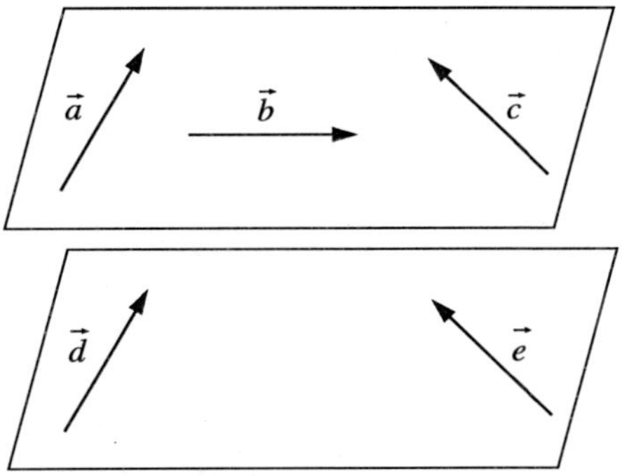

10. ***Negative of a vector.*** A vector having the same magnitude as that of the given vector but directed in the opposite sense is called the negative of the given vector.

    Thus if $\vec{a} = \overrightarrow{AB}$, then $\overrightarrow{BA} = -\vec{a}$.

11. ***Free vector.*** A vector is said to be free if it can be transferred to any position in its line of support or on a line parallel to its line of support, i.e. if its initial point can be shifted any where on its line of support or on a line parallel to its line of support. A free vector can therefore be shifted parallel to itself. Thus a free vector have no particular location in space.

12. ***Localized vector.*** A vector is said to be localized if its initial point is fixed and cannot be transferred to any other point on its line of support or on a line parallel to its line of support. Force is a localized vector.

5.5 MULTIPLICATION OF VECTORS BY SCALARS

Let $\vec{a}$ be a given vector and λ be a given scalar, then $\lambda\vec{a}$ is defined as a vector whose magnitude is λ times the modules of $\vec{a}$ and whose sense (direction) is same as that of $\vec{a}$ if λ is positive and opposite to that of $\vec{a}$ if λ is negative.

In particular, $\vec{a}$ and $\vec{b}$ are collinear iff $\vec{a} = \lambda\vec{b}$ for some scalar λ.

5.6 ADDITION OF VECTORS

Triangle law of addition of vectors. If two vectors $\vec{a}$ and $\vec{b}$ are represented (in magnitude and direction) by the two sides of a triangle taken in order, then their sum (resultant) $\vec{c}$ is represented by the third side taken in the apposite order.

Thus $\vec{c} = \vec{a} + \vec{b}$ i.e. $\overrightarrow{OQ} = \overrightarrow{OP} + \overrightarrow{PQ}$

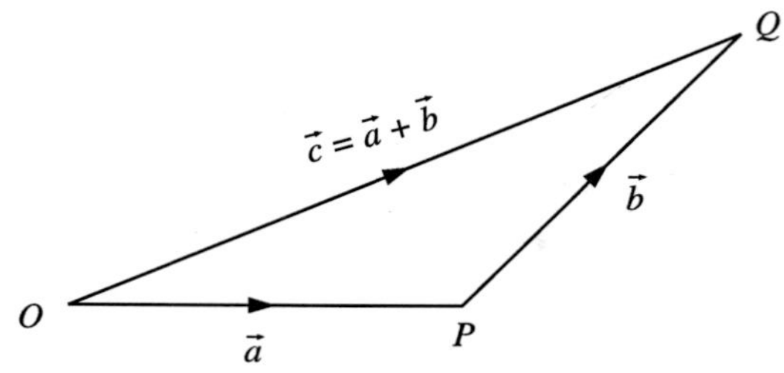

5.7 PARALLELOGRAM LAW OF ADDITION OF VECTORS

If two vectors $\vec{a}$ and $\vec{b}$ are represented (in magnitude and direction) by the two adjacent sides of a parallelogram, then their sum $\vec{c}$ is represented by the diagonal of the parallelogram which is co-initial with the given vectors.

Thus $\vec{a} + \vec{b} = \vec{c}$ or $\overrightarrow{OP} + \overrightarrow{OR} = \overrightarrow{OQ}$

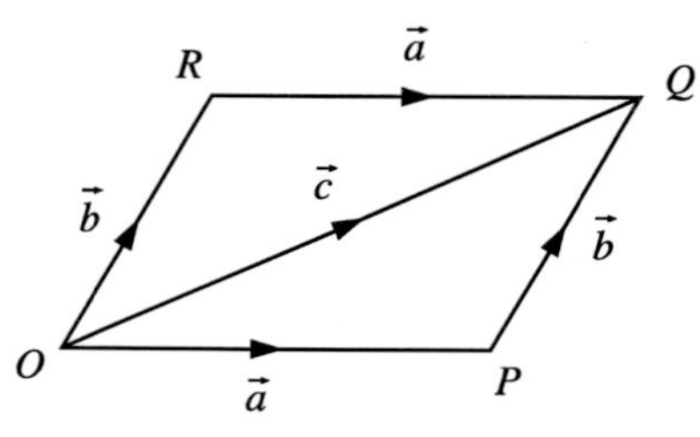

5.8 SUBTRACTION OF TWO VECTORS

For any two vectors, we define $\vec{a} - \vec{b} = \vec{a} + (-\vec{b})$ and call this operation subtraction (or difference) of vectors.

5.9 POSITION VECTORS (P.V.) OF A POINT

Let a fixed point O in space (or plane) be taken as origin. If P is any point then $\overrightarrow{OP}$ is called position vector (P.V.) of the point P with respect to O.

Let $\overrightarrow{PQ}$ be the given vector, then

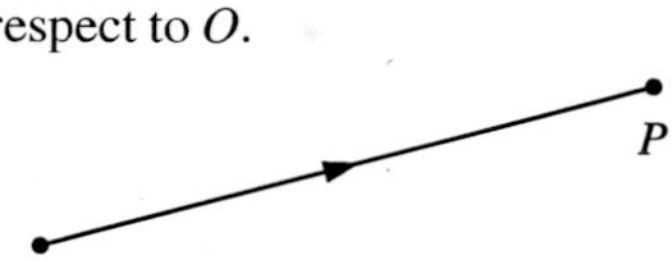

$$\overrightarrow{OP} + \overrightarrow{PQ} = \overrightarrow{OQ}$$

$$\Rightarrow \quad \overrightarrow{PQ} = \overrightarrow{OQ} - \overrightarrow{OP}$$

$$\Rightarrow \quad \overrightarrow{PQ} = PV \text{ of } Q - PV \text{ of } P$$

5.10 TRIANGLE INEQUALITY

If $\vec{a}, \vec{b}$ are two victors, then

$$|\vec{a}+\vec{b}| \le |\vec{a}| + |\vec{b}|$$

From triangle inequality, if follows that

$$|\vec{a}-\vec{b}| \le |\vec{a}| + |\vec{b}|$$

and $$|\vec{a}-\vec{b}| \ge \left||\vec{a}| - |\vec{b}|\right|$$

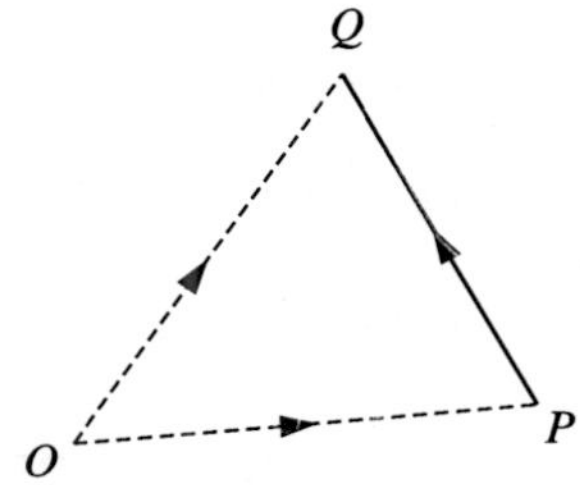

5.11 SECTION FORMULA

If $\vec{a}$ and $\vec{b}$ are the position vectors of the points A and B respectively, then the position vector of the point C dividing the segment AB in the ratio $m : n$ is $\vec{c} = \dfrac{m\vec{b}+n\vec{a}}{m+n}$. In particular, the position vector of the mid point of the segment AB is $\vec{c} = \dfrac{\vec{a}+\vec{b}}{2}$.

Note: If $\vec{a}, \vec{b}$ are two non-zero non-collinear vectors and α, β are two scalars then $\alpha\vec{a} + \beta\vec{b} = \vec{0}$ only if $\alpha = 0$ and $\beta = 0$.

5.12 COMPONENTS OF A VECTOR IN A PLANE

(*a*) Let $P(x, y)$ be a point in the coordinate plane XOY and $\hat{i}, \hat{j}$ be unit vectors along X-axis and Y-axis respectively. If $\vec{r}$ is the position vector of the point P, then

$$\vec{r} = \overrightarrow{OP} = \overrightarrow{OM} + \overrightarrow{MP} = \overrightarrow{OM} + \overrightarrow{ON} = x\hat{i} + y\hat{j}$$

$x\hat{i}, y\hat{j}$ are called component vectors of the vector $\vec{r}$ and x, y are called scalar components of $\vec{r}$ along X-axis and Y-axis respectively.

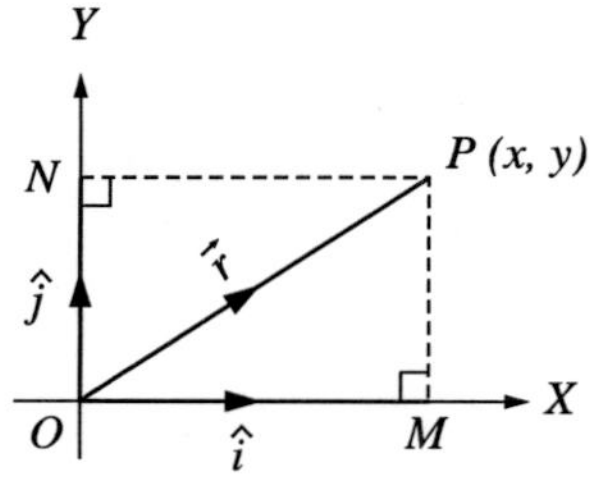

Also, $$|\overrightarrow{OP}| = |\vec{r}| = \sqrt{x^2 + y^2}$$

(*b*) If $P(x_1, y_1)$ and $Q(x_2, y_2)$ are any two points in XOY plane then

$$\overrightarrow{PQ} = \overrightarrow{OQ} - \overrightarrow{OP} = (x_2 - x_1)\hat{i} + (y_2 - y_1)\hat{j}$$

$(x_2 - x_1)\hat{i}$, $(y_2 - y_1)\hat{j}$ are called components vectors of $\overrightarrow{PQ}$ and $x_2 - x_1$, $y_2 - y_1$ are called components of $\overrightarrow{PQ}$ along X-axis and Y- axis respectively.

Also $|\overrightarrow{PQ}| = \sqrt{(x_2 - x_1)^2 + (y_2 - y_1)^2}$

5.13 COMPONENTS OF A VECTOR IN SPACE

(*a*) Let $P(x, y, z)$ be a point in space and $\hat{i}$, $\hat{j}$, $\hat{k}$ be unit vectors along X-axis, Y- axis and Z-axis respectively, If $\vec{r}$ is the position vector of the point P, then

$$\vec{r} = x\hat{i} + y\hat{j} + z\hat{k}$$

$x\hat{i}$, $y\hat{j}$, $z\hat{k}$ are called component vectors of the vector $\vec{r}$ and x, y, z are called scalar components of $\vec{r}$ along X-axis, Y-axis and Z-axis respectively.

Also $$|\overrightarrow{OP}| = |\vec{r}| = \sqrt{x^2 + y^2 + z^2}$$

(*b*) If $P(x_1, y_1, z_1)$ and $Q(x_2, y_2, z_2)$ are two points in space, then $\overrightarrow{PQ} = \overrightarrow{OQ} - \overrightarrow{QP}$ $= (x_2 - x_1)\hat{i} + (y_2 - y_1)\hat{j} + (z_2 - z_1)\hat{k}$; $(x_2 - x_1)\hat{i}$, $(y_2 - y_1)\hat{j}$, $(z_2 - z_1)\hat{k}$ are called component vectors of $\overrightarrow{PQ}$ and $x_2 - x_1$, $y_2 - y_1$, $z_2 - z_1$ are called components of $\overrightarrow{PQ}$ along X-axis, Y-axis and Z-axis respectively.

Also $$|\overrightarrow{PQ}| = |\vec{r}| = \sqrt{(x_2 - x_1)^2 + (y_2 - y_1)^2 + (z_2 - z_1)^2}$$

5.14 DIRECTION COSINES OF A GIVEN VECTOR

If a vector $\overrightarrow{OP}$ in space makes an angles α, β and γ with the three rectangular axes respectively, then $\cos\alpha$, $\cos\beta$, $\cos\gamma$ are called direction cosines of this vector. If a vector $\overrightarrow{AB} = \vec{a} = a_1\hat{i} + a_2\hat{j} + a_3\hat{k}$ makes angle α, β and γ with X-axis Y-axis and Z-axis respectively, then its direction cosines are given by

$$\cos\alpha = \frac{a_1}{|\vec{a}|} = \frac{a_1}{\sqrt{a_1^2 + a_2^2 + a_3^2}}, \cos\beta = \frac{a_2}{|\vec{a}|} = \frac{a_2}{\sqrt{a_1^2 + a_2^2 + a_3^2}}$$

and $$\cos\gamma = \frac{a_3}{|\vec{a}|} = \frac{a_3}{\sqrt{a_1^2 + a_2^2 + a_3^2}}$$

5.15 LINEAR DEPENDENCE, COLLINEARITY, COPLANARITY

Linear Combination

A vector $\hat{r}$ is said to be a liner combination of $\vec{a},\ \vec{b},\ \vec{c}$. If we can find scalars α, β, λ, such that $\vec{r} = \alpha\vec{a} + \beta\vec{b} + \gamma\vec{c}$.

Linearly Dependent & Independent Vectors

A set of vectors $\vec{a},\ \vec{b},\ \vec{c},...$ is called linearly dependent if we can find scalars α, β, γ, .. (not all zero) such that $\alpha\vec{a} + \beta\vec{b} + \gamma\vec{c} + ... = \vec{0}$, otherwise it is called linearly independent (set). In other words, a set of vectors $\vec{a},\ \vec{b},\ \vec{c},...$ is linearly independent iff $\alpha\vec{a} + \beta\vec{b} + \gamma\vec{c} + ... = \vec{0}$ implies

$$\alpha = \beta = \gamma = ... = 0$$

Note (*i*) Two vectors $\vec{a}$ and $\vec{b}$ are linearly dependent iff $\vec{b} = \alpha\vec{a}$ for some scalar a. i.e. iff $\vec{a}$ and $\vec{b}$ are collinear Thus, two vectors $\vec{a}$ and $\vec{b}$ are linearly independent if they are not collinear.

(*ii*) Two colinear vectors are linearly dependent.

(*iii*) Two non-zero, non colinear vectors are linearly independent.

5.16 THEOREM

The necessary and sufficient conditions for three points with position vectors $\vec{a},\ \vec{b},\ \vec{c}$ to be colinear is that three exists three scalars α, β, γ not all zero such that $\alpha\vec{a} + \beta\vec{b} + \gamma\vec{c} = \vec{0}$ where $\alpha + \beta + \gamma = 0$.

Proof (Necessary Condition) Suppose the points *A*, *B*, *C* with position vector $\vec{a},\ \vec{b},\ \vec{c}$ respectively are collinear. Then, surely the point C divides the segment *AB* in a ratio λ : 1 (say), so that

we get $$\vec{c} = \frac{\lambda\vec{b} + \vec{a}}{\lambda + 1}$$

$$\Rightarrow \quad \vec{a} + \lambda\vec{b} - \vec{c}(\lambda + 1) = 0$$

Thus, there exist non-zero scalars, α = 1, β = λ, γ = – λ – 1 such that $\alpha\vec{a} + \beta\vec{b} - \gamma\vec{c} = \vec{0}$ and $\alpha + \beta + \gamma = 0$.

(Sufficient Condition). Let $\vec{a},\ \vec{b},\ \vec{c}$ be related as $\alpha\vec{a} + \beta\vec{b} - \gamma\vec{c} = \vec{0}$, $\alpha + \beta + \gamma = 0$, without loss of generality, let $\gamma \neq 0$. then,

$$\vec{c} = \frac{\alpha\vec{a}+\beta\vec{b}}{-\gamma} = \frac{\alpha\vec{a}+\beta\vec{b}}{\alpha+\beta}$$

This implies that the point C lies on the joint of A and B, or in other words, the points A, B and C with position vector $\vec{a}, \vec{b}, \vec{c}$ are collinear.

Corollary 1. If $\vec{a}$ and $\vec{b}$ are two non-zero, non-collinear vectors, then every vector $\vec{r}$ coplanar with $\vec{a}$ and $\vec{b}$ can be uniquely expressed as $\vec{r} = \alpha\vec{a} + \beta\vec{b}$, for some scalars α and β.

Corollary 2. Three vectors are coplanar iff one of them is expressible as a linear combination of other two.

Corollary 3. If $\vec{a}, \vec{b}, \vec{c}$ are there non-zero, non-coplanar vectors, and α, β, γ are scalars, then $\alpha\vec{a}+\beta\vec{b}+\gamma\vec{c}=\vec{0}$ iff $\alpha=\beta=\gamma=0$

5.17 THEOREM

The necessary and sufficient condition for four points with position vector $\vec{a}, \vec{b}, \vec{c}$ and $\vec{d}$ to be coplanar is that three exists scalars $\alpha, \beta, \gamma, \delta$ not all zero such that

$$\alpha\vec{a} + \beta\vec{b} + \gamma\vec{c} + \delta\vec{d} = \vec{0} \text{ where } \alpha + \beta + \gamma + \delta = 0$$

Proof (Necessary condition). Suppose the four points A, B, C, D with position vectors $\vec{a}, \vec{b}, \vec{c}, \vec{d}$ respectively are coplanar. Then suppose that the line joining the points A and B intersect the line joining the points C and D. (In case the line AB is parallel to the line CD, we shall have to work with the lines AC and BD). The point of intersection will divide the line AB in some ratio, say $m : n$ and the line CD in some other ratio, say, $p : q$. Then we see that

$$\frac{m\vec{b}+n\vec{a}}{m+n} = \frac{q\vec{d}+p\vec{c}}{q+p}$$

$$\left(\frac{n}{m+n}\right)\vec{a}+\left(\frac{m}{m+n}\right)\vec{b}-\left(\frac{p}{q+p}\right)\vec{c}-\left(\frac{q}{q+p}\right)\vec{d}=0$$

Writing $\alpha = \dfrac{n}{m+n}, \beta = \dfrac{m}{m+n}, \gamma = \dfrac{-p}{p+q}, \delta = \dfrac{-q}{p+q}$

We see that,

$$\alpha\vec{a} + \beta\vec{b} + \gamma\vec{c} + \delta\vec{d} = \vec{0} \text{ and } \alpha + \beta + \gamma + \delta = 0$$

Sufficient Conditions

Suppose there exits scalars $\alpha, \beta, \gamma, \delta$ not all zero, such that

$$\alpha\vec{a} + \beta\vec{b} + \gamma\vec{c} + \delta\vec{d} = \vec{0} \text{ and } \alpha + \beta + \gamma + \delta = 0$$

Hence $$\frac{\alpha\vec{a}+\beta\vec{b}}{\alpha+\beta}=\frac{\gamma\vec{c}+\delta\vec{d}}{\gamma+\delta}$$

Which implies that the line joining the points *A* and *B* intersects the line joining the points *C* and *D*. This implies that all the four points *A*, *B*, *C* and *D* are coplanar.

If case $\gamma+\delta=0$ or $\alpha+\beta=0$, we choose the pairs *A* and *C*, and *B* and *D*. This completes the proof.

ILLUSTRATIVE EXAMPLES

Example 1. Show that the three points given by $\vec{a}-2\vec{b}+3\vec{c}$, $2\vec{a}+3\vec{b}-4\vec{c}$, $-7\vec{b}+10\vec{c}$ are collinear.

Solution: Let $\vec{p}=\vec{a}-2\vec{b}+3\vec{c}$, $\vec{q}=2\vec{a}+3\vec{b}-4\vec{c}$ and $\vec{\gamma}=-7\vec{b}+10\vec{c}$ ·

Clearly $2\vec{p}-\vec{q}-\vec{r}=\vec{0}$ and $2+(-1)+(-1)=0$. Hence $\vec{p}.\vec{q}.\vec{r}$ are collinear.

Example 2. Show that the points A, B, C whose position vectors are $-2\vec{a}+3\vec{b}+5\vec{c}$, $\vec{a}+2\vec{b}+3\vec{c}, 7\vec{a}-\vec{c}$ respectively are collinear, what ever be $\vec{a}, \vec{b}, \vec{c}$.

Solution:

$$\overrightarrow{AB} = P.V. \text{ of } B - P.V. \text{ of } A$$

$$= (\vec{a}+2\vec{b}+3\vec{c})-(-2\vec{a}+3\vec{b}+5\vec{c}) = 3\vec{a}-\vec{b}-2\vec{c}$$

$$\overrightarrow{AC} = P.V. \text{ of } C - P.V. \text{ of } A$$

$$\overrightarrow{AC} = (7\vec{a}-\vec{c})-(-2\vec{a}+3\vec{b}+5\vec{c}) = 9\vec{a}-3\vec{b}-6\vec{c}$$

$\Rightarrow$ $\overrightarrow{AC} = 3\overrightarrow{AB}$, which shows that $\overrightarrow{AC}$ and $\overrightarrow{AB}$ are collinear.

$\Rightarrow$ $\overrightarrow{AC}$ and $\overrightarrow{AB}$ have same or parallel supports.

But $\overrightarrow{AC}$ and $\overrightarrow{AB}$ have point A in common, therefore $\overrightarrow{AC}$ and $\overrightarrow{AB}$ have same support.

$\Rightarrow$ Points *A*, *B*, *C* are collinear.

Example 3. Show that the points (2, –1, 1), B(3, –5, 1) and (–1, 11, 9) are colinear.

Solution: Let A (2, –1, 3), B(3, –5, 1) and C(–1, 11, 9) and O be the origin. Then

$$\overrightarrow{OA} = 2\hat{i}-\hat{j}+3\hat{k},\ \overrightarrow{OB} = 3\hat{i}-5\hat{j}+\hat{k},\ \overrightarrow{OC} = -\hat{i}+11\hat{j}+9\hat{k}$$

$$\therefore \quad \overrightarrow{AB} = \overrightarrow{OB}-\overrightarrow{OA} = (3\hat{i}-5\hat{j}+\hat{k})-(2\hat{i}-\hat{j}+3\hat{k}) = \hat{i}-4\hat{j}-2\hat{k},$$

$$\overrightarrow{AC} = \overrightarrow{OC}-\overrightarrow{OA} = (-\hat{i}+11\hat{j}+9\hat{k})-(2\hat{i}-\hat{j}+3\hat{k}) = -3\hat{i}+12\hat{j}+6\hat{k}$$

$$= -3(\hat{i}-4\hat{j}-2\hat{k}) = -3\overrightarrow{AB}$$

$\Rightarrow$ the vectors $\overrightarrow{AB}$ and $\overrightarrow{AC}$ are collinear.

$\Rightarrow$ $\overrightarrow{AB}$ and $\overrightarrow{AC}$ have same or parallel supports.

But $\overrightarrow{AB}$ and $\overrightarrow{AC}$ are co-initial, therefore $\overrightarrow{AB}$ and $\overrightarrow{AC}$ have same support. Hence the points A, B, C are collinear.

Example 4. Find out whether the four points whose position vectors are $2\hat{i}+3\hat{j}-\hat{k}$, $\hat{i}-2\hat{j}+3\hat{k}$, $3\hat{i}+4\hat{j}-2\hat{k}$ and $\hat{i}-6\hat{j}+6\hat{k}$ are coplanar or not.

Solution: Let O be the origin and A, B, C, D be four given points.

Then $\overrightarrow{OA} = 2\hat{i}+3\hat{j}-\hat{k}, \overrightarrow{OB} = \hat{i}-2\hat{j}+3\hat{k}, \overrightarrow{OC} = 3\hat{i}+4\hat{j}-2\hat{k}$

$\overrightarrow{OD} = \hat{i}-6\hat{j}+6\hat{k}$

$\therefore$ $\overrightarrow{AB} = \overrightarrow{OB}-\overrightarrow{OA} = (\hat{i}-2\hat{j}+3\hat{k})-(2i+3\hat{j}-\hat{k}) = -\hat{i}-5\hat{j}+4\hat{k}$

$\overrightarrow{AC} = \hat{i}+\hat{j}-\hat{k}, \overrightarrow{AD} = -i-9\hat{j}+7\hat{k}$

Now A, B, C, D are coplanar if $\overrightarrow{AB}, \overrightarrow{AC}, \overrightarrow{AD}$ are coplanar i.e.

if $\overrightarrow{AB} = \alpha\overrightarrow{AC} + \beta\overrightarrow{AD}$ for some scalars α, β

$\Rightarrow$ $-\hat{i}-5\hat{j}+4\hat{k} = \alpha(\hat{i}+\hat{j}-\hat{k})+\beta(-\hat{i}-9\hat{j}+7\hat{k})$

$\Rightarrow$ $-\hat{i}-5\hat{j}+4\hat{k} = (\alpha-\beta)\hat{i}+(\alpha-9\beta)\hat{j}+(-\alpha+7\beta)\hat{k}$

$\Rightarrow$ $\alpha-\beta=-1, \quad \alpha-9\beta=-5, -\alpha+7\beta=4$

First two equations yield $\alpha = -\frac{1}{2}, \beta = \frac{1}{2}$ and there values satisfy third equation.

$\therefore$ $\overrightarrow{AB}, \overrightarrow{AC}, \overrightarrow{AD}$ coplanar.

Hence the given four points are coplanar.

Example 5. If $x_1\vec{a}+y_1\vec{b}+z_1\vec{c} = x_2\vec{a}+y_2\vec{b}+z_2\vec{c}$, where $\vec{a}, \vec{b}, \vec{c}$ are non-coplanar vectors, then show that $x_1 = x_2, y_1 = y_2, z_1 = z_2$.

Solution: Since $x_1\vec{a}+y_1\vec{b}+z_1\vec{c} = x_2\vec{a}+y_2\vec{b}+z_2\vec{c}$, that is $(x_1 - x_2)\vec{a}$ + $(y_1-y_2)\vec{b} + (z_1-z_2)\vec{c} = \vec{0}$

On using corollary 3, we have

$$x_1-x_2=0, y_1-y_2=0, z_1-z_2=0$$

$\Rightarrow$ $$x_1=x_2, y_1=y_2, z_1=z_2$$

Example 6. If $\vec{a}, \vec{b}, \vec{c}$ are linearly independent, prove that $\vec{a}-2\vec{b}+3\vec{c}$, $-2\vec{a}+3\vec{b}-4\vec{c}$ and $-\vec{b}+3\vec{c}$ are also linearly independent.

Solution: Let α, β, γ be scalars such that

$$\alpha(\vec{a}-2\vec{b}+3\vec{c})+\beta(-2\vec{a}+3\vec{b}-4\vec{c})+\gamma(-\vec{b}+3\vec{c})=\vec{0}$$

$$\Rightarrow \quad (\alpha-2\beta)\vec{a}+(-2\alpha+3\beta-\gamma)\vec{b}+(3\alpha-4\beta+3\gamma)\vec{c}=\vec{0} \quad ...(1)$$

Since $\vec{a}$, $\vec{b}$ and $\vec{c}$ are linearly independent, from (1), we get

$$\alpha-2\beta=0, -2\alpha+3\beta-\gamma=0, 3\alpha-4\beta+3\gamma=0$$

$$\Rightarrow \quad \alpha-2\beta=0, -4\beta+3\beta-\gamma=0, 6\beta-4\beta+3\gamma=0$$

$$\Rightarrow \quad \alpha=2\beta, \gamma=-\beta, \gamma=-\frac{2}{3}\beta, \Rightarrow -\beta=-\frac{2}{3}\beta \Rightarrow \beta=0$$

$$\Rightarrow \quad \alpha=\beta=\gamma=0$$

$\Rightarrow$ the given vectors are linearly independent.

Example 7. If $\vec{a}, \vec{b}, \vec{c}, \vec{d}$ are the position vectors of four points A, B, C, D respectively and if $\vec{b}-\vec{a}=\vec{c}-\vec{d}$, show that *ABCD* is a parallelogram.

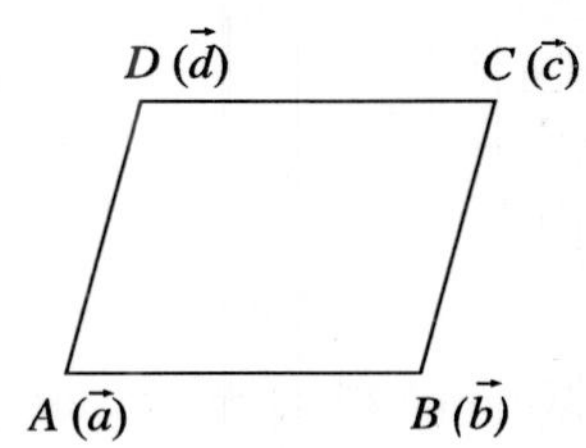

Solution: Given $\vec{b}-\vec{a}=\vec{c}-\vec{d}$

$$\Rightarrow \quad \overrightarrow{AB}=\overrightarrow{DC}$$

$\Rightarrow$ sides *AB* and *DC* of a quadrilateral are equal and parallel.

$\Rightarrow$ *ABCD* is a parallelogram.

Example 8. Show that a quadrilateral is a parallelogram if and only if its diagonals bisect each other.

Solution: Let *ABCD* be quadrilateral and O be the origin. Let $\vec{a}$, $\vec{b}$, $\vec{c}$ and $\vec{d}$ be the position vectors of the points *A*, *B*, *C* and *D* respectively with reference to origin *O*. Then the mid point of diagonal *AC* is $\frac{\vec{a}+\vec{c}}{2}$ and midpoint of diagonal BD is $\frac{\vec{b}+\vec{d}}{2}$.

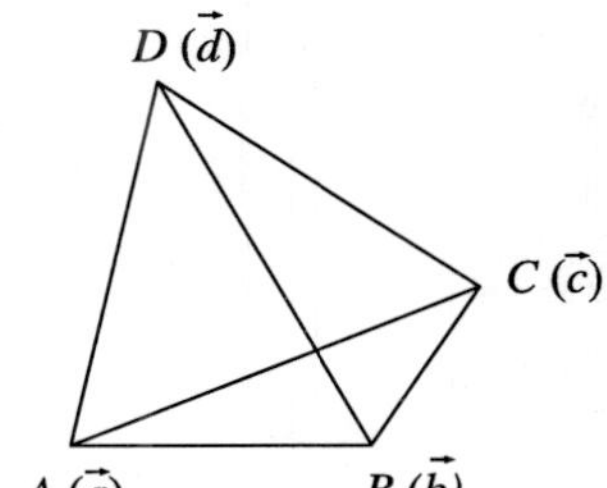

First, let the diagonals bisect each other

$\Rightarrow$ *AC* and *BD* have same midpoint.

$$\Rightarrow \quad \frac{\vec{a}+\vec{c}}{2}=\frac{\vec{b}+\vec{d}}{2} \Rightarrow \vec{b}-\vec{a}=\vec{c}-\vec{d}$$

$\Rightarrow \quad \overrightarrow{AB}=\overrightarrow{DC} \Rightarrow$ *AB* and *DC* are equal and parallel

$\Rightarrow$ *ABCD* is parallelogram.

Conversely, let *ABCD* be a parallelogram. Then *AB* and *DC* are equal and parallel

$$\Rightarrow \quad \overrightarrow{AB}=\overrightarrow{DC} \Rightarrow \vec{b}-\vec{a}=\vec{c}-\vec{d}$$

$$\Rightarrow \qquad \vec{b}+\vec{d} = \vec{a}+\vec{c} \Rightarrow \frac{\vec{a}+\vec{c}}{2} = \frac{\vec{b}+\vec{d}}{2}$$

$\Rightarrow$ diagonals AC and BD have same mid-point.

$\Rightarrow$ diagonals AC and BD bisect each other.

Example 9. Prove by vector method that the line segment joining the middle points of any two sides of a triangle is parallel to the third side and equal to half of it.

Solution: Let A, B, C be the vertices of a triangle and let $\vec{a}, \vec{b}, \vec{c}$ be their position vectors with reference to some arbitrary origin O.

Let D, E be mid-points of sides AB, AC respectively. As D is the mid point of side AB, its P.V. is $\frac{\vec{a}+\vec{b}}{2}$. Similarly as E is the mid-point of AC, its P.V. is $\frac{\vec{a}+\vec{c}}{2}$.

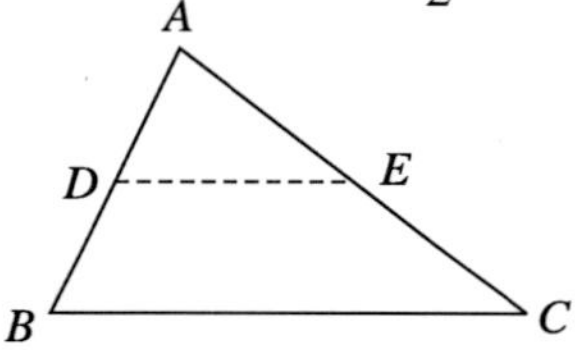

$$\therefore \quad \overrightarrow{DE} = \text{P.V. of E} - \text{P.V. of } D = \frac{\vec{a}+\vec{c}}{2} - \frac{\vec{a}+\vec{b}}{2}$$

$$= \frac{\vec{c}-\vec{b}}{2} = \frac{1}{2}(\vec{c}-\vec{b}) = \frac{1}{2}\overrightarrow{BC}$$

Hence DE is parallel to BC and one half of it in magnitude.

Example 10. Using vectors, show that the medians of a triangle meet at a point.

Solution: Let $\vec{a}$, $\vec{b}$, $\vec{c}$ be the position vectors of vertices A, B, C of a triangle ABC with reference to an arbitrary origin O. Let D be the mid-point of side BC so that AD is a median. By mid-point formula, the position vector of D is $\frac{\vec{b}+\vec{c}}{2}$.

Let G be a point which divides median AD internally in ratio 2 : 1. Using section formula, the position of point G is given by

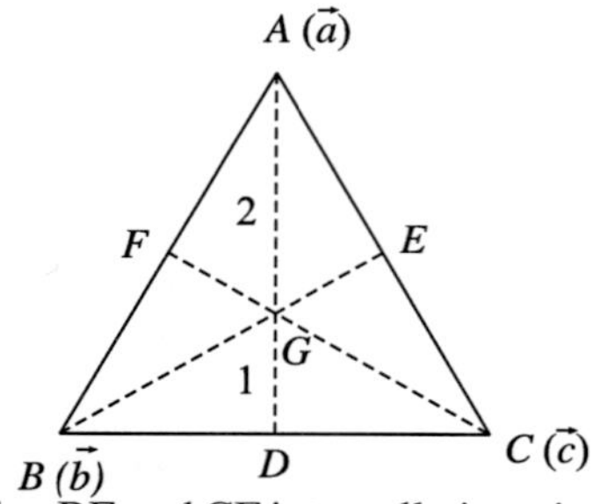

$$\overrightarrow{OG} = \frac{2.\left(\frac{\vec{b}+\vec{c}}{2}\right)+1.\vec{a}}{2+1} = \frac{\vec{a}+\vec{b}+\vec{c}}{3}$$

Similarly, we can show that the position vector of the point which divides either of the median BE and CF internally in ratio 2 : 1 is given by $\frac{\vec{a}+\vec{b}+\vec{c}}{3}$. Thus all the three medians pass through the point G with p.v. $\frac{\vec{a}+\vec{b}+\vec{c}}{3}$.

Hence the medians of a triangle are concurrent.

Note: Remember that G is called the **centroid** of the triangle and its position vector is $\frac{\vec{a}+\vec{b}+\vec{c}}{3}$.

Example 11. Prove that the internal bisectors of the angles of a triangle are concurrent.

Solution: Let ABC be any triangle and $\vec{x}, \vec{y}, \vec{z}$ be the position vectors of the points A, B, C respectively w.r.t. origin O. Let a, b, c be the lengths of the sides BC, CA, AB respectively. Let the internal bisector of angle A meet BC at D, then $BD : DC = AB : AC = c : b$.

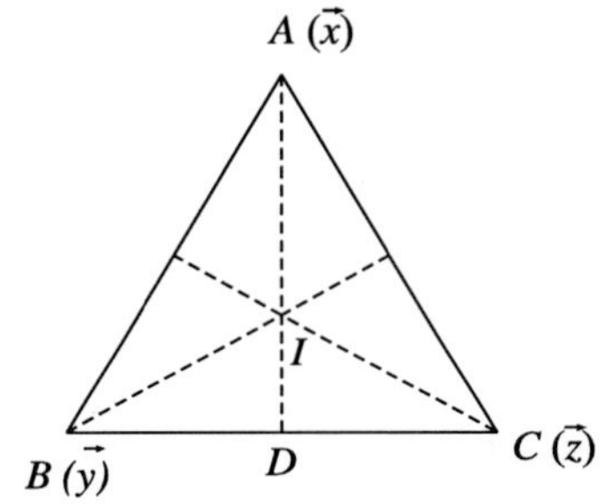

$\therefore$ By section formula, position vector of the point D is $\dfrac{b\vec{y}+c\vec{z}}{b+c}$

Let the internal bisector of angle B meet AD at I, then $DI : IA = BD : AB = \dfrac{ac}{b+c} : \text{c} = a : (b+c)$.

$$\left(\because \frac{BD}{DC}=\frac{c}{b} \Rightarrow \frac{BD}{BD+DC}=\frac{C}{b+c} \Rightarrow \frac{BD}{a}=\frac{c}{b+c} \Rightarrow BD=\frac{ac}{b+c}\right)$$

$\therefore$ the position vector of the point I is $\dfrac{(b+c).\dfrac{b\vec{y}+c\vec{z}}{(b+c)}+a\vec{x}}{(b+c)+a}$

i.e. $\dfrac{a\vec{x}+b\vec{y}+c\vec{z}}{a+b+c}$

The symmetry of the result shows that the bisector of $\angle C$ will also pass through I. Hence, the bisectors of the angle of a triangle are concurrent.

Example 12. The two adjacent sides of a parallelogram are $2\hat{i}+4\hat{j}-5\hat{k}$ and $\hat{i}+2\hat{j}+3\hat{k}$. Find the unit vectors along the diagonals of the parallelogram.

Solution: Let the two adjacent sides OA, OB of parallelogram $OACB$ be represented by $\vec{a}$ and $\vec{b}$ respectively.

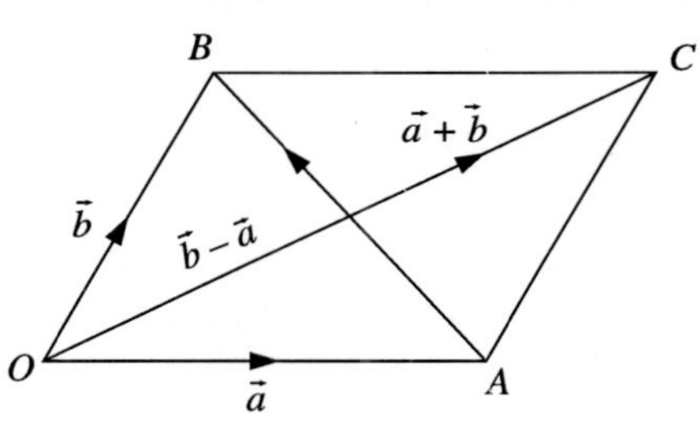

Then $\vec{a} = 2\hat{i}+4\hat{j}-5\hat{k}$

and $\vec{b} = \hat{i}+2\hat{j}+3\hat{k}$

The two diagonals are

$$\vec{d_1} = \vec{a}+\vec{b} = 3\hat{i}+6\hat{j}-2\hat{k}$$

and $\overrightarrow{d_2} = \vec{b} - \vec{a} = -\hat{i} - 2\hat{j} + 8\hat{k}$

The required unit vectors are

$$\overrightarrow{n_1} = \frac{\overrightarrow{d_1}}{|\overrightarrow{d_1}|} = \frac{3\hat{i} + 6\hat{j} - 2\hat{k}}{\sqrt{(3)^2 + (6)^2 + (-2)^2}} = \frac{3}{7}\hat{i} + \frac{6}{7}\hat{j} - \frac{2}{7}\hat{k}$$

and $$\overrightarrow{n_2} = \frac{\overrightarrow{d_2}}{|\overrightarrow{d_2}|} = \frac{-\hat{i} - 2\hat{j} + 8\hat{k}}{\sqrt{(-1)^2 + (-2)^2 + (8)^2}} = -\frac{1}{\sqrt{69}}\hat{i} - \frac{2}{\sqrt{69}}\hat{j} + \frac{8}{\sqrt{69}}\hat{k}$$

Example 13. The position vectors of two points A and B are $\hat{i} + \hat{j} + \hat{k}$ and $5\hat{i} - 3\hat{j} + \hat{k}$. Find a unit vector in direction of $\overrightarrow{AB}$, and also find the direction cosines of $\overrightarrow{AB}$. What angles does $\overrightarrow{AB}$ make with the three axes?

Solution: $\overrightarrow{AB} = \text{P.V. of B} - \text{P.V. of A} = (5\hat{i} - 3\hat{j} + \hat{k}) - (\hat{i} + \hat{j} + \hat{k}) = 4\hat{i} - 4\hat{j}$

$\therefore$ Unit vector in the direction of $\overrightarrow{AB}$ is

$$\hat{n} = \frac{\overrightarrow{AB}}{|\overrightarrow{AB}|} = \frac{4\hat{i} - 4\hat{j}}{\sqrt{(4)^2 + (-4)^2}} = \frac{1}{\sqrt{2}}\hat{i} - \frac{1}{\sqrt{2}}\hat{j}$$

$\therefore$ Direction cosines of $\overrightarrow{AB}$ are $\left\{\frac{1}{\sqrt{2}}, -\frac{1}{\sqrt{2}}, 0\right\}$

$$\Rightarrow \quad \cos\alpha = \frac{1}{\sqrt{2}}, \cos\beta = -\frac{1}{\sqrt{2}}, \cos\gamma = 0$$

$$\Rightarrow \quad \alpha = 45°, \beta = 135°, \gamma = 90°.$$

Therefore, $\overrightarrow{AB}$ makes angles of 45°, 135° and 90° with the three coordinate axes.

EXERCISE 5.1

1. If $A(2, 1, 3)$, $B(2, -3, 4)$ and $C(-1, 2, 7)$, find a point D such that $\overrightarrow{AB}$ and $\overrightarrow{CD}$ are parallel.
2. (*i*) Find a unit vector in the direction of the sum of the vectors

   $\vec{a} = -\hat{i} + \hat{j} + \hat{k}$ and $\vec{b} = 2\hat{i} + \hat{j} - 3\hat{k}$

   (*ii*) If $\vec{a} = \hat{i} + \hat{j} + \hat{k}, \vec{b} = \hat{j} + \hat{k}$ and $\vec{c} = \hat{i} + \hat{k}$ find a unit vector in the direction of $\vec{a} - 2\vec{b} + 3\vec{c}$
3. If the vectors $4\hat{i} + \alpha\hat{j} + \hat{k}$ and $\beta\hat{i} + 2\hat{j} + \hat{k}$ are parallel, find the values of α and β.

4. If the position vectors of the three vertices of a triangle are $2\hat{i}+3\hat{j}+4\hat{j}$, $3\hat{i}+4\hat{j}+2\hat{k}$ and $4\hat{i}+2\hat{j}+3k$. prove that ΔABC is equilateral.

5. O is the origin and $\overrightarrow{OA}=(1,2)$, $\overrightarrow{OB}=(3,4)$ and $\overrightarrow{OC}=(x,-2)$. Find by using vector method,

(*i*) x so that points A, B, C are collinear.

(*ii*) x so that points A, B, C form a triangle, right angled at B.

6. If $\vec{a}, \vec{b}$ are non-collinear and $\vec{c}=(x+4y)\vec{a}+(2x+y+1)\vec{b}$,

$$\vec{d}=(y-2x+2)\vec{a}+(2x-3y-1)\vec{b}\text{ , find } x, y \text{ such that } 3\vec{c}=2\vec{d}.$$

7. If $\vec{a}=\hat{i}+\hat{j}+\hat{k}$, $\vec{b}=2\hat{i}+3\hat{j}$, $\vec{c}=3\hat{i}+5\hat{j}-2\hat{k}$, $\vec{d}=\hat{k}-\hat{j}$ prove that the vectors $(\vec{b}-\vec{a})$ and $(\vec{d}-\vec{c})$ are parallel. Also find the ratio of their moduli.

8. If G is the centroid of a triangle ABC, prove that $\overrightarrow{GA}+\overrightarrow{GB}+\overrightarrow{GC}=\vec{0}$

[**Hint.** Let $\vec{a}, \vec{b}, \vec{c}$ be the position vectors of the vertices A, B, C respectively of ΔABC with reference to some origin O. Then the position vector of G is $\dfrac{\vec{a}+\vec{b}+\vec{c}}{3} \Rightarrow \overrightarrow{GA}=\dfrac{\vec{a}+\vec{b}+\vec{c}}{3}-\vec{a}$]

9. If ABC and $A'B'C'$ are two triangles and G, G' are their centroids, prove that $\overrightarrow{AA'}+\overrightarrow{BB'}+\overrightarrow{CC'}=3\overrightarrow{GG'}$ ·

[**Hint.** Let $\vec{a}, \vec{b}, \vec{c}$ be the position vectors of the verities A, B, C respectively of ΔABC with reference to some origin O and $\vec{a}', \vec{b}', \vec{c}'$ be the position vectors of the vertices A', B', C' respectively of $\Delta A'B'C'$ with reference to O. Then $\overrightarrow{AA'}=\vec{a}'-\vec{a}$.]

10. Show that the triangle ABC whose vertices A, B, C are $7\hat{j}+10\hat{k}, -\hat{i}+6\hat{j}+6\hat{k}$, $-4\hat{i}+9\hat{j}+6\hat{k}$ is isosceles and right angled. Also find the length of the median form A.

ANSWERS

1. $(-1,-2,8)$ **2.** (*i*) $\frac{1}{3}(\hat{i}+2\hat{j}-2\hat{k})$ (*ii*) $\frac{1}{3\sqrt{2}}(4\hat{i}-\hat{j}+\hat{k})$

3. $\alpha=2, \beta=4$ **5.** (*i*) -3 (*ii*) 9

6. $x=2, y=-1$. **7.** $1:3$

10. $\sqrt{\frac{45}{2}}$ Units.

5.18 SCALAR (OR DOT) PRODUCT OF VECTORS

Let $\vec{a}$ and $\vec{b}$ be two vectors and let Q be the angle between them. Then the scalar or dot produced of two (non-zero) vectors $\vec{a}$ and $\vec{b}$ denoted by $\vec{a} \cdot \vec{b}$ is defined as

$$\vec{a}.\vec{b} = |\vec{a}||\vec{b}| \cos\theta = ab\cos\theta$$

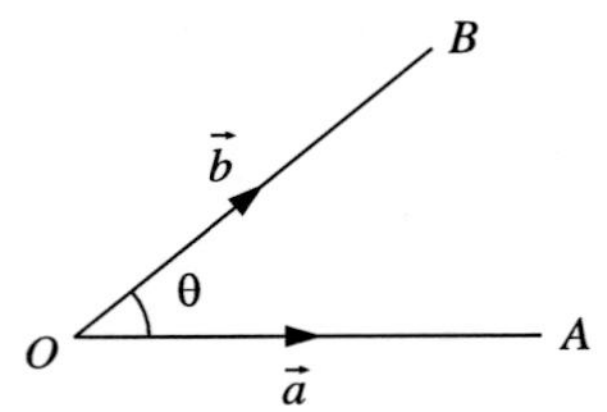

where $a = |\vec{a}|$, $b = |\vec{b}|$ and $\theta (0 \le \theta \le \pi)$ is the angle between $\vec{a}$ and $\vec{b}$.

Clearly; the scalar product of two vectors is a scalar.

5.19 ANGLE BETWEEN TWO VECTORS IN TERMS OF SCALAR PRODUCT

Let θ be the angle between two non-zero vectors $\vec{a}$ and $\vec{b}$, then,

$$\vec{a} \cdot \vec{b} = |\vec{a}||\vec{b}|\cos\theta$$

$$\therefore \qquad \cos\theta = \frac{\vec{a}.\vec{b}}{|\vec{a}||\vec{b}|} \Rightarrow \theta = \cos^{-1}\left(\frac{\vec{a}.\vec{b}}{|\vec{a}||\vec{b}|}\right)$$

5.20 LENGTH OF VECTOR

Let $\vec{a}$ be any vector. Then $\vec{a}.\vec{a} = |\vec{a}||\vec{a}|\cos 0^\circ = |\vec{a}|^2$

$\therefore |\vec{a}| = \sqrt{\vec{a}.\vec{a}}$

Remarks (*i*) If $\vec{a} = \vec{0}$ or $\vec{b} = \vec{0}$, we define $\vec{a}.\vec{b} = 0$.

(*ii*) If $\vec{a}$ and $\vec{b}$ are like vectors, we have $\theta = 0$.

$\therefore \vec{a}.\vec{b} = ab\cos 0^\circ = ab$

(*iii*) If $\vec{a}$ and $\vec{b}$ unlike vectors, we have $\theta = \pi$

$\therefore \vec{a}.\vec{b} = ab\cos\pi = -ab$

(*iv*) If $\vec{a}$ and $\vec{b}$ are orthogonal vectors, we have $\theta = \pi/2$

$\therefore \vec{a}.\vec{b} = ab\cos\pi/2 = 0$

5.21 ORTHONORMAL VECTOR TRIAD

Let $\hat{i}$, $\hat{j}$, $\hat{k}$ be unit vectors along three mutually perpendicular coordinate axes, the x-axis, y-axis and z-axis respectively. Then, these vectors are said to form an orthonormal triad of vectors.

Clearly, we have

(*i*) $\hat{i}.\hat{i} = |\hat{i}||\hat{i}| \cos 0° = 1$

Similarly $\hat{j}.\hat{j} = 1$, $\hat{k} \cdot \hat{k} = 1$

Thus $\hat{i} \cdot \hat{i} = \hat{j}.\hat{j} = \hat{k}.\hat{k} = 1$

(*ii*) $\hat{i}.\hat{j} = |\hat{i}||\hat{j}| \cos\frac{\pi}{2} = 0$

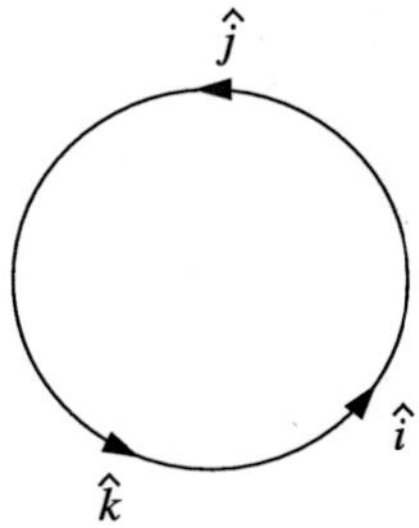

Similarly $\hat{i}.\hat{k} = 0$, $\hat{j}.\hat{k} = 0$, $\hat{j}.\hat{i} = 0$, $\hat{k}.\hat{i} = 0$ and $\hat{k}.\hat{j} = 0$

$\therefore\ \hat{i}.\hat{j} = \hat{i}.\hat{k} = \hat{j}.i = \hat{j}.\hat{k} = \hat{k}.\hat{i} = \hat{k}.\hat{j} = 0$

5.22 GEOMETRICAL INTERPRETATION OF SCALAR PRODUCT

Let $\overrightarrow{OA} = \vec{a}$, $\overrightarrow{OB} = \vec{b}$ and $\angle AOB = \theta$. Draw BM $\perp$ OA.

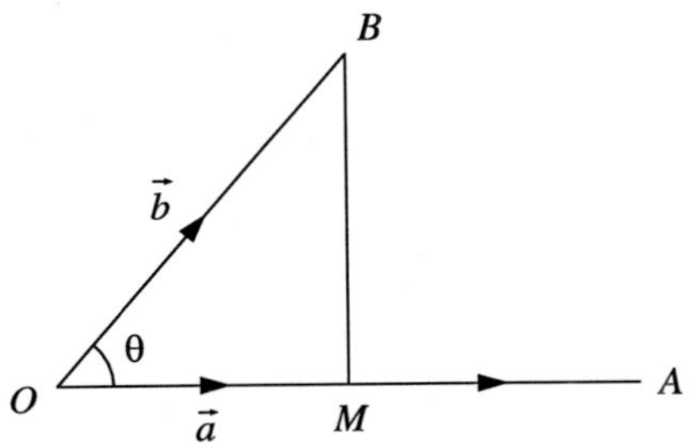

Then, $\vec{a}.\vec{b} = |\vec{a}||\vec{b}| \cos\theta = a\,(b\cos\theta)$

$= a(OB\cos\theta) = a(OM)$ $[\because OB\cos\theta = OM]$

$= a$ (*Projection* of $\vec{b}$ on $\vec{a}$)

Thus, $\vec{a}.\vec{b} = a$ (*Projection* of $\vec{b}$ on $\vec{a}$)

Similarly, $\vec{b}.\vec{a} = b$ (*Projection* of $\vec{a}$ on $\vec{b}$).

Remarks: As a consequence of the above results, we have

(*i*) Projection of $\vec{a}$ on $\vec{b} = \dfrac{\vec{a}.\vec{b}}{|\vec{b}|}$

(*ii*) Projection of $\vec{b}$ on $\vec{a} = \dfrac{\vec{a}.\vec{b}}{|\vec{a}|}$

5.23 PHYSICAL SIGNIFICANCE OF SCALAR PRODUCT

Let O be a particle and let $\overrightarrow{OA} = \vec{a}$ be a force of magnitude a. Let $\overrightarrow{OB} = \vec{b}$ be the displacement due to this force. Draw $BM \perp OA$.

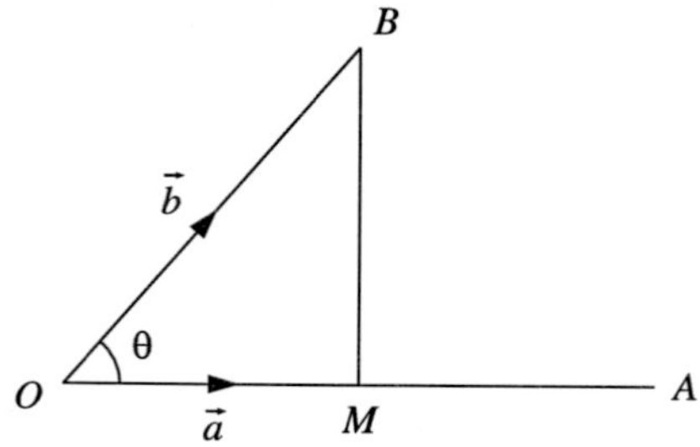

Then, $OM = OB \cos \theta = b \cos \theta$

$\therefore$ displacement in the direction of force $= OM = b \cos \theta$.

$\therefore$ Work done $= a(b \cos \theta) = \vec{a}.\vec{b}$

Thus, the work done by a force $\vec{a}$ acting on a particle in displacing it though $\vec{b}$ is $\vec{a}.\vec{b}$.

5.24 PROPERTIES OF SCALAR PRODUCT

Theorem 1. (Commutative law) For any two vectors $\vec{a}$ and $\vec{b}$, $\vec{a}.\vec{b} = \vec{b}.\vec{a}$

Proof: If $\vec{a}$ or $\vec{b}$ is a zero vector then $\vec{a}.\vec{b} = 0$ and $\vec{b}.\vec{a} = 0$. So, in this case, $\vec{a}.\vec{b} = \vec{b}.\vec{a}$

Now, let $\vec{a}$ and $\vec{b}$ be any two non-zero vectors, and let θ be the angle between them. Them,

$$\vec{a}.\vec{b} = |\vec{a}||\vec{b}| \cos \theta = ab \cos \theta, \text{ and}$$

$$\vec{b}.\vec{a} = |\vec{b}||\vec{a}| \cos (-\theta) = ba \cos \theta = ab \cos \theta$$

So, in this case also, $\vec{a}.\vec{b} = \vec{b}.\vec{a}$

Hence, $\vec{a}.\vec{b} = \vec{b}.\vec{a}$

Theorem 2. (Condition of perpendicularity of two vectors)

For any two vectors $\vec{a}$ and $\vec{b}$, $\vec{a}.\vec{b} = 0 \Leftrightarrow \vec{a} = \vec{0}$ or $\vec{b} = \vec{0}$ or $\vec{a} \perp \vec{b}$.

Proof: Let $\vec{a}.\vec{b} = 0$

Then, $\vec{a}.\vec{b} = 0 \Rightarrow ab \cos \theta = 0$

$$\Rightarrow a = 0 \text{ or } b = 0 \text{ or } \cos \theta = 0$$

$$\Rightarrow a = 0 \text{ or } b = 0 \text{ or } \cos\theta = \pi/2$$

$$\Rightarrow \vec{a} = \vec{0} \text{ or } \vec{b} = \vec{0} \text{ or } \vec{a} \perp \vec{b}$$

Thus $\vec{a}.\vec{b} = 0 \Rightarrow \vec{a} = 0$ or $\vec{b} = 0$ or $\vec{a} \perp \vec{b}$

Conversely, let $\vec{a} = \vec{0}$ or $\vec{b} = \vec{0}$ or $\vec{a} \perp \vec{b}$

If $\vec{a} = \vec{0}$ or $\vec{b} = \vec{0}$, then by defination, $\vec{a}.\vec{b} = 0$

If $\vec{a} \perp \vec{b}$ then $\vec{a}.\vec{b} = ab\cos\dfrac{\pi}{2} = 0$

$\therefore \vec{a} = \vec{0}$ or $\vec{b} = \vec{0}$ or $\vec{a} \perp \vec{b} \Rightarrow \vec{a}.\vec{b} = 0$

Hence $\vec{a}.\vec{b} = 0 \Leftrightarrow \vec{a} = \vec{0}$ or $\vec{b} = \vec{0}$ or $\vec{a} \perp \vec{b}$

Theorem 3. For any two vectors $\vec{a}$ and $\vec{b}$,

(*i*) $\vec{a}.(-\vec{b}) = -(\vec{a}.\vec{b}) = (-\vec{a}).\vec{b}$

(*ii*) $(-\vec{a}).(-\vec{b}) = \vec{a}.\vec{b}$

Theorem 4. (Distributive law) For any three vectors $\vec{a}, \vec{b}, \vec{c}$,

$$\vec{a}.(\vec{b} + \vec{c}) = \vec{a}.\vec{b} + \vec{a}.\vec{c}$$

Corollary 1. $\vec{a}.(\vec{b} - \vec{c}) = (\vec{a}.\vec{b} - \vec{a}.\vec{c})$

Theorem 5. If $\vec{a}, \vec{b}$ are any two vectors and m is any real number (scalar) then

$$(m\vec{a}).\vec{b} = m\cdot(\vec{a}.\vec{b}) = \vec{a}\cdot(m\vec{b})$$

Theorem 6. For any two vectors $\vec{a}$ and $\vec{b}$,

(*i*) $(\vec{a} + \vec{b})^2 = (\vec{a})^2 + (\vec{b})^2 + 2\vec{a}.\vec{b}$

(*ii*) $(\vec{a} - \vec{b})^2 = (\vec{a})^2 + (\vec{b})^2 - 2\vec{a}.\vec{b}$

(*iii*) $(\vec{a} + \vec{b}).(\vec{a} - \vec{b}) = (\vec{a})^2 - (\vec{b})^2$

(*iv*) $(\vec{a} + \vec{b})^2 + (\vec{a} - \vec{b})^2 = 2[(\vec{a})^2 + (\vec{b})^2]$

(*v*) $(\vec{a} + \vec{b})^2 - (\vec{a} - \vec{b})^2 = 4\vec{a}.\vec{b}$

Corollary 1. If $\vec{a}.\vec{b} = \vec{a}.\vec{c}$ then either $\vec{a} = 0$ or $\vec{b} = \vec{c}$ or $\vec{a}$ is perpendicular to $(\vec{b} - \vec{c})$

5.25 SCALAR PRODUCT IN TERMS OF RECTANGULAR COMPONENTS

Let $\vec{a} = a_1\hat{i} + a_2\hat{j} + a_3\hat{k}$ and $\vec{b} = b_1\hat{i} + b_2\hat{j} + b_3\hat{k}$, then

$\vec{a}.\vec{b} = a_1b_1 + a_2b_2 + a_3b_3$

Corollary 1. If $\vec{a} = a_1\hat{i} + a_2\hat{j} + a_3\hat{k}$, then $|\vec{a}| = \sqrt{a_1^2 + a_2^2 + a_3^2}$

Corollary 2. Let $\vec{a} = a_1\hat{i} + a_2\hat{j} + a_3\hat{k}$, $\vec{b} = b_1\hat{i} + b_2\hat{j} + b_3\hat{k}$, then $\vec{a}$ and $\vec{b}$ are perpendicular iff $\vec{a}.\vec{b} = \vec{0}$ i.e. iff $a_1b_1 + a_2b_2 + a_3b_3 = 0$

Thus $\vec{a}$ and $\vec{b}$ are perpendicular iff $a_1b_1 + a_2b_2 + a_3b_3 = 0$

Also note that $\vec{a}$ and $\vec{b}$ are parallel (collinear) iff

$$\frac{a_1}{b_1} = \frac{a_2}{b_2} = \frac{a_3}{b_3}$$

5.26 ANGLE BETWEEN TWO VECTORS

Let θ be the angle between two (non-zero) vectors $\vec{a}$ and $\vec{b}$ then $\vec{a}.\vec{b} = ab\cos\theta$ where $a = |\vec{a}|$ and $b = |\vec{b}|$

$$\Rightarrow \qquad \cos\theta = \frac{\vec{a}.\vec{b}}{ab} \qquad ...(1)$$

Let $\vec{a} = a_1\hat{i} + a_2\hat{j} + a_3\hat{k}$ and $\vec{b} = b_1\hat{i} + b_2\hat{j} + b_3\hat{k}$, then

$$\vec{a}.\vec{b} = a_1b_1 + a_2b_2 + a_3b_3$$

and
$$a = |\vec{a}| = \sqrt{a_1^2 + a_2^2 + a_3^2}, b = |\vec{b}| = \sqrt{b_1^2 + b_2^2 + b_3^2}$$

From (1) we get,
$$\cos\theta = \frac{a_1b_1 + a_2b_2 + a_3b_3}{\sqrt{a_1^2 + a_2^2 + a_3^3}\sqrt{b_1^2 + b_2^2 + b_3^2}}$$

$$\Rightarrow \qquad \theta = \cos^{-1}\left(\frac{a_1b_1 + a_2b_2 + a_3b_3}{\sqrt{a_1^2 + a_2^2 + a_3^2}\sqrt{b_1^2 + b_2^2 + b_3^2}}\right)$$

5.27 WORK DONE BY A FORCE

If a force $\vec{F}$ causes a displacement $\vec{d}$, then the work done by the force $\vec{F}$ is the product of the magnitude of displacement $\vec{d}$ and the magnitude of components of $\vec{F}$ along the displacement $\vec{d}$.

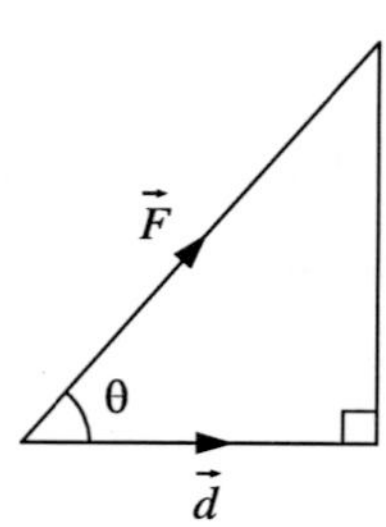

$\therefore$ work done by $\vec{F} = |\vec{d}||F\cos\theta|$

$$= |\vec{F}.\vec{d}|$$

Note: Force and displacement are vectors, whereas work done is a scalar quantity.

Remarks. If the dot product of the force vector and the displacement vector is 0, then it means that the displacement is at right angle to the force.

ILLUSTRATIVE EXAMPLES

Example 1. Let $\vec{a} = 2\hat{i} + 3\hat{j} - \hat{k}, \vec{b} = -\hat{i} + 3\hat{j} + 4\hat{k}$. Evolute

(*i*) $|\vec{a}|, |\vec{b}|$ (*ii*) $\vec{a}.\vec{b}$ (*iii*) the angle between the vectors $\vec{a}$ and $\vec{b}$

(*iv*) the projection of $\vec{a}$ on $\vec{b}$ (*v*) the projection of $\vec{b}$ on $\vec{a}$.

(*vi*) area of the ΔAOB, where O is the origin.

Solution. Given $\vec{a} = 2\hat{i} + 3\hat{j} - \hat{k}, \vec{b} = -\hat{i} + 3\hat{j} + 4\hat{k}$

(*i*) $|\vec{a}| = \sqrt{(2)^2 + (3)^2 + (-1)^2} = \sqrt{4+9+1} = \sqrt{14}$

$|\vec{b}| = \sqrt{(-1)^2 + (3)^2 + (4)^2} = \sqrt{1+9+16} = \sqrt{26}$

(*ii*) $\vec{a}.\vec{b} = (2)(-1) + (3)(3) + (-1)(4)$

$= -2 + 9 - 4 = 3$

(*iii*) The angle θ between the vectors $\vec{a}$ and $\vec{b}$ is given by

$$\cos\theta = \frac{\vec{a}.\vec{b}}{|\vec{a}||\vec{b}|} = \frac{3}{\sqrt{14}\sqrt{26}} = \frac{3}{2\sqrt{91}}$$

(*iv*) The Projection of $\vec{a}$ and $\vec{b}$ $= \dfrac{\vec{a}.\vec{b}}{|\vec{b}|} = \dfrac{3}{\sqrt{26}}$

(*v*) The Projection of $\vec{b}$ and $\vec{a}$ $= \dfrac{\vec{a}.\vec{b}}{|\vec{a}|} = \dfrac{3}{\sqrt{14}}$

(*vi*) Area of $\Delta AOB = \dfrac{1}{2}|\vec{a}||\vec{b}|\sin\theta$

$$\text{Now } \sin^2\theta = 1 - \cos^2\theta = 1 - \left(\frac{3}{2\sqrt{91}}\right)^2 = 1 - \frac{9}{364} = \frac{355}{364}$$

$$\therefore \text{Area of } \Delta AOB = \frac{1}{2}\sqrt{14}\sqrt{26}.\sqrt{\frac{355}{364}}$$

$$= \frac{1}{2}\sqrt{355} = 9.42 \text{ sq. units approx.}$$

Example 2. If $(\vec{a} - \vec{b}).(\vec{a} + \vec{b}) = 8$ and $|\vec{a}| = 8|\vec{b}|$, find $|\vec{a}|$ and $|\vec{b}|$.

Solution. Given $(\vec{a} - \vec{b}).(\vec{a} + \vec{b}) = 8$

$$\Rightarrow \quad (\vec{a})^2 - (\vec{b})^2 = 8 \Rightarrow |\vec{a}|^2 - |\vec{b}|^2 = 8$$

$$\Rightarrow \quad 64|\vec{b}|^2 - |\vec{b}|^2 = 8 \qquad (\because |\vec{a}| = 8|\vec{b}|, \text{ given})$$

$$63|\vec{b}|^2 = 8 \Rightarrow |\vec{b}| = \sqrt{\frac{8}{63}}$$

$\therefore \qquad |\vec{a}| = 8|\vec{b}| = 8\sqrt{\frac{8}{63}}$

Example 3. If the angle between two vectors $\vec{a}$ and $\vec{b}$ of equal magnitude is 30° and their scalar product is $2\sqrt{3}$, find their magnitudes.

Solution. Given $|\vec{a}| = |\vec{b}|$, θ (angle between $\vec{a}$ and $\vec{b}$) = 30°

and $\vec{a}.\vec{b} = 2\sqrt{3}$

$$\Rightarrow \qquad |\vec{a}||\vec{b}| \cos 30° = 2\sqrt{3} \Rightarrow |\vec{a}|^2 .\sqrt{\frac{3}{2}} = 2\sqrt{3}$$

$$\Rightarrow \qquad |\vec{a}|^2 = 4 \Rightarrow |\vec{a}| = 2$$

Hence $\qquad |\vec{a}| = 2 = |\vec{b}|$

Example 4. Find the value of λ for which the vectors $\vec{a} = 3\hat{i} + \hat{j} - 2\hat{k}$ and $\vec{b} = \hat{i} + \lambda\hat{j} - 3\hat{k}$ are perpendicular to each other.

Solution. Given $\vec{a} = 3\hat{i} + \hat{j} - 2\hat{k}$ and $\vec{b} = \hat{i} + \lambda\hat{j} - 3\hat{k}$

$$\vec{a} \perp \vec{b} \Rightarrow \vec{a}.\vec{b} = 0$$

$$\Rightarrow \qquad (3\hat{i} + \hat{j} - 2\hat{k}).(\hat{i} + \lambda\hat{j} - 3\hat{k}) = 0$$

$$\Rightarrow \qquad 3 \times 1 + 1 \times \lambda + (-2) \times (-3) = 0$$

$$\Rightarrow \qquad \lambda = -9$$

Hence the required value of λ is – 9.

Example 5. If $\vec{a} = \hat{i} + \hat{j} - 2\hat{k}$ and $\vec{b} = 3\hat{i} + 2\hat{j} - \hat{k}$, find the value of $(\vec{a} + 3\vec{b}).(2\vec{a} - \vec{b})$.

Solution. We have

$$(\vec{a} + 3\vec{b}) = (\hat{i} + \hat{j} + 2\hat{k}) + 3(3\hat{i} + 2\hat{j} - \hat{k})$$

$$= (\hat{i} + \hat{j} + 2\hat{k}) + (9\hat{i} + 6\hat{j} - 3\hat{k}) = 10\hat{i} + 7\hat{j} - \hat{k}$$

and

$$(2\vec{a} - \vec{b}) = 2(\hat{i} + \hat{j} + 2\hat{k}) - (3\hat{i} + 2\hat{j} - \hat{k})$$

$$= (2\hat{i} + 2\hat{j} + 4\hat{k}) - (3\hat{i} + 2\hat{j} - \hat{k})$$

$$= -\hat{i} + 5\hat{k}$$

$$\therefore \qquad (\vec{a} + 3\vec{b}).(2\vec{a} - \vec{b}) = (10\hat{i} + 7\hat{j} - \hat{k}).(-\hat{i} + 0\hat{j} + 5\hat{k})$$

$$= 10 \times (-1) + 7 \times 0 + (-1) \times 5 = -15$$

Hence $\qquad (\vec{a} + 3\vec{b}).(2\vec{a} - \vec{b}) = -15.$

Example 6. Find λ when the scalar projection of $\vec{a} = \lambda\hat{i} + \hat{j} + 4\hat{k}$ on $\vec{b} = 2\hat{i} + 6\hat{j} + 3\hat{k}$ is 4 units.

Solution. The scalar projection of $\vec{a}$ on $\vec{b}$ = $\dfrac{\vec{a}.\vec{b}}{|\vec{b}|} = 4$ (given) ...(1)

Here $\vec{a}.\vec{b} = \lambda.2 + 1.6 + 4.3 = 2\lambda + 18$ and $|\vec{b}| = \sqrt{2^2 + 6^2 + 3^2}$

$= \sqrt{49} = 7$

From (1), we get $\dfrac{2\lambda + 18}{7} = 4 \Rightarrow \lambda = 5.$

Example 7. If $\vec{a} = \hat{i} + 2\hat{j} - 3\hat{k}$ and $\vec{b} = 3\hat{i} - \hat{j} + 2\hat{k}$, show that $\vec{a} + \vec{b}$ is perpendicular to $\vec{a} - \vec{b}$.

Solution. Given $\vec{a} = \hat{i} + 2\hat{j} - 3\hat{k}$ and $\vec{b} = 3\hat{i} - \hat{j} + 2\hat{k}$

Now $\vec{a} + \vec{b} = 4\hat{i} + \hat{j} - \hat{k}$ and $\vec{a} - \vec{b} = -2\hat{i} + 3\hat{j} - 5\hat{k}$

$\therefore$ $(\vec{a} + \vec{b}).(\vec{a} - \vec{b}) = 4(-2) + 1.3 + (-1).(-5)$

$= -8 + 3 + 5 = 0$

$\Rightarrow \vec{a} + \vec{b}$ is perpendicular to $\vec{a} - \vec{b}$

Example 8. Find the value of λ so that the two vectors $2\hat{i} + 3\hat{j} - \hat{k}$ and $-4\hat{i} - 6\hat{j} + \lambda\hat{k}$ are (*i*) parallel (*ii*) perpendicular to each other.

Solution. Let $\vec{a} = 2\hat{i} + 3\hat{j} - \hat{k}$, $\vec{b} = -4\hat{i} - 6\hat{j} + \lambda\hat{k}$

(*i*) $\vec{a}$ and $\vec{b}$ are parallel to each other iff $\dfrac{a_1}{b_1} = \dfrac{a_2}{b_2} = \dfrac{a_3}{a_3}$

$\Leftrightarrow$ $\dfrac{2}{-4} = \dfrac{3}{-6} = \dfrac{-1}{\lambda}$

$\Leftrightarrow$ $\lambda = 2$

(*ii*) $\vec{a}$ and $\vec{b}$ are perpendicular to each other iff $\vec{a}.\vec{b} = 0$

i.e. if $(2)(-4) + (3)(-6) + \lambda\cdot(-1) = 0$

$\Rightarrow$ $\lambda = -8 - 18 = -26$

Example 9. The vectors $\vec{a} = 3\hat{i} + x\hat{j} - \hat{k}$ and $\vec{b} = 2\hat{i} + \hat{j} + y\hat{k}$ are mutually perpendicular. Given that $|\vec{a}| = |\vec{b}|$ find the values of x and y.

Solution. Since $\vec{a} \perp \vec{b}$, $\vec{a}.\vec{b} = \vec{0}$

$\Rightarrow$ $3.2 + x.1 + (-1)y = 0 \Rightarrow y - x = 6$ (1)

Given $|\vec{a}| = \vec{b} \Rightarrow |\vec{a}|^2 = |\vec{b}|^2$

$\Rightarrow$ $3^2 + x^2 + (-1)^2 = 2^2 + 1^2 + y^2 \Rightarrow y^2 - x^2 = 5$

$\Rightarrow$ $(y-x)(y+x) = 5$

$\Rightarrow$ $6(y+x) = 5$ (using (1))

$\Rightarrow$ $y + x = 5\backslash 6$ (2)

Solving (1) and (2), we get $x = -\dfrac{31}{12}$ and $y = \dfrac{41}{12}$.

Example 10. If $\overrightarrow{AB} = 2\hat{i} + \hat{j} - 3\hat{k}$ and the coordinate of A are (1, 2, –1) find the coordinate of B.

Solution. Let O be the origin. Then $\overrightarrow{OA} = \hat{i} + 2\hat{j} - \hat{k}$

$\therefore$ $\overrightarrow{AB} = \overrightarrow{OB} - \overrightarrow{OA}$

$\Rightarrow$ $\overrightarrow{OB} = \overrightarrow{AB} + \overrightarrow{OA}$

$= (2\hat{i} + \hat{j} - 3\hat{k}) + (\hat{i} + 2\hat{j} - \hat{k})$

$\overrightarrow{OB} = (3\hat{i} + 3\hat{j} - 4\hat{k})$

Hence the coordinates of B are (3, 3, –4)

Example 11. If $\vec{a}$ and $\vec{b}$ are vectors such that $|\vec{a}| = 2$, $|\vec{b}| = 3$ and $\vec{a}.\vec{b} = 4$, find $|\vec{a} - \vec{b}|$.

Solution. $|\vec{a} - \vec{b}|^2 = (\vec{a} - \vec{b}).(\vec{a} - \vec{b}) = |\vec{a}|^2 + |\vec{b}|^2 - 2\vec{a}.\vec{b} = 2^2 + 3^2 - 2 \times 4 = 5$

Hence, $|\vec{a} - \vec{b}| = \sqrt{5}$.

Example 12. If $\vec{a}$ and $\vec{b}$ are vectors such that $|\vec{a}| = |\vec{b}| = \sqrt{2}$ and $\vec{a}.\vec{b} = -1$, find the angle between $\vec{a}$ and $\vec{b}$

Solution. Let θ be the angle between $\vec{a}$ and $\vec{b}$.

$\vec{a}.\vec{b} = -1$ (given)

$\Rightarrow$ $|\vec{a}||\vec{b}|\cos\theta = -1$

$\Rightarrow$ $\sqrt{2}\sqrt{2} \times \cos\theta = -1$

$\Rightarrow$ $\cos\theta = -\dfrac{1}{2}$

$\Rightarrow$ $\theta = \dfrac{2\pi}{3}$ $[\because 0 \le \theta \le 2\pi]$

Hence, the angle between $\vec{a}$ and $\vec{b}$ is $\dfrac{2\pi}{3}$

Example 13. If $\vec{a}$ makes equal angles with the coordinate axes and has magnitude 3, then find the angle between $\vec{a}$ and each of the three coordinate axes.

Solution. Let $\vec{a} = a_1\hat{i} + a_2\hat{j} + a_3\hat{k}$ and θ be the angle between $\vec{a}$ and each of the three coordinate axes. Since $\hat{i}, \hat{j}, \hat{k}$ are unit vectors along the coordinate axes, we have

$$\cos\theta = \frac{\vec{a}.\hat{i}}{|\vec{a}||\hat{i}|} = \frac{a_1}{3} \qquad \left(\because |\vec{a}| = 3 \text{ and } \vec{a}.\hat{i} = a_1\right)$$

$\Rightarrow$ $a_1 = 3\cos\theta$

Similarly, $a_2 = 3\cos\theta$ and $a_3 = 3\cos\theta$

Now $|\vec{a}| = 3 \Rightarrow \sqrt{a_1^2 + a_2^2 + a_3^2} = 3$

$\Rightarrow \sqrt{9\cos^2\theta + 9\cos^2\theta + 9\cos^2\theta} = 3$

$\Rightarrow$ $3\sqrt{3}\cos\theta = 3$

$\Rightarrow$ $\cos\theta = \dfrac{1}{\sqrt{3}} \Rightarrow \theta = \cos^{-1}\left(\dfrac{1}{\sqrt{3}}\right).$

Example 14. Show that the vectors $\vec{a} = 3\hat{i} - 2\hat{j} + \hat{k}$, $\vec{b} = \hat{i} - 3\hat{j} + 5\hat{k}$, $\vec{c} = 2\hat{i} + \hat{j} - 4\hat{k}$ form a right angle triangle.

Solution. We have $\vec{b} + \vec{c} = (\hat{i} - 3\hat{j} + 5\hat{k}) + (2\hat{i} + \hat{j} - 4\hat{k}) = 3\hat{i} - 2\hat{j} + \hat{k} = \vec{a}$

$\Rightarrow \vec{a}, \vec{b}, \vec{c}$ are coplanar.

Also we observe that no two of these vectors are parallel, therefore, the given vectors form a triangle.

Further, $\vec{a}.\vec{c} = (3\hat{i} - 2\hat{j} + \hat{k}).(2\hat{i} + \hat{j} - 4\hat{k})$

$= (3)(2) + (-2)(1) + (1)(-4) = 0$

Thus, the dot product of the non-zero vectors, $\vec{a}$ and $\vec{c}$ is zero, therefore these vectors are perpendicular i.e. the sides represented by the vectors $\vec{a}$ and $\vec{c}$ are perpendicular. Hence the given vectors form a right angled triangle.

Example 15. If $\vec{a} \cdot \vec{b} = \vec{a} \cdot \vec{c}$, show that $\vec{a} = \vec{0}$ or $\vec{b} = \vec{c}$ or $\vec{a} \perp (\vec{b} - \vec{c})$.

Solution: $\vec{a} \cdot \vec{b} = \vec{a} \cdot \vec{c} \Rightarrow \vec{a} \cdot \vec{b} - \vec{a} \cdot \vec{c} = \vec{0}$

$\Rightarrow$ $\vec{a} \cdot (\vec{b} - \vec{c}) = \vec{0}$

$\Rightarrow$ $\vec{a} = \vec{0}$ or $(\vec{b} - \vec{c}) = \vec{0}$ or $\vec{a} \perp (\vec{b} - \vec{c})$

$$\Rightarrow \qquad \vec{a} = \vec{0} \text{ or } \vec{b} = \vec{c} \text{ or } \vec{a} \perp (\vec{b} - \vec{c})$$

Hence, $\vec{a} \cdot \vec{b} = \vec{a} \cdot \vec{c} \Rightarrow \vec{a} = \vec{0}$ or $\vec{b} = \vec{c}$ or $\vec{a} \perp (\vec{b} - \vec{c})$.

Example 16. If $|\vec{a} + \vec{b}| = |\vec{a} - \vec{b}|$, prove that $\vec{a}$ and $\vec{b}$ are perpendicular to each other.

Solution: Let $|\vec{a} + \vec{b}| = |\vec{a} - \vec{b}|$. then

$$|\vec{a} + \vec{b}|^2 = |\vec{a} - \vec{b}|^2$$

$$\Rightarrow \qquad (\vec{a} + \vec{b}) \cdot (\vec{a} + \vec{b}) = (\vec{a} - \vec{b}) \cdot (\vec{a} - \vec{b})$$

$$\Rightarrow \qquad \vec{a} \cdot \vec{a} + \vec{b} \cdot \vec{b} + 2\vec{a} \cdot \vec{b} = \vec{a} \cdot \vec{a} + \vec{b} \cdot \vec{b} - 2\vec{a} \cdot \vec{b} \quad [\because \vec{a} \cdot \vec{b} = \vec{b} \cdot \vec{a}\ ]$$

$$\Rightarrow \qquad 4\vec{a} \cdot \vec{b} = 0 \Rightarrow \vec{a} \cdot \vec{b} = 0$$

$$\Rightarrow \qquad \vec{a} \perp \vec{b}$$

Hence $|\vec{a} + \vec{b}| = |\vec{a} - \vec{b}| \Rightarrow \vec{a} \perp \vec{b}$.

Example 17. If $\vec{a}$ and $\vec{b}$ are unit vectors and θ is the angle between them, then show that $\sin\frac{\theta}{2} = \frac{1}{2}|\vec{a} - \vec{b}|$.

Solution: $\vec{a} \cdot \vec{b} = |\vec{a}||\vec{b}| \cos\theta = 1.1 \cdot \cos\theta = \cos\theta$

Now $(\vec{a} - \vec{b})^2 = |\vec{a}|^2 + |\vec{b}|^2 - 2\vec{a} \cdot \vec{b}$

$$= 1 + 1 - 2\cos\theta = 2(1 - \cos\theta) = 2\left(2\sin^2\frac{\theta}{2}\right)$$

Also $|\vec{a} - \vec{b}|^2 = (\vec{a} - \vec{b})^2$

$$\Rightarrow \qquad |\vec{a} - \vec{b}|^2 = 4\sin^2\frac{\theta}{2} \Rightarrow \sin\frac{\theta}{2} = \frac{1}{2}|\vec{a} - \vec{b}|.$$

EXERCISE 5.2

1. Find $\vec{a} \cdot \vec{b}$ when

   (*i*) $\vec{a} = \hat{i} - 2\hat{j} + \hat{k}$ and $\vec{b} = 3\hat{i} - 4\hat{j} - 2\hat{k}$

   (*ii*) $\vec{a} = \hat{i} + 2\hat{j} + 3\hat{k}$ and $\vec{b} = -2\hat{j} + 4\hat{k}$

   (*iii*) $\vec{a} = \hat{i} - \hat{j} + 5\hat{k}$ and $\vec{b} = 3\hat{i} - 2\hat{k}$

2. Evaluate the scalar product $(3\vec{a} - 5\vec{b}) \cdot (2\vec{a} + 7\vec{b})$.

3. If $|\vec{a}| = 3, |\vec{b}| = 4$ and $\vec{a} \cdot \vec{b} = 1$, find $(\vec{a} - \vec{b})^2$

4. If $\vec{a}$ is a unit vector and $(\vec{x} + \vec{a}) \cdot (\vec{x} + \vec{a}) = 12$, find $|\vec{x}|$.

5. Find the value of λ for which $\vec{a}$ and $\vec{b}$ are perpendicular, where

(*i*) $\vec{a} = 2\hat{i} + \lambda\hat{j} + \hat{k}$ and $\vec{b} = 4\hat{i} - 2\hat{j} + 2\hat{k}$

(*ii*) $\vec{a} = 3\hat{i} - \hat{j} + 4\hat{k}$ and $\vec{b} = -\lambda\hat{i} + 3\hat{j} + 3\hat{k}$

6. (*i*) if $\vec{a} = \hat{i} + 2\hat{j} - 3\hat{k}$ and $\vec{b} = 3\hat{i} - \hat{j} + 2\hat{k}$, show that $(\vec{a} + \vec{b})$ is perpendicular to $(\vec{a} - \vec{b})$.

(*ii*) If $\vec{a} = (5\hat{i} - \hat{j} - 3\hat{k})$ and $\vec{b} = (\hat{i} + 3\hat{j} - 5\hat{k})$ then show that $(\vec{a} + \vec{b})$ and $(\vec{a} - \vec{b})$ are orthogonal.

7. Show that the vector $\frac{1}{7}(2\hat{i} + 3\hat{j} + 6\hat{k}), \frac{1}{7}(3\hat{i} - 6\hat{j} + 2\hat{k})$ and $\frac{1}{7}(6\hat{i} + 2\hat{j} - 3\hat{k})$ are mutually perpendicular unit vectors.

8. Find the angle between the vectors $\vec{a}$ and $\vec{b}$, when

(*i*) $\vec{a} = \hat{i} - 2\hat{j} + 3\hat{k}$ and $\vec{b} = 3\hat{i} - 2\hat{j} + \hat{k}$,

(*ii*) $\vec{a} = 3\hat{i} + 2\hat{j} + 2\hat{k}$ and $\vec{b} = 2\hat{i} - 2\hat{j} + 4\hat{k}$

9. Find the angle between $\vec{a}$ and $\vec{b}$, when

(*i*) $|\vec{a}| = 2, |\vec{b}| = 1$ and $\vec{a} \cdot \vec{b} = \sqrt{3}$

(*ii*) $|\vec{a}| = \sqrt{3}, |\vec{b}| = 2$ and $\vec{a} \cdot \vec{b} = \sqrt{6}$

10. Let $\vec{a} = 2\hat{i} + 3\hat{j} + 2\hat{k}$ and $\vec{b} = \hat{i} + 2\hat{j} + \hat{k}$. Find the projection of (*i*) $\vec{a}$ on $\vec{b}$ (*ii*) $\vec{b}$ on $\vec{a}$.

11. Find the projection of $(\hat{i} - \hat{j})$ on $\hat{i} + \hat{j}$.

12. Find the projection of $(8\hat{i} + \hat{j})$ in the direction of $(\hat{i} + 2\hat{j} - 2\hat{k})$.

13. Show that the vector $\vec{a} = \hat{i} - 3\hat{j} - 5\hat{k}$, $\vec{b} = 2\hat{i} - \hat{j} + \hat{k}$ and $\vec{c} = \hat{i} + 2\hat{j} + 6\hat{k}$ form a right angled triangle.

14. If $\vec{a} = 2\hat{i} + 2\hat{j} + 3\hat{k}$, $\vec{b} = -\hat{i} + 2\hat{j} + \hat{k}$ and $\vec{c} = 3\hat{i} + \hat{j}$ such that $\vec{a} + \lambda\vec{b}$ is perpendicular to $\vec{c}$, then find the value of λ.

15. Show that vector $\hat{i} + \hat{j} + \hat{k}$ is equally inclined to the axes.

16. Find the work done by the force $\vec{F} = 2\hat{i} - \hat{j} - \hat{k}$ as its point of application moves from the origin to the point $P(2,-3, 1)$.

17. Using vectors, show that the angle between two diagonals of a cube is $\cos^{-1}\left(\frac{1}{3}\right)$.

18. Find the work done by the force $\vec{F} = \hat{i} + \hat{j} + \hat{k}$ acting on a particle if it is displaced from the point $A(3, 3, 3)$ to the point $B(4, 4, 4)$.

ANSWERS

1. (*i*) 9 (*ii*) 8 (*iii*) –7

2. $6|\vec{a}|^2 + 11\vec{a}\cdot\vec{b} - 35|\vec{b}|^2$ 3. 23 **4.** $\sqrt{13}$

5. (*i*) $\lambda = 5$ (*ii*) $\lambda = 3$ **8.** (*i*) $\cos^{-1}\frac{5}{7}$

(*ii*) $\cos^{-1}\sqrt{\frac{3}{7}}$ **9.** (*i*) 30° (*ii*) 45°

10. (*i*) $\frac{5\sqrt{6}}{3}$ (*ii*) $\frac{10\sqrt{17}}{17}$ **11.** 0

12. $\frac{10}{3}$ **14.** $\lambda = 8$ **16.** 6 units

18. 3 units,

5.28 VECTOR (OR CROSS) PRODUCT OF THE VECTORS

Let $\vec{a}$ and $\vec{b}$ be two nonzero, nonparallel vectors, and let θ be the angle between them such that $0 \le \theta \le \pi$.

Then, the vector product of $\vec{a}$ and $\vec{b}$ is defined as

$$\vec{a}\times\vec{b} = \left(|\vec{a}||\vec{b}|\sin\theta\right)\hat{n}$$

where $\hat{n}$ is a unit vectors perpendicular to both $\vec{a}$ and $\vec{b}$ such that $\vec{a}, \vec{b}, \hat{n}$ form a right handed system.

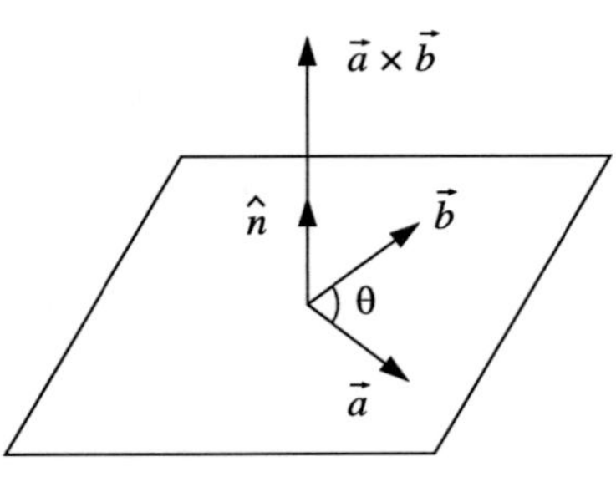

When a right handed screw is rotated from $\vec{a}$ to $\vec{b}$ and it advances along $\hat{n}$ then system is said to be right-handed.

Remarks:

1. If $\vec{a}$ and $\vec{b}$ are parallel or collinear. i.e, when $\theta = 0$ or $\theta = \pi$ then we define,

   $\vec{a}\times\vec{b} = \vec{0}$.

   In particular, $\vec{a}\times\vec{a} = \vec{0}$.

2. If $\vec{a} = \vec{0}$ or $\vec{b} = \vec{0}$, we define $\vec{a}\times\vec{b} = \vec{0}$.
3. For any vector $\vec{a}$, we have $\vec{a}\times\vec{a} = \left(|\vec{a}||\vec{a}|\sin\theta\right)\hat{n} = 0\hat{n} = \vec{0}$.
4. The cross product of two vectors is a vector quantity.
5. If $\vec{a}$ and $\vec{b}$ are unit vectors, then $\left|\vec{a}\times\vec{b}\right| = \sin\theta$. Thus, we find that the modulus of the vector product of two unit vector is the sine of the angle between their directions.
6. If $\vec{a}$ and $\vec{b}$ are perpendicular, then the three vectors $\vec{a}, \vec{b}$ and $\vec{a}\times\vec{b}$ form a right handed triad of mutually perpendicular vectors.

5.29 VECTORS PRODUCT OF UNIT VECTORS $\hat{i}, \hat{j}, \hat{k}$

Since $\hat{i}, \hat{j}, \hat{k}$ are unit vectors along three coordinate axes, these form a right-handed triad of mutually perpendicular unit vectors, we have

$\hat{i}\times\hat{j} = \hat{k} = -\hat{j}\times\hat{i}$, $\hat{j}\times\hat{k} = \hat{i} = -\hat{k}\times\hat{j}$ and $\hat{k}\times\hat{i} = \hat{j} = -\hat{i}\times\hat{k}$.

Also, as $\vec{a}\times\vec{a} = \vec{0}$ any vector $\vec{a}$, we have

$$\hat{i}\times\hat{i} = \hat{j}\times\hat{j} = \hat{k}\times\hat{k} = \vec{0}.$$

5.30 PROPERTIES OF VECTOR (OR CROSS) PRODUCT

Theorem 1. Vector product is not commutative

Proof: Let $\vec{a}$ and $\vec{b}$ be two nonzero, nonparallel vectors, and let θ be the angle between them, where $0 \le \theta \le \pi$.

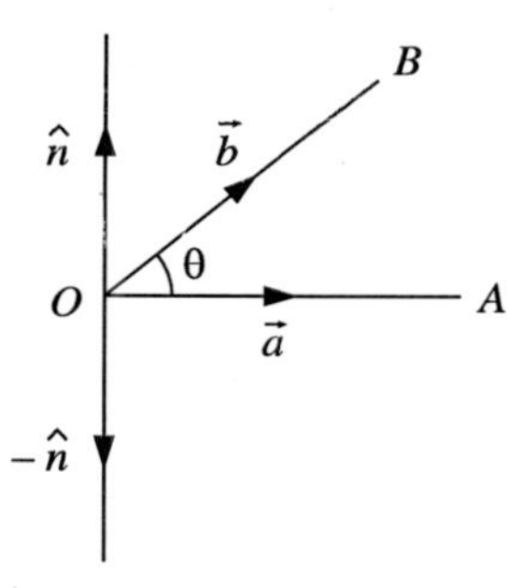

Let $\hat{n}$ be a unit vector perpendicular to both $\vec{a}$ and $\vec{b}$.

Then $\vec{a}$, $\vec{b}$, $\hat{n}$ form a right-handed system.

$$\therefore \quad \vec{a}\times\vec{b} = \left(|\vec{a}||\vec{b}|\sin\theta\right)\hat{n}$$

$$\Rightarrow \quad \vec{a}\times\vec{b} = (ab\sin\theta)\,\hat{n} \quad \text{...(1)}$$

Again, $\vec{b}, \vec{a}$ and $-\hat{n}$ form a right handed system.

$\therefore \quad \vec{b} \times \vec{a} = \left(|\vec{b}||\vec{a}|\sin\theta\right)(-\hat{n})$

$\Rightarrow \quad \vec{b} \times \vec{a} = -(ba \sin\theta)\,\hat{n} = -(ab \sin\theta)\,\hat{n}$

$\Rightarrow \quad \vec{b} \times \vec{a} = -(\vec{a} \times \vec{b})$ [using (1)]

Hence $\vec{b} \times \vec{a} \neq (\vec{a} \times \vec{b})$.

Theorem 2. For any two vectors $\vec{a}$ and $\vec{b}$,

$$(-\vec{b}) \times \vec{a} = \vec{a} \times \vec{b} = \vec{b} \times (-\vec{a})$$

Proof: Let θ be the angle between $\vec{a}$ and $\vec{b}$.

Then, the angle between $(-\vec{b})$ and $\vec{a}$ is $(\pi - \theta)$.

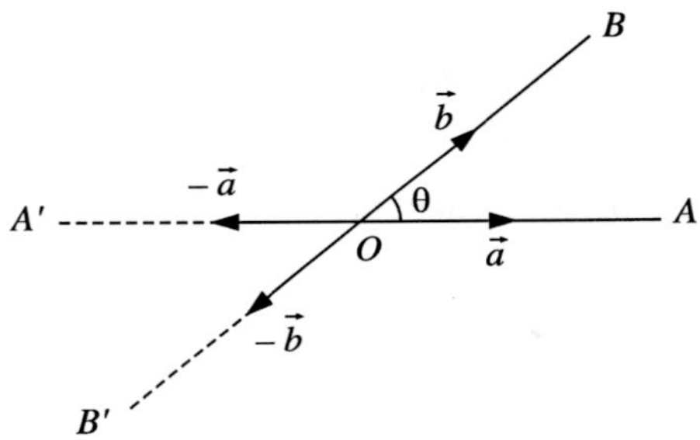

And, $(-\vec{b})$, $\vec{a}$, $\hat{n}$ form a right handed system.

$\therefore (-\vec{b}) \times \vec{a} = |-\vec{b}||\vec{a}| \sin(\pi - \theta)\,\hat{n}$

$= \left(|\vec{b}||\vec{a}|\sin\theta\right)\hat{n}$

$= (ab \sin\theta)\,\hat{n} = (\vec{a} \times \vec{b})$.

$\therefore \quad (-\vec{b}) \times \vec{a} = (\vec{a} \times \vec{b})$. ...(1)

Again, the rotation from $\vec{b}$ to $(-\vec{a})$ is the same as the rotation from $\vec{a}$ to $\vec{b}$, and the angle between $\vec{b}$ and $(-\vec{a})$ is $(\pi - \theta)$.

So, $\vec{b}$, $(-\vec{a})$ and $\hat{n}$ form a right-handed system.

$\therefore \quad \vec{b} \times (-\vec{a}) = \left(|\vec{b}||-\vec{a}|\sin\theta\right)\hat{n}$

$= (ba \sin\theta)\,\hat{n} = (ab \sin\theta)\,\hat{n} = \vec{a} \times \vec{b}$

$\therefore \quad \vec{b} \times (-\vec{a}) = \vec{a} \times \vec{b}$ ...(2)

Form (1) are (2), we get

$$(-\vec{b}) \times \vec{a} = (\vec{a} \times \vec{b}) = \vec{b} \times (-\vec{a}).$$

Theorem 3. For any scalar m, $(m\vec{a})\times\vec{b} = m(\vec{a}\times\vec{b}) = \vec{a}\times(m\vec{b})$.

Proof: Let θ be the angle between $\vec{a}$ and $\vec{b}$.

Let $\hat{n}$ be a unit vector perpendicular to both $\vec{a}$ and $\vec{b}$ such that $\vec{a}, \vec{b}, \hat{n}$ form a right-handed system.

Case I: When $m > 0$

In this case, $m\vec{a}$ and $\vec{a}$ are in the same direction.

∴ angle between $m\vec{a}$ and $\vec{b}$ = angle between $\vec{a}$ and $\vec{b}$ = θ.

Since $\vec{a}, \vec{b}, \hat{n}$ form a right-handed system, $m\vec{a}, \vec{b}, \hat{n}$ also form a right handed system.

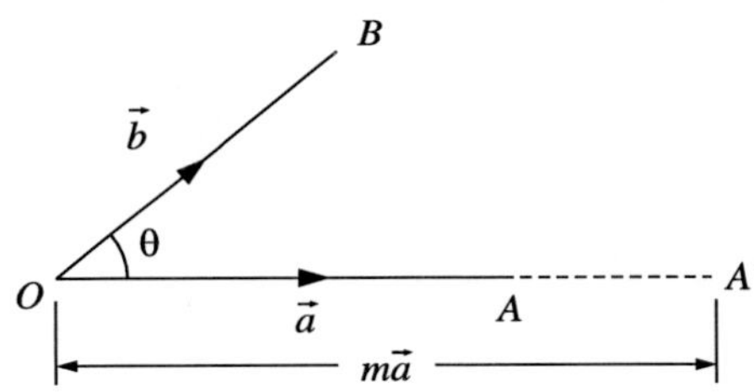

$$\therefore \quad (m\vec{a})\times\vec{b} = \left(|m\vec{a}|\,|\vec{b}|\sin\theta\right)\hat{n}$$

$$= \left(|m|\,|\vec{a}|\,|\vec{b}|\sin\theta\right)\hat{n}$$

$$= \left(m\,|\vec{a}|\,|\vec{b}|\sin\theta\right)\hat{n} \qquad [\because |m| = m]$$

$$= m\left\{\left(|\vec{a}|\,|\vec{b}|\sin\theta\right)\hat{n}\right\} = m(\vec{a}\times\vec{b}).$$

Again, $m\vec{b}$ and $\vec{b}$ are in the same direction.

∴ angle between $\vec{a}$ and $m\vec{b}$ = angle between $\vec{a}$ and $\vec{b}$ = θ.

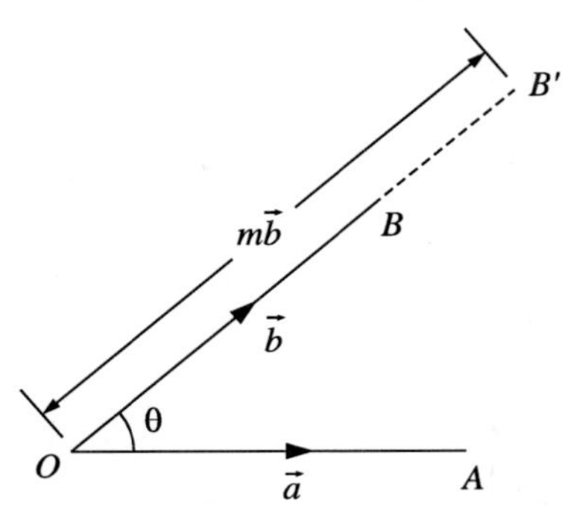

Since $\vec{a}, \vec{b}, \hat{n}$ form a right handed system, $\vec{a}$, $m\vec{b}$, $\hat{n}$ also form a right handed system.

$$\therefore \qquad \vec{a}\times(m\vec{b}) = \left(|\vec{a}|\,|m\vec{b}|\sin\theta\right)\hat{n}$$

$$= \left(|\vec{a}|\,|m|\,|\vec{b}|\sin\theta\right)\hat{n}$$

$$= \left(m|\vec{a}|\,|\vec{b}|\sin\theta\right)\hat{n} \qquad [\because \ |m| = m]$$

$$= m\left\{\left(|\vec{a}|\,|\vec{b}|\sin\theta\right)\hat{n}\right\} = m(\vec{a}\times\vec{b})$$

Thus, $(m\vec{a})\times\vec{b} = m(\vec{a}\times\vec{b}) = \vec{a}\times(m\vec{b})$, when $m > 0$.

Case II: When $m < 0$.

In this case, $m\vec{a}$ and $\vec{a}$ are unlike vectors.

Angle between $m\vec{a}$ and $\vec{b} = (\pi - \theta)$.

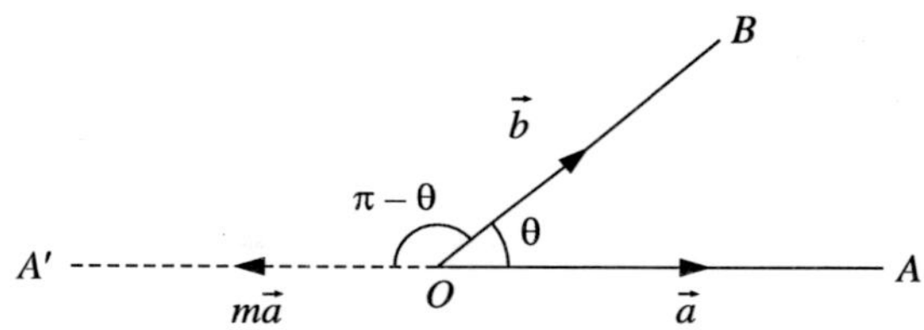

The rotation from $m\vec{a}$ to $\vec{b}$ is opposite to the rotation from $\vec{a}$ to $\vec{b}$.

$\therefore$ $m\vec{a}$, $\vec{b}$ and $-\hat{n}$ form a right-handed system.

$$\therefore \qquad (m\vec{a})\times\vec{b} = \left\{|m\vec{a}|\,|\vec{b}|\sin(\pi-\theta\right\}(-\hat{n})$$

$$= (|m|\,|\vec{a}|\,|\vec{b}|\sin\theta)(-\hat{n})$$

$$= \left\{(-m)|\vec{a}|\,|\vec{b}|\sin\theta\right\}(-\hat{n}) \qquad [\because m<0 \Rightarrow |m| = -m]$$

$$(m\vec{a})\times\vec{b} = m\left\{(|\vec{a}|\,|\vec{b}|\sin\theta)\hat{n}\right\} = m(\vec{a}\times\vec{b}).$$

Again, $m\vec{b}$ and $\vec{b}$ are unlike vectors. Angle between $\vec{a}$ and $m\vec{b}$ is $(\pi - \theta)$.

The rotation from $\vec{a}$ to $m\vec{b}$ is opposite to the rotation from $\vec{a}$ to $\vec{b}$.

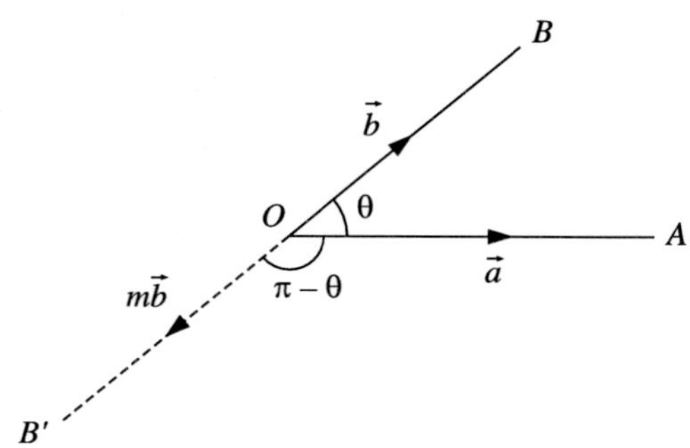

$\therefore\ \vec{a}$, $m\vec{b}$, $(-\hat{n})$ form a right-handed system.

$$\therefore \quad \vec{a}\times(m\vec{b}) = \left\{|\vec{a}||m\vec{b}|\sin(\pi-\theta)\right\}(-\hat{n})$$

$$= \left(|\vec{a}||m||\vec{b}|\sin\theta\right)(-\hat{n})$$

$$= -m\left(|\vec{a}||\vec{b}|\sin\theta\right)(-\hat{n}) \qquad [\because m<0 \Rightarrow |m| = -m]$$

$$= m\left\{\left(|\vec{a}||\vec{b}|\sin\theta\right)\hat{n}\right\} = m(\vec{a}\times\vec{b}).$$

Thus, $\quad (m\vec{a})\times\vec{b} = a\times(m\vec{b})$, when $m<0$.

Case III: When $m=0$. In this case, $(m\vec{a})\times\vec{b} = \vec{0}\times\vec{b} = \vec{0}$. $[\because\ m\vec{a} = \vec{0}\vec{a} = \vec{0}\,]$

$$m(\vec{a}\times\vec{b}) = 0(\vec{a}\times\vec{b}) = \vec{0}.$$

And, $\quad \vec{a}\times(m\vec{b}) = \vec{a}\times 0\vec{b} = \vec{a}\times\vec{0} = \vec{0}$.

$\therefore \quad (m\vec{a})\times\vec{b} = m(\vec{a}\times\vec{b}) = \vec{a}\times(m\vec{b})$, when $m=0$.

Hence, $\quad (m\vec{a})\times\vec{b} = m(\vec{a}\times\vec{b}) = \vec{a}\times(m\vec{b})$, for any scalar m.

Note: At a later stage, we shall prove that for any three vectors $\vec{a}, \vec{b}, \vec{c}$ we have $\vec{a}\times\vec{b}\cdot\vec{c} = \vec{a}\cdot\vec{b}\times\vec{c}$, i.e, the position of dot and cross can be interchanged. However, we shall use this fact in proving the following theorem.

Theorem 4: (Distributive law) For any vectors $\vec{a}, \vec{b}, \vec{c}$; $\vec{a}\times(\vec{b}+\vec{c}) = (\vec{a}\times\vec{b})+(\vec{a}\times\vec{c})$.

Proof: Let $\vec{p} = [\vec{a}\times(\vec{b}+\vec{c})-(\vec{a}\times\vec{b})-(\vec{a}\times\vec{c})]$, and let $\vec{d}$ be any arbitrary vector. Then,

$$\vec{d}\cdot\vec{p} = \vec{d}\cdot[\vec{a}\times(\vec{b}+\vec{c})-(\vec{a}\times\vec{b})-(\vec{a}\times\vec{c})]$$

$$= \vec{d}\cdot[\vec{a}\times(\vec{b}+\vec{c})]-\vec{d}\cdot(\vec{a}\times\vec{b})-\vec{d}\cdot(\vec{a}\times\vec{c})$$

$$= (\vec{d}\times\vec{a})\cdot(\vec{b}+\vec{c})-(\vec{d}\times\vec{a})\cdot\vec{b}-(\vec{d}\times\vec{a})\cdot\vec{c}$$

$$= (\vec{d}\times\vec{a})\cdot\vec{b}+(\vec{d}\times\vec{a})\cdot\vec{c}-(\vec{d}\times\vec{a})\cdot\vec{b}-(\vec{d}\times\vec{a})\cdot\vec{c}$$

$$=0$$

Now $\quad \vec{d}\cdot\vec{p} = 0 \Rightarrow \vec{d} = \vec{0}$ or $\vec{p} = \vec{0}$ or $(\vec{d}\perp\vec{p})$.

Since $\vec{d}$ is an arbitrary vector, we may choose it in such a way that $\vec{d}\neq 0$ and $\vec{d}$ is not perpendicular to $\vec{p}$. Then, $\vec{p} = \vec{0}$.

$\Rightarrow \vec{a}\times(\vec{b}+\vec{c})-(\vec{a}\times\vec{b})-(\vec{a}\times\vec{c}) = \vec{0}$

$\Rightarrow \qquad \vec{a}\times(\vec{b}+\vec{c}) = (\vec{a}\times\vec{b})+(\vec{a}\times\vec{c})$.

Hence, $\quad \vec{a}\times(\vec{b}+\vec{c}) = (\vec{a}\times\vec{b})+(\vec{a}\times\vec{c})$.

Corollary 1. $(\vec{a}\times\vec{b})\times\vec{c} = \vec{a}\times\vec{c}+\vec{b}\times\vec{c}$.

Proof: $(\vec{a}\times\vec{b})\times\vec{c} = -\vec{c}\times(\vec{a}+\vec{b}) = -(\vec{c}\times\vec{a}+\vec{c}\times\vec{b})$

$$= -\vec{c}\times\vec{a}-\vec{c}\times\vec{b} = \vec{a}\times\vec{c}+\vec{b}\times\vec{c}.$$

Corollary 2. $\vec{a}\times(\vec{b}+\vec{c}+\vec{d}) = \vec{a}\times\vec{b}+\vec{a}\times\vec{c}+\vec{a}\times\vec{d}$, and in general, $\vec{a}\times(\vec{b}+\vec{c}+\vec{d}+...) = \vec{a}\times\vec{b}+\vec{a}\times\vec{c}+\vec{a}\times\vec{d}+...$

Corollary 3. $\vec{a}\times(\vec{b}-\vec{c}) = \vec{a}\times\vec{b}-\vec{a}\times\vec{c}$.

Proof: $\vec{a}\times(\vec{b}-\vec{c}) = \vec{a}\times(\vec{b}+(-\vec{c})) = \vec{a}\times\vec{b}+\vec{a}\times(-\vec{c})$.

$$= \vec{a}\times\vec{b}+(-\vec{a}\times\vec{c}) = \vec{a}\times\vec{b}-\vec{a}\times\vec{c}.$$

Theorem 5. (Condition of parallelism of two vectors)

Two nonzero vectors $\vec{a}$ and $\vec{b}$ are parallel or collinear iff $\vec{a}\times\vec{b} = \vec{0}$.

Proof: Let $\vec{a}\neq\vec{0}$ and $\vec{b}\neq\vec{0}$, and let θ be angle between them. let us suppose that $\vec{a}$ and $\vec{b}$ are parallel or collinear. Then, $\theta = 0$ or $\theta = \pi$,

$\therefore \qquad \text{Sin}\,\theta = 0 \qquad [\therefore \sin 0 = 0 \text{ and } \sin\pi = 0]$.

$\therefore \qquad \vec{a}\times\vec{b} = (|\vec{a}||\vec{b}|\sin\theta)\hat{n}$

$= 0\hat{n} = \vec{0} \qquad [\therefore \sin\theta = 0]$

Thus, whenever two non zero vectors $\vec{a}$ and $\vec{b}$ are parallel or collinear

Then $\vec{a}\times\vec{b} = \vec{0}$.

Conversely, suppose $\vec{a}$ and $\vec{b}$ are nonzero vectors such that $\vec{a}\times\vec{b} = \vec{0}$.

Then, $\vec{a}\times\vec{b} = \vec{0} \Rightarrow |\vec{a}\times\vec{b}| = 0$

$\Rightarrow \qquad |\vec{a}||\vec{b}|\sin\theta = 0$

$\Rightarrow \qquad \sin\theta = 0 \qquad [\therefore \vec{a}\neq\vec{0} \text{ and } \vec{b}\neq\vec{0}]$

$\Rightarrow \qquad \theta = 0$

$\Rightarrow \qquad \theta = 0 \text{ or } \theta = \pi$

$\Rightarrow \vec{a}$ and $\vec{b}$ are parallel or collinear.

Thus, $\vec{a}\neq\vec{0}, \vec{b}\neq\vec{0}$ and $\vec{a}\times\vec{b} = \vec{0} \Rightarrow \vec{a}$ and $\vec{b}$ are parallel or collinear.

Theorem 6. For any two vectors $\vec{a}$ and $\vec{b}$, $\vec{a} \times \vec{b} = \vec{0} \Rightarrow \vec{a} = \vec{0}$ or

$\vec{b} = \vec{0}$ or $\vec{a}$ and $\vec{b}$ are parallel or collinear.

Proof: Let θ be the angle between $\vec{a}$ and $\vec{b}$, then,

$$\vec{a} \times \vec{b} = \vec{0} \Rightarrow (ab \sin\theta)\,\hat{n} = \vec{0}$$

$$\Rightarrow \quad ab \sin\theta = 0$$

$$\Rightarrow \quad a = 0, \text{ or } b = 0, \text{ or } \sin\theta = 0$$

$$\Rightarrow \quad \vec{a} = \vec{0}, \text{ or } \vec{b} = \vec{0}, \text{ or } \theta = 0 \text{ or } \theta = \pi$$

$$\Rightarrow \quad \vec{a} = \vec{0}, \text{ or } \vec{b} = \vec{0} \text{ or } \vec{a} \text{ and } \vec{b} \text{ are parallel or collinear}$$

Corollary 1. $\vec{a} \times \vec{b} = \vec{a} \times \vec{c}$ does not imply that $\vec{b} = \vec{c}$.

Proof: $\vec{a} \times \vec{b} = \vec{a} \times \vec{c} \Rightarrow \vec{a} \times \vec{b} - \vec{a} \times \vec{c} = \vec{0}$

$$\Rightarrow \quad \vec{a} \times (\vec{b} - \vec{c}) = \vec{0}$$

$$\Rightarrow \quad \vec{a} = \vec{0} \text{ or } (\vec{b} - \vec{c}) = 0 \text{ or } \vec{a} \parallel (\vec{b} - \vec{c})$$

$$\Rightarrow \quad \vec{a} = \vec{0} \text{ or } \vec{b} = \vec{c} \text{ or } \vec{a} \parallel (\vec{b} - \vec{c})$$

$$\therefore \quad \vec{a} \times \vec{b} = \vec{a} \times \vec{c} \text{ does not always mean that } \vec{b} = \vec{c}.$$

Theorem 7. (Geometrical Interpretation of vector product).

The area of a parallelogram whose adjacent sides are represented by $\vec{a}$ and $\vec{b}$ is $|\vec{a} \times \vec{b}|$.

Proof: Let $\overrightarrow{OA} = \vec{a}$, $\overrightarrow{OB} = \vec{b}$ and $\angle AOB = \theta$ complete the parallelogram $OACB$. Join AB. Draw $BD \perp OA$.

In ΔOAB, we have

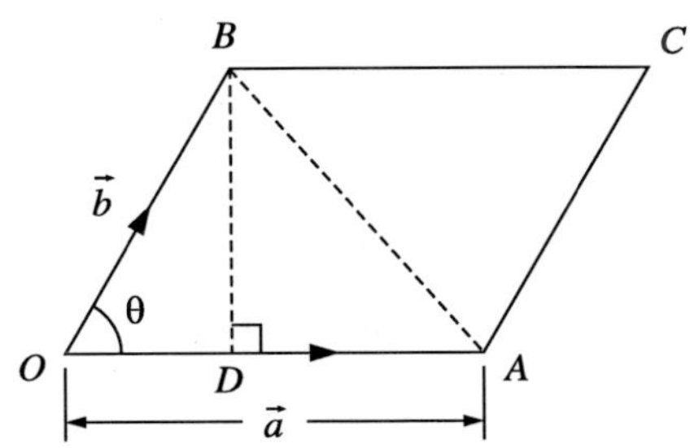

base $= OA = |\vec{a}|$, altitude $= BD = (OB)\sin\theta = |\vec{b}|\sin\theta$.

$\therefore$ Area of $\Delta OAB = \dfrac{1}{2} \times \text{base} \times \text{altitude}$

$$= \frac{1}{2}|\vec{a}||\vec{b}|\sin\theta = \frac{1}{2}ab\sin\theta = \frac{1}{2}|\vec{a}\times\vec{b}|.$$

$\therefore$ area of $\parallel$ gm $OACB = 2\times$(area of ΔOAB) $= |\vec{a}\times\vec{b}|$.

Hence, the area of a $\parallel$ gm whose adjacent sides are represented by $\vec{a}$ and $\vec{b}$ is given by $|\vec{a}\times\vec{b}|$.

Note: $(\vec{a}\times\vec{b})$ is called the vector area of the parallelogram.

5.31 VECTOR PRODUCT OF TWO VECTORS IN TERMS OF THEIR COMPONENTS.

Theorem 8: Let $\vec{a} = a_1\hat{i} + a_2\hat{j} + a_3\hat{k}$ and $\vec{b} = b_1\hat{i} + b_2\hat{j} + b_3\hat{k}$, then

$$\vec{a}\times\vec{b} = \begin{vmatrix} \hat{i} & \hat{j} & \hat{k} \\ a_1 & a_2 & a_3 \\ b_1 & b_2 & b_3 \end{vmatrix}$$

Proof: We have

$$\vec{a}\times\vec{b} = (a_1\hat{i} + a_2\hat{j} + a_3\hat{k})\times(b_1\hat{i} + b_2\hat{j} + b_3\hat{k})$$

$$= a_1b_1(\hat{i}\times\hat{i}) + a_1b_2(\hat{i}\times\hat{j}) + a_1b_3(\hat{i}\times\hat{k})$$

$$+ a_2b_1(\hat{j}\times\hat{i}) + a_2b_2(\hat{j}\times\hat{j})$$

$$+ a_2b_3(\hat{j}\times\hat{k}) + a_3b_1(\hat{k}\times\hat{i}) + a_3b_2(\hat{k}\times\hat{j}) + a_3b_3(\hat{k}\times\hat{k})$$

$$= (a_2b_3 - a_3b_2)\hat{i} + (a_3b_1 - a_1b_3)\hat{j} + (a_1b_2 - a_2b_1)\hat{k}$$

$$\vec{a}\times\vec{b} = \begin{vmatrix} \hat{i} & \hat{j} & \hat{k} \\ a_1 & a_2 & a_3 \\ b_1 & b_2 & b_3 \end{vmatrix}$$

5.32 ANGLE BETWEEN TWO VECTORS

Let θ be two angled between two (non-zero) vectors $\vec{a}, \vec{b}$ and $\hat{n}$ be a unit vector perpendicular to the plane of $\vec{a}, \vec{b}$ and positive for a right handed system from $\vec{a}$ to $\vec{b}$, then.

$$\vec{a}\times\vec{b} = |\vec{a}||\vec{b}|\sin\theta\,\hat{n} \Rightarrow |\vec{a}\times\vec{b}| = |\vec{a}||\vec{b}|\sin\theta$$

$$(\because |\hat{n}| = 1 \text{ and } 0 \le \theta \le \pi \Rightarrow \sin\theta \ge 0)$$

$$\Rightarrow \qquad \sin\theta = \frac{|\vec{a}\times\vec{b}|}{|\vec{a}||\vec{b}|} \qquad ...(1)$$

Let $\vec{a} = a_1\hat{i} + a_2\hat{j} + a_3\hat{k}$ and $\vec{b} = b_1\hat{i} + b_2\hat{j} + b_3\hat{k}$, then

$$\vec{a}\times\vec{b} = (a_2b_3 - a_3b_2)\hat{i} + (a_3b_1 - a_1b_3)\hat{j} + (a_1b_2 - a_2b_1)\hat{k}$$

$$\Rightarrow \quad |\vec{a}\times\vec{b}| = \sqrt{(a_2b_3 - a_3b_2)^2 + (a_3b_1 - a_1b_3)^2 + (a_1b_2 - a_2b_1)^2}$$

Then $|\vec{a}| = \sqrt{a_1^2 + a_2^2 + a_3^2}$ and $|\vec{b}| = \sqrt{b_1^2 + b_2^2 + b_3^2}$

Form (1), we get

$$\sin\theta = \frac{\sqrt{(a_2b_3 - a_3b_2)^2 + (a_3b_1 - a_1b_3)^2 + (a_1b_2 - a_2b_1)^2}}{\sqrt{a_1^2 + a_2^2 + a_3^2}\ \sqrt{b_1^2 + b_2^2 + b_3^2}}$$

5.33. UNIT VECTORS PERPENDICULAR TO TWO GIVEN VECTORS.

Let $\hat{n}$ be a unit vector perpendicular to two (non-zero) vectors $\vec{a}$, $\vec{b}$ and positive for a right handed system from $\vec{a}$ to $\vec{b}$ and θ be the angel between vectors $\vec{a}, \vec{b}$ then

$$\vec{a}\times\vec{b} = ab\sin\theta\ \hat{n}, \text{ where } a = |\vec{a}|, \text{b} = |\vec{b}|$$

$$\Rightarrow \quad |\vec{a}\times\vec{b}| = ab\sin\theta \qquad (\because |\hat{n}| = 1 \text{ and } 0 \le \theta \le \pi \Rightarrow \sin\theta \ge 0)$$

$$\text{Thus, } \vec{a}\times\vec{b} = |\vec{a}\times\vec{b}|\,\hat{n} \Rightarrow \hat{n} = \frac{\vec{a}\times\vec{b}}{|\vec{a}\times\vec{b}|}$$

Theorem: (Lagrange's identity)

For any vectors $\vec{a}$ and $\vec{b}$, prove that

$$|\vec{a}\times\vec{b}|^2 = |\vec{a}|^2|\vec{b}|^2 - (\vec{a}\cdot\vec{b})^2 = \begin{vmatrix} \vec{a}\cdot\vec{a} & \vec{a}\cdot\vec{b} \\ \vec{a}\cdot\vec{b} & \vec{b}\cdot\vec{b} \end{vmatrix}$$

Proof: Let θ be the angle between $\vec{a}$ and $\vec{b}$, then

$$\begin{aligned}
|\vec{a}\times\vec{b}|^2 &= (\vec{a}\times\vec{b})\cdot(\vec{a}\times\vec{b}) \\
&= (|\vec{a}||\vec{b}|\sin\theta)\hat{n}\cdot(|\vec{a}||\vec{b}|\sin\theta)\hat{n} \\
&= |\vec{a}|^2|\vec{b}|^2\sin^2\theta\,(\hat{n}\cdot\hat{n}) = |\vec{a}|^2|\vec{b}|^2\sin^2\theta \qquad [\because \hat{n}\cdot\hat{n} = 1] \\
&= |\vec{a}|^2|\vec{b}|^2(1 - \cos^2\theta) \\
&= |\vec{a}|^2|\vec{b}|^2 - |\vec{a}|^2|\vec{b}|^2\cos^2\theta \\
&= |\vec{a}|^2|\vec{b}|^2 - (\vec{a}\cdot\vec{b})^2
\end{aligned}$$

$$= (\vec{a}\cdot\vec{a})(\vec{b}\cdot\vec{b}) - (\vec{a}\cdot\vec{b})^2$$

$$= \begin{vmatrix} \vec{a}\cdot\vec{a} & \vec{a}\cdot\vec{b} \\ \vec{a}\cdot\vec{b} & \vec{b}\cdot\vec{b} \end{vmatrix}$$

Hence $|\vec{a}\times\vec{b}|^2 = |\vec{a}|^2|\vec{b}|^2 - (\vec{a}\cdot\vec{b})^2 = \begin{vmatrix} \vec{a}\cdot\vec{a} & \vec{a}\cdot\vec{b} \\ \vec{a}\cdot\vec{b} & \vec{b}\cdot\vec{b} \end{vmatrix}$

ILLUSTRATIVE EXAMPLE

Example 1. If $|\vec{a}| = \sqrt{26}$, $|\vec{b}| = 7$ and $|\vec{a}\times\vec{b}| = 35$, find $\vec{a}\cdot\vec{b}$

Solution: $|\vec{a}| = \sqrt{26}$, $|\vec{b}| = 7$ and $|\vec{a}\times\vec{b}| = 35$

$\therefore \quad |\vec{a}\times\vec{b}| = 35 \Rightarrow |\vec{a}||\vec{b}|\sin\theta = 35$

$\Rightarrow \quad \sin\theta = \dfrac{35}{|\vec{a}||\vec{b}|} = \dfrac{35}{(\sqrt{26})(7)} = \dfrac{5}{\sqrt{26}}$

$\cos\theta = \sqrt{1-\sin^2\theta} = \sqrt{1-\dfrac{25}{26}} = \dfrac{1}{\sqrt{26}}$

$\therefore \quad \vec{a}\cdot\vec{b} = |\vec{a}||\vec{b}|\cos\theta = \sqrt{26}\times 7\times\dfrac{1}{\sqrt{26}} = 7$

Hence, $\quad \vec{a}\cdot\vec{b} = 7$

Example 2. If $|\vec{a}| = 2$, $|\vec{b}| = 7$ and $\vec{a}\times\vec{b} = 3\hat{i} + 2\hat{j} + 6\hat{k}$, find the angle between $\vec{a}$ and $\vec{b}$.

Solution: Let θ be the angle between $\vec{a}$ and $\vec{b}$. Then, $\vec{a}\times\vec{b} = 3\hat{i} + 2\hat{j} + 6\hat{k}$

$\Rightarrow \quad |\vec{a}\times\vec{b}| = \sqrt{9+4+36} = \sqrt{49} = 7$

$\Rightarrow \quad |\vec{a}||\vec{b}|\sin\theta = 7 \qquad [\because |\vec{a}\times\vec{b}| = |\vec{a}||\vec{b}|\sin\theta]$

$\Rightarrow \quad \sin\theta = \dfrac{7}{|\vec{a}||\vec{b}|} = \dfrac{7}{2\times 7} = \dfrac{1}{2}$

$\Rightarrow \quad \theta = \dfrac{\pi}{6}.$

Hence, the angle between $\vec{a}$ and $\vec{b}$ is $\dfrac{\pi}{6}$.

Example 3. Find $\vec{a}\times\vec{b}$ and $\vec{b}\times\vec{a}$ if

(*i*) $\vec{a} = 3\hat{k} + 4\hat{j}$, $\vec{b} = \hat{i} + \hat{j} - \hat{k}$

(*ii*) $\vec{a} = (2, -1, 1)$, $\vec{b} = (3, 4, -1)$

Solution: (*i*) $\vec{a}\times\vec{b} = \begin{vmatrix} \hat{i} & \hat{j} & \hat{k} \\ 0 & 4 & 3 \\ 1 & 1 & -1 \end{vmatrix} = \hat{i}\,(-4-3) - \hat{j}\,(0-3) + \hat{k}(0-4)$

$= -7\hat{i} + 3\hat{j} - 4\hat{k}$

and $\quad \vec{b}\times\vec{a} = \begin{vmatrix} \hat{i} & \hat{j} & \hat{k} \\ 1 & 1 & -1 \\ 0 & 4 & 3 \end{vmatrix} = \hat{i}(3+4) - \hat{j}(3-0) + \hat{k}(4-0)$

$= 7\hat{i} - 3\hat{j} + 4\hat{k}$

(*ii*) $\quad \vec{a}\times\vec{b} = \begin{vmatrix} \hat{i} & \hat{j} & \hat{k} \\ 2 & -1 & 1 \\ 3 & 4 & -1 \end{vmatrix} = -3\hat{i} + 5\hat{j} + 11\hat{k}$ and

$\vec{b}\times\vec{a} = \begin{vmatrix} \hat{i} & \hat{j} & \hat{k} \\ 3 & 4 & -1 \\ 2 & -1 & 1 \end{vmatrix} = 3\hat{i} - 5\hat{j} - 11\hat{k}$.

Example 4. If $\vec{a} = 3\hat{i} + \hat{j} + 2\hat{k}$ and $\hat{b} = 2\hat{i} - 2\hat{j} + 4\hat{k}$, find

(*i*) The magnitude of $\vec{a}\times\vec{b}$.

(*ii*) a unit vector perpendicular to both $\vec{a}$ and $\vec{b}$.

(*iii*) the cosine and sine of the angle between the vectors $\vec{a}$ and $\vec{b}$.

Solution: (*i*) $\vec{a}\times\vec{b} = \begin{vmatrix} \hat{i} & \hat{j} & \hat{k} \\ 3 & 1 & 2 \\ 2 & -2 & 4 \end{vmatrix} = 8\hat{i} - 8\hat{j} - 8\hat{k}$

$\therefore$ Magnitude of $\vec{a}\times\vec{b} = |\vec{a}\times\vec{b}| = \sqrt{(8)^2 + (-8)^2 + (-8)^2} = 8\sqrt{3}$.

(*ii*) $\hat{n}$ (unit vector perpendicular to both $\vec{a}$ and $\vec{b}$.)

$$= \frac{\vec{a}\times\vec{b}}{|\vec{a}\times\vec{b}|} = \frac{8\hat{i} - 8\hat{j} - 8\hat{k}}{8\sqrt{3}} = \frac{1}{\sqrt{3}}(\hat{i} - \hat{j} - \hat{k}).$$

There are two unit vectors perpendicular to both $\vec{a}$ and $\vec{b}$. They are $\pm\hat{n} =$

$\pm\frac{1}{\sqrt{3}}(\hat{i} - \hat{j} - \hat{k})$

(*iii*) $\quad \vec{a}\cdot\vec{b} = ab\cos\theta = (3\hat{i} + \hat{j} + 2\hat{k})\cdot(2\hat{i} - 2\hat{j} + 4\hat{k})$

$= (3)(2) + (1)(-2) + (2)(4) = 12$

Now $\quad a = |\vec{a}| = \sqrt{(3)^2 + (1)^2 + (2)^2} = \sqrt{14}$

$$b = |\vec{b}| = \sqrt{(2)^2 + (-2)^2 + (4)^2} = \sqrt{24}$$

$\therefore \quad \cos\theta = \dfrac{\vec{a}\cdot\vec{b}}{ab} = \dfrac{12}{\sqrt{14}\sqrt{24}} = \sqrt{\dfrac{3}{7}}$

Also, $\quad \sin\theta = \dfrac{|\vec{a}\times\vec{b}|}{ab} = \dfrac{8\sqrt{3}}{\sqrt{14}\sqrt{24}} = \dfrac{2}{\sqrt{7}}.$

Example 5. Prove that $\vec{a}\times(\vec{b}+\vec{c}) + \vec{b}\times(\vec{c}+\vec{a}) + \vec{c}\times(\vec{a}+\vec{b}) = \vec{0}$.

Solution: $\vec{a}\times(\vec{b}+\vec{c}) + \vec{b}\times(\vec{c}+\vec{a}) + \vec{c}\times(\vec{a}+\vec{b})$

$$= \vec{a}\times\vec{b} + \vec{a}\times\vec{c} + \vec{b}\times\vec{c} + \vec{b}\times\vec{a} + \vec{c}\times\vec{a} + c\times\vec{b}$$

$$= \vec{a}\times\vec{b} - \vec{c}\times\vec{a} + \vec{b}\times\vec{c} - \vec{a}\times\vec{b} + \vec{c}\times\vec{a} - \vec{b}\times\vec{c} = \vec{0}.$$

Example 6. If $\vec{a}, \vec{b}, \vec{c}$, are vector such that $\vec{a}+\vec{b}+\vec{c}=\vec{0}$, then prove that $(\vec{a}\times\vec{b}) = (\vec{b}\times\vec{c}) = (\vec{c}\times\vec{a})$.

Solution. $\vec{a}+\vec{b}+\vec{c}=\vec{0} \Rightarrow \vec{a}+\vec{b} = -\vec{c}$

$\Rightarrow \quad (\vec{a}+\vec{b})\times\vec{b} = (-\vec{c})\times\vec{b}$

$\Rightarrow \quad (\vec{a}\times\vec{b}) + (\vec{b}\times\vec{b}) = (-\vec{c})\times\vec{b}$ [By distributive law]

$\Rightarrow \quad (\vec{a}\times\vec{b}) + \vec{0} = (\vec{b}\times\vec{c})$ [$\because \vec{b}\times\vec{b} = \vec{0}$ and $(-\vec{c})\times\vec{b} = \vec{b}\times\vec{c}$]

$\Rightarrow \quad \vec{a}\times\vec{b} = \vec{b}\times\vec{c}$. ...(1)

Also, $\quad \vec{a}+\vec{b}+\vec{c} = -\vec{0} \Rightarrow \vec{b}+\vec{c} = -\vec{a}$

$\Rightarrow \quad (\vec{b}+\vec{c})\times\vec{c} = (-\vec{a})\times\vec{c}$

$\Rightarrow \quad \vec{b}\times\vec{c} + \vec{c}\times\vec{c} = (-\vec{a})\times\vec{c}$ [By the distributive law]

$\Rightarrow \quad (\vec{b}\times\vec{c}) + \vec{0} = (\vec{c}\times\vec{a})$ [$\because \vec{c}\times\vec{c} = \vec{0}$ and $(-\vec{a})\times\vec{c} = \vec{c}\times\vec{a}$]

$\Rightarrow \quad \vec{b}\times\vec{c} = \vec{c}\times\vec{a}$ ...(2)

From (1) and (2) we get $\vec{a}\times\vec{b} = \vec{b}\times\vec{c} = \vec{c}\times\vec{a}$

Example 7. For any two vectors $\vec{a}$ and $\vec{b}$

$$(\vec{a}-\vec{b})\times(\vec{a}+\vec{b}) = 2(\vec{a}\times\vec{b}).$$

Solution. We have

$(\vec{a}-\vec{b})\times(\vec{a}+\vec{b})$

$= \vec{a}\times\vec{a} + \vec{a}\times\vec{b} - \vec{b}\times\vec{a} - \vec{b}\times\vec{b}$ [By the distributive law]

$= \vec{a}\times\vec{b} - \vec{b}\times\vec{a}$ $\quad [\because \vec{a}\times\vec{a} = \vec{0} \text{ and } \vec{b}\times\vec{b} = \vec{0}]$

$= (\vec{a}\times\vec{b}) + (\vec{a}\times\vec{b})$ $\quad [\because -\vec{b}\times\vec{a} = \vec{a}\times\vec{b}]$

$= 2(\vec{a}\times\vec{b})$

Example 8. When $\vec{a}\cdot\vec{b} = \vec{0}$ and $\vec{a}\times\vec{b} = \vec{0}$, then $\vec{a} = 0$ or $\vec{b} = 0$.

Solution. $\vec{a}, \vec{b} = 0$ and $\vec{a}\times\vec{b} = \vec{0}$

$\Rightarrow \vec{a} = \vec{0}$, or $\vec{b} = \vec{0}$, or $\vec{a} \perp \vec{b}$, and $\vec{a} = \vec{0}$, or $\vec{b} = \vec{0}$, or $\vec{a} \parallel \vec{b}$

$\Rightarrow \vec{a} = \vec{0}$, or $\vec{b} = \vec{0}$ $\quad$ [$\because \vec{a} \perp \vec{b}$, and $\vec{a} \parallel \vec{b}$ can never hold simultaneously].

Example 9. If $\vec{a}\cdot\vec{b} = \vec{a}\cdot\vec{c}, \vec{a}\times\vec{b}, = \vec{a}\times\vec{c}$ and $\vec{a} \neq \vec{0}$ then $\vec{b} = \vec{c}$.

Solution. $\vec{a}\cdot\vec{b} = \vec{a}\cdot\vec{c}$ and $\vec{a} \neq \vec{0}$

$\Rightarrow (\vec{a}\cdot\vec{b} - \vec{a}\cdot\vec{c}) = \vec{0}$ and $\vec{a} \neq \vec{0}$

$\Rightarrow \vec{a}.(\vec{b} - \vec{c}) = 0$ and $\vec{a} \neq \vec{0}$

$\Rightarrow \vec{b} - \vec{c} = \vec{0}$, and $\vec{a} \perp (\vec{b} - \vec{c})$

$\Rightarrow \vec{b} = \vec{c}$, or $\vec{a} \perp (\vec{b} - \vec{c})$ ...(1)

Again, $\vec{a}\times\vec{b} = \vec{a}\times\vec{c}$, and $\vec{a} \neq \vec{0}$

$\Rightarrow (\vec{a}\times\vec{b} - \vec{a}\times\vec{c}) = \vec{0}$, and $\vec{a} \neq \vec{0}$

$\Rightarrow \vec{b} - \vec{c} = \vec{0}$, or $\vec{a} \parallel (\vec{b} - \vec{c})$ ...(2)

Thus, from (1) and (2), it follows that

either $\vec{b} = \vec{c}$ or $\vec{a} \perp (\vec{b} - \vec{c})$ and $\vec{a} \parallel (\vec{b} - \vec{c})$

$\therefore \vec{b} = \vec{c}$ $\quad$ [$\because \vec{a}$ can not be both parallel or perpendicular to $(\vec{b} - \vec{c})$]

Example 10. If $\vec{a}\times\vec{b} = \vec{c}\times\vec{d}$ and $\vec{a}\times\vec{c} = \vec{b}\times\vec{d}$ show that $(\vec{a} - \vec{d})$ is parallel to $(\vec{b} - \vec{c})$, it being given that $\vec{a} \neq \vec{d}$ and $\vec{b} \neq \vec{c}$.

Solution. $\vec{a}\times\vec{b} = \vec{c}\times\vec{d}$ and $\vec{a}\times\vec{c} = \vec{b}\times\vec{d}$

$\Rightarrow \vec{a}\times\vec{b} - \vec{a}\times\vec{c} = \vec{c}\times\vec{d} - \vec{b}\times\vec{d}$

$\Rightarrow \vec{a}\times\vec{b} - \vec{a}\times\vec{c} + \vec{b}\times\vec{d} - \vec{c}\times\vec{d} = \vec{0}$

$\Rightarrow \vec{a}\times(\vec{b} - \vec{c}) + (\vec{b} - \vec{c})\times\vec{d} = \vec{0}$

$\Rightarrow \vec{a}\times(\vec{b} - \vec{c}) - \vec{d}\times(\vec{b}\times\vec{c}) = \vec{0}$

$\Rightarrow (\vec{a} - \vec{d})\times(\vec{b} - \vec{c}) = \vec{0}$

$\Rightarrow (\vec{a} - \vec{d}) \parallel (\vec{b} - \vec{c})$

Hence $(\vec{a} - \vec{d})$ is parallel to $(\vec{b} - \vec{c})$.

Example 11. If $\vec{a} = 3\hat{i} + 2\hat{j} + 2\hat{k}$ and $\vec{b} = \hat{i} + 2\hat{j} - 2\hat{k}$, then find a unit vector which is perpendicular to both the vectors $(\vec{a} + \vec{b})$ and $(\vec{a} - \vec{b})$.

Solution. $\vec{a} = 3\hat{i} + 2\hat{j} + 2\hat{k}$ and $\hat{b} = \hat{i} + 2\hat{j} - 2\hat{k}$

$\therefore \quad \vec{a} + \vec{b} = 4\hat{i} + 4\hat{j}$ and $\vec{a} - \vec{b} = 2\hat{i} + 4\hat{k}$

$$\therefore \quad (\vec{a} + \vec{b}) \times (\vec{a} - \vec{b}) = \begin{vmatrix} \hat{i} & \hat{j} & \hat{k} \\ 4 & 4 & 0 \\ 2 & 0 & 4 \end{vmatrix} = 16\hat{i} - 16\hat{j} - 8\hat{k}$$

$$\Rightarrow \quad \left|(\vec{a} + \vec{b}) \times (\vec{a} - \vec{b})\right| = \sqrt{(16)^2 + (-16)^2 + (-8)^2} = 24.$$

$\therefore \quad$ Unit vectors perpendicular to both the vectors $(\vec{a} + \vec{b})$ and $(\vec{a} - \vec{b})$

$$= \pm \frac{16\hat{i} - 16\hat{j} - 8\hat{k}}{24} = \pm \frac{1}{3}(2\hat{i} - 2\hat{j} - \hat{k})$$

Example 12. If $\vec{a} = \hat{i} + 4\hat{j} + 2\hat{k}, \vec{b} = 3\hat{i} - 2\hat{j} + 7\hat{k}$ and $\vec{c} = 2\hat{i} - \hat{j} + 4\hat{k}$, find a vector $\vec{d}$ which is perpendicular to both $\vec{a}$ and $\vec{b}$ and $\vec{c}.\vec{d} = 15$.

Solution.
$$\vec{a} \times \vec{b} = \begin{vmatrix} \hat{i} & \hat{j} & \hat{k} \\ 1 & 4 & 2 \\ 3 & -2 & 7 \end{vmatrix} = 32\hat{i} - \hat{j} - 14\hat{k}.$$

Since $\vec{d}$ is perpendicular to both $\vec{a}$ and $\vec{b}$, $\vec{d} = \lambda(\vec{a} \times \vec{b})$ for some scalar λ.

$\Rightarrow \quad \vec{d} = 32\lambda\hat{i} - \lambda\hat{j} - 14\lambda\hat{k}$

Given $\vec{c}.\vec{d} = 15 \Rightarrow 2.32\lambda + (-1).(-\lambda) + 4(-14\lambda) = 15$

$\Rightarrow 9\lambda = 15 \Rightarrow \lambda = 5/3$

$$\therefore \quad \vec{d} = \frac{5}{3}(32\hat{i} - \hat{j} - 14\hat{k}).$$

Example 13. Given $|\vec{a}| = 10, |\vec{b}| = 2$ and $\vec{a} \cdot \vec{b} = 12$, find $\left|\vec{a} \times \vec{b}\right|$.

Solution. By using Lagrange's identity, we have

$$(\vec{a} \times \vec{b})^2 = |\vec{a}|^2 |\vec{b}|^2 - (\vec{a} \cdot \vec{b})^2$$
$$= 10^2.2^2 - 12^2 = 400 - 144 = 256$$

$$\Rightarrow |\vec{a} \times \vec{b}|^2 = 256 \Rightarrow |\vec{a} \times \vec{b}| = 16.$$

Example 14. If $|\vec{a}| = \sqrt{26}, |\vec{b}| = 7$ and $|\vec{a} \times \vec{b}| = 35$, find $\vec{a} \cdot \vec{b}$

Solution: By Lagrange's identity, we have

$$(\vec{a} \times \vec{b}) = |\vec{a}|^2 |\vec{b}|^2 - (\vec{a} \cdot \vec{b})^2$$

$$\Rightarrow \quad |\vec{a} \times \vec{b}|^2 = |\vec{a}|^2 |\vec{b}|^2 - (\vec{a} \cdot \vec{b})^2$$

$$\Rightarrow \quad (35)^2 = \left(\sqrt{26}\right)^2 (7)^2 - (\vec{a}\cdot\vec{b})^2$$

$$\Rightarrow \quad (\vec{a}.\vec{b})^2 = (26)(49) - (35)^2 = 1274 - 1225 = 49$$

$$\Rightarrow \quad \vec{a}.\vec{b} = \pm 7.$$

Example 15. Find the value of λ for which the vectors $\vec{a} = \hat{i} + \lambda\hat{j} + 3\hat{k}$ and $\vec{b} = 3\hat{i} + 2\hat{j} + 9\hat{k}$ are parallel.

Solution. Clearly $\vec{a} \neq \vec{0}$ and $\vec{b} \neq \vec{0}$.

So, $\vec{a} \parallel \vec{b} \Leftrightarrow \vec{a}\times\vec{b} = \vec{0}$.

$$\text{Now, } \vec{a}\times\vec{b} = \begin{vmatrix} \hat{i} & \hat{j} & \hat{k} \\ 1 & \lambda & 3 \\ 3 & 2 & 9 \end{vmatrix}$$

$$= (9\lambda - 6)\hat{i} + (9-9)\hat{j} + (2 - 3\lambda)\hat{k}$$

$$= (9\lambda - 6)\hat{i} + (2 - 3\lambda)\hat{k}.$$

$$\therefore \quad \vec{a} \parallel \vec{b} \Leftrightarrow \vec{a}\times\vec{b} = \vec{0}.$$

$$\Leftrightarrow (9\lambda - 6)\hat{i} + (2 - 3\lambda)\,\hat{k} = \vec{0}$$

$$\Leftrightarrow 9\lambda - 6 = 0 \text{ and } 2 - 3\lambda = 0$$

$$\Leftrightarrow \lambda = 2/3$$

Hence, the value of λ is 2/3.

Example 16. In ΔOAB, $\overrightarrow{OA} = 3\hat{i} + 2\hat{j} - \hat{k}$ and $\overrightarrow{OB} = \hat{i} + 3\hat{j} + \hat{k}$. Find the area of the triangle.

Solution. Area of $\Delta OAB = \left|\frac{1}{2}(\overrightarrow{OA}\times\overrightarrow{OB})\right|$

$$\text{Now } \overrightarrow{OA}\times\overrightarrow{OB} = \begin{vmatrix} \hat{i} & \hat{j} & \hat{k} \\ 3 & 2 & -1 \\ 1 & 3 & 1 \end{vmatrix}$$

$$= (2+3)\,\hat{i} - (3+1)\hat{j} + (9-2)\hat{k}$$

$$= (5\hat{i} - 4\hat{j} + 7\hat{k})$$

$$\therefore \text{area of } \Delta\text{OAB} = \left|\frac{1}{2}(\overrightarrow{OA}\times\overrightarrow{OB})\right| = \left|\frac{5}{2}\hat{i} - 2\hat{j} + \frac{7}{2}\hat{k}\right|$$

$$= \sqrt{\frac{25}{4} + 4 + \frac{49}{4}} = \sqrt{\frac{90}{4}} = \frac{3}{2}\sqrt{10} \text{ sq. units.}$$

Hence the area of ΔOAB is $3\sqrt{\frac{5}{2}}$ sq. units.

Example 17. Let $\vec{a}, \vec{b}, \vec{c}$ be three non-zero vectors. If $\vec{a}.(\vec{b}\times\vec{c})=0$ and $\vec{b}, \vec{c}$ are not parallel, then prove that $\vec{a}=\lambda\vec{b}+\mu\vec{c}$ where λ, μ are some scalars.

Solution. $\vec{a}.(\vec{b}\times\vec{c})=0 \Rightarrow \vec{a}=\vec{0}$ or $\vec{b}\times\vec{c}=\vec{0}$ or $\vec{a}\perp(\vec{b}\times\vec{c})$

$\Rightarrow \quad \vec{b}\times\vec{c}=0$ or $\vec{a}\perp(\vec{b}\times\vec{c})$ $\qquad (\because \vec{a}\neq\vec{0}$, given)

$\Rightarrow \quad \vec{b}=\vec{0}$ or $\vec{c}=\vec{0}$ or $\vec{b}\,||\,\vec{c}$ or $\vec{a}\perp(\vec{b}\times\vec{c})$

$\Rightarrow \quad \vec{a}\perp(\vec{b}\times\vec{c})$ $\qquad (\because \vec{b}\neq\vec{0}, \vec{c}\neq\vec{0}$ and $\vec{b}, \vec{c}$ are not parallel given)

$\Rightarrow \quad \vec{a}$ is parallel to the plane of $\vec{b}$ and $\vec{c}$

$(\because \vec{b}\times\vec{c}$ is perpendicular to the plane of $\vec{b}$ and $\vec{c})$

$\Rightarrow \quad \vec{a}, \vec{b}, \vec{c}$ are coplanar $\Rightarrow \vec{a}=\lambda\vec{b}+\mu\vec{c}$ for some scalars λ and μ.

Example 18. If G is the centroid of ΔABC, prove that area of ΔAGB = area ΔBGC = area of $\Delta CGA = \frac{1}{3}$ area of ΔABC.

Solution. Taking A as origin, let $\vec{b}, \vec{c}$ be the position vectors of B, C respectively.

Then position vector of $G = \frac{\vec{0}+\vec{b}+\vec{c}}{3} = \frac{\vec{b}+\vec{c}}{3}$

Now area of $\Delta ABC = \frac{1}{2}|\vec{b}\times\vec{c}|$

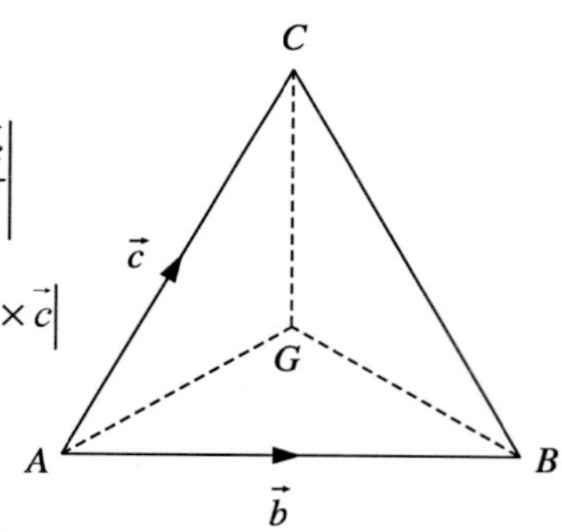

Area of $\Delta AGB = \frac{1}{2}|\overrightarrow{AB}\times\overrightarrow{AG}| = \frac{1}{2}\left|\vec{b}\times\frac{\vec{b}+\vec{c}}{3}\right|$

$= \frac{1}{6}|(\vec{b}\times\vec{b})+(\vec{b}\times\vec{c})| = \frac{1}{6}|\vec{b}\times\vec{c}|$

$= \frac{1}{3}$ area of ΔABC.

Similarly, the other two parts can be proved.

Example 19. Show that the points whose position vectors are $(5\hat{i}+6\hat{j}+7\hat{k})$, $(7\hat{i}-8\hat{j}+9\hat{k})$ and $(3\hat{i}+20\hat{j}+5\hat{k})$ are collinear.

Solution. Let $\vec{a}=5\hat{i}+6\hat{j}+7\hat{k}, \vec{b}=7\hat{i}-8\hat{j}+9\hat{k}$ and $\vec{c}=3\hat{i}+20\hat{j}+5\hat{k}$.

We know that the points A, B, C with position vectors $\vec{a}, \vec{b}, \vec{c}$ respectively are collinear iff $\vec{a}\times\vec{b}+\vec{b}\times\vec{c}+\vec{c}\times\vec{a}=\vec{0}$.

Now, $$\vec{a}\times\vec{b} = \begin{vmatrix} \hat{i} & \hat{j} & \hat{k} \\ 5 & 6 & 7 \\ 7 & -8 & 9 \end{vmatrix} = (110\hat{i}+4\hat{j}-82\hat{k})$$

$$\vec{b}\times\vec{c} = \begin{vmatrix} \hat{i} & \hat{j} & \hat{k} \\ 7 & -8 & 9 \\ 3 & 20 & 5 \end{vmatrix} = (-220\hat{i} - 8\hat{j} + 164\hat{k})$$

$$\vec{c}\times\vec{a} = \begin{vmatrix} \hat{i} & \hat{j} & \hat{k} \\ 3 & 20 & 5 \\ 5 & 6 & 7 \end{vmatrix} = (110\hat{i} + 4\hat{j} - 82\hat{k})$$

$\therefore \quad \vec{a}\times\vec{b} + \vec{b}\times\vec{c} + \vec{c}\times\vec{a} = \vec{0}.$

Hence the given points are collinear.

Example 20. Find the area of the parallelogram whose adjacent sides are represented by the vectors $(\hat{i} - 3\hat{j} + \hat{k})$ and $(\hat{i} + \hat{j} + \hat{k})$.

Solution. Let ABCD be the parallelogram whose adjacent sides are given by the vectors

$$\overrightarrow{AB} = (\hat{i} - 3\hat{j} + \hat{k}) \text{ and } \overrightarrow{AD} = (\hat{i} + \hat{j} + \hat{k})$$

$\therefore \quad$ Area of ||gm $ABCD = |\overrightarrow{AB}\times\overrightarrow{AD}|$

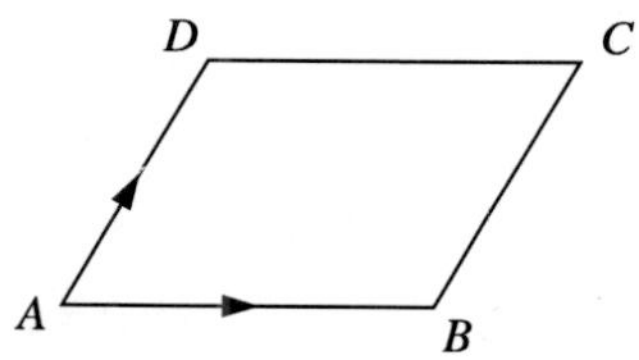

Now, $\quad \overrightarrow{AB}\times\overrightarrow{AD} = \begin{vmatrix} \hat{i} & \hat{j} & \hat{k} \\ 1 & -3 & 1 \\ 1 & 1 & 1 \end{vmatrix}$

$$= (-4\hat{i} + 4\hat{k})$$

$\therefore \quad$ Area of ||gm ABCD $= |-4\hat{i} + 0\hat{j} + 4\hat{k}|$

$$= \sqrt{(-4)^2 + 0^2 + 4^2} = \sqrt{32} = 4\sqrt{2} \text{ sq. units.}$$

Hence the area of the given ||gm is $4\sqrt{2}$ sq. Units.

Example 21. Find the area of the parallelogram whose diagonals are represented by the vectors $\vec{d_1} = (4\hat{i} - \hat{j} + 3\hat{k})$ and $\vec{d_2} = (-2\hat{i} + \hat{j} - 2\hat{k})$

Solution. Area of ||gm $= \dfrac{1}{2}|\vec{d_1}\times\vec{d_2}|$

Now, $\vec{d_1}\times\vec{d_2} = \begin{vmatrix} \hat{i} & \hat{j} & \hat{k} \\ 4 & -1 & -3 \\ -2 & 1 & -2 \end{vmatrix} = (5\hat{i} + 14\hat{j} + 2\hat{k}).$

$\therefore \quad$ Area of ||gm $= \dfrac{1}{2}|\vec{d_1}\times\vec{d_2}| = \dfrac{1}{2}\sqrt{(5)^2 + (14)^2 + (2)^2} = \dfrac{15}{2}$ sq. units

Hence, the area of the given ||gm is 7.5 sq. Units.

Example 22. (Sine rule) In any triangle, prove by vector method that $\dfrac{a}{\sin A} = \dfrac{b}{\sin B} = \dfrac{c}{\sin C}$.

Solution. In ΔABC, let $\overrightarrow{BC} = \vec{a}, \overrightarrow{CA} = \vec{b}$ and $\overrightarrow{AB} = \vec{c}$, then $a = \left|\overrightarrow{BC}\right|$, $b = \left|\overrightarrow{CA}\right|$ and $c = \left|\overrightarrow{AB}\right|$

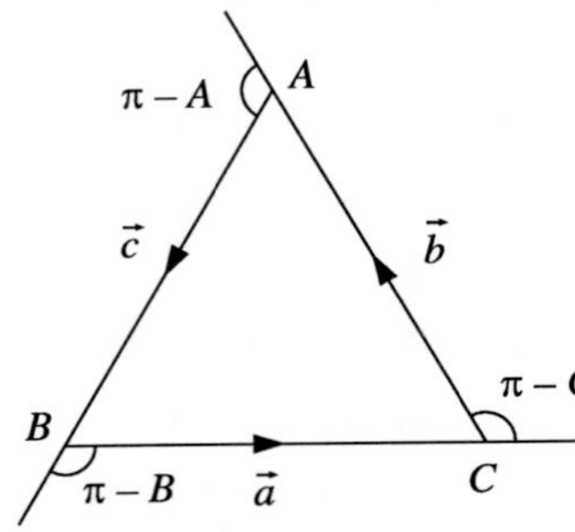

From ΔABC, we get

$$\overrightarrow{BC} + \overrightarrow{CA} = \overrightarrow{BA} \Rightarrow \overrightarrow{BC} + \overrightarrow{CA} + \overrightarrow{AB} = \vec{0}$$

$$\Rightarrow \vec{a} + \vec{b} + \vec{c} = \vec{0}$$

$$\Rightarrow \vec{a} \times (\vec{a} + \vec{b} + \vec{c}) = \vec{a} \times \vec{0}$$

$$\Rightarrow \vec{a} \times \vec{a} + \vec{a} \times \vec{b} + \vec{a} \times \vec{c} = \vec{0}$$

$$\Rightarrow \vec{a} \times \vec{b} = -\vec{a} \times \vec{c} \qquad \left[\because \vec{a} \times \vec{b} = \vec{0}\right]$$

$$\Rightarrow \vec{a} \times \vec{b} = \vec{c} \times \vec{a} \qquad ...(1)$$

Similarly, $\vec{a} \times \vec{b} = \vec{b} \times \vec{c}$ ...(2)

From (1) and (2), we get

$$\vec{a} \times \vec{b} = \vec{b} \times \vec{c} = \vec{c} \times \vec{a}$$

$$\Rightarrow \quad \left|\vec{a} \times \vec{b}\right| = \left|\vec{b} \times \vec{c}\right| = \left|\vec{c} \times \vec{a}\right|$$

$$\Rightarrow \quad ab \sin(\pi - C) = bc \sin(\pi - A) = ca \sin(\pi - B).$$

Dividing by abc, we get

$$\frac{\sin C}{c} = \frac{\sin A}{a} = \frac{\sin B}{b}$$

$$\Rightarrow \quad \frac{a}{\sin A} = \frac{b}{\sin B} = \frac{c}{\sin C}$$

Example 23. Prove by vector method that

(*i*) $\sin(\alpha - \beta) = \sin \alpha \cos \beta - \cos \alpha \sin \beta$

(*ii*) $\sin(\alpha + \beta) = \sin \alpha \cos \beta + \cos \alpha \sin \beta$

Solution. Let $\hat{i}, \hat{j}$ be unit vectors along two perpendicular line *OX*, *OY* in the plane of paper.

(*i*) Let $\overrightarrow{OA}, \overrightarrow{OB}$ be two unit vectors making angles α, β with *x*-axis, respectively, then $\angle AOB = \alpha - \beta$

Let $\hat{k}$ be a unit vector $\perp$ *XOY* plane such that $\hat{i}, \hat{j}, \hat{k}$ form a right handed triad.

$OP = |OA| \cos \alpha = 1.\cos \alpha = \cos \alpha,$

$PA = |OA| \sin \alpha = 1.\sin \alpha = \sin \alpha,$

$\therefore \quad \overrightarrow{OA} = \overrightarrow{OP} + \overrightarrow{PA} = \cos\alpha\,\hat{i} + \sin\alpha\,\hat{j}$

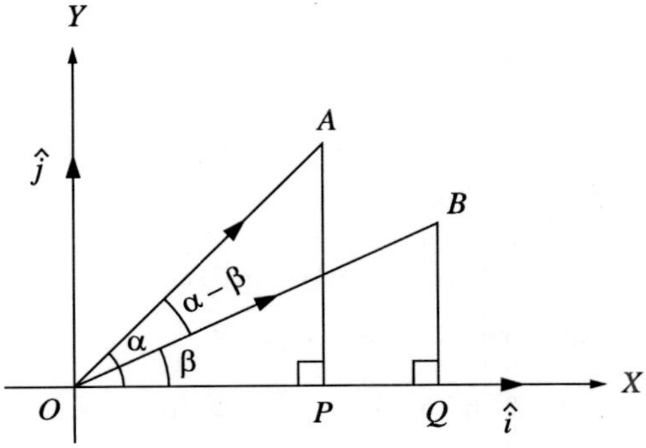

Similarly $\overrightarrow{OB} = \cos\beta\,\hat{i} + \sin\beta\,\hat{j}$

$$\therefore \quad \overrightarrow{OB} \times \overrightarrow{OA} = \left|\overrightarrow{OB}\right| \times \left|\overrightarrow{OA}\right| \sin(\alpha - \beta)\,\hat{k} = 1.1\sin(\alpha - \beta)\hat{k}$$

$$= \sin(\alpha - \beta)\,\hat{k}$$

$$\text{Also, } \overrightarrow{OB} \times \overrightarrow{OA} = \begin{vmatrix} \hat{i} & \hat{j} & \hat{k} \\ \cos\beta & \sin\beta & 0 \\ \cos\alpha & \sin\alpha & 0 \end{vmatrix} = (\sin\alpha\cos\beta - \cos\alpha\sin\beta)\hat{k}$$

$\Rightarrow \sin(\alpha - \beta)\,\hat{k} = (\sin\alpha\cos\beta - \cos\alpha\sin\beta)\,\hat{k}$

$\Rightarrow \sin(\alpha - \beta) = \sin\alpha\cos\beta - \cos\alpha\sin\beta.$

(*ii*) Let $\overrightarrow{OA}, \overrightarrow{OB}$ be unit vectors making angles α, $-\beta$ with the *x*-axis respectively then $\angle AOB = \alpha + \beta$.

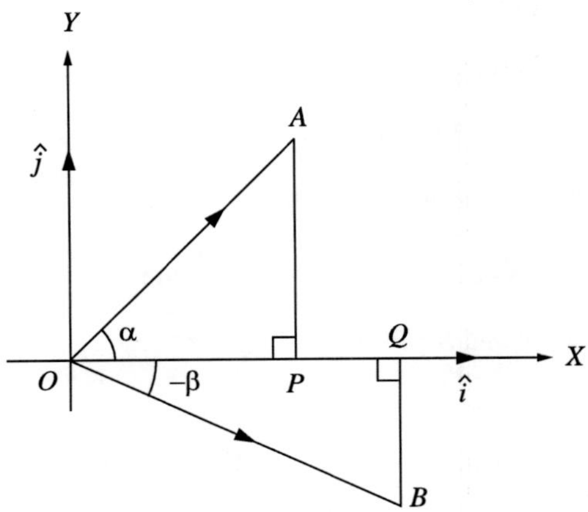

Let $\hat{k}$ be a unit vector $\perp$ *XOY* plane such that $\hat{i}, \hat{j}, \hat{k}$ form a right hand triad.

$\overrightarrow{OA} = \overrightarrow{OP} + \overrightarrow{PA} = \cos\alpha\,\hat{i} + \sin\alpha\,\hat{j}$

and $\overrightarrow{OB} = \overrightarrow{OQ} + \overrightarrow{QB} = \cos\beta\,\hat{i} - \sin\beta\,\hat{j}$

$$\therefore \quad \overrightarrow{OB} \times \overrightarrow{OA} = \left|\overrightarrow{OB}\right|\left|\overrightarrow{OA}\right| \sin(\alpha+\beta)\,\hat{k} = \sin(\alpha+\beta)\,\hat{k}.$$

$$\text{Also, } \overrightarrow{OB} \times \overrightarrow{OA} = \begin{vmatrix} \hat{i} & \hat{j} & \hat{k} \\ \cos\beta & -\sin\beta & 0 \\ \cos\alpha & \sin\alpha & 0 \end{vmatrix} = (\sin\alpha\cos\beta + \cos\alpha\sin\beta)\,\hat{k}$$

$$\Rightarrow \quad \sin(\alpha+\beta)\,\hat{k} = (\sin\alpha\cos\beta + \cos\alpha\sin\beta)\hat{k}$$

$$\Rightarrow \quad \sin(\alpha+\beta) = \sin\alpha\cos\beta + \cos\alpha + \sin\beta.$$

Example 24. If $\vec{a} = \hat{i} + \hat{j} + \hat{k}$ and $\vec{b} = 2\hat{i} + 3\hat{j} + 2\hat{k}$, find $\vec{a} \times \vec{b}$ and $\vec{a} \cdot \vec{b}$.

[I.U. 2007-2008]

Solution. $\vec{a} \cdot \vec{b} = (\hat{i} + \hat{j} + \hat{k}).(2\hat{i} + 3\hat{j} + 2\hat{k}) = 2 + 3 + 2 = 7$

$$\vec{a} \times \vec{b} = \begin{vmatrix} \hat{i} & \hat{j} & \hat{k} \\ 1 & 1 & 1 \\ 2 & 3 & 2 \end{vmatrix}$$

$$= \hat{i}\,(2-3) - \hat{j}\,(2-2) + \hat{k}\,(3-2)$$

$$= -\hat{i} - 0.\hat{j} + \hat{k} = -\hat{i} + \hat{k}$$

EXERCISE 5.3

1. Find $\vec{a} \times \vec{b}$, if $\vec{a} = 2\hat{i} + \hat{j} + \hat{k}$ and $\vec{b} = 3\hat{i} + \hat{j} + \hat{k}$.
2. Find $(\vec{a} \times \vec{b})$ and $|\vec{a} \times \vec{b}|$, when
   (*i*) $\vec{a} = \hat{i} - \hat{j} + 2\hat{k}$ and $\vec{b} = 2\hat{i} + 3\hat{j} - 4\hat{k}$
   (*ii*) $\vec{a} = 2\hat{i} + \hat{j} + 3\hat{k}$ and $\vec{b} = 3\hat{i} + 5\hat{j} - 2\hat{k}$
3. If $\vec{a} = -3\hat{i} + 4\hat{j} - 7\hat{k}$ and $\vec{b} = 6\hat{i} + 2\hat{j} - 3\hat{k}$ then find $(\vec{a} \times \vec{b})$. Verify that $\vec{b}$ and $(\vec{a} \times \vec{b})$ are perpendicular to each other.
4. Find the unit vectors perpendicular to both $\vec{a}$ and $\vec{b}$, when
   (*i*) $\vec{a} = 3\hat{i} + \hat{j} - 2\hat{k}$ and $\vec{b} = 2\hat{i} + 3\hat{j} - \hat{k}$
   (*ii*) $\vec{a} = 4\hat{i} + 2\hat{j} - \hat{k}$ and $\vec{b} = \hat{i} + 4\hat{j} - \hat{k}$
5. Find the magnitude of $\vec{a}$, where $\vec{a} = (\hat{i} + 3\hat{j} - 2\hat{k}) \times (-\hat{i} + 3\hat{k})$.
6. If $\vec{a} = 3\hat{i} - \hat{j} - 2\hat{k}$ and $\vec{b} = 2\hat{i} + 3\hat{j} + \hat{k}$, find $(\vec{a} + 2\vec{b}) \times (2\vec{a} - \vec{b})$.
7. Taking $\vec{a} = 2\hat{i} - 3\hat{j} - \hat{k}$ and $\vec{b} = \hat{i} + 4\hat{j} - 2\hat{k}$, verify that $\vec{a} \times \vec{b} \neq \vec{b} \times \vec{a}$.

8. If $\vec{a} = \hat{i} - 2\hat{j} + 3\hat{k}$ and $\vec{b} = 2\hat{i} + 3\hat{j} - 5\hat{k}$, find $\vec{a} \times \vec{b}$. Verify that $\vec{a}$ and $\vec{a} \times \vec{b}$ are perpendicular to each other.

9. If $|\vec{a}| = 2$, $|\vec{b}| = 7$ and $\vec{a} \times \vec{b} = 3\hat{i} + 2\hat{j} + 6\hat{k}$, find the angle between $\vec{a}$ and $\vec{b}$.

10. Find the angle between two vectors $\vec{a}$ and $\vec{b}$ if $|\vec{a} \times \vec{b}| = \vec{a}.\vec{b}$.

11. Define $\vec{a} \times \vec{b}$ and prove that $|\vec{a} \times \vec{b}| = (\vec{a} \times \vec{b}) \tan\theta$, where θ is the angle between $\vec{a}$ and $\vec{b}$.

12. Find $\vec{a} \times \vec{b}$ if $|\vec{a}| = 2, |\vec{b}| = 5$ and $|\vec{a} \times \vec{b}| = 8$.

13. Show that $\vec{a} \times \vec{b} = \vec{a} \times \vec{c}$ may not imply that $\vec{b} = \vec{c}$.

14. What conclusion can you draw when $\vec{a} \times \vec{b} = \vec{0}$ and $\vec{a} \cdot \vec{b} = 0$?

15. Find the area of the parallelogram whose adjacent sides are represented by the vectors.

(*i*) $\vec{a} = \hat{i} + 2\hat{j} + 3\hat{k}$ and $\vec{b} = 3\hat{i} + 2\hat{j} + \hat{k}$

(*ii*) $\vec{a} = 2\hat{i}$ and $\vec{b} = 5\hat{j}$

16. Find the area of the parallelogram whose diagonals are represented by the vectors $\vec{d_1} = 3\hat{i} + \hat{j} + 2\hat{k}$ and $\vec{d_2} = \hat{i} + \hat{j} + \hat{k}$.

17. Using vectors, find the area of ΔABC, whose vertices are

(*i*) $A(3, 1, 2), B(1, 1, 2)$ and $C(4, 3, 2)$.

(*ii*) $A(1, 2, 3), B(2, 5, -1)$ and $C(-1, 1, 2)$.

18. Show that the points having position vectors $\vec{a}, \vec{b}, (\overrightarrow{3a} - \overrightarrow{2b})$ are collinear, whatever be $\vec{a}, \vec{b}, \vec{c}$.

19. If three vectors $\vec{a}, \vec{b}, \vec{c}$ are such that $\vec{a} + \vec{b} + \vec{c} = \vec{0}$, prove that $\vec{a} \times \vec{b} = \vec{b} \times \vec{c} = \vec{c} \times \vec{a}$.

20. For any vectors $\vec{a}$ and $\vec{b}$, show that

$$(1 + |\vec{a}|^2)(1 + |\vec{b}|^2) = \{1 - (\vec{a}.\vec{b})\}^2 + |\vec{a} + \vec{b} + (\vec{a} \times \vec{b})|^2$$

21. Find λ and μ if $(2\hat{i} + 6\hat{j} + 27\hat{k}) \times (\hat{i} + \lambda\hat{j} + \mu\hat{k}) = \vec{0}$

ANSWERS

1. $\hat{j} - \hat{k}$

2. (*i*) $\vec{a} \times \vec{b} = (-2\hat{i} + 8\hat{j} + 5\hat{k})$ and $|\vec{a} \times \vec{b}| = \sqrt{93}$

(*ii*) $\vec{a} \times \vec{b} = (-17\hat{i} + 13\hat{j} + 7\hat{k})$ and $|\vec{a} \times \vec{b}| = 13\sqrt{3}$

3. $\vec{a}\times\vec{b}=(2\hat{i}-51\hat{j}-30\hat{k})$

4. (*i*) $\pm\dfrac{1}{5\sqrt{3}}(5\hat{i}-\hat{j}+7\hat{k})$ (*ii*) $\pm\dfrac{1}{\sqrt{209}}(2\hat{i}+3\hat{j}+14\hat{k})$

5. $|\vec{a}|=\sqrt{91}$ **6.** $-25\hat{i}+35\hat{j}-55\hat{k}$ **9.** $30°$ **10.** $\dfrac{\pi}{4}$ **12.** 6

15. (*i*) 9 sq. units (*ii*) 10 sq. Units **16.** $\dfrac{\sqrt{6}}{2}$ sq. units

17. (*i*) 2 sq. Units (*ii*) $\dfrac{\sqrt{155}}{2}$ sq. units **21.** $\lambda=3, \mu=-\dfrac{27}{2}$.

5.34 SCALAR TRIPLE PRODUCT

The scalar triple product of three vectors $\vec{a}, \vec{b}, \vec{c}$ is defined as $(\vec{a}\times\vec{b}).\vec{c}$, and it is denoted by $[\vec{a}\,\vec{b}\,\vec{c}]$.

Thus $[\vec{a}\,\vec{b}\,\vec{c}] = (\vec{a}\times\vec{b}).\vec{c}$.

Remark

The scalar triple product of three vectors is a scalar quantity.

5.35. GEOMETRICAL INTERPRETATION:

The scalar triple product $(\vec{a}\times\vec{b})\cdot\vec{c}$ represents the volume of a parallelopiped whose coterminous edges are represented by $\vec{a}, \vec{b}, \vec{c}$.

Proof: Let us consider a parallelpiped having coterminous edges OA, OB and OC respectively.

Let $\overrightarrow{OA}=\vec{a}, \overrightarrow{OB}=\vec{b}$ and $\overrightarrow{OC}=\vec{c}$.

Then $(\vec{a}\times\vec{b})$ is a vector perpendicular to the plane of $\vec{a}$ and $\vec{b}$. Let θ be the angle between $(\vec{a}\times\vec{b})$ and $\vec{c}$.

$$\begin{aligned}\therefore\ [\vec{a}\,\vec{b}\,\vec{c}] &= (\vec{a}\times\vec{b}).\vec{c}\\ &= |\vec{a}\times\vec{b}||\vec{c}|\cos\theta\\ &= (\text{area of }\|\text{gm OADB}).\ [\text{Projection of }\vec{c}\text{ along }(\vec{a}\times\vec{b})]\\ &= (\text{area of }\|\text{gm OADB}).\ \text{OL}\\ &= (\text{area of the base of the parallelopiped}).\ (\text{height})\\ &= \text{Volume of the parallelopiped with coterminous edges }\vec{a}, \vec{b}, \vec{c}.\end{aligned}$$

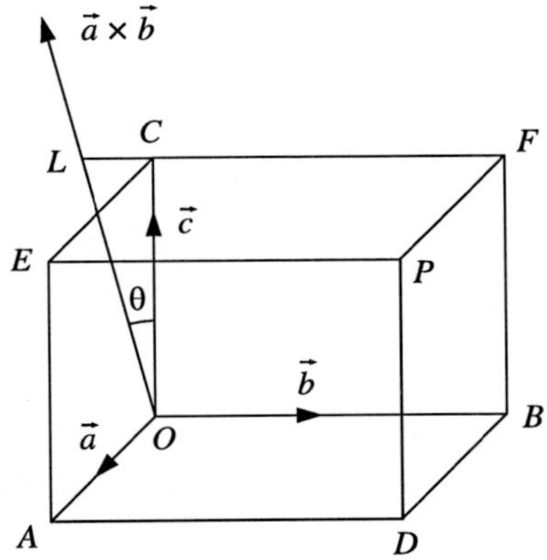

Hence $[\vec{a}\ \vec{b}\ \vec{c}]$ represents the volume of the parallelopiped with coterminous edges $\vec{a}, \vec{b}, \vec{c}$ forming a right handed system.

Theorem 1: For any three vectors $\vec{a}, \vec{b}, \vec{c}$, $(\vec{a}\times\vec{b})\cdot\vec{c} = \vec{a}.(\vec{b}\times\vec{c})$.

Proof: Let $\vec{a},\vec{b},\vec{c}$ form a right handed system representing the coterminous edges of a parallelepiped of volume V.

Then, V = $(\vec{a}\times\vec{b}).\vec{c}$

Again, $\vec{b}, \vec{c}, \vec{a}$ form a right handed system, representing the coterminous edges of the same parallelepiped.

$\therefore \quad V = (\vec{b}\times\vec{c}).\vec{a} = \vec{a}.(\vec{b}\times\vec{c})$

Hence $(\vec{a}\times\vec{b}).\vec{c} = \vec{a}.(\vec{b}\times\vec{c})$.

Theorem 2. The necessary and sufficient condition for three non zero, noncollinear vectors $\vec{a}, \vec{b}, \vec{c}$ to be coplanar to that $[\vec{a}\ \vec{b}\ \vec{c}] = 0$.

Proof: Let $\vec{a}, \vec{b}, \vec{c}$ be three nonzero, noncollinear and coplanar vectors. Then, $\vec{b}\times\vec{c}$ is perpendicular to the plane of $\vec{b}$ and $\vec{c}$.

$\Rightarrow \quad (\vec{b}\times\vec{c}) \perp \vec{a} \qquad [\because \vec{a}, \vec{b}, \vec{c}$ are coplaner]

$\Rightarrow \quad \vec{a}.(\vec{b}\times\vec{c}) = 0$

$\Rightarrow \quad [\vec{a}\,\vec{b}\,\vec{c}] = 0.$

Conversely, let $\vec{a}, \vec{b}, \vec{c}$ be three nonzero, noncollinear vector such that $[\vec{a}\,\vec{b}\,\vec{c}] = 0.$ Then,

$[\vec{a}\,\vec{b}\,\vec{c}] = 0 \Rightarrow \vec{a}.(\vec{b}\times\vec{c}) = 0$

$\Rightarrow \vec{a} = \vec{0}$, or $(\vec{b}\times\vec{c}) = \vec{0}$, or $(\vec{b}\times\vec{c}) \perp \vec{a}$

$\Rightarrow (\vec{b}\times\vec{c}) = \vec{0}$, or $(\vec{b}\times\vec{c}) \perp \vec{a}$ $\quad [\because \vec{a} \neq \vec{0}]$

$\Rightarrow (\vec{b}\times\vec{c}) \perp \vec{a}$ $\quad [\because \vec{b} \neq \vec{0}, \vec{c} \neq \vec{0}$ and $\vec{b}, \vec{c}$ are noncollinear $\Rightarrow \vec{b}\times\vec{c} \neq \vec{0}]$

$(\vec{b}\times\vec{c})$ Thus, is perpendicular to $\vec{a}$

But $(\vec{b}\times\vec{c})$ is perpendicular to the plane of $\vec{b}$ and $\vec{c}$.

$\therefore$ $\vec{a}$ must lie in the plane of $\vec{b}$ and $\vec{c}$.

Hence $\vec{a}, \vec{b}, \vec{c}$ must be coplanar.

5.36 SCALAR TRIPLE PRODUCT IN TERMS OF COMPONENTS

Let $\vec{a} = a_1\hat{i} + a_2\hat{j} + a_3\hat{k}, \vec{b} = b_1\hat{i} + b_2\hat{j} + b_3\hat{k}$ and $\vec{c} = c_1\hat{i} + c_2\hat{j} + c_3\hat{k}$

Then, $$[\vec{a}\,\vec{b}\,\vec{c}] = \begin{vmatrix} a_1 & a_2 & a_3 \\ b_1 & b_2 & b_3 \\ c_1 & c_2 & c_3 \end{vmatrix}$$

Proof: We have

$$\vec{a}\times\vec{b} = \begin{vmatrix} \hat{i} & \hat{j} & \hat{k} \\ a_1 & a_2 & a_3 \\ b_1 & b_2 & b_3 \end{vmatrix}$$

$$\vec{a}\times\vec{b} = (a_2b_3 - a_3b_3)\hat{i} + (a_3b_1 - a_1b_3)\,\hat{j} + (a_1b_2 - a_2b_1)\hat{k}$$

$$\therefore \quad [\vec{a}\,\vec{b}\,\vec{c}] = (\vec{a}\times\vec{b}).\vec{c}$$

$$\Rightarrow \quad (\vec{a}\times\vec{b}).\vec{c} = (a_2b_3 - a_3b_2)\,c_1 + (a_3\,b_1 - a_1b_3)c_2 + (a_1\,b_2 - a_2\,b_1)\,c_3$$

$$= a_1(b_2\,c_3 - b_3\,c_2) + a_2(b_3c_1 - b_1c_3) + a_3(b_1\,c_2 - b_2\,c_1)$$

$$\therefore (\vec{a}\times\vec{b}).\vec{c} = \begin{vmatrix} c_1 & c_2 & c_3 \\ a_1 & a_2 & a_3 \\ b_1 & b_2 & b_3 \end{vmatrix} = \begin{vmatrix} a_1 & a_2 & a_3 \\ b_1 & b_2 & b_3 \\ c_1 & c_2 & c_3 \end{vmatrix}. \text{ Hence } [\vec{a}\ \vec{b}\ \vec{c}] = \begin{vmatrix} a_1 & a_2 & a_3 \\ b_1 & b_2 & b_3 \\ c_1 & c_2 & c_3 \end{vmatrix}$$

Theorem 3. Vector product is distributive w.r.t., addition.

If $\vec{a}, \vec{b}, \vec{c}$ are any vectors, then $\vec{a}\times(\vec{b}+\vec{c}) = \vec{a}\times\vec{b} + \vec{a}\times\vec{c}$.

Proof: Let $\vec{r} = \vec{a}\times(\vec{b}+\vec{c}) - \vec{a}\times\vec{b} - \vec{a}\times\vec{c}$, and $\vec{p}$ be any arbitrary vector, then

$$\vec{p}.\vec{r} = \vec{p}.(\vec{a}\times(\vec{b}+\vec{c}) - \vec{a}\times\vec{b} - \vec{a}\times\vec{c})$$

$$= \vec{p}.(\vec{a}\times(\vec{b}+\vec{c})) - \vec{p}.(\vec{a}\times\vec{b}) - \vec{p}.(\vec{a}\times\vec{c})$$

[$\because$ dot product is distributive w.r.t. addition]

$$= (\vec{p}\times\vec{a}).(\vec{b}+\vec{c}) - (\vec{p}\times\vec{a}).\vec{b}) - (\vec{p}\times\vec{a}).\vec{c}$$

[$\because$ in a scalar triple product, the positions of dot and cross can be changed]

$$= (\vec{p}\times\vec{a}).\vec{b} + (\vec{p}\times\vec{a}).\vec{c} - (\vec{p}\times\vec{a}).\vec{b} - (\vec{p}\times\vec{a}).\vec{c}.$$

[$\because$ dot product is distributive w.r.t. addition]

$$= 0$$

$\therefore$ either $\vec{p} = \vec{0}$ or $\vec{r} = \vec{0}$ or $\vec{p}$ is $\perp \vec{r}$. But $\vec{p}$ being an arbitrary vector can be taken as non-zero vector and not $\perp$ to $\vec{r}$, it follows that $\vec{r} = \vec{0}$.

$$\Rightarrow \quad \vec{a}\times(\vec{b}+\vec{c}) - \vec{a}\times\vec{b} - \vec{a}\times\vec{c} = \vec{0}$$

$$\Rightarrow \quad \vec{a}\times(\vec{b}+\vec{c}) = \vec{a}\times\vec{b} + \vec{a}\times\vec{c}.$$

ILLUSTRATIVE EXAMPLES

Example 1. If $\vec{a} = 2\hat{i} + \hat{j} + 3\hat{k}, \vec{b} = -\hat{i} + 2\hat{j} + \hat{k}$ and $\vec{c} = -3\hat{i} + \hat{j} + 2\hat{k}$ find $[\vec{a}\,\vec{b}\,\vec{c}]$.

Solution. We have

$$[\vec{a}\,\vec{b}\,\vec{c}]. = \begin{vmatrix} 2 & 1 & 3 \\ -1 & 2 & 1 \\ -3 & 1 & 2 \end{vmatrix}$$

$$= 2(4-1) - 1(-2+3) + 3(-1+6)$$

$$= 2\times 3 - 1 + 15 = 6 - 1 + 15 = 20$$

Example 2. Find the volume of the parallelopiped whose edges are represented by $\vec{a} = 2\hat{i} - 3\hat{j} + 4\hat{k}, \vec{b} = \hat{i} + 2\hat{j} - \hat{k}$ and $\vec{c} = 3\hat{i} - \hat{j} - 2\hat{k}$.

Solution.

$$[\vec{a}\,\vec{b}\,\vec{c}] = \begin{vmatrix} 2 & -3 & 4 \\ 1 & 2 & -1 \\ 3 & -1 & 2 \end{vmatrix}$$

$$= 2(4-1) - (-3)(2+3) + 4(-1-6)$$

$$= 6 + 15 - 28 = -7$$

$\therefore$ The volume of the parallelopiped $= |-7| = 7$ cubic units.

Example 3. If $\vec{a} = -2\hat{i} - 2\hat{j} + 4\hat{k}, \vec{b} = -2\hat{i} + 4\hat{j} - 2\hat{k}, \vec{c} = 4\hat{i} - 2\hat{j} - 2\hat{k}$, prove that $\vec{a}, \vec{b}, \vec{c}$ are coplanar.

Solution.

Here $$[\vec{a}\,\vec{b}\,\vec{c}] = \begin{vmatrix} -2 & -2 & 4 \\ -2 & 4 & -2 \\ 4 & -2 & -2 \end{vmatrix}$$

$$= -2(-8-4) - (-2)(4+8) + 4(4-16)$$

$$= 24 + 24 - 48 = 0.$$

Also we note that the given vectors are non-zero, therefore the given vectors are coplanar.

Example 4. Find the value of λ so that the four points with position vectors $-\hat{j}+\hat{k}, 2\hat{i}-\hat{j}+\hat{k}, \hat{i}+\lambda\hat{j}+\hat{k}$ and $3\hat{j}+3\hat{k}$ are coplanar.

Solution. Let A, B, C and D be the given points whose position vectors are $-\hat{j}+\hat{k}, 2\hat{i}-\hat{j}+\hat{k}, \hat{i}+\lambda\hat{j}+\hat{k}$ and $3\hat{j}+3\hat{k}$ respectively.

$$\overrightarrow{AB} = 2\hat{i} - 2\hat{k}$$

$$\overrightarrow{AC} = \hat{i} + (\lambda+1)\hat{j} \text{ and } \overrightarrow{AD} = 4\hat{j} + 2\hat{k}$$

Since the points A, B, C and D are coplanar, vectors $\overrightarrow{AB}, \overrightarrow{AC}, \overrightarrow{AD}$ are coplanar.

$$\Rightarrow \left[\overrightarrow{AB}\ \overrightarrow{AC}\ \overrightarrow{AD}\right] = 0$$

$$\Rightarrow \begin{vmatrix} 2 & 0 & -2 \\ 1 & 1+\lambda & 0 \\ 0 & 4 & 2 \end{vmatrix} = 0$$

$$\Rightarrow \lambda = 1.$$

Example 5. Prove that $[\vec{a}+\vec{b}\ \ \vec{b}+\vec{c}\ \ \vec{c}+\vec{a}] = 2[\vec{a}\ \vec{b}\ \vec{c}]$

Solution. We have

$$[\vec{a}+\vec{b}\ \ \vec{b}+\vec{c}\ \ \vec{c}+\vec{a}] = (\vec{a}\times\vec{b}).(\vec{b}+\vec{c})\times(\vec{c}+\vec{a})]$$

$$= (\vec{a}\times\vec{b})\cdot[\vec{b}\times\vec{c}+\vec{b}\times\vec{a}+\vec{c}\times\vec{c}+\vec{c}\times\vec{a}]$$

$$= (\vec{a}\times\vec{b}).[(\vec{b}\times\vec{c})-(\vec{a}\times\vec{b})+(\vec{c}\times\vec{a})],$$

$$[\because \vec{c}\times\vec{c} = \vec{0} \text{ and } \vec{b}\times\vec{a} = -\vec{a}\times\vec{b}]$$

$$= \vec{a}.(\vec{b}\times\vec{c}) - \vec{a}.(\vec{a}\times\vec{b}) + \vec{a}.(\vec{c}\times\vec{a}) + \vec{b}.(\vec{b}\times\vec{c})$$
$$- \vec{b}.(\vec{a}\times\vec{b}) + \vec{b}.(\vec{c}\times\vec{a})$$

$$= [\vec{a}\ \vec{b}\ \vec{c}] - [\vec{a}\ \vec{a}\ \vec{b}] + [\vec{a}\ \vec{c}\ \vec{a}] + [\vec{b}\ \vec{b}\ \vec{c}] - [\vec{b}\ \vec{a}\ \vec{b}] + [\vec{b}\ \vec{c}\ \vec{a}]$$

$$= [\vec{a}\ \vec{b}\ \vec{c}] + [\vec{b}\ \vec{c}\ \vec{a}] = 2[\vec{a}\ \vec{b}\ \vec{c}]$$

Hence $[\vec{a}+\vec{b}\ \ \vec{b}+\vec{c}\ \ \vec{c}+\vec{a}] = 2[\vec{a}\ \vec{b}\ \vec{c}]$.

Example 6. Find λ if the vectors $\vec{a} = \hat{i}+3\hat{j}+\hat{k}, \vec{b} = 2\hat{i}-\hat{j}-\hat{k}$ and $\bar{c} = \lambda\hat{i}+7\hat{j}+3\hat{k}$ are coplanar.

Solution. Since $\vec{a}, \vec{b}$, and $\vec{c}$ are coplanar, we have $[\vec{a}\ \vec{b}\ \vec{c}] = 0$ *i.e.*,

$$\begin{vmatrix} 1 & 3 & 1 \\ 2 & -1 & -1 \\ \lambda & 7 & 3 \end{vmatrix} = 0$$

$$\Rightarrow \qquad \lambda = 0.$$

EXERCISE 5.4

1. Show that the vectors $\vec{a} = \hat{i} - 2\hat{j} + 3\hat{k}, \vec{b} = -2\hat{i} + 3\hat{j} - 4\hat{k}$ and $\vec{c} = \hat{i} - 3\hat{j} + 5\hat{k}$ are coplanar.
2. Find the volume of the parallelopiped whose edges are given by the vectors $2\hat{i} + 2\hat{j} - 3\hat{k},\ \hat{i} + \hat{j} - \hat{k}$, and $2\hat{i} - \hat{k}$.
3. Find $[\vec{a}\,\vec{b}\,\vec{c}]$, when $\vec{a} = 2\hat{i} + \hat{j} + 3\hat{k}, \vec{b} = -\hat{i} + 2\hat{j} + \hat{k}$ and $\vec{c} = 3\hat{i} + \hat{j} + 2\hat{k}$.
4. Show that the vectors $\vec{a}, \vec{b}, \vec{c}$ are coplanar, when $\vec{a} = \hat{i} - 2\hat{j} + 3\hat{k}$, $\vec{b} = -2\hat{i} + 3\hat{j} - 4\hat{k}$ and $\vec{c} = \hat{i} - 3\hat{j} + 5\hat{k}$.
5. Find the value of λ for which the vectors $\vec{a}, \vec{b}, \vec{c}$ are coplanar, where $\vec{a} = 2\hat{i} - \hat{j} + \hat{k}, \vec{b} = \hat{i} + 2\hat{j} + 3\hat{k}$ and $\vec{c} = 3\hat{i} + \lambda\hat{j} + 5\hat{k}$.
6. Show that the vectors $\vec{a} = \hat{i} + 3\hat{j} + \hat{k}, \vec{b} = 2\hat{i} - \hat{j} - \hat{k}$ and $\vec{c} = 7\hat{j} + 3\hat{k}$ are parallel to the same plane.

[Hint show that $[\vec{a}\,\vec{b}\,\vec{c}] = 0$]

ANSWERS

2. 2 **3.** -10 **5.** $\lambda = -4$

5.37. VECTOR TRIPLE PRODUCT

If $\vec{a}, \vec{b}, \vec{c}$ are any three vectors then $(\vec{a} \times \vec{b}) \times \vec{c}$ and $\vec{a} \times (\vec{b} \times \vec{c})$ etc. are known as vector triple products.

Remark:

The vector triple product is a vector quantity.

Theorem: If $\vec{a}, \vec{b}, \vec{c}$ are any three vectors, then

(*i*) $(\vec{a} \times \vec{b}) \times \vec{c} = (\vec{a}.\vec{c})\,\vec{b} - (\vec{b}.\vec{c})\,\vec{a}$ and

(*ii*) $\vec{a} \times (\vec{b} \times \vec{c}) = (\vec{a}.\vec{c})\,\vec{b} - (\vec{a}.\vec{b}) \cdot \vec{c}$

Proof: (*i*) Let $\vec{a} = a_1\hat{i} + a_2\hat{j} + a_3\hat{k}, \vec{b} = b_1\hat{i} + b_2\hat{j} + b_3\hat{k}, \vec{c} = c_1\hat{i} + c_2\hat{j} + c_3\hat{k}$

then, $\vec{a} \times \vec{b} = \begin{vmatrix} \hat{i} & \hat{j} & \hat{k} \\ a_1 & a_2 & a_3 \\ b_1 & b_2 & b_3 \end{vmatrix}$

$$= (a_2 b_3 - a_3 b_2)\hat{i} + (a_3 b_1 - a_1 b_3)\hat{j} + (a_1 b_2 - a_2 b_1)\hat{k}$$

$$\therefore \quad (\vec{a}\times\vec{b})\times\vec{c} = \begin{vmatrix} \hat{i} & \hat{j} & \hat{k} \\ a_2b_3 - a_3b_2 & a_3b_1 - a_1b_3 & a_1b_2 - a_2b_1 \\ c_1 & c_2 & c_3 \end{vmatrix} 2$$

$$= (a_3b_1c_3 - a_1b_3c_3 - a_1b_2c_2 + a_2b_1c_2)\hat{i}$$

$$+ \hat{j}(a_1b_2c_1 - a_2b_1c_1 - a_2b_3c_3 + a_3b_2c_3)$$

$$+ \hat{k}(a_2b_3c_2 - a_3b_2c_2 - a_3b_1c_1 + a_1b_3c_1)$$

Now, $\quad (\vec{a}.\vec{c})\,\vec{b} - (\vec{b}.\vec{c})\,\vec{a}$

$$= (a_1c_1 + a_2c_2 + a_3\,c_3)\,(b_1\hat{i} + b_2\hat{j} + b_3\hat{k})$$

$$- (b_1c_1 + b_2c_2 + b_3c_3)\,(a_1\hat{i} + a_2\hat{j} + a_3\hat{k})$$

$$= (a_1b_1c_1 + a_2b_1c_2 + a_3b_1c_3 - a_1b_1c_1 - a_1b_2c_2 - a_1b_3c_3)\hat{i} + ...$$

$$= (a_3b_1c_3 - a_1b_3c_3 - a_1b_2c_2 + a_2b_1c_2)\hat{i} + ...$$

The result follows

(ii) Proof yourself

ILLUSTRATIVE EXAMPLE

Example 1. If $\vec{a}$ and $\vec{b}$ are vectors such that $(\vec{a}\times\vec{b})\times\vec{b} = (\vec{b}\times\vec{a})\times\vec{a}$ then prove that $\vec{a}\times\vec{b} = \vec{0}$.

Solution. Given $(\vec{a}\times\vec{b})\times\vec{b} = (\vec{b}\times\vec{a})\times\vec{a}$

$$\Rightarrow (\vec{a}\times\vec{b})\times\vec{b} = -(\vec{a}\times\vec{b})\times\vec{a}$$

$$\Rightarrow (\vec{a}\times\vec{b})\times\vec{b} + (\vec{a}\times\vec{b})\times\vec{a} = \vec{0}$$

$$\Rightarrow (\vec{a}\times\vec{b})\times\vec{a} + (\vec{a}\times\vec{b})\times\vec{b} = \vec{0}$$

$$\Rightarrow (\vec{a}\times\vec{b})\times(\vec{a}+\vec{b}) = \vec{0} \qquad \text{[By distributive law]}$$

$$\Rightarrow (\vec{a}\times\vec{b}) = \vec{0} \text{ or } (\vec{a}+\vec{b}) = \vec{0} \text{ or } (\vec{a}\times\vec{b}) \parallel (\vec{a}+\vec{b}) \qquad ...(1)$$

But $(\vec{a}\times\vec{b}).(\vec{a}+\vec{b}) = (\vec{a}\times\vec{b})\cdot\vec{a} + (\vec{a}\times\vec{b}).\vec{b} = 0 + 0 = 0$

$$\Rightarrow (\vec{a}\times\vec{b}) \text{ is } \perp \text{ to } \vec{a}+\vec{b} \qquad ...(2)$$

So, we have either $\vec{a}\times\vec{b} = \vec{0}$ or $\vec{a}+\vec{b} = \vec{0}$.

If $\vec{a}+\vec{b} = \vec{0}$ then $\vec{a} = -\vec{b}$.

$$\therefore \quad \vec{a}\times\vec{b} = (-\vec{b})\times\vec{b} = -(\vec{b}\times\vec{b}) = \vec{0} \qquad ...(3)$$

From (1), (2) and (3), we get $\vec{a}\times\vec{b} = \vec{0}$.

Example 2. If $\vec{a}, \vec{b}, \vec{c}$ and $\vec{d}$ are any vectors, prove that

(*i*) $(\vec{b}\times\vec{c})\times(\vec{c}\times\vec{a}) = [\vec{a}\,\vec{b}\,\vec{c}]\,\vec{c}$

(*ii*) $(\vec{a}\times\vec{b})\times(\vec{c}\times\vec{d}) = [\vec{a}\,\vec{b}\,\vec{d}]\,\vec{c} - [\vec{a}\,\vec{b}\,\vec{c}]\,\vec{d}$

(*iii*) $(\vec{a}\times\vec{b}).(\vec{c}\times\vec{d}) = \begin{vmatrix} \vec{a}.\vec{c} & \vec{b}.\vec{c} \\ \vec{a}.\vec{d} & \vec{b}.\vec{d} \end{vmatrix}$

Solution. (*i*) $(\vec{b}\times\vec{c})\times(\vec{c}\times\vec{d}) = (\vec{b}\times\vec{c})\times\vec{r}$ where $\vec{r} = \vec{c}\times\vec{a}$

$$= (\vec{b}.\vec{r})\vec{c} - (\vec{c}.\vec{r})\vec{b}$$

$$= (\vec{b}\cdot\vec{c}\times\vec{a})\vec{c} - (\vec{c}\cdot\vec{c}\times\vec{a})\,\vec{b} \qquad [\because \vec{r} = \vec{c}\times\vec{a}]$$

$$= [\vec{b}\,\vec{c}\,\vec{a}]\vec{c} - O\vec{b} = [\vec{b}\,\vec{c}\,\vec{a}]\,\vec{c} - \vec{0}$$

$$= [\vec{a}\,\vec{b}\,\vec{c}]\vec{c}$$

(*ii*) $(\vec{a}\times\vec{b})\times(\vec{c}\times\vec{d}) = \vec{p}\times(\vec{c}\times\vec{d})$ when $\vec{p} = \vec{a}\times\vec{b}$

$$= (\vec{p}.\vec{d})\,\vec{c} - (\vec{p}.\vec{c})\,\vec{d}$$

$$= (\vec{a}\times\vec{b}.\vec{d})\,\vec{c} - (\vec{a}\times\vec{b}.\vec{c})\,\vec{d} \qquad [\because\ \vec{p} = \vec{a}\times\vec{b}]$$

$$= [\vec{a}\,\vec{b}\,\vec{d}]\,\vec{c} - [\vec{a}\,\vec{b}\,\vec{c}]\,\vec{d}$$

(*iii*) $(\vec{a}\times\vec{b}).(\vec{c}\times\vec{d}) = \vec{p}.(\vec{c}\times\vec{d})$ where $\vec{p} = \vec{a}\times\vec{b}$

$$= (\vec{p}\times\vec{c})\cdot\vec{d} \qquad [\because \text{dot and cross can be interchanged}]$$

$$= \left((\vec{a}\times\vec{b})\times\vec{c}\right).\vec{d} \quad \left[\because \vec{p} = \vec{a}\times\vec{b}\right]$$

$$= \left((\vec{a}.\vec{c})\vec{b} - (\vec{b}.\vec{c})\vec{a}\right)\vec{d}$$

$$= (\vec{a}.\vec{c})(\vec{b}.\vec{d}) - (\vec{b}.\vec{c})(\vec{a}.\vec{d})$$

$$= \begin{vmatrix} \vec{a}.\vec{c} & \vec{b}.\vec{c} \\ \vec{a}.\vec{d} & \vec{b}.\vec{d} \end{vmatrix}$$

Example 3. If $\vec{a}, \vec{b}$ and $\vec{c}$ are any vectors, prove that $[\vec{a}\times\vec{b}\ \ \vec{b}\times\vec{c}\ \ \vec{c}\times\vec{a}] = [\vec{a}\,\vec{b}\,\vec{c}]^2$

Solution. We have $(\vec{b}\times\vec{c})\times(\vec{c}\times\vec{a}) = \vec{r}\times(\vec{c}\times\vec{a})$ where $\vec{r} = \vec{b}\times\vec{c}$

$$= (\vec{r}.\vec{a})\,\vec{c} - (\vec{r}.\vec{c})\,\vec{a}$$

$$= \left\{(\vec{b}\times\vec{c}).\,\vec{a}\right\}\vec{c} - \left\{(\vec{b}\times\vec{c})\,\vec{c}\right\}\vec{a}$$

$$= \left[\vec{b}\,\vec{c}\,\vec{a}\right]\vec{c} - \left[\vec{b}\,\vec{c}\,\vec{c}\right]\vec{a}$$

$$= \left[\vec{b}\,\vec{c}\,\vec{a}\right]\vec{c} \qquad \left[\because [\vec{b}\,\vec{c}\,\vec{c}]\,\vec{a} = 0\,\vec{a} = 0\right]$$

$$= [\vec{a}\,\vec{b}\,\vec{c}]\,\vec{c} \qquad \left[\because [\vec{b}\,\vec{c}\,\vec{a}] = [\vec{a}\,\vec{b}\,\vec{c}]\right]$$

$$\Rightarrow (\vec{a}\times\vec{b}).\{(\vec{b}\times\vec{c})\times(\vec{c}\times\vec{a})\} = (\vec{a}\times\vec{b})\,[\vec{a}\,\vec{b}\,\vec{c}]\,\vec{c}$$

$$= [\vec{a}\,\vec{b}\,\vec{c}]\{(\vec{a}\times\vec{b}).\vec{c}\}$$

$$= [\vec{a}\,\vec{b}\,\vec{c}][\vec{a}\,\vec{b}\,\vec{c}] = [\vec{a}\,\vec{b}\,\vec{c}]^2$$

$$\Rightarrow [\vec{a}\times\vec{b}\;\; \vec{b}\times\vec{c}\;\; \vec{c}\times\vec{a}] = [\vec{a}\,\vec{b}\,\vec{c}]^2$$

Example 4. If $\vec{a} = \hat{i} + 2\hat{j} + 3\hat{k}, \vec{b} = 2\hat{i} - \hat{j} + \hat{k}$ and $\vec{c} = \hat{i} + \hat{j} - 2\hat{k},$ verify that $\vec{a}\times(\vec{b}\times\vec{c}) = (\vec{a}\cdot\vec{c})\,\vec{b} - (\vec{a}.\vec{b})\,\vec{c}$.

Solution. We have

$$\vec{b}\times\vec{c} = \begin{vmatrix} \hat{i} & \hat{j} & \hat{k} \\ 2 & -1 & 1 \\ 1 & 1 & -2 \end{vmatrix} = (\hat{i} + 5\hat{j} + 3\hat{k})$$

$$\therefore \quad \vec{a}\times(\vec{b}\times\vec{c}) = \begin{vmatrix} \hat{i} & \hat{j} & \hat{k} \\ 1 & 2 & 3 \\ 1 & 5 & 3 \end{vmatrix} = (-9\hat{i} + 3\hat{k})$$

Also, $(\vec{a}.\vec{c}) = (\hat{i} + 2\hat{j} + 3\hat{k}).(\hat{i} + \hat{j} - 2\hat{k}) = 1 + 2 - 6 = -3$

$$\Rightarrow (\vec{a}.\vec{c})\,\vec{b} = -3(2\hat{i} - \hat{j} + \hat{k}) = (-6\hat{i} + 3\hat{j} - 3\hat{k})$$

$$(\vec{a}.\vec{b}) = (\hat{i} + 2\hat{j} + 3\hat{k}).(2\hat{i} - \hat{j} + \hat{k}) = 2 - 2 + 3 = 3$$

$$\Rightarrow (\vec{a}.\vec{b})\vec{c} = 3(\hat{i} + \hat{j} - 2\hat{k}) = (3\hat{i} + 3\hat{j} - 6\hat{k})$$

$$\therefore \quad (\vec{a}.\vec{c})\vec{b} - (\vec{a}.\vec{b})\,\vec{c} = (-6\hat{i} + 3\hat{j} - 3\hat{k}) - (3\hat{i} + 3\hat{j} - 6\hat{k})$$

$$= (-9\hat{i} + 3\hat{k})$$

Hence $\vec{a}\times(\vec{b}\times\vec{c}) = (\vec{a}.\vec{c})\,\vec{b} - (\vec{a}.\vec{b})\,\vec{c}.$

5.38 RECIPROCAL SYSTEM OF VECTORS

Let $\vec{a}, \vec{b}, \vec{c}$ be three noncoplanar vector so that $[\vec{a}\,\vec{b}\,\vec{c}] \neq 0.$ Then the vectors,

$$\vec{a}' = \frac{(\vec{b}\times\vec{c})}{[\vec{a}\,\vec{b}\,\vec{c}]}, \vec{b}' = \frac{(\vec{c}\times\vec{a})}{[\vec{a}\,\vec{b}\,\vec{c}]} \text{ and } \vec{c}' = \frac{(\vec{a}\times\vec{b})}{[\vec{a}\,\vec{b}\,\vec{c}]}$$

are said to form a reciprocal system of vectors with the vectors $\vec{a}, \vec{b}, \vec{c}$.

ILLUSTRATIVE EXAMPLES

Example 1. If $\vec{a}, \vec{b}, \vec{c}$ and $\vec{a}', \vec{b}', \vec{c}'$ form a reciprocal system of vectors, then

(*i*) $\vec{a}.\vec{a}' = \vec{b}.\vec{b}' = \vec{c}.\vec{c}' = 1$

(*ii*) $[\vec{a}\,\vec{b}\,\vec{c}][\vec{a}'\,\vec{b}'\,\vec{c}'] = 1.$

Solution. We know that $\vec{a}' = \dfrac{(\vec{b}\times\vec{c})}{[\vec{a}\,\vec{b}\,\vec{c}]}, \vec{b}' = \dfrac{(\vec{c}\times\vec{a})}{[\vec{a}\,\vec{b}\,\vec{c}]}, \vec{c}' = \dfrac{(\vec{a}\times\vec{b})}{[\vec{a}\,\vec{b}\,\vec{c}]}$

(*i*) $\vec{a}.\vec{a}' = \vec{a}.\dfrac{(\vec{b}\times\vec{c})}{[\vec{a}\,\vec{b}\,\vec{c}]}, = \dfrac{\vec{a}.(\vec{b}\times\vec{c})}{[\vec{a}\,\vec{b}\,\vec{c}]} = \dfrac{[\vec{a}\,\vec{b}\,\vec{c}]}{[\vec{a}\,\vec{b}\,\vec{c}]} = 1$

$\vec{a}.\vec{a}' = 1$

Similarly $\vec{b}.\vec{b}' = 1$ and $\vec{c}.\vec{c}' = 1.$

Hence $\vec{a}.\vec{a}' = \vec{b}.\vec{b}' = \vec{c}.\vec{c}' = 1.$

$$(iii)\ [\vec{a}'\,\vec{b}'\,\vec{c}'] = \left\{\frac{(\vec{b}\times\vec{c})}{[\vec{a}\,\vec{b}\,\vec{c}]}.\frac{(\vec{b}\times\vec{a})}{[\vec{a}\,\vec{b}\,\vec{c}]}\times\frac{(\vec{a}\times\vec{b})}{[\vec{a}\,\vec{b}\,\vec{c}]}\right\}$$

$$= \frac{1}{[\vec{a}\,\vec{b}\,\vec{c}]^3}\cdot[\vec{b}\times\vec{c}\quad \vec{c}\times\vec{a}\quad \vec{a}\times\vec{b}]$$

$$= \frac{1}{[\vec{a}\,\vec{b}\,\vec{c}]^3}.[\vec{a}\,\vec{b}\,\vec{c}]^2 = \frac{1}{[\vec{a}\,\vec{b}\,\vec{c}]}\ \left\{\because [\vec{b}\times\vec{c}\quad \vec{c}\times\vec{a}\quad \vec{a}\times\vec{b}] = [\vec{a}\,\vec{b}\,\vec{c}]^2\right\}$$

$$\Rightarrow \qquad [\vec{a}'\,\vec{b}'\,\vec{c}']\cdot[\vec{a}\,\vec{b}\,\vec{c}] = 1.$$

Example 2. If $\vec{a}, \vec{b}, \vec{c}$ and $\vec{a}', \vec{b}', \vec{c}'$ be the reciprocal system of vectors then

$\vec{a}.\vec{a}' + \vec{b}.\vec{b}' + \vec{c}.\vec{c}' = 3$

Solution. We have $\vec{a}' = \dfrac{\vec{b}\times\vec{c}}{[\vec{a}\,\vec{b}\,\vec{c}]}, \vec{b}' = \dfrac{\vec{c}\times\vec{a}}{[\vec{a}\,\vec{b}\,\vec{c}]}, \vec{c}' = \dfrac{(\vec{a}\times\vec{b})}{[\vec{a}\,\vec{b}\,\vec{c}]}$, then

$$\vec{a}.\vec{a}' + \vec{b}.\vec{b}' + \vec{c}.\vec{c}'$$

$$= \vec{a}.\frac{(\vec{b}\times\vec{c})}{[\vec{a}\,\vec{b}\,\vec{c}]} + \vec{b}.\frac{(\vec{c}\times\vec{a})}{[\vec{a}\,\vec{b}\,\vec{c}]} + \vec{c}.\frac{(\vec{a}\times\vec{b})}{[\vec{a}\,\vec{b}\,\vec{c}]}$$

$$= \frac{\vec{a}.(\vec{b}\times\vec{c})}{[\vec{a}\,\vec{b}\,\vec{c}]} + \frac{\vec{b}(\vec{c}\times\vec{a})}{[\vec{a}\,\vec{b}\,\vec{c}]} + \frac{\vec{c}.(\vec{a}\times\vec{b})}{[\vec{a}\,\vec{b}\,\vec{c}]}$$

$$= \frac{[\vec{a}\,\vec{b}\,\vec{c}]}{[\vec{a}\,\vec{b}\,\vec{c}]} + \frac{[\vec{b}\,\vec{c}\,\vec{a}]}{[\vec{a}\,\vec{b}\,\vec{c}]} + \frac{[\vec{c}\,\vec{a}\,\vec{b}]}{[\vec{a}\,\vec{b}\,\vec{c}]} = \frac{[\vec{a}\,\vec{b}\,\vec{c}]}{[\vec{a}\,\vec{b}\,\vec{c}]} + \frac{[\vec{a}\,\vec{b}\,\vec{c}]}{[\vec{a}\,\vec{b}\,\vec{c}]} + \frac{[\vec{a}\,\vec{b}\,\vec{c}]}{[\vec{a}\,\vec{b}\,\vec{c}]}$$

Hence $\vec{a}.\vec{a}' + \vec{b}.\vec{b}' + \vec{c}.\vec{c}' = 3.$

EXERCISE 5.5

1. If $\vec{a} = 2\hat{i} - 3\hat{j} + \hat{k}, \vec{b} = \hat{i} - \hat{j} + 2\hat{k}, \vec{c} = 2\hat{i} + \hat{j} + \hat{k}$, then find (*i*) $(\vec{a} \times \vec{b}) \times \vec{c}$ (*ii*) $\vec{a} \times (\vec{b} \times \vec{c})$.

2. Show that the vectors $\vec{a} \times (\vec{b} \times \vec{c}), \vec{b} \times (\vec{c} \times \vec{a}), \vec{c} \times (\vec{a} \times \vec{b})$ are coplanar.

3. If $\vec{a}, \vec{b}, \vec{c}$ be three non coplanar vectors, ten show that $\vec{b} \times \vec{c}, \vec{c} \times \vec{a}, \vec{a} \times \vec{b}$ are also non-coplanar.

4. Prove that $\vec{a} \times (\vec{b} \times \vec{c}) = (\vec{a} \times \vec{b}) \times \vec{c}$ if and only if $(\vec{c} \times \vec{a}) \times \vec{b} = \vec{0}$.

5. Show that $\vec{a} \times (\vec{b} \times \vec{c}) + \vec{b} \times (\vec{c} \times \vec{a}) + \vec{c} \times (\vec{a} \times \vec{b}) = \vec{0}$.

6. Prove that $\vec{a} \times (\vec{a} \times (\vec{a} \times \vec{b}) = (\vec{a}.\vec{a})(\vec{b} \times \vec{a})$.

7. Prove that $(\vec{a} \times \vec{b}) \times (\vec{a} \times \vec{c}).\vec{d} = (\vec{a}.\vec{d})[\vec{a}\,\vec{b}\,\vec{c}]$

8. If $\vec{a} = 3\hat{i} - \hat{j} + 2\hat{k}, \vec{b} = 2\hat{i} + \hat{j} - \hat{k}$, and $\vec{c} = \hat{i} - 2\hat{j} + 2\hat{k}$ then show that $(\vec{a} \times \vec{b}) \times \vec{c} \neq \vec{a} \times (\vec{b} \times \vec{c})$.

9. If $\vec{a} = \hat{i} - 2\hat{j} + 3\hat{k}, \vec{b} = 2\hat{i} + \hat{j} + \hat{k}$, and $\vec{c} = \hat{i} + \hat{j} + 2\hat{k}$, show that $|(\vec{a} \times \vec{b}) \times \vec{c}| = 5\sqrt{14}$.

10. Verify that $\vec{a} \times (\vec{b} \times \vec{c}) = (\vec{a}.\vec{c})\vec{b} - (\vec{a}.\vec{b})\vec{c}$ when $\vec{a} = 2\hat{i} + \hat{j} + \hat{k}$, $\vec{b} = 3\hat{i} - 4\hat{j} + \hat{k}$ and $\vec{c} = 3\hat{i} + 7\hat{j} - 6\hat{k}$.

11. For any three vectors $\vec{a}, \vec{b}, \vec{c}$, show that $\vec{a} \times (\vec{b} \times \vec{c}) = (\vec{a} \times \vec{b}) \times \vec{c}$. iff $\vec{a}$ and $\vec{c}$ are collinear

12. Prove that the vectors $\hat{i}, \hat{j}, \hat{k}$ form a self orthogonal system.

13. Prove that $\vec{a}.(\vec{a} \times \vec{c}) = 0$.

14. Find the volume of the parallelopiped whose edges are given by the vectors $2\hat{i} + 2\hat{j} - 3\hat{k}, \hat{i} + \hat{j} - \hat{k}$ and $2\hat{i} - \hat{k}$.

15. If $\vec{a}, \vec{b}, \vec{c}, \vec{d}$ are four vector such that $[\vec{a}\,\vec{b}\,\vec{c}] \neq \vec{0}$, then show that

$$\vec{d} = \frac{[\vec{b}\,\vec{c}\,\vec{d}]\,\vec{a} + [\vec{c}\,\vec{a}\,\vec{d}]\,\vec{b} + [\vec{a}\,\vec{b}\,\vec{d}]\,\vec{c}}{[\vec{a}\,\vec{b}\,\vec{c}]}$$

ANSWERS

1. (*i*) $-4\hat{i} + 7\hat{j} + \hat{k}$ **2.** 2

(*ii*) $-3(4\hat{i} + 3\hat{j} + \hat{k})$

5.39. THE VECTOR DIFFERENTIAL OPERATOR DEL (∇)

The vector differential operator ∇ (read as del or nabla) is defined as

$$\nabla = \hat{i}\frac{\partial}{\partial x} + \hat{j}\frac{\partial}{\partial y} + \hat{k}\frac{\partial}{\partial z}$$

5.40. POINT FUNCTION

A variable quantity whose value at any point in a region of space depends upon the position of the point, is called a point function.

A point function is a function whose domain is a set of points.

There are two types of functions.

5.41. SCALAR POINT FUNCTION

If for each point P of a region R, there corresponds a scalar denoted by $f(P)$, then f is called a scalar point function for the region R.

The temperature $f(P)$ at any point P of a fluid occupying a certain region is an example of a scalar point function for that region.

5.42. VECTOR POINT FUNCTION

If for each point P of a region R there corresponds a vector $\vec{f}(P)$, then $\vec{f}$ is called vector point function for the region R.

If the velocity of a particle at a point P, at any given that t be $\vec{f}(P)$, then $\vec{f}$ is a vector point function for the region occupied by the particle at time t.

5.43. GRADIENT OR SLOPE OF SCALAR POINT FUNCTION

Let $f(x, y, z)$ be a function defining a scalar field, then the vector $\hat{i}\frac{\partial f}{\partial x} + \hat{j}\frac{\partial f}{\partial y} + \hat{k}\frac{\partial f}{\partial z}$ is called the gradient of the scalar field f and is denoted by grad f.

Thus, grad $f = \hat{i}\frac{\partial f}{\partial x} + \hat{j}\frac{\partial f}{\partial y} + \hat{k}\frac{\partial f}{\partial z}$.

Note: The gradient of a scalar field is a vector field.

5.44. GEOMETRICAL INTERPRETATION OF GRADIENT

Let $f(x, y, z)$ be a scalar valued function. Let $P(x_0, y_0, z_0)$ be any point. Let $f(x_0, y_0, z_0) = c$. Then the equation $f(x, y, z)$ represents a surface. Obviously (x_0, y_0, z_0) lies on this surface. Along this surface $df = 0$.

$$\Rightarrow \quad \frac{\partial f}{\partial x}dx + \frac{\partial f}{\partial y}dy + \frac{\partial f}{\partial z}dz = 0.$$

∴ ∇f is perpendicular to $\overrightarrow{dr}$ as long as $\overrightarrow{dr}$ represents a change from P to Q where Q remains on the surface $f(x, y, z) = c$.

$\therefore$ ∇f is normal to all the tangents to the surface at $P(x_0, y_0, z_0)$. Hence ∇f represents the normal to the surface $f(x, y, z) = c$.

Hence the unit normal vector $\hat{n}$ to the surface $f(x, y, z) = c$ is given by

$$\hat{n} = \frac{\nabla f}{|\nabla f|}$$

5.45 DIRECTIONAL DERIVATIVE

Let $\hat{n}$ be a unit vector. The component of the vector ∇f in the direction $\hat{n}$ is given by $\hat{n} \cdot \nabla f$ and is called the directional derivative of f in the direction $\hat{n}$.

In other words, we can say that the directional derivative of a scalar field f at a point $P(x, y, z)$ in the direction of unit vector $\hat{n}$ is given by $\frac{df}{ds} = (\text{grad} f) \cdot \hat{n}$.

5.46 PROPERTIES OF GRADIENT

(1) grad $(f \pm g)$ = grad $f \pm$ grad g

Proof: grad $(f \pm g) = \nabla(f \pm g) = \left(\hat{i}\frac{\partial}{\partial x} + \hat{j}\frac{\partial}{\partial y} + \hat{k}\frac{\partial}{\partial z}\right)(f \pm g)$

$$= \left(\hat{i}\frac{\partial f}{\partial x} + \hat{j}\frac{\partial f}{\partial y} + \hat{k}\frac{\partial f}{\partial z}\right) \pm \left(\hat{i}\frac{\partial g}{\partial x} + \hat{j}\frac{\partial g}{\partial y} + \hat{k}\frac{\partial g}{\partial z}\right)$$

$$= \nabla f \pm \nabla g$$

$$= \text{grad} f \pm \text{grad}\, g$$

(2) grad $(fg) = f$ grad $g + g$ grad f

Proof: grad $(fg) = \nabla(fg) = \sum \hat{i}\left(\frac{\partial}{\partial x}(fg)\right)$

$$= \sum \hat{i}\left(f\frac{\partial g}{\partial x} + g\frac{\partial f}{\partial x}\right)$$

$$= \sum \hat{i}\, f\frac{\partial g}{\partial x} + \sum \hat{i}\, g\frac{\partial f}{\partial x}$$

$$= f\left(\sum \hat{i}\frac{\partial g}{\partial x}\right) + g\left(\sum \hat{i}\frac{\partial f}{\partial x}\right)$$

$$= f(\nabla g) + g(\nabla f)$$

$$= f\, \text{grad}\, g + g\, \text{grad} f$$

(3) $\text{grad}\left(\frac{f}{g}\right) = \frac{g \text{ grad } f - f \text{ grad } g}{g^2}$

Proof. $\text{grad}\left(\frac{f}{g}\right) = \nabla\left(\frac{f}{g}\right) = \sum \hat{i}\frac{\partial}{\partial x}\left(\frac{f}{g}\right)$

$$= \sum \hat{i} \left(\frac{g \dfrac{\partial f}{\partial x} - f \dfrac{\partial g}{\partial x}}{g^2} \right)$$

$$= \frac{1}{g^2} \left(\sum \hat{i}\, g \frac{\partial f}{\partial x} - \sum \hat{i}\, f \frac{\partial g}{\partial x} \right)$$

$$= \frac{g \text{ grad } f - f \text{ grad } g}{g^2}$$

ILLUSTRATIVE EXAMPLES

Example 1. Find grad f at (1, 2, 3) if $f = 2x^2 + y^2 + z^2 - 3$

Solution. grad $f = \nabla f = \left(\hat{i} \frac{\partial}{\partial x} + \hat{j} \frac{\partial}{\partial y} + \hat{k} \frac{\partial}{\partial z} \right)(2x^2 + y^2 + z^2 - 3)$

$$= \hat{i}(4x) + \hat{j}(2y) + \hat{k}(2z)$$

$$\text{grad } f = 4x\hat{i} + 2y\,\hat{j} + 2z\hat{k}$$

$$\therefore \quad \text{grad } f \text{ at } (1, 2, 3) = 4\hat{i} + 4\hat{j} + 6\hat{k}$$

Example 2. If $f = x^2y - y^3 z^2 + 2$ find ∇f at the point (1, 1, – 2)

Solution. $\nabla f = \left(\hat{i} \frac{\partial}{\partial x} + \hat{j} \frac{\partial}{\partial y} + \hat{k} \frac{\partial}{\partial z} \right)(x^2 y - y^3 z^2 + 2)$

$$= \hat{i}\,(2xy) + \hat{j}\,(x^2 - 3y^2 z^2) + \hat{k}(-2y^3 z)$$

$$\nabla f = 2\hat{i} - 11\hat{j} + 4\hat{k} \text{ at } (1, 1, -2)$$

Example 3. Show that $\nabla r^2 = 2\vec{r}$ where $r^2 = \vec{r}.\vec{r}$

Solution. Let $\vec{r} = x\hat{i} + y\hat{j} + z\hat{k}$

$$\therefore r^2 = \vec{r}.\vec{r} = x^2 + y^2 + z^2$$

Now, $\nabla r^2 = \left(\hat{i} \frac{\partial}{\partial x} + \hat{j} \frac{\partial}{\partial y} + \hat{k} \frac{\partial}{\partial z} \right)(x^2 + y^2 + z^2)$

$$= 2x\hat{i} + 2y\hat{j} + 2z\hat{k} = 2(x\hat{i} + y\hat{j} + z\hat{k}) = 2\vec{r}.$$

Example 4. If $u = x + y + z$, $y = x^2 + y^2 + z^2$, $w = yz + zx + xy$, prove that grad u, grad v and grad w are coplanar vectors.

Solution. grad $u = \left(\hat{i} \frac{\partial}{\partial x} + \hat{j} \frac{\partial}{\partial y} + \hat{k} \frac{\partial}{\partial z} \right)(x + y + z) = \hat{i} + \hat{j} + \hat{k}$

$$\text{grad } v = \left(\hat{i}\frac{\partial}{\partial x} + \hat{j}\frac{\partial}{\partial y} + \hat{k}\frac{\partial}{\partial z}\right)(x^2 + y^2 + z^2) = 2x\hat{i} + 2y\hat{j} + 2z\hat{k}$$

$$\text{grad } w = \left(\hat{i}\frac{\partial}{\partial x} + \hat{j}\frac{\partial}{\partial y} + \hat{k}\frac{\partial}{\partial z}\right)(yz + zx + xy)$$

$$= \hat{i}(z + y) + \hat{j}(z + x) + \hat{k}(y + x)$$

$$\text{grad } u . (\text{grad } v \times \text{grad } w) = \begin{vmatrix} 1 & 1 & 1 \\ 2x & 2y & 2z \\ z+y & z+x & y+x \end{vmatrix}$$

$$= 2\begin{vmatrix} 1 & 1 & 1 \\ x & y & z \\ z+y & z+x & y+x \end{vmatrix}$$

$$= 2\begin{vmatrix} 1 & 1 & 1 \\ x+y+z & y+z+x & z+y+x \\ z+y & z+x & y+x \end{vmatrix}, R_2 \to R_2 + R_3$$

$$2(x+y+z)\begin{vmatrix} 1 & 1 & 1 \\ 1 & 1 & 1 \\ z+y & z+x & y+z \end{vmatrix} = 0$$

Hence grad u, grad v and grad w are coplanar vectors.

Example 5. Show that $\nabla r^n = nr^{n-2}\vec{r}$ and hence evaluate $\nabla\frac{1}{r^2}$ where $\vec{r} = x\hat{i} + y\hat{j} + z\hat{k}$

Solution. $\nabla r^n = \left(\hat{i}\frac{\partial}{\partial x} + \hat{j}\frac{\partial}{\partial y} + \hat{k}\frac{\partial}{\partial z}\right)r^n$

$$= \hat{i}\left(nr^{n-1}\frac{\partial r}{\partial x}\right) + \hat{j}\left(nr^{n-1}\frac{\partial r}{\partial y}\right) + \hat{k}\left(nr^{n-1}\frac{\partial r}{\partial z}\right) \quad ...(1)$$

$\because \quad \vec{r} = x\hat{i} + y\hat{j} + z\hat{k}$

$\therefore \quad r = |\vec{r}| = \sqrt{x^2 + y^2 + z^2}$ or $r^2 = x^2 + y^2 + z^2$ ...(2)

Differentiating (2) partially w.r.t. x, we get

$$2r\frac{\partial r}{\partial x} = 2x$$

$$\frac{\partial r}{\partial x} = \frac{x}{r}$$

similarly $\frac{\partial r}{\partial y} = \frac{y}{r}$ and $\frac{\partial r}{\partial z} = \frac{z}{r}$

$\therefore$ From equation (1),

$$\nabla r^n = \hat{i}\left(nr^{n-1}\frac{x}{r}\right) + \hat{j}\left(nr^{n-1}\frac{y}{r}\right) + \hat{k}\left(nr^{n-1}\frac{z}{r}\right)$$

$$= nr^{n-2}(x\hat{i} + y\hat{j} + z\hat{k})$$

$$\Rightarrow \nabla r^n = nr^{n-2}\vec{r} \quad \text{...(3)}$$

Putting $n = -2$ in equation (3), we get

$$\nabla\left(\frac{1}{r^2}\right) = -2\frac{\vec{r}}{r^4}$$

Example 6. Find the directional derivative of $\phi = x^2yz + 4x\,z^2$ at (1, 2, 3) in the direction of $2\hat{i} + \hat{j} + \hat{k}$.

Solution. $\nabla\phi = \left(\hat{i}\frac{\partial}{\partial x} + \hat{j}\frac{\partial}{\partial y} + \hat{k}\frac{\partial}{\partial z}\right)(x^2yz + 4xz^2)$

$$= (2xyz + 4z^2)\hat{i} + x^2z\hat{j} + (x^2y + 8xz)\hat{k}$$

$$\nabla\phi = 48\hat{i} + 3\hat{j} + 26\hat{k} \text{ at } (1, 2, 3)$$

The unit vector $\hat{n}$ along the given direction $= \dfrac{2\hat{i} + \hat{j} + \hat{k}}{\sqrt{6}}$

$\therefore$ The required directional derivative is

$$\hat{n}.\nabla\phi = \left(\frac{2\hat{i} + \hat{j} + \hat{k}}{\sqrt{6}}\right).(48\hat{i} + 3\hat{j} + 26\hat{k})$$

$$= \frac{96 + 3 + 26}{\sqrt{6}} = \frac{125}{\sqrt{6}}$$

Example 7. Find the unit normal to the surface $x^2 + y^2 + z^2 = 1$ at (1, 1, 2).

Solution. Let $\phi = x^2 + y^2 + z^2 - 1$

Let $\hat{n}$ denote the unit normal to the surface.

$$\therefore \quad \hat{n} = \frac{\nabla\phi}{|\nabla\phi|}$$

Now, $\nabla\phi = 2x\hat{i} + 2y\hat{j} + 2z\hat{k}$

$\therefore$ $\nabla\phi$ at $(1, 1, 2) = 2\hat{i} + 2\hat{j} + 2\hat{k}$

$$\therefore \hat{n} = \frac{2\hat{i} + 2\hat{j} + 4\hat{k}}{\sqrt{2^2 + 2^2 + 4^2}} = \frac{2(\hat{i} + \hat{j} + 2\hat{k})}{\sqrt{24}} = \frac{\hat{i} + \hat{j} + 2\hat{k}}{\sqrt{6}}$$

Example 8. Prove that $\nabla f(r) = \left(\frac{f'(r)}{r}\right)\vec{r}$. where $\vec{r} = x\hat{i} + y\hat{j} + z\hat{k}$

Solution. Since $\vec{r} = x\hat{i} + y\hat{j} + z\hat{k}$, we have $r^2 = x^2 + y^2 + z^2$.

$$\therefore \quad \frac{\partial r}{\partial x} = \frac{x}{r}, \frac{\partial r}{\partial y} = \frac{y}{r}, \frac{\partial r}{\partial z} = \frac{z}{r}$$

$$\nabla f(r) = \left(\hat{i}\frac{\partial}{\partial x} + \hat{j}\frac{\partial}{\partial y} + \hat{k}\frac{\partial}{\partial z}\right) f(r)$$

$$= \left\{\hat{i} f'(r)\frac{\partial r}{\partial x} + \hat{j}\, f'(r)\frac{\partial r}{\partial y} + \hat{k}\, f'(r)\frac{\partial r}{\partial z}\right\}$$

$$= f'(r)\left[\hat{i}\left(\frac{x}{r}\right) + \hat{j}\left(\frac{y}{r}\right) + \hat{k}\left(\frac{z}{r}\right)\right]$$

$$= \frac{f'(r)}{r}(x\hat{i} + y\hat{j} + z\hat{k}) = \left(\frac{f'(r)}{r}\right)\vec{r}.$$

Example 9. Find the directional derivative of $f = x^2y - 2y^2z + 5z^2x$ at the point (1, 1, 2) in the direction of the line $\frac{x-1}{2} = \frac{y-5}{2} = \frac{z}{1}$

Solution. $f = x^2y - 2y^2z + 5z^2x$

$$\text{grad} f = \hat{i}\frac{\partial f}{\partial x} + \hat{j}\frac{\partial f}{\partial y} + \hat{k}\frac{\partial f}{\partial z}$$

$$= (2xy + 5z^2)\hat{i} + (x^2 - 4yz)\,\hat{j} + (-2y^2 + 10zx)\,\hat{k}$$

$$\text{grad} f = 22\hat{i} - 7\hat{j} + 18\hat{k} \text{ at } (1, 1, 2)$$

Here $\hat{n} = \frac{2\hat{i} + 2\hat{j} + \hat{k}}{3}$

$\therefore$ Directional derivative = (grad f) . $\hat{n}$

$$= (22\hat{i} - 7\hat{j} + 18\hat{k}).\left(\frac{2\hat{i} + 2\hat{j} + \hat{k}}{3}\right)$$

$$= 16.$$

EXERCISE 5.6

1. Find grad ϕ for the following at the points indicated.

(*i*) $\phi(x, y, z) = x^2 + y^2 + z^2 - 25$ at (1, 1, 3)

(*ii*) $\phi(x, y, z) = \log(x^2 + y^2 + z^2)$ at (1, 2, 4)

(*iii*) $\phi(x, y, z) = x^3 + y^3 + 3xyz$ at (1, 1, 1)

2. Find the unit vector normal to the surface $x^2 - y^2 + z = 2$ at the point $(1, -1, 2)$.

3. Find a unit vector normal to the surface $xy^3 z^2 = 4$ at the point $(-1, -1, 2)$.

4. Find the directional derivative of $\frac{1}{r}$ in the direction of $\vec{r}$, where $\vec{r} = x\hat{i} + y\hat{j} + z\hat{k}$.

5. Prove that the directional derivative of $\phi = x^3y^2z$ at $(1, 2, 3)$ is maximum along the direction $9\hat{i} + 3\hat{j} + \hat{k}$. Also find the maximum directional derivative.

6. Find the directional derivative of $f(x, y, z) = x^2yz + 4xz^2$ at the point $(1, -2, -1)$ in the direction of the vector $2\hat{i} - \hat{j} - 2\hat{k}$

7. If $r = |\vec{r}|$ where $\vec{r} = x\hat{i} + y\hat{j} + z\hat{k}$, prove that

(*i*) $\nabla\left(\frac{1}{r}\right) = -\frac{\vec{r}}{r^3}$ (*ii*) $\nabla \log|\vec{r}| = \frac{\vec{r}}{r^2}$

8. Find the directional derivative of the function $f = x^2 - y^2 + z^2$ at the point P(1, 2, 3) in the direction of the line PQ where Q is the point (5, 0, 4).

9. Show that grad $f(r) \times \vec{r} = \vec{0}$.

10. If $\phi(x, y) = \log\sqrt{x^2 + y^2}$, show that

$$\text{grad } \phi = \frac{\vec{r} - \vec{k}\,(\vec{k}.\vec{r})}{\{\vec{r} - \vec{k}(\vec{k}.\vec{r})\}.\{\vec{r} - \vec{k}(\vec{k}.\vec{r})\}}$$

ANSWERS

1. (*i*) $2\hat{i} + 2\hat{j} + 6\hat{k}$ (*ii*) $\frac{2}{21}(\hat{i} + 2\hat{j} + 4\hat{k})$ (*iii*) $6\hat{i} + 6\hat{j} + 3\hat{k}$

2. $\frac{1}{\sqrt{3}}(\hat{i} + \hat{j} + \hat{k})$ **3.** $-\frac{1}{\sqrt{11}}(\hat{i} + 3\hat{j} - \hat{k})$ **4.** $-\frac{1}{x^2 + y^2 + z^2}$

5. Max. directional derivative $= 4\sqrt{91}$ **6.** $\frac{37}{3}$ **7.** $\frac{4}{3}\sqrt{21}$

5.47. DIVERGENCE OF VECTOR POINT FUNCTION.

Let $\vec{V} = V_1\hat{i} + V_2\hat{j} + V_3\hat{k}$ be a vector valued function. The divergence of $\vec{V}$ denoted by $\nabla\vec{V}$ or div. $\vec{V}$ is defined by

$$\nabla\vec{V} = \frac{\partial V_1}{\partial x} + \frac{\partial V_2}{\partial y} + \frac{\partial V_3}{\partial z}$$

Note: (1) The divergence of a vector valued function is a scalar valued function.

(2) If div $\vec{V} = 0$ then $\vec{V}$ is called **solenoidal** vector point function.

(3) For a constant vector $\vec{a}$, div $\vec{a} = 0$.

5.48. CURL OF A VECTOR POINT FUNCTION

The curl of a vector valued function $\vec{V}$ is denoted by curl $\vec{V}$ or $\nabla \times \vec{V}$ and is defined as

$$\text{curl } \vec{V} = \nabla \times \vec{V} = \left(\hat{i}\frac{\partial}{\partial x} + \hat{j}\frac{\partial}{\partial y} + \hat{k}\frac{\partial}{\partial z} \right) \times \vec{V} = \hat{i} \times \frac{\partial \vec{V}}{\partial x} + \hat{j} \times \frac{\partial \vec{V}}{\partial y} + \hat{k}\frac{\partial \vec{V}}{\partial z}$$

If $\vec{V} = V_1\hat{i} + V_2\hat{j} + V_3\hat{k}$ then

$$\text{curl } \vec{V} = \nabla \times \vec{V} = \left(\hat{i}\frac{\partial}{\partial x} + \hat{j}\frac{\partial}{\partial y} + \hat{k}\frac{\partial}{\partial z} \right) \times [V_1\hat{i} + V_2\hat{j} + V_3\hat{k}]$$

$$\text{curl } \vec{V} = \begin{vmatrix} \hat{i} & \hat{j} & \hat{k} \\ \frac{\partial}{\partial x} & \frac{\partial}{\partial y} & \frac{\partial}{\partial z} \\ V_1 & V_2 & V_3 \end{vmatrix}$$

$$\text{or curl } \vec{V} = \hat{i}\left(\frac{\partial V_3}{\partial y} - \frac{\partial V_2}{\partial z} \right) + \hat{j}\left(\frac{\partial V_1}{\partial z} - \frac{\partial V_3}{\partial x} \right) + \hat{k}\left(\frac{\partial V_2}{\partial x} - \frac{\partial V_1}{\partial y} \right).$$

Note: (1) The curl of a vector valued function is a vector valued function.

(2) If curl $\vec{V} = \vec{0}$, then $\vec{V}$ is said to be an **irrotational** vector.

(3) For a constant vector $\vec{a}$, curl $\vec{a} = \vec{0}$.

5.49 PHYSICAL INTERPRETATION OF DIVERGENCE

Consider a rectangular parallelopiped of dimensions δx, δy, δz parallel to x, y and z respectively.

Let $\vec{V} = V_x\hat{i} + V_y\hat{j} + V_z\hat{k}$ be the velocity of the fluid at $P(x, y, z)$.

The mass of the fluid flowing through the face ABCD per unit time is $V_y\,\delta x\,\delta z$
Mass of the fluid flowing out through the face PQRS per unit time

$$= \left(V_y + \frac{\partial V_y}{\partial y}\delta y \right)(\delta x\,\delta z)$$

Net decrease in the mass of fluid flowing across y-axis per unit time

$$= \left(V_y + \frac{\partial V_y}{\partial y}\delta y \right)\delta x\,\delta z - V_y\,\delta x\,\delta z = \frac{\partial V_y}{\partial y}\delta x\,\delta y\,\delta z$$

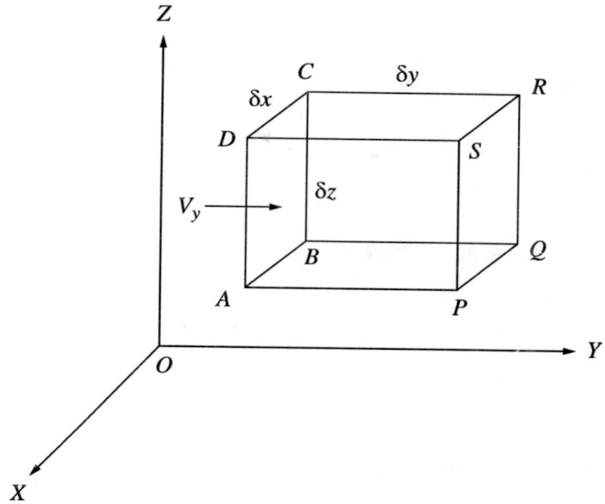

Similarly, the decrease in mass of fluid to the flow along x-axis and z-axis are

$$\frac{\partial V_x}{\partial x}\,\delta x\,\delta y\,\delta z \text{ and } \frac{\partial V_z}{\partial z}\,\delta x\,\delta y\,\delta z \text{ respectively.}$$

Total decrease of the fluid per unit time

$$= \left(\frac{\partial Vx}{\partial x}+\frac{\partial V_y}{\partial y}+\frac{\partial V_z}{\partial z}\right)\delta x\,\delta y\,\delta z \qquad ...(1)$$

Dividing equation (1) by volume $\delta x\,\delta y\,\delta z$ of parallelopiped, we get the rate of

decrease of fluid per unit time $= \dfrac{\partial Vx}{\partial x}+\dfrac{\partial V_y}{\partial y}+\dfrac{\partial V_z}{\partial z} = \nabla.\vec{V} = \text{div}\,\vec{V}$

Hence div $\vec{V}$ gives the rate at which fluid is originating at a point per unit volume.

Note: If the fluid is incompressible, there can be no gain or loss in the volume element and hence div $\vec{V} = 0$, then $\vec{V}$ is called a solenoidal vector function.

5.50. PHYSICAL INTERPRETATION OF CURL

Consider the motion of a rigid body rotating with angular velocity $\vec{W} = w_1\hat{i} + w_2\hat{j} + w_3\hat{k}$ about a fixed axis through 0.

We know that $\vec{V} = \vec{W}\times\vec{r}$, $\vec{V}$ is the linear velocity and $\vec{r} = x\hat{i} + y\hat{j} + z\hat{k}$ is the position vector of a point on the rotating body.

$$\therefore \quad \vec{V} = \vec{W}\times\vec{r} = \begin{vmatrix} \hat{i} & \hat{j} & \hat{k} \\ w_1 & w_2 & w_3 \\ x & y & z \end{vmatrix}$$

$$= (w_2 z - w_3 y)\hat{i} + (w_3 x - w_1 z)\,\hat{j} + (w_1 y - w_2 x)\hat{k}$$

$$\text{Curl } \vec{V} = \nabla \times \vec{V} = \begin{vmatrix} \hat{i} & \hat{j} & \hat{k} \\ \frac{\partial}{\partial x} & \frac{\partial}{\partial y} & \frac{\partial}{\partial z} \\ w_2 z - w_3 y & w_3 x - w_1 z & w_1 y - w_2 x \end{vmatrix}$$

$$= (w_1 + w_1)\,\hat{i} - (-w_2 - w_2)\hat{j} + (w_3 + w_3)\hat{k}$$

$$= 2w_1\hat{i} + 2w_2\hat{j} + 2w_3\hat{k} = 2(w_1\hat{i} + w_2\hat{j} + w_3\hat{k})$$

$$\text{Curl } \vec{V} = 2\vec{W}$$

$$\Rightarrow \quad \vec{W} = \frac{1}{2}\text{Curl } \vec{V}$$

Hence the angular velocity at any point is equal to half of the curl of the linear velocity $\vec{V}$ at that point of the body.

5.51. IMPORTANT VECTOR INDEMNITIES

1. $\text{div}(\vec{a} + \vec{b}) = \text{div } \vec{a} + \text{div } \vec{b}$

or $\nabla.(\vec{a} + \vec{b}) = \nabla.\vec{a} + \nabla.\vec{b}$

Proof. $\text{div } (\vec{a} + \vec{b}) = \nabla.(\vec{a} + \vec{b})$

$$= \left(\hat{i}\frac{\partial}{\partial x} + \hat{j}\frac{\partial}{\partial y} + \hat{k}\frac{\partial}{\partial z}\right).(\vec{a} + \vec{b})$$

$$= \hat{i}\frac{\partial}{\partial x}(\vec{a} + \vec{b}) + \hat{j}\frac{\partial}{\partial y}(\vec{a} + \vec{b}) + \hat{k}\frac{\partial}{\partial z}.(\vec{a} + \vec{b})$$

$$= \hat{i}\left(\frac{\partial \vec{a}}{\partial x} + \frac{\partial \vec{b}}{\partial x}\right) + \hat{j}\left(\frac{\partial \vec{a}}{\partial y} + \frac{\partial \vec{b}}{\partial y}\right) + \hat{k}\left(\frac{\partial \vec{a}}{\partial z} + \frac{\partial \vec{b}}{\partial z}\right)$$

$$= \left(\hat{i}\frac{\partial \vec{a}}{\partial x} + \hat{j}\frac{\partial \vec{a}}{\partial y} + \hat{k}\frac{\partial \vec{a}}{\partial z}\right) + \left(\hat{i}\frac{\partial \vec{b}}{\partial x} + \hat{j}\frac{\partial \vec{b}}{\partial y} + \hat{k}\frac{\partial \vec{b}}{\partial z}\right)$$

$$= \nabla.\vec{a} + \nabla.\vec{b} = \text{div } \vec{a} + \text{div } \vec{b}.$$

2. $\text{div } (\vec{a} \times \vec{b}) = \vec{b} \text{ curl } \vec{a} - \vec{a} \text{ curl } \vec{b}$

Proof. $\text{div } (\vec{a} \times \vec{b}) = \sum \hat{i}.\frac{\partial}{\partial x}(\vec{a} \times \vec{b}) = \sum \hat{i}.\left(\frac{\partial \vec{a}}{\partial x} \times \vec{b} + \vec{a} \times \frac{\partial \vec{b}}{\partial x}\right)$

$$= \sum \hat{i}.\left(\frac{\partial \vec{a}}{\partial x} \times \vec{b}\right) - \sum \hat{i}.\left(\frac{\partial \vec{b}}{\partial x} \times \vec{a}\right)$$

$$= \sum\left(\hat{i} \times \frac{\partial \vec{a}}{\partial x}\right).\vec{b} - \sum\left(\hat{i} \times \frac{\partial \vec{b}}{\partial x}\right).\vec{a}$$

$$= (\text{curl } \vec{a}).\vec{b} - (\text{curl } \vec{b}).\vec{a} = \vec{b}.\text{curl } \vec{a} - \vec{a}.\text{curl } \vec{b}$$

3. If $\vec{a}$ is a differentiable vector function and α is a differentiable scalar function, then

$$\text{div}\,(\alpha\,\vec{a}) = \alpha\,\text{div}\,\vec{a} + (\text{grad}\,\alpha).\,\vec{a}$$

Proof. $\text{div}\,(\alpha\,\vec{a}) = \nabla.(\alpha\,\vec{a}) = \left(\hat{i}\frac{\partial}{\partial x} + \hat{j}\frac{\partial}{\partial y} + \hat{k}\frac{\partial}{\partial z}\right).(\alpha\,\vec{a})$

$$= \hat{i}.\frac{\partial}{\partial x}(\alpha\,\vec{a}) + \hat{j}.\frac{\partial}{\partial y}(\alpha\,\vec{a}) + \hat{k}.\frac{\partial}{\partial z}(\alpha\,\vec{a})$$

$$= \sum\left\{\hat{i}.\frac{\partial}{\partial x}(\alpha\,\vec{a})\right\} = \sum\hat{i}.\left\{\frac{\partial\alpha}{\partial x}\vec{a} + \alpha\frac{\partial\vec{a}}{\partial x}\right\}$$

$$= \sum\left\{\hat{i}.\left(\frac{\partial\alpha}{\partial x}\cdot\vec{a}\right)\right\} + \sum\left\{\hat{i}.\left(\alpha\cdot\frac{\partial\vec{a}}{\partial x}\right)\right\}$$

$$= \sum\left\{\left(\frac{\partial\alpha}{\partial x}\hat{i}\right)\cdot\vec{a}\right\} + \sum\left\{\alpha\left(\hat{i}\cdot\frac{\partial\vec{a}}{\partial x}\right)\right\}$$

$$= \left\{\sum\frac{\partial\alpha}{\partial x}\hat{i}\right\}.\,\vec{a} + \alpha\sum\left(\hat{i}\;\frac{\partial\vec{a}}{\partial x}\right)$$

$$= (\nabla\alpha).\,\vec{a} + \alpha(\nabla\vec{a}) = (\text{grad}\,\alpha).\,\vec{a} + \alpha\,\text{div}\,\vec{a}$$

4. $\text{div}\,(\text{grad}\,f) = \nabla^2 f$

Proof. $\text{div}\,(\text{grad}\,f) = \nabla.\,(\nabla f) = \left(\hat{i}\frac{\partial}{\partial x} + \hat{j}\frac{\partial}{\partial y} + \hat{k}\frac{\partial}{\partial z}\right).\left(\hat{i}\frac{\partial f}{\partial x} + \hat{j}\frac{\partial f}{\partial y} + \hat{k}\frac{\partial f}{\partial z}\right)$

$$= \frac{\partial}{\partial x}\left(\frac{\partial f}{\partial x}\right) + \frac{\partial}{\partial y}\left(\frac{\partial f}{\partial y}\right) + \frac{\partial}{\partial z}\left(\frac{\partial f}{\partial z}\right)$$

$$= \frac{\partial^2 f}{\partial x^2} + \frac{\partial^2 f}{\partial^2 y} + \frac{\partial^2 f}{\partial z^2} = \left(\frac{\partial^2}{\partial x^2} + \frac{\partial^2}{\partial y^2} + \frac{\partial^2}{\partial z^2}\right)f$$

$$= \nabla^2 f \quad \text{where } \nabla^2 = \frac{\partial^2}{\partial x^2} + \frac{\partial^2}{\partial y^2} + \frac{\partial^2}{\partial z^2}$$

$\nabla^2 f$ is called laplacian operator and if $\boldsymbol{\nabla^2 f = 0}$ then it is called **Laplace's equation.**

5. $\text{div}\,(\text{curl}\,\vec{V}) = \nabla.\,(\nabla\times\vec{V}) = 0$

Proof. Let $\vec{V} = V_1\hat{i} + V_2\hat{j} + V_3\hat{k}$, then

$$\text{curl}\,\vec{V} = \begin{vmatrix} \hat{i} & \hat{j} & \hat{k} \\ \frac{\partial}{\partial x} & \frac{\partial}{\partial y} & \frac{\partial}{\partial z} \\ V_1 & V_2 & V_3 \end{vmatrix}$$

$$= \hat{i}\left(\frac{\partial V_3}{\partial y} - \frac{\partial V_2}{\partial z}\right) + \hat{j}\left(\frac{\partial V_1}{\partial z} - \frac{\partial V_3}{\partial x}\right) + \hat{k}\left(\frac{\partial V_2}{\partial x} - \frac{\partial V_1}{\partial y}\right)$$

Now, div (curl $\vec{V}$) = $\nabla.(\nabla \times \vec{V})$

$$= \frac{\partial}{\partial x}\left(\frac{\partial V_3}{\partial y} - \frac{\partial V_2}{\partial z}\right) + \frac{\partial}{\partial y}\left(\frac{\partial V_1}{\partial z} - \frac{\partial V_3}{\partial x}\right) + \frac{\partial}{\partial z}\left(\frac{\partial V_2}{\partial x} - \frac{\partial V_1}{\partial y}\right)$$

$$= \left(\frac{\partial^2 V_3}{\partial x \partial y} - \frac{\partial^2 V_2}{\partial x \partial z}\right) + \left(\frac{\partial^2 V_1}{\partial y \partial z} - \frac{\partial^2 V_3}{\partial y \partial x}\right) + \left(\frac{\partial^2 V_2}{\partial z \partial x} - \frac{\partial^2 V_1}{\partial z \partial y}\right) = 0$$

6. curl $(\vec{a} + \vec{b})$ = curl $\vec{a}$ + curl $\vec{b}$

or $\nabla \times (\vec{a} + \vec{b}) = \nabla \times \vec{a} + \nabla \times \vec{b}$

Proof. curl $(\vec{a} + \vec{b}) = \nabla \times (\vec{a} + \vec{b})$

$$= \left(\hat{i}\frac{\partial}{\partial x} + \hat{j}\frac{\partial}{\partial x} + \hat{k}\frac{\partial}{\partial x}\right) \times (\vec{a} + \vec{b}) = \sum \hat{i} \times \frac{\partial}{\partial x}(\vec{a} + \vec{b})$$

$$= \sum \hat{i} \times \left(\frac{\partial \vec{a}}{\partial x} + \frac{\partial \vec{b}}{\partial x}\right)$$

$$= \sum \hat{i} \times \frac{\partial \vec{a}}{\partial x} + \sum \hat{i} \times \frac{\partial \vec{b}}{\partial x} = \text{curl } \vec{a} + \text{curl } \vec{b}$$

7. $\vec{a}$ div $\vec{b} - \vec{b}$ div $\vec{a} + (\vec{b}.\nabla)\vec{a} - (\vec{a}.\nabla)\vec{b}$ = curl $(\vec{a} \times \vec{b})$

Proof. curl $(\vec{a} \times \vec{b}) = \sum \hat{i} \times \frac{\partial}{\partial x}(\vec{a} \times \vec{b})$

$$= \sum \hat{i} \times \left(\frac{\partial \vec{a}}{\partial x} \times \vec{b} + \vec{a} \times \frac{\partial \vec{b}}{\partial x}\right)$$

$$= \sum \hat{i} \times \left(\frac{\partial \vec{a}}{\partial x} \times \vec{b}\right) + \sum \hat{i} \times \left(\vec{a} \times \frac{\partial \vec{b}}{\partial x}\right)$$

$$= \sum \left[(\hat{i} \cdot \vec{b})\frac{\partial \vec{a}}{\partial x} - \left(\hat{i} \cdot \frac{\partial \vec{a}}{\partial x}\right)\vec{b}\right] + \sum \left[\left(\hat{i} \cdot \frac{\partial \vec{b}}{\partial x}\right) \cdot \vec{a} - (\hat{i} \cdot \vec{a})\frac{\partial \vec{b}}{\partial x}\right]$$

$$= \sum (\vec{b}.\hat{i})\frac{\partial \vec{a}}{\partial x} - \left(\sum \hat{i}. \times \frac{\partial \vec{a}}{\partial x}\right)\vec{b} \sum \left(\hat{i}.\frac{\partial \vec{b}}{\partial x}\right)\vec{a} - \sum (\vec{a}.\hat{i})\frac{\partial \vec{b}}{\partial x}$$

$$= \left(\vec{b}. \sum \hat{i}\frac{\partial}{\partial x}\right)\vec{a} - \vec{b} \text{ div } \vec{a} + \vec{a} \text{ div } \vec{b} - \left(\vec{a}. \sum \hat{i}\frac{\partial}{\partial x}\right)\vec{b}$$

$$= (\vec{b}.\nabla)\vec{a} - \vec{b} \text{ div } \vec{a} + \vec{a} \text{ div } \vec{b} - (\vec{a}.\nabla)\vec{b}$$

$$= \vec{a} \text{ div } \vec{b} - \vec{b} \text{ div } \vec{a} + (\vec{b}.\nabla)\vec{a} - (\vec{a}.\nabla)\vec{b}$$

8. curl (grad f) = $\nabla \times \nabla f = \vec{0}$.

$$= \begin{vmatrix} \hat{i} & \hat{j} & \hat{k} \\ \dfrac{\partial}{\partial x} & \dfrac{\partial}{\partial y} & \dfrac{\partial}{\partial z} \\ \dfrac{\partial f}{\partial x} & \dfrac{\partial f}{\partial y} & \dfrac{\partial f}{\partial z} \end{vmatrix}$$

$$= \hat{i}\left(\frac{\partial^2 f}{\partial y \partial z} - \frac{\partial^2 f}{\partial y \partial z}\right) + \hat{j}\left(\frac{\partial^2 f}{\partial x \partial z} - \frac{\partial^2 f}{\partial x \partial z}\right) + \hat{k}\left(\frac{\partial^2 f}{\partial x \partial y} - \frac{\partial^2 f}{\partial x \partial y}\right)$$

$$= \vec{0}.$$

ILLUSTRATIVE EXAMPLE

Example 1. If $f(x, y, z) = 3x^2y + y^3z^2$, find grad f at the point (1, 2, 3).

Solution. grad $f = \nabla f = \left(\hat{i}\dfrac{\partial}{\partial x} + \hat{j}\dfrac{\partial}{\partial y} + \hat{k}\dfrac{\partial}{\partial z}\right)(3x^2y + y^3z^2)$

$$= \hat{i}\frac{\partial}{\partial x}(3x^2y + y^3z^2) + \hat{j}\frac{\partial}{\partial y}(3x^2y + y^3z^2) + \hat{k}\frac{\partial}{\partial z}(3x^2y + y^3z^2)$$

$$= \hat{i}(6xy) + \hat{j}(3x^2 + 3y^2z^2) + \hat{k}(2y^3z)$$

grad $f = 6xy\,\hat{i} + (3x^2 + 3y^2z^2)\,\hat{j} + 2y^3z\,\hat{k}$

$\therefore$ Putting $x = 1, y = 2, z = 3$, we get

grad $f = \nabla f = 12\hat{i} + 111\hat{j} + 48\hat{k}$

Example 2. If $\vec{r} = x\hat{i} + y\hat{j} + z\hat{k}$ and $r = |\vec{r}|$, show that

(*i*) div $(\vec{r}f) = 3f + \vec{r}$. grad f

(*ii*) div $(\hat{r}) = \dfrac{2}{r}$.

Solution. We know that div $(\alpha\,\vec{a}) = \alpha$ div $\vec{a} + \vec{a}$. grad α

$\therefore$ div $(f\,\vec{r}) = f$ div $\vec{r} + \vec{r}$. grad f

Now, div $\vec{r} = \nabla.\vec{r} = \left(\hat{i}\dfrac{\partial}{\partial x} + \hat{j}\dfrac{\partial}{\partial y} + \hat{k}\dfrac{\partial}{\partial z}\right).(x\hat{i} + y\hat{j} + z\hat{k})$

div $\vec{r} = 1 + 1 + 1 = 3$

$\therefore$ div $(f\vec{r}) = 3f + \vec{r}$. grad f

and div $(\hat{r})$ = div$\left(\dfrac{1}{r}\vec{r}\right)$ $\qquad \left[\because \hat{r} = \dfrac{\vec{r}}{r}\right]$

$$= \frac{1}{r}\text{div}\,\vec{r} + \vec{r}.\,\text{grad}\,\frac{1}{r} = \frac{3}{r} + \vec{r}\left(-\frac{1}{r^2}\hat{r}\right)$$

$$= \frac{3}{r} + \vec{r}.\left(-\frac{\vec{r}}{r^3}\right) = \frac{3}{r} - \frac{r^2}{r^3} = \frac{3}{r} - \frac{1}{r} = \frac{2}{r}$$

Example. If $\vec{V} = e^{xyz}(\hat{i} + \hat{j} + \hat{k})$, find curl $\vec{V}$.

Solution. We have

$$\text{curl}\,\vec{V} = \begin{vmatrix} \hat{i} & \hat{j} & \hat{k} \\ \frac{\partial}{\partial x} & \frac{\partial}{\partial y} & \frac{\partial}{\partial z} \\ e^{xyz} & e^{xyz} & e^{xyz} \end{vmatrix} = \left[\frac{\partial}{\partial y}(e^{xyz}) - \frac{\partial}{\partial z}(e^{xyz})\right]\hat{i}$$

$$-\left[\frac{\partial}{\partial x}(e^{xyz}) - \frac{\partial}{\partial z}(e^{xyz})\right]\hat{j} + \left[\frac{\partial}{\partial x}(e^{xyz}) - \frac{\partial}{\partial y}(e^{xyz})\right]\hat{k}$$

$$= e^{xyz}(xz - xy)\,\hat{i} + e^{xyz}(xy - yz)\,\hat{j} + e^{xyz}(yz - xz)\,\hat{k}$$

Example 4. Find ∇f and $|\nabla f|$ if $f = (x^2 + y^2 + z^2) + e^{-(x^2 + y^2 + z^2)^{\frac{1}{2}}}$

Solution. Let $\vec{r} = x\hat{i} + y\hat{j} + z\hat{k}$, then $r^2 = \vec{r}\cdot\vec{r} = x^2 + y^2 + z^2$

$$\therefore f = r^2 e^{-r}$$

Now, $\nabla f = \frac{\partial f}{\partial x}\hat{i} + \frac{\partial f}{\partial y}\hat{j} + \frac{\partial f}{\partial z}\hat{k}$

$$\frac{\partial f}{\partial x} = \frac{\partial f}{\partial r}.\frac{\partial f}{\partial x} = (2r\,e^{-r} - r^2 e^{-r})\frac{\partial r}{\partial x}$$

But $r^2 = x^2 + y^2 + z^2$

$$\therefore \quad 2r\frac{\partial r}{\partial x} = 2x \Rightarrow \frac{\partial r}{\partial x} = \frac{x}{r} \quad \therefore \quad \frac{\partial f}{\partial x} = re^{-r}(2 - r)\frac{x}{r} = (2 - r)e^{-r}x$$

Similarly, $\frac{\partial f}{\partial y} = (2 - r)\,e^{-r}\,y$ and $\frac{\partial f}{\partial z} = (2 - r)\,e^{-r}.z$

$$\because \quad \nabla f = (2 - r)e^{-r}(x\hat{i} + y\hat{j} + z\hat{k}) = (2 - r)e^{-r}\vec{r}$$

Also, $|\nabla f| = |(2 - r)\,e^{-r}\vec{r}| = (2 - r)\,e^{-r}\,|\vec{r}| = (2 - r)\,e^{-r}.r$

Example 5. Prove that div $(r^n\,\vec{r}) = (n + 3)\,r^n$.

Solution. Let $\vec{r} = x\hat{i} + y\hat{j} + z\hat{k}$ then $r^2 = \vec{r}.\vec{r} = x^2 + y^2 + z^2$

$$\therefore \quad \text{div}\,(r^n\,\vec{r}) = \nabla.\,r^n\,\vec{r} = \nabla.\,(x^2 + y^2 + z^2)^{\frac{n}{2}}\,(x\hat{i} + y\hat{j} + z\hat{k})$$

$$= \left(\hat{i}\frac{\partial}{\partial x} + \hat{j}\frac{\partial}{\partial y} + \hat{k}\frac{\partial}{\partial z}\right).\left[(x^2 + y^2 + z^2)^{n/2}\cdot x\hat{i}\right.$$

$$\left. + (x^2 + y^2 + z^2)^{n/2}\cdot y\hat{j} + (x^2 + y^2 + z^2)^{n/2}\cdot z\hat{k}\right]$$

$$= \frac{n}{2}(x^2 + y^2 + z^2)^{\frac{n-2}{2}}(2x^2) + (x^2 + y^2 + z^2)^{\frac{n}{2}}$$

$$+ \frac{n}{2}(x^2 + y^2 + z^2)^{\frac{n-2}{2}}(2y^2) + (x^2 + y^2 + z^2)^{n/2}$$

$$+ \frac{n}{2}(x^2 + y^2 + z^2)^{\frac{n-2}{2}}(2z^2) + (x^2 + y^2 + z^2)^{n/2}$$

$$= n(x^2 + y^2 + z^2)^{\frac{n}{2}-1} + (x^2 + y^2 + z^2) + 3(x^2 + y^2 + z^2)^{n/2}$$

$$= n(x^2 + y^2 + z^2)^{n/2} + 3(x^2 + y^2 + z^2)^{n/2}$$

$$= (n + 3)(x^2 + y^2 + z^2)^{n/2} = (n + 3)r^n.$$

Example 6. Show that $\text{div}(\text{grad } r^n) = \nabla^2(r^n) = n(n + 1)\, r^{n-2}$, where $\vec{r} = x\hat{i} + y\hat{j} + z\hat{k}$ and hence find $\nabla^2\left(\frac{1}{r}\right)$.

Solution. Since $\vec{r} = x\hat{i} + y\hat{j} + z\hat{k}$ then $r^2 = \vec{r}.\vec{r} = x^2 + y^2 + z^2$

$$\therefore \frac{\partial r}{\partial x} = \frac{x}{r}, \frac{\partial r}{\partial y} = \frac{y}{r}, \frac{\partial r}{\partial z} = \frac{z}{r}$$

$$\text{grad } r^n = nr^{n-1}\hat{r} = nr^{n-1}\frac{\vec{r}}{r} = nr^{n-2}\vec{r} \qquad \left[\because \hat{r} = \frac{\vec{r}}{r}\right]$$

$$\therefore \quad \text{div}(\text{grad } r^n) = \text{div}(nr^{n-2}\vec{r})$$

$$= nr^{n-2}\,\text{div}\,\vec{r} + \text{grad}(nr^{n-2}).\vec{r}$$

$$= 3nr^{n-2} + n(n-2)\, r^{n-3}\hat{r}.\vec{r} \qquad [\because \text{div}\,\vec{r} = 3]$$

$$= 3nr^{n-2} + n(n-2)\, r^{n-4}(\vec{r}.\vec{r})$$

$$\text{div}(\text{grad } r^n) = 3nr^{n-2} + n(n-2)r^{n-4}\, r^2 = 3nr^{n-2} + n\,(n-2)\, r^{n-2}$$

$$\Rightarrow \quad \nabla^2(r^n) = (3n + n^2 - 2n)r^{n-2} = n(n + 1)\, r^{n-2}$$

Putting $n = -1$, we get

$$\nabla^2\left(\frac{1}{r}\right) = 0.$$

Example 7. If $\vec{r} = x\hat{i} + y\hat{j} + z\hat{k}$, show that

(*i*) $\text{div}\,\vec{r} = 3$

(*ii*) $\text{curl}\,\vec{r} = 0$ **[I.U. 2007–2008]**

Solution. (*i*) $\text{div}\,\vec{r} = \nabla.\vec{r} = \left(\hat{i}\frac{\partial}{\partial x} + \hat{j}\frac{\partial}{\partial y} + \hat{k}\frac{\partial}{\partial z}\right).(x\hat{i} + y\hat{j} + z\hat{k})$

$$\text{div}\,\vec{r} = 1 + 1 + 1 = 3$$

(*ii*) $\text{curl}\,\vec{r} = \nabla\times\vec{r} = \left(\hat{i}\frac{\partial}{\partial x} + \hat{j}\frac{\partial}{\partial y} + \hat{k}\frac{\partial}{\partial z}\right).(x\hat{i} + y\hat{j} + z\hat{k})$

$$= \begin{vmatrix} \hat{i} & \hat{j} & \hat{k} \\ \frac{\partial}{\partial x} & \frac{\partial}{\partial y} & \frac{\partial}{\partial z} \\ x & y & z \end{vmatrix}$$

$$= \hat{i}\,(0-0) - \hat{j}\,(0-0) + \hat{k}\,(0-0)$$

$$\text{curl}\,\vec{r} = 0\hat{i} - 0\hat{j} + 0\hat{k} = \vec{0}.$$

Example 8. Show that $\nabla^2 f(r) = f''(r) + \frac{2}{r} f'(r)$

Solution. Let $\vec{r} = x\hat{i} + y\hat{j} + z\hat{k}$

$$\text{grad}\, f(r) = f'(r)\,\hat{r} = \frac{1}{r} f'(r)\,\vec{r}$$

$$\text{div}\{\text{grad}\, f(r)\} = \nabla^2 f(r)$$

$$= \text{div}\left\{\frac{f'(r)\vec{r}}{r}\right\} = \frac{f'(r)}{r}\,\text{div}\,\vec{r} + \text{grad}\left\{\frac{f'(r)}{r}\right\}\vec{r}$$

$$= \frac{3}{r} f'(r) + \left\{\frac{r\,f''(r) - f'(r)}{r^2}\right\}\hat{r}.\vec{r} \qquad [\because \text{div}\,\vec{r} = 3]$$

$$= \frac{3}{r} f'(r) + \left\{\frac{r\,f''(r) - f'(r)}{r^3}\right\}(\vec{r}.\vec{r})$$

$$= \frac{3}{r} f'(r) + \left\{\frac{r\,f''(r) - f'(r)}{r}\right\} \qquad [\because \vec{r}.\vec{r} = r^2]$$

$$= f''(r) + \frac{2}{r} f'(r)$$

Example 9. If $\vec{V} = \frac{x\hat{i} + y\hat{j} + z\hat{k}}{\sqrt{x^2+y^2+z^2}}$, show that $\nabla.\vec{V} = \frac{2}{\sqrt{x^2+y^2+z^2}}$ and $\nabla\times\vec{V} = \vec{0}$.

Solution. $\nabla\times\vec{V} = \left(\hat{i}\frac{\partial}{\partial x} + \hat{j}\frac{\partial}{\partial y} + \hat{k}\frac{\partial}{\partial z}\right).\left(\frac{x\hat{i} + y\hat{j} + z\hat{k}}{\sqrt{x^2+y^2+z^2}}\right)$

$$= \frac{\partial}{\partial x}\left(\frac{x}{\sqrt{x^2+y^2+z^2}}\right) + \frac{\partial}{\partial y}\left(\frac{y}{\sqrt{x^2+y^2+z^2}}\right) + \frac{\partial}{\partial z}\left(\frac{z}{\sqrt{x^2+y^2+z^2}}\right) \quad ...(1)$$

$$\text{Now, } \frac{\partial}{\partial x}\left(\frac{x}{\sqrt{x^2+y^2+z^2}}\right) = \frac{y^2+z^2}{(x^2+y^2+z^2)^{3/2}},$$

$$\frac{\partial}{\partial y}\left(\frac{y}{\sqrt{x^2+y^2+z^2}}\right) = \frac{x^2+z^2}{(x^2+y^2+z^2)^{3/2}}$$

and $$\frac{\partial}{\partial z}\left(\frac{z}{\sqrt{x^2+y^2+z^2}}\right) = \frac{x^2+y^2}{(x^2+y^2+z^2)^{3/2}}$$

$$\therefore \quad \nabla.\vec{V} = \frac{2(x^2+y^2+z^2)}{(x^2+y^2+z^2)^{3/2}} = \frac{2}{\sqrt{x^2+y^2+z^2}}$$

$$\nabla\times\vec{V} = \left(\hat{i}\frac{\partial}{\partial x}+\hat{j}\frac{\partial}{\partial y}+\hat{k}\frac{\partial}{\partial z}\right)\times\left(\frac{x\hat{i}+y\hat{j}+z\hat{k}}{\sqrt{x^2+y^2+z^2}}\right)$$

$$= \begin{vmatrix} \hat{i} & \hat{j} & \hat{k} \\ \dfrac{\partial}{\partial x} & \dfrac{\partial}{\partial y} & \dfrac{\partial}{\partial z} \\ \dfrac{x}{\sqrt{x^2+y^2+z^2}} & \dfrac{y}{\sqrt{x^2+y^2+z^2}} & \dfrac{z}{\sqrt{x^2+y^2+z^2}} \end{vmatrix}$$

$$\nabla\times\vec{V} = \vec{0}.$$

EXERCISE 5.7

1. Show that the vector field $\vec{V} = (\sin y + z)\hat{i} + (x\cos y - z)\hat{j} + (x-y)\hat{k}$ is irrotational.
2. Prove that $\text{div}\,(r^n\,\vec{r}) = (n+3)\,r^n$, and hence show that $r^n\,\vec{r}$ is solenoidal if $n = -3$.
3. Prove that $\text{div}\,(\nabla\phi\times\nabla\psi) = 0$.
4. If $f(x, y, z) = x^2y + y^2x + z^2$, find ∇f at the point (1, 1, 1).
5. Prove that curl grad $r^n = \vec{0}$.
6. If $f = 2x^3y^2z^4$, find div (grad f).
7. If $\phi = x^2 - y^2 + 4z$, show that $\nabla^2\phi = 0$
8. Find the divergence and curl of the vector field $V = (x^2-y^2)\hat{i} + 2xy\,\hat{j} + (y^2-xy)\,\hat{k}$.
9. Show that the vector $V = (x+3y)\hat{i} + (y-3z)\hat{j} + (x-2z)\,\hat{k}$ is solenoidal.
10. Find div V and curl V, where $V = \text{grad}\,(x^3+y^2+z^3-3xyz)$.

11. Prove that curl grad $f = \vec{0}$, if $f = x^2y + 2xy + z^2$.

12. Prove that curl grad $r^n = \vec{0}$.

13. If a is constant vector and $\vec{r}$ is the radius vector, prove that

(*i*) div $(\vec{r} \times \vec{a}) = 0$ (*ii*) curl $(\vec{r} \times \vec{a}) = -2\vec{a}$.

where $\vec{r} = x\hat{i} + y\hat{j} + z\hat{k}$ and $\vec{a} = a_1\hat{i} + a_2\hat{j} + a_3\hat{k}$

ANSWERS

4. $3\hat{i} + 3\hat{j} + 2\hat{k}$

6. $12xy^2z^4 + 4x^3z^4 + 24x^3y^2z^2$

8. div $\vec{V} = 4x$, curl $\vec{V} = (2y - x)\hat{i} + y\hat{j} + 4\hat{k}$

10. div $\vec{V} = 6(x + y + z)$, curl $\vec{V} = \vec{0}$.